Developments in Applied Phycology

Volume 14

Series Editor

Michael A. Borowitzka, Algae R&D Centre, School of Veterinary and Life Sciences
Murdoch University
Murdoch, Australia

Aims and Scope

Applied Phycology, the practical use of algae, encompasses a diverse range of fields including algal culture and seaweed farming, the use of algae to produce commercial products such as hydrocolloids, carotenoids and pharmaceuticals, algae as biofertilizers and soil conditioners, the application of algae in wastewater treatment, renewable energy production, algae as environmental indicators, environmental bioremediation and the management of algal blooms. The commercial production of seaweeds and microalgae and products derived there from is a large and well established industry and new algal species, products and processes are being continuously developed.

The aim of this book series, Developments in Applied Phycology, is to present state-of-the-art syntheses of research and development in the field. Volumes of the series will consist of reference books, subject-specific monographs, peer reviewed contributions from conferences, comprehensive evaluations of large-scale projects, and other book-length contributions to the science and practice of applied phycology.

Prospective authors and/or editors should consult the Series Editor or Publishing Editor for more details. Series Editor: Michael A. Borowitzka - M.Borowitzka@murdoch.edu.au Publishing Editor: Éva Loerinczi - eva.loerinczi@springer.com

David Ulises Santos Ballardo • Sergio Rossi
Editors

Microalgae as Promising Source of Commercial Bioproducts

Advances and Emerging Technologies in Bioactive Compounds, Biofuels, Food and Biorefineries

Springer

Editors
David Ulises Santos Ballardo
Ecotoxicology and Bioprocesses Laboratory
Research Center in Food and Development
Mazatlán, Sinaloa, Mexico

Sergio Rossi
DiSTeBA
University of Salento
Lecce, Lecce, Italy

ISSN 2543-0599 ISSN 2543-0602 (electronic)
Developments in Applied Phycology
ISBN 978-3-031-86432-2 ISBN 978-3-031-86433-9 (eBook)
https://doi.org/10.1007/978-3-031-86433-9

This Springer imprint is published by the registered company Springer Nature Switzerland AG
The registered company address is: Gewerbestrasse 11, 6330 Cham, Switzerland

Santos Ballardo David Ulises dedicates this book to Marcos and Lupita (his parents) and especially to Dulce, Alana, and Ulises (the motor of his life)

Sergio Rossi dedicates this book to his family, always there for him

Preface

This book contains a wide compendium of information related to microalgae production and its use as commercial resources and also includes microalgal culture basics and the newest scientific and industrial advances in microalgal applications. The book is divided into seven parts (fundamentals of microalgae cultures, microalgae as bioenergy products, microalgae as source of edible products, health products developed from microalgal biomass, cosmetic applications from microalgae, alternative high value application for microalgae biomass, techno-economical approaches applied to microalgal technologies and products), which enclose traditional uses such as renewable energy biofuels, wastewater treatment, edible products, microalgae harvesting techniques. It will also present emerging perspectives of microalgae valorization, for example, health/cosmetic products, bioactive proteins and peptides, pigments, polyunsaturated fatty acids, bioactive polysaccharides, biofertilizers, bioplastics, nanotechnology, and enzymes bio-factories; the importance of the techno-economical analysis is also discussed.

Moreover, the book provides a comprehensive overview for those readers who are initiating the study about the microalgae potential as a renewable source of valuable commercial products and highlights the state of the art on innovative advances for scientific researchers, engineers, students, and microalgae commercial producers, who already focus on developing microalgae biotechnology for bioenergy, food, and other applications. Also, the role of microalgae technology in the blue economy is analyzed, an aspect with high relevance in the present and the future time, but which is regularly understudied.

Mazatlán, Sinaloa, Mexico — David Ulises Santos Ballardo
Lecce, Lecce, Italy — Sergio Rossi

Contents

Editors and Contributors

About the Editors

David Ulises Santos Ballardo, is a research scientist at the Research Center of Food and Development in Mexico, is interested in using second-generation and microalgae biomass to develop biorefineries, including biofuels and other valuable products. He has more than 14 years of field and laboratory experience on this topic, having published more than 27 publications (articles, chapters, books, and divulgation publications, published, in press, or submitted) of research and scientific outreach.

Sergio Rossi is a research scientist specialized in marine natural resources and biological oceanography, focused on health indicators of coastal benthic suspension feeder populations, flows of matter and energy in the benthic-pelagic coupling, the animal forests as carbon sinks, biofuels from algal cultures and the management/conservation of marine wildlife, with also a wide experience in articles publications, and in editing and publishing books in Springer-Nature. He has more than 180 scientific publications (published, in press, or submitted).

Contributors

Ernesto Aguilar-Palazuelos Posgrado en Ciencia y Tecnología de Alimentos, Facultad de Ciencias Químico-Biológicas, Universidad Autónoma de Sinaloa, Culiacan, Sinaloa, Mexico

Liliana Alzate-Gaviria Renewable Energy Division, The Yucatan Scientific Research Center, Merida, Yucatan, Mexico

Dulce L. Ambriz-Pérez Universidad Politécnica de Sinaloa, Mazatlán, Mexico

Estancias Postdoctorales por México SECIHTI, Mexico City, Mexico

Jesús Mateo Amillano-Cisneros Ingeniería en Agrotecnología, Universidad Politécnica del Mar y la Sierra, La Cruz, Sinaloa, Mexico

Andréia Anschau Department of Biotechnology and Bioprocess Engineering, Universidade Federal do Paraná – UTFPR, Paraná, Brazil

Maíra Bueno Ariede Department of Pharmacy, Faculty of Pharmaceutical Sciences, University of São Paulo, São Paulo, SP, Brazil

André Rolim Baby Department of Pharmacy, Faculty of Pharmaceutical Sciences, University of São Paulo, São Paulo, SP, Brazil

Raquel Barrena Research Group on Environmental (Bio)Remediation and Valorization of Effluents (BRAVE) Department of Chemical, Biological and Environmental Engineering, Escola d'Enginyeria, Universitat Autònoma de Barcalona, Barcelona, Spain

Israel Benítez-García Maestría en Ciencias Aplicadas, Unidad Académica de Ingeniería en Biotecnología, Universidad Politécnica de Sinaloa (UPSIN), Mazatlán, Sinaloa, Mexico

Magdalena Berges-Tiznado Universidad Politécnica de Sinaloa, Mazatlán, Mexico

Manuel de Jesús Bernal-Millán Centro de Investigación en Alimentación y Desarrollo, Culiacan, Sinaloa, Mexico

Miguel Betancourt-Lozano Laboratorio de Ecotoxicología, Centro de Investigación en Alimentación y Desarrollo Unidad Mazatlán, Mazatlán, Sinaloa, Mexico

Paqui Blánquez Research Group on Environmental (Bio)Remediation and Valorization of Effluents (BRAVE) Department of Chemical, Biological and Environmental Engineering, Escola d'Enginyeria, Universitat Autònoma de Barcalona, Barcelona, Spain

Carolina Bojórquez-Sánchez Universidad Politécnica de Sinaloa, Mazatlán, Mexico

Luis Angel Cabanillas-Bojórquez Postdoc CONAHCYT-Centro de Investigación en Alimentación y Desarrollo, Culiacan, Sinaloa, Mexico

Centro de Investigación en Alimentación y Desarrollo, Culiacan, Sinaloa, Mexico

Blondy Canto-Canche Biotechnology Division, The Yucatan Scientific Research Center, Merida, Yucatan, Mexico

Feliznando Isidro Cárdenas-Torres Posgrado de Ciencias de la Nutrición y Alimentos Medicinales, Facultad de Ciencias de la Nutrición y Gastronomía, Universidad Autónoma de Sinaloa, Culiacan, Sinaloa, Mexico

Karen Itzel Carrillo-Cañedo Posgrado en Ciencia y Tecnología de Alimentos, Facultad de Ciencias Químico-Biológicas, Universidad Autónoma de Sinaloa, Culiacan, Sinaloa, Mexico

Maritza E. Cervantes-Gaxiola Department of Chemical Engineering, Facultad de Ciencias Químico-Biológicas, Universidad Autónoma de Sinaloa, Culiacán, Sinaloa, Mexico

Laura Aracely Contreras-Angulo Centro de Investigación en Alimentación y Desarrollo, Culiacan, Sinaloa, Mexico

Ada Rebeca Contreras-Rodríguez Unidad Académica de Ciencia y Tecnología de la Luz y la Materia, Universidad Autónoma de Zacatecas, Zacatecas, Mexico

Jorge Alberto Vieira Costa School of Chemistry and Food, Federal University of Rio Grande, Rio Grande, RS, Brazil

Ana Priscila Centeno da Rosa School of Chemistry and Food, Federal University of Rio Grande, Rio Grande, RS, Brazil

Gabriela Moura Kurowiski de Brito School of Chemistry and Food, Federal University of Rio Grande, Rio Grande, RS, Brazil

João Carlos Monteiro de Carvalho Biochemical and Pharmaceutical Technology Department, Faculty of Pharmaceutical Sciences, University of São Paulo, São Paulo, SP, Brazil

Débora Bohrer de Lima Department of Biotechnology and Bioprocess Engineering, Universidade Federal do Paraná – UTFPR, Paraná, Brazil

Michele Greque de Morais School of Chemistry and Food, Federal University of Rio Grande, Rio Grande, RS, Brazil

Catalina Lugo de Ossa Bioprocess Research Group, Department of Chemical Engineering, Faculty of Engineering, Universidad de Antioquia UdeA, Medellin, Colombia

Sofia de Souza Oliveira Department of Biotechnology and Bioprocess Engineering, Universidade Federal do Paraná – UTFPR, Paraná, Brazil

Carolina G. Delgado-Álvarez Universidad Politécnica de Sinaloa, Mazatlán, Mexico

Estancias Postdoctorales por México SECIHTI, Mexico City, Mexico

Mariany Costa Deprá Bioprocess Intensification Group, Federal University of Santa Maria, UFSM, Santa Maria, RS, Brazil

Rosangela Rodrigues Dias Bioprocess Intensification Group, Federal University of Santa Maria, UFSM, Santa Maria, RS, Brazil

Roberto Gutiérrez Dorado Food Science and Technology Graduate Program Faculty of Chemical and Biological Sciences (FCQB) of the Autonomous University of Sinaloa (UAS), University City, Culiacan, Sinaloa, Mexico

Biotechnology Graduate Integral Program, Faculty of Chemical and Biological Sciences (FCQB) of the Autonomous University of Sinaloa (UAS), University City, Culiacan, Sinaloa, Mexico

Darissa Alves Dutra Bioprocess Intensification Group, Federal University of Santa Maria, UFSM, Santa Maria, RS, Brazil

Elda Espana-Gamboa Renewable Energy Division, The Yucatan Scientific Research Center, Merida, Yucatan, Mexico

Ana F. Esteves LEPABE – Laboratory for Process Engineering, Environment, Biotechnology and Energy, Faculty of Engineering, University of Porto, Porto, Portugal

ALiCE – Associate Laboratory in Chemical Engineering, Faculty of Engineering, University of Porto, Porto, Portugal

Perla Rosa Fitch-Vargas Facultad de Ciencias del Mar, Universidad Autónoma de Sinaloa, Mazatlán, Sinaloa, Mexico

Xavier Font Research Group on Environmental (Bio)Remediation and Valorization of Effluents (BRAVE) Department of Chemical, Biological and Environmental Engineering, Escola d'Enginyeria, Universitat Autònoma de Barcalona, Barcelona, Spain

Laura Marina Franco-Herrera Ministry of Science, Humanities, Technology, and Innovation-SEICHTI, Jalisco State Center for Research and Assistance in Technology and Design-CIATEJ, Industrial Biotechnology, Zapopan, Mexico

Miguel Angel Franco-Nava National Technological Institute of Mexico, Technological Institute of Mazatlan (ITMAZ-TecNM), Bioengineering Laboratory, Mazatlán, Mexico

María Anel Fuentes-Valencia Ingeniería en Agrotecnología, Universidad Politécnica del Mar y la Sierra, La Cruz, Sinaloa, Mexico

Andrea de Jesús Gárate-Osuna Programa Regional de Posgrado en Biotecnología, Facultad de Ciencias Químico-Biológicas, Universidad Autónoma de Sinaloa, Culiacan, Sinaloa, Mexico

Melissa García-Carrasco Centro de Investigación en Alimentación y Desarrollo, Culiacan, Sinaloa, Mexico

Jose Manuel García-Padilla Laboratorio de Biotecnología e Ingeniería Genética, Facultad de Ciencias Químico-Biológicas, Universidad Autónoma de Sinaloa, Culiacan, Sinaloa, Mexico

Lourdes Janeth Germán-Báez Laboratorio de Biotecnología e Ingeniería Genética, Facultad de Ciencias Químico-Biológicas, Universidad Autónoma de Sinaloa, Culiacan, Sinaloa, Mexico

Amol D. Gholap Department of Pharmaceutics St. John Institute of Pharmacy and Research, Palghar, Maharashtra, India

Mario Armando Gómez-Favela Programa de Posgrado en Ciencia y Tecnología de Alimentos, Facultad de Ciencias Químico Biológicas, Universidad Autónoma de Sinaloa, Culiacan, Sinaloa, Mexico

Posgrado de Ciencias de la Nutrición y Alimentos Medicinales, Facultad de Ciencias de la Nutrición y Gastronomía, Universidad Autónoma de Sinaloa, Culiacan, Sinaloa, Mexico

Salvador Gómez-Jiménez Maestría en Ciencia e Ingeniería de Materiales de la Unidad Académica de Ingeniería I, Universidad Autónoma de Zacatecas, Zacatecas, Mexico

Ana L. Gonçalves LEPABE – Laboratory for Process Engineering, Environment, Biotechnology and Energy, Faculty of Engineering, University of Porto, Porto, Portugal

ALiCE – Associate Laboratory in Chemical Engineering, Faculty of Engineering, University of Porto, Porto, Portugal

Julián González-Trinidad Unidad Académica de Ciencia y Tecnología de la Luz y la Materia, Universidad Autónoma de Zacatecas, Zacatecas, Mexico

Marcos D. González-Llanes Department of Chemical Engineering, Facultad de Ciencias Químico-Biológicas, Universidad Autónoma de Sinaloa, Culiacán, Sinaloa, Mexico

Miriam Domingues Guimarães Department of Biotechnology and Bioprocess Engineering, Universidade Federal do Paraná – UTFPR, Paraná, Brazil

Erick Paúl Gutiérrez-Grijalva Programa Investigadoras e investigadores por México SECIHTI-Centro de Investigación en Alimentación y Desarrollo, Culiacan, Sinaloa, Mexico

José Basilio Heredia Centro de Investigación en Alimentación y Desarrollo, Culiacan, Sinaloa, Mexico

Oscar M. Hernández-Calderón Department of Chemical Engineering, Facultad de Ciencias Químico-Biológicas, Universidad Autónoma de Sinaloa, Culiacán, Sinaloa, Mexico

Eduardo Jacob-Lopes Bioprocess Intensification Group, Federal University of Santa Maria, UFSM, Santa Maria, RS, Brazil

Jennifer Ortiz-Letechipia Unidad Académica de Ciencias Químicas, Universidad Autónoma de Zacatecas, Zacatecas, Mexico

María G. Laguna-Martínez Department of Chemical Engineering, Facultad de Ciencias Químico-Biológicas, Universidad Autónoma de Sinaloa, Culiacán, Sinaloa, Mexico

Nayely Leyva-López Postdoc CONAHCYT-Centro de Investigación en Alimentación y Desarrollo, Culiacan, Sinaloa, Mexico

Centro de Investigación en Alimentación y Desarrollo, Culiacan, Sinaloa, Mexico

José Belisario Leyva-Morales Área Académica de Química, Instituto de Ciencias Básicas e Ingeniería, Universidad Autónoma del Estado de Hidalgo, Mineral de la Reforma, Hidalgo, Mexico

Nidia Araiza Lizarde Ingeniería en Biotecnología Universidad Politécnica de Sinaloa, Mazatlán, Sinaloa, Mexico

Karla Denisse Luna-Avelar Estancias posdoctorales por México, SECIHTI, Mexico City, Mexico

Laboratorio de Ecotoxicología y Bioprocesos, Centro de Investigación en Alimentación y Desarrollo Unidad Mazatlán, Mazatlan, Sinaloa, Mexico

Bruno Roswag Machado School of Chemistry and Food, Federal University of Rio Grande, Rio Grande, RS, Brazil

Cláudia A. Machado LEPABE – Laboratory for Process Engineering, Environment, Biotechnology and Energy, Faculty of Engineering, University of Porto, Porto, Portugal

ALiCE – Associate Laboratory in Chemical Engineering, Faculty of Engineering, University of Porto, Porto, Portugal

Carolina Maia LEPABE – Laboratory for Process Engineering, Environment, Biotechnology and Energy, Faculty of Engineering, University of Porto, Porto, Portugal

ALiCE – Associate Laboratory in Chemical Engineering, Faculty of Engineering, University of Porto, Porto, Portugal

Henri Márquez-Pacheco Maestría en Biotecnología Agropecuaria, Universidad Politécnica del Mar y la Sierra, La Cruz, Elota, Sinaloa, Mexico

Emmanuel Martínez-Montaño Maestría en Ciencias Aplicadas, Unidad Académica de Ingeniería en Biotecnología, Universidad Politécnica de Sinaloa (UPSIN), Mazatlán, Sinaloa, Mexico

Investigadores por México CONAHCyT, Consejo Nacional de Humanidades, Ciencia y Tecnología, Ciudad de México, Mexico

Renata Miliani Martinez Department of Pharmacy, Faculty of Pharmaceutical Sciences, University of São Paulo, São Paulo, SP, Brazil

Jennifer Viancy Félix Medina CONACYT-Agroindustrial Engineering, Polytechnic University of Sea and the Sierra, La Cruz, Elota, Sinaloa, Mexico

National Technological of Mexico (TecNM), Technological Institute of Culiacan (ITC), Culiacan, Sinaloa, Mexico

Karelia Arisbeth Meza-Ayala Laboratorio de Biotecnología e Ingeniería Genética, Facultad de Ciencias Químico-Biológicas, Universidad Autónoma de Sinaloa, Culiacan, Sinaloa, Mexico

Paula Fernandes Montanher Department of Biotechnology and Bioprocess Engineering, Universidade Federal do Paraná – UTFPR, Paraná, Brazil

Omar Nateras-Ramírez Instituto Tecnológico de Mazatlán, Instituto Tecnológico de Mexico, Mazatlán, Sinaloa, Mexico

Luz Adriana Ontiveros-García Ingeniería en Agrotecnología, Universidad Politécnica del Mar y la Sierra, La Cruz, Sinaloa, Mexico

Jacqueline Stephanie Ortiz-Letechipia Doctorado en Ingeniería para la Innovación Tecnológica, Unidad Académica de Ingeniería Eléctrica, Universidad Autónoma de Zacatecas, Zacatecas, Mexico

Jesús R. Ortiz-del-Castillo Department of Chemical Engineering, Facultad de Ciencias Químico-Biológicas, Universidad Autónoma de Sinaloa, Culiacán, Sinaloa, Mexico

Catalina Quevedo Ospina Bioprocess Research Group, Department of Chemical Engineering, Faculty of Engineering, Universidad de Antioquia UdeA, Medellin, Colombia

Carmen C. Osuna-Martínez Facultad de Ciencias del Mar, Universidad Autónoma de Sinaloa, Mazatlán, Mexico

Idalia Osuna-Ruiz Maestría en Ciencias Aplicadas, Unidad Académica de Ingeniería en Biotecnología, Universidad Politécnica de Sinaloa (UPSIN), Mazatlán, Sinaloa, Mexico

Rodrigo T. Patino-Diaz The Center for Research and Advanced Studies of the National Polytechnic Institute, Cinvestav – Unidad Merida, Merida, Yucatan, Mexico

Xiomara Patricia Perea-Domínguez Doctorado en Sustentabilidad, Universidad Autónoma de Occidente, Unidad Regional Guasave, Guasave, Sinaloa, Mexico

Janitzio Xiomara Korina Perales Sánchez Food Science and Technology Graduate Program Faculty of Chemical and Biological Sciences (FCQB) of the Autonomous University of Sinaloa (UAS), University City, Culiacan, Sinaloa, Mexico

Biotechnology Graduate Integral Program, Faculty of Chemical and Biological Sciences (FCQB) of the Autonomous University of Sinaloa (UAS), University City, Culiacan, Sinaloa, Mexico

Hugo Virgilio Perales-Vela Laboratorio de Bioquimica, Unidad de Morfologia y Funcion, Facultad de Estudios Superiores Iztacala, Universidad Nacional Autonoma de Mexico, Tlalnepantla de Baz, Estado de Mexico, Mexico

Fabiana Vieira Lima Solino Pessoa Department of Health Sciences, Federal University of Espírito Santo, São Mateus, Espírito Santo, ES, Brazil

Mariana Santos Pinheiro Department of Health Sciences, Federal University of Espírito Santo, São Mateus, Espírito Santo, ES, Brazil

José C. M. Pires LEPABE – Laboratory for Process Engineering, Environment, Biotechnology and Energy, Faculty of Engineering, University of Porto, Porto, Portugal

ALiCE – Associate Laboratory in Chemical Engineering, Faculty of Engineering, University of Porto, Porto, Portugal

Evelyn Pereira Ribeiro Garcia Department of Biotechnology and Bioprocess Engineering, Universidade Federal do Paraná – UTFPR, Paraná, Brazil

Gissel Daniela Ríos-Herrera Estancias Posdoctorales por México CONAHCyT, Consejo Nacional de Humanidades, Ciencia y Tecnología, Ciudad de México, México/Universidad Politécnica de Sinaloa (UPSIN), Mazatlán, Mexico

Universidad Politécnica del Mar y la Sierra, La Cruz de Elota, Sinaloa, Mexico

Erika Y. Rios-Iribe Programa de Posgrado Integral en Biotecnología, Facultad de Ciencias Químico Biológicas, Universidad Autónoma de Sinaloa, Culiacán, Sinaloa, México

Sergio Rossi DiSTeBA, University of Salento, Lecce, Italy

Institute of Marine Sciences (LABOMAR), Federal University of Ceará, Fortaleza, CE, Brazil

Underwater Gardens International, Barcelona, Spain

Eusiel Rubio-Castro Department of Chemical Engineering, Facultad de Ciencias Químico-Biológicas, Universidad Autónoma de Sinaloa, Culiacán, Sinaloa, Mexico

Xóchitl Ariadna Ruiz-Armenta Programa de Posgrado en Ciencia y Tecnología de Alimentos, Facultad de Ciencias Químico Biológicas, Universidad Autónoma de Sinaloa, Culiacan, Sinaloa, Mexico

Jesús Aaron Salazar-Leyva Maestría en Ciencias Aplicadas, Unidad Académica de Ingeniería en Biotecnología, Universidad Politécnica de Sinaloa (UPSIN), Mazatlán, Sinaloa, Mexico

Eva M. Salgado LEPABE – Laboratory for Process Engineering, Environment, Biotechnology and Energy, Faculty of Engineering, University of Porto, Porto, Portugal

ALiCE – Associate Laboratory in Chemical Engineering, Faculty of Engineering, University of Porto, Porto, Portugal

Ana Luisa Félix Sámano Food Science and Technology Graduate Program Faculty of Chemical and Biological Sciences (FCQB) of the Autonomous University of Sinaloa (UAS), University City, Culiacan, Sinaloa, Mexico

Antoni Sánchez Research Group on Environmental (Bio)Remediation and Valorization of Effluents (BRAVE) Department of Chemical, Biological and Environmental Engineering, Escola d'Enginyeria, Universitat Autònoma de Barcalona, Barcelona, Spain

Jorge Manuel Sandoval-Gallardo Facultad de Ingeniería y Tecnología de Mazatlán, Universidad Autónoma de Sinaloa, Mazatlán, Sinaloa, Mexico

David Ulises Santos-Ballardo Laboratorio de Ecotoxicología, Centro de Investigación en Alimentación y Desarrollo A.C, Mazatlán, Mexico

Lucielen Oliveira Santos School of Chemistry and Food, Federal University of Rio Grande, Rio Grande, RS, Brazil

Ana Margarita Silva-Benavides Centro de Investigacion en Ciencias Del Mar y Limnología, Universidad de Costa Rica, San Pedro, San Jose, Costa Rica

Izabela Queiroz Silva School of Chemistry and Food, Federal University of Rio Grande, Rio Grande, RS, Brazil

Pedro Garcia Pereira Silva School of Chemistry and Food, Federal University of Rio Grande, Rio Grande, RS, Brazil

Giuseppe Torzillo Institute of BioEconomy, IBE, CNR, Florence, Italy

Centro de Investigacion en Ciencias Del Mar y Limnología, Universidad de Costa Rica, San Pedro, San Jose, Costa Rica

Ruby Alejandra Valdez-Ojeda Renewable Energy Division, The Yucatan Scientific Research Center, Merida, Yucatan, Mexico

Angel Valdez-Ortiz Laboratorio de Biotecnología e Ingeniería Genética, Facultad de Ciencias Químico-Biológicas, Universidad Autónoma de Sinaloa, Culiacán, Sinaloa, Mexico

Rafael Valdez-Ortiz Service of Nephrology, General Hospital of Mexico, Dr. Eduardo Liceaga, Mexico City, Mexico

Mariana Peñuela Vásquez Bioprocess Research Group, Department of Chemical Engineering, Faculty of Engineering, Universidad de Antioquia UdeA, Medellin, Colombia

Claudia Amezcua Vega Ingeniería en Biotecnología Universidad Politécnica de Sinaloa, Mazatlán, Sinaloa, Mexico

Munise Zaparoli Bioprocess Engineering and Biotechnology Department, Federal University of Paraná, Curitiba, PR, Brazil

Leila Queiroz Zepka Bioprocess Intensification Group, Federal University of Santa Maria, UFSM, Santa Maria, RS, Brazil

Part I

Fundamentals of Microalgae Cultures

1 The Beginning to Integration: Evolution and Context of Microalgae Biotechnology

Darissa Alves Dutra, Rosangela Rodrigues Dias, Mariany Costa Deprá, Leila Queiroz Zepka, and Eduardo Jacob-Lopes

Abstract

Microalgae biotechnology is an emerging field of research and a viable option for developing environmentally friendly technologies. The immense potential of the biotechnological applications based on different microalgae has fostered a wave of research on their potential as a sustainable resource for products and solutions for human and animal nutrition, aquaculture, agriculture, biofuels, and emergent areas such as cosmetics and pharmaceuticals. "In this context", the market is still developing for most of these applications, and intensive research is necessary to unlock microalgae biotechnology's full potential. In light of it all, we present a retrospective on microalgae biotechnology in this chapter. Microalgae-based products and solutions that have overcome the valley of death and those that have not yet crossed it are addressed. An overview of microalgae biotechnology's various promises and limitations is also presented.

Keywords

Algae · Biomass · Bioproducts

1.1 Introduction

In a world increasingly concerned about sustainability, microalgae emerge as stars, offering hope and concrete solutions for a greener future. These photosynthetic microorganisms are the most numerous in aquatic environments; around 72,500 species are consistently cataloged. Microalgae biotechnology proves that minor living things can have the most significant positive impact, providing us with powerful tools to solve global challenges ecologically consciously (Borowitzka 2015; Queiroz et al. 2020).

If we take some time to analyze the world as it is today, much still needs to be done so that the effects of anthropogenic actions are not irreversible; with the environmental crisis, the word sustainability emerges as a significant agenda of discussions in the pillars of society, academia, industry, and politics. It is becoming increasingly urgent to create sustainable plans. On the face of it, finding economically viable resources to overcome traditional models based on fossil inputs raises the attention of the pillars of society to microalgae as an escape valve. As an area on the rise, microalgae biotechnology is increasingly gaining commercial space due to its potential for exploration. This green gold is a significant and valuable source of biologically active compounds and applies to several industrial sectors (Chisti 2020).

The constituents of the raw material of microalgae are proteins, lipids, and carbohydrates, which promote a biomass of expansive usefulness. Their main application is in the chemical, pharmaceutical, cosmetic, and food sectors (Deprá et al. 2022). All these market possibilities occur due to their high yield per area, giving these microorganisms one of the most attractive characteristics compared to other traditional crops—their cultivation does not compete with arable land to produce food, biofuels, or waste treatment (Severo et al. 2017). It is worth highlighting that even though microalgae are considered a technological bomb, there are still several obstacles to be overcome, mainly in the upstream and downstream processes, with emphasis on the high energy demand required in the production of these microorganisms, which means that there is still a great struggle to compete with existing traditional practices (Agbede et al. 2020).

The issue of climate change is a highly pressing concern that has far-reaching implications for the future of our planet. In light of this, microalgae biotechnology must form part of humanity's reckoning with the Earth, making the necessary adjustments to meet the demands of producing emission-neutral products included in net zero. The current chapter

D. A. Dutra · R. R. Dias · M. C. Deprá · L. Q. Zepka
E. Jacob-Lopes (✉)
Bioprocess Intensification Group, Federal University of Santa Maria, UFSM, Santa Maria, RS, Brazil

D. U. Santos Ballardo, S. Rossi (eds.), *Microalgae as Promising Source of Commercial Bioproducts*, Developments in Applied Phycology 14, https://doi.org/10.1007/978-3-031-86433-9_1

provides an overview of the advancements achieved in microalgae biotechnology.

1.2 A Retrospective on Microalgae Biotechnology

Biotechnology has a rich history of serving multiple purposes. Microalgae cultivation is one of the emerging forms of biotechnology that stands out as a promising option for the future (Mozafari et al. 2020). Microalgae—known as the most primitive form of life on Earth—have incredible biodiversity, and their biomass results from the variability in the biochemical composition of microalgae cultures. When combined with genetic improvement and the establishment of large-scale cultivation technology, certain species are allowed to be commercially used (Rodrigues et al. 2018). The historical timeline of algae biomass production has existed for more than 100 years; initially, indigenous peoples extracted the biomass without commercialization and used it only for food consumption (Jeevanandam et al. 2020).

Historically, the kickoff in the commercial culture of microalgae began in the 1960s with *Chlorella*, and after a decade, *Arthrospira platensis* was cultivated. During the early stages of microalgae culture exploration, the industry was primarily interested in protein as a potential ingredient for preventative food. Going forward a little in time, in 1980, the commercial production of *Dunaliella* and *Haematococcus* became significant in the industrial trade of microalgae; their manufacture was directed toward the production of pigments, with a focus on beta (β)-carotene and astaxanthin, used for additive food and animal feed. At the beginning of the 1990s, the production of polyunsaturated fatty acids was consolidated, especially docosahexaenoic acid (DHA) and eicosapentaenoic acid (EPA), intended for use in aquaculture feed and, in some cases, as an improvement in dietary food products. *Nannochloropsis*, *Nitzschia*, *Schizochytrium*, and *Crypthecodinium* were commonly used (Jacob-Lopes et al. 2019; Chisti 2020).

We have been learning about the physiology, growth, and production of microalgae for over 50 years (Hlavova et al. 2015). Nowadays, the leading producers of microalgae-based products are Taiwan, Japan, the USA, China, Brazil, Spain, Israel, Germany, and Myanmar. Giving rise to approximately US$ 6 billion, the annual production of microalgae dry biomass can reach 20,000 tons. Due to their remarkable metabolic plasticity and several chemical compounds within their cells, microalgae have been the commercial box office burst in the biotechnology segment. With their rise, these green microorganisms became a valuable solution to provide continuity to global ecosystems and enable applications in different technological fields (De Carvalho Silvello et al. 2022). From this perspective, studies have started to trace the path with the objective of applying microalgae as raw materials in different technological fields (Min et al. 2022).

The idea of clean technologies has extended to microalgae biotechnology, mainly because these microorganisms are effective tools for environmental processes such as wastewater bioremediation and their ability to mitigate gaseous pollutants—the capture of CO_2 (Carraro et al. 2022).

After many bottlenecks emerged during commercial applications of microalgae, research changed the purpose of the applications. To improve microalgae production, it has been necessary to study its biological principles, seeking to improve the production stages. Since then, the academy has joined forces focused on advances in the genetic engineering of microalgae, trying to promote better development upstream and downstream. The great challenge observed was the energy requirement in these processes (Deprá et al. 2020; Muthukrishnan 2022).

There is no doubt that microalgae biotechnology is well-established today. Nor can we refute their particularity as products that are less harmful to human health, which have been in increasing demand in recent years, thanks to the awakening of the population in search of more beneficial products. Unfortunately, it must be said that most production routes for microalgae-based products are only for a single product. This action increases the difficulty of reaching the splendor of microalgae biotechnology. Adopting a multi-component approach in microalgae biorefinery can economically valorize the offered biological resources (Jacob-Lopes and Zepka 2023). The application of this technology united through efforts may, 1 day, transform microalgae into a commodity.

1.3 From the Petri Dish to the High-Value Commercial Products

Currently, the market for microalgae-based products is mainly dominated by *Chlorella* and *Arthrospira platensis*, sold as whole dry biomass, which is a type of unicellular protein. These microalgae generate bioactive products that are widely used in several applications, including natural dyes, food additives, and supplements. The market for bio-products generated by microalgae is poised for significant growth in the future, given the wide range of applications and the immense health benefits offered by these bioproducts. The expansion of microalgae-based products is demonstrated in Table 1.1. In recent times, the focus on producing sustainable and commercially viable biofuels has increased significantly. Consequently, there was an increasing interest in microalgae, considered abundant sources of valuable metabolites. Scientists are working to extract valuable substances from microalgae to help cover the production expenses. Microalgae are also valuable for creating bioprod-

Table 1.1 A short list of microalgae-based products that have production potential in the market

Products	Microalgae	Description	Application	Reference
Algae protein bars	*Chlorella vulgaris*	Nutrient-rich protein bars made from microalgae	Nutritional supplements, snacks	Fanari et al. (2023)
Algae biofuel	*A. platensis*	Sustainable biofuels produced from microalgae	Transportation, energy sector	Li et al. (2023)
Algae-based cosmetics	*Haematococcus pluvialis*	Skin and hair care products with microalgae extracts	Cosmetics and personal care	Indarto et al. (2023)
Algae-based fertilizers	*Chlorella* sp.	Organic fertilizers enriched with microalgae nutrients	Packaging, manufacturing	Ammar et al. (2022)
Animal feed	*Aurantiochytrium limacinum*	Animal supplementation	Enhancement of the lipid profile in swine tissues with DHA enrichment	Moran et al. (2018)

ucts and kick-starting their circular economy (Dias et al. 2020; Ramírez-Mérida 2023).

In addition to generating dry biomass that serves as an abundant source of unicellular protein, it also produces eicosapentaenoic acid (EPA), docosahexaenoic acid (DHA), beta-carotene, astaxanthin, and phycocyanin. These natural substances are commonly used as organic pigments, dietary supplements, and food additives, contributing to the improvement of various industries (Jacob-Lopes et al. 2019). Typically, the production costs of metabolite compounds from microalgae surpass those of traditional sources. This is what makes the process less attractive to investors' pockets. However, the exceptional benefits of these metabolites make them more advantageous in an industrial context. An excellent example is phycocyanin, the only natural blue dye suitable for use (Dias et al. 2020).

The cost of producing microalgae-based products is often high because most of these products are intracellular, which require costly extraction and purification processes. Polyunsaturated fatty acids and pigments are high-value products that require more complicated processing, resulting in higher costs. These elevated prices are due to intracellular products requiring expensive extraction and purification procedures. In terms of outlay, the downstream processes to produce total dry biomass involve only around 2%, while for polyunsaturated fatty acids and pigment extractions, the percentages are around 70% and 85%, respectively (Jacob-Lopes et al. 2019).

Ensuring product quality can increase the chances of microalgae-based products establishing themselves in the market, ensuring future investments. Therefore, companies should pay attention to the quality of the products they choose to produce—whole dry biomass or intracellular products. As these products are so sensitive, attention is essential to every part of their production. It is worrying that microalgae biotechnology requires more attention to improve metabolic processes and modernize extraction methods in the age of technology. It is essential to highlight that several microalgae are currently under review and are advancing through different stages of development, in addition to those already used for commercial purposes (Dias et al. 2020). However, the clock is ticking in relation to the range of applications that microalgae biotechnology can achieve. On the other hand, an awakening to this bottleneck needs to receive the attention it deserves so that we will be taking baby steps with microalgae biotechnology.

For example, when unicellular protein produced from microalgae is mentioned, it is consolidated that *Chlorella* and *Arthrospira platensis* (*Spirulina*) are champions for production in the market. This product, which humans usually consume as a dietary supplement, is not subjected to any additional processing besides drying.

Another example of a microalgae-based product consolidated on the market is recombinant protein, which is used more in the production of vaccines. Due to the cellular structure of microalgae, which have chloroplasts, these microorganisms are more susceptible to producing complex recombinant proteins (Camacho et al. 2019).

Studies show that *Chlamydomonas reinhardtii* and *Schizochytrium* sp. are potential sources for producing pharmaceutical proteins and vaccines against *Staphylococcus aureus* and Zika virus. Moreover, microalgae have been proven to trigger mucosal and systemic humoral immune responses. Besides, recombinant proteins have a variety of applications, including cost-effective Human Papillomavirus (HPV) vaccines. Functional antibodies are more expensive to produce in mammalian cells, while plants have lower estimated costs. Therefore, using microalgae becomes more attractive due to the cost–benefit ratio of recombinant proteins applied in the industrial vaccine sector (Dias et al. 2020; Calijuri et al. 2022).

Continuing with products already in advanced development based on microalgae, phycoerythrin is a pigment belonging to phycobiliproteins, commonly extracted from red microalgae. It is considered one of the brightest natural fluorophores. As a natural pigment, phycoerythrin has bioactive compounds that are beneficial to human health. Some studies have recently shown that β-phycoerythrin from *Porphyridium cruentum* species has a standard application as an indicator in quantitative assays and can also promote the reduction of tumor cell myeloid in vitro.

Furthermore, fucoxanthin is part of the group of carotenoids—not provitamin A, which gives brown microalgae their characteristic color. These molecules are related to fucoxanthin antioxidant, anti-inflammatory, anticancer, and anti-obesity activities; these compounds also present interesting configurations for human health mainly because they do not produce harmful actions. Currently, the microalgae most used to obtain fucoxanthin is *Phaeodactylum tricornutum*. The main application of this substance is as a dietary supplement for humans (Wang et al. 2021; Ji et al. 2023).

In products containing carotenoids, lutein provides a lemon-yellow color. Since the human body does not produce this compound, it must be consumed through the diet. Some studies have shown that lutein can be a carrier of antioxidant and anti-inflammatory properties, helping prevent atherosclerosis and reducing the chances of macular degeneration and cardiovascular diseases. The global lutein market, valued at USD 308 million in 2018, reached USD 396.4 million in 2024 due to the demand for its various applications. Lutein is commonly used as a colorant, food additive, and eye health supplement.

Currently, lutein is mainly obtained from marigold flowers, but it is already confirmed that some microalgae strains are also interesting lutein sources, such as *Chlorella vulgaris, Muriellopsis sp., Scenedesmus sp., Desmodesmus sp., Chromochloris zofingiensis, and Chlorella protothecoides* (Global 2019; Bhatia et al. 2020).

The market currently offers various microalgae-based products, with even more in development. Although some of these products are still in the early stages, they hold great potential to become commercially available in the coming years.

1.4 Application Areas of Microalgae Biotechnology

What exciting developments are being made in microalgae biotechnology? While already established commercial applications exist, more potential remains to be unlocked. These commercial applications include biofuels, polyhydroxyalkanoates (PHAs), enzymes, zeaxanthin, prebiotics, phenolics, and other compounds. Recently, there has been a growing interest in producing biofuels as alternatives to fossil fuels. Microalgae have gained attention as potential sources of biofuels due to their high lipid content in the oil extracted from their biomass. This oil can be used to produce biodiesel and biohydrogen. Recent research has highlighted the production of biodiesel and biomethane from the microalgae *Chlorella sorokiniana* (Constantino et al. 2021). The production of biofuels is quite challenging, especially after going through an in-depth analysis of their current state. Their production has difficulties, making it a complex task to carry out. This is especially true when considering the various factors that come into play in the market (Deprá et al. 2018; Adeniyi et al. 2018). Exciting advancements in biotechnology have made microalgae promising alternatives to oil as a raw material, providing much-needed hope for a sustainable future.

Another perspective for the application of microalgae biotechnology is the growing concern about the plastic pollution crisis around the world, which has resulted in a renewed effort to discover biodegradable plastics. Polyhydroxyalkanoates (PHAs) are promising alternatives to petroleum-based plastics that can be regenerative and have the ability to decompose fully in soil and marine environments. The exorbitant cost of PHAs compared to conventional plastics serves as a significant obstacle to their widespread application. Recent research has shown that microalgae biomass is an economical option for use as a raw material in the synthesis of PHAs. Microalgae biomass has been studied as a carbon source for producing PHAs due to its high carbohydrate yield and lack of lignin. Microalgal technology makes it simpler to recover fermentable sugars at a reduced cost. Consequently, using microalgae biomass as a raw material for synthesizing polyhydroxyalkanoates is an ideal economic choice. A study with the *Scenedesmus* sp. strain revealed an accumulation of 8.61% of PHA (Tan et al. 2022). Besides, García et al. (2020) reported that this compound has a concentration of 29.92% in dry mass, using *Scenedesmus* sp., while its lowest concentration was 0.83%.

On the other hand, microalgae biotechnology is also a valuable source for producing commercially important enzymes, which can serve as potent biocatalysts in various industrial applications. Cellulases, galactosidases, proteases, lipases, phytases, laccases, and amylases are among the enzymes that can be derived from microalgae. Despite the absence of any commercial production of microalgae-based enzymes, laboratory studies have demonstrated their remarkable potential to synthesize these substances (García and de Llasera 2021; De Llasera et al. 2022). In order to minimize operating costs and enhance enzyme concentration, new and inventive strategies are currently being developed. The biorefinery concept is being explored as a practical approach to overcome current limitations and soon pave the way for enzyme production (Dias et al. 2020).

Macular pigment, also known as zeaxanthin, is produced by microalgae such as *Dunaliella*, *Arthrospira platensis*, *Synechococcus*, *Chlorella*, *Prochlorococcus*, and *Prochlorothrix*. Zeaxanthin is a pigment that belongs to the xanthophylls group of carotenoids. It can be found in many fruits and vegetables and is commonly found in the skin and eyes of humans and animals. Zeaxanthin is known to have antioxidant and anti-inflammatory effects, which help it perform important physiological functions. Numerous studies have explored its potential health benefits, such as its ability to help combat severe disorders like neurological diseases, allergies, and cancer. Producing zeaxanthin from microalgae is a promising alternative for applications such as dyes and food supplements. In this way, biotechnology can work with the help of genetic engineering to make the increase and accumulation of zeaxanthin in microalgae possible, making them commercially viable sources (Mrowicka et al. 2022).

Did you know that microalgae can be a great source of prebiotics? Prebiotics are polysaccharides, including xylo-oligosaccharides, galactooligosaccharides, galactans, and β-glucans. Additionally, microalgae-derived prebiotics have been found to promote the growth of beneficial bacteria in the gut after consumption. Microalgae-based nutritional supplements containing prebiotics like *Arthrospira platensis* and *Chlorella vulgaris* are now commercially available at affordable prices. To fully utilize microalgae as alternative sources of prebiotics, it is necessary to improve extraction methods and analytical techniques used to measure their chemical composition. The enhancement of existing biotechnology can aid in identifying fresh bioactive prebiotics. Improving existing biotechnology can help identify new bioactive prebiotics for use in different areas, such as pharmaceuticals, functional foods, and other scientific applications (Gouda et al. 2022).

Moreover, phenolics are important antioxidants in plants and microalgae. Recent research shows that microalgae extracts contain phenolics contributing to their antioxidant activity. Many food products contain functional ingredients and preservatives that are high in antioxidants. These components are crucial in improving the taste, texture, and shelf life of different foods. Additionally, antioxidants are known for their health benefits as they help protect the body against harmful free radicals (Coulombier et al. 2021). Advancing studies on phenolics through microalgae is taking a step beyond what has already been established about these microorganisms. This is an interesting development that can still be made in microalgae biotechnology.

Microalgae have the potential to become sustainable sources of biomass due to their renewability, ability to mitigate greenhouse gases, and the possibility of being converted into high-value bioproducts to meet different application needs. An overview of the application of products obtained through microalgae biotechnology is presented in Fig. 1.1. In several countries, algae-based products have been promoted as a nutritional supplement to combat malnutrition. Numerous companies in developing countries have implemented measures to ensure the well-being of their inhabitants. Despite the promising potential of microalgae, their commercialization remains constrained by several factors, including inadequate research funding, small-scale operations, and challenges in implementing policies. It is time to turn the corner and reach the splendor of this green gold that is applicable in different industrial scenarios.

1.5 The Promises and Challenges of Microalgae Biotechnology

While there is fascinating scientific research with promising results and potential applications in pharmaceuticals, nutritional supplements, aquaculture food products, and bioremediation solutions, this is not the only factor in realizing commercial applications. High-quality research is crucial to harness the vast diversity of microalgae and innovate in the field of biotechnology. Unfortunately, the microalgal industry is still taking baby steps to achieve more excellent biotechnology development. In fact, the economic factor still carries the most significant weight, mainly because the upstream and downstream processes are still extravagantly expensive. To assist in the majority achievement of microalgae potential, some disruptive technologies are entering the scene, such as the Internet of Things (IoT) and genetic engineering (Dias et al. 2020).

Most microalgae cultivation still relies on manually monitoring growth in open ponds or photobioreactors, but it is time to embrace advances in research and technology. Operators must collect samples from open ponds or photobioreactors daily to measure biomass concentration. This process is time-consuming and can be affected by various environmental factors, such as temperature and pH changes, which can impede microalgae growth. However, these limitations can be solved through IoT integration. IoT integration can monitor and improve microalgae growth and harvest. This is essential to reduce labor and operating costs in upstream bioprocessing (Fabris et al. 2020).

According to recent studies, implementing IoT technology in microalgae cultivation has significantly reduced production costs by up to 30% while increasing overall production by 20% (Zambon et al. 2019). In 2018, Giannino et al. observed that when the Internet of Things was applied to the system, it showed an increase in production of ~9% compared to a system without monitoring. Recent studies have found that integrating IoT technology into microalgae pond systems has the potential to decrease energy use by a staggering 60% (Hermadi et al. 2021).

It is interesting to remember that the energy issue is one of the biggest bottlenecks in producing microalgae, and the high level of energy demand and its reduction in consumption positively impact environmental sustainability. As a result, implementing the Internet of Things in microalgae cultivation successfully achieves more productive biotechnology. Embracing this idea raises the level of biotechnology already established in society. It aims at greater resourcefulness in generating products based on microalgae, reducing

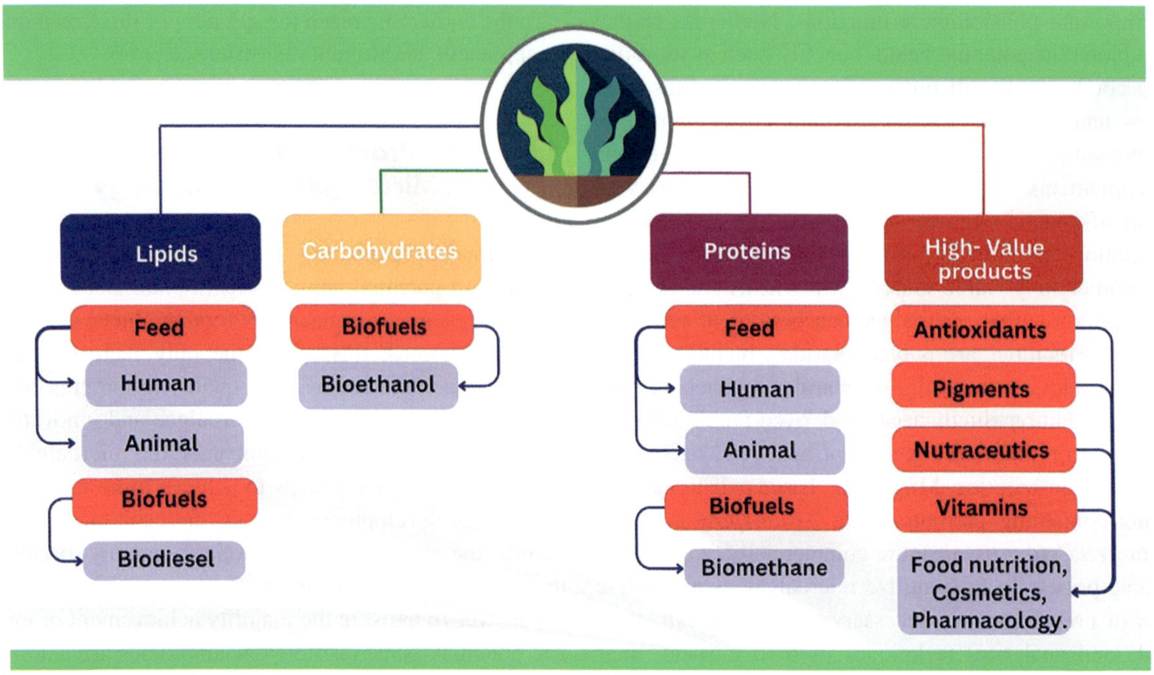

Fig. 1.1 Application of products obtained through microalgae biotechnology

the environmental impact, and making the economic sector more pleasant (Muñoz et al. 2022).

Another way to thrive with microalgae biotechnology is to advance genetic engineering. However, optimizing the cellular metabolism of microalgae is a challenge for biotechnology. Scientists and researchers have made enormous efforts to improve these biological processes (Thanigaivel et al. 2022). Biotechnology has genetic engineering as one of its main pillars. Its set includes molecular analyses, generally through recombination, with the primary objective of synthesizing improved microorganisms, occurring in the improvement or genetic structure of the species under analysis, according to the needs of the research. Carrying out genetic improvement in microalgae cells further expands its applicability in several sectors. Genetic engineering of microalgae can assist in producing pharmaceutical proteins and bioactive metabolites due to their metabolic diversity and ease of genetic modification. They can quickly be grown in bioreactors at rates comparable to Chinese hamster ovary (CHO) cells—commonly used to produce recombinant proteins. The microalgae *Chlamydomonas reinhardtii* is a well-studied species of algae and can be used to produce recombinant proteins (Ramos-Martinez et al. 2017).

It is possible to increase biofuel productivity by 33% using a microalgae strain with a high lipid content. Thanks to recent advances in genetic engineering tools like CRISPR-Cas9 and their optimization for microalgae, targeting specific genes can be done more efficiently. A study on *Chlamydomonas reinhardtii* strains demonstrated a successful increase in the overall lipid productivity of microalgae using CRISPR-Cas9 (Ng et al. 2017; Shin et al. 2019).

Moreover, the role of genetic engineering in the production of carotenoids and biohydrogen also already exists. There are indications in the literature that its application in the microalgae *Chlorella zofingiensis* was able to increase their global production of carotenoids by 32.1%. Furthermore, astaxanthin production increased by 54.1%. Current research explores genetic engineering to increase the efficiency of H_2 photoproduction in microalgae, promoting a better scope of biotechnology in biohydrogen production through microalgae. The result of genetic alteration in *Synechococcus 7002* promoted a fivefold increase in H_2 production compared to wild-type cells (Grama et al. 2022).

The development of safe and efficient genetically modified strains is essential. Production scalability and cost savings are areas for improvement. However, the future is promising. Genetic engineering will continue to drive the production of valuable products, contributing to a more sustainable world. As research advances, the vision is that microalgae will become a core source of numerous biotechnological applications, promoting a greener and more prosperous future.

1.6 Conclusion and Future Perspectives

As we look to the horizon of microalgae biotechnology, we cannot help but marvel at the incredible progress we have witnessed. Over the past few decades, research in this field has expanded our horizons and redefined our expectations regarding the potential of microalgae. The increasing use of these versatile single-celled organisms has opened the door to many innovative applications, from producing biofuels and pharmaceuticals to removing pollutants and contributing to food security. Generally, researchers are optimistic about the future of microalgae biotechnology. We are just beginning to unravel the vast spectrum of possibilities that these tiny organisms can offer. As we continue to deepen our understanding of the genetics, metabolism, and ecology of microalgae, the possibilities of developing sustainable and practical solutions to many of the most pressing global challenges become increasingly accurate.

However, we must maintain a realistic view. Important scientific and technological challenges still need to be overcome to improve the cultivation, harvesting, and processing of microalgae on a large scale. As microalgae biotechnology advances, it offers an exciting glimpse into a more sustainable future, where we can harness natural resources more efficiently, reduce our dependence on fossil fuels, and address environmental and health challenges with increasingly sophisticated solutions. It is a journey that will continue to captivate the mind and inspire the imagination of researchers, offering hope for a better, more balanced world.

References

Adeniyi OM, Azimov U, Burluka A (2018) Algae biofuel: current status and future applications. Renew Sust Energ Rev 90:316–335

Agbede OO, Oke EO, Akinfenwa SI, Wahab KT, Ogundipe S, Aworanti OA et al (2020) Thin layer drying of green microalgae (*Chlorella* sp.) paste biomass: drying characteristics, energy requirement and mathematical modeling. Bioresour Technol Rep 11:100467

Ammar EE, Aioub AA, Elesawy AE, Karkour AM, Mouhamed MS, Amer AA, El-Shershaby NA (2022) Algae as Bio-fertilizers: between current situation and future prospective. Saudi J Biol Sci 29(5):3083–3096

Bhatia L, Bachheti RK, Garlapati VK, Chandel AK (2020) Third-generation biorefineries: a sustainable platform for food, clean energy, and nutraceuticals production. Biomass Convers Biorefinery:1–16

Borowitzka MA (2015) Algal biotechnology. In: Sahoo D, Seckbach J (eds) The algae world. Cellular origin, life in extreme habitats and astrobiology, vol 26. Springer, Dordrecht, pp 319–338

Calijuri ML, Silva TA, Magalhães IB, de Paula Pereira ASA, Marangon BB, de Assis LR, Lorentz JF (2022) Bioproducts from microalgae biomass: technology, sustainability, challenges and opportunities. Chemosphere 305:135508

Camacho F, Macedo A, Malcata F (2019) Potential industrial applications and commercialization of microalgae in the functional food and feed industries: a short review. Mar Drugs 17(6):312

Carraro CDFF, Loures CCA, de Castro JA (2022) Microalgae bioremediation and CO2 fixation of industrial wastewater. Clean Eng Technol 8:100466

Chisti Y (2020) Microalgae biotechnology: a brief introduction. In: Handbook of microalgae-based processes and products. Academic Press, pp 3–23

Constantino A, Rodrigues B, Leon R, Barros R, Raposo S (2021) Alternative chemo-enzymatic hydrolysis strategy applied to different microalgae species for bioethanol production. Algal Res 56:102329

Coulombier N, Jauffrais T, Lebouvier N (2021) Antioxidant compounds from microalgae: a review. Mar Drugs 19(10):549

De Carvalho Silvello MA, Gonçalves IS, Azambuja SPH, Costa SS, Silva PGP, Santos LO, Goldbeck R (2022) Microalgae-based carbohydrates: a green innovative source of bioenergy. Bioresour Technol 344:126304

De Llasera MG, Pérez AF, Marín GP, Calva EB (2022) First evidence of extracellular enzymatic degradation of benzo (a) pyrene by the phytoplankton species Selenastrum capricornutum and the influence of temperature. Environ Adv 8:100246

Deprá MC, dos Santos AM, Severo IA, Santos AB, Zepka LQ, Jacob-Lopes E (2018) Microalgal biorefineries for bioenergy production: can we move from concept to industrial reality? Bioenergy Res 11:727–747

Deprá MC, Jacob-Lopes E, Zepka LQ (2022) Introductory chapter: Microalgal biotechnology-the gold-green in modern sustainable world. In: Progress in microalgae research-a path for shaping sustainable futures. IntechOpen

Depra MC, Severo IA, dos Santos AM, Zepka LQ, Jacob-Lopes E (2020) Environmental impacts on commercial microalgae-based products: Sustainability metrics and indicators. Algal Research, 51, 102056

Dias RR, Severo IA, Deprá MC, Maroneze MM, Zepka LQ, Jacob-Lopes E (2020) The next-generation of microalgae-based products. In: Microbial enzymes and biotechniques: interdisciplinary perspectives. Springer Nature, pp 15–42

Fabris M, Abbriano RM, Pernice M, Sutherland DL, Commault AS, Hall CC et al (2020) Emerging technologies in algal biotechnology: toward the establishment of a sustainable, algae-based bioeconomy. Front Plant Sci 11:279

Fanari F, Comaposada J, Boukid F, Climent E, Coma AC, Guerrero L, Castellari M (2023) Enhancing energy bars with microalgae: a study on nutritional, physicochemical and sensory properties. J Funct Foods 109:105768

García MM, de Llasera MPG (2021) A review on the enzymes and metabolites identified by mass spectrometry from bacteria and microalgae involved in the degradation of high molecular weight PAHs. Sci Total Environ 797:149035

García G, Sosa-Hernández JE, Rodas-Zuluaga LI, Castillo-Zacarías C, Iqbal H, Parra-Saldívar R (2020) Accumulation of PHA in the microalgae Scenedesmus sp. under nutrient-deficient conditions. Polymers 13(1):131

Global (2019) Global lutein market to witness a CAGR of 5.2% during 2018–2024. Available in https://www.globenewswire.com. Accessed 30 July 23

Gouda M, Tadda MA, Zhao Y, Farmanullah F, Chu B, Li X, He Y (2022) Microalgae bioactive carbohydrates as a novel sustainable and eco-friendly source of prebiotics: emerging health functionality and recent technologies for extraction and detection. Front Nutr 9:806692

Grama SB, Liu Z, Li J (2022) Emerging trends in genetic engineering of microalgae for commercial applications. Mar Drugs 20(5):285

Hermadi I, Setiadianto IR, Al Zahran DFI, Simbolon MN, Saefurahman G, Wibawa DS, Arkeman Y (2021, May) Development of smart algae pond system for microalgae biomass production. In: IOP conference series: earth and environmental science, vol 749, no 1. IOP Publishing, p 012068

Hlavova M, Turoczy Z, Bisova K (2015) Improving microalgae for biotechnology—From genetics to synthetic biology. Biotechnol Adv 33(6):1194–1203

Indarto I, Rahmah GF, Sari W, Kafiana V, Rudini M (2023) Sunscreen cream base on combination of Curcuma longa and Haematococcus

Pluvialis extracts: formulation and in vivo test. In: AIP conference proceedings, vol 2595, no 1. AIP Publishing
Jacob-Lopes E, Zepka LQ (2023) Food and feed from microalgae: a historical perspective to future directions. In: Handbook of food and feed from microalgae. Academic Press, pp 3–7
Jacob-Lopes E, Maroneze MM, Deprá MC, Sartori RB, Dias RR, Zepka LQ (2019) Bioactive food compounds from microalgae: an innovative framework on industrial biorefineries. Curr Opin Food Sci 25:1–7
Jeevanandam J, Harun MR, Lau SY, Sewu DD, Danquah MK (2020) Microalgal biomass generation via electroflotation: a cost-effective dewatering technology. Appl Sci 10(24):9053
Ji L, Qiu S, Wang Z, Zhao C, Tang B, Gao Z, Fan J (2023) Phycobiliproteins from algae: current updates in sustainable production and applications in food and health. Food Res Int:112737
Li S, Chang H, Zhang S, Ho SH (2023) Production of sustainable biofuels from microalgae with CO_2 bio-sequestration and life cycle assessment. Environ Res:115730
Min KH, Kim DH, Ki MR, Pack SP (2022) Recent progress in flocculation, dewatering, and drying technologies for microalgae utilization: scalable and low-cost harvesting process development. Bioresour Technol 344:126404
Moran CA, Morlacchini M, Keegan JD, Delles R, Fusconi G (2018) Effects of a DHA-rich unextracted microalgae as a dietary supplement on performance, carcass traits and meat fatty acid profile in growing-finishing pigs. J Anim Physiol Anim Nutr 102(4):1026–1038
Mozafari M, Tariverdian T, Beynaghi A (2020) Trends in biotechnology at the turn of the millennium. Recent Pat Biotechnol 14(1):78–82
Mrowicka M, Mrowicki J, Kucharska E, Majsterek I (2022) Lutein and zeaxanthin and their roles in age-related macular degeneration—neurodegenerative disease. Nutrients 14(4):827
Muñoz M, Guzmán JL, Torres M, Acién FG (2022) An IoT platform for data management in an industrial-scale microalgae cultivation plant. IEEE Access 10:127128–127139
Muthukrishnan L (2022) Bio-engineering of microalgae: challenges and future prospects toward industrial and environmental applications. J Basic Microbiol 62(3–4):310–329
Ng IS, Tan SI, Kao PH, Chang YK, Chang JS (2017) Recent developments on genetic engineering of microalgae for biofuels and bio-based chemicals. Biotechnol J 12(10):1600644
Queiroz MI, Vieira JG, Maroneze MM (2020) Morphophysiological, structural, and metabolic aspects of microalgae. In: Handbook of microalgae-based processes and products. Academic Press, pp 25–48
Ramírez-Mérida LG (2023) Commercial facilities of microalgae-based products around the world. In: Handbook of food and feed from microalgae. Academic Press, pp 33–40
Ramos-Martinez EM, Fimognari L, Sakuragi Y (2017) High-yield secretion of recombinant proteins from the microalga Chlamydomonas reinhardtii. Plant Biotechnol J 15(9):1214–1224
Rodrigues LS, do Valle AF, D'Elia E (2018) Biomass of microalgae spirulina maxima as a corrosion inhibitor for 1020 carbon steel in acidic solution. Int J Electrochem Sci 13:6169–6189
Severo IA, Zepka LQ, Lopes EJ (2017) Microalgae biotechnology in integrated processes. J Chem Eng Process Technol 8(366):2
Shin YS, Jeong J, Nguyen THT, Kim JYH, Jin E, Sim SJ (2019) Targeted knockout of phospholipase A2 to increase lipid productivity in Chlamydomonas reinhardtii for biodiesel production. Bioresour Technol 271:368–374
Tan FHP, Nadir N, Sudesh K (2022) Microalgal biomass as feedstock for bacterial production of PHA: advances and future prospects. Front Bioeng Biotechnol 10:879476
Thanigaivel S, Priya AK, Dutta K, Rajendran S, Vasseghian Y (2022) Engineering strategies and opportunities of next generation biofuel from microalgae: a perspective review on the potential bioenergy feedstock. Fuel 312:122827
Wang S, Wu S, Yang G, Pan K, Wang L, Hu Z (2021) A review on the progress, challenges and prospects in commercializing microalgal fucoxanthin. Biotechnol Adv 53:107865
Zambon I, Cecchini M, Egidi G, Saporito MG, Colantoni A (2019) Revolution 4.0: industry vs. agriculture in a future development for SMEs. Processes 7(1):36

Microalgae Production Systems and Photobioreactor Designs: Recent Trends and Future Prospects

2

Miguel Angel Franco-Nava
and Laura Marina Franco-Herrera

Abstract

Microalgae and cyanobacteria convert light energy and inorganic nutrients into biomass that can be used as a source of commercial bioproducts like pharmaceuticals, biofertilizers, biofuels, biopolymers, cosmetics, and food products for humans and animals. However, practical issues need to be addressed before scaling up from laboratory to mass production of algal biomass. The main bioengineering items that determine the performance of microalgae production systems and photobioreactor designs are addressed in this chapter. Trends in microalgae mass production facilities are examined by analyzing each component's specialized functions. Recently proposed technologies, not only for photobioreactors but also for their complementary and service devices, are concisely referred to. Therefore, a multidisciplinary team of engineers and scientists should analyze novel microalgae production systems or photobioreactors to design, calculate, and implement the best solution for a specific bioprocess or application.

Keywords

Microalgae production systems · Algal bioengineering · Aquaculture-controlled systems · Algal biotechnology

M. A. Franco-Nava (✉)
National Technological Institute of Mexico, Technological Institute of Mazatlan (ITMAZ-TecNM), Bioengineering Laboratory, Mazatlán, Mexico
e-mail: miguel.fn@mazatlan.tecnm.mx

L. M. Franco-Herrera
Ministry of Science, Humanities, Technology, and Innovation-SEICHTI, Jalisco State Center for Research and Assistance in Technology and Design-CIATEJ, Industrial Biotechnology, Zapopan, Mexico

2.1 Introduction

The production of single-cell protein was the primary approach to start the development of microalgae's mass culture during the past century. Throughout the years, the practice of this biotechnology demonstrated that the microalgae's biomass could provide a wide range of valuable chemical substances. Because the microalgae and cyanobacteria convert light energy into biomass, this could be used as a source of commercial valuable metabolites to produce pharmaceuticals (Borowitzka 2013), biofertilizers (Benemann et al. 2018), biofuels (Gonzalez-Fernandez et al. 2018), biopolymers (Adamu Ugya et al. 2023), and cosmetics (Ventura et al. 2017). Another way to valorize this biomass type is in food products for humans and animals (Liang et al. 2018; Spolaore et al. 2006).

Modern aquaculture systems have been used to produce microalgae and cyanobacteria since the past century when commercial production systems were developed. Efforts to research and develop vigorous microalgae propagation in large-scale facilities date back to the 1940s and 1950s (Öncel et al. 2015). However, practical issues need to be addressed before scaling up from laboratory to mass production of algal biomass. These issues were classified for convenience analysis as biological or engineering factors.

Biological factors include adequate quantification of microalgal growth, defined as the increase of a microalgae population. The specific growth rate (μ, in day^{-1} or h^{-1}) is the main parameter relative to the performance of microalgae culture systems. Nowadays, modeling the growth kinetics for any mass-produced microalgae is essential to designing, controlling, estimating, and optimizing the microalgae-based bioprocess (Muñoz-Tamayo et al. 2013). The selection of microalgae species depends on the main product, by-products, or environmental services to be provided by the bioprocess (Salama et al. 2017). Besides, contamination by other organisms could be a problem, especially in pilot or

D. U. Santos Ballardo, S. Rossi (eds.), *Microalgae as Promising Source of Commercial Bioproducts*, Developments in Applied Phycology 14, https://doi.org/10.1007/978-3-031-86433-9_2

large-scale installations. In these cases, coculture or polyculture are other options that are applied today (Watanabe and Isdepsky 2021).

The environmental culture parameters that affect biological performance and modify the bioprocess are the concentration and quality of nutrients, gas concentrations, light intensity, temperature, pH, and salinity, among others (Bermejo et al. 2021). Regarding nutrient concentration, recycling of old culture medium has been widely proposed since the past century as an essential practice in continuous algal cultivation (Chakraborty et al. 2023). However, the possible presence of growth-inhibiting substances that may affect the performance of the bioprocess must be considered (Schädler et al. 2020). Regardless, the environmental culture parameters must be established and maintained for optimal growth, and then a pilot plant or large-scale mass culture systems must monitor and/or control them. However, as the scale of production systems increases, monitoring and control become more complex. This change in the steady state of the bioprocess requires many engineering proposals for solutions.

Rezvani and Rostami (2023) published a review paper discussing the basic principles of closed photobioreactor (PBR) design, focusing on the mass transfer of gases and the performance of the microalgae-based bioprocess. Thanks to advanced automation and standardized control systems, it is now possible to control various growth factors like temperature, pH, and light intensity. This technology can manage these factors to create precise climate simulations for the study of the bioprocess. Hence, these simulations are crucial for scaling up production processes toward outdoor cultivation at an industrial level (Fuchs et al. 2021).

Engineering factors that determine the performance of microalgae production systems are directly related to applying biological knowledge of the species in culture to design and improve technological facilities. Furthermore, installations replicating the environmental and biological conditions and allowing the cultivated species to perform the proposed bioprocess are preferred. In this way, the engineering facilities should be able to replicate natural or controlled stress conditions to enhance the bioprocess.

One main objective of microalgae production systems is to provide the means for a maximum and continuous growth rate (Sánchez-Zurano et al. 2021) while allowing continuous harvest without disturbing the bioprocess (Peter et al. 2022). This should occur by providing the best yield ratio per unit area, making this measurement a fundamental parameter for designing microalgae biomass production facilities (Kumar and Bharadvaja 2020).

The yield per unit area is primarily influenced by the light intensity from the exposed area to the maximum depth or distance where the aquatic medium completely absorbs the light radiation (Song et al. 2023). The attenuation of light with depth is determined not only by its absorption and conversion through microalgal metabolism or transformation into other forms of energy, such as heat, but also by scattering, reflection, and absorption by microorganisms and dissolved molecules. Additionally, the intrinsic absorption coefficient of the aquatic environment plays a crucial role, varying according to the nature of the incident light, particularly its wavelength composition (Sánchez Mirón et al. 2000). The design of a microalgal mass production system could be divided into zones with specific functions, facilitating analyses for performance optimization and facility redesign, while also allowing partial shutdowns for maintenance or in response to system failures.

Therefore, the design of an industrial facility for pilot or massive production of microalgae must consider the following as principal components (Fig. 2.1):

(a) Containers (tanks, pools, etc.) that must allow light to penetrate the culture medium.
(b) A lighting source, which can be natural, artificial, or a combination of both.
(c) An agitation device that prevents sedimentation of the cultured microalgae and maintains homogeneous access of the microalgae to the dissolved gas and nutrients, as well as to the light source.
(d) A gas control system that introduces and/or purges gases to and/or from the culture medium as needed by the bioprocess.
(e) Devices for continuous or batch addition of dissolved inorganic nutrients.
(f) A temperature control system that maintains the culture medium closest to the optimal growth temperature or that allows thermal shocks to be applied as stress factors if required by the bioprocess.
(g) Instrumentation for the evaluation and control of the bioprocess.
(h) A microalgae harvesting system that provides a continuous or discontinuous biomass yield, in the best case, with the possibility of recirculating culture medium.
(i) A processing plant for treatment/preservation until the use or sale of the harvested biomass.
(j) An emergency power plant, which is always required for an eventual power outage.

2.2 Design of Photobioreactors

The photobioreactor (PBR) is the core of a microalgae production system. Hence, its performance is essential for the success of the bioprocess because it plays an essential role in maintaining environmental culture parameters that affect the growth rate of microalgae. The characteristics of the PBR depend on several factors that must be considered during the

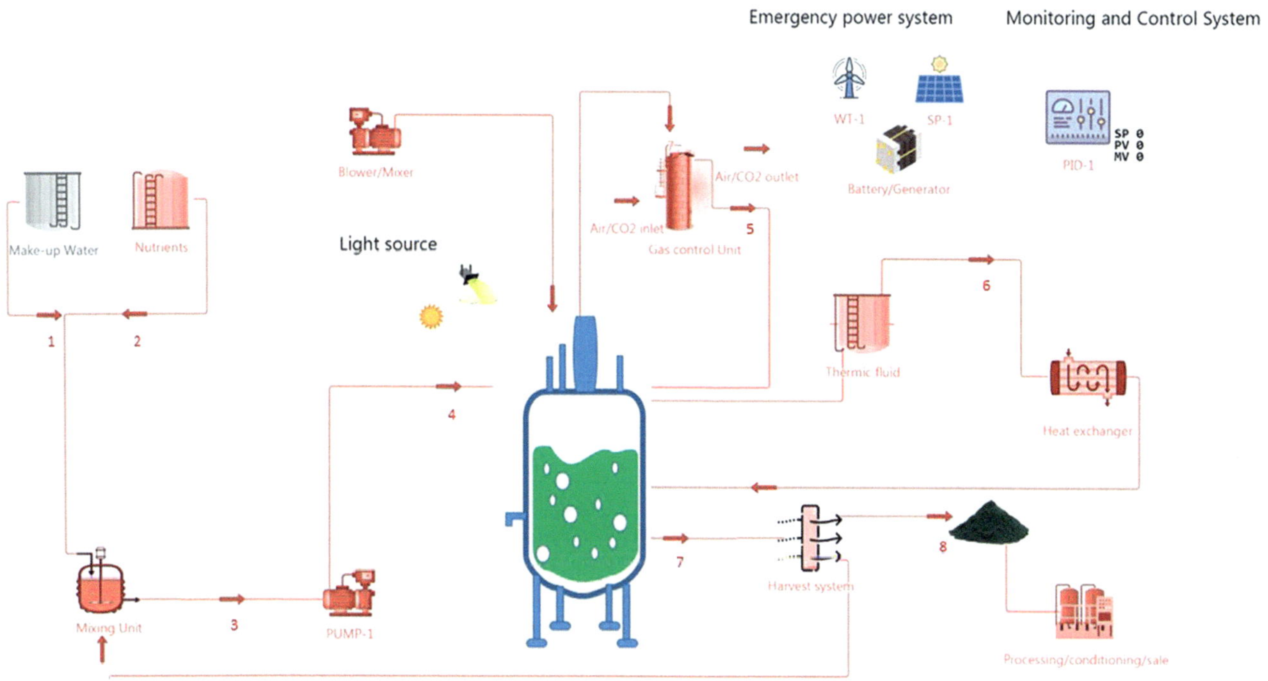

Fig. 2.1 Microalgal mass production system

creative design process. This process includes an iterative and dynamic exercise that starts with a requirement and is followed by some solution proposals; then, they are analyzed, and the most favorable proposal is implemented. The design process allows the identification of each subsystem's main functions and elements that may be necessary to structure the PBR. The process continues with an evaluation that generates feedback information at any time for continuous improvement. There are many examples and methodologies of this process in the literature that bioengineers have applied for the classic design of PBRs and the "nonconventional" proposals (Clagnan et al. 2023; Legrand et al. 2021; Mustapa et al. 2020; Sathinathan et al. 2023; Solimeno et al. 2017; Vajdi and Aslani 2023).

The mass balances between the components of the microalgae production system are very important steps for designing the PBR and for sizing the capacity of all the necessary devices for the bioprocess. Based on this information, all necessary materials, electrical energy, labor, daily production, and coproduct generation are calculated (Borella et al. 2021).

According to the main components of the previously defined microalgal mass production system, the PBR is essentially the container where the microalgae are grown. Hydrodynamic characteristics, such as adequate mixing, bulk rheology, and shear stress, affect other factors vital to microalgae, such as mass and heat transfer in all PBRs.

Furthermore, PBRs can be classified into open and closed reactors, depending on their exposure to environmental elements (Zhou et al. 2020). The design of the microalgae production system must consider this characteristic, which has technical and economic relevance for the bioprocess. Container geometry varies significantly between all PBR types, but the main design criteria described above should be applied in all cases.

Tredici et al. (2010) suggest that the main design criteria for PBR include geometry and structure, construction materials, surface-to-volume (S/V) ratio, orientation and inclination, mixing and degassing devices, and systems for cleaning and temperature regulation. On the other hand, in commercial PBRs, ease of operation, construction, and operating costs are critical.

2.3 Bioprocess Strategy

As a particular type of bioreactor, the PBR also has three operation modes called bioprocess strategies. These bioprocess strategies depend on the input and output flow rates of the PBR. They are the batch bioprocess, the fed-batch (or semicontinuous) bioprocess, and the continuous bioprocess (Rezvani and Rostami 2023). The main features of this strategy are explained below.

Batch bioprocess: Microalgae cultivation is developed in a discontinuous bioprocess until the nutrients are completely depleted; this is considered a suitable cultivation method. However, this method may not be ideal for commercial purposes (Solís-Salinas et al. 2021). According to Acién

Fernández et al. (2013), the main inconvenience found in the discontinuous (batch) bioprocess is the need for a longer time period between each batch to obtain an adequate biomass concentration for harvest, as well as the high operating expenses associated with starting the process each time.

Continuous and fed-batch bioprocesses: Continuous and closed PBR is the main way to achieve higher microalgal biomass production for industrial applications. This bioprocess strategy is executed by feeding fresh medium at a constant rate of 10–100% of the daily PBR working volume. Continuous or semicontinuous processes are more efficient for long-term biomass production, costs, and operation. Shekh et al. (2022) introduced a literature review on using PBRs to cultivate *Dunaliella salina*, where a comparison between the continuous and semicontinuous modes of operation of PBRs is discussed regarding biomass productivity.

2.4 Geometry of the Container

2.4.1 Natural Lagoons, Large Ponds, and Raceways

The microalgae industry started with natural ponds and artificial lagoons with circular, large rectangular, or raceway geometries (Öncel et al. 2015, 2016). The main production systems in the 1980s were extensive, shallow, unmixed open ponds (extensive method) and open raceways (intensive method). Due to better mixing, biomass productivity in raceways is much higher than in extensive open ponds. However, some efforts are being made today to improve mixing in large ponds (Ghaffar et al. 2023).

The area of each pond varies; there are production units larger than 200 ha, achieving a total production area of around 700 ha (Borowitzka 2013). The average depth of the water column in these ponds can be 0.15–1.50 m because sunlight can only penetrate a limited depth depending on the turbidity of the culture medium and other local environmental factors. One of the most important factors regarding the depth:area ratio is salinity, which can be affected by evaporation and dilution phenomena. It is not practical to use large ponds to maintain low-to-medium-salinity cultures due to their size and high evaporation rate in the summer. On the other hand, it is essential to consider the rainy season as it can affect low-salinity waters. Kusmayadi et al. (2020) published a review on computational fluid dynamics (CFD) as a tool for designing raceways for the cultivation of microalgae while highlighting that there are still significant challenges in the application of CFD for the design and analysis of this type of PBRs.

Low-cost maintenance and operation are the benefits of using large pond or raceway geometries (Tan et al. 2020). However, culture conditions depend on seasonal changes, so productivity varies throughout the year. There, the cultivation process involves continuous harvesting of a portion of the pond and returning clarified medium culture to recycle the remaining nutrients.

Besides, PBRs based on the raceway geometry can be provided with elongated baffles to direct the flow of the microalgae culture medium and prevent short circuits. The main disadvantage of open ponds is the high risk of contamination and the lake's control over the environmental and culture conditions. Finally, the great advantage of this type of installation is the low cost of energy and materials (Jorquera et al. 2010).

2.4.2 High-Rate Algal Ponds (HRAPs)

High-rate algal ponds (HRAPs) have a channel geometry with a circular change of direction for a serpentine water flow. Hydraulic retention time (HRT) is one of the main design parameters that must be calculated as $HRT = V \cdot Q^{-1}$, where V is the total volume of water medium in the HRAP and Q is the volume of fluid introduced daily. The dimensions of a typical HRAP vary depending on the biological and technical factors discussed above. However, it is common to find designs with a depth of 20 cm working at an HRT of 3 days with a water medium circulating at a speed of 0.3 m s^{-1} (Oswald and Golueke 1960).

2.4.3 Flat Panel

One of the most common geometries of PBRs is the flat panel, with advantages described by Vo et al. (2019). Regarding this geometry, the high surface-to-volume ratio provides better efficiency for the photosynthetic bioprocess, thanks to the considerably smaller width compared to the height and length dimensions. An application of this geometry is the photobioreactors integrated into the façade of the Bio Intelligent Quotient (BIQ) building in Germany, manufactured for research purposes on biomass production and heat transfer. There, 129 modules, 270 cm × 70 cm × 8 cm, were used as active elements of the southeast and southwest facades of the building (IBA Hamburg GmbH 2013). Vajdi and Aslani (2023) proposed a flat panel for the center of Teheran, Iran, with modules of 200 cm × 100 cm × 10 cm. A series of deflector sheets were proposed within the flat panel designs to prevent gravitational sedimentation. Notably, the baffles within the flat panel considerably change the flow dynamics and all implied aquatic medium and cell movement. Ardiani et al. (2019) have proposed this geometry for architectural applications, but this is not yet widely used.

2.4.4 Columns, Airlift, and Cylindrical Tank

Column is another geometry used for PBRs applied in mass cultivation of microalgae, where a hollow cylinder is the main characteristic of this container. Three main configurations have been reported using this geometry: bubble column, airlift, and stirred tank (Cheng et al. 2002; Gautam and Vinu 2020).

Bubble column photobioreactors are low-cost and have no internal structure or moving parts. However, when an impeller is added, it becomes a stirred tank. Besides, airlift configuration consists of a combination of columns that can be either concentric or connected by pipes. One column generates an ascending flow due to naturally rising bubbles, while the other facilitates a descending flow driven by mass liquid transfer between them. The volume in the gas–liquid section of the airlift PBR should be minimized to have a better cost–benefit ratio of investment in this kind of facility. Furthermore, many phenomena could affect the performance of this kind of PBR, but one of the most important is the gas–liquid mass transfer. This phenomenon is a limiting factor for a high-density microalgal culture, where the degassing of produced oxygen, and thus an equilibrium between the atmospheric and dissolved gases in the culture medium, is required. Even more, it is estimated that 15–20% of the total operating costs of this kind of PBR depend on the performance of this gas–liquid mass transfer (Cheng et al. 2002). Behin and Amiri (2023) summarize the leading designs and factors that affect this geometry.

Here, the volumetric mass transfer coefficient (k_{La}) is the parameter that has been recently used as a fixed design parameter through the scale-up or the hydrodynamic characterization of PBRs (Mohamadnia et al. 2022). One of the most important design criteria for this geometry and configuration is the distance between the entrance of air (or gases) and the liquid medium's surface. This distance should be long enough to allow the smaller bubbles to disengage before the liquid flows back down (Chisti and Moo-young 1993).

The core component of a stirred cylindrical tank is the mechanical agitation, which is provided by agitators or impellers of various shapes and sizes. In addition to mixing, they contribute other functions, such as aeration, degassing, heat transfer, and mass transfer. So, the main design parameters of this geometry are stirrer rate, momentum factor, impeller tip, power consumption, oxygen transfer rate, drop size distribution, speed, and mixing rate (Aslanbay Guler et al. 2019; de Souza Kirnev et al. 2022; Deniz 2020; Mohamadnia et al. 2022). During their scale-up process, these parameters must remain constant to improve product and biomass productivity compared to small-scale results. The baffles are equidistant vertical plates or bars that can reduce vortex and improve mixing. The practical operating volume of a stirred tank photobioreactor is typically 70–80% of the total volume. This design feature allows adequate bioreactor headspace for accommodating foam formation and detachment of liquid droplets from rising gas, which is essential for optimal performance (Assunção and Malcata 2020).

2.4.5 Tubular Geometry

Tubular geometry is commonly used as a transparent circuit of coiling tubes in a horizontal configuration, but some examples can be found using helical or diagonal configurations. The diameters of the pipes or tubes may be around 60 mm or less. For the PBRs with tubular geometry, the light path length (ϕ_{eq}) that affects the average irradiance (I_{av}) is related to the diameter of the tube (ϕ) and the solar zenith angle (θ), measured from the vertical to the sun's position in the sky as proposed by Acién Fernández et al. (1997); thus, Eq. 2.1 is as follows:

$$\phi_{eq} = \frac{\phi}{\cos\theta} \tag{2.1}$$

The other important design parameter for tubular PBRs is the spacing between adjacent parallel rungs of the looped tube (separation). It is expressed as a function of the tube's diameter as follows.

$$separation = n_T \cdot \phi \tag{2.2}$$

where (n_T) is the number of separation diameters between the adjacent parallel tubes.

2.4.6 Hybrid Geometry

A hybrid geometry of tubular and column geometries has been constructed for pilot-scale outdoor mass production systems (Acién Fernández et al. 2001). This design includes a serpentine-shaped tubing loop exposed to sunlight and coupled to an airlift configuration unit for degassing purposes. Dogaris et al. (2015) proposed a horizontal PBR (HBR) coupled to a pair of airlift devices with rectangular geometry. The airlift was used to mix and drive the flow through the flat panel, which had baffles to direct the flux of the culture medium. All these structures were tested under indoor conditions, submerged in a tub, and outdoors, floating in a pond. Like any other new configuration, this proposal is for small production volumes and is still in the early stages of development.

On the other hand, plastic bags without any rigid structure are often used as containers; therefore, the bag adopts a semiround shape and functions as a flat panel or bubble column. Some authors have reported other small-scale geometric and hydrodynamic configurations (Assunção and Malcata 2020; Cui et al. 2022; Johnson et al. 2018).

2.5 Container Materials

Any material used to manufacture a PBR must meet these characteristics:

(a) innocuous to the bioprocess, (b) resistant to ultraviolet (UV) degradation for outdoor applications, (c) resistant to working temperatures that are within the optimal range for microalgae or the required environmental conditions, (d) resistant to a wide pH range, and (e) resistant to high pressures, which is another benefit that is required by engineering design (Tredici et al. 2010).

The most common materials used to construct PBRs are (a) hydraulic concrete, (b) sheets or tubes of a wide variety of plastic polymers, (c) composed materials, (d) corrosion-resistant steel, (e) glass, and (f) acrylic (Dasgupta et al. 2010; Ugwu et al. 2008).

Hydraulic concrete is a very common material for constructing raceway channels, high-rate algae ponds (HRAPs), and large rectangular or circular pool geometries of PBRs (Ahmad et al. 2021).

Moreover, polyvinyl chloride (PVC) sheets or other durable food-grade plastic materials are commonly used as lining material on soils or other reinforced structures because they can adapt to a wide variety of tanks, pools, or artificial lagoons (Sirohi et al. 2023). Transparent PVC pipes are another suitable material for constructing a tubular PBR. This material features UV-blocking technology that reduces harmful wavelengths and allows beneficial wavelengths to penetrate. In the same context, transparent polyethylene (PE) or translucent low-density polyethylene (LDPE) and Linear LDPE (LLDPE) sheet of 0.15–0.2 mm thickness has been used as a material for the construction of a large number of PBR geometries (Dogaris et al. 2015). Transparent waterproof polyurethane (PU) with a thickness of 0.2 mm has also been used to construct PBRs (Harris et al. 2013). Besides, fiberglass is a more expensive but preferred material in some cases for particular geometries (Collao et al. 2021).

Stainless-steel framing is the most common material used in flat panel geometries for structural and fastening purposes where its corrosion is adequate for exposure to water (Kossalbayev et al. 2024).

Laminated safety glass (LSG) was applied in the photobioreactor of the "BQI" building (IBA Hamburg GmbH 2013). There, a 1.7-cm-thick LSG was the transparent element on both sides of a flat panel photobioreactor. Transparent polycarbonate sheet is also used as a wall for flat panel geometries with 1.27 cm thickness for a panel up to 300 cm × 150 cm (Kim 2014).

In the case of tubular PBRs, the tubes can be made of glass, but the main disadvantages of using this material are their fragility and the risk of breakage. Glass containers are common but only in small-scale PBRs due to the same risk.

Acrylic, also called plexiglass, is often used to construct closed systems in a greenhouse. This transparent or translucent material functions as a mechanical barrier to prevent chemical and biological contamination and controls light intensity, temperature, and evaporation (Singh and Sharma 2012). Acrylic tubes with 18.4 cm inner diameter, 203.3 cm height, and 5 mm wall thickness are commonly used for PBR columns. A variation of this material is the acrylic sheets, which are largely used for flat panel and rectangular airlift PBRs.

Stirred tanks and column PBRs are primarily constructed of translucent or transparent materials such as cylindrical plexiglass, borosilicate glass, and fiberglass.

2.6 Lighting

Microalgal biomass production systems are based on photosynthetic metabolism; due to this, illumination is one of the most important factors determining this bioprocess's success. Two main aspects should be considered regarding the natural illumination or artificial lighting design for a PBR. First, the photosynthetically active radiation (PAR) corresponds to the 400–700-nm wavelength range of light radiation that photosynthetic microorganisms can use. Second, light intensities, which have been proposed, range from 10 to 1000 μmol m^{-2} s^{-1} for photobioreactors when an LED lamp provides the illumination.

Furthermore, the quality and quantity of light vary widely from one required product to another, from one bioreactor geometry to another, from one scale to another, from one source of light to another, and from one species of algae to another. The quality and quantity of light in a photosynthetic bioprocess depends on the optical dynamic caused by the interactions of aquatic medium with individual cells, particles, and molecules suspended or dissolved in the water. A complex fluid dynamic and other physical interactions between different PBR biotic and abiotic elements generate specific optical proprieties for each PBR geometry that must be considered in PBR designs.

Regarding the outdoor production systems, one of the factors to consider is the geographical location of production facilities. This affects the timing, intensity, and orientation of daily solar irradiation and seasonality, three lighting elements that directly affect the efficiency of light utilization by microalgae in the outdoor PBRs. Some published works expose this complexity and propose modeling strategies for specific bioprocess assessment and simulation (Huesemann et al. 2016; Song et al. 2023; Vélez-Landa et al. 2021).

It is common for stirred tank PBRs to use external artificial light sources, such as fluorescent or halogen lamps. Immersed optical fibers have been tested as alternatives for profound illumination; however, their use is not widely recommended because they hinder the mixing pattern within the PBR. Another published proposal suggests a lighting system with a light-emitting tube connected to an optical ring via an optical fiber (Csogor et al. 2001).

Nwoba et al. (2019) published a review article on light management technologies in photobioreactors, where they suggest that the feasibility of implementing light management approaches on large-scale photobioreactors depends on economic viability, product value, system life cycle, and ease of implementation. These authors conclude that the most promising approach to advancing microalgal culture on PBRs is a synergistic combination of existing and emerging light management technologies to improve biological productivity, reduce production costs, and minimize environmental impact.

2.7 Stirring Devices

Agitation is a crucial engineering factor in the design of microalgae production facilities. All types of PBRs require maintaining cells in suspension and minimizing light and nutrient gradients. The homogeneity of the aquatic medium is guaranteed by agitation devices that must consider cellular fragility. These devices can cause problems derived from high shear forces that can cause cell lysis. Therefore, the correct design or selection of stirring devices is essential for a successful bioprocess.

Mixing within open ponds can be achieved through various methods such as paddlewheels, rotating arms, mechanic tools (e.g., pumping), or gas transfer systems (e.g., pipes, propellers, and jet aerators) (Ali et al. 2023). Paddlewheels are the most common mixing and flow generator devices within the raceway channels and the HRAPs.

For the continuous agitation of PBRs, the "air lift" technique is one of the most viable and friendly techniques for the cultivation of microalgae. This technique consists of a continuous supply of compressed air from the bottom of the container to prevent the sedimentation of microalgae. This water movement must reach sufficient flow rates to prevent the deposition of microalgae and biofouling along the internal surfaces of the PBR. Using a diffuser at the bottom of the PBR is very common to obtain homogeneous agitation due to the dispersion of microbubbles.

The circulator pumping system is usually composed of centrifugal pumps that provide a mechanical momentum that generates a flow of aquatic medium to circulate throughout the production units. This mechanical impulse is useful for mixing or circulating the medium through all the complementary devices or tanks that make up the massive microalgae production system.

The power and number of pumps depend on several factors, such as the total length and design of the piping system, the required water level, the hydraulic head specified in the production system's design, the specific pumping application, and the operating regimen. Pumps with a power range of 3–4 hp. were utilized by Vajdi and Aslani (2023).

2.8 Gas Control System

The gas supply and the degassing process are two aspects that directly affect the performance of photosynthesis, where both the efficacy of CO_2 solubilization and the purge of produced oxygen depend on the mass transfer capacity of PBRs. Nitrogen has been used to purge the O_2 produced through photosynthesis; however, air is usually used due to its low cost.

A PBR gas purge outlet is always required for gas control because the air injected for agitation and the oxygen produced must be removed from the liquid phase and, therefore, from the PBR. For this purpose, an aeration and degassing device should be considered; in certain cases, an oxygen recovery device is recommended. Some oxygen recovery devices have been proposed particularly for live support systems in space (Hendrickx et al. 2006; Oguchi et al. 1989).

On the other hand, a CO_2 injection system is also required to control the pH and as a nutritional factor for microalgae. In the case of CO_2 injection into water, it is essential to consider its chemical interactions. Once the CO_2 is dissolved in the liquid phase, the carbonate–bicarbonate chemical equilibrium shifts, initiating a complex physicochemical interaction of ions involving different chemical species.

It is essential to consider this in seawater systems where there is a diversity of ions and higher concentrations. CO_2 injection requires a microbubble generator that maximizes the contact surface between the gas–liquid phases to allow the dissolution of CO_2 within the aquatic medium. Gas usage measurement is performed using a flowmeter at a wide range of flow rates (e.g., 10 L min^{-1}), and a solenoid valve is used to control CO_2 injection, where an electronic control system manages this valve.

Many aeration and degassing technologies can be used to achieve gas–liquid equilibrium between air and water. One of the most popular are the degassing towers, or packed degassing columns as they are also called. There, the water medium falls in countercurrent against a large volume of air that maximizes the gas–liquid ratio to around 10:1 (Skov 2019).

The mass transfer coefficient (k_La) is a very important design criterion that is a function of the contact between gas and liquid. This phenomenon is mainly affected by the mixing within PBRs or gas–liquid contactors, the size of the bubbles, and the gas flow rates. Acién Fernández et al. (2013) propose an equation (Eq. 2.3) that establishes the relationship between mass transfer coefficients and the mass transfer volume for oxygen and carbon dioxide to avoid excessive oxygen accumulation or carbon dioxide limitation. There, $K_l a_l,_{O_2}$ and $K_l a_l,_{CO_2}$ are the respective volumetric mass transfer coefficients of oxygen and carbon dioxide; $[CO_2]$ and $\left[CO_2^*\right]$ are the concentration of CO_2 in the culture and the concentration in equilibrium with the gas phase in contact with the liquid, normally pure CO_2. $[O_2]$ and $\left[O_2^*\right]$ are the same for oxygen, but in the case of O_2^*, it is normally in the air. V_{mt,CO_2} and V_{mt,O_2} are the volumes of the mass transfer unit for CO_2 and O_2, respectively.

$$\frac{K_l a_l,_{O_2}}{K_l a_l,_{CO_2}} = -\frac{\left([CO_2]-\left[CO_2^*\right]\right)V_{mt,CO_2}}{\left([O_2]-\left[O_2^*\right]\right)V_{mt,O_2}} \quad (2.3)$$

2.9 Feeding Devices

Feeding devices are all the necessary equipment to store, prepare, and incorporate the inorganic nutrients, the supply of water, and required stock biomass into the microalgal culture. These devices include dissolution vessels fitted with agitators for rapid and homogeneous dissolution of the salts. A mixing unit sometimes functions as a water collection unit in the recirculating system where it is applied; there, the mass of aquatic medium can compensate for the variations in water levels and flows derived from the clogging of filters, inlets, or outlets of the aquatic medium from or to the recirculating system. This device is sometimes used for general temperature and pH monitoring and corrections.

Pumps are the primary devices used to feed a continuous culture of microalgae and transfer culture media between tanks through the production facilities. Peristaltic pumps are preferred for media recycling due to their physical mechanism, which minimizes the shear stress and takes care of cell integrity. However, their cost is higher than that of centrifugal pumps, so they are mainly used in laboratory or pilot-scale facilities.

A recent proposal for a feeding device uses anion exchange membranes. The proposal aims to provide nutrients to microalgae at a constant rate of consumption using organic polymer membranes coated with quaternary amine functional groups (Fu et al. 2016). Other similar feeding strategies have been proposed by Shao et al. (2022).

2.10 Instrumentation for the Control and Evaluation of the Bioprocess

Various sensors are used to monitor physicochemical changes during the bioprocess, which are increasingly common in microalgae production facilities. The sensors are connected to electronic devices for data recording and processing. These electronic devices, called controllers, could activate the actuators (pumps, solenoid valves, blowers, etc.) to control pH, light, and temperature.

2.10.1 Temperature Control System

Closed systems tend to heat up rapidly, which can affect productivity. Therefore, it is important to ensure that these systems are adequately cooled for optimal performance. For this, various heat exchanger technologies are used to cool and heat aquatic media depending on specific circumstances. In photobioreactors, heat exchangers are commonly used to maintain temperatures between 25 and 30°C. When a heat exchanger is installed, the heat transferred can be profitable inside or outside the production system.

The light source usually provides light and heat in highly controlled indoor cultivation systems. In these cases, the climatic conditions of the place determine the characteristics of temperature control systems. For a cold place, the heat from the light source is favorable and works as an actuator of the temperature control system, while in tropical latitudes, this is inconvenient, and an additional cooling device is required.

2.10.2 pH Control System

pH control in microalgae production systems is mainly achieved by dossing CO_2 gas as required by the biological and physicochemical carbon cycle within the PBRs. The pH value of water is determined by the CO_2–carbonate–bicarbonate equilibrium, which is defined by the following series of chemical reactions 2.4–2.5.

$$H_2O + CO_{2(aq)} \leftrightarrow H_2CO_{3(aq)} \leftrightarrow HCO_3^-(aq) + H^+_{(aq)} \quad (2.4)$$

$$HCO_3^-(aq) \leftrightarrow CO_3^{-2}{}_{(aq)} + H^+_{(aq)} \quad (2.5)$$

The reactions 2.4 and 2.5 provide protons to the aquatic environment in the PBR, and the pH decreases, while the added CO_2 dissociates into carbonic acid, bicarbonate, and carbonate. This is because protons affect the water equilib-

... the reaction 2.6. Therefore, as pH decreases, ...al amount of dissolved inorganic carbon (DIC) increases.

$$H_2O \leftrightarrow H^+_{(aq)} + OH^-_{(aq)} \quad (2.6)$$

In contrast, when the equilibrium of the reactions shifts toward the aqueous carbon dioxide by photosynthesis, the pH increases. These equilibriums depend on salinity and temperature; hence, the monitoring and control of the bioprocess should also be considered. Gas contactors are the primary engineering devices widely used for pH control.

2.11 The Harvest System

The harvest system is the first step in product recovery. Methods currently used in commercial production facilities include centrifugation, coagulation, filtration, flocculation, flotation, screening, and sedimentation. However, electrophoresis, electroflotation, and ultrasound are occasionally used on smaller scales. The choice of the harvest system should be made considering price and effectiveness (Assad and Kumar 2024). Additionally, the desired product and subsequent separation or extraction procedures must be considered, as they may be affected by the harvesting method applied.

Furthermore, flotation and settling methods exploit the salinity-dependent buoyancy properties of microalgae cells in stationary and moving gradients. Settling is another harvesting technology in which the biomass is separated at the bottom of the sedimentation unit, and a portion of the microalgae is recycled into the production system with most culture medium. An example of this technology is the in-ground gravity settler where a concentration of microalgae from 0.5 g L^{-1} to 4.5 g L^{-1} can be concentrated with an efficiency of 90% (Vajdi and Aslani 2023). The salinity-dependent hydrophobic adhesion properties of microalgae cells have also been used where flocculation and bubble columns are the most commonly used devices.

Continuous centrifuge (including spiral plate or disc stack separation technologies) is another technical solution for microalgae harvesting. This technology can produce pastes from 31.5 to 34% of dry solids (GEA 2013). Filtration is applied as a harvest method, with many commercial systems available that allow the extraction of various products. This method includes high-pressure filtration using diatomaceous earth as a filter medium or deep-bed filtration. Continuous filtration of biomass is achieved with drum filters.

Dewatering units reduce the humidity from the paste, reducing its volume and facilitating its handling, packaging, and storage. The centrifugal separator can be used as a harvesting or dewatering unit. This technology achieves dewatering efficiencies of up to 98% when used as a secondary solid–liquid separation unit. Concentrations of up to 200 g L^{-1} can be achieved with centrifugal separators (Vajdi and Aslani 2023). The extracted culture media can then be recirculated. When heat needs to be recovered from the aquatic medium, a heat exchanger is required. The recovered heat is then transferred to a practical application, typically for heating, and preheating water. The culture media is then recirculated into the bioprocess if necessary.

2.12 Prospects (Ongoing Technologies)

In the near future, advanced engineering techniques, such as computational fluid dynamics (CFD) simulation and chlorophyll fluorescence measurements, will be used to design new reactors and improve the efficiency of existing ones. This could enhance the microalgae's ability to use light, thus improving photobioreactor performance.

Gas exchangers are becoming very important for new gas sequestration and collection applications. New technologies include gas exchangers for using oxygen in spacecraft and submarines during CO_2 control in artificial atmospheres (Maestrup et al. 2023). Similarly, gas exchangers are being developed for CO_2 sequestration, offering potential solutions for mitigating global warming. As an example, a large-scale PBR in an open-air pond in Akhfenir, Morocco, is cultivating microalgae to capture CO_2 in one of the world's most extensive facilities within a 30,000-square meter site (Page 2023), while in New Zealand, there is one of the largest ponds to produce biofuels from microalgae, and an area of 1.25 ha is currently in operation.

A novel algal biofilm membrane photobioreactor (BMPBR) equipped with solid carriers and a submerged membrane module was constructed to grow attached microalgae and remove nutrients from the secondary effluent (Gao et al. 2015). Another novel gas-permeable bag photobioreactor was developed using an inexpensive microporous film with high oxygen permeability, as reported by Kishi et al. (2021). According to the authors, if additional improvements are made, productivity can be increased without consuming a significant amount of energy for aeration.

The thin-layer cascade is a new technique for producing microalgae in an open system and is proposed as an alternative to raceway ponds. There are promising published results that suggest that their performance, design guidelines, and technical details are in the process of improvement (Clagnan et al. 2023).

A proposed electromagnetic photobioreactor (E-PBR) integrates electromagnets to assess microalgae's growth under the magnetic field's effect. Li et al. (2007) suggested that the electromagnetic field may improve the assimilation of C, N, P, and minerals by *Spirulina platensis* culture.

Deamici et al. (2023) applied an electromagnetic field to *Chlorella fusca* and suggested that this is a nontoxic and low-cost alternative to enhance intracellular carbohydrate production.

Regarding the lab-scale photobioreactors, several papers describe the incorporation of novel technologies, mainly those involving their instrumentation and new control process devices (Benner et al. 2022; Bonnanfant et al. 2021; Le Gouic et al. 2021). Mustapa et al. (2020) reported a particularly innovative PBR design, which has a torus shape and is fully instrumented and controlled. This system has previously demonstrated its feasibility in controlling with high accuracy all culture parameters influencing the photosynthesis and growth of cultured microalgae.

Internally illuminated PBR (IIPBR) is an emerging technology under study and development (Alonso-Bastida et al. 2022; Jiménez-González et al. 2017). Erbland et al. (2020) have shown the feasibility of constructing an IIPBR that is both cost-effective and easy to use. They demonstrate that internal lighting, automated CO_2 injection, and pH management make it possible to obtain high cell densities and manage energy efficiency. In addition, the same authors suggest that on a larger scale, these kinds of PBRs could make it financially viable to install monitoring and control systems, which could further enhance cell densities and reduce its overall footprint. Amaral et al. (2020) proposed using an internal light-integrated photobioreactor and an internal light bubble column photobioreactor to assess specific growth rates by the microalgae *Chlorella minutissima.*

Wave-driven floating photobioreactors use wave energy for culture mixing. Dogaris et al. (2015) suggested that they can be constructed with low capital costs, are easy to scale up, and require low energy inputs, and it is possible to obtain high biomass productivity. Zhao et al. (2023) studied the fluid dynamics in this kind of PBR without mixing devices. They examined how wave conditions and the shape of the PBR (cylindrical, square, or rectangular) impact the process. Their findings suggest that further investigation is needed, specifically on utilizing weaker waves to improve mixing efficiency.

The mesh ultrathin layer photobioreactor (MUTL-PBR) was briefly described by Pulz et al. (2013), and Sandmann et al. (2021) recently demonstrated that it had a biomass yield that was about three times higher and, specifically, antioxidant capacity that was about tenfold greater than the traditional tubular photobioreactor. Assunção and Malcata (2020) describe and discuss “nonconventional” photobioreactor configurations, including a qualitative comparison of their performance. They first introduce parameters that affect the photobioreactor performance and then discuss the most common designs that they propose as suitable for the microalgae in culture. Recently, HRAP is an established wastewater treatment technology that recovers resources like microalgal biomass. This biomass can be used to produce a variety of products, such as fertilizers, feedsto… and bioplastics (Adamu Ugya et al. 2023; Assad and Kum… 2024; Nishshanka et al. 2022; de Paula Pereira et al. 2023).

2.13 Conclusions

There is no best geometry or configuration of photobioreactor for any microalgae species or by-product. An interdisciplinary team must analyze the case to design, calculate, and implement the best solution for a specific bioprocess or application.

The use of open ponds in monocultures is often avoided due to the possibility of contamination by other microorganisms. In addition, open systems require precise control over factors such as temperature and lighting, which can hinder the success of the bioprocess. As a result, closed systems are preferred, as they offer better control over key culture conditions and minimize de risk of contamination (Banerjee and Ramaswamy 2019).

Closed photobioreactor designs may meet the requirements of the pharmaceutical industry, making it possible to meet the good manufacturing practice (GMP) for this and other industries to produce biomass for food, feed, additives, and a variety of bioactive molecules.

The design of PBRs should be focused on achieving the operating conditions closest to the optimal cultivation conditions required by the species of microalgae. This design can be created while considering the reduction of costs and energy consumption.

References

Acién Fernández FG, García Camacho F, Sánchez Pérez JA, Fernández Sevilla JM, Molina Grima E (1997) A model for light distribution and average solar irradiance inside outdoor tubular photobioreactors for the microalgal mass culture. Biotechnol Bioeng 55(5):701–714. https://doi.org/10.1002/(SICI)1097-0290(19970905)55:5<701::AID-BIT1>3.0.CO;2-F

Acién Fernández FG, Fernández Sevilla JM, Sánchez Pérez JA, Molina Grima E, Chisti Y (2001) Airlift-driven external-loop tubular photobioreactors for outdoor production of microalgae: assessment of design and performance. Chem Eng Sci 56(8):2721–2732. https://doi.org/10.1016/S0009-2509(00)00521-2

Acién Fernández FG, Fernández Sevilla JM, Molina Grima E (2013) Photobioreactors for the production of microalgae. Rev Environ Sci Biotechnol 12(2):131–151. https://doi.org/10.1007/s11157-012-9307-6

Adamu Ugya Y, Chen H, Sheng Y, Ajibade FO, Wang Q (2023) A review of microalgae biofilm as an eco-friendly approach to bioplastics, promoting environmental sustainability. Environ Res 236(P2):116833. https://doi.org/10.1016/j.envres.2023.116833

Ahmad I, Abdullah N, Koji I, Yuzir A, Eva Muhammad S (2021) Evolution of photobioreactors: a review based on microalgal perspective. IOP Conf Ser Mater Sci Eng 1142(1):012004. https://doi.org/10.1088/1757-899x/1142/1/012004

Ali EM, Elshobary M, El-Sheekh MM (2023) Algal bioenergy: the fuel for tomorrow. In: Shah MP (ed) Green approach to alternative fuel for a sustainable future. Elsevier, pp 409–432. https://doi.org/10.1016/B978-0-12-824318-3.00021-7

Alonso-Bastida A, Franco-Nava MA, Adam-Medina M, López-Zapata B, Álvarez-Gutiérrez PE, García-Morales J (2022) Mathematical modeling of thermal interactions in a self-cooling pilot-scale photobioreactor. Case Stud Therm Eng 31(September 2021):101825. https://doi.org/10.1016/j.csite.2022.101825

Amaral MS, Loures CCA, Naves FL, Baeta BEL, Silva MB, Prata AMR (2020) Evaluation of cell growth performance of microalgae Chlorella minutissima using an internal light integrated photobioreactor. J Environ Chem Eng 8(5):104200. https://doi.org/10.1016/j.jece.2020.104200

Ardiani NA, Koerniawan MD, Martokusumo W, Suyono EA, Poerbo HW (2019) Feasibility of Algae Photobioreactor as Façe in the Office Building in Indonesia. IOP Conf Ser Earth Environ Sci 322(1):012020. https://doi.org/10.1088/1755-1315/322/1/012020

Aslanbay Guler B, Deniz I, Demirel Z, Oncel SS, Imamoglu E (2019) Transition from start-up to scale-up for fucoxanthin production in flat plate photobioreactor. J Appl Phycol 31(3):1525–1533. https://doi.org/10.1007/s10811-018-1696-7

Assad H, Kumar A (2024) Biofuels from microalgae: production, processing, and extraction technologies. In: Jeguirim M, Zorpas AA (eds) Advances in biofuels production, optimization and applications. Elsevier Inc., pp 145–163. https://doi.org/10.1016/B978-0-323-95076-3.00014-4

Assunção J, Malcata FX (2020) Enclosed "non-conventional" photobioreactors for microalga production: a review. Algal Res 52(October):102107. https://doi.org/10.1016/j.algal.2020.102107

Banerjee S, Ramaswamy S (2019) Comparison of productivity and economic analysis of microalgae cultivation in open raceways and flat panel photobioreactor. Bioresour Technol Rep 8(September):100328. https://doi.org/10.1016/j.biteb.2019.100328

Behin J, Amiri P (2023) A review of recent advances in airlift reactors technology with emphasis on environmental remediation. J Environ Manag 335(March):117560. https://doi.org/10.1016/j.jenvman.2023.117560

Benemann JR, Woertz I, Lundquist T (2018) Autotrophic microalgae biomass production: from niche markets to commodities. Ind Biotechnol 14(1):3–10. https://doi.org/10.1089/ind.2018.29118.jrb

Benner P, Meier L, Pfeffer A, Krüger K, Oropeza Vargas JE, Weuster-Botz D (2022) Lab-scale photobioreactor systems: principles, applications, and scalability. Bioprocess Biosyst Eng 45(5):791–813. https://doi.org/10.1007/s00449-022-02711-1

Bermejo E, Rayen F, Taidi B (2021) Microalgae culture quality indicators: a review. Crit Rev Biotechnol 41(4):457–473. https://doi.org/10.1080/07388551.2020.1854672

Bonnanfant M, Marec H, Jesus B, Mouget JL, Pruvost J (2021) Investigation of the photosynthetic response of Chlorella vulgaris to light changes in a torus-shape photobioreactor. Appl Microbiol Biotechnol 105(23):8689–8701. https://doi.org/10.1007/s00253-021-11636-w

Borella L, Sforza E, Bertucco A (2021) Effect of residence time in continuous photobioreactor on mass and energy balance of microalgal protein production. New Biotechnol 64(June):46–53. https://doi.org/10.1016/j.nbt.2021.05.006

Borowitzka MA (2013) Dunaliella : biology , production , and markets. In: Richmond A, Hu Q (eds) Handbook of microalgal culture: applied phycology and biotechnology, 2nd edn. Wiley, pp 359–368. https://doi.org/10.1002/9781118567166.ch18

Chakraborty B, Gayen K, Bhowmick TK (2023) Transition from synthetic to alternative media for microalgae cultivation: a critical review. Sci Total Environ 897(July):165412. https://doi.org/10.1016/j.scitotenv.2023.165412

Cheng H-P, Wang P-M, Chen J-W, Wu W-T (2002) Cultivation of Acetobacter xylinum for bacterial cellulose production in a modified airlift reactor. Biotechnol Appl Biochem 35(2):125–132. https://doi.org/10.1111/j.1470-8744.2002.tb01180.x

Chisti Y, Moo-young M (1993) Improve the performance of airlift reactors. Chem Eng Prog 89:38–45. https://api.semanticscholar.org/CorpusID:114398830

Clagnan E, Dell'Orto M, Štěrbová K, Grivalský T, Artur Câmara Manoel J, Masojídek J, D'Imporzano G, Gabriel Acién-Fernández F, Adani F (2023) Impact of photobioreactor design on microalgae-bacteria communities grown on wastewater: differences between thin-layer cascade and thin-layer raceway ponds. Bioresour Technol 374(January):128781. https://doi.org/10.1016/j.biortech.2023.128781

Collao J, del Mar Morales-Amaral M, Acién-Fernández FG, Bolado-Rodríguez S, Fernandez-Gonzalez N (2021) Effect of operational parameters, environmental conditions, and biotic interactions on bacterial communities present in urban wastewater treatment photobioreactors. Chemosphere 284(November 2020):131271. https://doi.org/10.1016/j.chemosphere.2021.131271

Csogor Z, Herrenbauer M, Schmidt K, Posten C (2001) Light distribution in a novel photobioreactor-modeling for optimization. J Appl Phycol 13:325–333

Cui J, Purton S, Baganz F (2022) Characterisation of a simple 'hanging bag' photobioreactor for low-cost cultivation of microalgae. J Chem Technol Biotechnol 97(3):608–619. https://doi.org/10.1002/jctb.6985

Dasgupta CN, Jose Gilbert J, Lindblad P, Heidorn T, Borgvang SA, Skjanes K, Das D (2010) Recent trends on the development of photobiological processes and photobioreactors for the improvement of hydrogen production. Int J Hydrog Energy 35(19):10218–10238. https://doi.org/10.1016/j.ijhydene.2010.06.029

de Souza Kirnev PC, de Souza Vandenberghe LP, Soccol CR, de Carvalho JC (2022) Mixing and agitation in photobioreactors. In: Innovations in fermentation and phytopharmaceutical technologies. Elsevier Inc. https://doi.org/10.1016/B978-0-12-821877-8.00005-1

Deamici KM, Silva PGP, Costa JAV, Santos LO (2023) Low electromagnetic fields applied to chlorella fusca cultivation to increase production of microalga-based carbohydrates. Bioenergy Res 16(3):1548–1555. https://doi.org/10.1007/s12155-022-10562-7

Deniz I (2020) Scaling-up of Haematococcus pluvialis production in stirred tank photobioreactor. Bioresour Technol 310(April):123434. https://doi.org/10.1016/j.biortech.2020.123434

Dogaris I, Welch M, Meiser A, Walmsley L, Philippidis G (2015) A novel horizontal photobioreactor for high-density cultivation of microalgae. Bioresour Technol 198:316–324. https://doi.org/10.1016/j.biortech.2015.09.030

Erbland P, Caron S, Peterson M, Alyokhin A (2020) Design and performance of a low-cost, automated, large-scale photobioreactor for microalgae production. Aquac Eng 90:102103. https://doi.org/10.1016/j.aquaeng.2020.102103

de Paula Pereira ASA, Magalhães IB, Ferreira J, de Siqueira Castro J, Calijuri ML (2023) Microalgae organomineral fertilizer production: a life cycle approach. Algal Res 71(February):103035. https://doi.org/10.1016/j.algal.2023.103035

Fu Q, Chang HX, Huang Y, Liao Q, Zhu X, Xia A, Sun YH (2016) A novel self-adaptive microalgae photobioreactor using anion exchange membranes for continuous supply of nutrients. Bioresour Technol 214:629–636. https://doi.org/10.1016/j.biortech.2016.04.081

Fuchs T, Arnold ND, Garbe D, Deimel S, Lorenzen J, Masri M, Mehlmer N, Weuster-Botz D, Brück TB (2021) A newly designed automatically controlled, sterilizable flat panel photobioreactor for axenic algae culture. Front Bioeng Biotechnol 9. https://doi.org/10.3389/fbioe.2021.697354

Gao F, Yang ZH, Li C, Zeng GM, Ma DH, Zhou L (2015) A novel algal biofilm membrane photobioreactor for attached microalgae growth and nutrients removal from secondary effluent. Bioresour Technol 179:8–12. https://doi.org/10.1016/j.biortech.2014.11.108

Gautam R, Vinu R (2020) Reaction engineering and kinetics of algae conversion to biofuels and chemicals: via pyrolysis and

hydrothermal liquefaction. React Chem Eng 5(8). https://doi.org/10.1039/d0re00084a
GEA (2013) Setting the record straight on algae separation. Biomass Magazine 4. https://issuu.com/bbiinternational/docs/september_13-bmm-outlines
Ghaffar I, Deepanraj B, Sundar LS, Vo DVN, Saikumar A, Hussain A (2023) A review on the sustainable procurement of microalgal biomass from wastewaters for the production of biofuels. Chemosphere 311(P2):137094. https://doi.org/10.1016/j.chemosphere.2022.137094
Gonzalez-Fernandez C, Barreiro-Vescovo S, De Godos I, Fernandez M, Zouhayr A, Ballesteros M (2018) Biochemical methane potential of microalgae biomass using different microbial inocula. Biotechnol Biofuels 11(1):1–11. https://doi.org/10.1186/s13068-018-1188-7
Harris L, Tozzi S, Wiley P, Young C, Richardson TMJ, Clark K, Trent JD (2013) Potential impact of biofouling on the photobioreactors of the Offshore Membrane Enclosures for Growing Algae (OMEGA) system. Bioresour Technol 144:420–428. https://doi.org/10.1016/j.biortech.2013.06.125
Hendrickx L, De Wever H, Hermans V, Mastroleo F, Morin N, Wilmotte A, Janssen P, Mergeay M (2006) Microbial ecology of the closed artificial ecosystem MELiSSA (Micro-Ecological Life Support System Alternative): reinventing and compartmentalizing the Earth's food and oxygen regeneration system for long-haul space exploration missions. Res Microbiol 157(1):77–86. https://doi.org/10.1016/j.resmic.2005.06.014
Huesemann M, Crowe B, Waller P, Chavis A, Hobbs S, Edmundson S, Wigmosta M (2016) A validated model to predict microalgae growth in outdoor pond cultures subjected to fluctuating light intensities and water temperatures. Algal Res 13:195–206. https://doi.org/10.1016/j.algal.2015.11.008
IBA Hamburg GmbH (2013) Smart material house BIQ. International Building Exhibition Hamburg, July, 1–22. www.iba-hamburg.de
Jiménez-González A, Adam-Medina M, Franco-Nava MA, Guerrero-Ramírez GV (2017) Grey-box model identification of temperature dynamics in a photobioreactor. Chem Eng Res Des 121:125. https://doi.org/10.1016/j.cherd.2017.03.004
Johnson TJ, Katuwal S, Anderson GA, Gu L, Zhou R, Gibbons WR (2018) Photobioreactor cultivation strategies for microalgae and cyanobacteria. Biotechnol Prog 34(4):811–827. https://doi.org/10.1002/btpr.2628
Jorquera O, Kiperstok A, Sales EA, Embiruçu M, Ghirardi ML (2010) Comparative energy life-cycle analyses of microalgal biomass production in open ponds and photobioreactors. Bioresour Technol 101(4):1406–1413. https://doi.org/10.1016/j.biortech.2009.09.038
Kim K-H (2014) International conference on sustainable building Asia a feasibility study of an algae façade system. International conference on sustainable building Asia, Doe 2010, 333–341. https://www.semanticscholar.org/paper/A-Feasibility-Study-of-an-Algae-Façade-System-Kim/2f7a6d33ba10cf1eda2a9ebc320cbe54543418b1
Kishi M, Tanaka K, Akizuki S, Toda T (2021) Development of a gas-permeable bag photobioreactor for energy-efficient oxygen removal from algal culture. Algal Res 60:102543. https://doi.org/10.1016/j.algal.2021.102543
Kossalbayev BD, Yilmaz G, Sadvakasova AK, Zayadan BK, Belkozhayev AM, Kamshybayeva GK, Sainova GA, Bozieva AM, Alharby HF, Tomo T, Allakhverdiev SI (2024) Biotechnological production of hydrogen: design features of photobioreactors and improvement of conditions for cultivating cyanobacteria. Int J Hydrog Energy 49:413–432. https://doi.org/10.1016/j.ijhydene.2023.09.001
Kumar L, Bharadvaja N (2020) A review on microalgae biofuel and biorefinery: challenges and way forward. In: Energy sources, part A: recovery, utilization and environmental effects. Bellwether Publishing, Ltd. https://doi.org/10.1080/15567036.2020.1836084
Kusmayadi A, Suyono EA, Nagarajan D, Chang JS, Yen HW (2020) Application of computational fluid dynamics (CFD) on the raceway design for the cultivation of microalgae: a review. J Ind Microbiol Biotechnol 47(4–5):373–382. https://doi.org/10.1007/s10295-020-02273-9
Le Gouic B, Marec H, Pruvost J, Cornet JF (2021) Investigation of growth limitation by CO_2 mass transfer and inorganic carbon source for the microalga Chlorella vulgaris in a dedicated photobioreactor. Chem Eng Sci 233:116388. https://doi.org/10.1016/j.ces.2020.116388
Legrand J, Artu A, Pruvost J (2021) A review on photobioreactor design and modelling for microalgae production. React Chem Eng 6(7):1134–1151. https://doi.org/10.1039/d0re00450b
Li ZY, Guo SY, Li L, Cai MY (2007) Effects of electromagnetic field on the batch cultivation and nutritional composition of Spirulina platensis in an air-lift photobioreactor. Bioresour Technol 98(3):700–705. https://doi.org/10.1016/j.biortech.2006.01.024
Liang MH, Wang L, Wang Q, Zhu J, Jiang JG (2018) High-value bioproducts from microalgae: strategies and progress. Crit Rev Food Sci Nutr 59(15):2423–2441. https://doi.org/10.1080/10408398.2018.1455030
Maestrup PG, Appelganc K, Band S, Stechmann F, Hagemann V, Förster A, Verseux C, Heinicke C (2023) Integration of a photobioreactor into the MaMBA facility as part of a human-centered life support system. 52nd International conference on environmental systems, July, ICES-2023-258
Mohamadnia S, Tavakoli O, Faramarzi MA (2022) Production of fucoxanthin from the microalga Tisochrysis lutea in the bubble column photobioreactor applying mass transfer coefficient. J Biotechnol 348(December 2021):47–54. https://doi.org/10.1016/j.jbiotec.2022.03.009
Muñoz-Tamayo R, Martinon P, Bougaran G, Mairet F, Bernard O (2013) Design of optimal experiments for parameter estimation of microalgae growth models. IFAC Proc. Volumes (IFAC-PapersOnline) 12(Part 1):315–320. https://doi.org/10.3182/20131216-3-IN-2044.00043
Mustapa NS, Abu Mansor MS, Serri NA (2020) Design and development of centred-light photobioreactor for microalgae cultivation system. IOP Conf Ser: Mater Sci Eng 716(1):1–7. https://doi.org/10.1088/1757-899X/716/1/012009
Nishshanka GKSH, Anthonio RADP, Nimarshana PHV, Ariyadasa TU, Chang JS (2022) Marine microalgae as sustainable feedstock for multi-product biorefineries. Biochem Eng J 187(June):108593. https://doi.org/10.1016/j.bej.2022.108593
Nwoba EG, Parlevliet DA, Laird DW, Alameh K, Moheimani NR (2019) Light management technologies for increasing algal photobioreactor efficiency. Algal Res 39(January):101433. https://doi.org/10.1016/j.algal.2019.101433
Oguchi M, Otsubo K, Nitta K, Shimada A, Fujii S, Koyano T, Miki K (1989) Closed and continuous algae cultivation system for food production and gas exchange in celss. Adv Space Res 9(8):169–177. https://doi.org/10.1016/0273-1177(89)90042-2
Öncel S, Kose A, Vardar F, Torzillo G (2015) From the ancient tribes to modern societies, microalgae evolution from a simple food to an alternative fuel source. In: Kim S-K (ed) Handbook of marine microalgae: biotechnology advances. Elsevier Inc., pp 127–144. https://doi.org/10.1016/B978-0-12-800776-1.00009-1
Öncel S, Köse A, Öncel D (2016) Façade integrated photobioreactors for building energy efficiency. In: Fernando Pacheco-Torgal SM, Rasmussen E, Granqvist C-G, Ivanov V, Kaklauskas A (eds) Start-up creation: the smart eco-efficient built environment. Elsevier Ltd., pp 237–299. https://doi.org/10.1016/B978-0-08-100546-0.00011-X
Oswald W, Golueke C (1960) Biological transformation of solar energy. Adv Appl Microbiol 2:223–262. https://doi.org/10.1016/S0065-2164(08)70127-8

Page T (2023) Growing and burying algae in the Sahara is the latest solution for the climate crisis. CNN Published 4:36 AM EDT, Fri August 18, 2023. https://edition.cnn.com/2023/08/18/africa/brilliant-planet-algae-carbon-removal-climate-morocco-scn-spc-intl/index.html

Peter AP, Koyande AK, Chew KW, Ho SH, Chen WH, Chang JS, Krishnamoorthy R, Banat F, Show PL (2022) Continuous cultivation of microalgae in photobioreactors as a source of renewable energy: current status and future challenges. Renew Sust Energ Rev 154(September 2021):111852. https://doi.org/10.1016/j.rser.2021.111852

Pulz O, Broneske J, Waldeck P (2013) IGV GmbH experience report, industrial production of microalgae under controlled conditions: innovative prospects. In: Richmond A, Hu Q (eds) Handbook of microalgal culture: applied phycology and biotechnology, 2nd edn. Wiley, pp 445–460. https://doi.org/10.1002/9781118567166.ch24

Rezvani F, Rostami K (2023) Photobioreactors for utility-scale applications: effect of gas–liquid mass transfer coefficient and other critical parameters. Environ Sci Pollut Res 30(31):76263–76282. https://doi.org/10.1007/s11356-023-27644-4

Salama ES, Kurade MB, Abou-Shanab RAI, El-Dalatony MM, Yang IS, Min B, Jeon BH (2017) Recent progress in microalgal biomass production coupled with wastewater treatment for biofuel generation. Renew Sust Energ Rev 79(June):1189–1211. https://doi.org/10.1016/j.rser.2017.05.091

Sánchez Mirón A, Molina Grima E, Fernández Sevilla JM, Chisti Y, García Camacho F (2000) Assessment of the photosynthetically active incident radiation in outdoor photobioreactors using oxalic acid/uranyl sulfate chemical actinometer. J Appl Phycol 12(3–5):385–394. https://doi.org/10.1023/a:1008153630822

Sánchez-Zurano A, Rodríguez-miranda E, Guzmán JL, Acién-fernández FG, Fernández-sevilla JM, Grima EM (2021) Abaco: a new model of microalgae-bacteria consortia for biological treatment of wastewaters. Appl Sci 11(3):1–24. https://doi.org/10.3390/app11030998

Sandmann M, Smetana S, Heinz V, Rohn S (2021) Comparative life cycle assessment of a mesh ultra-thin layer photobioreactor and a tubular glass photobioreactor for the production of bioactive algae extracts. Bioresour Technol 340(June):125657. https://doi.org/10.1016/j.biortech.2021.125657

Sathinathan P, Parab HM, Yusoff R, Ibrahim S, Vello V, Ngoh GC (2023) Photobioreactor design and parameters essential for algal cultivation using industrial wastewater: a review. Renew Sust Energ Rev 173(November 2022):113096. https://doi.org/10.1016/j.rser.2022.113096

Schädler T, Neumann-Cip AC, Wieland K, Glöckler D, Haisch C, Brück T, Weuster-Botz D (2020) High-density microalgae cultivation in open thin-layer cascade photobioreactors with water recycling. Appl Sci 10(11):1–13. https://doi.org/10.3390/app10113883

Shao H, Sun Y, Jiang X, Hu J, Guo C, Lu C, Guo F, Sun C, Wang Y, Dai C (2022) Towards biomass production and wastewater treatment by enhancing the microalgae-based nutrients recovery from liquid digestate in an innovative photobioreactor integrated with dialysis bag. J Environ Manag 317(May):115337. https://doi.org/10.1016/j.jenvman.2022.115337

Shekh A, Sharma A, Schenk PM, Kumar G, Mudliar S (2022) Microalgae cultivation: photobioreactors, CO_2 utilization, and value-added products of industrial importance. J Chem Technol Biotechnol 97(5):1064–1085. https://doi.org/10.1002/jctb.6902

Singh RN, Sharma S (2012) Development of suitable photobioreactor for algae production – a review. Renew Sust Energ Rev 16(4):2347–2353. https://doi.org/10.1016/j.rser.2012.01.026

Sirohi R, Sim SJ, Pandey A (2023) Photobioreactors: an introduction. In: Sirohi R, Pandey A, Sim S, Chang J-S, Lee D-J (eds) Current developments in biotechnology and bioengineering. photobioreactors : design and applications. Susan Dennis, pp 3–10. https://doi.org/10.1016/B978-0-323-99911-3.00008-7

Skov PV (2019) CO2 in aquaculture. Fish Physiology 37:287–321. https://doi.org/10.1016/bs.fp.2019.07.004

Solimeno A, Gabriel F, García J (2017) Mechanistic model for design, analysis, operation and control of microalgae cultures: calibration and application to tubular photobioreactors. Algal Res 21:236–246. https://doi.org/10.1016/j.algal.2016.11.023

Solís-Salinas CE, Patlán-Juárez G, Okoye PU, Guillén-Garcés A, Sebastian PJ, Arias DM (2021) Long-term semi-continuous production of carbohydrate-enriched microalgae biomass cultivated in low-loaded domestic wastewater. Sci Total Environ 798:149227. https://doi.org/10.1016/j.scitotenv.2021.149227

Song Y, Cheng J, Yang Y, Gao Z (2023) Irradiance penetration distribution and flashing light frequency simultaneously affected with microalgal cell absorption and CO_2 bubble scattering in a raceway pond. Sci Total Environ 864(December 2022):160988. https://doi.org/10.1016/j.scitotenv.2022.160988

Spolaore P, Joannis-Cassan C, Duran E, Isambert A (2006) Commercial applications of microalgae. J Biosci Bioeng 101(2):87–96. https://doi.org/10.1263/jbb.101.87

Tan JS, Lee SY, Chew KW, Lam MK, Lim JW, Ho SH, Show PL (2020) A review on microalgae cultivation and harvesting, and their biomass extraction processing using ionic liquids. Bioengineered 11(1):116–129. https://doi.org/10.1080/21655979.2020.1711626

Tredici MR, Chini Zittelli G, Rodolfi L (2010) Photobioreactors. Encycl Ind Biotechnol 1:1–15. https://doi.org/10.1002/9780470054581.eib479

Ugwu CU, Aoyagi H, Uchiyama H (2008) Photobioreactors for mass cultivation of algae. Bioresour Technol 99(10):4021–4028. https://doi.org/10.1016/j.biortech.2007.01.046

Vajdi S, Aslani A (2023) Design and techno-economic analysis of direct CO_2 capturing with integrated photobioreactors as a building façade. Sustain Energy Technol Assess 56(March 2022):103068. https://doi.org/10.1016/j.seta.2023.103068

Vélez-Landa L, Hernández-De León HR, Pérez-Luna YDC, Velázquez-Trujillo S, Moreira-Acosta J, Berrones-Hernández R, Sánchez-Roque Y (2021) Influence of light intensity and photoperiod on the photoautotrophic growth and lipid content of the microalgae verrucodesmus verrucosus in a photobioreactor. Sustainability (Switzerland) 13(12):6606. https://doi.org/10.3390/su13126606

Ventura SPM, Nobre BP, Ertekin F, Hayes M, Garciá-Vaquero M, Vieira F, Koc M, Gouveia L, Aires-Barros MR, Palavra AMF (2017) Extraction of value-added compounds from microalgae. In: Microalgae-based biofuels and bioproducts: from feedstock cultivation to end-products, pp 461–483. https://doi.org/10.1016/B978-0-08-101023-5.00019-4

Vo HNP, Ngo HH, Guo W, Nguyen TMH, Liu Y, Liu Y, Nguyen DD, Chang SW (2019) A critical review on designs and applications of microalgae-based photobioreactors for pollutants treatment. Sci Total Environ 651:1549–1568. https://doi.org/10.1016/j.scitotenv.2018.09.282

Watanabe MM, Isdepsky A (2021) Biocrude oil production by integrating microalgae polyculture and wastewater treatment: novel proposal on the use of deep water-depth polyculture of mixotrophic microalgae. Energies 14(21):6992. https://doi.org/10.3390/en14216992

Zhao Y, Jia G, Cheng Y, Zhu H, Chi Z, Shen H, Zhu C (2023) Numerical study on the internal fluid mixing and its influencing mechanisms of the wave-driven floating photobioreactor for microalgae production. Front Mar Sci 10(January):1095590. https://doi.org/10.3389/fmars.2023.1095590

Zhou W, Lu Q, Han P, Li J (2020) Microalgae cultivation and photobioreactor design. In: Yousuf A (ed) Microalgae cultivation for biofuels production. Elsevier Inc., pp 31–50. https://doi.org/10.1016/B978-0-12-817536-1.00003-5

Dewatering and Microalgal Harvesting Techniques: The Bottleneck to Industrial Microalgae Processing

3

Carolina Maia, Ana F. Esteves, Cláudia A. Machado, Ana L. Gonçalves, Eva M. Salgado, and José C. M. Pires

Abstract

Microalgae have attracted increasing interest due to their composition, which is rich in proteins, lipids, carbohydrates, and pigments, and can be used in several applications. Despite the enormous potential, the harvesting procedure makes the commercialization of microalgal biomass economically unfeasible. This process represents almost 30% of the total cost of microalgal production. Harvesting these microorganisms is a significant challenge because they grow in diluted suspensions and have a small size, negative surface charge, and low density. Centrifugation, gravity sedimentation, flocculation, flotation, and filtration are the standard methods for microalgal harvesting. When a combination of these processes is used, the harvesting procedure involves two steps: thickening and dehydration. There is no standard harvesting method as it varies with the microalgal species and the media composition. Regardless of the method chosen for harvesting, it must not contaminate the biomass or change its quality. Furthermore, it should have a low operation time, a high recovery rate, and the potential for large-scale use. This chapter overviews all the main methods currently used to harvest microalgae. Furthermore, the advantages and limitations of each technique are addressed.

C. Maia · A. F. Esteves · C. A. Machado · A. L. Gonçalves
E. M. Salgado · J. C. M. Pires (✉)
LEPABE – Laboratory for Process Engineering, Environment, Biotechnology and Energy, Faculty of Engineering, University of Porto, Porto, Portugal

ALiCE – Associate Laboratory in Chemical Engineering, Faculty of Engineering, University of Porto, Porto, Portugal
e-mail: jcpires@fe.up.pt

Keywords

Centrifugation · Dewatering · Filtration · Flocculation · Flotation · Gravity sedimentation · Microalgal harvesting · Thickening

3.1 Introduction

Microalgae come under a broad category of photosynthetic microorganisms that can accumulate biomass rich in proteins, lipids, carbohydrates, and pigments. The huge taxonomic diversity and wide environmental distribution of microalgae make them singular microorganisms with great potential to be explored as sustainable feedstock in several areas, such as environment, energy, pharmaceutical and cosmetics industries, aquaculture, human food, and animal feed (Esteves et al. 2021, 2022; Silva et al. 2021; Liu et al. 2023). Besides the biotechnological potential of microalgal biomass, these microorganisms present several advantages over other photosynthetic organisms (e.g., terrestrial crops) (Alishah Aratboni et al. 2019; Hosseinkhani et al. 2022; Liu et al. 2023): (i) higher biomass productivities; (ii) shorter reproduction cycles; (iii) higher lipid contents and nutritional values; (iv) production in nonarable land, thus not competing with the one intended for food production; and (v) use of low-quality waters, thus reducing biomass production costs (associated to nutrients supply) and environmental impact (associated to freshwater use). Although there is much potential, producing and processing microalgae on a large scale still needs to be economically viable. The harvesting process is the most significant barrier to industrial microalgal processing, and it is important to note that there is still no standard method for harvesting microalgae (Liu et al. 2023; Laamanen et al. 2021).

D. U. Santos Ballardo, S. Rossi (eds.), *Microalgae as Promising Source of Commercial Bioproducts*, Developments in Applied Phycology 14, https://doi.org/10.1007/978-3-031-86433-9_3

The biomass harvesting step accounts for 20–30% of the total biomass production costs because microalgae have specific characteristics that make the harvesting process complex, expensive, and time-consuming (Figueiredo et al. 2022). These characteristics include (Kratzer and Murkovic 2021; Laamanen et al. 2021) (i) the small size of microalgal cells (2–30 μm), (ii) the obtention of dilute cultures (0.2–5 g L^{-1}), (iii) the low density of microalgal cells (water-like density), and (iv) the negatively charged microalgal cell surface, which makes microalgae remain stable in suspension. Furthermore, the volume to be processed is large; the microalgal species, media composition, and culture conditions vary; and the need for harvesting is frequent (Laamanen et al. 2021; Kumar et al. 2023). Although several harvesting techniques have been developed, most present high energy demand, time consumption, and biomass contamination limitations. Therefore, before the scale-up of microalgae production, it is crucial to evaluate current harvesting technologies and find cost-effective, rapid alternatives that do not contaminate the biomass or the environment (Laamanen et al. 2021; Liu et al. 2023). This chapter presents an overview of all the techniques currently employed in the microalgal harvesting process and discusses their advantages and limitations.

3.2 Harvesting Methods

Microalgal harvesting consists of recovering microalgal biomass from the culture medium using several techniques summarized in Fig. 3.1. These methods can be classified as physical, chemical, biological, or electrical-based. Also, it is possible to combine methods to obtain higher recovery rates (Tork et al. 2017). When combined methods are used, the harvesting process comprises two stages (Tan et al. 2018): (i) thickening, where the microalgal slurry is concentrated to 2–7% of total suspended solids (TSSs); and (ii) dewatering, where the slurry is processed into a cake with 15–25% of TSS. The most common methods used in the thickening stage include sedimentation, flocculation, and flotation. High energy–demanding methods, such as centrifugation and filtration, are used in the dewatering stage (Esteves et al. 2020).

For a harvesting method to be considered efficient, it must fulfill specific requirements (Tang et al. 2020): (i) low operation time, (ii) capability of maintaining the quality of the compounds after harvesting, (iii) low operation and maintenance costs, (iv) potential for large-scale use, (v) high recovery rate, (vi) suitability for different microalgal species, and (vii) pollutant-free. Furthermore, to reduce the water footprint of microalgal production, the harvesting method must allow the reuse of the culture medium (Muylaert et al. 2017). The selection of the harvesting method is also influenced by the species chosen and the desired end product (Tan et al. 2018). For example, to produce products for human consumption, it is also crucial to bear in mind that not all harvesting methods can be used, especially those involving the addition of chemical substances, as they can lead to biomass contamination with toxic compounds or alter the composition of the desired product. Additionally, neither the target products nor the microalgal cells should be damaged during harvesting (Laamanen et al. 2021). An overview of published studies on microalgal harvesting methods can be found in Table 3.1. At the industrial scale, the most commonly used

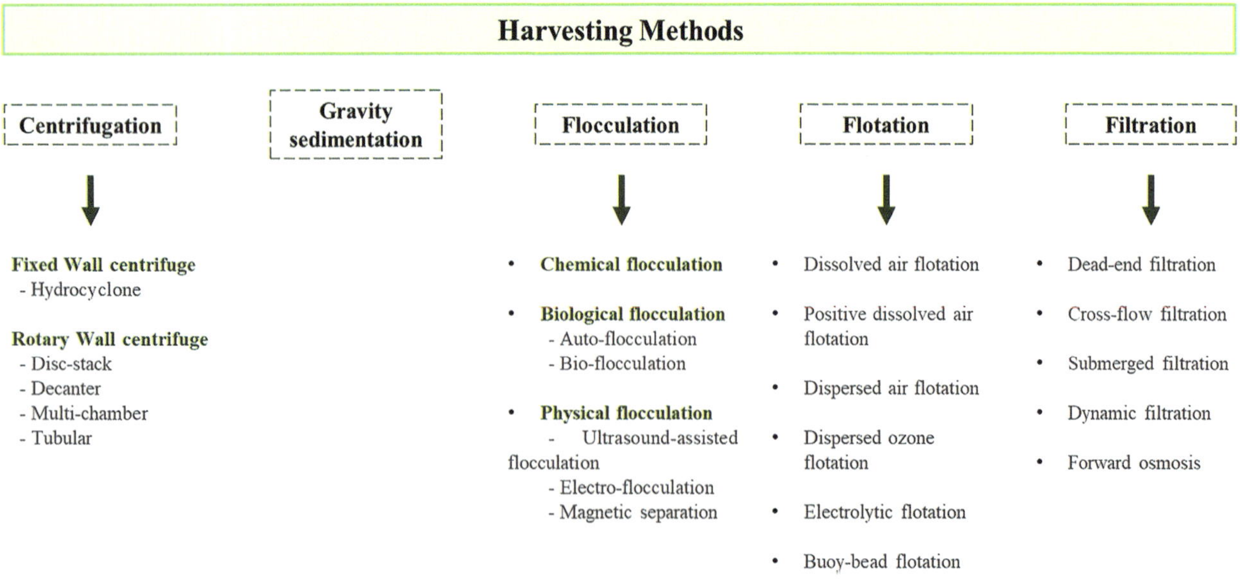

Fig. 3.1 Schematic representation of microalgal harvesting techniques

Table 3.1 Current techniques used for microalgal harvesting

Harvesting method	Microalgae	Operational conditions	Energy consumption	Results	References
Centrifugation	*Aurantiochytrium* sp.	*X*: 25 g L^{-1}; PT: 30 min; *v*: 9000 rpm	–	HE = 84.6%	Kim et al. (2015a)
Centrifugation	*Aurantiochytrium* sp.	*X*: 25 g L^{-1}; PT: 30 min; *v*: 1000 rpm	1.94 kWh kg^{-1}	HE = 87.2%	Kim et al. (2015a)
Centrifugation	*Arthrospira platensis*	*V*: 0.005 L; *X*: 0.41 g L^{-1}; PT: 10 min; *v*: 3000 *g*	–	HE = 85%	Cuellar-Bermudez et al. (2020)
Centrifugation	*Scenedesmus obliquus*	*V*: 70 L; *X*: 0.75 g L^{-1}; PT: 50 min	120.4 kWh kg^{-1}	HE = 99.3%	Wang et al. (2019b)
Centrifugation	*Scenedesmus spinosus*	FR: 14 L min^{-1}; *X*: 0.4 g L^{-1}; PT: 10 min; *v*: 5500 rpm	–	HE = 82.1%	Reyes and Labra (2016)
Sedimentation	*Chlorella vulgaris*	*V*: 0.1 L; *T*: 25 °C; PT: 180 min	–	HE = 25%	Salim et al. (2012)
	Neochloris oleoabundans	*V*: 0.1 L; *T*: 25 °C; PT: 180 min	–	HE = 40%	
Chemical flocculation	*Nannochloropsis oculata*	*A*: 10 mg L^{-1} Tanfloc; *X*: 400 mg L^{-1}; *V*: 250 mL; PT: 30 min; pH: 6	–	HE = 99%	Roselet et al. (2016)
Chemical flocculation	*Neochloris oleoabundans*	*A*: 16 mg L^{-1} LMW chitosan; *V*: 50 mL; PT: 30 min; pH: 6	–	HE >95%	Kirnev et al. (2018)
Auto-flocculation	*Chlorococcum* sp.	*A*: 40 mg L^{-1} NaOH; *V*: 100 mL; pH: 12	–	HE = 94%	Ummalyma et al. (2016)
Auto-flocculation	*Scenedesmus quadricauda*	*X*: 0.54 g L^{-1}; *V*: 25 mL; PT: 30 min; pH:11.6	–	HE = 95%	Huo et al. (2014)
Auto-flocculation	*Chaetoceros muelleri*	*X*: 0.42 g L^{-1}; *V*: 25 mL; PT: 30 min; pH: 11.5	–	HE = 100%	Huo et al. (2014)
Bio-flocculation (extracellular product of *Bacillus subtilis*)	*Chlorella vulgaris*	*A*: 22.03 mg L^{-1} Poly (γ -glutamic acid); *X*: 0.57 g L^{-1}; PT: 120 min	–	HE = 91%	Zheng et al. (2012)
Bio-flocculation (fungal pelletization)	*Chlorella vulgaris*	*A*: 1:300 ratio of *Aspergillus niger* to microalgae; 2 g L^{-1} glucose	–	HE >90%	Gultom et al. (2014)
Electro flocculation (non-sacrificial graphite electrodes)	*Scenedesmus* sp.	*X*: 341.9 mg L^{-1}; pH: 3; *T*: 25 °C; PT: 40 min; Voltage: 3 V; CD: 2 mA cm^{-2}	0.88 kWh kg^{-1}	HE = 23%	Liu et al. (2018)
Electro-flocculation (stainless steel cathode and carbon steel anode)	*Chlorella vulgaris*	*X*: 0.6–1.8 g L^{-1}; pH: 3; *V*: 500 mL; pH: 6–7.5; *T*: 20–22 °C; PT: 26 min; CD: 0.02 A; voltage: 5 V	0.10 kWh kg^{-1}	HE >95%	Lucakova et al. (2021)
Magnetic separation (iron-based nanoparticles)	*Chlorella vulgaris*	*A*: 25 mg L^{-1} Fe-MNP; *v*: 120 rpm; PT: 10 min	–	HE >95%	Almomani (2020)
Magnetic separation	*Nannocloropsis oculata*	*A*: 120 mg L^{-1} Fe_3O_4/PS; *X*: 500 mg L^{-1}; pH: 4; PT: 180 min; *V*: 300 mL	–	HE = 96%	Chu et al. (2020)
Magnetic separation	*Chlorella vulgaris*	*A*: 0.199 g Fe_3O_4/g of dry algae; *X*: 1.683 g L^{-1}; pH: 4	–	HE = 70.8%	Gerulová et al. (2022)
Magnetic separation (coated MNP)	*Chlorella vulgaris*	*A*: 0.199 g Fe_3O_4-PEI/g of dry algae; *X*: 1.683 g L^{-1}; pH: 4	–	HE = 87.8%	Gerulová et al. (2022)
Flotation (with biological extract)	*Desmodesmus brasiliensis*	*A*: 0.035 g L^{-1} CAPB; *X*: 1.35 g L^{-1}; pH: 3; PT: 15 min	–	HE = 93.6%	Shao et al. (2018)
Electrolytic flotation (stainless steel cathode and carbon anode)	*Chlorella vulgaris*	*X*: 0.55 g L^{-1}; PT: 60 min; *V*: 250 L; electric current of 3 A; current density of 0.01 A cm^{-2}	2.73 kWh kg^{-1}	HE = 52.2%	Luo et al. (2017)

(continued)

Table 3.1 (continued)

Harvesting method	Microalgae	Operational conditions	Energy consumption	Results	References
Dissolved air flotation	*Dunaliella salina*	*A*: 0.0085 mol L^{-1} NaOH; *X*: 0.4–0.6 g L^{-1}; *V*: 600 mL; PT: 7 min	–	HE = 80%	Besson and Guiraud (2013)
Dissolved air flotation	*Chlorella vulgaris*	Initial OD_{750} = 0.8; *V*: 100 mL; pH = 6; bubble volume = 50 mL	–	HE = 30%	Demir-Yilmaz et al. (2023)
Positive dissolved air flotation (coated bubbles)	*Chlorella vulgaris*	*A*: 25 mg L^{-1} polyoctyl chitosan; Initial OD_{750} = 0.8; *V*: 100 mL; pH = 6; bubble volume = 50 mL	–	HE = 60%	Demir-Yilmaz et al. (2023)
Flotation (buoyant bead)	*Scenedesmus obliquus*	*A*: 30 mg L^{-1} chitosan, 0.55 mg L^{-1} sodium borosilicate microspheres; *X*: 9.4 × 10^6 cells mL^{-1}; pH: 7.5; *V*: 2 L; PT: 2 min	–	HE = 92.5%	Zou et al. (2018)
Flotation (buoyant bead)	*Chlorella vulgaris*	*A*: 20 mg L^{-1} chitosan, 0.55 mg L^{-1} sodium borosilicate microspheres; *X*: 7.6 × 10^6 cells mL^{-1}; pH: 7.5; *V*: 2 L; PT: 2 min	–	HE = 83.8%	Zou et al. (2018)
Microfiltration (membrane ceramic)	*Arthrospira* sp.	*X*: 0.66 g L^{-1}; *F*: 230 L m^{-2} h^{-1}; PT: 150 min	0.91 kWh m^{-3}	CF = 12.5	Jana et al. (2018)
Microfiltration (membrane PVDF)	*Aurantiochytrium* sp.	*X*: 25 g L^{-1}; FR: 8 L min^{-1}; PT: 240 min; TMP: 100 kPa	–	HE = 97.3%	Kim et al. (2015a)
Ultrafiltration (membrane PVDF)	*Aurantiochytrium* sp.	*X*: 25 g L^{-1}; FR: 8 L min^{-1}; PT: 240 min; TMP: 100 kPa	–	HE = 99.9%	Kim et al. (2015a)
Ultrafiltration (membrane PES)	*Aurantiochytrium* sp.	*X*: 25 g L^{-1}; FR: 8 L min^{-1}; PT: 180 min; *v*: 1600 rpm; TMP: 100 kPa	0.39 kWh kg^{-1}	HE = 99.8%	Kim et al. (2015a)
Ultrafiltration (membrane PES)	*Dunaliella salina*	*X*: 2–3 × 10^5 cells mL^{-1}; *F*: 31 L m^{-2} h^{-1}	0.87 kWh m^{-3}	CF = 5.9	Monte et al. (2018)
Ultrafiltration (membrane PES)	*Dunaliella salina*	*X*: 4 × 10^5 cells mL^{-1}; *F*: 22 L m^{-2} h^{-1}	–	CF = 16.4	Monte et al. (2018)
Ultrafiltration	*Scenedesmus acuminatus*	*X*: 1 g L^{-1}; *V*: 10000 L; *F*: 53.6 L m^{-2} h^{-1}	–	HE = 93%	Wang et al. (2019a)
Submerged microfiltration (PVDF)	*Chlorella vulgaris*	*X*: 0.41 g L^{-1}; *V*: 25 L; *F*: 38.3 L m^{-2} h^{-1}	0.27 kWh m^{-3}	HE = 98%	Bilad et al. (2012)
Forward osmosis (FO) (membrane CTA)	*Chlorella vulgaris*	Draw solution: 750 mL of artificial seawater (35 g L^{-1} NaCl); *X*: 0.5–2 g L^{-1}; PT: 4 h; *V*: 250 mL	–	DW rate = 2 L m^{-2} h^{-1}	Buckwalter et al. (2013)
Forward osmosis (membrane CTA)	*Chlorella* sp. ADE4	Draw solution: SWRO concentrate; *X*: 0.93 g L^{-1}; FR: 0.85 L min^{-1}; PT: 10 h	–	DW rate = 4.8 L m^{-2} h^{-1}	Kim et al. (2015b)

A additives, CAPB cocamidopropyl betaine, *CD* current density, *CF* concentration factor, *CTA* cellulose triacetate, *DW* rate: dewatering rate, *F* flux, *FR* flow rate, *HE* harvesting efficiency, *MNP* magnetic nanoparticle, *PEI* polyethyleneimine, *PES* polyether sulfone, *PT* processing time, *PVDF* polyvinylidene fluoride, *SWRO* seawater reverse osmosis, *TMP* transmembrane pressure, *V* volume, *v* velocity, *X* biomass concentration

methods are centrifugation and filtration, where, despite their high energy consumption, no toxic substances are added, the biomass maintains its properties during the process, and high recovery rates are obtained (Figueiredo et al. 2022).

3.2.1 Centrifugation

Centrifugation uses centrifugal forces to separate microalgal biomass from the culture medium (Kowthaman et al. 2022). It is the most commonly used method since it achieves high separation efficiencies (95–99%) and can be used in microalgal production for high-value products (nutraceuticals, cosmetics, and pharmaceuticals) (Mkpuma et al. 2022). Centrifugation is not viable for low-value products (biofuels) due to the high energy inputs needed.

Industrial centrifuges have two main configurations: (i) with a fixed wall (hydrocyclone) and (ii) with a rotary wall, also known as sedimentary centrifuges (disc-stack centrifuge, decanter centrifuge, multichamber centrifuge, and tubular centrifuge) (Najjar and Abu-Shamleh 2020; Ghazvini et al. 2022). Within these designs, the most widely used to harvest microalgae are the disc-stack and decanter centrifuges, as they achieve better results than other types of centrifuges (Laraib et al. 2021; Liber et al. 2020). In disc-stack centrifugation, microalgae are separated based on their density. In this process, the less dense particles flow to the interior of the discs, toward the center, whereas the denser particles migrate to the outside of the discs, close to the walls (Najjar and Abu-Shamleh 2020; Laraib et al. 2021). Although these centrifuges can achieve high harvesting efficiencies and can harvest small-sized particles present in a low-density culture, they require high energy consumption and can cause cell rupture (Najjar and Abu-Shamleh 2020; Kumar et al. 2023; Singh and Patidar 2018). The decanter centrifuge consists of a horizontal conical bowl with a screw conveyor inside that rotates in the same direction but at different speeds. It is the screw conveyor that transports the deposited solids out of the centrifuge (Najjar and Abu-Shamleh 2020). This centrifuge works continuously, has low maintenance costs, and can handle large volumes of liquids and high solid concentrations (Kumar et al. 2023; Roy and Mohanty 2019). However, the decanter centrifuge has high energy consumption and cannot separate small particles (<15 μm) (Laraib et al. 2021). The multichamber centrifuge comprises multiple tubular containers placed coaxially so that each chamber collects particles of a specific size due to the application of different centrifugal forces in the chambers. A tubular centrifuge consists of a long tube and rotating bearings at each end. Due to the strong centrifugal force to which the suspension is subjected, the liquid will flow to the top of the centrifuge, while the solid particles migrate out of the container (Najjar and Abu-Shamleh 2020; Ghazvini et al. 2022).

Although multichamber and tubular centrifuges have high recovery efficiencies, they cannot work continuously, requiring periodic manual cleaning (Esteves et al. 2020). Finally, hydrocyclones consist of a cylindrical section attached to a conical base. In this process, a radial centrifugal force is applied to the fluid as it is injected tangentially into the upper part of the cylindrical region, allowing the fluid's particles to be removed. Even though hydrocyclones can function continuously, their efficiency fluctuates depending on the fluid dynamics, leading to incomplete solid–liquid separation (Ghazvini et al. 2022; Singh and Patidar 2018).

Compared with other harvesting methods, centrifugation offers additional advantages (Siddiki et al. 2022; Esteves et al. 2020): (i) it is suitable for all types of microalgal species; (ii) it is efficient and fast, preventing the deterioration of biomass quality; (iii) it can process large volumes of slurry; and (iv) it is a chemical-free technique. However, this method has disadvantages, such as the risk of cell damage (which may occur when using microalgal species sensitive to high g-force) and high energy costs. To decrease energy consumption, a pre-concentration of the biomass should be done through flocculation, flotation, or filtration, for example. By pre-concentrating the biomass, the culture volume that needs to be harvested is significantly lowered, reducing the energy costs of the process (Muylaert et al. 2017). Fasaei et al. (2018) performed a techno-economic assessment of microalgal harvesting systems. They concluded that combining chitosan flocculation with centrifugation required 0.07 kWh kg^{-1}, whereas centrifugation alone required 1.94 kWh kg^{-1}. However, the use of flocculants can further increase the overall process costs. Badvipour et al. (2016) also evaluated the harvesting costs of different methods. In this study, the authors found that when only centrifugation was used, the cost of the process was 1184 \$ $mton^{-1}$, but when using membrane filtration followed by centrifugation, the cost of the process was reduced by 875 \$ $mton^{-1}$.

3.2.2 Gravity Sedimentation

Gravity sedimentation is the simplest mechanical separation method to harvest microalgae (Ghazvini et al. 2022). Settling tanks can be used for sedimentation, although they have a big footprint. Lamella separators, also known as inclined settlers, are receiving more attention. These settlers are composed of inclined plates that increase the effective area available for sedimentation and minimize the footprint. They can also be modified to improve the sedimentation process, for example, by adding vibration or altering the surface to prevent microalgae from adhering (Smith et al. 2013; Muylaert et al. 2017).

The sedimentation process can be explained by Stokes' law, according to which the settling velocity of a microalgal cell is directly affected by the mass density difference between the microalgal cells and the culture medium and by the square of the cells' radius (Roselet et al. 2019; Roy and Mohanty 2019). In the cases where the cells have a higher density than the culture medium, they settle due to gravitational forces. Besides density, the size and shape of microalgal cells also influence the settling velocity (Farfan-Cabrera et al. 2022). Theoretically, according to Stokes' law, a spherical cell, like *Chlorella,* has an average settling rate of 0.1 m d^{-1} (Ghazvini et al. 2022; Roy and Mohanty 2019). However, microalgal cells have different complex morphologies, affecting their sedimentation rate. Settling velocity has been reported to be between 0.4 and 2.2 m d^{-1} for microalgae sizes 10–1000 μm, but no direct correlation was observed between these factors (Peperzak et al. 2003).

Unlike other mechanical methods, the sedimentation process is simple, does not damage the microalgal structure, and has negligible energy and operational costs (Roselet et al. 2019; Ghazvini et al. 2022). Nevertheless, microalgae are microscopic, and cultures achieve low biomass concentrations, with densities close to the water, making sedimentation difficult (Roselet et al. 2019; Alam and Wang 2019). Microalgae also present carboxylic (–COOH) and amine (–NH_2) groups on the cell surface, with a negative surface charge at typical culture pH because these groups tend to be deprotonated. The negatively charged surface combined with the allogenic organic matter (AOM) carried by the cells does not allow microalgae to easily aggregate, remaining in stable suspensions instead (Roselet et al. 2019). Therefore, sedimentation of most microalgal species is only reliable with a previous flocculation process to increase cell density (through flocs' formation). With flocculation, settling velocities higher than 52 m d^{-1} have been reported (Ghazvini et al. 2022).

3.2.3 Flocculation

Flocculation is an efficient, low-energy process that works with various microalgal species and large culture volumes (Kumar et al. 2023; Singh and Patidar 2018). By aggregating the microalgal cells, flocculation can facilitate subsequent separation by sedimentation, filtration, flotation, or centrifugation. Briefly, flocculation starts after destabilizing suspensions or solutions under rapid mixing (in a process called coagulation). It corresponds to the slow formation of large agglomerates from the previously destabilized particles (Branyikova et al. 2018; Laamanen et al. 2021). Most authors do not distinguish between these processes when referring to microalgal harvesting. Thus, this differentiation will not be made in this chapter either.

To reduce the zone of influence of the repulsion forces between microalgal cells, it is necessary to minimize the zeta potential, which can be done through the compression of the electrical double layer by increasing the ionic strength of the culture medium (Davis 2010; Sawyer et al. 2003). Adding cationic flocculants or altering the pH of the medium can induce flocculation (Kumar et al. 2023). In fact, according to Roy and Mohanty (2019), this technique stimulates explicitly ionic interaction between the cells, flocculants, and salts already present in the growth media. Selecting the right flocculant is essential to avoid contamination of the microalgal biomass. Therefore, the chosen flocculant should (i) be ecological, (ii) be low-cost and nontoxic, (iii) not change the biomass' quality, (iv) be efficient at low concentrations, and (v) allow the culture medium to be reused (Singh and Patidar 2018; Roy and Mohanty 2019).

The flocculation process is typically explained by four mechanisms that can co-occur (Davis 2010; Sawyer et al. 2003; Kumar et al. 2023): (i) charge neutralization, (ii) sweep mechanism (enmeshment in a precipitate), (iii) charge patching, and (iv) bridging mechanism. Charge neutralization occurs when flocculants with a different charge from microalgae cells are added to the suspension. The surface charge of the microalgal cells is neutralized, which reduces electrostatic repulsion, resulting in cell aggregation and floc formation. Sweeping happens when high amounts of inorganic flocculants are added to the culture suspension. Their presence results in large amorphous metal hydroxide precipitates that capture or enmesh microalgal cells. Charge patching is when a polymer with a high charge density binds to microalgal cells. This results in an irregular polarity of the microalgae's surface charge, which causes particle aggregation and the subsequent formation of large flocs. Furthermore, when polymeric flocculants are used, the bridging mechanism occurs. In this mechanism, the polymer or colloid bonds to the surfaces of two different particles, forming flocs (Kumar et al. 2023; Yin et al. 2020; Branyikova et al. 2018). Flocculation for microalgal harvesting can be performed by chemical, biological, or physical methods, detailed in the following sections.

3.2.4 Chemical Flocculation

Chemical flocculation is the primary approach for considerably improving the economic and energetic balance of harvesting different microalgal species (Branyikova et al. 2018). In this process, floc formation is achieved by adding organic or inorganic compounds with flocculant activity. Chemical flocculants used in microalgal harvesting are classified into metallic salts (inorganic flocculants) and synthetic organic polymers (polyelectrolytes) (Roy and Mohanty 2019). The most commonly used metallic salts include aluminum or iron-based salts due to their high efficiency, availability, and low cost. The high recovery rates obtained using these compounds are related to their ability to form multi-charged polynuclear complexes in solution (Min et al. 2022). Pre-hydrolyzed metal salts such as polyaluminum chloride are gaining increased interest, as their development enables the production of the correct hydrolysis species, regardless of the treatment process conditions (Bratby 2016). The negative environmental impacts (i.e., non-biodegradability and the production of large amounts of contaminated sludge that require a complex and expensive treatment) are the major drawbacks of using inorganic flocculants (Al-Sahari et al. 2020). Polyelectrolyte, in turn, is a class of polymers that contain ionizable functional groups in their structure (Bratby 2016). Polyelectrolytes produce lower volumes of contaminated sludge than inorganic flocculants, and the harvested biomass is free from metallic residues (Teh et al. 2016). However, they are more expensive, difficult to biodegrade, and the monomers, subunits from which they are derived, are usually toxic (Roy and Mohanty 2019).

Due to the environmental problems associated with chemical flocculants, they are not recommended for food and feed applications. Considering their expected greener production and usage, natural organic flocculants have emerged as alternatives (Kirnev et al. 2018; Matter et al. 2019). These flocculants are abundant, cheaper, environmentally friendly, biodegradable, and nontoxic. In addition, these compounds are not affected by changes in the pH, thus having a wide range of uses and resulting in high harvesting efficiencies (Al-Sahari et al. 2020; Ang and Mohammad 2020). Since natural flocculants have a polymeric structure with charged functional groups, the main mechanisms involved in the destabilization process are charge neutralization and the formation of bridges between cells (Min et al. 2022). Despite the commercialization challenges of natural flocculants, including regulatory approval, market awareness, support, recognition, and local competition, natural tannin-based flocculants (Kirnev et al. 2018; Roselet et al. 2016) and chitosan derived from crustaceans have been successfully applied in microalgal harvesting as alternatives to chemical coagulants (Japar et al. 2021; Xu et al. 2013; Kirnev et al. 2018).

3.2.5 Biological Flocculation

The terms "auto-flocculation" and "bio-flocculation" are sometimes used in an undifferentiated way. Nevertheless, they describe different mechanisms. Auto-flocculation refers to the spontaneous formation of flocs by microalgae without adding flocculants (Min et al. 2022; Roy and Mohanty 2019). Some microalgae flocculate in response to environmental changes in pH, nutrient content (magnesium and calcium ions), and dissolved oxygen concentration. Culture aging can also cause auto-flocculation of the microalgae *Scenedesmus obliquus* and *Botryococcus braunii* (Min et al. 2022). On a laboratory scale, it was found that at basic pH (>9), auto-flocculation can occur due to cell charge neutralization with divalent cations like Ca^{2+} and Mg^{2+} (Roy and Mohanty 2019). For instance, at high pH, positively charged calcium phosphate precipitates can be formed in phosphate-rich waters (Roselet et al. 2019; Alam and Wang 2019). These precipitates interact with the negatively charged microalgal cells, forming flocs due to charge neutralization or sweep flocculation. Auto-flocculation can also be induced by magnesium hydroxide or calcium carbonate precipitates in waters with low phosphate concentrations. Since these precipitates are not as harmful as metals, there are fewer issues with biomass contamination (Muylaert et al. 2017). Moreover, some researchers have discovered that dissolved oxygen increases the number of binding sites on microalgae cells' surface, facilitating flocs' formation (Laraib et al. 2021). The auto-flocculation process is simple, inexpensive, and environmentally friendly and allows the reuse of culture medium (because chemical compounds are not used) (Laraib et al. 2021; Min et al. 2022). Although this process may be advantageous at the laboratory scale, it is still not suitable for large-scale harvesting since (Roselet et al. 2019; Roy and Mohanty 2019): (i) large quantities of acids and bases are needed to adjust the pH, (ii) it is time-consuming, and (iii) it is only applicable for some microalgal species. Hence, more research is required regarding the energy consumption of auto-flocculation processes.

The flocculation of microalgae cells can also be performed through microorganisms (fungi, bacteria, and algae) or their extracellular biopolymers in a process known as bio-flocculation (Laraib et al. 2021; Matter et al. 2019). In bio-flocculation, large flocs are formed through the interaction of cells with microorganisms without the addition of chemical flocculants. The main mechanisms involved in bio-flocculation are charge neutralization, electrostatic patch, and interparticle bridging (Roy and Mohanty 2019). In this technique, the flocculating microorganisms can be cocultured with microalgae in the presence of an organic carbon source or added just before the harvesting step (Esteves et al. 2020). Bacteria are the most common microorganisms used in this procedure (Yin et al. 2020). Microalgae–bacteria flocs occur when bacteria release biopolymers, like extracellular polymeric substances (EPSs), that alter the functional groups on the surface of microalgal cells and cause them to aggregate. Although bacterial bio-flocculation is receiving much attention, the potential for bacterial contamination, which affects the quality of the biomass produced, must be considered before its industrial use (Liu et al. 2023). On the other hand, when filamentous fungi are used, flocculation occurs due to the algae–fungi co-pelletization process through the interaction of the microalgal surface with the fungal mycelia, which is positively charged (Zhao et al. 2019). Even though this method is efficient, environmental contamination remains a concern (Matter et al. 2019). Algal–algal bio-flocculation happens when non-flocculating microalgae are harvested with the addition of self-flocculating microalgae (or their EPSs) (Matter et al. 2019). Increasing flocculation through bridging and patching mechanisms promotes the harvesting of non-flocculating microalgae (Alam et al. 2016). In bio-flocculation, algae can provide valuable advantages over bacteria and fungi regarding microbial contamination. Since this method does not require different growing conditions, it is more affordable and avoids contamination. To increase the sustainability of this method, it is still important to research the biochemical and physiological mechanisms that make flocculation happen (Matter et al. 2019; Nazari et al. 2020).

Moreover, bio-flocculation is an environmentally friendly and energy-efficient process that does not affect the cell's integrity (Branyikova et al. 2018). However, bio-flocculation is not considered safe for human food or animal feed applications due to potential microbial contamination. In addition, coculturing requires significant amounts of organic carbon sources and high energy costs (Esteves et al. 2020; Roy and Mohanty 2019).

3.2.6 Physical Flocculation

Flocculation can also be achieved through physical forces, such as ultrasound-assisted flocculation, electro-flocculation, and magnetic separation. Since this method does not use chemical compounds, it is possible to avoid biomass contamination (Li et al. 2020; Branyikova et al. 2018). In ultrasound-assisted flocculation, high frequency, in the order of MHz, and low amplitude ultrasounds allow microalgal cells to aggregate while maintaining their integrity. On the contrary, low frequency (kHz) and high amplitude can cause cell lysis (Branyikova et al. 2018). The microalgae cells tend to form agglomerates in the ultrasonic field by aggregating into the knots of the ultrasonic wave. Due to gravity, larger agglomerates quickly sediment from the fluid (Li et al. 2020). Bosma et al. (2003) studied ultrasound (resonance frequency of 2.1 MHz) with enhanced sedimentation to harvest *Monodus subterraneus*. At 4–6 L d^{-1} flow rates, removal efficiencies higher than 90% were achieved at the laboratory scale. Although ultrasound-assisted flocculation is suitable for continuous operation, the process scale-up is challenging and requires special equipment and high energy consumption (Vandamme et al. 2013; Li et al. 2020).

In electro-flocculation, negatively charged microalgal cells are destabilized using electrodes forming an electric field. The electrodes can be sacrificial electrodes, with the electrolytic release of metal ions, or non-sacrifice electrodes, with nonreactive cathodes and anodes (Matter et al. 2019). When a sacrificial metal anode is used, positively charged metal ions are released. These metal cations can generate flocs by electrostatically attracting negatively charged microalgae (Matter et al. 2019; Li et al. 2020). Although the probability of biomass contamination with sacrificial electrodes is low, non-sacrificial electrodes are preferable to harvest microalgae for food and feed applications (Silva et al. 2021). When non-sacrificial electrodes are used under direct current conditions, the negatively charged microalgal cells move toward the anode electrode. When they get there, they become neutralized and flocculating (Matter et al. 2019). The chemical nature of the electrodes strongly influences the process, with aluminum electrodes being the most effective, although toxic to human and animal health (Liu et al. 2018). Current density is also an important parameter. Higher current density allows for a more efficient and faster harvesting process but increases energy consumption (Roy and Mohanty 2019). Therefore, it is crucial to balance this parameter and energy consumption. At a laboratory scale, electro-flocculation allows microalgae harvesting with low energy costs (Liu et al. 2017; Lucakova et al. 2021). However, on an industrial scale, the cost of electrodes and their maintenance is high, making the process unappealing (Min et al. 2022; Silva et al. 2021).

In magnetic separation, flocculation and separation occur simultaneously (Branyikova et al. 2018). In this method, micro- or nano-sized magnetic particles are used to mark the microalgal cells, allowing them to be easily separated from the culture medium by applying an external magnetic field (Yin et al. 2020; Roy and Mohanty 2019). Nanoparticles have a larger surface area than microparticles, increasing their biomass recovery capacity (Mkpuma et al. 2022). The ideal magnetic particle should have high adsorption potential to ensure efficient harvesting and be low-cost, nontoxic, stable, and reusable (Roy and Mohanty 2019). The most commonly used particles are iron oxide (Fe_3O_4) nanoparticles, also known as magnetite, due to their specific surface area and superparamagnetism (Matter et al. 2019; Han et al. 2020). However, when naked magnetite is used, the pH value needs to be controlled for the harvesting process to be successful. Increasing harvesting efficiency and avoiding pH value adjustment can be achieved by modifying the surface of magnetic nanoparticles with cationic functional groups (Han et al. 2020). There are two approaches for adding the functional groups: "immobilized on" or "attached-to" strategies. In the "immobilized on" approach, the naked magnetic particles are functionalized with cationic polyelectrolytes that help in binding with the microalgae. In the "attached-to" strategy, microalgal cells are coated with polymeric ligands to help attach bare magnetic particles. Without coating, magnetic particles are more unstable and are more likely to aggregate into large clusters, which decreases their dispersion and harvesting efficiency. As a result, the "immobilized" strategy is more popular among these two approaches because it enables greater harvesting efficiency under the same conditions (Roy and Mohanty 2019; Han et al. 2020).

By adding coating, magnetic nanoparticles can be multifunctional. In other words, in addition to harvesting microalgae, they can cause cell disruption, facilitating the extraction of lipids used in the biorefinery (Han et al. 2020; Seo et al. 2016). Seo et al. (2016) studied using Fe_3O_4 nanoparticles decorated with cetrimonium bromide (CTAB), a cationic surfactant, to harvest microalgae and disrupt cells simultaneously. The authors demonstrated that nanoparticles functionalized with CTAB could disrupt cells and effectively harvest microalgae, reaching an efficiency of 95%. Furthermore, they managed to regenerate the nanoparticles by adding an anionic surfactant, sodium dodecyl sulfate (SDS), and using sonication, achieving a reuse efficiency of around 80%. The ability to regenerate and reuse flocculants is one of the main challenges that needs to be resolved to reduce operational costs (Matter et al. 2019). Although biomass harvesting using magnetic separation is fast and efficient, the high cost of magnetic particles has to be considered when applying this process on a large scale (Roy and Mohanty 2019; Matter et al. 2019).

3.3 Flotation

Flotation is a harvesting technique where air bubbles are pressurized in the culture medium to attach and carry the microalgal cells to the surface of the liquid to be collected by skimming processes (Junior et al. 2020). This method can reach high concentration factors because the water is drained by gravity, making it more efficient under the same conditions than sedimentation (Laamanen et al. 2016). Furthermore, flotation is viable on a large scale, requires a low initial investment in equipment, and has a small footprint and a short operating time (Luo et al. 2017; Laamanen et al. 2016). Nevertheless, this method still presents high energy consumption that needs to be optimized to become economically viable on a large scale (Roy and Mohanty 2019).

Three stages comprise the flotation process (Laamanen et al. 2021): (i) generation of bubbles or addition of low-density microspheres, (ii) interaction and adhesion of microalgae with bubbles or microspheres, and (iii) rise and development of a concentrated layer of microalgae on the liquid's surface. Depending on the techniques used for bubble formation and bubble size, there are different types of flotation: dissolved air flotation, dispersed air flotation, dispersed ozone flotation, and electrolytic flotation (Singh and Patidar 2018). Furthermore, buoy-bead flotation can also be used for microalgal harvesting, in which low-density microspheres are used instead of bubbles (Zou et al. 2018).

Dissolved air flotation is the most commonly used method. Supersaturated air is applied, producing bubbles with sizes between 10 and 100 μm, leading to high energy consumption. Dispersed air flotation is a cheaper and less efficient technique; the air is injected through a mechanical agitator or a porous material, and the bubble size ranges between 700 and 1500 μm (Singh and Patidar 2018). In dispersed ozone flotation, ozone is used instead of air. This method can be applied to harvest and disrupt microalgal cells, but it is expensive and can affect the quality of the harvested biomass (Kowthaman et al. 2022). In electrolytic flotation, unlike traditional methods, hydrogen and oxygen bubbles are generated in the medium by pairs of electrodes instead of being injected by external sources. The electrodes can be made from various materials, such as iron and aluminum. To avoid metal contamination in biomass, non-sacrificial anodes (e.g., carbon) are being studied and used (Luo et al. 2017; Laamanen et al. 2016). Moreover, electrolytic flotation is mainly used to harvest marine microalgae since seawater has a higher electrical conductivity than freshwater (Esteves et al. 2020). The release of hydrogen bubbles, which raises health and safety concerns, remains the method's primary drawback (Laamanen et al. 2016).

In flotation, biomass separation is greatly influenced by the size of the bubbles and their adherence to the microalgal cells (Kumar et al. 2023). To increase the adhesion probability between the air bubbles and the microalgae, surfactants or flocculants are added (Singh and Patidar 2018). Therefore, flocculation commonly precedes flotation to facilitate the harvesting process and increase recovery rates (Luo et al. 2017, Laamanen et al. 2016). Another possibility is to modify the surface of the bubbles with polymers or surfactants by adding them to the saturator. When bubbles form, they become coated with these compounds, changing their surface charge, increasing the swept area, and, consequently, the capture efficiency. This method is called positive dissolved air flotation and avoids the flocculation step, which shortens the harvesting process and lowers costs (Laamanen et al. 2016; Demir-Yilmaz et al. 2023). Demir-Yilmaz et al. (2023) studied the impact of bubble functionalization on microalgal harvesting. They conducted dissolved air flotation experiments using clean and functionalized bubbles. The bubbles were coated with polyoctyl chitosan (PO-chitosan) in the second scenario. The authors demonstrated that the functionalization of the bubbles increased the separation efficiency from 30% with clean bubbles to almost 60% with coated bubbles. These results proved that it is possible to increase the efficiency of the flotation process by coating the bubbles.

Flotation is a low-cost and fast method, suitable for some microalgal species, with moderate recovery rates (between 50% and 99%) (Esteves et al. 2020). However, possible metal contamination during electrolytic flotation or the presence of contaminants due to the use of flocculants makes this method unsuitable for food applications (Barros et al. 2015). For this reason, this process is mainly used for low-added-value target products such as biofuels (Tang et al. 2020). Nevertheless, some alternatives are being developed to overcome the contamination issues, such as using non-sacrificial anodes and bio-flocculants (Laamanen et al. 2021). Zou et al. (2018) tested a buoy-bead flotation using a pre-flocculation with the bio-flocculant chitosan as an alternative to the chemical flocculant ferric chloride (which has high toxicity and is nonbiodegradable). The authors demonstrated that the highest harvesting efficiencies were achieved when chitosan was used as a pre-flocculant (92%), compared with 88% of efficiency obtained with ferric chloride. These results demonstrated that chitosan is a suitable alternative to ferric chloride and can be potentially used in the food industry due to its low toxicity.

3.4 Filtration

Filtration is a harvesting method that uses a filter or a membrane to recover microalgal biomass from the culture medium (Junior et al. 2020). Filtration methods can be classified according to (i) the membrane pore size (macrofiltration, microfiltration, ultrafiltration, nanofiltration, and reverse osmosis) (Tan et al. 2018), (ii) the configuration (dead-end, cross-flow, submerged, forward osmosis, and dynamic filtration) (Ennaceri et al. 2023), and (iii) the type of pressure applied (gravitational, pressure, and vacuum) (Muylaert et al. 2017). The membrane pores used in macrofiltration are greater than 10 μm. Hence, this type of filtration is mainly applied to large microalgae, such as filamentous microalgae. Microfiltration occurs when the pore size is within the range of 0.1–10 μm, while ultrafiltration, a process used to separate organic molecules, is achieved with a pore size between 0.01 and 0.1 μm. Typically, microfiltration and ultrafiltration are the primary techniques used in the microalgal industry. Regarding this, Jana et al. (2018) compared the use of microfiltration and ultrafiltration for harvesting *Arthrospira* sp. The authors concluded that microfiltration led to a higher concentration factor, but the ultrafiltration membrane showed less fouling potential. About the energy consumption (associated with pumps and air backwash), microfiltration achieved lower consumption values (0.91 kWh m^{-3}) than ultrafiltration (2.62 kWh m^{-3}). The authors also tested the effect of bio-flocculant addition, observing that fouling decreased in both filtration techniques.

In terms of configuration, in dead-end filtration, the liquid flow is perpendicular to the membrane, forcing microalgal cells to settle on the surface of the membrane. Dead-end filtration is mainly used with low-concentrated cultures and is highly prone to fouling, so it is not economically viable for microalgal harvesting. In cross-flow filtration, the permeate passes perpendicularly through the filter, and the culture flows parallel to the membrane, creating shear stress that reduces fouling. High shear pressure, however, can cause cell damage in some microalgae and induce their rupture, aggravating the fouling problem (Muylaert et al. 2017). When submerged filtration is used, the membrane modules are immersed directly in the culture medium. In this system, air bubbles interact with the membrane surface, creating tangential shear, which reduces fouling (Zhao et al. 2023). This process does not damage the microalgal cells, but its efficacy depends on the aeration rate (Ennaceri et al. 2023; Zhao et al. 2023; Mkpuma et al. 2022). To overcome this limitation, turbulence was analyzed, and the dynamic membrane system was created. Unlike the submerged system, the vibrating membrane prevents foulants from adhering to its surface, thereby controlling fouling (Zhao et al. 2023; Mkpuma et al. 2022). Although this method increases membrane flux and does not cause damage to microalgae, its mechanical complexity limits its expansion (Ennaceri et al. 2023). More recently, forward osmosis (FO), a passive filtration configuration, has gained interest as a technique for harvesting microalgae due to its resistance to fouling (Mkpuma et al. 2022; Ennaceri et al. 2023). This method dehydrates the microalgae culture using an osmotic gradient by passing water from the feed solution to the draw solution (Kim et al. 2015b). This occurs due to the high osmotic pressure in the draw solution, which forces water to pass the semipermeable membrane (Buckwalter et al. 2013). Since FO does not use hydraulic pressure, it is ideal for harvesting microalgae sensitive to shear stress (Ennaceri et al. 2023). However, this approach has some drawbacks, including a low final microalgal concentration and the need to regenerate the draw solution, which raises energy costs (Zhao et al. 2023).

The major limitation of filtration methods is, indeed, the membrane fouling. This phenomenon occurs because microalgae release a great variety of AOM and tend to form biofilms. The surface charge and the molecular weight of AOM, the membrane characteristics, and the operating conditions affect the membrane fouling potential, negatively impacting the process efficiency (Novoa et al. 2021). Due to this, antifouling strategies are being studied and developed (Mkpuma et al. 2022; Zhao et al. 2021). Nonetheless, filtration is a cost-effective and chemical-free method, has low energy consumption (pressure and gravitational filtration) and high recovery efficiencies (70–90%) with low shear stress, and enables water recycling. However, frequent membrane replacement and cleaning due to membrane fouling and pumping costs increase operational and maintenance costs (Siddiki et al. 2022). Also, the process is time-consuming because it entails using pressure or a vacuum. It is not suitable for processing large culture volumes. In the case of vacuum filtration, the energy consumption is high (Kowthaman et al. 2022).

3.5 Conclusions and Future Perspectives

The desire to commercialize microalgal biomass on a large scale is growing, but the harvesting process is preventing it from happening. Since microalgae are such a diverse group of microorganisms, there is no standard approach for harvesting them. When deciding which process to use, it is important to consider the final use of the biomass as well as the cost, danger of contamination, harvesting efficiency, suitability for industrial usage, and length of the process. All methods covered in this chapter can be used to concentrate microalgae, although the efficiency of the process varies. One of the most used techniques is centrifugation since it can achieve high harvesting efficiencies. However, because of the energy requirements, it is expensive and not viable for low-value products. Sedimentation is the most straightforward and affordable method. Yet, since microalgae are so small and have low density, this approach only works if the flocculation process is carried out first. However, when chemical flocculants are used, the biomass could be contaminated. More recently, the application of natural organic flocculants has aroused interest as they do not cause contamination. Although these substances have already proven to be effective, more research is still required to address marketing issues. When biological flocculants are used, there is a risk of microbiological contamination. In physical flocculation, the only viable option is magnetic separation, which allows microalgae to be harvested and facilitates lipids extraction. However, optimizing nanoparticle recovery methods to reduce operating costs before large-scale use is necessary. Flotation can obtain high concentration factors; however, energy consumption needs to be reduced before employing it extensively. For the filtration process, it is crucial to research methods that minimize membrane fouling. Therefore, the future of the harvesting process involves improving the techniques described here or combining them. More research needs to be done to develop an economically sustainable harvesting process.

Acknowledgments This work was financially supported by (i) LA/P/0045/2020 (ALiCE) and UIDB/00511/2020-UIDP/00511/2020 (LEPABE), funded by national funds through FCT/MCTES (PIDDAC), and (ii) Project PhotoBioValue (ref. PTDC/BTA-BTA/2902/2021), funded by FEDER funds through COMPETE2020—Programa Operacional Competitividade e Internacionalização (POCI) and by national funds (PIDDAC) through FCT/MCTES. C. Maia, A.F. Esteves, and E.M. Salgado thank FCT for the financial support of their work through the FCT PhD Research Scholarships 2023.04096.BD, 2020.05477.BD, and 2021.07412.BD, respectively.

References

Al-Sahari M, Al-Gheethi AAS, Radin Mohamed RMS (2020) Natural coagulates for wastewater treatment; a review for application and mechanism. In: Prospects of fresh market wastes management in developing countries. Springer, pp 17–31

Alam MA, Vandamme D, Chun W, Zhao X, Foubert I, Wang Z, Muylaert K, Yuan Z (2016) Bioflocculation as an innovative harvesting strategy for microalgae. Rev Environ Sci Biotechnol 15:573–583

Alam MA, Wang Z (2019) Microalgae biotechnology for development of biofuel and wastewater treatment. Springer

Alishah Aratboni H, Rafiei N, Garcia-Granados R, Alemzadeh A, Morones-Ramírez JR (2019) Biomass and lipid induction strategies in microalgae for biofuel production and other applications. Microb Cell Factories 18:1–17

Almomani F (2020) Algal cells harvesting using cost-effective magnetic nanoparticles. Sci Total Environ 720:137621

Ang WL, Mohammad AW (2020) State of the art and sustainability of natural coagulants in water and wastewater treatment. J Clean Prod 262:121267

Badvipour S, Eustance E, Sommerfeld MR (2016) Process evaluation of energy requirements for feed production using dairy wastewater for algal cultivation: theoretical approach. Algal Res 19:207–214

Barros AI, Gonçalves AL, Simões M, Pires JC (2015) Harvesting techniques applied to microalgae: a review. Renew Sust Energ Rev 41:1489–1500

Besson A, Guiraud P (2013) High-pH-induced flocculation–flotation of the hypersaline microalga Dunaliella salina. Bioresour Technol 147:464–470

Bilad MR, Vandamme D, Foubert I, Muylaert K, Vankelecom IF (2012) Harvesting microalgal biomass using submerged microfiltration membranes. Bioresour Technol 111:343–352

Bosma R, Van Spronsen WA, Tramper J, Wijffels RH (2003) Ultrasound, a new separation technique to harvest microalgae. J Appl Phycol 15:143–153

Branyikova I, Prochazkova G, Potocar T, Jezkova Z, Branyik T (2018) Harvesting of microalgae by flocculation. Fermentation 4:93

Bratby J (2016) Coagulation and flocculation in water and wastewater treatment. IWA publishing

Buckwalter P, Embaye T, Gormly S, Trent JD (2013) Dewatering microalgae by forward osmosis. Desalination 312:19–22

Chu F-J, Wan T-J, Chen H, Wu C-H, Kao P-M (2020) Magnetophoretic harvesting of Nannochloropsis oculata using iron oxide immobilized beads. Water 12:236

Cuellar-Bermudez SP, Kilimtzidi E, Devaere J, Goiris K, Gonzalez-Fernandez C, Wattiez R, Muylaert K (2020) Harvesting of Arthrospira platensis with helicoidal and straight trichomes using filtration and centrifugation. Sep Sci Technol 55:2381–2390

Davis ML (2010) Water and wastewater engineering: design principles and practice. McGraw-Hill Education

Demir-Yilmaz I, Ftouhi MS, Balayssac S, Guiraud P, Coudret C, Formosa-Dague C (2023) Bubble functionalization in flotation process improve microalgae harvesting. Chem Eng J 452:139349

Ennaceri H, Mkpuma VO, Moheimani NR (2023) Nano-clay modified membranes: a promising green strategy for microalgal antifouling filtration. Sci Total Environ 902:166479

Esteves AF, Almeida CJ, Gonçalves AL, Pires JC (2020) Microalgae harvesting techniques. In: Handbook of microalgae-based processes and products. Elsevier

Esteves AF, Pires JC, Gonçalves AL (2021) Current utilization of microalgae in the food industry beyond direct human consumption. In: Cultured microalgae for the food industry. Elsevier

Esteves AF, Soares SM, Salgado EM, Boaventura RA, Pires JC (2022) Microalgal growth in aquaculture effluent: coupling biomass valorisation with nutrients removal. Appl Sci 12:12608

Farfan-Cabrera LI, Franco-Morgado M, González-Sánchez A, Pérez-González J, Marín-Santibáñez BM (2022) Microalgae biomass as a new potential source of sustainable green lubricants. Molecules 27:1205

Fasaei F, Bitter J, Slegers P, Van Boxtel A (2018) Techno-economic evaluation of microalgae harvesting and dewatering systems. Algal Res 31:347–362

Figueiredo D, Ferreira A, Quelhas P, Schulze PS, Gouveia L (2022) Nannochloropsis oceanica harvested using electrocoagulation with alternative electrodes–an innovative approach on potential biomass applications. Bioresour Technol 344:126222

Gerulová K, Kucmanová A, Sanny Z, Garaiová Z, Seiler E, Čaplovičová M, Čaplovič Ľ, Palcut M (2022) Fe3O4-PEI nanocomposites for magnetic harvesting of chlorella vulgaris, chlorella ellipsoidea, microcystis aeruginosa, and auxenochlorella protothecoides. Nano 12:1786

Ghazvini M, Kavosi M, Sharma R, Kim M (2022) A review on mechanical-based microalgae harvesting methods for biofuel production. Biomass Bioenergy 158:106348

Gultom SO, Zamalloa C, Hu B (2014) Microalgae harvest through fungal pelletization—co-culture of Chlorella vulgaris and Aspergillus Niger. Energies 7:4417–4429

Han S-F, Jin W, Tu R, Gao S-H, Zhou X (2020) Microalgae harvesting by magnetic flocculation for biodiesel production: current status and potential. World J Microbiol Biotechnol 36:1–10

Hosseinkhani N, Mccauley JI, Ralph PJ (2022) Key challenges for the commercial expansion of ingredients from algae into human food products. Algal Res 64:102696

Huo S, Wang Z, Zhu S, Cui F, Zou B, You W, Yuan Z, Dong R (2014) Optimization of Alkaline Flocculation for Harvesting of

Scenedesmus quadricauda# 507 and Chaetoceros muelleri# 862. Energies 7:6186–6195

Jana A, Ghosh S, Majumdar S (2018) Energy efficient harvesting of Arthrospira sp. using ceramic membranes: analyzing the effect of membrane pore size and incorporation of flocculant as fouling control strategy. J Chem Technol Biotechnol 93:1085–1096

Japar AS, Takriff MS, Mohd Yasin NH, Mahmod SS (2021) Optimization of Chlorella biomass harvesting by flocculation and its potential for biofuel production. J Appl Phycol 33:1621–1629

Junior WGM, Gorgich M, Corrêa PS, Martins AA, Mata TM, Caetano NS (2020) Microalgae for biotechnological applications: cultivation, harvesting and biomass processing. Aquaculture 528:735562

Kim K, Shin H, Moon M, Ryu B-G, Han J-I, Yang J-W, Chang YK (2015a) Evaluation of various harvesting methods for high-density microalgae, Aurantiochytrium sp. KRS101. Bioresour Technol 198:828–835

Kim S-B, Paudel S, Seo GT (2015b) Forward osmosis membrane filtration for microalgae harvesting cultivated in sewage effluent. Environ Eng Res 20:99–104

Kirnev P, De Carvalho J, Miyaoka J, Cartas L, Vandenberghe L, Soccol C (2018) Harvesting Neochloris oleoabundans using commercial organic flocculants. J Appl Phycol 30:2317–2324

Kowthaman C, Kumar PS, Selvan VAM, Ganesh D (2022) A comprehensive insight from microalgae production process to characterization of biofuel for the sustainable energy. Fuel 310:122320

Kratzer R, Murkovic M (2021) Food ingredients and nutraceuticals from microalgae: main product classes and biotechnological production. Food Secur 10:1626

Kumar N, Banerjee C, Negi S, Shukla P (2023) Microalgae harvesting techniques: updates and recent technological interventions. Crit Rev Biotechnol 43:342–368

Laamanen C, Desjardins S, Senhorinho G, Scott J (2021) Harvesting microalgae for health beneficial dietary supplements. Algal Res 54:102189

Laamanen CA, Ross GM, Scott JA (2016) Flotation harvesting of microalgae. Renew Sust Energ Rev 58:75–86

Laraib N, Hussain A, Javid A, Noor T, Ahmad Q-U-A, Chaudhary A, Manzoor M, Akmal M, Bukhari SM, Ali W (2021) Recent trends in microalgal harvesting: an overview. Environ Dev Sustain 24:1–31

Li S, Hu T, Xu Y, Wang J, Chu R, Yin Z, Mo F, Zhu L (2020) A review on flocculation as an efficient method to harvest energy microalgae: mechanisms, performances, influencing factors and perspectives. Renew Sust Energ Rev 131:110005

Liber JA, Bryson AE, Bonito G, Du ZY (2020) Harvesting microalgae for food and energy products. Small Methods 4:2000349

Liu Q, Zhang M, Lv T, Chen H, Chika AO, Xiang C, Guo M, Wu M, Li J, Jia L (2017) Energy-producing electro-flocculation for harvest of Dunaliella salina. Bioresour Technol 241:1022–1026

Liu S, Hajar HAA, Riefler G, Stuart BJ (2018) Investigation of electrolytic flocculation for microalga Scenedesmus sp. using aluminum and graphite electrodes. RSC Adv 8:38808–38817

Liu Z, Hao N, Hou Y, Wang Q, Liu Q, Yan S, Chen F, Zhao L (2023) Technologies for harvesting the microalgae for industrial applications: current trends and perspectives. Bioresour Technol 387:129631

Lucakova S, Branyikova I, Kovacikova S, Pivokonsky M, Filipenska M, Branyik T, Ruzicka MC (2021) Electrocoagulation reduces harvesting costs for microalgae. Bioresour Technol 323:124606

Luo S, Griffith R, Li W, Peng P, ChenG Y, Chen P, Addy MM, Liu Y, Ruan R (2017) A continuous flocculants-free electrolytic flotation system for microalgae harvesting. Bioresour Technol 238:439–449

Matter IA, Bui VKH, Jung M, Seo JY, Kim Y-E, Lee Y-C, Oh Y-K (2019) Flocculation harvesting techniques for microalgae: a review. Appl Sci 9:3069

Min KH, Kim DH, Ki M-R, Pack SP (2022) Recent progress in flocculation, dewatering, and drying technologies for microalgae utilization: scalable and low-cost harvesting process development. Bioresour Technol 344:126404

Mkpuma VO, Moheimani NR, Ennaceri H (2022) Microalgal dewatering with focus on filtration and antifouling strategies: a review. Algal Res 61:102588

Monte J, Sá M, Galinha CF, Costa L, Hoekstra H, Brazinha C, Crespo JG (2018) Harvesting of Dunaliella salina by membrane filtration at pilot scale. Sep Purif Technol 190:252–260

Muylaert K, Bastiaens L, Vandamme D, Gouveia L (2017) Harvesting of microalgae: overview of process options and their strengths and drawbacks. In: Microalgae-based biofuels and bioproducts. Woodhead Publishing, pp 113–132

Najjar YS, Abu-Shamleh A (2020) Harvesting of microalgae by centrifugation for biodiesel production: a review. Algal Res 51:102046

Nazari MT, Freitag JF, Cavanhi VAF, Colla LM (2020) Microalgae harvesting by fungal-assisted bioflocculation. Rev Environ Sci Biotechnol 19:369–388

Novoa AF, Vrouwenvelder JS, Fortunato L (2021) Membrane fouling in algal separation processes: a review of influencing factors and mechanisms. Front Chem Eng 3:21

Peperzak L, Colijn F, Koeman R, Gieskes W, Joordens J (2003) Phytoplankton sinking rates in the Rhine region of freshwater influence. J Plankton Res 25:365–383

Reyes JF, Labra C (2016) Biomass harvesting and concentration of microalgae Scenedesmus sp. cultivated in a pilot phobioreactor. Biomass Bioenergy 87:78–83

Roselet F, Burkert J, Abreu PC (2016) Flocculation of Nannochloropsis oculata using a tannin-based polymer: bench scale optimization and pilot scale reproducibility. Biomass Bioenergy 87:55–60

Roselet F, Vandamme D, Muylaert K, Abreu PC (2019) Harvesting of microalgae for biomass production. In: Microalgae biotechnology for development of biofuel and wastewater treatment. Springer

Roy M, Mohanty K (2019) A comprehensive review on microalgal harvesting strategies: current status and future prospects. Algal Res 44:101683

Salim S, Vermuë M, Wijffels R (2012) Ratio between autoflocculating and target microalgae affects the energy-efficient harvesting by bioflocculation. Bioresour Technol 118:49–55

Sawyer CN, Mccarty PL, Parkin GF (2003) Chemistry for environmental engineering and science. McGraw-Hill

Seo JY, Praveenkumar R, Kim B, Seo J-C, Park J-Y, Na J-G, Jeon SG, Park SB, Lee K, Oh Y-K (2016) Downstream integration of microalgae harvesting and cell disruption by means of cationic surfactant-decorated Fe_3O_4 nanoparticles. Green Chem 18:3981–3989

Shao W, Zhang J, Wang K, Liu C, Cui S (2018) Cocamidopropyl betaine-assisted foam separation of freshwater microalgae Desmodesmus brasiliensis. Biochem Eng J 140:38–46

Siddiki SYA, Mofijur M, Kumar PS, Ahmed SF, Inayat A, Kusumo F, Badruddin IA, Khan TY, Nghiem L, Ong HC (2022) Microalgae biomass as a sustainable source for biofuel, biochemical and bio-based value-added products: an integrated biorefinery concept. Fuel 307:121782

Silva MI, Esteves AF, Pires JC, Gonçalves AL (2021) Microalgal biorefineries: key processes and main challenges. In: Microalgal biotechnology. https://doi.org/10.1039/9781839162473-00036

Singh G, Patidar S (2018) Microalgae harvesting techniques: a review. J Environ Manag 217:499–508

Smith BT, Halperin J, Darzins A, Davis RH (2013) Enhanced sediment flow in inclined settlers via surface modification or applied vibration for harvesting microalgae. Algal Res 2:369–377

Tan XB, Lam MK, Uemura Y, Lim JW, Wong CY, Lee KT (2018) Cultivation of microalgae for biodiesel production: a review on upstream and downstream processing. Chin J Chem Eng 26:17–30

Tang DYY, Khoo KS, Chew KW, Tao Y, Ho S-H, Show PL (2020) Potential utilization of bioproducts from microalgae for the quality enhancement of natural products. Bioresour Technol 304:122997

Teh CY, Budiman PM, Shak KPY, Wu TY (2016) Recent advancement of coagulation–flocculation and its application in wastewater treatment. Ind Eng Chem Res 55:4363–4389

Tork MB, Khalilzadeh R, Kouchakzadeh H (2017) Efficient harvesting of marine Chlorella vulgaris microalgae utilizing cationic starch nanoparticles by response surface methodology. Bioresour Technol 243:583–588

Ummalyma SB, Mathew AK, Pandey A, Sukumaran RK (2016) Harvesting of microalgal biomass: efficient method for flocculation through pH modulation. Bioresour Technol 213:216–221

Vandamme D, Foubert I, Muylaert K (2013) Flocculation as a low-cost method for harvesting microalgae for bulk biomass production. Trends Biotechnol 31:233–239

Wang L, Pan B, Gao Y, Li C, Ye J, Yang L, Chen Y, Hu Q, Zhang X (2019a) Efficient membrane microalgal harvesting: pilot-scale performance and techno-economic analysis. J Clean Prod 218:83–95

Wang S, Yerkebulan M, Abomohra AE-F, El-Khodary S, Wang Q (2019b) Microalgae harvest influences the energy recovery: a case study on chemical flocculation of Scenedesmus obliquus for biodiesel and crude bio-oil production. Bioresour Technol 286:121371

Xu Y, Purton S, Baganz F (2013) Chitosan flocculation to aid the harvesting of the microalga Chlorella sorokiniana. Bioresour Technol 129:296–301

Yin Z, Zhu L, Li S, Hu T, Chu R, Mo F, Hu D, Liu C, Li B (2020) A comprehensive review on cultivation and harvesting of microalgae for biodiesel production: environmental pollution control and future directions. Bioresour Technol 301:122804

Zhao Y, Guo G, Sun S, Hu C, Liu J (2019) Co-pelletization of microalgae and fungi for efficient nutrient purification and biogas upgrading. Bioresour Technol 289:121656

Zhao Z, Liu B, Ilyas A, Vanierschot M, Muylaert K, Vankelecom IF (2021) Harvesting microalgae using vibrating, negatively charged, patterned polysulfone membranes. J Membr Sci 618:118617

Zhao Z, Muylaert K, Vankelecom IF (2023) Applying membrane technology in microalgae industry: a comprehensive review. Renew Sust Energ Rev 172:113041

Zheng H, Gao Z, Yin J, Tang X, Ji X, Huang H (2012) Harvesting of microalgae by flocculation with poly (γ-glutamic acid). Bioresour Technol 112:212–220

Zou X, Li Y, Xu K, Wen H, Shen Z, Ren X (2018) Microalgae harvesting by buoy-bead flotation process using Bioflocculant as alternative to chemical Flocculant. Algal Res 32:233–240

Wastewater Treatment Based on Microalgae Bioremediation

4

Carmen C. Osuna-Martínez, Magdalena Berges-Tiznado, Carolina G. Delgado-Álvarez, Carolina Bojórquez-Sánchez, David Ulises Santos-Ballardo, and Dulce L. Ambriz-Pérez

Abstract

Wastewater is a combination of organic and inorganic components; some of them can be considered nutrients, while others have toxic potential; therefore, the elimination of hazardous substances is crucial to guarantee the safety of living beings and the environment, as well as complying with international regulations in this matter. There are several methods for treating wastewater; among them, microalgae emerge as ecological and effective options. This chapter discusses the nature of wastewater and conventional treatment methods compared to microalgae systems. In addition, some applications are presented, and the advantages and challenges of microalgae bioremediation are also considered.

Keywords

Phycoremediation · Biotreatments · Effluents · Environmental pollutants · Environmental applications

C. C. Osuna-Martínez
Facultad de Ciencias del Mar, Universidad Autónoma de Sinaloa, Mazatlán, Mexico

M. Berges-Tiznado · C. Bojórquez-Sánchez
Universidad Politécnica de Sinaloa, Mazatlán, Mexico

C. G. Delgado-Álvarez · D. L. Ambriz-Pérez (✉)
Universidad Politécnica de Sinaloa, Mazatlán, Mexico

Estancias Postdoctorales por México SECIHTI, Mexico City, Mexico
e-mail: dambriz@upsin.edu.mx

D. U. Santos-Ballardo
Laboratorio de Ecotoxicología, Centro de Investigación en Alimentación y Desarrollo A.C, Mazatlán, Mexico

4.1 Introduction

Despite water being one of the most valuable resources on the planet, water bodies are contaminated daily with the discharge of wastewater from diverse sources (Amaro et al. 2023). Wastewaters consist of a combination of many components, such as organic and inorganic compounds, heavy metals, pharmaceutical residues, pesticides, pathogen microorganisms, oil, and grease, among others; this combination depends on the source, which includes municipal, industrial, and agricultural wastewaters and landfill leachate, among others. All the components present in wastewater represent several problems in their proper management, which is essential to ensure the health and wellness of human beings and the environment. This is why many worldwide regulations exist (Jais et al. 2017).

Organic and inorganic substances affect and modify the pH in the medium, its biochemical oxygen demand (BOD) and chemical oxygen demand (COD), and the BOD/COD ratio; if BOD increases, there is a change in pH value, depletion of oxygen level required for various living organisms in the water body, and rapid bacterial growth; also, changes in the dissolved oxygen quantity for extended periods may limit sensitive fish and aquatic lives (Jais et al. 2017). Likewise, oil and grease are highly hydrophobic. Therefore, they are usually located on the wastewater surface. Besides, they have very low biodegradability, affecting aquatic life by decreasing both light penetration and the oxygen transfer between air and water. Additionally, even the thinnest oily layer could contain toxic micropollutants, especially hydrophobic ones, such as benzene, toluene, xylene, and hydrocarbons, among others (Pintor et al. 2016).

Another component of concern is antibiotics; their presence in water bodies where wastewater is disposed of might increase microbial antibiotic resistance (Jais et al. 2017).

D. U. Santos Ballardo, S. Rossi (eds.), *Microalgae as Promising Source of Commercial Bioproducts*, Developments in Applied Phycology 14, https://doi.org/10.1007/978-3-031-86433-9_4

Finally, wastewater could contain toxic substances to human beings, animals, and the ecosystem, such as heavy metals or pesticides.

For the reasons mentioned above, wastewater should be treated appropriately; for this, physicochemical or biological methods can be applied; however, biological systems have advantages mainly in terms of cost-effectiveness, eco-friendly approach, easy and safe operation (Andrade and Andrade 2018). Among biological systems, microalgae have been seen as environmentally friendly solutions to detoxify conventional pollutants, antibiotics, heavy metals, pesticides, pathogens, and antibiotic-resistant genes (Sutherland and Ralph 2019; Amaro et al. 2023). Several authors have evaluated the use of microalgal systems for the treatment of different types of wastewaters, such as agricultural, municipal, and industrial, observing high removal efficiencies of (1) conventional pollutants, up to 99%, 99%, and 90% of total nitrogen (N), total phosphorus (P), and/or organic carbon ©, respectively; (2) antibiotics frequently found in wastewaters, by about 42%–93%, depending on the antibiotic studied; and (3) heavy metals, up to 90% of Pb, 97% of Hg, and 100% of Cr (VI), just to mention a few (Goswami et al. 2022; Amaro et al. 2023).

In this sense, microalgae could manage pollutants through different mechanisms, mostly biosorption, bioaccumulation, and biodegradation.

4.2 Main Types of Polluting Wastes and Associated Problems

The continuous generation of waste is one of the leading environmental problems affecting the world in recent years. The domestic wastes generated by the population are added to the wastes generated by different industries during the production of food, clothing, transportation, electronic devices, and all the services and needs of human beings.

Due to the multiple characteristics of pollutants, their classification is very diverse; it can be by size, composition, environmental persistence, toxicity, or emission source. Table 4.1 shows the groups of pollutants and their characteristics.

This section will explore the main types of polluting wastes and the problems associated with their generation and management. These wastes range from toxic and chemical wastes to nonbiodegradable materials, and their proper management is essential for preventing adverse impacts on human health and the environment.

In conclusion, it is acknowledged that these discharges and wastes contaminate water resources, jeopardizing the quality and availability of water. Water pollution is a global issue whose origins lie in various human activities essential for economic and social development.

Table 4.1 Groups of pollutants and their main characteristics

Group	Description	Reference
Atmospheric pollutants	Atmospheric deposits of pollutants that fall into bodies of water, such as acid rain, can acidify and damage aquatic ecosystems	EPA (2023).
Biological contaminants	Bacteria: Microorganisms such as *Escherichia coli* (*E. coli*) and fecal coliforms indicate fecal contamination and can cause waterborne diseases Viruses: Infectious agents, including norovirus and rotavirus, can cause gastrointestinal illnesses	Plummer et al. (2014) and Some et al. (2021).
Chemical contaminants	Organic substances: These include compounds such as pesticides, herbicides, pharmaceuticals, and industrial chemicals that can be toxic or endocrine disruptors Heavy metals: Lead, mercury, cadmium, and chromium are toxic and can accumulate in aquatic organisms and biomagnify in the food chain Nutrients: Nitrogen and phosphorus can cause eutrophication of water bodies and harmful algal blooms	Mandal et al. (2016).
Emerging pollutants	They include chemicals and pharmaceutical compounds that are being studied due to their potential impact on health and the environment	Geissen et al. (2015).
Physical pollutants	Temperature: Changes in water temperature can negatively affect aquatic organisms. Power plants and factories release hot water into water bodies, which can negatively affect aquatic life	Råman Vinnå et al. (2017).
Radioactive contaminants	Nuclear activities can release radioactive substances and have a long-term impact on health and the environment	Ma et al. (2023).
Sedimentary pollutants	Suspended solids: These are fine solid particles that can cloud water and damage aquatic habitats Sediment and particles carried by rain runoff can increase turbidity and affect aquatic life	Rügner et al. (2013).
Solid waste	Garbage and solid waste dumped in water bodies, such as plastics and other materials, can cause pollution and harm aquatic life	Haward (2018).

Contaminants that affect water quality can be classified into several categories according to their origin and nature. The following is a general classification of these pollutants:

4.2.1 Municipal

Municipal waste includes a large amount of waste generated by the inhabitants of a place when carrying out their daily activities. This includes waste generated in common areas, such as parks, schools, squares, and beaches, as well as waste generated in the homes of each citizen, also known as household waste. These can be divided into two main categories: solid urban waste and household wastewater.

Solid waste refers to food scraps, batteries, paper, cardboard, clothing, plastic bottles, and others. Once urban solid waste is generated in households, it is collected by municipal services and transferred for final disposal in a sanitary landfill (52%), a controlled landfill (11.5%), or an uncontrolled landfill (32.9%) (SEMARNAT 2017). Once there, the waste generates leachates, infiltrating the soil and subsoil and reaching water bodies.

On the other hand, wastewater involves everything that goes down the drains of homes, schools, offices, etc., daily, i.e., waste from toilets, food waste such as fats and oils and water from washing machines. Once the different sources emit these, the municipality must treat them before dumping them into a body of water such as a river, stream, or directly into the sea.

4.2.2 Agricultural

Agriculture is considered the world's primary food source and one of the activities with the most significant presence in Mexico. Mexico is the leading exporter of various products resulting from agriculture. After the food crisis in the 1940s, Mexico began to invest more in research and technology, which increased pesticide and fertilizer use.

The use of chemicals in agriculture mainly focuses on controlling pests that can affect crops; these chemicals include herbicides, insecticides, acaricides, nematicides, molluscicides, rodenticides, and fungicides. Herbicides are the most widely used, causing an impact on the environment, mainly on the soil where they are applied and on the water bodies where they leach.

Using highly hazardous pesticides in agricultural fields triggers various effects on human health, wildlife, and the environment. Pesticides are spread through agricultural soils and reach irrigation canals, animals, rural populations, etc. As a result, native species may disappear from the area, and ecosystems may be altered by the disappearance of beneficial insects (Del Puerto Rodríguez et al. 2014).

Fertilizer is another type of pollution triggered by agriculture, affecting water bodies due to its high concentration of elements such as nitrogen, phosphorus, and potassium. It is known that about 20% of these fertilizers emitted into the environment end up accumulating in soils, while 35% end up in the oceans, causing problems such as eutrophication (Mateo-Sagasta et al. 2017).

Furthermore, oxygen depletion (hypoxia) results from the nutrient overloading of water bodies, causing what is commonly known as eutrophication. Nearshore water bodies with high nutrient loads increase the amount of organic matter, thus increasing their primary productivity, and this is how algae and green plants cover the water surface, causing a decrease in oxygen and reducing the passage of light, thus creating a dead zone. This generates unpleasant odors; a loss of biodiversity in the area; problems for human health, mainly the respiratory system; and even sanitary problems for people near the area (Rabalais 2019).

Moreover, industrial activity has increased over the years; this is an embracing necessity of today's society for its high generation of jobs, the services it provides, and its generation of tools and instruments to facilitate the life of man in all aspects. This activity results in multiple types of industries, such as automotive, food, commerce, construction, and hospitality, which impact the environment through the waste they generate.

4.2.3 Industrial

Process wastewater generated by various industries must be treated before being discharged into a receiving water body, and this treatment focuses on the pollutants that may be present depending on the process activity carried out. The waters may be enriched with biodegradable or nonbiodegradable organic matter and contaminated with fats and oils, heavy metals, salts, or excess minerals, among other liquid industrial wastes (Ahmed et al. 2021).

On the other hand, atmospheric emissions generated by different industries, either wet or dry, eventually tend to be deposited in water bodies. These emissions may contain particulate matter, which, once precipitated, contaminates the water. Even the term "acid rain" originates from pollutants emitted into the atmosphere, which eventually precipitate with rain in the form of acids, causing corrosion in structures and vehicles.

4.2.4 Mining

The exploitation of mineral resources is an activity with a significant economic impact in Mexico; it is known that Mexico is the main producer of silver, thereby creating a substantial direct and indirect economic flow. In mining, water resources are used in the process of exploration, exploitation, concentration, and mining leaching, and water is a resource affected by pollution in this industry.

The main threats in terms of water contamination are the tailings, which are accumulations of wastewater from mining, mainly loaded with acids and heavy metals. Despite being regulated under Mexican regulations, due to natural phenomena or human accidents, these tailings can overflow and reach rivers and streams in mountainous areas and flow downstream until they reach the oceans, contaminating hundreds of kilometers of bodies of water.

Among these mining leachate contaminants are cyanide, sulfuric acid, arsenic, lead, and mercury, which are highly polluting and harmful to aquatic ecosystems, wildlife, and humans.

4.2.5 Others

Many sources can contaminate water; however, a significant one is the oil spill caused by the extraction platforms of this finite natural resource. Once in the water, the oil creates a surface layer that impedes the passage of sunlight, drastically affecting the life cycle in these places.

On the other hand, livestock farming is a crucial food production activity that has increased its output to meet the current food demand. This implies higher concentrations of contaminants, such as organic matter, antibiotics, vaccines, and hormonal growth promoters; these travel through water from farms to ecosystems, causing damage to ecosystem health.

4.3 Polluting Waste Regulations

The appropriate management of pollutants directly affects the lifestyle and health of people and the ecosystem. Therefore, the issue is reflected in international standards. In this sense, the Universal Declaration of Human Rights declares in Article 25, "Everyone has the right to a standard of living adequate for the health and well-being of himself and his family, including food, clothing, housing and medical care and necessary social services." In order to achieve the right above-mentioned, the International Covenant on Economic, Social and Cultural Rights declares in Article 11: "The States Parties to the present Covenant will take appropriate steps to ensure the realization of this right, recognizing to this effect the essential importance of international cooperation based on free consent"; meanwhile, Article 12 declares: "The States Parties recognize the right of everyone to the enjoyment of the highest attainable standard of physical and mental health. The steps to be taken by the States Parties to the present Covenant to achieve the full realization of this right shall include those necessary for: The improvement of all aspects of environmental and industrial hygiene and the prevention, treatment, and control of epidemic, endemic, occupational and other diseases" (OHCHR). The articles mentioned before include proper waste and pollutant management to avoid detriment to human health and/or lifestyle.

All of these have led to the Basel Convention on the Control of Transboundary Movements of Hazardous Wastes and their Disposal, which declares in Article 4 that "each Party shall take the appropriate measures to ensure that the generation of hazardous wastes and other wastes within it is reduced to a minimum, ensure that persons involved in the management of hazardous wastes or other wastes within it take such steps as are necessary to prevent pollution due to hazardous wastes and other wastes arising from such management and, if such pollution occurs, to minimize the consequences thereof for human health and the environment" (OHCHR).

The above-mentioned standards have led to the need to establish regulations worldwide, such as those applicable to North American countries. The United States counts on Screening Quick Reference Tables (SQuiRTs), where the National Oceanic and Atmospheric Administration (NOAA) proposes a series of limit concentrations tables for various organic and inorganic pollutants in various environmental compartments. Its objective is to identify potential impacts on coastal resources and habitats likely to be affected by hazardous waste (Buchman 2008). Also, in the United States, there is the Clean Water Act (CWA), which is the main tool for protecting the quality of surface waters in that country (USEPA 2021).

For its part, Canada has the Canadian Council of Ministers of the Environment (CCME), which is the principal ministerial-led intergovernmental forum for collective action on environmental issues of national and international concern (CCM). This council proposes the Canadian Environmental Quality Guidelines, which list a wide repertoire of organic and inorganic contaminants in soils, sediments, groundwater, wastewater, and water for agriculture and aquatic life (CCM).

In Mexico, the so-called Official Mexican Standards (NOM, by its acronym in Spanish) establish regulations about what hazardous waste is, in what concentration it is necessary to apply a remediation method, how its final disposal should be, and the limited concentrations it can have when it is available, among other aspects. The NOMs involved in the topic are NOM-052-SEMARNAT-2005, which establishes the characteristics of the hazardous waste, the list of these, and the limits that make the waste dangerous due to its environmental toxicity (SEMARNAT 2005a). NOM-001-SEMARNAT-2021 establishes the maximum permissible limits of contaminants in wastewater discharges into national waters and assets (SEMARNAT 2022).

NOM-002-SEMARNAT-1996 establishes the maximum permissible limits of contaminants in wastewater discharges to urban or municipal sewage systems (SEMARNAT 1998a). Also, Official Mexican Standard NOM-003-SEMARNAT-1997 establishes the maximum permissible limits of contaminants for treated wastewater reused in public services (SEMARNAT 1998b). Additionally, NOM-147-SEMARNAT/SSA1–2004, establishes the criteria to determine the remediation concentrations of soils contaminated by arsenic, barium, beryllium, cadmium, hexavalent chromium, mercury, nickel, silver, lead, selenium, thallium, and/or vanadium (SEMARNAT 2007). Besides, NOM-138-SEMARNAT/ss-2003 establishes the maximum permissible limits of hydrocarbons in soils and specifications for their characterization and remediation (SEMARNAT 2005b).

4.4 Current Conventional Treatments and Perspectives

Since the beginning of the implementation of wastewater treatment strategies, numerous technologies and approaches have been developed to return water to its initial state. These technologies are usually approached from multiple perspectives, combining different methods and levels involving physical, chemical, and biological processes. Under environmental management, wastewater treatment plays a fundamental role in eliminating, regulating, or mitigating pollutants through returning high-quality treated water to the environment or its reuse in various applications.In this section, conventional wastewater treatment technologies will be explored, which have been developed and refined over time and are the basis of wastewater management worldwide. The key processes used in wastewater treatment will be explained, from coarse solid removal to disinfection and nutrient removal. In addition, the importance of these technologies in protecting public health and conserving water resources will be discussed.

As environmental awareness and demands for wastewater treatment grow, it is important to understand the basis of conventional technologies and how they are being adapted and complemented by more advanced and sustainable approaches. Table 4.2 presents some of the conventional technologies in wastewater treatment when considering some types of pollutants.

These conventional wastewater treatment technologies are essential to protecting public health and the environment by reducing pollution and ensuring that treated water meets the required quality standards before release or reuse. However, it is important to note that advanced and sustainable technologies are also gaining relevance to address environmental challenges and improve wastewater treatment efficiency.

In wastewater treatment science, technologies are designed from a pragmatic and versatile perspective. As mentioned at the beginning, different stepwise methods are combined in this process.

Table 4.2 Conventional technologies in wastewater treatment according to the type of contaminant

Pollutant	Treatment used
Biodegradable organic	Activated sludge Trickling filter Rotating biological discs Oxidation ponds Sand filtration Physicochemical
Inorganic solids	Ion exchange Inverse osmosis Electrodialysis
Metals	Chemical precipitation Ion exchange
Nutrients (N, P)	Nitrification and denitrification with suspended biomass Nitrification and denitrification with fixed biomass Drag with ammonia Ion exchange Chlorination at the breaking point Coagulation/sedimentation with metal salts Coagulation/sedimentation with lime Biochemical removal
Pathogens	Chlorination Ozonation
Suspended solids	Screening and shredding Sedimentation Flotation Filtration Coagulation/sedimentation

Adapted from Moeller Chávez et al. (2011)

Figure 4.1 shows a diagram of the urban wastewater treatment process, where the water flows through several stages, each intended to treat a specific type of pollutant and give it a new purpose.

According to the environmental regulations of each country, treated water can follow two routes: reuse or release. Treated water can be reused for various applications, such as agricultural irrigation or industrial use; it can also be released into natural water bodies if it complies with regulations to avoid negative impacts on the environment.

Thus, wastewater treatment technologies play a fundamental role in protecting public health and preserving the environment. These treatments allow the effective removal of pollutants present in wastewater, which, at the same time, can prevent the release of pollutants into natural water bodies. This contributes to properly managing waste generated by various human activities, reducing the negative environmental impact.

Ensuring that wastewater is treated effectively and complies with environmental regulations is necessary to maintain water quality and ecosystem health and prevent the spread of waterborne diseases.

Several stages are carried out in the wastewater treatment process, each with a specific purpose. This process is shown in Fig. 4.2.

Despite the importance of wastewater treatment technologies, it cannot be denied that there are several limitations to their widespread implementation. One of the most prominent limitations is the availability of financial resources, as the investment to develop and maintain water treatment infrastructure is often costly and may be beyond the financial capacity of many communities and countries. In addition, a lack of participation on the part of the population can lead to inadequate wastewater management, exacerbating the problem. As the demand for water treatment services increases due to population growth and urbanization, there is an additional challenge to providing adequate and sustainable solutions. Overcoming these constraints requires government and community efforts, strategic investments, and a more significant commitment to environmental preservation, public health, and economic development.

Bioremediation using microalgae offers a promising approach to establishing more sustainable wastewater treatment systems. These microalgae can absorb and remove nutrients, such as nitrogen and phosphorus, from wastewater, avoiding eutrophication, which helps reduce pollution. In addition, the microalgae bioremediation process can generate biomass that, in turn, can have valuable applications. This

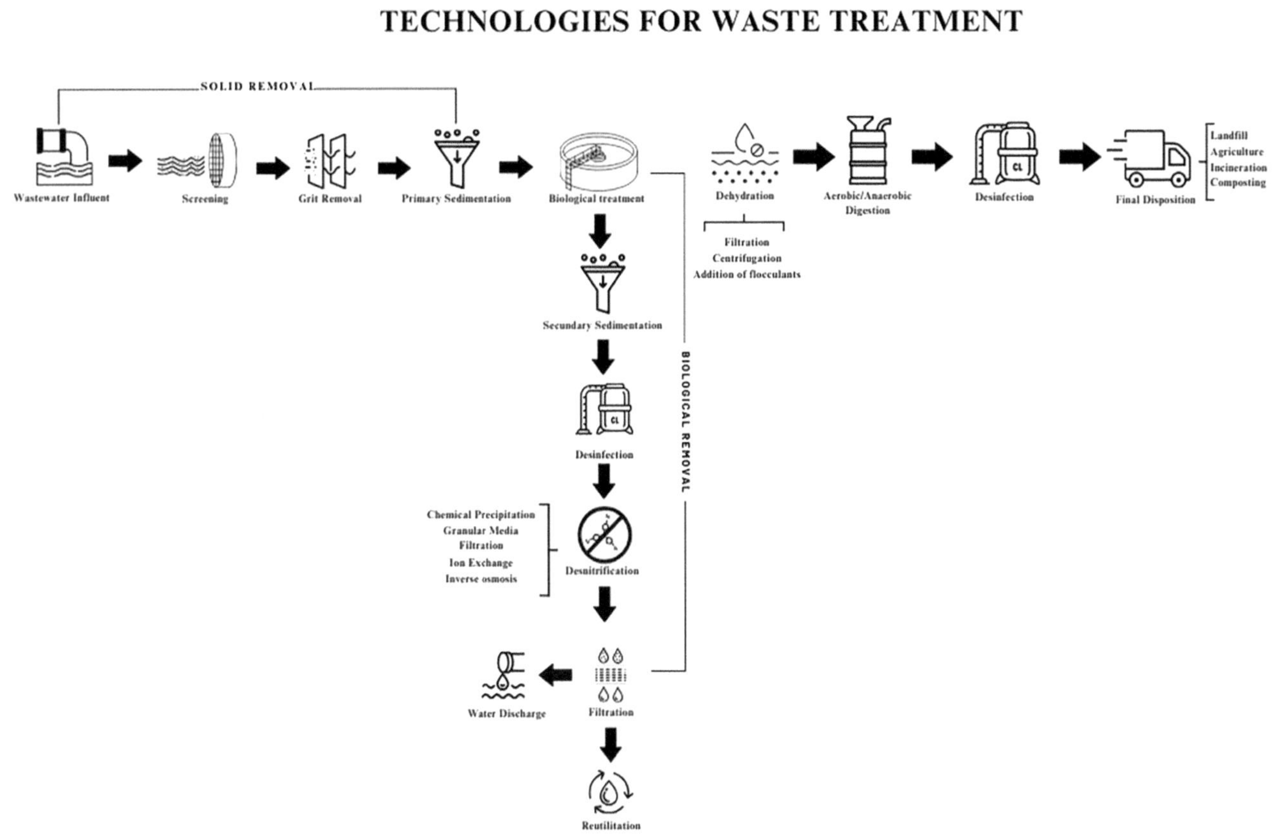

Fig. 4.1 Technologies for waste treatment

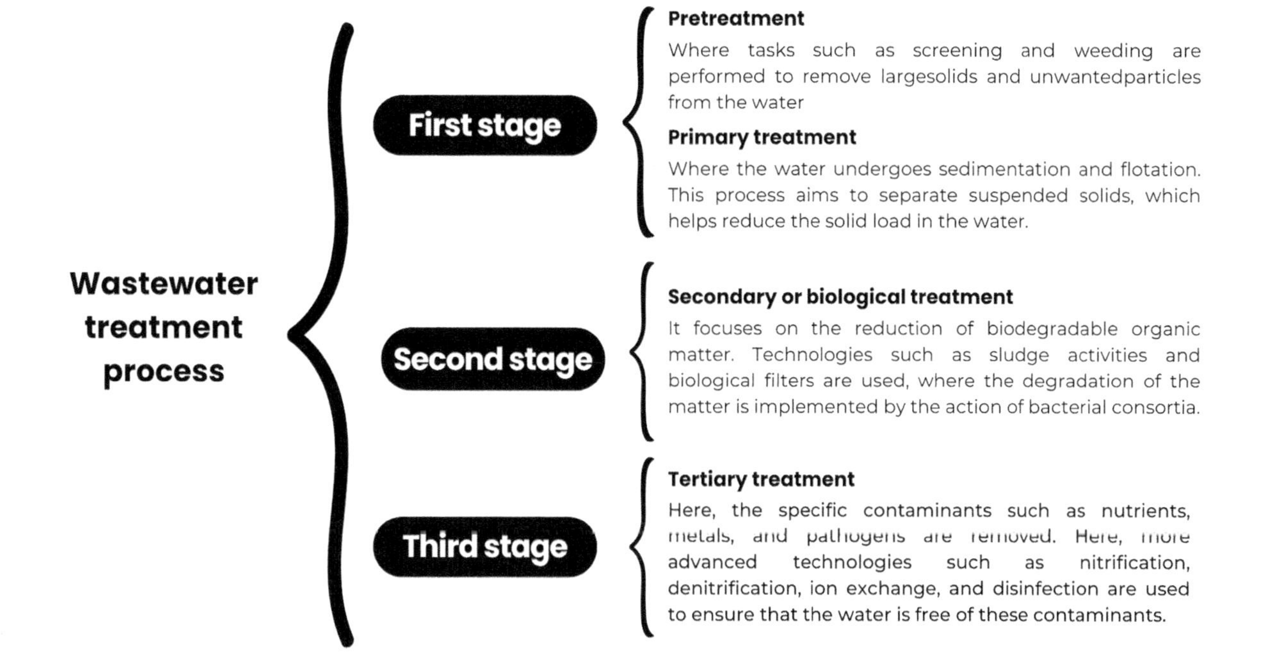

Fig. 4.2 Wastewater treatment process

approach treats wastewater effectively and offers a way to recover resources and promote sustainability in wastewater management.

4.5 Microalgal Bioremediation

Microalgal biomass has been widely used to remove some common environmental pollutants, representing a biological solution to reduce anthropogenic negative effects and the associated wastewater impacts on the environment, and in some scenarios, value-added biomass products (Carvalho Nogueira da Gama et al. 2023). Also, microalgae have a high capacity to produce large biomass and lipid products for biofuel production as well as the ability to capture carbon dioxide through photosynthesis (Singh et al. 2023). Algae bioremediation experiments began with Caldwell's (1946) and Oswald and Gotaas's (1957) works on sewage oxidation ponds; they studied the symbiotic relationships among microalgae, bacteria, and oxidation tanks for wastewater treatment in both open and closed ponds (Sahu et al. 2019). This sustainable approach to domestic and industry wastewater treatments implies not only environmental benefits but also economic profits; some of the applications are described below.

4.5.1 Alimentary Industry

Carvalho Nogueira da Gama et al. (2023) evaluated the removal of total C, N, and P by an algae culture, mostly of *Chlorella vulgaris* (98.17–99.73%) and others (*Limnothrix planctonica*, *Pseudanabaena catenata,* and *Monoraphidium tortile*), from the wastewater effluents of meat and brewing industries and different mixes of both industries' wastewater on optimal, deficit, and excess conditions. There was no optimal treatment removal of C, N, or P, and algae mix growth was not affected by the differences in nutritional conditions of every culture, but the bioremediation ability was entirely influenced by nutrition; the algae mix removed P and N under C-deficit conditions, whereas the best C removal was in wastewater media with optimal availability and excess C.

4.5.2 Domestic Wastewaters

Usually, large amounts of ammonia can be found in domestic wastewater, and despite several treatment methods, microalgae consortiums can be a green-sustainable approach to remove this hazardous nutrient as well as carbon, phosphorus, and other N species from wastewater streams. Likewise, in the experiment where a consortium mainly composed of the microalgae *Tetradesmus obliquus* and *Desmodesmus* sp. and bacteria were used to reduce the mentioned components from toilet wastewater, ammonium microalgae absorption was 8.5 ± 0.4%, and carbon was reduced by 34.2% (Hu et al. 2023).

4.5.3 Tannery Industry

Tannery is a billion-dollar industry, but its wastewater has organic and inorganic pollutants toxic to water bodies, animals, and humans, including high chemical oxygen demand, total dissolved and suspended solids, phosphates, sulfates, nitrogen, and elevated chromium (Devi et al. 2023). An experiment that was designed using the microalgae *Nannochloropsis oculata*, *Chlorella vulgaris,* and *Chlorella sorokiniana* proved their interactions with temperature, light intensity, inoculum density, and light period to explore the bioremediation potential in nutrient removal from tannery wastewater previously treated with ozone (Saranya and Shanthakumar 2020). Based on the bioremediation potential, *C. sorokiniana* > *C. vulgaris* > *N. oculata* had removal percentages of C of 90, 87, and 85%, and ammonium had reductions of 90, 83, and 80% in the stationary phase, respectively; for each microalga, 30–46% of the total N and P removal was on the first day.

The microalgae genus *Tetraselmis* sp. was used successfully to remove phosphates, ammonia, and total N under wastewater conditions from purely tannery effluents, tannery effluents after primary treatment, and tannery effluents after primary and secondary treatment (all of them at 50% water dilution); 99.9% of ammonia was removed by the three treatments, phosphates' bioremediation was higher in the third treatment (91.75%) and lower in the pure tannery untreated effluent (84.72%), and surprisingly, total N remotion was higher in the first treatment (82.22%) of pure raw wastewater (De Cassia Campos Pena et al. 2019). By using the marine microalgae *Chlorella* sp. and *Phormidium* sp., cultured individually as well as in consortiums together, considerable reductions in total N and P were reached, ranging from 88 to 91.16% from tannery wastewater (Das et al. 2018). Other studies on tannery wastewater treatment have used the microalgae *Chlorella vulgaris* (grown for 21 days) with very promising removal rates of 100% of nitrates after 6 days and 99% of phosphates after day 21; after 21 days sulfate content was reduced by 67.4% showing high efficiency on reduce tannery wastewater load of nutrients (Das et al. 2017).

4.5.4 Aquaculture

Aquaculture is an intensive economic activity in which effluents represent a serious environmental issue because of eutrophication problems in receptor water bodies (Thomsen et al. 2020). A novel bio-alternative to aquaculture wastewater treatment rich in ammonia is to use a bio-composite from a coculture of the microalgae *Chlorella sorokiniana* and the bacteria *Komagataeibacter europaeus* with the ability to produce cellulose; the excellent ammonia removal was 91.64 ± 1.27%, and the microalgae demonstrated the capability to utilize ammonia as an N source to enhance biomass production by the exchange action of the bacterial cellulose membrane (Tseng et al. 2023). Nevertheless, aquaculture farms have widely used antibiotics like florfenicol to prevent culture organism diseases. However, the untransformed fraction remains in the water, adding a treatment plus eutrophication by its effluents (Gao et al. 2018). Oliveira et al. (2023) demonstrated the ability of a microalgae community (genera *Chloroidium* and *Nitzschia*) to bioremediate aquaculture C-, P-, and N-rich effluents with occasional loads of florfenicol. In the experiments, microalgae community exhibited high and stable organic carbon removal among 83–100% Otherwise, ammonium removal percentage was 6–94% all through the 148 days of the experiment, but in the presence of the antibiotic, the removal efficiency dropped to 70%; nitrate reached its maximum removal from days 22 to 43 with 73%, and it reduced in the presence of florfenicol, but nitrite remained in the aquaculture wastewater. The total N removal during the first 21 days was 93–94%; during the next 21 days, the N bioremediation achieved was between 18 and 79%, and the minimum N removal was during the next 144 days (6–62%) with the lowest N recovery when florfenicol was in the mix. Finally, on day 49, the removal of phosphates had a maximum efficiency (46%), and the overall system could remove low amounts of florfenicol (5.3–11.4 mg).

4.6 Mechanisms of Action of Microalgae in Bioremediation

Microalgae have specific mechanisms for each conventional pollutant removal, which may entail pathways to store nutrients in the biomass or assimilate them into essential molecules for microalgae. The microalgal mechanisms for pollutant removal mostly include biosorption, bioaccumulation, and biodegradation. These mechanisms may occur as follows: (1) fast passive adsorption due to the physicochemical properties of the cell surface and pollutants, followed by (2) fairly slow mass transfer through the cell membrane, and (3) culminates either with bioaccumulation, biodegradation, or both (Amaro et al. 2023).

Bio-adsorption is a non-metabolic process, binding pollutant molecules onto the microalgal surface; this occurs on both living and nonliving microalgal cell surfaces because most of the cell receptors for the contaminant remain viable even after the cell has died (Sutherland and Ralph 2019). The use of nonliving microalgal biomass has several advantages, as follows: there are no contaminant toxicity limitations; the contaminants can be desorbed, and the microalgal biomass can be reused; increased algal sustainability and ecological acceptability; and substantially lower operational costs (Sutherland and Ralph 2019; Amaro et al. 2023).

4.6.1 Inorganic Carbon and Nitrogen

It is well known that carbon (C), nitrogen (N), and phosphorus (P) are elements that limit organisms' growth. Microalgae particularly need N and P because they are essential for their growth and can be available from wastewater effluents; there is a dependence on the removal of P, and it has proved to be limited when N is low (Beuckels et al. 2015). Microalgae consortiums, sometimes in coculture with bacteria, have been used to remove nutrients and other compounds from various wastewater-type effluents, such as from meat processing, brewing, tannery and winery industries, aquaculture, and domestic waste..

Microalgae can use inorganic carbon, nitrogen, and phosphorus from the environment through biotransformation. Most microalgal species have carbon-concentrating mechanisms that involve the reversible intracellular or extracellular conversion of HCO_3^- to CO_2, which allows them to secure a carbon source, even when the concentration of atmospheric CO_2 is low. Inorganic carbon can enter the microalgal cell through three different processes: (1) direct passive diffusion of CO_2; (2) transformation of HCO_3^- to CO_2, catalyzed by an extracellular carbonic anhydrase, followed by passive diffusion of CO_2; and (3) direct active transport of HCO_3^-, aided by inorganic carbon uptake transporters.

Regarding inorganic nitrogen, microalgae can assimilate this nutrient from wastewaters as NH_4^+, NO_2^-, and NO_3^-. The inorganic nitrogen species are actively transported into the microalgal cell, and NH_4^+ can be directly incorporated into 2-oxoglutarate, leading to glutamate production and amino acid synthesis through the glutamine synthetase/glutamate synthase cycle. However, NO_3^- and NO_2^- are reduced to NH_4^+ before assimilation: NO_3^- is primarily converted to NO_2^- in the cytosol by the action of nitrate reductase, and NO_2^- is then reduced to NH_4^+ in the chloroplast in a reaction catalyzed by nitrite reductase. Therefore, NH_4^+ is the preferred form of nitrogen by most microalgae as it has lower energy demand than other nitrogen sources.

About inorganic phosphorus, the primary source for microalgae is inorganic phosphate in the forms of $H_2PO_4^-$, HPO_4^{2-}, or PO_4^{3-}, which are transferred through the cell membrane by active transport using inorganic phosphorus transporters. Once inside the cell, they can be used directly to synthesize nucleic acids, phospholipids, and adenosine triphosphate (ATP) through phosphorylation. Besides, in the presence of light, phosphates can be converted into acid-soluble polyphosphates in a reaction catalyzed by polyphosphate kinase and used for deoxyribonucleic acid (DNA) and protein synthesis (Amaro et al. 2023).

4.6.2 Heavy Metals

Microalgae consume heavy metals such as boron (B), cobalt (Co), copper (Cu), iron (Fe), molybdenum (Mo), manganese (Mn), and zinc (Zn) as trace elements for enzymatic process and cell metabolism. However, other heavy metals such as arsenic (As), cadmium (Cd), chromium (Cr), lead (Pb), and mercury (Hg) are toxic to microalgae. Due to the hormesis phenomenon, low-toxicity heavy metal concentrations can stimulate the growth and metabolism of microalgae, which is the case of *Anabaena* sp., *Oscillatoria* sp., *Phormidium* sp., and *Spirogyra* sp. (Leong and Chang 2020).

Microalgae have many strategies to deal with heavy metals, such as heavy metal immobilization, gene regulation, exclusion, chelation, and antioxidants or reducing enzymes, which reduce heavy metals via redox reactions. Microalgae form complexes with heavy metals; these complexes are separated inside vacuoles to help regulate the concentration of heavy metal ions in the cytoplasm and, thereby, their toxicity (Leong and Chang 2020).

Microalgae remove heavy metals by a two-stage mechanism: (1) rapid extracellular passive adsorption (biosorption) and (2) slow intracellular positive diffusion and accumulation (bioaccumulation) (Leong and Chang 2020).

Additionally, as a response to stress, microalgae excrete extracellular polymeric substances (EPSs) into the surrounding environment; these are high-molecular-weight polymers, such as polysaccharides, lipids, proteins, nucleic acids, and humic substances, whose function is to protect cells from severe environments. Microalgal EPSs can be attached to the cell surface or are soluble when excreted by microalgae, and they can bind heavy metals (Sutherland and Ralph 2019; Leong and Chang 2020; Amaro et al. 2023).

The absorption of heavy metals can also take place on the surface of the microalgae cell; this is a rapid process that consists of (1) the formation of covalent bonds between ionized cell walls with heavy metals, (2) ionic exchange of heavy metal ions with cell wall cation, and (3) binding of heavy metal cations with negatively charged uronic acids of microalgae exopolysaccharides. While the process of heavy metal accumulation inside the cell is much slower, here, (1) the heavy metals are actively transported across the cell membrane and into the cytoplasm, (2) followed by diffusion, and (3) subsequent binding with internal binding sites of proteins and peptides, metal transporters, oxidative stress–reducing agents, and phytochelatins (Leong and Chang 2020).

4.6.3 Bigger Pollutant Molecules

Microalgae have different absorption interactions depending on the molecule's hydrophilic or hydrophobic nature. Hydrophobic molecules can be absorbed into the microalgal cell walls or organic molecules discharged by microalgae into their adjacent environments, such as EPSs. Additionally, the charge of pollutants is an important factor since microalgal cell walls and EPSs generally have a negative charge due to hydroxyl, carboxyl, and phosphoryl functional groups; hence, hydrophobic, cationic pollutants are actively attracted toward the microalgal cell surface through electrostatic interactions, whereas hydrophilic pollutants are repelled, but this directly depends on the EPS composition (Sutherland and Ralph 2019; Amaro et al. 2023).

Microalgae absorb pollutants as metabolic intracellular active pathways, primarily due to the binding of molecules to intracellular proteins. It is observed that pollutant accumulation in microalgae can lead to an increase in reactive oxygen species (ROS) (which enhances the lipid content of microalgae), or it can also be a pre-step for biodegradation (Leong and Chang 2020; Amaro et al. 2023). Biodegradation involves the transformation of complex compounds into simpler breakdown molecules through catalytic metabolic degradation (Sutherland and Ralph 2019). Moreover, pollutants are broken down within or outside the algal cells, and algae consume the originated derivatives. In intracellular degradation, the molecule is first adsorbed on algae, then gradually transmitted through the algal cell walls, and last, fragmented by algal enzymes inside the cell. In extracellular degradation processes, the molecules are fragmented due to the action of extracellular enzymes, and the intermediates/end products are metabolized inside the algal cells (Amaro et al. 2023).

4.7 Culture Systems for Microalgal Bioremediation

There are many microalgae cultivation systems; this section will describe the open pond and closed photobioreactor systems.

4.7.1 Open Pond Systems for Microalgae Cultivation

Open ponds include natural systems (e.g., lakes and lagoons), artificial ponds, or containers. They are the oldest and simplest systems for microalgal production. A typical open pond is made from poured concrete or dug into the earth and lined with a plastic liner to prevent the ground from soaking up the liquid (Shen et al. 2009; Sun et al. 2016). However, open systems have limiting factors such as the need for large areas to scale up and susceptibility to contamination (e.g., introduced by birds) and adverse weather conditions.

Additionally, contamination with other microalgae, high bacterial loads, and grazers is common in large-scale open pond cultivation systems (Narala et al. 2016). These reactors are recommended for species with high resistance that are grown in open systems, such as *Dunaliella* sp. (resistant to high salinity), *Spirulina* sp. (grown in high alkalinity), and *Chlorella* sp. (grown with high nutrient concentrations); however, open ponds have been used on an industrial scale for the microalgae mentioned above and *Haematococcus pluvialis* as well (Costa and De Morais 2014). Commonly used forms include raceway ponds, circular ponds, and unstirred ponds, of which raceway ponds have received the most attention (Sun et al. 2016).

4.7.1.1 Raceway Pond System

The open raceway pond system has a very simple structure consisting of a closed raceway and a paddle wheel. This minimal design can be constructed using low-cost materials, including compacted earth, cement, and plastics (Fig. 4.3a) (Lu et al. 2021; Amaro et al. 2023). Microalgae are inoculated and cultivated in the raceway, and the paddle wheel promotes the circulation of culture media or wastewater. The water-flow velocity required to prevent algae cells from deposition and settling depends on the sinking rate of the cells. A velocity of 10–20 cm s^{-1} has been found effective, and higher velocities are preferred, but the velocity must remain under 30 cm s^{-1} to be energetically viable (Shen et al. 2009; Lu et al. 2021). This system can be operated outdoors, reducing the operation cost by using sunlight to perform photosynthesis (Lu et al. 2021). The authors recommend this type of system to produce robust, fast-growing strains tolerant to extreme conditions; here, the biomass productivity is highly dependent on the strain used, climatic conditions, and culture depth; this last one oscillates between 15 and 30 cm (Shen et al. 2009; Acién et al. 2017). The low investment and operation costs make the raceway pond system one of the most used for microalgae growth worldwide.

4.7.1.2 Circular Pond System

Circular ponds have a design like raceway ponds and are normally up to 45 m in diameter and 30–70 cm in depth with a centrally pivoted agitator. The circular ponds have size as a limiting factor since a long rotating arm leads to poor mixing efficiency (e.g., >50 m in diameter) (Fig. 4.3b) (Shen et al. 2009). Moreover, this design has the disadvantages of high energy demand in the agitation process and high construction costs. This cultivation system has been used in Japan and Taiwan to culture *Chlorella* sp. (Tan et al. 2020).

4.7.1.3 Unstirred Pond System

Unstirred ponds represent the most economical and least technical of all commercial culture methods since they can be simply natural lakes or constructed from natural water ponds with uncovered beds, usually less than 50 cm deep (Fig. 4.3c). Despite their simplicity, unstirred ponds are limited by the used strain, using only microalgae that can grow in poor conditions or have a competitive advantage that allows them to outgrow contaminants such as protozoa, other microalgae, viruses, and bacteria (Shen et al. 2009).

4.7.2 Closed Photobioreactor Systems

Closed photobioreactors are more efficient as they can be operated at highly controlled conditions and, therefore, can overcome the disadvantages of an open cultivation system. Photobioreactors can be designed and optimized by the specific strain used; their design is covered with a transparent material or contained within transparent tubing. Most designs consist of tubes of various shapes, sizes, and lengths, constructed of transparent materials such as glass and plastic. Photobioreactors are also widely designed as plate (flat panel) types to maximize light exposure. This closed system utilizes relatively little space while increasing light availability and significantly decreasing contamination issues (Shen et al. 2009; Narala et al. 2016) (Fig. 4.4).

4.8 Factors Affecting Biomass Yield

After a phycoremediation process with microalgae, the biomass obtained has several applications, including aquaculture, human nutrition, pharmaceuticals, and biofuels, among others. However, several factors could affect the yield and quality of the recovered biomass (Jais et al. 2017). Below, two of the most important factors are described.

4.8.1 Wastewater Characteristics

The production of microalgae biomass in wastewater depends on the physical and chemical characteristics of the wastes, such as the presence of nutrients necessary for microalgae growth and/or toxic pollutant compounds. The physical and chemical characteristics of wastewater directly depend on its source; wastewater often consists of a combination of organic and inorganic compounds; it might have

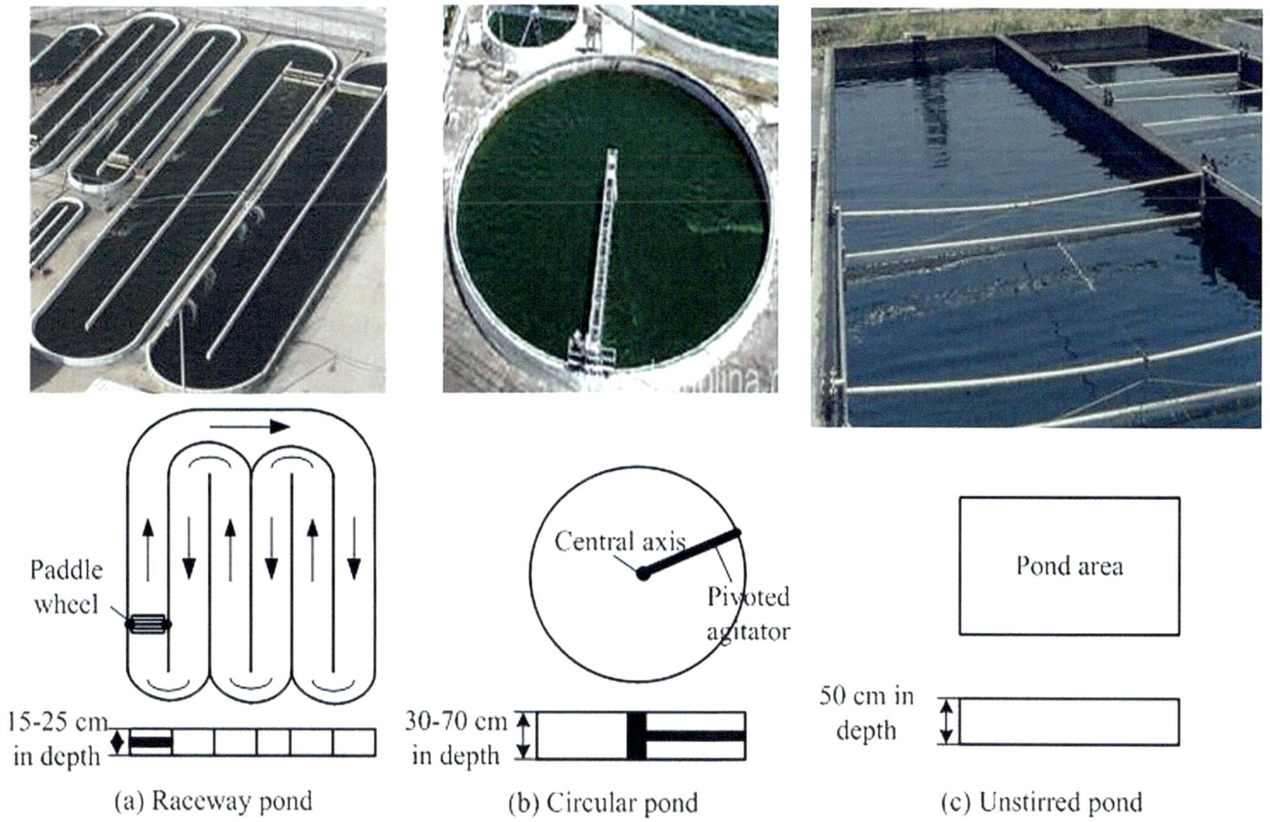

Fig. 4.3 Open pond systems for microalgae cultivation. (**a**) Raceway pond, (**b**) circular pond, and (**c**) unstirred pond. (Taken from Shen et al. (2009))

Fig. 4.4 Example of a flat panel photobioreactor system (maximum working volume 1300 L); it is composed of 20 vertical parallel plates of 50 L, 1150 × 1150 × 50 mm (W × H × D), in direct communication with each other. (Adapted from Lindblad et al. (2019))

the presence of heavy metals, pesticides, pharmaceutical residues, pathogenic microorganisms, oil, and grease, among other undesired contaminants. To overcome the toxicity of wastewater components, it is crucial to use resistant microalgae strains, which can survive in different types of wastewaters and efficiently remove nutrients and heavy metals. In this sense, the strains most used are indigenous microalgae, which have the potential to produce high-quality wastewater and microalgae biomass; the species most frequently found are *Chlamydomonas reinhardtii*, *Euglena pascheri*, *Lepocinclis steinii*, *Lyngbya pseudospirulina*, *Oscillatoria chalybaea*, and *Oscillatoria terebriformis*. However, this also has limitations; for example, these microalgae strains may be effective in one region but not necessarily in others (Jais et al. 2017).

4.8.2 C/N and P/N Relationships

There is a great influence of C/N ratios on microalgae biomass production; low ratios lead to low nutrient removal, and high levels of ammonia interfere with the photosynthesis process as well as biomass production. Dang et al. (2022) determined an optimal C/N ratio to enhance biomass production and improve the bioremediation capacity of *Chlorella* sp. in coculture with activated sludge for pollutants. Also, ammonia tolerance was tested in synthetic wastewater (a mix of acetic acid, ammonium chloride, potassium phosphate, iron sulfate, magnesium sulfate, calcium chloride, Ethylenediaminetetraacetic acid (EDTA) disodium salt, and sodium carbonate) prepared at different concentrations of nutrients. The optimal C/N = 5:1 removed 89% of C and had a higher biomass production as well as a remotion of 98.6% of N on the tenth day, but from days 6 to 10, total depletion of ammonia caused cell death; P removal was 98% at C/N 5:1 but was also high at C/N = 10:1.

On the other hand, pork meat has typically used its wastewater to produce biogas, but the effluent remains rich in N, P, and other nutrients; microalgae bioremediation has been used in diluted wastewater, with no previous expensive treatments, from the pork industry with low removal percentage of nutrients due to the low C/N ratios, which inhibits the microalgae growth (Wang et al. 2016; Wang et al. 2017). A solution to this low C/N was to mix this pig meat effluent with brewery wastewater (C/N > 60) to achieve an optimal C/N ratio to improve nutrients and enhance microalgae biomass production. The goal was to develop a low-cost microalgae bioremediation strategy using *Chlorella vulgaris* in a yield composed of piggery wastewater mixed with effluents from three different stages of the process of brewing wastewater. The most successful wastewater mix was 1:5 using effluents from the packaging process of brewing wastewater, reaching an optimal C/N value of 7.9 with the removal of 100% of ammonia, 96% of N, 95% of P, and 93% of C and a biomass of 276 g/L (Zheng et al. 2018).

4.9 Bioremediation Applications

Bioremediation possesses an advantage over physical and chemical methods to mitigate contaminants due to its efficiency and quality to benefit the environment. Bioremediation is a natural process that involves microorganisms to reduce pollutant concentration, including bacteria, yeasts, fungi, microalgae, and their metabolites (Oindrila et al. 2021). Among these, microalgae have risen in prominence, primarily due to their potential for simultaneous bioremediation, CO_2 mitigation, and the production of high added-value molecules such as long-chain polyunsaturated fatty acids, phenolic compounds, volatile compounds, sterols, proteins, amino acids, peptides, vitamins, polysaccharides, and pigments. Additionally, approximately 50,000 microalgae species are known, representing countless possible strategies (Andrade and Andrade 2018). Microalgae bioremediation processes can occur at the contaminated site (in situ) or under controlled conditions (ex situ).

4.9.1 Controlled Conditions (Ex Situ)

Ex situ bioremediation involves removing contaminants from the original site using biological treatment by microorganisms, such as microalgae, in acclimated bioreactors. This process requires environmental adaptations (salinity, pH, temperature, and light intensity), nutritional adjustments (macro- and micronutrient and contaminant concentration to avoid nutrient limitation or substrate inhibition, mainly caused by hydrocarbons), and operational modifications (type of bioreactor, treatment time, and mode of operation: batch or continuous) to maximize the treatment performance. These adaptations lead to achieving high removal rates of pollutants. Among these conditions, pollutant concentration is crucial, as many contaminants may be toxic to the microalgae species used (Mclellan et al. 2019; da Silva Gonçalves et al. 2023).

4.9.2 Environment (In Situ)

In situ bioremediation implies reducing or removing pollutants inside the contaminated ecosystem. In situ bioremediation might be performed in different environments, such as water and soils, for various contaminants. For example, microalgae are used in the in situ bioremediation of water bodies contaminated with heavy metals, as some microalgae are heavy metal accumulators and play a key role in bioremediation in their native habitats. Algae adopt different strategies for survival and growth when exposed to higher metal concentrations; one is the formation of algal mats. A microalgal mat remediates nutrients and pollutants from the aquatic body during its growth. Patidar et al. (2015) explored the bioaccumulation of metals and lipids in a microalgae mat; they observed that the season affects the metal's bioaccumulation. Pre-monsoon season nurtured the growth of the *Cyanophyceae* and diatom communities, which exhibited better metal endurance and could be utilized for the bioremediation of heavy metals.

4.10 Challenges

In a search for published information on challenges related to wastewater treatment based on microalgae bioremediation (Scopus 2023), a total of 842 documents were recovered up to the 2023; most of them (810) have been written in the past 10 years (2013–2023). While the potential use of microalgae as environmental bioremediation agents is widely recognized, there are still some challenges to be considered, namely, efficiency, toxicity, nutrient deficiency, the multiplicity of contaminants, water quality, light and shading effect, condition control, and costs, among others.

4.10.1 Toxicity

This kind of study has been conducted in two senses: (1) some effluents with high pollutant levels are toxic to microalgae and (2) the reduction of the toxicity levels of the wastewater pollutants by microalgae. In the first case, studies are necessary to determine which species are more resistant to certain pollutants since wastewater could be toxic to some of the microalgal strains; this is one of the most significant challenges in algal bioremediation (Gondi et al. 2022). Some species can be more tolerant than others; for instance, Xiong et al. (2016) evaluated the toxicity and cellular stresses of carbamazepine on *Chlamydomonas mexicana* and *Scenedesmus obliquus*. The authors demonstrated that *C. mexicana* was more tolerant to carbamazepine and could be used to treat contaminated wastewater. Besides, Renuka et al. (2015) emphasized the importance of nutrient ratio as a limiting factor inducing algal blooms, which, in turn, can also stimulate the production of allelochemicals in certain microalgal species and exert an adverse effect on other algae, thereby inhibiting their growth. Such conditions could lead to the dominance of certain microalgal species, particularly cyanobacteria, which produce allelochemicals and toxins. In the second case, in a bioremediation study of bisphenol A, Zhao et al. (2022) verified using a phytohormone that regulates the resistance of *Chlorella* to bisphenol. Toxicity stimulates microalgal growth. The addition of precise compounds could improve the removal of pollutants. So studying what kind of compounds could be used is necessary.

4.10.2 Nutrient Deficiency

According to Paskuliakova et al. (2018), landfill leachate is a type of wastewater that is difficult to treat; the authors found that phosphate addition promotes microalgal growth of proven strains and total ammonia nitrogen removal. However, when strong leachates are diluted to acceptable total ammonia nitrogen levels before remediation with microalgae, the reduced concentration of other co-occurring nutrients contributes to slow microalgal growth and may cause secondary nutrient limitations. This study enhanced microalgal growth rate and biomass production by adding some minerals in leachate samples. In another study, Xu et al. (2020) mentioned that carbon deficiency (among other factors) is a problem jeopardizing microalgae-based wastewater remediation; they proved that adding glycerol is a good carbon source for microalgae growth in the wastewater.

4.10.3 Multiple Pollutants

It is necessary to know better about potential interactions among pollutants (e.g., additivity, synergism, antagonism, inhibition, potentiation, and masking) in wastewater and the function of microalgae as bioremediators. In the case of nutrients, for instance, understanding the mechanisms of nutrient uptake and their interaction/interference with other contaminants, such as heavy metals, is required (Renuka et al. 2015). On the other hand, in some published papers related to antibiotic bioremediation, the authors reported better results when they tested different pollutants at once. Xiong et al. (2019) found that the removal rate of sulfamethazine by *Scenedesmus obliquus* increased with the addition of sulfamethoxazole.

Besides, Kiki et al. (2022) reported higher removal rates of sulfamethoxazole, trimethoprim, and levofloxacin in combined than in individual treatments. Liu et al. (2023) found that the combined treatment enhanced the antibiotic removal ability of *Chlorella pyrenoidosa* under the dual impacts of erythromycin and roxithromycin. More bioassays with multiple pollutants (from the same or different groups) should be performed to understand the bioremediation process better.

4.10.4 Water Quality

Microalgae could be used as bioindicators of water quality with several potential advantages, such as (1) biological impacts can be determined (2) potential synergetic and antagonistic impacts of various combined pollutants on ecosystems can be exhibited, (3) early-stage diagnosis of putative harmful effects of toxins on human and animal health can be monitored and (4) they can be considered as potentially viable economic alternatives to the use of conventional sophisticated methods (O'neill and Rowan 2022).

Furthermore, diatom assemblages could be used as water quality bioindicators for pollution incidences; in fact, over 20 diatom-based indices exist for water quality monitoring; however, most traditional indices vastly underutilize the full potential for diatoms to act as bioindicators (Kock et al. 2023). Diatoms are excellent options since they are more directly affected by climate change and can, therefore, respond in a more immediate way, being early bioindicators. Green microalgae could be used for this purpose as well; for example, *Pseudokirchneriella subcapitata* has been suggested as a potential early warning indicator for altering issues associated with aquaculture processing due to environmental variances, including climate change, presenting advantages such as low-cost and both highly reliable and reproducible culture (O'Neill and Rowan 2022).

4.10.5 Light and Shading Effect

Among all the factors that affect microalgae productivity and growth, light could be the most important since it provides all the energy required for metabolism through photosynthesis (Saccardo et al. 2022). Researchers must deal with both extremes of this factor: high and low light intensity. Recently, Min et al. (2022) conducted field experiments on sewage inflow river water to investigate the effect (among others) of light intensity. They observed that in an open pond without a light shielding facility, the productivity of *Hydrodictyon reticulatum* was very low, resulting in photoinhibition. However, the authors mentioned that if the light intensity is adjusted by installing a shading facility, the water temperature decreases, and the productivity of *H. reticulatum* could be improved. Then, light intensity was moderately reduced using a transparent film, and adequate water temperature could be maintained, increasing nutrient removal efficiency and biomass productivity. On the other hand, there are situations where shading is a factor to be faced. For instance, high algal density would lead to self-shading, which is a light attenuation by microalgae cell absorption that reduces light availability in the inner parts of the photobioreactors; this leads to autoinhibition and a reduction in photosynthetic efficiency (Abdel-Raouf et al. 2012; Saccardo et al. 2022). Also, microalgae are exposed to light–dark cycles because of the light gradient induced by the self-shading and the turbulent mixing in the reactor; due to these conditions, cells receive light intermittently (Saccardo et al. 2022). Emparan et al. (2021) observed the reduction in photosynthetic efficiency as an inhibitory effect of self-shading, resulting in the slow growth rate of microalgae cells responsible for removing low chemical oxygen demand.

4.10.6 Condition Control

Several abiotic factors influence the effectiveness of microalgal systems on pollutant removal (pH, temperature, light intensity, and agitation, among others) (Amaro et al. 2023), which must be controlled. The effect of temperature can be very different according to the algal species, phyla, and even strain. Therefore, its correct modeling in microalgae growth is crucial (Rossi et al. 2023). In the case of pH, it is an important parameter that can impact the mechanism of algal bioremediation since pH has a significant impact on microalgae metabolism and growth, which could affect uptake of ions, enzymatic activity, phosphorus, and inorganic carbon availability, and pollutant tolerance (Sousa et al. 2021). Microalgal cultivation systems are exposed to fluctuating environmental conditions on an industrial scale. Thus, further research should focus on the optimization of the operating conditions, considering this real scenario at a large scale (Amaro et al. 2023). Moreover, Abdelfattah et al. (2023) pointed out that the main challenge facing large-scale wastewater treatment plants is the complicated operation and monitoring of the whole process, which requires developing new technologies such as online monitoring and remote control.

4.10.7 Costs

Wastewater generated from different anthropogenic activities usually undergoes cost-effective treatments before being discharged into natural water bodies. Therefore, it is crucial to reduce the raw material costs of production for photosynthetic microalgae cultivation (e.g., nutrients, carbon dioxide, and light energy). For instance, the continuous use of artificial light increases electricity costs (Maroneze et al. 2019); thus, evaluating the influence of different photoperiods would reduce costs. About carbon dioxide, Rossi et al. (2023) mentioned that microalgae can remove it from gaseous streams and provide oxygen to aerobic bacteria in algae–bacteria systems; this implies cost reduction for oxygenation, which is the main drawback of conventional bioremediation processes. In this sense, using consortiums of microalgae with bacteria or fungi has been studied as a good strategy to reduce costs (Renuka et al. 2015).

4.10.8 Efficiency

Microalgal bio-absorption technologies have been well-documented in the scientific literature over the past decade. The pollutant adsorption rate (efficiency) has been reported as variable; this could be due to the pollutant solubility; hydrophilic compounds are anionic (negatively charged) and have low adsorption affinity values with microalgal cells due to the cells also being negatively charged. In contrast, lipophilic compounds, which are cationic, have high bio-adsorption affinity values with microalgae (De Wilt et al. 2016).

Another factor that affects the efficiency is the microalgae species utilized; there exist species that have a high affinity for the target pollutants, preferentially adsorbing them onto the cell surface, essentially acting as "hyper-adsorbents" or "hyper-accumulators"; among them, we find species belonging to the genera *Anabaena* (*Dolichospermum*), *Chlorella, Cladophora, Oscillatoria,* and *Scenedesmus* (Sutherland and Ralph 2019).

In most of the literature reviewed, the aspects mentioned above are related to the efficiency of microalgae cultures and their subsequent use as bioremediators; to the extent that these challenges are addressed, the efficiency of these processes will also improve.

4.11 Conclusion

Conventional pollutants in wastewater are considered potential threats to living beings and the environment. Therefore, it is necessary to look for effective strategies to remove them. Orthodox wastewater treatment systems entail several environmental issues, such as high energy consumption, high CO_2 emissions, and the use of chemicals or the generation of harmful by-products. Hence, microalgae have emerged as promising and eco-friendly solutions to treat wastewater, showing numerous advantages over other methods. Microalgae deal with contaminant substances through different mechanisms, the most outstanding being biodegradation, which could allow effluents to comply with established regulations. Although microalgae bioremediation is an effective process, there are still challenges to overcome. However, the bioremediation scheme with microalgae and the subsequent use of the biomass generated during the process have gained popularity since the biomass can be used for aquaculture, agriculture, pharmaceutical, cosmetic, and food industries, among others, allowing a circular economy approach.

Acknowledgments The authors, as collaborators of the Estancias Postdoctorales por México 2022 program, thank CONAHCYT, especially D. Ambriz as a postdoc researcher.

References

Abdelfattah A, Ali SS, Ramadan H, El-Aswar EI, Eltawab R, Ho SH, Elsamahy T, Li S, El-Sheekh MM, Schagerl M, Kornaros M, Sun J (2023) Microalgae-based wastewater treatment: mechanisms, challenges, recent advances, and future prospects. Environ Sci Ecotechnol 13:100205–100205

Abdel-Raouf N, Al-Homaidan AA, Ibraheem IBM (2012) Microalgae and wastewater treatment. Saudi J Biol Sci 19:257–275

Acién FG, Molina E, Reis A, Torzillo G, Zittelli GC, Sepúlveda C, Masojídek J (2017) Photobioreactors for the production of microalgae. In: Microalgae-based biofuels and bioproducts: from feedstock cultivation to end-products, pp 1–44

Ahmed J, Thakur A, Goyal A (2021) Industrial wastewater and its toxic effects. In: Biological Treatment of Industrial Wastewater, pp 1–14
Amaro HM, Salgado EM, Nunes OC, Pires JCM, Esteves AF (2023) Microalgae systems – environmental agents for wastewater treatment and further potential biomass valorisation. J Environ Manag 337:117678–117678
Andrade CJD, Andrade LMD (2018) Microalgae for bioremediation of textile wastewater: an overview. MOJ Food Process Technol 6
Beuckels A, Smolders E, Muylaert K (2015) Nitrogen availability influences phosphorus removal in microalgae-based wastewater treatment. Water Res 77:98–106
Buchman MF (2008) NOAA screening quick reference tables. Office of Response and Restoration Division, Seattle
Caldwell DH (1946) Sewage oxidation ponds: performance, operation and design. Sew Work J 18:433–458
Carvalho Nogueira da Gama R, Peixoto-Assemany P, Rodrigues de Assis L, Vital-Oliveira L, Cecon PR, Calijuri ML (2023) Influence of C/N ratio on microalgae-bacteria joint culture: treatment performance and phytoplankton dynamics in mixed wastewaters. Bioresour Technol Rep 23:101516–101516
Costa J, De Morais MG (2014) An open pond system for microalgal cultivation. Biofuels Algae 1–22. https://doi.org/10.1016/B978-0-444-59558-4.00001-2
da Silva Gonçalves AH, de Andrade FP, de Farias Silva CE, Medeiros JA, Santos GKS, do Nascimento MAA, Tonholo J, Almeida RMRG (2023) Biological treatment of petroleum produced water ex situ using microorganisms: an overview, main developments and challenges. Energy Ecol Environ 8:289–303
Dang BT, Nguyen TT, Ngo HH, Pham MDT, Le LT, Nguyen NKQ, Vo TDH, Varjani S, You SJ, Lin KA, Huynh KPH, Bui XT (2022) influence of c/n ratios on treatment performance and biomass production during co-culture of microalgae and activated sludge. Sci Total Environ 837:155832–155832
Das C, Naseera K, Ram A, Meena RM, Ramaiah N (2017) Bioremediation of tannery wastewater by a salt-tolerant strain of Chlorella vulgaris. J Appl Phycol 29:235–243
Das C, Ramaiah N, Pereira E, Naseera K (2018) Efficient bioremediation of tannery wastewater by monostrains and consortium of marine chlorella sp. and Phormidium sp. Int J Phytoremediation 20:284–292
De Cassia Campos Pena A, Bertoldi CF, Da Fontoura JT, Trierweiler LF, Gutterres M (2019) Consortium of microalgae for tannery effluent treatment. Braz Arch Biol Technol 62:e19170518–e19170518
De Wilt A, Butkovskyi A, Tuantet K (2016) Micropollutant removal in an algal treatment system fed with source separated wastewater streams. J Hazard Mater 304:84–92. https://doi.org/10.1016/J.JHAZMAT.2015.10.033
Del Puerto Rodríguez AM, Suárez Tamayo S, Palacio Estrada DE (2014) Efectos de los plaguicidas sobre el ambiente y la salud. Rev Cubana de Hig y Epidemiol 52:372–387
Devi A, Verma M, Saratale GD, Saratale RG, Ferreira LFR, Mulla SI, Bharagava RN, Devi A, Verma M, Saratale GD, Saratale RG, Ferreira LFR, Mulla SI, Bharagava RN (2023) Microalgae: a green eco-friendly agents for bioremediation of tannery wastewater with simultaneous production of value-added products. Chmsp 336:139192–139192
Emparan Q, Harun R, Sing Jye Y (2021) Efficiency of pollutants removal in treated palm oil mill effluent (TPOME) using different concentrations of sodium alginate-immobilized Nannochloropsis sp. cells. Int J Phytoremediation 23:454–461
EPA (2023) Effects of Acid Rain | US EPA
Gao F, LI Z, Chang Q, Gao M, She Z, Wu J, Jin C, Zheng D, Guo L, Zhao Y, Wang S (2018) Effect of florfenicol on performance and microbial community of a sequencing batch biofilm reactor treating mariculture wastewater. Environ Technol 39:363–372
Geissen V, Mol H, Klumpp E, Umlauf G, Nadal M, Van Der Ploeg M, van de Zee SEATM, Ritsema CJ (2015) Emerging pollutants in the environment: A challenge for water resource management. Int Soil Water Conserv Res 3:57–65
Gondi R, Kavitha S, Yukesh Kannah R, Parthiba Karthikeyan O, Kumar G, Kumar Tyagi V, Rajesh Banu J (2022) Algal-based system for removal of emerging pollutants from wastewater: a review. Bioresour Technol 344:126245–126245
Goswami RK, Agrawal K, Shah MP, Verma P (2022) Bioremediation of heavy metals from wastewater: a current perspective on microalgae-based future. Lett Appl Microbiol 75:701–717
Haward M (2018) Plastic pollution of the world's seas and oceans as a contemporary challenge in ocean governance. Nat Commun 9(1):667
Hu Y, Qiu S, Bi Q, Chen Z, Zhang X, Ge S (2023) Start-up and maintenance of indigenous microalgae-bacteria consortium treating toilet wastewater through partial nitrification and nitrite-type denitrification. Water Res 239:120029–120029
Jais NM, Mohamed RMSR, Al-Gheethi AA, Hashim MKA (2017) The dual roles of phycoremediation of wet market wastewater for nutrients and heavy metals removal and microalgae biomass production. Clean Techn Environ Policy 19:37–52
Kiki C, Rashid A, Zhang Y, LI X, Chen TY, Eloise Adéoye AB, Peter PO, Sun Q (2022) Microalgal mediated antibiotic co-metabolism: kinetics, transformation products and pathways. Chemosphere 292:133438–133438
Kock A, Glanville HC, Law AC, Stanton T, Carter LJ, Taylor JC (2023) Emerging challenges of the impacts of pharmaceuticals on aquatic ecosystems: a diatom perspective. Sci Total Environ 878:162939–162939
Leong YK, Chang JS (2020) Bioremediation of heavy metals using microalgae: recent advances and mechanisms. Bioresour Technol 303:122886–122886
Lindblad P, Fuente D, Borbe F, Cicchi B, Conejero JA, Couto N, Čelešnik H, Diano MM, Dolinar M, Esposito S, Evans C, Ferreira EA, Keller J, Khanna N, Kind G, Landels A, Lemus L, Noirel J, Ocklenburg S, Oliveira P, Pacheco CC, Parker JL, Pereira J, Pham TK, Pinto F, Rexroth S, Rögner M, Schmitz HJ, Benavides AMS, Siurana M, Tamagnini P, Touloupakis E, Torzillo G, Urchueguía JF, Wegelius A, Wiegand K, Wright PC, Wutschel M, Wünschiers R (2019) CyanoFactory, a European consortium to develop technologies needed to advance cyanobacteria as chassis for production of chemicals and fuels. Algal Res 41:101510
Liu K, LI J, Zhou Y, Li W, Cheng H, Han J (2023) Combined toxicity of erythromycin and roxithromycin and their removal by Chlorella pyrenoidosa. Ecotoxicol Environ Saf 257:114929–114929
Lu Q, Xiao Y, Chen F (2021) Application of microalgae biotechnology for the sustainable development of aquaculture. Adv Bioenergy 6:117–163
Ma H, Shen M, Tong Y, Wang X (2023) Radioactive wastewater treatment technologies: a review. Molecules *28*:1935
Mandal S, Kunhikrishnan A, Bolan NS, Wijesekara H, Naidu R (2016) Application of biochar produced from biowaste materials for environmental protection and sustainable agriculture production. In: Environmental materials and waste: resource recovery and pollution prevention, pp 73–89
Maroneze MM, Deprá MC, Zepka LQ, Jacob-Lopes E (2019) Artificial lighting strategies in photobioreactors for bioenergy production by Scenedesmus obliquus CPCC05. SN Appl Sci 1:1–12
Mateo-Sagasta J, Zadeh SM, Turral H, Burke J (2017) Water pollution from agriculture: a global review executive summary. Food and Agriculture Organization of the United Nations and the International Water Management Institute on behalf of the Water Land and Ecosystems research program

Mclellan J, Gupta SK, Kumar M (2019) Feasibility of using bacterial-microalgal consortium for the bioremediation of organic pesticides: application constraints and future prospects. In: Gupta SK, Bux F (eds) Application of microalgae in wastewater treatment. Springer, Cham, pp 341–362

Min KJ, OH DY, Park KY (2022) Field test of water-net based wastewater treatment for nutrient removal and bioethanol production. Chemosphere 301:134791–134791

Moeller Chávez M, Sandoval Yoval L, Ramírez González A, Ramírez Camperos E, Ramírez Camperos E, Escalante Estrada V, Tomasini Ortiz C, Miranda Mandujano V, Mijaylova Nacheva P, Ortíz Rendón G (2011) Tratamiento de aguas residuales. Instituto Mexicano de Tecnología del Agua

Narala RR, Garg S, Sharma KK, Thomas-Hall SR, Deme M, LI Y, Schenk PM (2016) Comparison of microalgae cultivation in photobioreactor, open raceway pond, and a two-stage hybrid system. Front Energy Res 4:208116–208116

O'neill EA, Rowan NJ (2022) Microalgae as a natural ecological bioindicator for the simple real-time monitoring of aquaculture wastewater quality including provision for assessing impact of extremes in climate variance—A comparative case study from the Republic of Ireland. Sci Total Environ 802:149800. https://doi.org/10.1016/J.SCITOTENV.2021.149800

Oindrila P, Amrita J, Dibyajit L, Moupriya N, Rina Rani R (2021) In situ and ex situ bioremediation of heavy metals: the present scenario. J Environ Eng Landsc Manag 29:454–469

Oliveira AS, Alves M, Leitão F, Tacão M, Henriques I, Castro PML, Amorim CL (2023) Bioremediation of coastal aquaculture Effluents Spiked with florfenicol using microalgae-based granular sludge—a promising solution for recirculating aquaculture systems. Water Res 233:119733–119733

Oswald WJ, Gotaas HB (1957) Photosynthesis in sewage treatment. Trans Am Soc Civ Eng 122:73–97

Paskuliakova A, Mcgowan T, Tonry S, Touzet N (2018) Microalgal bioremediation of nitrogenous compounds in landfill leachate—the importance of micronutrient balance in the treatment of leachates of variable composition. Algal Res 32:162–171

Patidar SK, Mishra SK, Bhattacharya S, Ghosh T, Paliwal C, Goel S, Mishra S (2015) Naturally floating microalgal mat for in situ bioremediation and potential for biofuel production. Algal Res 9:275–282

Pintor AMA, Vilar VJP, Botelho CMS, Boaventura RAR (2016) Oil and grease removal from wastewaters: sorption treatment as an alternative to state-of-the-art technologies. A critical review. Chem Eng J 297:229–255

Plummer JD, Long SC, Charest AJ, Roop DO (2014) Bacterial and viral indicators of fecal contamination in drinking water. J Am Water Works Assoc 106:E200–E211

Rabalais NN (2019) Ocean deoxygenation from eutrophication (human nutrient inputs). In: Baxter JM, Laffoley D (eds) Ocean deoxygenation: everyone's problem. IUCN

Råman Vinnå L, Wüest A, Bouffard D (2017) Physical effects of thermal pollution in lakes. Water Resour Res 53:3968–3987

Renuka N, Sood A, Prasanna R, Ahluwalia AS (2015) Phycoremediation of wastewaters: a synergistic approach using microalgae for bioremediation and biomass generation. Int J Environ Sci Technol 12:1443–1460

Rossi S, Carecci D, Ficara E (2023) Thermal response Analysis and compilation of cardinal temperatures for 424 strains of microalgae, cyanobacteria, diatoms and other species. Sci Total Environ 873:162275–162275

Rügner H, Schwientek M, Beckingham B, Kuch B, Grathwohl P (2013) Turbidity as a proxy for total suspended solids (TSS) and particle facilitated pollutant transport in catchments. Environ Earth Sci 69:373–380

Saccardo A, Bezzo F, Sforza E (2022) Microalgae growth in ultra-thin steady-state continuous photobioreactors: assessing self-shading effects. Front Bioeng Biotechnol 10:977429

Sahu SN, Sahoo NK, Naik SN (2019) Phycoremediation technology: a global prospective. In: Gupta SK, Bux F (eds) Application of microalgae in wastewater treatment. Springer, Cham

Saranya D, Shanthakumar S (2020) Effect of culture conditions on biomass yield of acclimatized microalgae in ozone pre-treated tannery effluent: a simultaneous exploration of bioremediation and lipid accumulation potential. J Environ Manag 273:111129

Scopus (2023) *Scopus – sources* [Online]. Available: https://www.scopus.com/sources.uri?zone=TopNavBar&origin=searchauthorfreelookup [Accessed]

Semarnat (1998a) NOM-002-SEMARNAT-1996. DOF (Diario Oficial de la Federación). DOF Diario Oficial de la Federación

Semarnat (1998b) NOM-003-SEMARNAT-1997. DOF (Diario Oficial de la Federación). DOF Diario Oficial de la Federación

Semarnat (2005a) NOM-052-SEMARNAT-2005. DOF (Diario Oficial de la Federación). DOF Diario Oficial de la Federación

Semarnat (2005b) NOM-138-SEMARNAT/ss-2003. OF (Diario Oficial de la Federación). DOF Diario Oficial de la Federación

Semarnat (2007) NOM-147-SEMARNAT/SSA1–2004. DOF (Diario Oficial de la Federación). DOF Diario Oficial de la Federación

Semarnat (2017) Residuos sólidos urbanos. Semarnat

Semarnat (2022) NOM-001-SEMARNAT-2021. DOF (Diario oficial de la federación). DOF (Diario Oficial de la Federación)

Shen Y, Yuan W, Pei ZJ, Wu Q, Mao E (2009) Microalgae mass production methods. Trans ASABE 52:1275–1287

Singh K, Ansari FA, Ingle KN, Gupta SK, Ahirwal J, Dhyani S, Singh S, Abhilash PC, Rawat I, Byun C, Bux F (2023) Microalgae from wastewaters to wastelands: leveraging microalgal research conducive to achieve the UN sustainable development goals. Renew Sust Energ Rev 188:113773–113773

Some S, Mondal R, Mitra D, Jain D, Verma D, Das S (2021) Microbial pollution of water with special reference to coliform bacteria and their nexus with environment. Energy Nexus 1:100008–100008

Sousa CA, Sousa H, Vale F, Simões M (2021) Microalgae-based bioremediation of wastewaters—influencing parameters and mathematical growth modelling. Chem Eng J 425:131412. https://doi.org/10.1016/J.CEJ.2021.131412

Sun Z, Liu J, Zhou ZG (2016) Algae for biofuels: an emerging feedstock. In: Handbook of biofuels production: processes and technologies, 2nd edn. Elsevier, pp 673–698

Sutherland DL, Ralph PJ (2019) Microalgal bioremediation of emerging contaminants – opportunities and challenges. Water Res 164:114921–114921

Tan J, Sen Lee SY, Chew KW (2020) A review on microalgae cultivation and harvesting, and their biomass extraction processing using ionic liquids. Bioengineered 11:116–129. https://doi.org/10.1080/21655979.2020.1711626

Thomsen E, Herbeck LS, Jennerjahn TC (2020) The end of resilience: surpassed nitrogen thresholds in coastal waters led to severe seagrass loss after decades of exposure to aquaculture effluents. Mar Environ Res 160:104986–104986

Tseng YS, Patel AK, Haldar D, Chen CW, Dong CD, Singhania RR (2023) Microalgae and nano-cellulose composite produced via a co-culturing strategy for ammonia removal from the aqueous phase. Bioresour Technol 389:129801–129801

USEPA (2021) Effluent limitations guidelines and standards (ELG) database

Wang M, Yang Y, Chen Z, Chen Y, Wen Y, Chen B (2016) Removal of nutrients from undiluted anaerobically treated piggery wastewater by improved microalgae. Bioresour Technol 222:130–138

Wang S, Wang L, Deng L, Zheng D, Zhang Y, Jiang Y, Yang H, Lei Y (2017) Performance of autotrophic nitrogen removal from digested piggery wastewater. Bioresour Technol 241:465–472

Xiong JQ, Kurade MB, Abou-Shanab RAI, Ji MK, Choi J, Kim JO, Jeon BH (2016) Biodegradation of carbamazepine using freshwater microalgae Chlamydomonas mexicana and Scenedesmus obliquus and the determination of its metabolic fate. Bioresour Technol 205:183–190

Xiong JQ, Kim SJ, Kurade MB, Govindwar S, Abou-Shanab RAI, Kim JR, Roh HS, Khan MA, Jeon BH (2019) Combined effects of sulfamethazine and sulfamethoxazole on a freshwater microalga, Scenedesmus obliquus: toxicity, biodegradation, and metabolic fate. J Hazard Mater 370:138–146

Xu M, Zeng Q, Li H, Zhong Y, Tong L, Ruan R, Liu H (2020) Contribution of glycerol addition and algal–bacterial cooperation to nutrients recovery: a study on the mechanisms of microalgae-based wastewater remediation. J Chem Technol Biotechnol 95:1717–1728

Zhao Z, Yang H, Feng Z, Huo Y, Fu L, Zhou D (2022) Role of naphthaleneacetic acid in the degradation of bisphenol A and wastewater treatment by microalgae: enhancement and signaling. Chemosphere 307:135829

Zheng H, Liu M, Lu Q, Wu X, Ma Y, Cheng Y, Addy M, Liu Y, Ruan R (2018) Balancing carbon/nitrogen ratio to improve nutrients removal and algal biomass production in piggery and brewery wastewaters. Bioresour Technol 249:479–486

Part II

Microalgae as Bioenergy Products

Thermochemical Conversion of Microalgal Biomass: A Potential Source of Bio-oil, Biochar, and Syngas

5

Lucielen Oliveira Santos, Pedro Garcia Pereira Silva, Bruno Roswag Machado, Izabela Queiroz Silva, Munise Zaparoli, Gabriela Moura Kurowiski de Brito, Michele Greque de Morais, Jorge Alberto Vieira Costa, and Ana Priscila Centeno da Rosa

Abstract

Thermochemical conversion of microalgae biomass is a promising area of renewable energy and sustainability research when integrated into the biorefinery context. Microalgae are single-celled microorganisms that use photosynthesis, converting sunlight into chemical energy. As a result, they may accumulate a significant amount of biomass rich in lipids, carbohydrates, and proteins, making them valuable sources of raw material for energy production. Thus, this chapter describes thermochemical conversion methods that may be applied to microalgal biomass, such as pyrolysis, torrefaction, hydrothermal conversion, and gasification. These conversion methods may reduce the dependence on fossil fuels and decrease greenhouse gas emissions. In addition, the main challenges of using microalgae biomass to obtain products of interest, such as bio-oil, biochar, and synthesis gas (syngas), were also addressed.

Keywords

Microalgae · Biomass · Sustainability · Conversion methods · Bioenergy

L. O. Santos (✉) · P. G. P. Silva · B. R. Machado · I. Q. Silva
G. M. K. de Brito · M. G. de Morais · J. A. V. Costa
A. P. C. da Rosa
School of Chemistry and Food, Federal University of Rio Grande, Rio Grande, RS, Brazil

M. Zaparoli
Bioprocess Engineering and Biotechnology Department, Federal University of Paraná, Curitiba, PR, Brazil

5.1 Introduction

Renewable energy sources have emerged as a viable solution to replace fossil fuels, primarily due to the environmental impacts caused by the latter. Furthermore, the continuous growth of the global population drives the global energy demand. Therefore, a practical approach to address these challenges is the pursuit of ecologically sustainable and environmentally friendly fuel alternatives (Aliyu et al. 2021; Azizi et al. 2020).

The search for sustainable energy sources and bioproducts within the context of biorefineries has driven extensive research into alternative biomass conversion technologies. In this context, microalgae biomass has garnered interest due to its CO_2 fixation capacity, rapid growth, production of biofuels, and other high-value-added products. Furthermore, unlike conventional terrestrial biomass sources, microalgae offer advantages, such as the potential for cultivation in wastewater, significantly reducing biomass production costs (Costa et al. 2023; Yu et al. 2017).

Microalgae biofuels may be produced by chemical, biochemical, or thermochemical techniques. Thermochemical conversion methods, such as pyrolysis, torrefaction, gasification, and hydrothermal conversion of microalgal biomass, are considered environmentally sustainable. These processes subject microalgae biomass to high temperatures and controlled chemical reactions, producing bio-oil, biochar, and synthesis gas (syngas) (Aliyu et al. 2021; Sekar et al. 2021). These products have versatile applications, including their use as fertilizers and soil conditioners and in environmental remediation. They also play a crucial role in producing biofuels, bioenergy, and carbon storage (Costa et al. 2023; Sekar et al. 2021; Wang et al. 2013).

Thermochemical conversion processes may also be integrated into a biorefinery context. Due to their photosynthetic characteristics, microalgae may produce various products

D. U. Santos Ballardo, S. Rossi (eds.), *Microalgae as Promising Source of Commercial Bioproducts*, Developments in Applied Phycology 14, https://doi.org/10.1007/978-3-031-86433-9_5

depending on cultivation. Among these products, biodiesel is noteworthy. However, lipid extraction from microalgae results in a significant amount of residual biomass containing proteins, carbohydrates, and small quantities of lipids. Some thermochemical conversion techniques, such as pyrolysis, have the potential to convert microalgae residues into bio-oil, biochar, and syngas, reducing process costs and maximizing the use of biomass (Wang et al. 2013).

There are studies demonstrating the production of these products derived from microalgae biomass and their potential applications. However, there are few reports on the conversion processes of whole microalgae biomass or extracted microalgae biomass into bioproducts in comparison with lignocellulosic biomass (Wang et al. 2013). Therefore, several challenges remain to be overcome to achieve profitable and competitive processes in the market (Andrade et al. 2018).

Therefore, this chapter aims to clarify the fundamental role of microalgae in the sustainable bioeconomy using thermochemical conversion techniques. Furthermore, it addresses future challenges associated with thermochemical conversion technologies such as pyrolysis, torrefaction, hydrothermal conversion, gasification processes, and their respective variations. In addition, the main challenges of using microalgae biomass to obtain products of interest, such as bio-oil, biochar, and synthesis gas, are also addressed.

5.2 Microalgae Biomass

Microalgae are heterogeneous microorganisms, usually microscopic, unicellular, colonial or filamentous, photoautotrophic, and phylogenetic; they may be prokaryotic or eukaryotic (Ayub et al. 2022). Microalgae biomass production fundamentally depends on some conditions established for cell growth. The interaction between physical, chemical, and biological factors directly influences microalgae cultivation, which may promote or inhibit their growth, impacting the maximization of biomass and other compounds of commercial interest, such as pigments, lipids, carbohydrates, and proteins (Aliyu et al. 2021). Also, the chemical composition of microalgae may change according to each species due to the nutritional and environmental conditions to which they are subjected during their life cycle (Krishnan et al. 2022).

5.2.1 Technological Advances of Microalgae Biomass as Feedstock

The use of microalgae biomass as feedstock in industrial chemical processes has become increasingly frequent. The interest in new alternatives that enable biomass production to produce bioenergy has gained prominence in the past few years (Chen et al. 2015; Das et al. 2021; Hussin et al. 2023).

Developing efficient large-scale cultivation processes that achieve high cell density is crucial to commercializing microalgal products, which may help reduce the environmental impact of bioenergy production (Razzak et al. 2022). The predominant approach for large-scale microalgae cultivation is using outdoor cultivation systems. In contrast, indoor microalgae cultivation, often called a closed cultivation system, is conducted within tubular photobioreactors. Within these enclosed setups, stringent monitoring and control measures are implemented to manage factors such as excessive light exposure, overheating, parasitic contaminations, and proliferation of predatory microalgae species (Hussin et al. 2023; Razzak et al. 2022). Thus, the cost of producing microalgae biomass may be minimized using new environmentally friendly technologies that favor its growth, such as magnetic fields (Santos et al. 2022) and genetic modification (Choudhary et al. 2022).

5.3 Thermochemical Conversion

Thermochemical conversion methods stand out as highly efficient approaches for converting microalgae biomass into bioenergy. These methods offer distinct advantages over the biochemical route, primarily owing to their versatility in using microalgae biomass for a direct conversion process. During this thermal process, organic materials decompose at exceptionally high temperatures, producing bio-oils, biochar, or syngas. These conversion methods, as displayed in Fig. 5.1, may reduce the dependence on fossil fuels and decrease greenhouse gas emissions, as explained below.

5.3.1 Pyrolysis

Pyrolysis is a thermochemical conversion process carried out at high temperatures and without oxygen. This process has received attention due to its simplicity of operation, speed, and versatility. The main products obtained from this process are bio-oil, biochar, and non-condensable gases. Getting these products depends on process conditions, such as temperature, residence time, raw material used, and other reactor conditions. Different pyrolysis techniques may be employed to obtain liquid, solid, and gaseous products, including fast pyrolysis, slow pyrolysis, catalytic pyrolysis, or microwave pyrolysis (Sekar et al. 2021). Some examples are elucidated in Table 5.1.

5.3.1.1 Fast Pyrolysis

Fast pyrolysis operates at high temperatures (850–1250 °C) and high heating rates, ranging from 10 to 200 °C in short periods, from 1 to 10 s (Tripathi et al. 2016). This technique is widely used for bio-oil production due to its low secondary cracking, reduced polymerization, and poor condensation

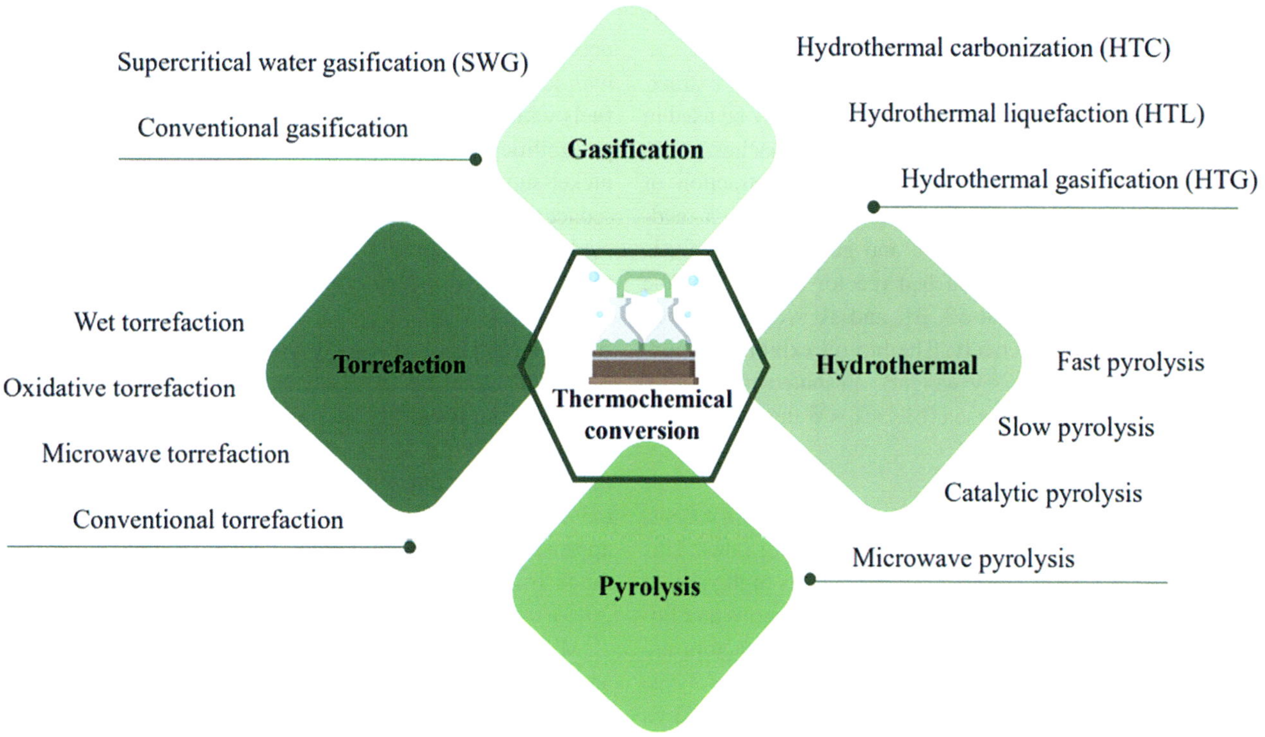

Fig. 5.1 Schematic diagram of thermochemical conversion methods and their variations

Table 5.1 Thermochemical conversion methods of microalgae biomass and its main products

Microalgae strain	Thermochemical conversion method	Bio-oil yield (%)	Biochar yield (%)	Syngas yield (%)	Reference
Tetraselmis sp.	Hydrothermal liquefaction	21.4	–	–	Vo et al. (2017)
Phaeodactylum tricornutum	Hydrothermal liquefaction	38.8	61	CH_4 (3.7); CO (8.8)	Christensen et al. (2014)
Chlorella sp.	Hydrothermal liquefaction	32.5	–	–	Reddy et al. (2016)
Chlorella vulgaris	Catalytic pyrolysis	10.4	–	–	Zainan et al. (2018)
Scenedesmus sp.	Fast pyrolysis	18.4	4.6	–	Grierson et al. (2011)
Porphyra tenera	Slow pyrolysis	48	68	H_2 (3); CO (14.1); CO_2 (38)	Bae et al. (2011)
Scenedesmus almeriensis	Microwave pyrolysis	41.1	44.5	–	Beneroso et al. (2013)
Chlorella vulgaris	Conventional gasification	–	–	H_2 (42.3); CO (31.9); CO_2 (8); CH_4 (12.7)	Khani et al. (2023)
Chlorella vulgaris	Supercritical water gasification	–	–	H_2 (55); CO (25); CH_4 (5)	Nurcahyani et al. (2020)
Nannochloropsis sp.	Conventional gasification	–	–	H_2 (45); CO (31); CO_2 (12); CH_4 (2)	Sanchez-Silva et al. (2013)

(Sekar et al. 2021). The high heating rate employed in fast pyrolysis transforms biomass into a liquid product before the reaction and formation of biochar. Depending on process conditions, this technique may produce approximately 70 wt% bio-oil, 10 wt% biochar, and 20 wt% gas (Yu et al. 2017).

The quantity obtained from each bioproduct (bio-oil, biochar, and syngas) also depends on the microalgae used. In a study conducted by Azizi et al. (2020), bio-oil, biogas, and biochar yields from different microalgae were evaluated using fast pyrolysis at 500 °C. The microalgae evaluated exhibited varied bioproduct results. Similar behaviors were observed when pyrolyzing *Nannochloropsis* and *Tetraselmis* biomasses, producing 50% of bio-oil, 30% of biochar, and 13% of biogas. In the case of *Isochrysis galbana* biomass, the highest bio-oil content was 66%, while biochar content was lower by 21%. Furthermore, it was reported that the biochar formed had higher nitrogen and phosphorus contents than other biomass sources.

Another way of using biomass may be done after extracting compounds of interest, such as lipids that may be used in biodiesel production. Through biorefinery approaches, it was found that the residual biomass from lipid extraction of *Chlorella vulgaris* showed potential when used as a feedstock to produce bio-oil, biochar, and gas in fast pyrolysis experiments using a fluidized bed reactor at 500 °C. The pyrolyzed biomass yielded 53, 31, and 10 wt% of bio-oil, biochar, and gas, respectively. The bio-oil exhibited a high nitrogen content (12.8% by weight), attributed to the high protein content in the feedstock (61.24%) (Wang et al. 2013).

5.3.1.2 Slow Pyrolysis

Slow pyrolysis, also known as conventional pyrolysis, employs long residence times and slow heating rates. This technique primarily produces biochar due to its high yield of solid residues. However, small quantities of gaseous and liquid products may also be formed. Fixed-bed reactors are commonly used for biomass conversion by slow pyrolysis, primarily due to their low cost and ease of operation (Aliyu et al. 2021).

Microalgae are introduced into a reactor and slowly heated to the desired temperature for the slow pyrolysis process, with residence times ranging from minutes to hours. The temperature used in slow pyrolysis may vary from 300 to 1000 °C. This parameter is crucial in slow pyrolysis, particularly in producing biochar with a higher surface area and porosity (Costa et al. 2023). Furthermore, the extended contact times in this technique and the slower heating rates are sufficient for completing secondary reactions and allow for removing vapors produced during these reactions (Sekar et al. 2021; Tripathi et al. 2016).

5.3.1.3 Catalytic Pyrolysis

Catalytic pyrolysis is a technique in which catalysts are added to improve the pyrolysis process. This results in a more stable oil and prevents polymerization, condensation reactions, and oxygen content in the bio-oil. In this process, tubular fixed-bed reactors may be used, operating at temperatures ranging from 300 to 600 °C (Sekar et al. 2021).

This thermochemical conversion technique is essential when obtaining bio-oil from highly proteinaceous microalgae. Microalgae with these characteristics tend to produce bio-oil with lower oxygen content and higher nitrogen content than lignocellulosic biomass. The presence of oxygen-containing compounds may reduce the energy content and stability of the bio-oil. In contrast, nitrogen-containing compounds may lead to the emission of NOx (a pollutant gas) if the bio-oil is used as a transportation fuel (Andrade et al. 2018).

Therefore, using catalysts may enhance bio-oil quality by deoxygenation and denitrogenation reactions, increasing the hydrocarbon content and yielding environmentally friendly fuels with a high calorific value (Li et al. 2019). Among the possibilities for catalyst utilization, catalytic pyrolysis with nickel-supported zeolites stands out. This catalyst effectively reduces nitrogen-containing compounds, oxygenates, and acids, thereby improving the quality of the hydrocarbons in the bio-oil (Aliyu et al. 2021).

5.3.1.4 Microwave Pyrolysis

Microwave-assisted pyrolysis processes are intermediate-speed, falling between fast and slow pyrolysis. Microwaves generated by electrical energy take the form of electromagnetic radiation, operating in the radio frequency range of 0.3–300 GHz. The corresponding wavelength varies from approximately 1000 to 1 mM, and temperatures typically range from 400 to 800 °C (Sekar et al. 2021; Yang et al. 2019).

Microwave pyrolysis processes are considered alternative and promising methods for pyrolysis. This method has proven to be an excellent source for producing renewable fuels from raw materials, such as microalgae biomass, primarily due to the acceleration and intensification of chemical reactions. Due to the excellent calorific value of this method, higher yields of bio-syngas and bio-oil may be obtained in comparison with conventional processes. Furthermore, this method has received attention for its ability to process large volumes of biomass with uniform heating and without agitation. However, microalgae biomass has the disadvantage of poor absorption of irradiation. This problem may be solved by mixing the microalgae with an absorbent, such as activated carbons or metal oxides, to reach a high temperature during the process (Sekar et al. 2021).

In this context, this technique was conducted on *Chlorella* sp. to produce biofuels. The highest bio-oil yields from biomass reached 28.6%, obtained from applying a microwave power of 750 W. The resulting bio-oil had a higher heating value (HHV) of 30.7 MJ kg^{-1}, density of 0.98 kg L^{-1}, and viscosity of 61.2 cSt (Du et al. 2011).

5.3.2 Torrefaction

5.3.2.1 Conventional Torrefaction

Conventional torrefaction is the process of controlled heating of biomass with limited oxygen, inducing physical and chemical changes (Huang et al. 2020). The thermal texture leads to the volatilization of organic compounds, while stable products, such as coal and biochar, are formed (Tripathi et al. 2016).

The advantages of torrefaction using microalgae biomass are remarkable. The increased carbon concentration in the torrefied biomass makes it a promising candidate for biofuel production (Aliyu et al. 2021). At the same time, the fragmentation of cell structures during torrefaction simplifies access to internal components, boosting efficiency processes (Das et al. 2021). However, torrefaction also presents challenges that need to be considered. Precise temperature control is vital to obtain desired products, while process optimization is essential to maximize biological and environmental benefits (Chen et al. 2021).

Biochar production from this biomass provides a carbonaceous base with numerous uses, from improving soil fertility to treating industrial effluents and removing environmental contaminants (Costa et al. 2023). Research and innovation, together with advances in technology and a better understanding of roasting processes, have the potential to diversify its applications (Doddapaneni and Kikas 2023).

Thus, microalgae biomass conventional torrefaction stands out as a multifaceted strategy with expressive use in several industrial sectors, including renewable energy, agriculture, and environmental management. Its versatile role, which encompasses biofuel production and biochar generation, reinforces its potential to contribute to a more sustainable future.

5.3.2.2 Oxidative Torrefaction

Conventional torrefaction focuses mainly on the thermal decomposition of biomass. In contrast, in oxidative torrefaction, thermal decomposition and oxidation occur due to the active involvement of oxygen in the torrefaction reactions. Oxygen significantly modulates the chemical responses during torrefaction, resulting in end products distinct from those obtained in conventional torrefaction (Zhang et al. 2022). This process influences the molecular structure and alters the organic compounds formed (Doddapaneni and Kikas 2023).

However, the oxidative torrefaction process requires careful control due to the complex interaction of oxygen in the chemical reactions. Oxygen may accelerate oxidation reactions, resulting in unwanted effects or excessive biomass consumption (Chen et al. 2021). Finding the balance between desired oxidation reactions and preventing over-oxidation requires fine regulation of factors, such as temperature, oxygen concentration, and residence time (Zhang et al. 2022). The intrinsic variability of biomass feedstocks also plays a role, requiring careful adjustment of operating conditions to obtain desired and predictable products (Huang et al. 2020).

Therefore, oxidative torrefaction is a distinct approach in comparison with conventional torrefaction. The ability to incorporate oxygen into reactions creates a scenario of complex transformations with great potential for customized products. However, this complexity requires strict control of process conditions, ensuring predictable and desired results.

5.3.2.3 Microwave Torrefaction

Microwave torrefaction is an innovative process that uses electromagnetic radiation to heat and transform coproducts, such as microalgae biomass, under a controlled environment (Chen et al. 2021; Siddique et al. 2022). During this process, the microalgae are exposed to a microwave source, which leads to the evaporation of water and the activation of chemical reactions, such as breaking bonds and eliminating functional groups (Zhang et al. 2022).

The speed and selectivity of microwaves result in shorter processing times and precise temperature control, which may better preserve the valuable compounds in microalgae (Ayub et al. 2022). In addition, as microwave torrefaction takes place in a closed environment and with lower energy consumption, it may reduce polluting gas emissions compared to conventional torrefaction processes (Siddique et al. 2022). In conclusion, microwave roasting of microalgae is a promising technique that combines the advantages of microwaves with the benefits of roasting, offering an efficient and controlled approach to transforming microalgae into value-added products.

5.3.2.4 Wet Torrefaction

Wet torrefaction is an innovative process that uses water or water vapor as an integral part of the heat treatment of coproducts, such as microalgae biomass (Aliyu et al. 2021). During this process, microalgae biomass is heated in a controlled manner in the presence of water, which leads to water evaporation and activation of chemical reactions, such as breaking bonds and eliminating functional groups (Doddapaneni and Kikas 2023).

The presence of water contributes to the regulation of thermal reactions, giving them more excellent smoothness and control, resulting in shorter processing times and more precise temperature control, which may more effectively preserve the valuable compounds present in microalgae biomass (He et al. 2018). In this context, wet torrefaction of microalgae is a promising technique that uses the presence of water as part of the torrefaction process. This combines the advantages of precise temperature control and preserving valuable compounds in microalgae, making it an efficient and controlled approach to transforming microalgae into high-value-added products.

5.3.3 Hydrothermal Conversion

5.3.3.1 Hydrothermal Carbonization (HTC)

Hydrothermal carbonization (HTC) is a thermochemical process that facilitates the conversion of microalgae biomass into carbon-rich biochar. HTC process is conducted in aqueous medium at low temperatures (180–250 °C) and pressures (2–10 MPa). Besides, using moderate pressure and temperature conditions, the biochar maintains consistent characteristics in terms of porosity, chemical composition, and surface properties (Costa et al. 2023).

The reactions involved in HTC encompass hydrolysis, dehydration, decarboxylation, condensation, polymerization, and aromatization. Therefore, the process is well suited for biomasses with high water content. This occurs because water serves as a solvent and aids catalyze reactions. Catalysis is enabled by the dissolution of gases produced during HTC (CO_2, NO_X, and SO_X) by forming salts and acids (Djandja et al. 2023).

The composition of microalgae exerts a fundamental role in the HTC process. Variation in water content among different microalgae species requires the implementation of pretreatments to adjust levels. Chemical composition is also a critical factor, as essential nutrients and minerals may directly impact the quality of the biochar produced. Therefore, it is necessary to consider these aspects to optimize the HTC process and obtain high-quality biochar (Venkata et al. 2023).

In this context, the study of different microalgae has been carried out to refine the HTC process that has been conducted. Various temperatures and reaction times are observed in the studies, which are assessed based on the composition of the microalgae biomass. For example, Liu et al. (2019) conducted HTC on *Scenedesmus* biomass, evaluating temperatures from 180 to 260 °C at 2 MPa, with a residence time of 4 h. Efficiency of 44.7% was achieved in ash removal, a carbon content of 51.9%, and a high heating value. Moreover, Sztancs et al. (2021) used biomass from *Chlorella vulgaris* at 220 °C with a reaction time of 45 min, obtaining hydrochar with a heating value of 18.87% and an energy yield of 33.5%.

Regarding yield, HTC is influenced by temperature, reaction time, and solid–liquid ratio parameters. Biomass degradation for achieving higher biochar yields is contingent on temperature. Although higher temperatures increase biochar aromaticity, improving its thermal stability and energy content, they also reduce yield. Additionally, extended HTC duration leads to a decrease in oxygenated chemical groups on the surface, also resulting in lower process yields (Ebhodaghe et al. 2022).

A higher solid–liquid ratio with increased water content enables better biomass dissolution in the HTC process. However, conversions may be facilitated by including catalysts, which must be selected in a manner that does not cause corrosion to the reactor. Another possibility is the simultaneous use of homogeneous and heterogeneous catalysts. Nevertheless, satisfactory biochar yields hinge on optimizing the addressed HTC parameters (Das et al. 2021).

Regarding the advantages of conducting HTC, the process stands out due to the absence of the need for biomass pre-drying and the use of lower temperatures and pressures. As a result, operational energy costs are significantly reduced in comparison with conventional thermal technologies. However, the main drawback of employing the HTC technique on microalgae biomass is the low calorific value of the primary product (Aliyu et al. 2021).

5.3.3.2 Hydrothermal Liquefaction (HTL)

Hydrothermal liquefaction (HTL) is a process in which the water present in wet biomass is subjected to subcritical conditions. At high pressures (10–25 MPa) and temperatures (250–380 °C), microalgae biomass is transformed into liquid biofuel in the form of bio-oil due to the action of water as a solvent. Additionally, catalysts such as sodium carbonate (Na_2CO_3), nickel (Ni), and ruthenium (Ru) may be used to facilitate the process (Razzak et al. 2022).

HTL process consists of three chemical stages: depolymerization, decomposition, and recombination. In this manner, hydrolysis of bonds in solid biomass occurs, forming liquid compounds. The outcome is the production of microalgae bio-oil with a composition rich in lipophilic macromolecular compounds of low molecular weight (Krishnan et al. 2022). However, during the HTL process, solid residues are generated from microalgae biomass. The depolymerization process produces biochar from these residues, characterized by its high molecular mass. The composition of the biochar produced exhibits a high concentration of reactive oxygen species and an absence of hydrogen (He et al. 2023).

Studies involving the use of microalgae for bio-oil production may lead to variations in yields due to the chemical differences between species. Pongsiriyakul et al. (2021) investigated HTL in *Nannochloropsis* sp. using a temperature of 320 °C with a reaction time of 30 min; 50% of bio-oil yield was obtained, and 64.1% of carbon content was reported.

The main challenge in using the HTL technique is related to scaling up. Using continuous reactors to obtain large-scale products may lead to corrosion issues. Therefore, it is necessary to construct reactors with high resistance that may withstand corrosion and other chemical reactions. Another limitation of HTL is related to biomass viscosity, resulting in pressure drops and reactor blockages (Hussin et al. 2023).

Regarding advantages, HTL applied to wet microalgae biomass is considered an energetically efficient process in comparison with pyrolysis. This is because it results in the generation of a stable liquid product with low oxygen content. A detailed understanding of biomass degradation mechanisms is crucial for optimizing process conditions and ensuring high-quality products. Therefore, despite the scaling challenges, HTL stands out as a versatile and promising technique for the efficient conversion of microalgae biomass into valuable products, contributing to environmental sustainability and diversification of the energy matrix (Aliyu et al. 2021; Hussin et al. 2023).

5.3.3.3 Hydrothermal Gasification (HTG)

Hydrothermal gasification (HTG) is a process employed in the conversion of biomass with high water content into synthesis gases, also known as syngas, which are formed mainly by carbon monoxide (CO), carbon dioxide (CO_2), methane (CH_4), and hydrogen (H_2). This method involves subjecting wet biomass to high temperatures and pressures, usually above the critical point of water (TC = 373 °C, PC = 22.1 MPa) (Shahbeik et al. 2022).

In this manner, its activation occurs when wet microalgae biomass is subjected to water's supercritical state. This causes the water to act as a solvent and break the biomass component's intermolecular bonds. However, water may also be subjected to subcritical conditions to execute the HTG process, providing a means to selectively generate syngas (Razzak et al. 2022).

As a result, HTG may be conducted in three distinct ways depending on the choice of synthesis gases. The first method is called aqueous phase reforming (APR), where biomass is gasified under subcritical conditions (230–250 °C) using heterogeneous catalysts, such as platinum (Pt) and palladium (Pd). APR yields multiple combustible gases and CO_2 as products. The second is catalytic gasification in a near-critical state (350–400 °C), which employs the same catalysts, but thermodynamic conditions primarily favor the formation of CH_4 and CO_2 (Aliyu et al. 2021; Hussin et al. 2023).

Supercritical water gasification (SWG) is the third and final mode of performing HTG, exhibiting higher calorific efficiency when compared to the other methods. This method is employed for biomasses with water content exceeding 30% and can be conducted in the presence or absence of catalysts. Catalysts utilized in supercritical gasification are nickel-based as they are cost-effective and participate in the hydrogenation reaction. The primary synthesis gases formed are H_2 and CO_2 (Krishnan et al. 2022).

However, the primary advantage of employing the HTG technique is the production of high-quality syngas, virtually free from impurities, such as tar and char. Consequently, production costs are reduced due to the elimination of additional purification steps (Das et al. 2021). Compared to traditional gasification methods, HTG implies lower alloy investments due to low operating temperatures. HTG faces significant challenges, such as the need for large quantities of water, which may be problematic in regions with scarce water resources. Additionally, operation at high pressure, especially in supercritical gasification, increases costs due to the requirement for unique materials to withstand the extreme conditions (Zhou et al. 2023).

The effectiveness of HTG in producing gases from microalgal biomass is strongly correlated with temperature. Low temperatures favor CH_4 production, while high temperatures promote H_2 formation. The ratio of biomass to water, along with the use of a catalyst, may also influence process yield. Catalysts drive desired reactions and inhibit undesired reactions, increasing selectivity in producing valuable gases. Therefore, to ensure maximum results and energy efficiency in the process, optimizing operational conditions and selecting appropriate catalysts is crucial (Hussin et al. 2023).

5.3.4 Gasification

This thermochemical process involves the transformation of the carbon-based material present in microalgae into a combustible gas known as syngas, such as CO, CO_2, H_2, and CH_4 (Das et al. 2021; Raheem et al. 2015). Gasification occurs in a controlled environment of high temperature and established pressure where the biomass is subjected to complex chemical reactions (Ayub et al. 2022).

One of the main benefits of gasification is the versatility of the products generated. The syngas produced may be used for different purposes, from generating electricity and heat to synthesizing valuable chemicals. This makes it possible to maximize the value of microalgae biomass by harnessing its carbon-rich composition to create various valuable products (Ong et al. 2019).

Technological advances are driving research in the area, with the development of more efficient reactors and catalysts and the integration of syngas purification processes. As the gasification of microalgae biomass continues to evolve, it is expected to play a vital role in the transition to a low-carbon economy, contributing to the reduction of greenhouse gas emissions and the diversification of energy and chemical sources (Qi et al. 2023).

In addition, gasification offers environmental advantages. Unlike direct burning, where air pollutants can be released, gasification is an alternative route, possibly capturing polluting by-products before they are released into the atmosphere. This contributes to reducing negative impacts on air and climate, aligning with the sustainability goals (Khani et al. 2023). However, gasification also presents challenges as the complex nature of the chemical reactions involved requires precise control of process conditions to optimize syngas production and minimize the formation of unwanted by-products.

5.3.4.1 Conventional Gasification

The conventional gasification of microalgae biomass is converting carbon-based material into a synthesis gas through heating at high temperatures, typically in an environment with low oxygen content (Raheem et al. 2015). This method has been commonly used to produce synthesis gas rich in H_2 and CO, which may be employed in various applications. Conventional gasification offers syngas production for electricity generation, heat, and raw materials for chemical synthesis (Chang et al. 2023).

This method is well known and has existing infrastructure in some industries, which may facilitate its implementation. If carbon capture and storage (CCS) techniques are implemented, gasification may reduce carbon emissions. However, conventional gasification can release greenhouse gases, such as CO_2 and other pollutants, if emission control measures are not adopted (Ayub et al. 2022). Furthermore, the process requires high temperatures, leading to product degradation issues and reduced energy efficiency. The generated synthesis gas often requires additional treatments for impurity removal before being used in industrial applications (Qi et al. 2023).

5.3.4.2 Supercritical Water Gasification (SWG)

SWG is an innovative approach that involves the conversion of microalgae biomass into synthesis gas using water in a supercritical state, an intermediate state between liquid and vapor, achieved at high temperatures and pressures (Ayub et al. 2022). SWG may be more energy-efficient, as supercritical water possesses unique properties as a solvent and reactant that may enhance reaction rates. Moreover, using supercritical water may help reduce emissions of pollutant gases in comparison with conventional gasification (Chang et al. 2023).

Due to special reaction conditions, the process may also lead to the formation of high-value products, such as hydrocarbons and chemicals. On the other hand, SWG requires precise temperature and pressure control, making the process more complex and potentially less costly. Furthermore, this technology is in its early stages of development, requiring further research and optimization to become commercially viable. Implementing SWG may be more expensive due to specific operational conditions and technologies in demand (Khani et al. 2023).

Thus, SWG has been studied to improve process parameters and optimization. Caputo et al. (2016) investigated the SWG of the *Nannochloropsis gaditana* biomass using parameters of 663 °C, 24 MPa, and a reaction time of 2.13 min. The study demonstrated that 97.4% of the microalgae biomass was gasified under these conditions. Another study by Tiong and Komiyama (2019) carried out the gasification of *Chlorella vulgaris* with supercritical water (385 °C and 26 MPa). Utilizing Ru as a catalyst, 87% of the *C. vulgaris* biomass was gasified in a reaction time of 30 min. This indicates that microalgae species influence process parameters and gasification rates.

Both conventional gasification and SWG have advantages and disadvantages in converting microalgae biomass into syngas. In addition, the overall conversion efficiency and economic viability depend on the proper selection of biomass, pretreatment process, and gasification technologies used (Das et al. 2021). Both approaches have the potential to significantly contribute to the transition to a more sustainable and low-carbon economy (Das et al. 2021; Qi et al. 2023).

5.4 Bioenergy Main Products

5.4.1 Bio-oil

Bio-oil is a complex mixture of oxygen-rich compounds and various organic substances, including sugar monomers derived from cellulose/hemicellulose, sugar oligomers, and their derivatives, and accompanied by a considerable proportion of water. This mixture originates from both the intrinsic humidity of biomass and the reactions involved (Sekar et al. 2021).

Bio-oil from microalgae has an amplified stability and exhibits a higher calorific value in comparison with others obtained from organic materials due to its reduced oxygen/carbon ratio, consequently decreasing the concentration of sulfur and releasing less SO_2 in comparison with traditional fuels (Li et al. 2023). Furthermore, bio-oil has a wide range of applications, including use in heating systems and power generation through boilers, as well as in engines and turbines for electricity production. It also manufactures various chemical products, such as phenols, organic acids, and oxygenated compounds (Li et al. 2019).

The transformation of biomass into bio-oil may be performed by two main processes (Table 5.1): flash pyrolysis and hydrothermal liquefaction. Flash pyrolysis is characterized by the rapid thermal decomposition of organic components, occurring in the absence of oxygen. This process results in the production of liquids, gases, and char. On the other hand, HTL involves the reaction of biomass in water under elevated temperatures and pressures. This process can occur with or without a catalyst (Li et al. 2023). Research related to the pyrolysis of microalgae biomass is extensively significant and has produced encouraging results, which have the potential to open the path to commercial applications (Aliyu et al. 2023).

Biomass composition also has a crucial role in bio-oil's physical and chemical properties. This composition can diversify during the microalgae pyrolysis process, which is the focus of investigation in different studies exploring the analysis of the thermal decomposition of microalgae to produce bio-oil within a wide range of temperatures, heating rates, and exposure times. Carpio et al. (2021) evaluated

HTL of demineralized wastewater microalgae biomass with a mixture of various species of microalgae under different reaction temperatures (260–320 °C) and times (0–90 min) in bio-oil content. The highest oil yield (26%) was obtained at 280 °C and 60 min.

Aliyu et al. (2023) studied the pyrolysis process of *Chlorella vulgaris* to produce bio-oil and biochar at various temperatures to investigate the trade-off between energy used in pyrolysis. They observed that the highest efficiencies achieved were 42.4% for bio-oil at 500 °C and 45.2% for biochar at 400 °C. According to Li et al. (2023), this phenomenon is expected in the pyrolysis of microalgae. A lower pyrolysis temperature produces a higher biochar yield and a lower bio-oil crop, so the process conditions depend mainly on the product. This shows the importance of adapting pyrolysis parameters to the required finished product, emphasizing the versatility of microalgae as a raw material for bioenergy production and showing the necessity of further research to optimize these processes for sustainable and efficient production of biofuel and biochar.

5.4.2 Biochar

Biochar is a high-carbon content charcoal, which may be obtained from the thermal decomposition of microalgae biomass under oxygen-limited conditions (Table 5.1). The main processes to obtain microalgal biochar are slow pyrolysis, torrefaction, and hydrothermal carbonization (Costa et al. 2023). Slow pyrolysis is generally considered a viable technique for biochar production, yielding solid products in a range of 15–43% by weight for microalgae biomass (Yang et al. 2019). Biochar may have high porosity, surface area, functional groups, aromatic structure, and mineralogical composition. These characteristics enable diverse applications, including its use as a soil fertilizer and fuel, in the reduction of atmospheric CO_2, and in the treatment of various pollutants in wastewater (Yu et al. 2017).

Biochar production may be influenced by the technique employed and various parameters. Factors such as biomass characteristics (biomass type, particle size, and moisture content), reaction conditions (temperature, duration, and heating rate), surrounding environment (type of gas used for transport and gas flow rate), and other elements (such as reactor type and the presence of catalysts) all play a critical role in determining the yield and properties of produced biochar (Tripathi et al. 2016).

Pyrolysis processes using high residence times at low temperatures that allow polymerization reactions generally yield high solids (Yu et al. 2017). For instance, the higher the temperature, the lower the biochar yield. The heating rate also significantly influences biochar production. Low heating rates are favored for biochar formation. Under these conditions, biochar production is better since it ensures that thermal cracking of biomass does not occur and the potential for secondary pyrolysis reactions is significantly reduced (Tripathi et al. 2016).

Furthermore, biochar production from microalgae may be integrated into a biorefinery context, reducing process costs and increasing the yield of bioproducts. This becomes feasible when utilizing the remaining microalgae biomass, as exemplified in the study conducted by Wang et al. (2013); after lipid extraction, residual biomass produced biochar enriched in nitrogen and minerals. Moreover, it exhibited high concentrations of trace elements, such as nitrogen, phosphorus, potassium, calcium, and magnesium, compared to the biochar produced from lignocellulosic biomass. These characteristics indicate significant potential for the application of biochar as a fertilizer.

5.4.3 Syngas

The syngas process, also known as synthesis gasification, plays a crucial role in pursuing sustainable energy solutions and maximizing the value of microalgae biomass (Ebhodaghe et al. 2022). This innovative method involves the conversion of biomass into a synthesis gas rich in H_2 and CO by complex thermochemical reactions (Table 5.1). Syngas is obtained by gasifying microalgae biomass in a controlled environment of high temperature and low oxygen, resulting in a gas that may be further utilized (Chang et al. 2023).

Syngas offers several advantages, and its high versatility allows it to be employed in electricity generation, heat production, and even in the synthesis of valuable chemical products. Additionally, it contributes to waste reduction by transforming microalgae biomass into a valuable resource and can help decrease greenhouse gas emissions due to controlled gasification (Ayub et al. 2022).

However, there are challenges and opportunities associated with the syngas process. The complexity of the process requires precise control of gasification conditions, while continuous research is needed to enhance efficiency and explore new approaches. Economic viability is also critical, depending on system optimization and infrastructure availability (Qi et al. 2023).

Syngas holds promising future applications, including producing advanced biofuels, such as biogas and biodiesel, and using the resulting hydrogen in fuel cells or industrial processes. Furthermore, syngas may be a raw material for sustainable chemical products (Chen et al. 2015; Krishnan et al. 2022).

The syngas process offers an innovative approach to converting microalgae biomass into clean energy and high-value products. Despite technical and economic challenges, syngas has significant potential to contribute to energy and environmental sustainability. With ongoing advances in research and development, syngas may play a pivotal role in transitioning toward a more sustainable and low-carbon economy, harnessing the full potential of microalgae biomass.

5.5 Conclusion and Final Remarks

Thermochemical conversion of microalgae biomass represents a promising alternative to renewable energy and sustainability, mostly when integrated into the biorefinery context. The main thermochemical conversion methods, including pyrolysis, torrefaction, hydrothermal conversion, and gasification, may reduce the reliance on fossil fuels and contribute to reducing greenhouse gas emissions. Additionally, challenges are associated with harnessing microalgae biomass to obtain valuable products, such as bio-oil, biochar, and syngas. The research field continues advancing and is pivotal in steering us toward a cleaner, more sustainable energy future. However, it is essential to highlight that research in this area is ongoing to optimize processes, increase efficiency, and reduce the costs involved in the thermochemical conversion of microalgae biomass. With continued advances, this technology has the potential to play a significant role in the transition to a cleaner and more sustainable energy matrix.

References

Aliyu A, Lee JGM, Harvey AP (2021) Microalgae for biofuels: a review of thermochemical conversion processes and associated opportunities and challenges. Bioresour Technol Rep 15:100694. https://doi.org/10.1016/j.biteb.2021.100694

Aliyu A, Lee JGM, Harvey AP (2023) Microalgae for biofuel: isothermal pyrolysis of a fresh and a marine microalga with mass and energy assessment. Chem Eng J Adv 14:100474. https://doi.org/10.1016/j.ceja.2023.100474

Andrade LA, Batista FRX, Lira TS, Barrozo MAS, Vieira LGM (2018) Characterization and product formation during the catalytic and non-catalytic pyrolysis of the green microalgae *Chlamydomonas reinhardtii*. Renew Energy 119:731–740. https://doi.org/10.1016/j.renene.2017.12.056

Ayub HMU, Ahmed A, Lam SS, Lee J, Show PL, Park YK (2022) Sustainable valorization of algae biomass via thermochemical processing route: an overview. Bioresour Technol 344:126399. https://doi.org/10.1016/j.biortech.2021.126399

Azizi K, Keshavarz Moraveji M, Arregi A, Amutio M, Lopez G, Olazar M (2020) On the pyrolysis of different microalgae species in a conical spouted bed reactor: bio-fuel yields and characterization. Bioresour Technol 311:123561. https://doi.org/10.1016/j.biortech.2020.123561

Bae YJ, Ryu C, Jeon JK, Park J, Suh DJ, Suh YW, Chang D, Park YK (2011) The characteristics of bio-oil produced from the pyrolysis of three marine macroalgae. Bioresour Technol 102:3512–3520. https://doi.org/10.1016/j.biortech.2010.11.023

Beneroso D, Bermúdez JM, Arenillas A, Menéndez JA (2013) Microwave pyrolysis of microalgae for high syngas production. Bioresour Technol 144:240–246. https://doi.org/10.1016/j.biortech.2013.06.102

Caputo G, Dispenza M, Rubio P, Scargiali F, Marotta G, Brucato A (2016) Supercritical water gasification of microalgae and their constituents in a continuous reactor. J Supercrit Fluids 118:163–170. https://doi.org/10.1016/j.supflu.2016.08.007

Carpio RB, Zhang Y, Kuo CT, Chen WT, Schideman LC, de Leon R (2021) Effects of reaction temperature and reaction time on the hydrothermal liquefaction of demineralized wastewater algal biomass. Bioresour Technol Rep 14:100679. https://doi.org/10.1016/j.biteb.2021.100679

Chang YJ, Chang JS, Lee DJ (2023) Gasification of biomass for syngas production: research update and stoichiometry diagram presentation. Bioresour Technol 387:129535. https://doi.org/10.1016/j.biortech.2023.129535

Chen WH, Lin BJ, Huang MY, Chang JS (2015) Thermochemical conversion of microalgal biomass into biofuels: a review. Bioresour Technol 184:314–327. https://doi.org/10.1016/j.biortech.2014.11.050

Chen WH, Lin BJ, Lin YY, Chu YS, Ubando AT, Show PL, Ong HC, Chang JS, Ho SH, Culaba AB, Pétrissans A, Pétrissans M (2021) Progress in biomass torrefaction: principles, applications and challenges. Prog Energy Combust Sci 82:100887. https://doi.org/10.1016/j.pecs.2020.100887

Choudhary S, Tripathi S, Poluri KM (2022) Microalgal-based bioenergy: strategies, prospects, and sustainability. Energy Fuels 36:14584–14612. https://doi.org/10.1021/acs.energyfuels.2c02922

Christensen PS, Peng G, Vogel F, Iversen BB (2014) Hydrothermal liquefaction of the microalgae *Phaeodactylum tricornutum*: impact of reaction conditions on product and elemental distribution. Energy Fuels 28:5792–5803. https://doi.org/10.1021/ef5012808

Costa JAV, Zaparoli M, Cassuriaga APA, Cardias BB, Vaz BS, Morais MG, Moreira JB (2023) Biochar production from microalgae: a new sustainable approach to wastewater treatment based on a circular economy. Enzym Microb Technol 169:110281. https://doi.org/10.1016/j.enzmictec.2023.110281

Das P, Chandramohan VP, Mathimani T, Pugazhendhi A (2021) A comprehensive review on the factors affecting thermochemical conversion efficiency of algal biomass to energy. Sci Total Environ 766:144213. https://doi.org/10.1016/j.scitotenv.2020.144213

Djandja OS, Liew RK, Liu C, Liang J, Yuan H, He W, Feng Y, Lougou BG, Duan PG, Lu X, Kang S (2023) Catalytic hydrothermal carbonization of wet organic solid waste: a review. Sci Total Environ 873:162119. https://doi.org/10.1016/j.scitotenv.2023.162119

Doddapaneni TRKC, Kikas T (2023) Advanced applications of torrefied biomass: a perspective view. Energies 16. https://doi.org/10.3390/en16041635

Du Z, Li Y, Wang X, Wan Y, Chen Q, Wang C, Lin X, Liu Y, Chen P, Ruan R (2011) Microwave-assisted pyrolysis of microalgae for biofuel production. Bioresour Technol 102:4890–4896. https://doi.org/10.1016/j.biortech.2011.01.055

Ebhodaghe SO, Imanah OE, Ndibe H (2022) Biofuels from microalgae biomass: a review of conversion processes and procedures. Arab J Chem 15:103591. https://doi.org/10.1016/j.arabjc.2021.103591

Grierson S, Strezov V, Shah P (2011) Properties of oil and char derived from slow pyrolysis of *Tetraselmis chui*. Bioresour Technol 102:8232–8240. https://doi.org/10.1016/j.biortech.2011.06.010

He C, Tang C, Li C, Yuan J, Tran KQ, Bach QV, Qiu R, Yang Y (2018) Wet torrefaction of biomass for high quality solid fuel production: a review. Renew Sust Energ Rev 91:259–271. https://doi.org/10.1016/j.rser.2018.03.097

He S, Barati B, Hu X, Wang S (2023) Carbon migration of microalgae from cultivation towards biofuel production by hydrothermal technology: a review. Fuel Process Technol 240:107563. https://doi.org/10.1016/j.fuproc.2022.107563

Huang Y, Li B, Liu D, Xie X, Zhang H, Sun H, Hu X, Zhang S (2020) Fundamental advances in biomass autothermal/oxidative pyrolysis: a review. ACS Sustain Chem Eng 8:11888–11905. https://doi.org/10.1021/acssuschemeng.0c04196

Hussin F, Hazani NN, Khalil M, Aroua MK (2023) Environmental life cycle assessment of biomass conversion using hydrothermal technology: a review. Fuel Process Technol 246:107747. https://doi.org/10.1016/j.fuproc.2023.107747

Khani Y, Valizadeh S, Yim H, Chai S, Chang D, Valizadeh B, Farooq A, Ko CH, Park YK (2023) Production of hydrogen via microalgae air gasification over acid mine sludge waste coated on monolith. Chem Eng J 465:1–12. https://doi.org/10.1016/j.cej.2023.142797

Krishnan RY, Manikandan S, Subbaiya R, Kim W, Karmegam N, Govarthanan M (2022) Advanced thermochemical conversion of algal biomass to liquid and gaseous biofuels: a comprehensive review of recent advances. Sustain Energy Technol Assess 52:102211. https://doi.org/10.1016/j.seta.2022.102211

Li F, Srivatsa SC, Bhattacharya S (2019) A review on catalytic pyrolysis of microalgae to high-quality bio-oil with low oxygeneous and nitrogenous compounds. Renew Sust Energ Rev 108:481–497. https://doi.org/10.1016/j.rser.2019.03.026

Li H, Dong Z, Wang B, Wu W, Cao M, Zhang Y, Liu Z (2023) High-quality bio-oil production via catalytic pyrolysis of biocrude oil from hydrothermal liquefaction of microalgae Spirulina. J Anal Appl Pyrolysis 173:106057. https://doi.org/10.1016/j.jaap.2023.106057

Liu H, Chen Y, Yang H, Gentili FG, Söderlind U, Wang X, Zhang W, Chen H (2019) Hydrothermal carbonization of natural microalgae containing a high ash content. Fuel 249, 441–448. https://doi.org/10.1016/j.fuel.2019.03.004

Nurcahyani PR, Hashimoto S, Matsumura Y (2020) Supercritical water gasification of microalgae with and without oil extraction. J Supercrit Fluids 165:104936. https://doi.org/10.1016/j.supflu.2020.104936

Ong HC, Chen WH, Farooq A, Gan YY, Lee KT, Ashokkumar V (2019) Catalytic thermochemical conversion of biomass for biofuel production: a comprehensive review. Renew Sust Energ Rev 113:109266. https://doi.org/10.1016/j.rser.2019.109266

Pongsiriyakul K, Kiatkittipong W, Adhikari S, Lim JW, Lam SS, Kiatkittipong K, Dankeaw A, Reubroycharoen P, Laosiripojana N, Faungnawakij K, Assabumrungrat S (2021) Effective Cu/Re promoted Ni-supported γ-Al2O3 catalyst for upgrading algae bio-crude oil produced by hydrothermal liquefaction. Fuel Process Technol 216:106670. https://doi.org/10.1016/j.fuproc.2020.106670

Qi H, Xu H, Zhang J, Xu Z, Zhong L, Cui P, Zhu Z, Wang Y (2023) Thermodynamic and techno-economic analyses of hydrogen production from different algae biomass by plasma gasification. Int J Hydrog Energy 48:35895. https://doi.org/10.1016/j.ijhydene.2023.06.038

Raheem A, Wan Azlina WAKG, Taufiq Yap YH, Danquah MK, Harun R (2015) Thermochemical conversion of microalgal biomass for biofuel production. Renew Sust Energ Rev 49:990–999. https://doi.org/10.1016/j.rser.2015.04.186

Razzak SA, Lucky RA, Hossain MM, deLasa H (2022) Valorization of microalgae biomass to biofuel production: a review. Energy Nexus 7:100139. https://doi.org/10.1016/j.nexus.2022.100139

Reddy HK, Muppaneni T, Ponnusamy S, Sudasinghe N, Pegallapati A, Selvaratnam T, Seger M, Dungan B, Nirmalakhandan N, Schaub T, Holguin FO, Lammers P, Voorhies W, Deng S (2016) Temperature effect on hydrothermal liquefaction of *Nannochloropsis gaditana* and *Chlorella* sp. Appl Energy 165:943–951. https://doi.org/10.1016/j.apenergy.2015.11.067

Sanchez-Silva L, López-González D, Garcia-Minguillan AM, Valverde JL (2013) Pyrolysis, combustion and gasification characteristics of *Nannochloropsis gaditana* microalgae. Bioresour Technol 130:321–331. https://doi.org/10.1016/j.biortech.2012.12.002

Santos LO, Silva PGP, Machado BR, Sala L, Deamici KM (2022) Update on the application of magnetic fields to microalgal cultures. World J Microbiol Biotechnol 38:1–10. https://doi.org/10.1007/s11274-022-03398-y

Sekar M, Mathimani T, Alagumalai A, Chi NTL, Duc PA, Bhatia SK, Brindhadevi K, Pugazhendhi A (2021) A review on the pyrolysis of algal biomass for biochar and bio-oil – bottlenecks and scope. Fuel 283:119190. https://doi.org/10.1016/j.fuel.2020.119190

Shahbeik H, Peng W, Kazemi Shariat Panahi H, Dehhaghi M, Guillemin GJ, Fallahi A, Amiri H, Rehan M, Raikwar D, Latine H, Pandalone B, Khoshnevisan B, Sonne C, Vaccaro L, Nizami AS, Gupta VK, Lam SS, Pan J, Luque R, Sels B, Tabatabaei M, Aghbashlo M (2022) Synthesis of liquid biofuels from biomass by hydrothermal gasification: a critical review. Renew Sust Energ Rev 167:112833. https://doi.org/10.1016/j.rser.2022.112833

Siddique IJ, Salema AA, Antunes E, Vinu R (2022) Technical challenges in scaling up the microwave technology for biomass processing. Renew Sust Energ Rev 153:111767. https://doi.org/10.1016/j.rser.2021.111767

Sztancs G, Kovacs A, Toth AJ, Mizsey P, Billen P, Fozer D (2021) Catalytic hydrothermal carbonization of microalgae biomass for low-carbon emission power generation: the environmental impacts of hydrochar co-firing. Fuel 300:120927. https://doi.org/10.1016/j.fuel.2021.120927

Tiong L, Komiyama M (2019) Supercritical water gasification of microalga Chlorella vulgaris over supported Ru. J. Supercrit. Fluids 144, 1–7. https://doi.org/10.1016/j.supflu.2018.05.011

Tripathi M, Sahu JN, Ganesan P (2016) Effect of process parameters on production of biochar from biomass waste through pyrolysis: a review. Renew Sust Energ Rev 55:467–481. https://doi.org/10.1016/j.rser.2015.10.122

Venkata K, Rama T, Doddapaneni KC, Kumar P, Kaushal P, Ziauddin S, Pollmann K, Jain R (2023) Critical review on production, characterization and applications of microalgal hydrochar: insights on circular bioeconomy through hydrothermal carbonization. Chem Eng J 473:145059. https://doi.org/10.1016/j.cej.2023.145059

Vo TK, Kim SS, Ly HV, Lee EY, Lee CG, Kim J (2017) A general reaction network and kinetic model of the hydrothermal liquefaction of microalgae *Tetraselmis* sp. Bioresour Technol 241:610–619. https://doi.org/10.1016/j.biortech.2017.05.186

Wang K, Brown RC, Homsy S, Martinez L, Sidhu SS (2013) Fast pyrolysis of microalgae remnants in a fluidized bed reactor for bio-oil and biochar production. Bioresour Technol 127:494–499. https://doi.org/10.1016/j.biortech.2012.08.016

Yang C, Li R, Zhang B, Qiu Q, Wang B, Yang H, Ding Y, Wang C (2019) Pyrolysis of microalgae: a critical review. Fuel Process Technol 186:53–72. https://doi.org/10.1016/j.fuproc.2018.12.012

Yu KL, Show PL, Ong HC, Ling TC, Chi-Wei Lan J, Chen WH, Chang JS (2017) Microalgae from wastewater treatment to biochar – feedstock preparation and conversion technologies. Energy Convers Manag 150:1–13. https://doi.org/10.1016/j.enconman.2017.07.060

Zainan NH, Srivatsa SC, Li F, Bhattacharya S (2018) Quality of bio-oil from catalytic pyrolysis of microalgae *Chlorella vulgaris*. Fuel 223:12–19. https://doi.org/10.1016/j.fuel.2018.02.166

Zhang C, Wang M, Chen WH, Pétrissans A, Pétrissans M, Ho SH (2022) A comparison of conventional and oxidative torrefaction of microalga *Nannochloropsis oceanica* through energy efficiency analysis and life cycle assessment. J Clean Prod 369:133236. https://doi.org/10.1016/j.jclepro.2022.133236

Zhou Y, Remón J, Pang X, Jiang Z, Liu H, Ding W (2023) Hydrothermal conversion of biomass to fuels, chemicals and materials: a review holistically connecting product properties and marketable applications. Sci Total Environ 886:163920. https://doi.org/10.1016/j.scitotenv.2023.163920

Biodiesel Production from Microalgae Biomass

6

Mariana Peñuela Vásquez, Catalina Quevedo Ospina, and Catalina Lugo de Ossa

Abstract

The growing demand for alternative energy sources to fossil fuels that can partially or fully replace them has driven a significant increase in research aimed at generating sustainable long-term energy technologies that are environmentally friendly and capable of reversing or halting the environmental impacts caused by conventional fuels. One of the areas of study is the production of bio-oil from microalgae. These microorganisms can produce lipids as free fatty acids (FFAs) or triacylglycerols (TAGs) through their cellular metabolism. These molecules can, in turn, be transformed into biodiesel through various chemical processes. The liquid fuels obtained from microalgae offer significant advantages over other fuels, especially from an environmental perspective, as their production is based on the fixation and consumption of atmospheric CO_2 or combustion products, generating oxygen (O_2) as a metabolic byproduct. These cultivation systems are cost-effective because they do not require large land areas, and the cultivation media can consist of industrial waste or contaminated water. Additionally, they can utilize natural light as an energy source for photosynthesis. Simultaneously with biodiesel production, other value-added products such as proteins, pigments, and antioxidants can be obtained. All of these characteristics make microalgae a promising source of biofuels. Due to this, this chapter aims to present the most relevant aspects of biodiesel production from microalgae, organized into four main stages of the process: (1) microorganism cultivation and growth, (2) cell disruption, (3) lipid extraction, and (4) chemical or thermochemical conversion of lipids to biodiesel.

M. Peñuela Vásquez (✉) · C. Quevedo Ospina · C. Lugo de Ossa
Bioprocess Research Group, Department of Chemical Engineering, Faculty of Engineering, Universidad de Antioquia UdeA, Medellín, Colombia
e-mail: mariana.penuela@udea.edu.co

Keywords

Biodiesel · Microalgae · Cell disruption · Lipid extraction · Biorefineries

6.1 Introduction

Microalgae are unicellular, mainly aquatic organisms found in freshwater or saltwater sources. They are primary producers in aquatic ecosystems like rivers, lakes, and oceans. Algae are broadly classified into macroalgae, microalgae, and cyanobacteria, which vary by species and genus, making this group quite diverse (Levasseur et al. 2020). Like plants, algae can perform photosynthesis by capturing CO_2, nutrients, and light energy to convert them into organic matter—that is, more algae or biomass. This biomass is rich in carbohydrates, proteins, lipids, and other valuable substances like photosynthetic pigments and antioxidants (Abdelfattah et al. 2023; Liu et al. 2024). From a biochemical standpoint, macroalgae, with a higher carbohydrate content, are suitable for the food sector, while microalgae are preferred for oil production due to their higher lipid content. Macroalgae, microalgae, and cyanobacteria have been studied as lipid sources for biodiesel production (Aniruddha et al. 2022).

Biodiesel production from microalgae is not yet widely implemented due to the variability in lipid productivity of different strains, which is influenced by both genotype and cultivation conditions. This causes the lipid percentage in biomass to vary significantly, from 20% to 60% of cell weight, even within the same species, as shown in Table 6.1 (Balouch et al. 2023; Krishnamoorthy et al. 2023).

The lipid content and biodiesel yield from microalgae oil are 15–300 times higher than those of oil crops like corn, soybeans, sunflower, and castor beans (Aniruddha et al. 2022). These agricultural products are considered first-generation fuel sources but negatively impact food security and biodiversity, require large tracts of land, are often mono-

D. U. Santos Ballardo, S. Rossi (eds.), *Microalgae as Promising Source of Commercial Bioproducts*, Developments in Applied Phycology 14, https://doi.org/10.1007/978-3-031-86433-9_6

Table 6.1 Microalgae used for lipid production

Microalgae	Rendimiento de Biomass yield	Contenido de Lipid content (% of dry weight)	Reference
Chlorella vulgaris	0.29 (g/L) 0.69–1.05 (g/L)	30 24.28–31.04*	Razzak et al. (2022) *Aniruddha et al. (2022)
Chlorella pyrenoidosa	1.41 g/L	30.9	Zhang et al. (2022)
Nannochloropsis oculata *Nannochloropsis* sp.		17.91 31–68**	Krishnamoorthy et al. (2023) **Li et al. (2023)
Nephrochlamys subsolitaria	1.21 (g/Ld)	12.71	Balouch et al. (2023)
Dunaliella tertiolecta	1.5 (g/L)	38.8	Yang et al. (2024)
Chlamydomonas reinhardtii	5.47 (g/L)	21–51	Aniruddha et al. (2022)
Scenedesmus bijuga *Scenedesmus dimorphus* *Scenedesmus obliquus*	3.65–6.07 (g/L) 0.28–5.52 (g/L) 0.31 (g/L)*	34.10–35.24 33.70–40.33 27*	Aniruddha et al. (2022) *Razzak et al. (2022)
Desmodesmus pannonicus *Desmodesmus abundans*	1.35 (g/Ld) 18.4 (g/L)*	14.51 40.4*	Balouch et al. (2023) *Yang et al. (2024)
Anabaena cylindrica	–	4–7	Li et al. (2023)
Synechococcus	–	11	Li et al. (2023)
Monoraphidium griffithii *Monoraphidium dybowskii*	1.12 (g/Ld) 42.12–45.92 (mg/L/d)	13.82 30.12–31.78*	Balouch et al. (2023) *Aniruddha et al. (2022)
Ankistrodesmus falcatus	1.09 (g/Ld)	13.41	Balouch et al. (2023)
Parachlorella kessleri	1.42 (g/Ld)	16.46	Balouch et al. (2023)
Neochloris oleoabundans	1.050 (g/L)	38	Razzak et al. (2022)

cultures that use significant amounts of pesticides and fertilizers, and are highly dependent on climate conditions (Siddiki et al. 2022).

On the other hand, microalgae are not significant human food sources, do not require extensive land for cultivation, and are not dependent on climatic conditions, allowing for year-round cultivation. They can also be grown in contaminated water containing various chemicals, including organic matter, organic and inorganic compounds, and heavy metals, among others. Additionally, they are natural CO_2 fixers and produce O_2 through photosynthesis. These characteristics make biodiesel production from microalgae attractive for the fuel industry due to its lower economic and energy costs and bioremediation potential, significantly benefiting the environment (Aniruddha et al. 2022). The fuel produced from microalgae is classified as third-generation biodiesel and is considered biodegradable, environmentally friendly, and highly energetic. It can replace diesel, jet fuel, and fuels used for transportation (Siddiki et al. 2022).

Several critical factors ensure maximum productivity in lipid production processes from microalgae. These include evaluating microalgae species with potential as feedstock for biodiesel production, cultivation conditions, harvesting strategies, cell disruption techniques, and the most suitable technologies for lipid extraction and biofuel production (Krishnamoorthy et al. 2023). As shown in Fig. 6.1, biodiesel production involves multiple stages, regardless of the feedstock type. These stages include (1) microorganism cultivation and growth, (2) cell disruption, (3) lipid extraction, and (4) chemical or thermochemical conversion of lipids to biodiesel (Bharti et al. 2024).

Technological advancements currently focus on finding techniques or methods that reduce production costs by minimizing energy consumption, reducing the number of stages required to obtain biodiesel, and increasing conversion efficiency to biodiesel through chemical and thermochemical methods (Al-Rabiah et al. 2022; da Costa Magalhães et al. 2023; Deepika et al. 2024; Gómez et al. 2022; Marangon et al. 2024; Silva et al. 2024). There is also a growing interest in maximizing the number of value-added products derived from the same cultivation system through the implementation of the biorefinery concept in microalgae production systems, which not only enables the generation of marketable products but also significantly reduces environmental impacts (Siddiki et al. 2022; Silva et al. 2024; Zarei and Zamani 2024; Zhao et al. 2024).

This chapter presents advances in biodiesel production from microalgae divided into four fundamental topics: (1) cultivation systems, from bioreactors used for microalgae cultivation to lipid metabolism and abiotic factors affecting metabolite production; (2) techniques and methods for conventional and non-conventional lipid separation; (3) biodiesel production using chemical and thermochemical technologies; and (4) biorefineries.

6.2 Cultivation Systems

6.2.1 Photobioreactors

Microalgae cultivation systems vary significantly and are distinct from those typically used for other microorganisms.

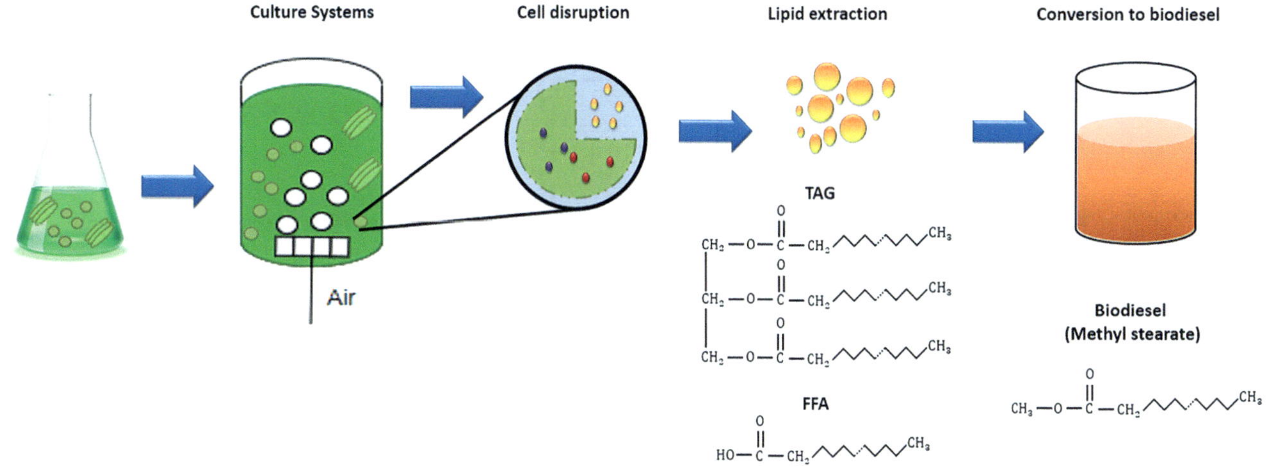

Fig. 6.1 General description of the biodiesel production process from microalgae

Generally, algae are grown in open pond systems or closed photobioreactors (PBRs) (Penloglou et al. 2024), as shown in Fig. 6.2.

6.2.1.1 Open Systems

Open systems are the most commercially explored due to their low operational costs. They are commonly used in cultivation for human food, aquaculture, and wastewater treatment. Essential aspects to consider in these systems include energy, land cost, water, nutrients, algae biology, local climate conditions, and the final product. These systems may be slanted, mixed ponds, shallow lagoon-type ponds, channel ponds, and central-pivot circular ponds. They use sunlight as the primary energy source and atmospheric CO_2 (Tomar et al. 2023). Open systems are generally more economical due to their low construction costs, ease of operation, and scalability. However, they have several limitations, such as low productivity due to low CO_2 transfer rates, high evaporation rates, and contamination risks (Casanova et al. 2023).

6.2.1.2 Closed Systems

In these systems, algae are cultivated in a closed environment, allowing for controlled environmental conditions, accelerated growth, and biomass accumulation over the short term and all year round, making them more efficient for cultivation. Photobioreactors are highly versatile, with configurations such as vertical tubes (including bubble columns, airlift, and concentric tubes) and horizontal designs like flat panels and stirred tanks (Tomar et al. 2023). These systems can use natural or artificial light, and their mass transfer rate is much higher than in open systems, improving cell exposure to light. The main limitations are the high operational and energy costs when operated at a large scale (Casanova et al. 2023).

Given their specific energy and nutrient requirements, it is essential to determine cultivation conditions and the type of photobioreactor that will optimize growth and lipid production for each microalgae species. This diversity is why a variety of cultivation systems for lipid production are explored in literature, evaluating variables specific to the microalgae of interest. For instance, Pereira et al. cultivated the marine microalga *Chlorella minutissima* in 18 and 44 L bubble columns, optimizing biomass and lipids by considering factors such as CO_2 percentage, nutrients, temperature, and salinity (Pereira et al. 2018). Pierre et al. Growth *Nannochloropsis* sp. in a raceway system under nutritional stress conditions to accumulate lipids to produce biodiesel (Perrier et al. 2015). Besides, Chen et al. cultivated *Chlorella vulgaris* in a tubular photobioreactor using a two-stage approach to accumulate lipids for biodiesel production (Chen et al. 2014). Huang et al. cultured *Chlorella pyrenoidosa* under different lighting gradients in a flat-panel photobioreactor (Huang et al. 2014). Bácsi et al. grew the green microalgae *Monoraphidium pusillum* in a conventional Erlenmeyer flask (air bubbled in a tank-type photobioreactor) and in a hybrid (fermenter type + helical tubular type) photobioreactor of the same volume (2.8 L) (Bácsi et al. 2022).

6.3 Metabolism

The metabolism of microalgae involves a complex network of enzymatic reactions, including metabolic pathways such as photosynthesis, glycolysis, the Calvin cycle, carbon concentration mechanisms (CCMs), the tricarboxylic acid (TCA) cycle, beta (β)-oxidation, and other physiological and biochemical processes associated with lipid synthesis, carbon chain growth, liposome transport, and the formation of storage molecules, among others (Priyadharsini et al. 2022; Wu et al. 2023). Due to their versatility with carbon sources, which can be organic and/or inorganic, and their adaptability to the presence or absence of light, microalgae

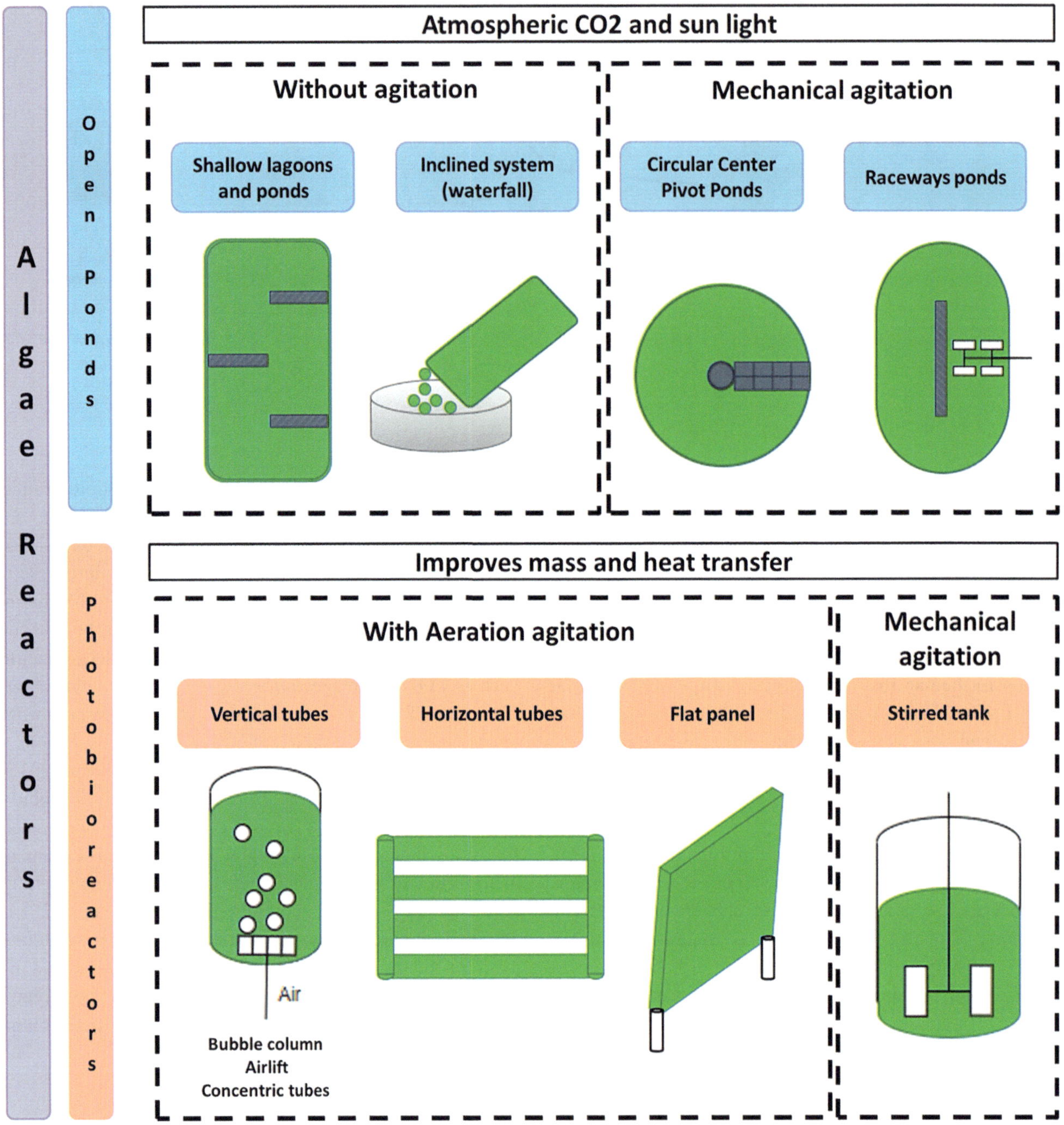

Fig. 6.2 Types of photobioreactors for the culture of microalgae

can exhibit three basic cultivation modes: photoautotrophic, heterotrophic, and mixotrophic, as shown in Fig. 6.3 (Cao et al. 2023; Khoo et al. 2023). These cultivation systems are directly related to the type of metabolism the microalgae will use for energy production, reproduction, growth, and cellular maintenance, as well as the quantity and quality of metabolites of industrial interest, such as proteins, pigments, and lipids (Udaypal et al. 2024; Wu et al. 2023).

6.3.1 Photoautotrophic

This is the most commonly used cultivation system due to its simplicity and low cost. The essential components are CO_2 or HCO_3 and light energy, making it the most environmentally friendly system. However, its biomass productivity is lower, so this system's choice depends on the microalgae's nutritional needs, the desired carbon source, and the target metabolite. It is important to note that phototrophy allows for

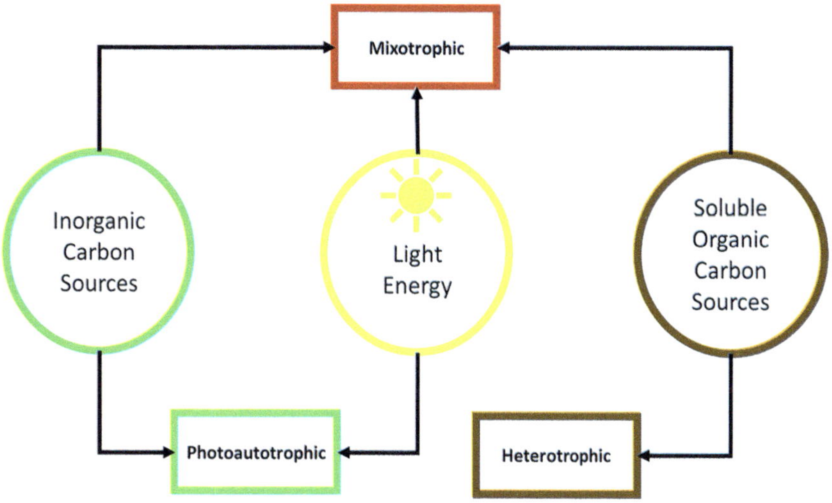

Fig. 6.3 Basic forms for the culture of microalgae

the production of photosynthetic pigments and antioxidants, while other systems may not be suitable for this purpose (Abdelfattah et al. 2023; Zhang et al. 2022; Liu et al. 2024).

6.3.2 Heterotrophic

This cultivation mode occurs in the absence of light, where cellular growth mainly depends on an organic carbon source. Without photosynthetic metabolic pathways, the biochemical composition of biomass exhibits higher percentages of carbohydrates and lipids. Additionally, this system is prone to bacterial contamination, although biomass production yields are much higher than in an autotrophic system (Pandey et al. 2024).

6.3.3 Mixotrophic

Mixotrophy combines photoautotrophic and heterotrophic systems. In this mode, microalgae consume both organic and inorganic carbon sources, improving microalgal biomass productivity while allowing for the production of metabolites associated with photosynthesis (Zhang et al. 2022).

Most microalgae can adapt and grow under various CO_2 concentrations, a trait associated with the CCM. This mechanism involves inorganic carbon transporters, such as Adenosine triphosphastase (ATPases), which enable the active transport of HCO_3, along with one or more carbonic anhydrases that transform CO_2 into bicarbonate (Priyadharsini et al. 2022). Simultaneously, cells use photosynthesis to assimilate light energy and convert it into chemical energy.

Photosynthesis occurs in two phases: in the first phase, known as the light-dependent phase, cells use light to break down water into oxygen, protons, and electrons. In the second phase, the light-independent (or dark) phase, protons and electrons help reduce CO_2 to starch via the Calvin cycle. Although these phases differ, they occur simultaneously. Later, the starch produced by the Calvin cycle can be broken down and metabolized to form fatty acids and TAGs (Khoo et al. 2023).

In the dark phase, an enzyme called RuBisCO is involved in carbon fixation in the first stage of the Calvin cycle (Fig. 6.4). However, RuBisCO can sometimes fix O_2 instead of CO_2, a process known as photorespiration, which can occur due to temperature and light intensity changes. Under intense light, oxygen (O_2) accumulation increases the local O_2/CO_2 ratio, thereby reducing carboxylase activity and increasing oxygenase activity. In photosystem II, excess electrons react with the oxygen produced by photosynthesis, forming reactive oxygen species (ROS) that negatively affect microalgal growth and lipid accumulation (Sousa et al. 2012).

6.4 Lipid Metabolism

Microalgal lipids can be categorized as polar and nonpolar. Polar lipids primarily serve structural functions, such as phospholipids and sterols, as part of cell membranes. These lipids are also crucial in cellular signaling as intermediates or precursors. Nonpolar lipids, on the other hand, are primarily storage molecules that can be used as energy sources. Most are triacylglycerols (TAGs) in microalgae (Khoo et al. 2023; Liu et al. 2024; Mulgund 2022; Sibi et al. 2016).

TAG synthesis in microalgae is believed to occur in the chloroplast membrane and endoplasmic reticulum primarily (Fig. 6.5). Free fatty acids (FFAs) are synthesized in the chloroplast and then transported to the cytoplasm to synthesize additional TAGs in the endoplasmic reticulum. In chloroplasts, acetyl-coenzyme A (CoA) carboxylase (ACC) is a key rate-limiting enzyme in fatty acid biosynthesis, catalyzing the conversion of acetyl-CoA to malonyl-CoA, which is subsequently transferred to the fatty acid synthesis (FAS)

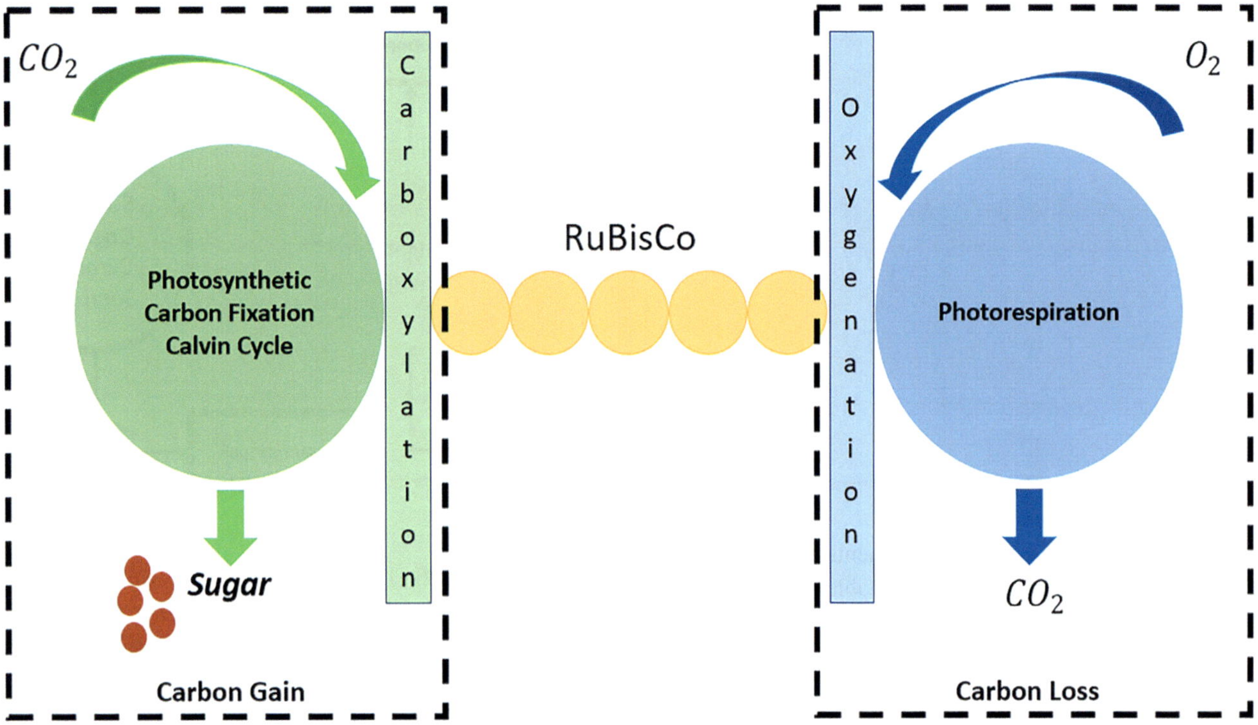

Fig. 6.4 Metabolic RuBisCO pathway

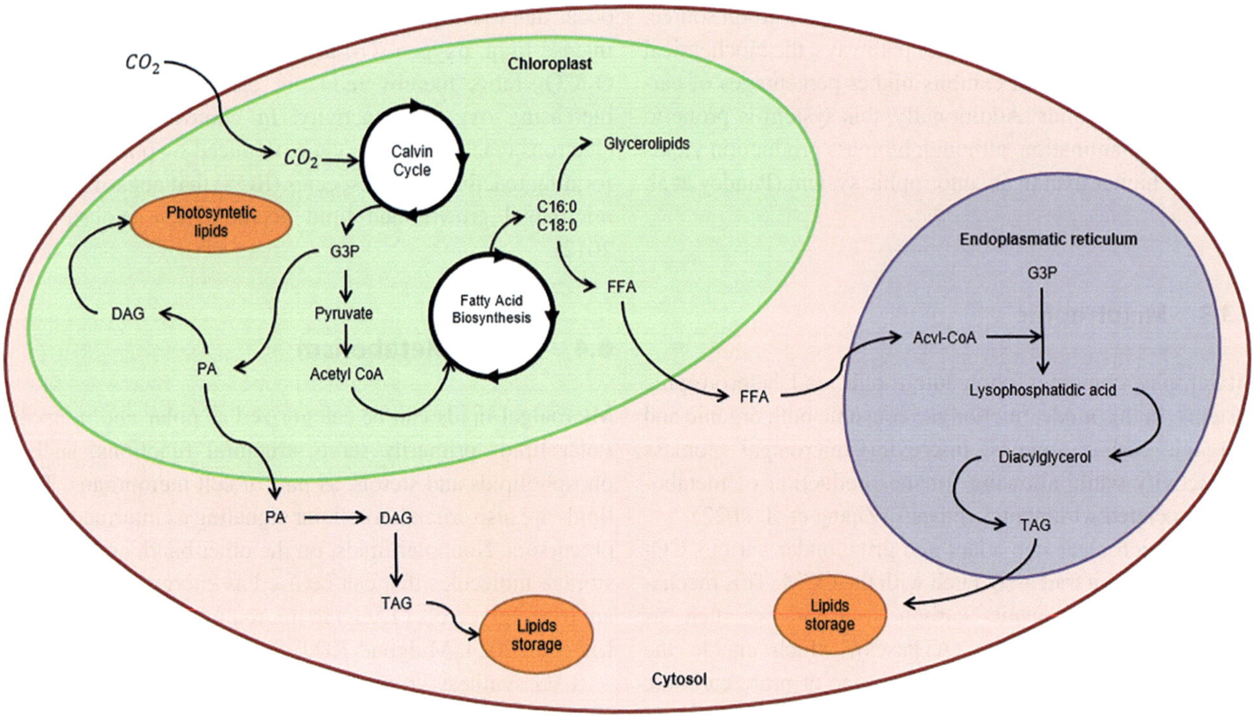

Fig. 6.5 Lipidic metabolism in microalgae (Khoo et al. 2023; Liu et al. 2024; Mulgund 2022; Tibocha-Bonilla et al. 2018)

pathway to form long carbon chains of C_{16} and C_{18} (Khoo et al. 2023; Liu et al. 2024). Acyl chains are then cleaved to release fatty acids, which are transported to the endoplasmic reticulum for further processing. Acetyl-CoA production and transfer are key regulatory steps in the fatty acid synthesis and oxidation. Under nutrient restrictions or environmental stress, half of the total lipid content is synthesized de novo, while the remaining lipids are derived from recycling existing fatty acids. The main TAG synthesis site is the endoplasmic reticulum, and TAG synthesis primarily follows the Kennedy pathway. The carbon chain of acyl-coenzyme A (acyl-CoA) and glycerol-3-phosphate acyltransferase (GPAT) undergo a series of transformations to form TAGs (Khoo et al. 2023; Mulgund 2022; Tibocha-Bonilla et al. 2018).

6.4.1 Genetics and Metabolic Regulation

Genetic modification and metabolic regulation engineering enable precise control over metabolic pathways. Genetic modification in algae requires cloning vectors, meaning extrachromosomal DNA molecules that self-replicate independently and correspond to the gene intended for expression or insertion. These vectors are usually chromosomal segments from the organism of interest and must include a promoter region, transcription regulator, and selection marker genes. Meanwhile, metabolic regulation and metabolic flux analysis help evaluate the direction of a metabolic pathway, enabling the identification of bottlenecks and genetic regulation mechanisms (Priyadharsini et al. 2022; Udaypal et al. 2024). These techniques facilitate the overexpression of rate-limiting enzymes in lipid synthesis pathways. This overexpression induces higher lipid accumulation without affecting other physiological and biochemical processes (Khoo et al. 2023; Mulgund 2022; Sundaram et al. 2023). Some bioengineering techniques applicable to lipid production in microalgae for genetic modification include clustered regularly interspaced short palindromic repeats (CRISPR) associated with protein 9, random mutagenesis, and zinc finger nuclease (ZFN), primarily used for genetic sequence alteration. For gene expression repression and activation, microRNAs (miRNAs) and short-interfering RNAs (siRNAs) are commonly used (Tomar et al. 2023).

6.4.2 Abiotic Factors Affecting Lipid Production

Lipid production in microalgae requires suitable conditions for cell growth and oil production, often specific to the cultivated microalgal species. Environmental stress and metabolic regulation can promote TAG production and accumulation, causing an imbalance that inhibits most fundamental metabolic processes and activates lipid synthesis pathways. Environmental stress factors that can enhance lipid production and accumulation include nutrient limitation, light intensity, temperature, pH, and CO_2 levels, among others (Cao et al. 2023; Liu et al. 2024).

Thus, defining the cultivation system, understanding optimal growth conditions, and identifying the stress conditions that force microalgae to divert their metabolism toward generating storage molecules like TAGs and carotenoids are important. This accumulation of reserves and the degradation of nitrogen-containing compounds, such as chlorophyll and proteins, enable microalgae to cope with adverse environments by increasing the availability of carbon and energy in the form of neutral lipids, which can also reduce oxidative damage caused by cellular stress (Tomar et al. 2023).

6.4.2.1 Nutrient Limitation

Nutrient limitation is considered an effective strategy for lipid production. The presence of carbon, nitrogen (N), phosphorus (P), and ions like Mg and Ca in the culture medium is crucial for microalgal biochemistry and physiology, participating in cell division, photosynthesis, and lipid accumulation (Pandey et al. 2024). Macronutrients like nitrogen (N) and phosphorus (P) enable the formation of molecules such as proteins and DNA. Additionally, deficiency in these two nutrients can severely impact intracellular metabolism. Phosphorus, in particular, affects the carbon flow between starch and lipid synthesis pathways by inhibiting the activity of Adenosine diphosphate-glucose pyrophosphorylase (AGPase), a key enzyme in the first step of starch synthesis (Shahid et al. 2020). On the other hand, micronutrients like Mg are essential for forming metalloproteins with enzymatic functions involved in carbon fixation and photosynthesis, and they also act as enzyme activators, such as Acetil-CoA carboxilase (ACC), which promotes fatty acid biosynthesis (Khoo et al. 2023; Liu et al. 2024). Calcium (Ca) is an integral part of cell wall and cell membrane structures, promoting cell communication and environmental stimuli by acting as an intracellular messenger (Yang et al. 2022).

Notably, nitrogen limitation can increase lipid content more efficiently than phosphorus limitation. This behavior was reported by Srinuanpan et al. (2018), who found lipid percentages of 40.3% in *Scenedesmus obliquus* SIT06 and 30.5% in *Micractinium reisseri* SIT04 under phosphorus limitation (Srinuanpan et al. 2018). Under nitrogen limitation, lipid content significantly increased, reaching 56.8% and 36.6%, respectively. Similarly, Omar Faruque et al. (2020) found that increasing the C/N ratio in cultures under nitrogen stress reduces protein synthesis, with carbon stored as intracellular carbohydrates and lipids. Lipid accumulation increased the activity of acetyl-CoA and acetyl-CoA carboxylase to recalibrate the distribution of starch and lipids (Faruque et al. 2020). Meanwhile, Dong et al. (2020) observed that cells preferentially synthesize starch and, over prolonged stress periods, degrade starch, using the degraded carbon skeleton primarily for lipid synthesis (Dong et al. 2020).

6.4.2.2 CO_2 as a Carbon Source

In microalgae, the lipid accumulation mechanism involves converting CO_2 into biomass through photosynthesis, where fixed carbon atoms are used to form lipids via biochemical reactions. Each microalgal species has an optimal CO_2 concentration that benefits metabolic activity, growth, and lipid accumulation (Mulgund 2022). Reducing CO_2 concentration will decrease lipid content due to reduced metabolic rates, while increasing CO_2 concentration above its optimal level may cause medium oversaturation and potential toxicity, adversely affecting biomass and lipid production. This phenomenon may relate to the chemical carbon equilibrium, where unused CO_2 converts into carbonic acid (H_2CO_3), lowering the culture pH, acidifying the medium, and altering physiological mechanisms associated with cell reproduction, growth, and metabolism (Sibi et al. 2016).

Because carbon fixation is directly linked to photosynthesis, CO_2 and light are often evaluated simultaneously for growth and lipid production. For example, Mohsenpour and Willoughby (2016) found that increasing the CO_2 concentration from 5% to 15% under blue light increased lipid content by sixfold on certain microalgae species. Similarly, Khan et al. (2022) found that the total lipid content rose from 8.0% to 22.0% of dry biomass when CO_2 supply was 15%. Other researchers also consider the importance of macronutrient sources, such as Faruque et al. (2020), who examined lipid composition in *Scenedesmus dimorphus* by adjusting CO_2 concentration and the N/P ratio, concluding that lipid accumulation increases with a favorable N/P ratio and adequate CO_2 supplementation.

6.4.2.3 Light Energy

As photosynthetic autotrophs, light plays a critical role in microalgae growth and lipid accumulation. Light intensity, wavelength, and photoperiod are important cultivation parameters closely related to light energy supply. The optimal light intensity for growth in most microalgae is between 200 and 400 μmol/m^{2}s (Maltsev et al. 2021). However, each microalga has an optimal irradiance level for growth, and levels exceeding this value can harm cells, causing photoinhibition and increased heat dissipation (Abdur Razzak et al. 2024; Lisondro et al. 2022).

Light's effect on lipid production is well-documented, as it can influence the quantity and type of lipids produced. Regarding this, Chin et al. (2023) observed an optimal light intensity of 135 μmol/m^{2}s in *Chaetoceros muelleri* and *Isochrysis galbana*, with the lipid content reaching up to 43.4% dry weight, primarily composed of saturated and monounsaturated fatty acids, the highest proportion being oleic acid (27.7%), a good option for algal biodiesel production. Different light sources and photoperiods can also positively affect growth and lipid production. For instance, Ren et al. (2020) evaluated the effects of different light sources and photoperiods on the growth and lipid accumulation of *Scenedesmus* sp., achieving biomass concentrations of 3.2 g/L and lipid content of 40.3%. Wavelength is another important factor that can enhance lipid accumulation. Moreover, Sirisuk et al. (2018) reported increased lipid production using green LED light in *Nannochloropsis salina*, *Isochrysis galbana*, and *Phaeodactylum tricornutum*, with values of 49.4%, 63.3%, and 62.0% dry weight, respectively.

6.4.2.4 Temperature

Temperature can affect the rate of photosynthesis, cell growth, and lipid accumulation in microalgae because it directly influences enzyme activity (Abdelkarim et al. 2024). Increasing temperature in cultures below the optimal growth temperature positively affects photosynthesis and cell division. However, the temperature exceeding the optimal range may lead to excessive growth, inhibition, or cell death. Like other organisms, microalgae have different temperature tolerance ranges, though most can survive between 20 and 30 °C (Abdur Razzak et al. 2024). This does not preclude the existence of thermophilic microalgae, which can tolerate temperatures as high as 40 °C due to heat-resistant structural molecules and thermostable enzymes (Covarrubias et al. 2016). In general, higher temperatures can enhance biomass production and metabolites like carbohydrates and lipids, possibly due to the activity of thermophilic enzyme systems, whereas lower temperatures usually reduce carboxylase activity, directly impacting carbon fixation and the synthesis of carbon-based molecules (Abdur Razzak et al. 2024; Cao et al. 2023).

The effect of temperature on lipid content can vary by species. For instance, Converti et al. (2009) reported an increase in lipid content from 7.9% to 14.9% in *Nannochloropsis oculata* with a temperature change from 20 to 25 °C, while *Chlorella vulgaris* showed a decrease in lipid content from 14.7% to 5.9% with a temperature change from 25 to 30 °C. Similarly, Chaisutyakorn et al. (2018) found that lipid content decreased as temperature increased from 25 to 40 °C in *Chaetoceros* sp. *FIKU035*, *Tetraselmis suecica FIKU032*, and *Nannochloropsis* sp. *FIKU036*.

6.4.2.5 pH

This factor can impact many physiological mechanisms related to growth, metabolism, lipid accumulation, and ion transport. pH determines the solubility and availability of CO_2 and other essential nutrients, as well as the activities of intracellular enzymes and those associated with the cell wall. Acidic pH interferes with nutrient absorption and cellular metabolism, while alkaline pH reduces the cell's CO_2 affinity, increases cell wall flexibility, delays cell cycle completion, and interferes with cell division (Udaypal et al. 2024). Due to the sensitivity of photosynthetic structures and enzymes to extreme pH conditions, maintaining the culture pH within an appropriate range for each microalga is crucial. However, in cultures where the carbon source is bicarbonate or CO_2, medium pH can vary depending on the photoperiod, becoming basic during light periods when photosynthesis is more active and acidic during dark periods when cells tend to respire. For this reason, pH fluctuations are often observed throughout day and night cycles (Liu et al. 2024).

Most microalgal species grow best at a nearly neutral pH between 7.0 and 7.6. However, some species, like *Chlamydomonas reinhardtii*, show a higher lipid production at a slightly basic pH of 7.8 (Ochoa-Alfaro et al. 2019). Others, such as *Chlorella*, can survive in a broader pH range. For example, *Parachlorella kessleri R-3* can accumulate substantial lipids in a pH range of 3–11 (Song et al. 2023a).

6.4.2.6 Salt Stress

Salt plays a role in the physiological and biochemical pathways of growth, reproduction, and fatty acid metabolism, increasing lipid content in microalgae (Cao et al. 2023; Liu et al. 2024). It often promotes the production of saturated fatty acids, ideal for biodiesel production (Tomar et al. 2023). Saline stress creates osmotic pressure differences within cells, forcing them to adapt by shifting their metabolism, inducing oxidative stress, and stimulating fatty acid metabolism. This occurs because NaCl can influence carbon distribution, causing it to shift from starch to lipid synthesis through positive regulation of ACCase (the key enzyme responsible for lipid synthesis) and negative regulation of Adenosine diphosphate glucan pyrophosphorylase (AGPase) (the key enzyme responsible for starch synthesis) (Khoo et al. 2023).

6.4.2.7 Heavy Metal Stress

Microalgae exhibit a strong ability to assimilate metal ions. Heavy metals such as Cu, Zn, Fe, Mn, Ni, and Cr are essential micronutrients for biological processes, acting as precursors to vitamins, catalytic cofactors in several metalloenzymes, and structural proteins in cell membranes (Tomar et al. 2023). Because these ions play vital roles in these organisms, essential metal ions can influence growth rate and lipid content, such as Fe^{3+}, which participates in photosynthesis; Mg^{2+}, an important component of chlorophyll structure; and Cu^{2+} and Zn^{2+}, which can increase lipid content in microalgae like *Chlorella sorokiniana* and *Scenedesmus acuminatus* (Hamed et al. 2022). However, high concentrations of essential heavy metals or low concentrations of nonessential heavy metals can induce the formation of reactive oxygen species (ROS), resulting in oxidative damage to proteins, peptides, and lipids and activating the antioxidant system (Song et al. 2023b). This system is a protective mechanism in microalgae that activates oxygen-eliminating enzymes, promotes secondary metabolites, and increases TAG synthesis to protect cells (Khoo et al. 2023; Liu et al. 2024). In the microalga *Parachlorella kessleri R-3*, low Cr^{6+} concentrations (0.5–2 mg/L) promoted growth, but concentrations above 5 mg/L inhibited growth, although a lipid content of 61.0% was achieved at 5 mg/L Cr^{6+} (Song et al. 2023b). Heavy metal mixtures can also be used to induce lipid accumulation, as in *Tetraselmis* sp., which was subjected to binary mixtures of Ni, Cr, and Co to overstimulate lipid metabolism (Dammak et al. 2022). The main advantage of using such elicitors is that heavy metals are abundant in industrial wastewater and currently represent a significant environmental problem. These metals can be remediated by microalgae and used as raw materials for biodiesel production (Song et al. 2023b).

6.5 Lipid Separation

6.5.1 Cell Disruption

Since most microalgae have thick, rigid cell walls and lipids are intracellular metabolites, extracting these compounds can be challenging. Breaking one or more cellular barriers may be necessary, depending on the cultivated species. The internal cell wall of microalgae consists of a hemicellulose and cellulose microfibril matrix, while the outer cell wall is enriched with a rigid polysaccharide matrix (Tomar et al. 2023). Each microalgal species has unique characteristics in its cell wall; some, such as the *Scenedesmus* genus, possess a coenobial wall—a structure allowing colony formation made of polysaccharides, providing additional rigidity (Baudelet et al. 2017). Consequently, cell disruption can become a limiting step in lipid extraction from microalgae, making finding techniques that maximize lipid recovery essential. This step is typically carried out using mechanical and nonmechanical techniques (Unterlander et al. 2017).

6.5.1.1 Mechanical Methods

Mechanical methods include milling, high-pressure homogenization, supercritical fluid extraction, pressurized liquid extraction, ultrasound-assisted extraction, microwave-assisted extraction, and pulsed electric field techniques. These methods can have drawbacks, such as high energy consumption, equipment corrosion, and the inability to effectively fractionate microalgal biomass due to the extremely small size of some microalgae (3–10 μm) (Yu et al. 2024).

6.5.1.2 Nonmechanical Methods

Nonmechanical methods include chemical treatments with acids, alkalis, organic solvents, oxidants, and biological pretreatment (e.g., enzymes and microorganisms). Biological methods are less frequently used than mechanical extraction methods, even if they are considered cleaner and more environmentally friendly. They avoid using toxic solvents, reduce waste generation, and lower energy consumption (Unterlander et al. 2017).

An important factor in cell disruption, regardless of the technique, is biomass hydration. Using wet biomass may affect the efficiency and effectiveness of this process and subsequent processes like lipid extraction and oil transesterification. The negative impact of water content is directly related to the nonpolar, hydrophobic nature of lipids. Therefore, it is recommended to dry the biomass beforehand to improve disruption and lipid extraction (Krishnamoorthy et al. 2023). Various drying methods are available, including rotary, spray, solar, cross-flow air, vacuum-shelf, and flash drying (Unterlander et al. 2017).

6.5.2 Lipid Extraction

Lipid extraction constitutes 30–40% of microalgae's total biodiesel production cost. Traditional methods for this process typically use nonpolar organic solvents with high lipid affinity. However, alternative methods using eco-friendly solvents have been developed due to the significant waste generation associated with these methods.

6.5.2.1 Conventional Methods

In the Soxhlet method, the extraction solvent must be explicitly chosen, as different solvents will yield various natural compounds from the same substance. Extract composition will vary by solvent. For *Chlorella vulgaris*, lipid yields up to 20% by weight of extract/dry biomass have been reported with hexane Soxhlet extraction (Wetterwald et al. 2023). Meanwhile, methanol:chloroform (2:1) solution yielded 18.6% lipids in a freshwater microalga (Kumar et al. 2019).

Soxhlet extraction is straightforward, requiring minimal effort, and the lipid can be used immediately for conversion; however, it could be time-consuming and energy-intensive (Tomar et al. 2023).

Besides, the Folch method is a popular technique for lipid separation in biological samples, using a chloroform:methanol:water solvent system in an 8:4:3 (v/v/v) ratio. It is a simple and effective method for large sample volumes, but the drawback is its reliance on hazardous chemicals, increasing operational costs (Kumar et al. 2019; Tomar et al. 2023).

The Bligh and Dyer method uses a monophasic mixture of methanol, chloroform, and water for extraction. It involves a two-step process, where lipids are measured in the chloroform phase after separation. This method is often used for lipid extraction from microalgae. Furthermore, in dried *Nanochloropsis* cells, the order of solvent addition can affect lipid extraction efficiency. Adding solvents in increasing polarity order increased extraction yield from 185 mg (18.5%) to 210 mg (21.0%) (Tomar et al. 2023).

6.5.2.2 Eco-Friendly Solvents

Historically, organic solvents have been used to recover lipids from microalgae. However, due to the environmental impact of these compounds, various unconventional technologies using eco-friendly solvents have been developed based on green chemistry principles. These solvents are generally inert, have low volatility, and are minimally toxic or nontoxic. This group includes ionic liquids (ILs), liquid biphasic systems (LBSs), deep eutectic solvents (DESs), and switchable solvents (SSs) (Tomar et al. 2023; Yu et al. 2024).

Ionic liquids (ILs) consist of cations and anions, forming an organic salt that is liquid at room temperature, primarily composed of zinc, aluminum, tin, and iron chlorides with quaternary ammonium salts. ILs can serve as both solvents and catalysts for biodiesel production from microalgae, making them environmentally friendly production methods. The catalytic reaction with ILs has advantages, including high reactivity and simple separation steps under mild conditions (Ong et al. 2021).

Liquid biphasic systems (LBSs) are used for pretreating microalgal biomass, with cost-saving benefits, high separation efficiency, and protein stability. They are nonflammable, nonvolatile, and recyclable and exhibit low vapor pressure and melting point, among other properties. However, their practical implementation requires prior cell wall disruption of the microalgae (Yu et al. 2024).

Deep eutectic solvents (DESs) are characterized by a significant reduction in melting point, generally exceeding 150 °C. DESs effectively break down the hydrogen-bond network of the microalgal cell wall. They are hydrophilic, and adding adequate water reduces viscosity and enhances performance. DESs have a high affinity for proteins, allowing for protein dissolution and extraction during microalgal biomass treatment. After extraction, lipids are converted into biodiesel via transesterification, and carbohydrates can be converted into biofuels through fermentation. Compared to ILs, DESs are lower-cost, easy to prepare, nontoxic, and biodegradable, making them very promising (Yu et al. 2024).

Switchable solvents (SSs) can reversibly change physical properties like polarity and hydrophobicity, allowing them to function as extractors. SSs can be used for cell disruption in their polar state, while in their nonpolar state, they have a high affinity for lipids, facilitating their extraction (Ismail and Al-Zuhair 2020).

6.6 Biodiesel Production

Biodiesel is one of the most important biofuels because it can replace nonrenewable fuels used in industry and transportation. Biodiesel is a liquid fuel considered less toxic, renewable, and reusable compared to liquid petroleum-based fuels. It can also be used in standard diesel engines without modifications or additional adaptations. A typical pathway for converting microalgal biomass into liquid biofuels involves the production of intracellular lipids, subsequent extraction, and finally, transformation processes that can be chemical, like transesterification, or thermochemical, like gasification, pyrolysis, or liquefaction, which produce liquid fuel as shown in Fig. 6.6 (Siddiki et al. 2022).

6.6.1 Transesterification

Transesterification is the most effective and least expensive method for producing biodiesel from algae. It comprises four stages: (1) mixing alcohol and catalyst, (2) reaction, (3) separation, and (4) washing the methyl ester (Alibas et al. 2023). In this reaction, fatty acids are hydrolyzed and methylated into fatty acid methyl esters (or biodiesel). There are two methods for the transesterification of oils: the traditional two-step method, which is considered costly due to high energy consumption, and the single-step method, also called in situ transesterification, where lipid extraction and catalytic transesterification occur simultaneously (Tien Thanh et al. 2022).

Most biodiesel production catalysts are homogeneous, attracting interest due to their ability to produce high biodiesel yields at low cost and in a short time (Neag et al. 2023). However, this type of catalyst usually mixes with the product, making separation difficult, which can lead to equipment corrosion and water contamination (Gupta et al. 2023). Heterogeneous catalysts are viable substitutes, offering high efficiency and environmental friendliness, and they are suitable for raw materials with high free fatty acid concentrations.

To minimize pollution from toxic chemicals, bioenzyme-catalyzed transesterification processes, particularly lipase-based transesterification, have been evaluated. However, enzymatic transesterification is costly, has a slow reaction rate, and limits catalyst use and regeneration. Alternatively, supercritical transesterification requires no catalyst addition, raw material pretreatment, or lipid extraction from the microalgal biomass. Extraction and reaction can occur simultaneously in a single step, producing biodiesel within minutes (Pandey et al. 2024).

6.6.2 Gasification

Gasification is a versatile thermal conversion technology where biomass is transformed into various gaseous compounds, including H_2 (10–50%), CO (7–52%), and CH_4, with a solid residue that can be used as charcoal or fertilizer or for other applications. This process is carried out at temperatures between 700 and 1000 °C, requiring that the residual biomass have a moisture content below 15% and controlled conditions of flow, oxygen, and CO_2/air (Al-Rabiah et al. 2022; Raheem et al. 2021).

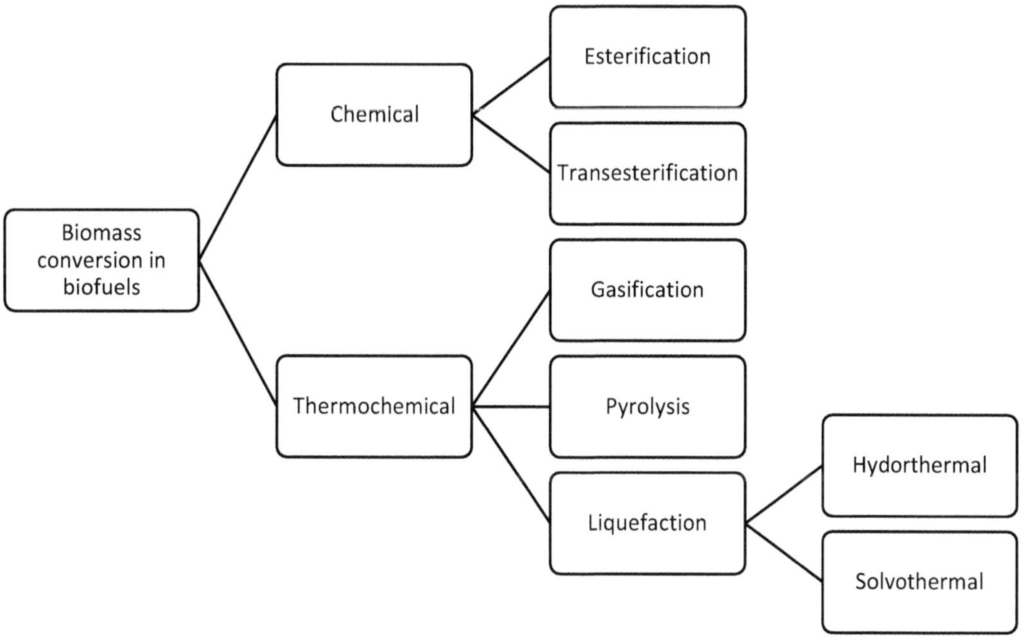

Fig. 6.6 Common bio-oil production techniques for microalgae

6.6.3 Liquefaction

6.6.3.1 Hydrothermal Liquefaction (HTL)

HTL is a thermochemical conversion method where wet microalgal biomass is subjected to high temperatures and pressures in the presence of water. During the process, microalgal biomass is converted into four products: gases, aqueous products, organic products (bio-oil or bio-crude), and solids. The advantage of this technology lies in reduced operational costs because it can use wet biomass, avoiding the high economic and energy costs associated with biomass drying (da Costa Magalhães et al. 2023; Gómez et al. 2022; Silva et al. 2024). HTL has a smaller carbon footprint, a better energy return on investment, and more significant economic potential than other methods. However, bio-oil composition highly depends on the type of microalga used, mainly due to the high nitrogen content typically found in these organisms, resulting in higher nitrogen and oxygen content than crude oil (Deepika et al. 2024).

6.6.3.2 Solvothermal Liquefaction (STL)

STL is a solvolysis reaction that cleaves various chemical bonds in the biomass using organic solvents like polyhydroxy alcohols. The entire biomass, not just lipids, is converted into a liquid product. The advantage of this technique is that it does not require biomass drying, allowing it to be processed wet, thus eliminating a high-cost economic and energy-intensive drying stage. Bio-crude oil obtained from microalgae contains chemical compounds such as linear hydrocarbons, aromatics, phenol derivatives, and other oxygen- and nitrogen-containing organic compounds (Ha Tran and Lee 2020; Kim et al. 2015). While bio-crude oil has physicochemical properties similar to crude oil, its chemical composition may vary depending on the moisture, ash, and elemental composition (e.g., O, N, S, and P) of the biomass used. It also has a lower viscosity and H ratio than petroleum oils (Marangon et al. 2024).

6.6.4 Pyrolysis

Pyrolysis is a technology that converts organic matter into biofuels in an oxygen-free environment at temperatures between 300 and 800 °C. This process produces solid (biochar) and liquid (bio-oil) products, as well as non-condensable gases (NCGs). Pyrolysis can be classified as either slow or fast, depending on the reactor configuration and the predominant final product. Slow pyrolysis maximizes biochar production, while fast pyrolysis enhances the production of high-quality bio-oil in large quantities. However, the high nitrogen content in biomass used as a raw material may reduce bio-oil yield and increase biochar production, inhibiting organic matter decomposition and promoting secondary carbonization reactions, which poses a disadvantage compared to fossil fuel production and other bio-oil production techniques (Aliyu et al. 2021; Li et al. 2022; Marinič et al. 2023; Mohd Nasir et al. 2023).

6.7 Biorefinery

Currently, two main strategies are employed for compound production in biorefineries. The first strategy focuses on producing biofuels from renewable and sustainable sources to replace conventional petroleum-derived sources. The second approach aims to establish a sustainable and robust industry that maximizes the generation of value-added products from the microorganism of interest (Zarei and Zamani 2024). Biorefineries involving the production of biofuels and other high-value bioproducts from microalgae integrate multiple production processes that operate simultaneously. Microalgae are increasingly considered renewable and sustainable options for biorefineries due to their ability to produce biomass, bioproducts, and high-energy compounds at low economic costs.

The economic efficiency of these processes is mainly because microalgal cultivation does not require large tracts of land. Additionally, pollutants from industrial and domestic wastewater can be used as nutrient sources, and industrial gaseous waste containing CO_2 can be integrated, using photoreaction systems that harness natural light, significantly reducing production costs.

One of the advantages of this type of process is that it is possible to reduce or eliminate waste generation, as each component of the microalgal biomass can be used as raw material in various industries, such as pharmaceuticals, cosmetics, food for humans and animals, and biofertilizers. Therefore, these biorefineries allow for the design of closed production cycles under the concept of circular economy (Siddiki et al. 2022; Silva et al. 2024; Zhao et al. 2023).

6.8 Concluding Remarks

Biodiesel production from microalgae represents a profitable solution to sustainable energy generation and environmental preservation challenges. Microalgae's high lipid content, rapid growth rate, and ability to be grown in nonarable lands using wastewater and industrial CO_2 make them attractive alternatives to traditional biofuel sources.

However, several challenges remain in scaling this technology. Key barriers include high production costs, primarily driven by energy-intensive processes like cultivation, cell disruption, and lipid extraction. Besides, the variability in lipid productivity among strains and their sensitivity to environmental factors further complicate commercial feasibility.

Additionally, the economic and energy demands of biomass drying and the chemical treatments used in extraction hinder broader adoption.

Technological advances have significantly addressed some of these challenges. Innovations in photobioreactor designs (including optimized configurations for enhanced light penetration and CO_2 transfer) have improved biomass productivity. Moreover, alternative eco-friendly lipid extraction techniques, such as ionic liquids and deep eutectic solvents, present environmentally sustainable choices to conventional methods. Furthermore, genetic engineering and metabolic regulation strategies enable the development of microalgae strains with improved lipid yields and resilience under stress conditions. Moreover, their integration of microalgae into biorefinery schemes enhances their economic viability by generating high-value coproducts alongside biodiesel.

In synthesis, while the pathway to widespread biodiesel production from microalgae is complex, ongoing research and technological progress continue to enhance its viability. A multidisciplinary approach that combines advances in engineering, biology, and chemistry is essential for overcoming the current constraints. Future efforts should focus on cost reduction, process integration, and maximization of the economic potential of biorefineries to realize the full promise of microalgae as cornerstones of sustainable energy systems.

References

Abdelfattah A, Ali SS, Ramadan H, El-Aswar EI, Eltawab R, Ho SH, Elsamahy T, Li S, El-Sheekh MM, Schagerl M, Kornaros M, Sun J (2023) Microalgae-based wastewater treatment: mechanisms, challenges, recent advances, and future prospects. Environ Sci Ecotechnol 13(October 2022):100205. https://doi.org/10.1016/j.csc.2022.100205

Abdelkarim OH, Wijffels RH, Barbosa MJ (2024) Microalgal lipid production: A comparative analysis of Nannochloropsis and Microchloropsis strains. J Appl Phycol. https://doi.org/10.1007/s10811-024-03318-7

Abdur Razzak S, Bahar K, Islam KMO, Haniffa AK, Faruque MO, Hossain SMZ, Hossain MM (2024) Microalgae cultivation in photobioreactors: sustainable solutions for a greener future. In: Green chemical engineering, vol 5, issue 4. KeAi Communications Co, pp 418–439. https://doi.org/10.1016/j.gce.2023.10.004

Alibas I, Yilmaz A, Alibas K, Arslan M, Koc C (2023) The effect of different drying methods and microalgae species on the quality parameters of biodiesel obtained by transesterification technique. Biomass Bioenergy 168(October 2022):106688. https://doi.org/10.1016/j.biombioe.2022.106688

Aliyu A, Lee JGM, Harvey AP (2021) Microalgae for biofuels: a review of thermochemical conversion processes and associated opportunities and challenges. Bioresour Technol Rep 15(March):100694. https://doi.org/10.1016/j.biteb.2021.100694

Al-Rabiah AA, Al-Dawsari JN, Ajbar AM, Al Darwish RK, Abdelaziz OY (2022) Development of a biomass gasification process for the coproduction of methanol and power from red sea microalgae. Energies 15(21). https://doi.org/10.3390/en15217890

Aniruddha R, Rajendran A, Sindhu S (2022) A study on biofuel generation from microalgae species. Mater Today Proc 57:1660–1665. https://doi.org/10.1016/j.matpr.2021.12.269

Bácsi I, Tóthfalusi F, Márton K, B-Béres V, Gonda S (2022) The effects of photobioreactor type on biomass and lipid production of the green microalga Monoraphidium pusillum in laboratory scale. Appl Sci (Switz) 12(4). https://doi.org/10.3390/app12042196

Balouch H, Zayadan BK, Sadvakasova AK, Kossalbayev BD, Bolatkhan K, Gencer D, Civelek D, Demirbag Z, Alharby HF, Allakhverdiev SI (2023) Prospecting the biofuel potential of new microalgae isolates. Int J Hydrog Energy 48(50):19060–19073. https://doi.org/10.1016/j.ijhydene.2023.02.028

Baudelet PH, Ricochon G, Linder M, Muniglia L (2017) A new insight into cell walls of Chlorophyta. Algal Res 25(April):333–371. https://doi.org/10.1016/j.algal.2017.04.008

Bharti RK, Kaushal C, Singh A, Dhar DW, Babu R, Kaushik A (2024) Evaluation of fuel properties for possible biodiesel output based on the fatty acid composition of oleaginous plants and microalgae. Sci Total Environ 918(January):170448. https://doi.org/10.1016/j.scitotenv.2024.170448

Cao Y, Yang S, Wang J, Kong W, Guo B, Xi Y, Zhang A, Yue B (2023) Metabolomic exploration of the physiological regulatory mechanism of the growth and metabolism characteristics of Chlorella vulgaris under photoautotrophic, mixotrophic, and heterotrophic cultivation conditions. Biomass Bioenergy 173:106775. https://doi.org/10.1016/j.biombioe.2023.106775

Casanova LM, Mendes LBB, Corrêa TDS, da Silva RB, Joao RR, Macrae A, Vermelho AB (2023) Development of microalgae biodiesel: current status and perspectives. Microorganisms 11(1):1–20. https://doi.org/10.3390/microorganisms11010034

Chaisutyakorn P, Praiboon J, Kaewsuralikhit C (2018) The effect of temperature on growth and lipid and fatty acid composition on marine microalgae used for biodiesel production. J Appl Phycol 30(1):37–45. https://doi.org/10.1007/s10811-017-1186-3

Chen CY, Yeh KL, Chang HY, Chang JS (2014) Strategies to improve oil/lipid production of microalgae in outdoor cultivation using vertical tubular-Type photobioreactors. Energy Procedia 61:2755–2758. https://doi.org/10.1016/j.egypro.2014.12.298

Chin GJWL, Andrew AR, Abdul-Sani ER, Yong WTL, Misson M, Anton A (2023) The effects of light intensity and nitrogen concentration to enhance lipid production in four tropical microalgae. Biocatal Agric Biotechnol 48(February):102660. https://doi.org/10.1016/j.bcab.2023.102660

Converti A, Casazza AA, Ortiz EY, Perego P, Del Borghi M (2009) Effect of temperature and nitrogen concentration on the growth and lipid content of Nannochloropsis oculata and Chlorella vulgaris for biodiesel production. Chem Eng Process Process Intensif 48(6):1146–1151. https://doi.org/10.1016/j.cep.2009.03.006

Covarrubias Y, Cantoral-Uriza EA, Casas-Flores JS, García-Meza JV (2016) Thermophile mats of microalgae growing on the woody structure of a cooling tower of a thermoelectric power plant in Central Mexico. Rev Mex Biodivers 87(2):277–287. https://doi.org/10.1016/j.rmb.2016.04.001

da Costa Magalhães B, Matricon L, Ramirez Romero LA, Checa R, Lorentz C, Chambonniere P, Delrue F, Roubaud A, Afanasiev P, Laurenti D, Geantet C (2023) Catalytic hydrotreatment of bio-oil from continuous HTL of Chlorella sorokiniana and Chlorella vulgaris microalgae for biofuel production. Biomass Bioenergy 173(February):106798. https://doi.org/10.1016/j.biombioe.2023.106798

Dammak M, Hlima HB, Tounsi L, Michaud P, Fendri I, Abdelkafi S (2022) Effect of heavy metals mixture on the growth and physiology of Tetraselmis sp.: applications to lipid production and bioremediation. Bioresour Technol 360:127584

Deepika C, Mrinal, Pohrmen CB, Jaiswal KS, Sangmesh B, Jaiswal KK, Ramasamy AP, Jaiswal AK (2024) Hydrothermal liquefaction of wet microalgal biomass for biofuels and platform chemicals: advances and future prospects. Discov Appl Sci 6(5). https://doi.org/10.1007/s42452-024-05911-4

Dong L, Li D, Li C (2020) Characteristics of lipid biosynthesis of Chlorella pyrenoidosa under stress conditions. Bioprocess Biosyst Eng 43(5):877–884. https://doi.org/10.1007/s00449-020-02284-x

Faruque MO, Razzak SA, Hossain MM (2020) Application of heterogeneous catalysts for biodiesel production from microalgal oil—a review. In: Catalysts, vol 10, issue 9. MDPI, pp 1–25. https://doi.org/10.3390/catal10091025

Gómez EA, Ocampo D, Rios LA (2022) Reúso de residuos líquidos generados en la licuefacción hidrotérmica de microalgas provenientes de la industria cementera. Inf Tecnol 33(6):103–112. https://doi.org/10.4067/s0718-07642022000600103

Gupta P, Rajak U, Verma TN, Arya M, Singh TS (2023) Impact of fuel injection pressure on the common rail direct fuel injection engine powered by microalgae, kapok oil, and soybean biodiesel blend. Ind Crop Prod 194(February):116332. https://doi.org/10.1016/j.indcrop.2023.116332

Ha Tran M, Lee EY (2020) Development and optimization of solvothermal liquefaction of marine macroalgae Saccharina japonica biomass for biopolyol and biopolyurethane production. J Ind Eng Chem 81:167–177. https://doi.org/10.1016/j.jiec.2019.09.005

Hamed SM, El Shimi HI, Van Dijk JR, Osman AI, Korany SM, Abdelgawad H (2022) A novel integrated system for heavy metals removal and biodiesel production via green microalgae: a techno-economic feasibility assessment. J Environ Chem Eng 10(6):108804. https://doi.org/10.1016/j.jece.2022.108804

Huang J, Li Y, Wan M, Yan Y, Feng F, Qu X, Wang J, Shen G, Li W, Fan J, Wang W (2014) Novel flat-plate photobioreactors for microalgae cultivation with special mixers to promote mixing along the light gradient. Bioresour Technol 159:8–16. https://doi.org/10.1016/j.biortech.2014.01.134

Ismail M, Al-Zuhair S (2020) Thermo-responsive switchable solvents for simultaneous microalgae cell disruption, oil extraction-reaction, and product separation for biodiesel production. Biocatal Agric Biotechnol 26(May):101667. https://doi.org/10.1016/j.bcab.2020.101667

Khan AA, Mondal M, Halder GN, Saha AK (2022) Bioconversion of chemically captured carbon dioxide into microalgal lipids, a potential source of biodiesel: an integrated technique. Fuel 311(July 2021):122549. https://doi.org/10.1016/j.fuel.2021.122549

Khoo KS, Ahmad I, Chew KW, Iwamoto K, Bhatnagar A, Show PL (2023) Enhanced microalgal lipid production for biofuel using different strategies including genetic modification of microalgae: a review. In: Progress in energy and combustion science, vol 96. Elsevier Ltd. https://doi.org/10.1016/j.pecs.2023.101071

Kim KH, Jo YJ, Lee CG, Lee EY (2015) Solvothermal liquefaction of microalgal Tetraselmis sp. biomass to prepare biopolyols by using PEG#400-blended glycerol. Algal Res 12:539–544. https://doi.org/10.1016/j.algal.2015.08.007

Krishnamoorthy A, Rodriguez C, Durrant A (2023) Optimisation of ultrasonication pretreatment on microalgae Chlorella Vulgaris & Nannochloropsis Oculata for lipid extraction in biodiesel production. Energy 278(May):128026. https://doi.org/10.1016/j.energy.2023.128026

Kumar V, Kumar S, Chauhan PK, Verma M, Bahuguna V, Joshi HC, Ahmad W, Negi P, Sharma N, Ramola B, Rautela I, Nanda M, Vlaskin MS (2019) Low-temperature catalyst based hydrothermal liquefaction of harmful macroalgal blooms, and aqueous phase nutrient recycling by microalgae. Sci Rep 9(1):1–9. https://doi.org/10.1038/s41598-019-47664-w

Levasseur W, Perré P, Pozzobon V (2020) A review of high value-added molecules production by microalgae in light of the classification. Biotechnol Adv 41(January):107545. https://doi.org/10.1016/j.biotechadv.2020.107545

Li G, Hu R, Wang N, Yang T, Xu F, Li J, Wu J, Huang Z, Pan M, Lyu T (2022) Cultivation of microalgae in adjusted wastewater to enhance biofuel production and reduce environmental impact: pyrolysis performances and life cycle assessment. J Clean Prod 355(December 2021):131768. https://doi.org/10.1016/j.jclepro.2022.131768

Li S, Chang H, Zhang S, Ho SH (2023) Production of sustainable biofuels from microalgae with CO2 bio-sequestration and life cycle assessment. Environ Res 227(March):115730. https://doi.org/10.1016/j.envres.2023.115730

Lisondro I, Gómez Serrano C, Sepúlveda C, Batista Ceballos AI, Acién Fernández FG (2022) Influence of irradiance on the growth and biochemical composition of Nitzschia aff. pellucida. J Appl Phycol 34(1):19–30. https://doi.org/10.1007/s10811-021-02605-x

Liu H, Liu TJ, Guo HW, Wang YJ, Ji R, Le Kang L, Wang YT, Guo X, Li JG, Jiang LQ, Fang Z (2024) A review of the strategy to promote microalgae value in CO2 conversion-lipid enrichment-biodiesel production. J Clean Prod 436(January):140538. https://doi.org/10.1016/j.jclepro.2023.140538

Maltsev Y, Maltseva K, Kulikovskiy M, Maltseva S (2021) Influence of light conditions on microalgae growth and content of lipids, carotenoids, and fatty acid composition. In: Biology, vol 10, issue 10. MDPI. https://doi.org/10.3390/biology10101060

Marangon BB, de Siqueira Castro J, Calijuri ML (2024) Aviation fuel based on wastewater-grown microalgae: challenges and opportunities of hydrothermal liquefaction and hydrotreatment. J Environ Manag 354(February):120418. https://doi.org/10.1016/j.jenvman.2024.120418

Marinič D, Grilc M, Hočevar B, Delrue F, Likozar B (2023) Liquefaction, cracking and hydrogenation of microalgae biomass resources to CO2 negative advanced biofuels: mechanisms, reaction microkinetics and modelling. Renew Energy 203(October 2022):382–393. https://doi.org/10.1016/j.renene.2022.12.055

Mohd Nasir AS, Mohamed B, Li LY (2023) Comparative life cycle assessment of co-pyrolysing sewage sludge and wastewater-grown microalgae for biofuel production. Resour Conserv Recycl 190(May 2022). https://doi.org/10.1016/j.resconrec.2022.106780

Mohsenpour SF, Willoughby N (2016) Effect of CO2 aeration on cultivation of microalgae in luminescent photobioreactors. Biomass Bioenergy 85:168–177. https://doi.org/10.1016/j.biombioe.2015.12.002

Mulgund A (2022) Increasing lipid accumulation in microalgae through environmental manipulation, metabolic and genetic engineering: a review in the energy NEXUS framework. In: Energy nexus, vol 5. Elsevier Ltd. https://doi.org/10.1016/j.nexus.2022.100054

Neag E, Stupar Z, Maicaneanu SA, Roman C (2023) Advances in biodiesel production from microalgae. In: Energies, vol 16, issue 3. MDPI. https://doi.org/10.3390/en16031129

Ochoa-Alfaro AE, Gaytán-Luna DE, González-Ortega O, Zavala-Arias KG, Paz-Maldonado LMT, Rocha-Uribe A, Soria-Guerra RE (2019) pH effects on the lipid and fatty acids accumulation in Chlamydomonas reinhardtii. Biotechnol Prog 35(6):1–8. https://doi.org/10.1002/btpr.2891

Omar Faruque M, Ilyas M, Mozahar Hossain M, Abdur Razzak S (2020) Influence of nitrogen to phosphorus ratio and CO2 concentration on lipids accumulation of Scenedesmus dimorphus for bioenergy production and CO2 biofixation. Chem Asian J 15(24):4307–4320. https://doi.org/10.1002/asia.202001063

Ong HC, Tiong YW, Goh BHH, Gan YY, Mofijur M, Fattah IMR, Chong CT, Alam MA, Lee HV, Silitonga AS, Mahlia TMI (2021) Recent advances in biodiesel production from agricultural products and microalgae using ionic liquids: opportunities and challenges. In: Energy conversion and management, vol 228. Elsevier Ltd. https://doi.org/10.1016/j.enconman.2020.113647

Pandey S, Narayanan I, Selvaraj R, Varadavenkatesan T, Vinayagam R (2024) Biodiesel production from microalgae: a comprehensive review on influential factors, transesterification processes, and challenges. In: Fuel, vol 367. Elsevier Ltd. https://doi.org/10.1016/j.fuel.2024.131547

Penloglou G, Pavlou A, Kiparissides C (2024) Recent advancements in Photobioreactors for microalgae cultivation: a brief overview. In: Processes, vol 12, issue 6. Multidisciplinary Digital Publishing Institute (MDPI). https://doi.org/10.3390/pr12061104

Pereira FM, Loures CCA, Amaral MS, Gomes FM, Pedro GA, Machado MAG, Reis CER, Silva MB (2018) Evaluation of fatty acids production by Chlorella minutissima in batch bubble-column photobioreactor. Fuel 230(April):155–162. https://doi.org/10.1016/j.fuel.2018.04.170

Perrier B, Crampon C, Guézet O, Simon C, Maire F, Lépine O, Pruvost J, Lozano P, Bernard O, Badens E (2015) Production of a methyl ester from the microalgae Nannochloropsis grown in raceways on the French west coast. Fuel 153:640–649. https://doi.org/10.1016/j.fuel.2015.03.011

Priyadharsini P, Nirmala N, Dawn SS, Baskaran A, SundarRajan P, Gopinath KP, Arun J (2022) Genetic improvement of microalgae for enhanced carbon dioxide sequestration and enriched biomass productivity: review on CO2 bio-fixation pathways modifications. Algal Res 66:102810. https://doi.org/10.1016/j.algal.2022.102810

Raheem A, Abbasi SA, Mangi FH, Ahmed S, He Q, Ding L, Memon AA, Zhao M, Yu G (2021) Gasification of algal residue for synthesis gas production. Algal Res 58(July):102411. https://doi.org/10.1016/j.algal.2021.102411

Razzak SA, Lucky RA, Hossain MM, deLasa H (2022) Valorization of microalgae biomass to biofuel production: a review. Energy Nexus 7(August):100139. https://doi.org/10.1016/j.nexus.2022.100139

Ren HY, Dai YQ, Kong F, Xing D, Zhao L, Ren NQ, Ma J, Liu BF (2020) Enhanced microalgal growth and lipid accumulation by addition of different nanoparticles under xenon lamp illumination. Bioresour Technol 297(November 2019):122409. https://doi.org/10.1016/j.biortech.2019.122409

Shahid A, Rehman AU, Usman M, Ashraf MUF, Javed MR, Khan AZ, Gill SS, Mehmood MA (2020) Engineering the metabolic pathways of lipid biosynthesis to develop robust microalgal strains for biodiesel production. Biotechnol Appl Biochem 67(1):41–51. https://doi.org/10.1002/bab.1812

Sibi G, Shetty V, Mokashi K (2016) Enhanced lipid productivity approaches in microalgae as an alternate for fossil fuels – a review. In: Journal of the energy institute, vol 89, issue 3. Elsevier B.V., pp 330–334. https://doi.org/10.1016/j.joei.2015.03.008

Siddiki SYA, Mofijur M, Kumar PS, Ahmed SF, Inayat A, Kusumo F, Badruddin IA, Khan TMY, Nghiem LD, Ong HC, Mahlia TMI (2022) Microalgae biomass as a sustainable source for biofuel, biochemical and biobased value-added products: an integrated biorefinery concept. Fuel 307(August 2021):121782. https://doi.org/10.1016/j.fuel.2021.121782

Silva TA, do Couto EDA, Assemany PP, Costa PAC, Marques PASS, Paradela F, dos Reis AJD, Calijuri ML (2024) Biofuel from wastewater-grown microalgae: a biorefinery approach using hydrothermal liquefaction and catalyst upgrading. J Environ Manag 368(July):122091. https://doi.org/10.1016/j.jenvman.2024.122091

Sirisuk P, Ra CH, Jeong GT, Kim SK (2018) Effects of wavelength mixing ratio and photoperiod on microalgal biomass and lipid production in a two-phase culture system using LED illumination. Bioresour Technol 253(January):175–181. https://doi.org/10.1016/j.biortech.2018.01.020

Song X, Liu BF, Kong F, Song Q, Ren NQ, Ren HY (2023a) Lipid accumulation by a novel microalga Parachlorella kessleri R-3 with wide pH tolerance for promising biodiesel production. Algal Res 69:102925. https://doi.org/10.1016/j.algal.2022.102925

Song X, Liu BF, Kong F, Song Q, Ren NQ, Ren HY (2023b) Simultaneous chromium removal and lipid accumulation by microalgae under acidic and low temperature conditions for promising biodiesel production. Bioresour Technol 370(November 2022):128515. https://doi.org/10.1016/j.biortech.2022.128515

Sousa C, de Winter L, Janssen M, Vermuë MH, Wijffels RH (2012) Growth of the microalgae Neochloris oleoabundans at high partial oxygen pressures and sub-saturating light intensity. Bioresour Technol 104:565–570. https://doi.org/10.1016/j.biortech.2011.10.048

Srinuanpan S, Cheirsilp B, Prasertsan P, Kato Y, Asano Y (2018) Strategies to increase the potential use of oleaginous microalgae as biodiesel feedstocks: nutrient starvations and cost-effective harvesting process. Renew Energy 122:507–516. https://doi.org/10.1016/j.renene.2018.01.121

Sundaram T, Rajendran S, Gnanasekaran L, Rachmadona N, Jiang JJ, Khoo KS, Show PL (2023) Bioengineering strategies of microalgae biomass for biofuel production: recent advancement and insight. In: Bioengineered, vol 14, issue 1. NLM (Medline), p 2252228. https://doi.org/10.1080/21655979.2023.2252228

Tibocha-Bonilla JD, Zuñiga C, Godoy-Silva RD, Zengler K (2018) Advances in metabolic modeling of oleaginous microalgae Mike Himmel. In: Biotechnology for biofuels, vol 11, issue 1. BioMed Central Ltd. https://doi.org/10.1186/s13068-018-1244-3

Tien Thanh N, Mostapha M, Lam MK, Ishak S, Kanna Dasan Y, Lim JW, Tan IS, Lau SY, Chin BLF, Hadibarata T (2022) Fundamental understanding of in-situ transesterification of microalgae biomass to biodiesel: a critical review. Energy Convers Manag 270(May):116212. https://doi.org/10.1016/j.enconman.2022.116212

Tomar S, Agarwal S, Singh H, Kumar R, Qureshi KA, Jaremko M, Emwas AH, Rai PK (2023) Microalgae: a promising source for biofuel production. Biocatal Agric Biotechnol 53(August):102877. https://doi.org/10.1016/j.bcab.2023.102877

Udaypal, Goswami RK, Mehariya S, Verma P (2024) Advances in microalgae-based carbon sequestration: current status and future perspectives. In: Environmental research, vol 249. Academic Press Inc. https://doi.org/10.1016/j.envres.2024.118397

Unterlander N, Champagne P, Plaxton WC (2017) Lyophilization pretreatment facilitates extraction of soluble proteins and active enzymes from the oil-accumulating microalga Chlorella vulgaris. Algal Res 25(November 2016):439–444. https://doi.org/10.1016/j.algal.2017.06.010

Wetterwald L, Leybros A, Fleury G, Delrue F, Dimitriades-Lemaire A, Chambonniere P, Hertz A (2023) Supercritical CO2 extraction of neutral lipids from dry and wet Chlorella vulgaris NIES 227 microalgae for biodiesel production. J Environ Chem Eng 11(5):110628. https://doi.org/10.1016/j.jece.2023.110628

Wu W, Tan L, Chang H, Zhang C, Tan X, Liao Q, Zhong N, Zhang X, Zhang Y, Ho SH (2023) Advancements on process regulation for microalgae-based carbon neutrality and biodiesel production. Renew Sust Energ Rev 171:112969. https://doi.org/10.1016/j.rser.2022.112969

Yang J, Li W, Xing C, Xing G, Guo Y, Yuan H (2022) Ca2+ participates in the regulation of microalgae triacylglycerol metabolism under heat stress. Environ Res 208:112696. https://doi.org/10.1016/j.envres.2022.112696

Yang Z, Chen J, Tang B, Lu Y, Ho SH, Wang Y, Chen C, Shen L (2024) Metabolic interpretation of NaCl stress-induced lipid accumulation in microalgae for promising biodiesel production with saline wastewater. Chem Eng Sci 284(July 2023):119447. https://doi.org/10.1016/j.ces.2023.119447

Yu D, Luo H, Tan J, Yuan J, Chen J, Liu X, Li H (2024) Efficient fractionation of microalgae biomass as a whole for biodiesel production enabled by deep eutectic solvent and water. Fuel 363(January):130892. https://doi.org/10.1016/j.fuel.2024.130892

Zarei Z, Zamani H (2024) Biorefinery approach to stimulate astaxanthin and biofuel generation in microalga Haematococcus pluvialis under different light irradiance. Clean Techn Environ Policy 0123456789:3333. https://doi.org/10.1007/s10098-024-02803-4

Zhang S, Zhang L, Xu G, Li F, Li X (2022) A review on biodiesel production from microalgae: influencing parameters and recent advanced technologies. Front Microbiol 13(July):1–20. https://doi.org/10.3389/fmicb.2022.970028

Zhao Q, Han F, You Z, Huang Y, She X (2023) Evaluation of the relationship of wastewater treatment and biodiesel production by microalgae cultivated in the photobioreactor. Fuel 350(May):128750. https://doi.org/10.1016/j.fuel.2023.128750

Zhao Q, Han F, You Z, Huang Y, She X, Shi X, Han P (2024) Evaluation of microalgae biodiesel for carbon neutrality based on the waste treatment by the autotrophic and heterotrophic combination. Energy 291(July 2023):130314. https://doi.org/10.1016/j.energy.2024.130314

7 Biogas Production from Anaerobic Digestion of Microalgae

Raquel Barrena, Paqui Blánquez, Xavier Font, and Antoni Sánchez

Abstract

Microalgal biomass is a promising source for biogas production. However, biomass pretreatment is generally a crucial step in disintegrating cell wall structure and allowing it to exploit its potential fully. As a result, methane yield from microalgal biomass is influenced by various factors, including the microalgae species, biomass pretreatment, process operational parameters (such as pH, temperature, and nutrient availability), and the presence or absence of inhibitors. Besides, the co-digestion of microalgal biomass with carbon-rich residues is a suitable strategy to improve biogas production yields. The anaerobic digestion (AD) of microalgae and its integration into biorefineries play a crucial role in closing loops within the current circular economy context. Typically, this integration involves the anaerobic digestion of either microalgal biomass or residual exhaust microalgal biomass (such as that obtained after lipid extraction). However, ensuring the proper integration of both materials involves a rigorous assessment of their environmental impact and economic performance.

Keywords

Microalgae · Anaerobic digestion · Bioenergy · Biomethane

R. Barrena (✉) · P. Blánquez · X. Font · A. Sánchez
Research Group on Environmental (Bio)Remediation and Valorization of Effluents (BRAVE) Department of Chemical, Biological and Environmental Engineering, Escola d'Enginyeria, Universitat Autònoma de Barcalona, Barcelona, Spain
e-mail: raquel.barrena@uab.cat

7.1 Introduction

7.1.1 Current Framework and Benefits of Anaerobic Digestion

Anaerobic digestion (AD) is a biological process in which organic matter is degraded in the strict absence of oxygen by the action of specific microorganisms. As a result of this process, a gas with high energy content called biogas (a mixture mainly composed of methane (CH_4) and carbon dioxide (CO_2), together with other trace compounds) and an organic waste sludge with high moisture content ("digestate") are obtained. The anaerobic process is a process that is found spontaneously in nature, such as in swamps, underground deposits, and in the stomachs of ruminants. Although well known in the past two centuries, it was not until the end of the twentieth century when, because of the international oil crisis and the increase in the cost of treatment systems of the growing food industry, this technology was considered a cost-effective and adequate treatment with industrial applications worldwide.

Although the fermentative processes in the anaerobic degradation of organic matter are widely studied, the industrial process admits a large number of operational configurations, working scales, digested waste, etc., which results in great complexity when choosing the most appropriate treatment system. In recent years, anaerobic digestion has gained significant relevance due to a worldwide increasing demand for energy. Besides, energy from anaerobic digestion has three characteristics that make it extraordinarily attractive:

- It is renewable.
- It is obtained from waste.
- It is locally available.

D. U. Santos Ballardo, S. Rossi (eds.), *Microalgae as Promising Source of Commercial Bioproducts*, Developments in Applied Phycology 14, https://doi.org/10.1007/978-3-031-86433-9_7

7.1.2 Fundamentals of Anaerobic Digestion

Anaerobic digestion relies on a complex biological process involving several steps, as presented in Fig. 7.1. Biologically, the anaerobic degradation pathway of organic matter is a multistep process of series and parallel reactions, which makes it challenging to define a unique limiting step, especially when using complex substrates such as organic waste (Gerardi 2003).

Apart from the biology issues, anaerobic digestion can be classified into several categories according to some specific criteria:

(a) Operation temperature: Mesophilic (within 35–40 °C) and thermophilic (around 55 °C) are the most typical ranges. Although, even if theoretically mesophilic anaerobic digestion presents the lowest biodegradation rates, it is usually more implemented than thermophilic conditions, mainly because the process is more stable and robust. Psychrophilic or ambient temperatures are very interesting, but the corresponding process is still in a preliminary stage of development (Wang and Zhou 2023).

(b) Operation mode: Anaerobic digestion at full scale is practically always performed under the continuous or semicontinuous regime, with reactors in the magnitude of thousand cubic meters of capacity (Lora Grando et al. 2017). Notwithstanding this, batch low-scale anaerobic tests (known as biochemical methane potential (BMP)) are very popular as the first assay to know the feasibility of a substrate being anaerobically digested (Luna-Avelar et al. 2021). However, these tests must be carefully interpreted, as some critical aspects of the process (inoculation need, mixing regime, or microbial communities' adaptation to a specific substrate) are missed.

(c) Stages: Following the biological steps of anaerobic digestion (Fig. 7.1), several attempts have separated these steps in a configuration similar to stirred tank reactors in series, typical of those from chemical reactors' theory. Each reactor intends to reproduce the most favorable conditions (pH, temperature, redox potential, or organic loading rate). Although most of these configurations have been mainly developed at the pilot scale (Mata-Alvarez et al. 2000), a configuration based on a hydrolysis reactor (not strictly anaerobic) followed by a one-pot system including acidogenesis and methanogenesis under low redox potential to reduce organic matter to biogas has gained popularity at full scale. This configuration is recommended for plants treating complex substrates such as biowaste, where hydrolysis can be the limiting biological step (Llabrés-Luengo and Mata-Alvarez 1988; Abad et al. 2019).

(d) Moisture content: It is typical to classify anaerobic digestion systems into "wet" and "dry" bioreactors (USEPA, 2023). These terms refer to the working moisture of the reactor or, preferably, its dry matter content. Thus, wet anaerobic digestion includes systems with dry matter content under 10–15%, while in dry mode, it is usually over 15%. This results in wet reactors, similar to stirred homogenous tanks, which are suitable for substrates such as wastewater sludge or liquid manures. In contrast, dry reactors include several configurations based on mechanical devices, which are especially adequate for biowaste (Angelonidi and Smith 2015). In

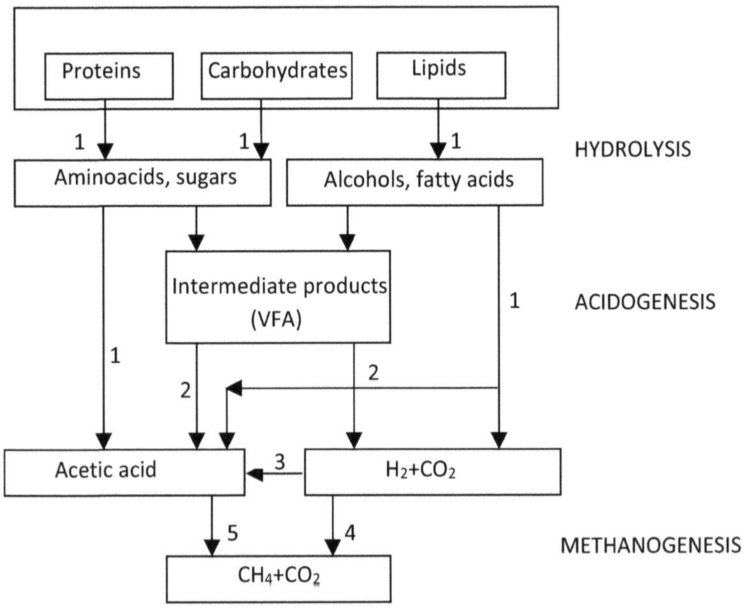

Fig. 7.1 Biological pathways of anaerobic digestion from a complex mixture of organic matter: (1) hydrolytic–acidogenic bacteria, (2) acetogenic bacteria, (3) homoacetogenic bacteria, (4) methanogenic hydrogen–consuming bacteria, and (5) methanogenic acetoclastic bacteria. (*VFA* volatile fatty acid)

summary, the selection of one of these two configurations is closely related to the moisture content of the organic waste to be treated.

Considering all these previous items and as a general rule, modern anaerobic digestion is performed in medium- to big-size one-pot reactors (except for hydrolysis) under mesophilic conditions and continuous or semicontinuous operation mode.

7.1.3 Biogas: Composition, Upgrading, and Uses

Table 7.1 shows the main composition of biogas. It can be observed that methane is practically always the main compound, carbon dioxide is also one at high percentages, and the rest of the compounds can be considered trace gases, with their content highly dependent on the chemical composition of the organic waste that is anaerobically digested (Calbry-Muzyka et al. 2022).

More specifically, biogas composition has important implications for using biogas as an alternative renewable energy source, a trend that is exponentially increasing in the current situation of lack of energy and irregular distribution of fossil fuels in the world (Lanni et al. 2023). Particularly, the implications are different according to the objective of substituting natural gas by the so-called biomethane from biogas:

7.1.3.1 Main Compounds

According to Table 7.1, the main biogas compounds are methane and carbon dioxide. However, the ratio of both gases strongly depends on the organic waste used as substrate and the addition of certain additives, such as trace metals or nanomaterials (Barrena et al. 2021). Biogas has been traditionally burned in cogeneration units for the combined production of electricity and thermal energy in the form of water vapor that is typically used to maintain the temperature of the anaerobic digester (Abad et al. 2019). Although economically attractive, this option is being overcome by upgrading biogas, which can be defined as separating methane from carbon dioxide and other gases from biogas (especially those with undesirable properties, as explained later). The resulting biogas, close to 100% methane, is called biomethane or renewable natural gas, which can be used as a vehicle fuel or injected into the gas distribution network. In this case, which has recently gained extraordinary interest, the key point is the separation of carbon dioxide, which is the secondary compound present at the highest percentage. Several approaches have been presented in the literature, with some of them available at full scale and in operation in waste or wastewater treatment plants (WWTP) or landfills:

- Membrane separation: Developed with several materials and other technologies such as membrane bioreactors, membrane electrolysis, or gas–liquid absorption (Gkotsis et al. 2023).
- Biomethanation: Where biogas is supplemented with hydrogen gas and biologically combined with carbon dioxide with methanogenic microorganisms to generate more methane, usually in trickle-bed reactors (Thapa et al. 2023).
- Heterogeneous catalysis: Using solid catalysts and packed-bed plug-flow reactors. Recently, advanced materials such as biochar or metal-organic frameworks in combination with nanomaterials have presented high yields and selectivity compared to traditional catalysts such as zeolite or activated carbon (Su et al. 2016). These catalysts have also shown high performance when obtaining other interesting products, such as methanol, an easily manageable form of energy, and a chemical platform (García et al. 2022).

7.1.3.2 Undesirable Impurities

Hydrogen Sulfide (H_2S) This specific compound has been the object of numerous studies as it is a highly corrosive product for most of the equipment used for biogas or biomethane recovery in the form of energy. The resulting products of its combustion (mainly SO_2) are also corrosive. For

Table 7.1 Typical properties and average values of the composition of biogas from several organic wastes

Characteristics	CH_4	CO_2	H_2	H_2S	Biogas properties[a]
Fraction (% v/v)	55–70	27–44	1	3	100
Calorific value (MJ/m^3)	35.8	–	10.8	22.8	21.5
Ignition range (% volume in air)	5–15	–	4–80	4–45	6–12
Ignition temperature (°C)	650–750	–	585	–	650–750
Critical pressure (MPa)	4.7	7.5	1.2	8.9	7.5–8.9
Relative density to air	0.55	2.5	0.07	1.2	0.83

[a]Assuming a biogas composition of 60% CH_4 and 40% CO_2

this reason, different approaches have been reported for its removal. As hydrogen sulfide is inherently produced by sulfate-reducing bacteria during the anaerobic digestion of any organic waste (Moestedt et al. 2013), it can be removed using two completely different strategies:

1. From the mixed liquor in the anaerobic digester, where the most typical option is using iron salts to precipitate sulfur inside the reactor (Moestedt et al. 2013; Maharaj et al. 2018).
2. From the biogas stream, where several options can be found, such as heterogeneous catalysis to transform hydrogen sulfide into elemental sulfur (Tarek et al. 2024), adsorption with several materials (Choudhury and Lansing 2021), and biological reduction to elemental sulfur or even sulfate (Montebello et al. 2014).

Siloxanes This group refers to two main types of siloxanes: poly-dimethyl siloxanes and methyl-siloxanes, which are volatile and can be easily volatilized into biogas. Although the concentration of siloxane in biogas does not exceed 0.4–0.5 mg/L (Dewil et al. 2006), a large number of attempts have been proposed and implemented for its removal due to its adverse effects on the equipment, which imposes high costs of maintenance and frequent equipment replacement (Ruiling et al. 2017). Among these technologies, adsorption on solid materials is the most used strategy to remove siloxanes from biogas streams. Typical adsorbents such as activated carbon or silica gel or, more recently, advanced materials such as graphene oxide, nanomaterials, and metal-organic frameworks have been tested (Vali et al. 2023).

7.1.4 The Role of Anaerobic Digestion in Circular Bioeconomy: Biorefineries

In recent years, microalgae-based treatments have been represented as complements in waste and, more predominantly, in wastewater treatment plants (WWTPs). Microalgae can grow in aquatic environments and use light (photosynthesis) and carbon dioxide to create biomass. Microalgae contain lipids and fatty acids as membrane components, storage products, metabolites, and energy sources. The potential of microalgae as alternative biofuel sources has been widely reviewed (Gaurav et al. 2024). Besides, their role in the final stages of cleaning wastewater by accumulating nutrients such as nitrogen and phosphorus is also proposed, as explained in this book.

Considering this fact, the first link between microalgae and anaerobic digestion consists of obtaining biogas from excess microalgae or exhaust microalgae once the produced biofuels have been extracted. Both approaches have been widely studied, as explained later in this chapter (Santos-Ballardo et al. 2015).

In addition to this straightforward relationship, waste and WWTPs are evolving to more complex facilities denominated biorefineries in which, apart from treating a polluted stream, another complementary goal is to obtain bioproducts and bioenergy. Therefore, biorefineries are industrial facilities that convert biomass into energy, chemicals, and materials. In this context, anaerobic digestion stands out as the primary process that provides energy to the other processes occurring in a biorefinery, playing an important role in the environmental and economic sustainability of these facilities and thus making them feasible (Font and Sánchez 2021).

Regarding microalgae, their presence in biorefineries has also been highlighted (Choi et al. 2021). Some examples include their use as producers of very specific biomaterials (Morais et al. 2022; Vasistha et al. 2023; Karan et al. 2023), although most of the published works are related to lipid or biofuel production (Li et al. 2015). Other studies focus on the uptake of atmospheric carbon dioxide to mitigate its greenhouse effect (Fal et al. 2023). In most cases, anaerobic co-digestion of excess microalgae-derived sludge with other co-substrates is a typical way to manage this residual matter to obtain a surplus of energy (Vassalle et al. 2022). A high-complexity scheme of a potential biorefinery, including anaerobic digestion and microalgae treatment, is presented in Fig. 7.2, which tries to squeeze the potential of both processes. The reader can figure out many alternative configurations.

Before entering the specific points to be considered when implementing anaerobic co-digestion processes of microalgae-derived materials, such as pretreatments, process performance, or techno-economic analysis, among others, it is important to highlight a specific problem that microalgae present regarding their separation and concentration from aqueous streams. This is related to microalgae's low sedimentability and dewaterability, which hamper their separation and further use. In this sense, advanced strategies have been proposed in the literature beyond gravity separation, although most have been presented at a laboratory scale. This is the case of using magnetic iron nanoparticles (Wang et al. 2015; Abo Markeb et al. 2019).

7.2 Biogas Potential Production from Microalgae

AD is considered a promising process for recovering energy from microalgal biomass, which presents several advantages over other biofuel feedstocks (Chia et al. 2018). Microalgal biomass composition is strain-specific, but generally, its macromolecular content varies between 6.7–65% of carbohydrates, 15–84% of proteins, and 1–63% of lipids

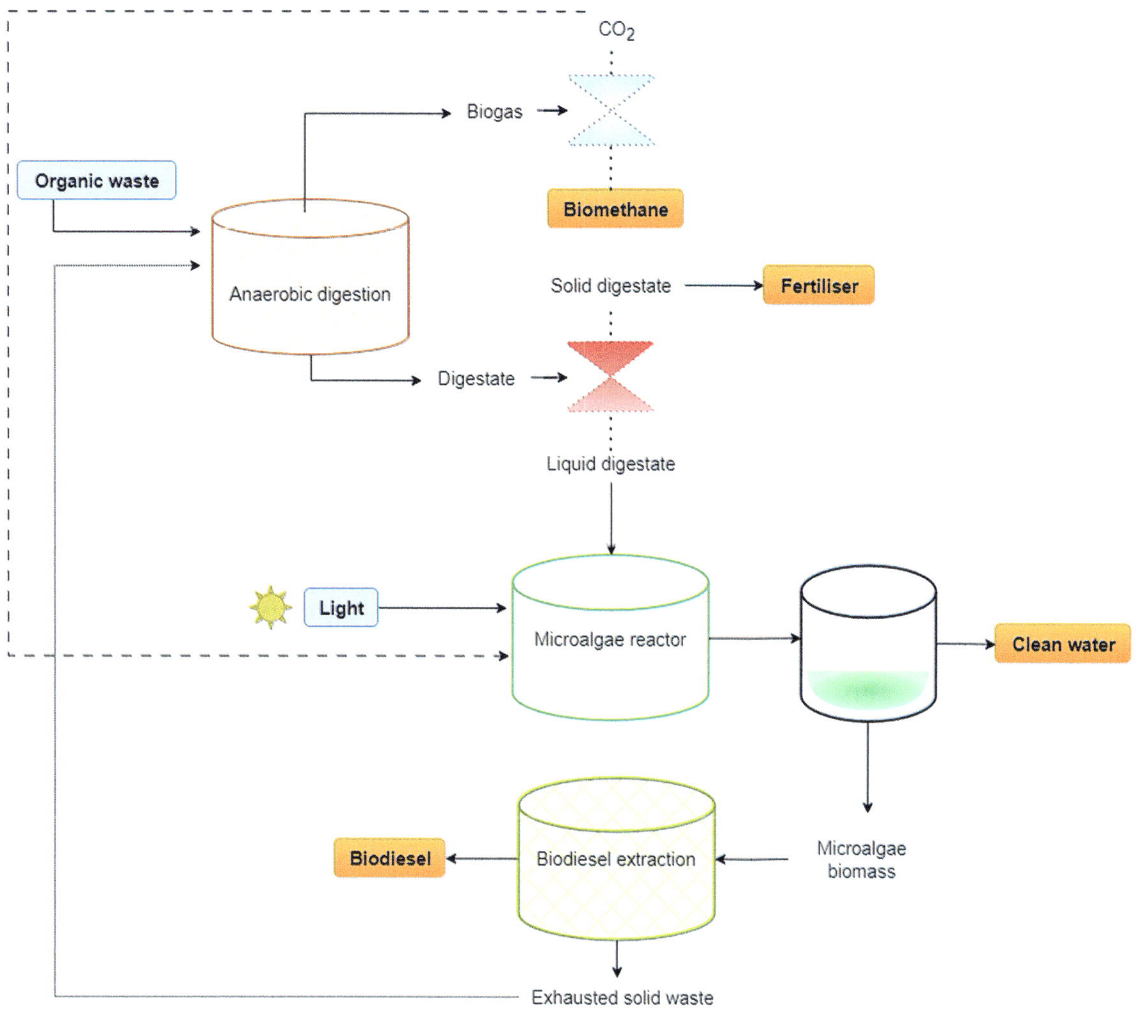

Fig. 7.2 Possible integration of anaerobic digestion and microalgae in a complex biorefinery. Blue represents biorefinery inputs, orange represents biorefinery outputs, and black represents intermediate products

based on volatile solid (VS) basis (Xia et al. 2015). When microalgal biomass is used for AD, all macromolecules are employed for biogas production, and carbohydrates and lipid extraction are not required (Avila et al. 2021).

In other cases, anaerobic digestion could be applied to residual algal biomass, for example, after lipid extraction to produce other biofuels. Moreover, microalgae biomass does not contain lignin, making it a more easily hydrolyzable substrate in comparison with crops and lignocellulosic biomasses (Singh et al. 2015). Methane yield from algal biomass depends on several factors, such as the microalgae species, pretreatment of the biomass, process operational parameters (pH, temperature, or nutrient availability, among others), and the presence or absence of inhibitors, among other factors that can be critical (Mussgnug et al. 2010). This variability can make it difficult to compare different microalgae species, which are also influenced by other factors like the pretreatments applied, lipid extraction methods, and co-digestion with other residues.

Biochemical methane potential (BMP) tests are commonly applied to analyze various simultaneous effects on anaerobic digestion, aiming to determine the ultimate methane yield under specific conditions. However, it is important to consider that the impact of pretreatments in continuous full-scale biogas plants may vary depending on factors such as hydraulic retention time (HRT) and the mixing regime.

Generally, the microalgae cell wall is a tough and resistant structure with an inner and outer layer (Yamada and Sakaguchi 1982). For instance, green algae contain a well-defined membrane and a complex layered cell wall structure (or the absence of it, depending on the microalgae species) typically formed by polymers such as cellulose, hemicellulose, pectin, and starch, while in cyanobacteria, glucan is the main polysaccharide (Hallenbeck and Benemann 2002; Nagarajan et al. 2020). Microalgae cell walls act as a protective layer that regulates environmental interactions, providing rigidity, mechanical strength, and chemical resistance to the microalgae (Lari et al. 2019). However, the microalgae cell wall is a rigid structure resistant to hydrolysis, representing a boundary for microalgae organic matter release (Soto-Sierra et al., 2018). Moreover, some microalgae cell walls contain two biopolymers (namely sporopollenin and algaenan), which are recalcitrant macromolecules that resist acetolysis and hamper the release of organic matter from microalgae through bacterial degradation and chemical hydrolysis (de Leeuw et al. 2005; González-Fernández et al. 2012b; Ometto et al. 2014). Some algae species, such as *Chlamydomonas reinhardtii*, *Dunaliella salina*, and *Asthrospira platensis*, showed a higher anaerobic biodegradability and higher biogas production in comparison with *Chlorella kessleri* and *Scenedesmus* sp. (Mussgnug et al. 2010). When the microalgal cell wall is not affected by the disruption process and intracellular content is not released, enzymes from the anaerobic consortium will have less access to algal biomass, thus limiting biogas production (Sialve et al. 2009).

7.2.1 Algal Biomass Pretreatment Technologies

Considering the recalcitrancy of microalgae biomass due to its cell wall structure and composition, microalgae biomass pretreatment is generally a crucial step to disintegrate the cell wall structure (Avila et al. 2021). Those pretreatments pursue cell membrane disintegration, the release of intracellular macromolecules followed by their solubilization, making them bioavailable for microorganisms' degradation and further conversion into biofuels and bioproducts (González-Fernández et al. 2012a; Mendez et al. 2013). Methods for microalgae pretreatment are categorized according to the effect on the biomass and the nature of the disruption force such as mechanical, thermal, chemical, and biological pretreatments.

Mechanical pretreatments of microalgae biomass involve physical methods targeted at either size reduction through milling or high-pressure homogenization or the disruption of cell walls by inducing physical damage through techniques like ultrasound and microwave treatments. Ultrasound pretreatment is the most common mechanical pretreatment applied to algal biomass (de Oliveira et al. 2022); the cavitation generated by high-frequency sonic waves breaks the cell wall from the inside out. In contrast, microwaves can disrupt cells through athermal and thermal effects. Athermically, microwaves polarize and realign macromolecules along the electromagnetic field, breaking down the hydrogen bonds, altering their structures and causing damage to the cell wall. However, thermically, heat is generated inside the cells by absorbing microwave energy through cellular mass or surrounding media, damaging the cells (Rodriguez et al. 2015; Jankowska et al. 2017).

Generally, higher specific power or extended exposure time favors the mechanical pretreatments, resulting in an increased biomass solubilization and methane yield compared to untreated biomass, as presented in Table 7.2, an example using *Chlorella* sp. microalgae. For certain species, pretreatment primarily facilitates the rupture of the cell wall, enhancing the microorganisms' access to intracellular material. Despite seemingly low solubilization degrees, the released intracellular content significantly impacts the methane yields of treated versus untreated biomass (Scholz et al. 2014). However, physical pretreatments are inefficient from the energy perspective since the energy required by mechanical pretreatments is often higher than the energy generated from anaerobic digestion (Passos et al. 2014; Córdova et al. 2018a).

On the other hand, cell walls are solubilized by *thermal pretreatments*, which involves subjecting the biomass to temperatures between 50 and 240 °C (Agarwalla et al. 2023). Thermal hydrolysis is the term for thermal pretreatment carried out at temperatures below 100 °C, whereas hydrothermal pretreatment occurs at temperatures between 100 and 180 °C with a slow pressure decrease (<2 MPa). Some studies have shown that thermal hydrolysis pretreatment can increase biomethane production by up to 18% (Córdova et al. 2018a). Other works compare thermal and hydrothermal treatments, obtaining better results for the hydrothermal

Table 7.2 Effect of pretreatment in methane yield on some *Chlorella* sp. microalgae and mixed microalgae consortium

Microalgae species	Methane yield (L CH_4 kg^{-1} VS)	Pretreatment	Increase (%)	Reference
Chlorella sorokiniana	318 ± 1	Physical (mechanical)	39	Córdova et al. (2018a)
		Physical (thermal)	6–18	
Chlorella vulgaris	120 ± 15	Biological	53–162	Ferreira et al. (2022)
Chlorella pyrenoidosa	222	Physical (thermal)	57	Xiao et al. (2019)
Chlorella pyrenoidosa	218	Chemical	7–133	Fu et al. (2023)
Mixed microalgae consortium	260 ± 8	Chemical	18–66	Cardeña et al. (2016)
Mixed microalgae consortium	300	Biological	100	Aydin et al. (2017)

treatments (Wang et al. 2017). However, the high energy required to heat the biomass to the desired pretreatment temperature is a matter of concern. Solar-driven hydrothermal pretreatment methods have been proposed to reduce energy expenditure. Thus, Xiao et al. (2019) used parabolic trough collectors to concentrate solar radiations for hydrothermal pretreatment of microalgae biomass, achieving 57% higher biomethane production. The solar-driven hydrothermal system achieved the highest exergy efficiency of 41%, compared to 36 and 26% for hydrothermal and control systems. Despite these promising results, solar-driven hydrothermal systems are still in development and require further engineering innovations to reduce capital costs and better integrate into existing bioenergy infrastructures.

Furthermore, chemical pretreatments include using compounds such as acids, bases, and solvents, among others, to produce chemical reactions on biomass structure, resulting in the solubilization of the cell wall components and intracellular organic matter (lipids, proteins, and carbohydrates). Acidic pretreatment converts cellulosic and hemicellulosic matrices in the cell wall into simple sugars by hydrolyzing them with sulfuric, hydrochloric, and free nitrous acids (Agarwalla et al. 2023). Furfural and 5-hydroxymethylfurfural can be produced by pretreatment involving concentrated acid and/or high heat reacting with solubilized sugars (Amin et al. 2017; Agarwalla et al. 2023). Besides, furfural or 5-hydroxymethylfurfural at 2 g/L concentration inhibits microbial growth and methanogenic activity during anaerobic digestion (mesophilic and thermophilic), according to Ghasimi et al. (2016).

Unlike acidic pretreatments, alkaline pretreatments permeabilize cell walls by saponifying uronic acids and esters in the wall using alkaline compounds like sodium, potassium, or calcium hydroxide. This causes swelling and increases the specific surface areas available for protein solubilization and microbial degradation (de Oliveira et al. 2022).

Moreover, divalent alkalis only partially ionize in aqueous conditions; hence, using monovalent alkalis instead of divalent ones improves biomass solubilization (Penaud et al. 1999). Mild temperature and pressure conditions allow for oxidative pretreatments without the need for concentrated acids or bases. As a result, oxidative pretreatment is more cost-effective and environmentally friendly than conventional chemical procedures (Kendir and Ugurlu 2018). However, research in this field is scarce, and the results of assessing the oxidative pretreatment cost-effectiveness based on chemical and energy input are still preliminary.

Furthermore, chemical concentration, biomass concentration, temperature, and time are the most important parameters influencing chemical pretreatment, which has been the subject of many research works (M'Arimi et al. 2020; Choudhary et al. 2022). On one hand, the main advantage of chemical pretreatments is the high degree of solubilization obtained with low energy use. On the other hand, the main drawbacks are the cost of the chemicals, the pH imbalance, the corrosion of the reactors, and the inhibition of the methanogenic bacteria as a result of the accumulation of toxic substances produced during pretreatment (Rodriguez et al. 2015; Xiao et al. 2020).

Biological pretreatments include using natural microorganisms (fungal strains, bacterial strains, and microbial consortia) or enzymes for biomass hydrolysis to simpler molecules under controlled temperature and pH conditions (Zabed et al. 2019). It has been demonstrated that using specific enzymes for algal cell wall disruption and biomass solubilization increases the methane yield (Ehimen et al. 2009; Avila et al. 2021). Hydrolytic enzymes, such as cellulase, hemicellulase, protease, carbohydrase, and amylase, are the most used for this purpose. Biological pretreatments are gaining interest over other methods since they require minimal energy input, are simple and safe, present high selectivity and low downstream processing costs, and imply the absence or reduced formation of toxic compounds (Zabed et al. 2019). On the negative side, biological pretreatment is time-consuming, the enzymes are substrate-specific, and the production cost is still considerable (Bohutskyi and Bouwer 2013), hindering its large-scale implementation for improving biomethane production. Selecting an enzyme requires considering the microalgae's chemical composition and cell wall, requiring research on the best enzyme or combination of enzymes for a particular microalgae species (Bohutskyi and Bouwer 2013; Avila et al. 2021).

Notwithstanding this, enzymes can be generated using inexpensive substrates, such as organic waste, to lower their high cost (Sakhuja et al. 2021), or they can be immobilized to enable recovery and reuse in several pretreatment cycles (Bolivar et al. 2022). Nonetheless, the production of enzymes using inexpensive substrates and their immobilization for later application continue to be highly researched topics. A further method to reduce the expense of biological enzymatic pretreatments is the in situ production of crude enzymes by extracellular enzyme-secreting bacteria or fungi to increase methane output more effectively than commercial enzymes. This is due to a synergistic effect between multiple enzymes secreted by the bacteria or fungi and enzymes secreted by the hydrolytic bacteria that also participate in anaerobic digestion (Hom-Diaz et al. 2016; Córdova et al. 2019). The in situ production of crude enzymes is cheaper and more efficient. However, selecting the proper microorganism, growing the culture, and obtaining the optimal inoculum/algae ratio are required to obtain the desired results, which is a time-consuming process.

Many pretreatment techniques are usually combined with other methods to enhance the advantages of a single method and improve cell wall disruption and hydrolysis of the organic matter released (Yang and Wang 2019). Generally, chemical and mechanical pretreatments can be simultaneously applied; on the other hand, physical pretreatments are applied, followed by biological or chemical pretreatments (Atelge et al. 2020).

Considering the microalgae differences from strain to strain, the best pretreatment method or combination of pretreatments depends on microalgal-specific features and pretreatment conditions. To improve microalgae's anaerobic digestibility, pretreatments chosen should preserve its organic matter content and avoid generating inhibitory products that could imbalance the anaerobic digestion process (Córdova et al. 2018b). Furthermore, the pretreatment efficiency is determined by the improvement in the energy or product yield and the cost-effectiveness of the pretreatment.

7.2.2 Methane Production from Residual Microalgae Biomass

The valorization of residual biomass after the extraction of valuable compounds from microalgae gains strength in a circular economy context. The anaerobic digestion of residual biomass after lipid extraction from microalgae is highlighted as a key step for the economic and energy balance in several techno-economic studies on the microalgae biodiesel process (see Sect. 7.3).

The extraction of some compounds, like the lipid fraction, is expected to reduce the potential for methane production. However, lipid extraction from microalgae can be considered a pretreatment Fe, sometimes improving methane production (Neves et al. 2016). For instance, methane production of residual biomass of *Chlorella* sp. substantially increased after the extraction of valuable compounds from the cell wall disruption and organic matter solubilization (Markou et al. 2022). Different techniques can be used, including thermal, chemical, or physical processes affecting different microalgae configuration aspects. While the effect of lipid extraction is the most reported, residual microalgae biomass after the extraction of other valuable compounds, such as amino acids, is also reported. Ramos-Suárez and Carreras (2014) observed a positive hydrology effect on biomass after amino acid rather than lipid extraction, consequently leading to higher methane production. Table 7.3 compares the methane production for both raw and residual microalgal biomasses from different authors. As observed, the extraction could positively or negatively affect biomethane production. Nevertheless, choosing an appropriate extraction method might be important, mainly because some solvents could inhibit methane production (Ehimen et al. 2009).

7.2.3 Microalgal Biomass Co-digestion

The co-digestion of both raw and residual microalgal biomass has been extensively studied to enhance their methane yields. As previously mentioned, microalgae can have a high protein content (ranging from 15% to 84%), resulting in a low carbon-to-nitrogen (C/N) ratio. Typically, microalgae present a C/N ratio below 10, which can potentially hamper the anaerobic digestion process due to ammonia toxicity. An imbalance between carbon and nitrogen leads to a nitrogen release, inhibiting methanogens and causing volatile fatty acid (VFA) accumulation, resulting in anaerobic digestion failure. This issue becomes particularly critical in long-term digestion processes. To overcome these problems, co-digestion has been used to balance the carbon-to-nitrogen ratio, typically recommended in the range of 20–30.

Various carbon-rich substrates have been employed in co-digestion studies, particularly in batch tests, to determine the optimal proportions for further investigation. Table 7.4 provides typical substrates used in co-digestion experiments

Table 7.3 Methane production comparison of whole microalgae and residual biomass

Microalgae species	Methane yield: whole microalgae (L CH_4 kg^{-1} VS)	Extracted product	Effect on methane yield: decrease or increase (%)	Reference
Chlorella sp.	237 ± 7	Lipids	−7.6	Markou et al. (2022)
		Chlorophyll	−11.4	
		Proteins	−12.7	
Chlorella sp.	337	Lipids	−6.8	Zhao et al. (2014)
Nannochloropsis sp.	557 ± 5	Lipids	28.4	
Nanofrustulum sp.	507 ± 5	Lipids	40.0	
Phaeodactylum tricornutum	337 ± 15	Lipids	−0.3	
Scenedesmus sp.	140 ± 29	Lipids	−51.3	Ramos-Suárez and Carreras (2014)
		Amino acids	−94.4	
Tetraselmis sp.	160	Lipids	47.5	Hernández et al. (2014)
Cyanobacterial microalgal consortium	239 ± 32	Chlorophyll	97.5	Águila Carricondo et al. (2023)

Table 7.4 Some examples of microalgae co-digestion in batch reactors using different microalgae: co-substrate ratios

Microalgae species	Co-substrate	Ratio microalgae co-substrate (SV)	Best ratio	Reference
Dunaliella salina	Olive mill solid waste	3:1		Fernández-Rodríguez et al. (2014)
		1:1		
		0.3:1	✓	
Dunaliella tertiolecta (after lipid extraction)	Mango residues (peel and seed)	0.1:1		Luna-Avelar et al. (2021)
		0.3:1	✓	
		0.4:1		
		0.7:1		
Chlorella vulgaris	Potato processing waste	3:1		Zhang et al. (2018)
		1:1		
		0.3:1	✓	
Mixed microalgae consortium	Sewage sludge	0.1:1		Olsson et al. (2014)
		0.3:1		
		0.6:1	✓	
Mixed microalgae consortium	Primary sludge	3:1		Solé-Bundó et al. (2018)
		1:1		
		0.3:1	✓	
Mixed microalgae consortium	Food waste	1:1		Ferreira et al. (2022)
		0.5:1	✓	

with microalgae biomass. As observed, the most effective options typically involve low microalgae proportions in the initial mixture. In this sense, balancing the mixtures by considering microalgae and other possible co-substrate availability is important. In some studies, the improvements are not significant; however, co digestion has demonstrated an improvement in the kinetics of the process (Ramos-Suárez and Carreras 2014).

Figure 7.3 summarizes different suitable co-substrates studied in different scenarios for anaerobic co-digestion using microalgae. These studies pointed out the use of glycerol (a by-product of biodiesel production) as an easy and economic co-substrate to increase methane production (Ehimen et al. 2009; Santos-Ballardo et al. 2015).

Different agricultural wastes have also been explored as co-substrates due to their high carbon content. Additionally, other carbon-rich substrates, such as paper or food waste, have also been tested. However, one of the most used co-substrates is sewage sludge. As mentioned above, microalgae can effectively grow in a nutrient-rich environment like wastewater. Microalgae-based wastewater treatment is, in fact, a promising technology for removing nutrients from wastewater while generating biomass that can be processed into fuels or value-added products (Fig. 7.2).

Simultaneous treatment of microalgal biomass (which potentially requires pretreatment, as explained in Sect. 7.2.1) with sewage sludge can positively impact the sustainability of WWTPs. For instance, anaerobic–aerobic treatment solutions, such as up-flow anaerobic sludge blanket (UASB) reactors followed by high-rate algal ponds (HRAPs), have already proven to be efficient for water treatment, as well as for energy recovery from the co-digestion of raw sewage and microalgal biomass (Vassalle et al. 2022). In this last example, the energy balance was positive, with the total energy produced being twice the energy consumed in the system.

7.3 Environmental Assessment and Techno-Economic Analysis

Anaerobic digestion of microalgae and its integration into biorefineries is crucial for closing loops in the current circular economy framework. In most cases, this integration includes the anaerobic digestion of microalgae or lipid-extracted microalgae. Assessing its environmental impact and conducting an economic analysis is essential to ensure proper integration. In this context, there are different aspects to consider when conducting environmental and techno-economic analyses of algae anaerobic digestion, which should be addressed. These factors encompass significant variability in the flowcharts of algae anaerobic digestion, making it challenging to provide a simple response to whether the anaerobic digestion of algae is environmentally and economically viable or to compare different studies.

Figure 7.4 represents a schematic algae production and anaerobic digestion process, presenting the variability in the

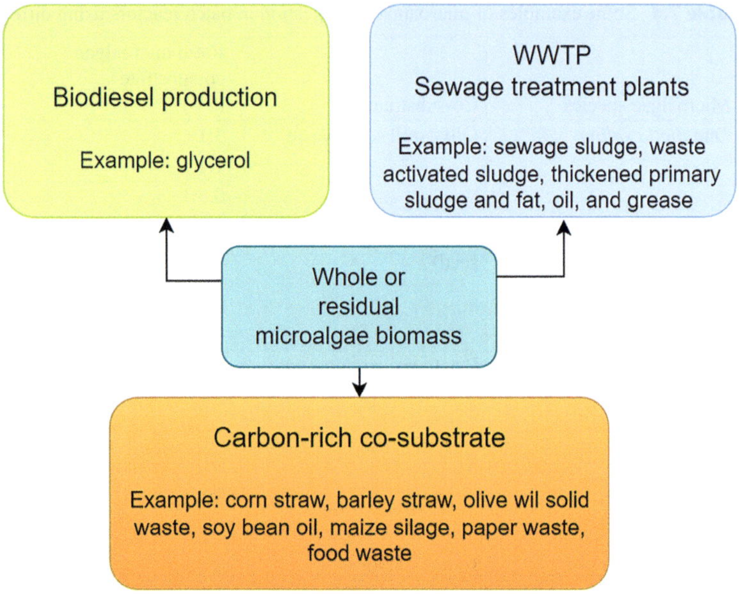

Fig. 7.3 Suitable co-substrates for anaerobic co-digestion of microalgae (raw and residual biomass) from different approaches

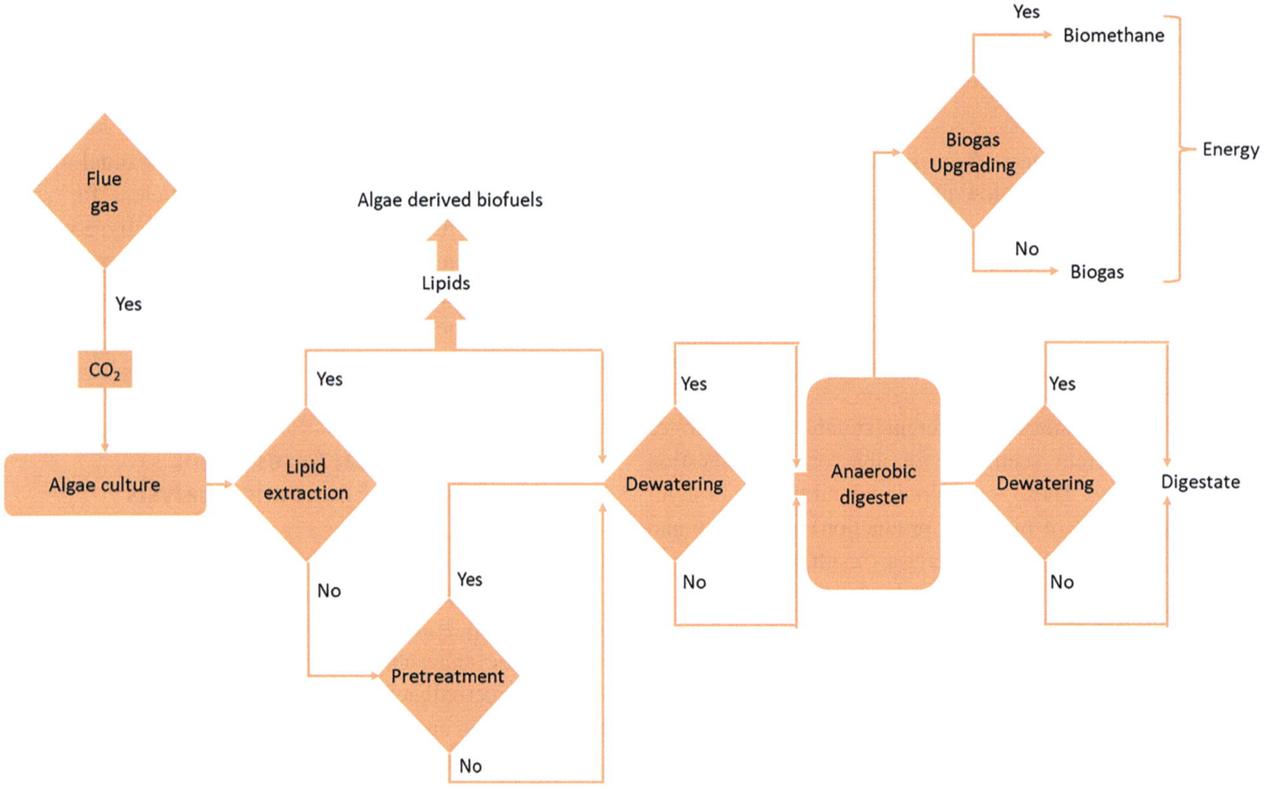

Fig. 7.4 Representation of an algae production and anaerobic digestion integrated process, considering some steps that can be conducted or not

process flow. Differences in operating and capital costs are shown in Table 7.5, presenting a high variability related to different flow-diagram processes.

The facility's geographical location is another parameter to consider when comparing the anaerobic digestion of microalgae systems. Ambient temperature will promote or hinder algae production and facilitate anaerobic digestion, reducing the costs of digester heating. Solar radiation is also critical, making those places near the equator more suitable (Wu et al. 2019). When producing microalgae, some studies propose obtaining carbon dioxide from flue gases (Xu et al. 2023). Besides, the plant location is an important issue,

Table 7.5 Economic performance of different processes combining algae production and anaerobic digestion

Algae	Pretreatment	Reactor	Algae productivity (g/m² per day)	Methane yield (m³ CH_4/kg VS)	Capex (MM $)	Opex (MM $)	Payback (years)	Other	Reference
Mixed	No	ORP	19–30	0.09–0.34	21–24	3.4–3.9	5.5–10	Settling+DAF high-rate AD at 37 °C	Zamalloa et al. (2011)
Chlorella sp.	No	ORP	25	0.174	6.7	0.324	22.6	Settling+centrifugation AD at 35 °C, HRT = 28 d	Xiao et al. (2020)
Chlorella sp.	Hydrothermal	ORP	25	0.239	6.7	0.325	16.6	Settling+centrifugation AD at 35 °C, HRT = 28 d	Xiao et al. (2020)
Chlorella sp.	Solar-driven hydrothermal	ORP	25	0.313	7.2	0.345	12.3	Settling+centrifugation AD at 35 °C, HRT = 28 d	Xiao et al. (2020)
Cyanothece BG0011	No	ORP	22	300 mL/g DM	478	96		AD at 37 °C, HRT = 25 d	Wu et al. (2019)
Cyanothece BG0011	No	ORP	22	300 mL/g DM	513	59		AD at 20 °C, HRT = 50 d	Wu et al. (2019)
Cyanothece BG0011	No	ORP	22	300 mL/g DM	419	59		AD in covered lagoon, HRT = 50 d	Wu et al. (2019)
Isochrysis sp.	Cell disruption	ORP		5.5 kW/m³	87	32		AD at 35 °C, HRT = 19 d	Bravo-Fritz et al. (2016)
Tetraselmis sp.	Cell disruption	ORP		5.5 kW/m³	117.5	34.3		AD at 35 °C, HRT = 19 d	Bravo-Fritz et al. (2016)

ORPs open raceway ponds, *DAF* dissolved air flotation

mainly because a large land extension is necessary to facilitate microalgae production and posterior anaerobic digestion. Even more, in most cases, wastewater or manure are used as substrates for algae production. In this case, a combination of the necessary free space, the presence of flue gas, and wastewater must converge in the same location.

The origin of the microalgae biomass is also important as it may influence its biogas potential. For example, in some cases, microalgae can specifically be produced to be anaerobically digested to obtain biogas. However, the microalgae could sometimes be produced to extract their lipids and produce other microalgae-derived biofuels. In the last case, the biomethane potential will be lower. For example, in the case of *Nannochloropsis salina* lipid-extracted biomass, it was reported that three times less methane was produced when compared to the raw microalgal biomass (Quinn et al. 2014), although these results could be improved (Table 7.3). Besides, when digesting the raw algae, depending on the type of algae cultivated, the biogas potential will be different (see Sect. 7.2), thus affecting the economics and environmental sustainability of the process.

Some authors (Collet et al. 2011) propose a combined process consisting of a lipid recovery process from a fraction of the algae biomass produced and an anaerobic digestion process (methane production) from both raw algae and the residue from algae after lipid extraction.

The microalgae cultivation step is the major cost in microalgae production, which is influenced by several factors, including microalgae productivity (Richardson et al. 2014; Bravo-Fritz et al. 2016). A wide range of production costs from 150 to 6000 $/t is reported in the literature (Wu et al. 2019), influenced by parameters such as algae productivity, type of bioreactor, and harvesting process, among others.

When the raw microalgae are fed to the anaerobic digestion process pretreatments (see Sect. 7.2.1), which could even include a dewatering step, it may be necessary to increase biodegradability by breaking the external cell wall. The capital cost, energy cost, and environmental impact of the pretreatment may influence the economic or environmental assessment of the entire process. Besides, microalgae biomass usually presents a low carbon-to-nitrogen ratio, which may hinder anaerobic digestion or cause ammonia toxicity issues. Co-digestion, to balance the carbon-to-nitrogen ratio, may be necessary. From a theoretical point of view, however, co-digestion complicates the comparison between studies.

On the other hand, the dewatering step can be necessary before or after the anaerobic digestion. Energy and reagent consumption during this process stage can significantly impact the results of environmental and techno-economic analyses. However, dewatering intensity is lower when raw algae are used as a source for anaerobic digestion than when algae are intended to produce biodiesel, as in this latter case, more intensity is needed (Wu et al. 2019).

Furthermore, anaerobic digestion itself may also influence the economics of the process. The type of anaerobic digester selected (e.g., high-rate anaerobic digesters that can operate at five to ten times higher volumetric loading rates compared to conventional anaerobic processes), process temperature (psychrophilic, mesophilic, or thermophilic), hydraulic residence time (HRT), mixing frequency, and the presence of additives, among others, are important parameters to be studied in detail. Thus, to process the same microalgae biomass, the digester volume will change depending on the election of these parameters, influencing capital costs. Wu et al. (2019) performed a techno-economic analysis comparing psychrophilic and mesophilic anaerobic digestion, reporting an increase in the fixed capital cost of the anaerobic digester from 67 million $ (mesophilic) to 102 million $ (psychrophilic) but a reduction in running costs from 21 to 2.3 million $, respectively.

In most studies, biogas is burned in combined heat and power (CHP) cogeneration systems, producing electricity and heat, generally in a proportion of 35–40% of electric energy generation and 40–45% of thermal energy. However, other more beneficial uses are possible, such as upgrading biogas to obtain biomethane to be used as a substitute for natural gas.

Moreover, biogas upgrading investment costs may range from 1000 to 2000 $/m^3 and maintenance costs from 15 to 59 $/year/m^3 (Ahmed et al. 2021), influencing the process economics. Besides, governments may subsidize feed-in tariffs to increase renewable energy production from locally available sources, which may impact process economics. Feed-in tariff is another parameter related to the geographical location of the plant that should be taken into account when developing economic analysis (Figueroa-Torres and Theodoropoulos 2023; López-Sánchez et al. 2023).

Finally, the use/fate of the digestate obtained from the anaerobic digestion process is important. This nutrient-rich flow should be used as a fertilizer or soil enhancer (Bravo-Fritz et al. 2016). Nowadays, the destination or further treatment of digestate is a topic of high interest, although out of the scope of this book, and classical strategies such as direct application to soil or composting coexist with new approaches to revalorize this product of increasing generation (Cerda et al. 2019).

7.4 Conclusions

Besides all the variables mentioned before, it should be noted that many studies rely on data (methane yield, algae yield, energy consumption, etc.) obtained from laboratory and pilot plant experiments, and there are few environmental and techno-economic impact studies conducted at full scale, which hampers the implementation of these technologies.

The current research in the anaerobic digestion of algae cannot be separated from the biorefinery concept. Thus, the most recent techno-economic and environmental impact studies encompass both anaerobic digestion and bioproduct production, analyzing the biorefinery's overall scope.

References

Abad V, Avila R, Vicent T, Font X (2019) Promoting circular economy in the surroundings of an organic fraction of municipal solid waste anaerobic digestion treatment plant: biogas production impact and economic factors. Bioresour Technol 283:10–17. https://doi.org/10.1016/J.BIORTECH.2019.03.064

Abo Markeb A, Llimós-Turet J, Ferrer I et al (2019) The use of magnetic iron oxide based nanoparticles to improve microalgae harvesting in real wastewater. Water Res 159:490–500. https://doi.org/10.1016/J.WATRES.2019.05.023

Agarwalla A, Komandur J, Mohanty K (2023) Current trends in the pretreatment of microalgal biomass for efficient and enhanced bioenergy production. Bioresour Technol 369:128330. https://doi.org/10.1016/J.BIORTECH.2022.128330

Águila-Carricondo P, de la Roche Cadavid JP, Galán PL et al (2023) New green biorefineries from cyanobacterial-microalgal consortia: production of chlorophyll-rich extracts for the cosmetic industry and sustainable biogas. J Clean Prod 429:139652. https://doi.org/10.1016/j.jclepro.2023.139652

Ahmed SF, Mofijur M, Tarannum K et al (2021) Biogas upgrading, economy and utilization: a review. Environ Chem Letter 19(6):4137–4164. https://doi.org/10.1007/S10311-021-01292-X

Amin FR, Khalid H, Zhang H et al (2017) Pretreatment methods of lignocellulosic biomass for anaerobic digestion. AMB Expr 7:72. https://doi.org/10.1186/s13568-017-0375-4

Angelonidi E, Smith SR (2015) A comparison of wet and dry anaerobic digestion processes for the treatment of municipal solid waste and food waste. Water Environ J 29:549–557. https://doi.org/10.1111/WEJ.12130

Atelge MR, Atabani AE, Banu JR et al (2020) A critical review of pretreatment technologies to enhance anaerobic digestion and energy recovery. Fuel 270:117494. https://doi.org/10.1016/J.FUEL.2020.117494

Avila CR, Vicent I Huguet T, Blánquez Cano P (2021) Microalgae-based systems for micropollutants removal, resource recovery and bioenergy production towards a circular bioeconomy approach. Universitat Autònoma de Barcelona. Programa de Doctorat en Ciència i Tecnologia Ambientals. https://ddd.uab.cat/record/257730

Aydin S, Yıldırım E, Ince O, Ince B (2017) Rumen anaerobic fungi create new opportunities for enhanced methane production from microalgae biomass. Algal Res 23:150–160. https://doi.org/10.1016/J.ALGAL.2016.12.016

Barrena R, del Carmen Vargas-García M, Capell G et al (2021) Sustained effect of zero-valent iron nanoparticles under semi-continuous anaerobic digestion of sewage sludge: evolution of nanoparticles and microbial community dynamics. Sci Total Environ 145969. https://doi.org/10.1016/j.scitotenv.2021.145969

Bohutskyi P, Bouwer E (2013) Biogas production from algae and cyanobacteria through anaerobic digestion: a review, analysis, and research needs. In: Lee JW (ed) Advanced biofuels and bioproducts. Springer, New York, pp 873–975

Bolivar JM, Woodley JM, Fernandez-Lafuente R (2022) Is enzyme immobilization a mature discipline? Some critical considerations to capitalize on the benefits of immobilization. Chem Soc Rev 51:6251–6290. https://doi.org/10.1039/D2CS00083K

Bravo-Fritz CP, Sáez-Navarrete CA, Herrera-Zeppelin LA, Varas-Concha F (2016) Multi-scenario energy-economic evaluation for a biorefinery based on microalgae biomass with application of anaerobic digestion. Algal Res 16:292–307. https://doi.org/10.1016/J.ALGAL.2016.03.028

Calbry-Muzyka A, Madi H, Rüsch-Pfund F et al (2022) Biogas composition from agricultural sources and organic fraction of municipal solid waste. Renew Energy 181:1000–1007. https://doi.org/10.1016/J.RENENE.2021.09.100

Cardeña R, Moreno G, Bakonyi P, Buitrón G (2016) Enhancement of methane production from various microalgae cultures via novel ozonation pretreatment. Chem Eng J. https://doi.org/10.1016/j.cej.2016.09.016

Cerda A, Mejias L, Rodríguez P et al (2019) Valorisation of digestate from biowaste through solid-state fermentation to obtain value added bioproducts: a first approach. Bioresour Technol 271:409–416. https://doi.org/10.1016/J.BIORTECH.2018.09.131

Chia SR, Ong HC, Chew KW et al (2018) Sustainable approaches for algae utilisation in bioenergy production. Renew Energy 129:838–852. https://doi.org/10.1016/J.RENENE.2017.04.001

Choi HI, Hwang SW, Patel AK, Sim SJ (2021) Microalgal biorefinery: a sustainable technology toward circular bioeconomy and microalgal biomass valorization. Biomass Biofuel Biochem 323–350. https://doi.org/10.1016/B978-0-12-821878-5.00023-4

Choudhary S, Tripathi S, Poluri KM (2022) Microalgal-based bioenergy: strategies, prospects, and sustainability. Energy Fuels 36:14584–14612. https://doi.org/10.1021/ACS.ENERGYFUELS.2C02922/ASSET/IMAGES/MEDIUM/EF2C02922_0006.GIF

Choudhury A, Lansing S (2021) Adsorption of hydrogen sulfide in biogas using a novel iron-impregnated biochar scrubbing system. J Environ Chem Eng 9:104837. https://doi.org/10.1016/J.JECE.2020.104837

Collet P, Hélias Arnaud A, Lardon L et al (2011) Life-cycle assessment of microalgae culture coupled to biogas production. Bioresour Technol 102:207–214. https://doi.org/10.1016/J.BIORTECH.2010.06.154

Córdova O, Passos F, Chamy R (2018a) Physical pretreatment methods for improving microalgae anaerobic biodegradability. Appl Biochem Biotechnol 185:114–126. https://doi.org/10.1007/S12010-017-2646-6/FIGURES/2

Córdova O, Santis J, Ruiz-Fillipi G et al (2018b) Microalgae digestive pretreatment for increasing biogas production. Renew Sust Energ Rev 82:2806–2813. https://doi.org/10.1016/J.RSER.2017.10.005

Córdova O, Passos F, Chamy R (2019) Enzymatic pretreatment of microalgae: cell wall disruption, biomass solubilisation and methane yield increase. Appl Biochem Biotechnol 189:787–797. https://doi.org/10.1007/S12010-019-03044-8/FIGURES/3

de Leeuw JW, Versteegh GJM, van Bergen PF (2005) Biomacromolecules of algae and plants and their fossil analogues. Plant Climate Change 209–233. https://doi.org/10.1007/978-1-4020-4443-4_14

de Oliveira MC, Bassin ID, Cammarota MC (2022) Microalgae and cyanobacteria biomass pretreatment methods: a comparative analysis of chemical and thermochemical pretreatment methods aimed at methane production. Fermentation 8:497. https://doi.org/10.3390/FERMENTATION8100497

Dewil R, Appels L, Baeyens J (2006) Energy use of biogas hampered by the presence of siloxanes. Energy Convers Manag 47:1711–1722. https://doi.org/10.1016/J.ENCONMAN.2005.10.016

Ehimen EA, Connaugton S, Sun Z, Carrington GC (2009) Energy recovery from lipid extracted, transesterified and glycerol codigested microalgae biomass. GCB Bioenergy 1:371–381. https://doi.org/10.1111/j.1757-1707.2009.01029.x

Fal S, Smouni A, Arroussi HE (2023) Integrated microalgae-based biorefinery for wastewater treatment, industrial CO_2 sequestration and microalgal biomass valorization: a circular bioeconomy approach. Environ Adv 12:100365. https://doi.org/10.1016/J.ENVADV.2023.100365

Fernández-Rodríguez MJ, Rincón B, Fermoso FG et al (2014) Assessment of two-phase olive mill solid waste and microalgae co-digestion to improve methane production and process kinetics. Bioresour Technol 157:263–269. https://doi.org/10.1016/J.BIORTECH.2014.01.096

Ferreira LO, Astals S, Passos F (2022) Anaerobic co-digestion of food waste and microalgae in an integrated treatment plant. J Chem Technol Biotechnol 97:1545–1554. https://doi.org/10.1002/JCTB.6900

Figueroa-Torres GM, Theodoropoulos C (2023) Techno-economic analysis of a microalgae-based biorefinery network for biofuels and value-added products. Bioresour Technol Rep 23:101524. https://doi.org/10.1016/J.BITEB.2023.101524

Font X, Sánchez A (2021) Significance of anaerobic digestion in circular bioeconomy. Biomass Biofuels Biochemical 269–289. https://doi.org/10.1016/B978-0-12-821878-5.00020-9

Fu J, Yan B, Gui S et al (2023) Anaerobic co-digestion of thermo-alkaline pretreated microalgae and sewage sludge: methane potential and microbial community. J Environ Sci 127:133–142. https://doi.org/10.1016/J.JES.2022.04.020

García AC, Moral-Vico J, Markeb AA, Sánchez A (2022) Conversion of carbon dioxide into methanol using Cu–Zn nanostructured materials as catalysts. Nanomaterials 12:999. https://doi.org/10.3390/NANO12060999

Gaurav K, Neeti K, Singh R (2024) Microalgae-based biodiesel production and its challenges and future opportunities: a review. Green Technol Sustain 2:100060. https://doi.org/10.1016/J.GRETS.2023.100060

Gerardi MH (2003) The microbiology of anaerobic digesters. Wiley

Ghasimi DS, Aboudi K, de Kreuk M, Zandvoort MH, van Lier JB (2016) Impact of lignocellulosic-waste intermediates on hydrolysis and methanogenesis under thermophilic and mesophilic conditions. Chem. Eng. J. 2016, 295, 181–191. https://doi.org/10.1016/j.cej.2016.03.045

Gkotsis P, Kougias P, Mitrakas M, Zouboulis A (2023) Biogas upgrading technologies—recent advances in membrane-based processes. Int J Hydrog Energy 48:3965–3993. https://doi.org/10.1016/J.IJHYDENE.2022.10.228

González-Fernández C, Sialve B, Bernet N, Steyer JP (2012a) Comparison of ultrasound and thermal pretreatment of Scenedesmus biomass on methane production. Bioresour Technol 110:610–616. https://doi.org/10.1016/J.BIORTECH.2012.01.043

González-Fernández C, Sialve B, Bernet N, Steyer JP (2012b) Impact of microalgae characteristics on their conversion to biofuel. Part II: focus on biomethane production. Biofuels Bioprod Biorefin 6:205–218. https://doi.org/10.1002/BBB.337

Hallenbeck PC, Benemann JR (2002) Biological hydrogen production; fundamentals and limiting processes. Int J Hydrog Energy 27:1185–1193. https://doi.org/10.1016/S0360-3199(02)00131-3

Hernández D, Solana M, Riaño B et al (2014) Biofuels from microalgae: lipid extraction and methane production from the residual biomass in a biorefinery approach. Bioresour Technol 170:370–378. https://doi.org/10.1016/J.BIORTECH.2014.07.109

Hom-Diaz A, Passos F, Ferrer I et al (2016) Enzymatic pretreatment of microalgae using fungal broth from Trametes versicolor and commercial laccase for improved biogas production. Algal Res 19:184–188. https://doi.org/10.1016/J.ALGAL.2016.08.006

Jankowska E, Sahu AK, Oleskowicz-Popiel P (2017) Biogas from microalgae: review on microalgae's cultivation, harvesting and pretreatment for anaerobic digestion. Renew Sust Energ Rev 75:692–709. https://doi.org/10.1016/j.rser.2016.11.045

Karan H, Roles J, Hankamer B, Ross IL (2023) Targeting greens and yellows: a solar biorefinery analysis for the microalgae-based co-production of pigments, proteins, and fuel. Algal Res 74:103187. https://doi.org/10.1016/J.ALGAL.2023.103187

Kendir E, Ugurlu A (2018) A comprehensive review on pretreatment of microalgae for biogas production. Int J Energy Res 42:3711–3731. https://doi.org/10.1002/ER.4100

Lanni D, Minutillo M, Cigolotti V, Perna A (2023) Biomethane production through the power to gas concept: a strategy for increasing the renewable sources exploitation and promoting the green energy transition. Energy Convers Manag 293:117538. https://doi.org/10.1016/J.ENCONMAN.2023.117538

Lari Z, Ahmadzadeh H, Hosseini M (2019) Cell wall disruption: a critical upstream process for biofuel production. Adv Feed Convers Technol Altern Fuels Bioprod:21–35. https://doi.org/10.1016/B978-0-12-817937-6.00002-3

Li J, Liu Y, Cheng JJ et al (2015) Biological potential of microalgae in China for biorefinery-based production of biofuels and high value compounds. New Biotechnol 32:588–596. https://doi.org/10.1016/J.NBT.2015.02.001

Llabrés-Luengo P, Mata-Alvarez J (1988) The hydrolytic step in a dry digestion system. Biol Wastes 23:25–37. https://doi.org/10.1016/0269-7483(88)90042-0

López-Sánchez A, Silva-Gálvez AL, González-López ME et al (2023) Valorization of livestock waste through combined anaerobic digestion and microalgae-based treatment in México: a techno-economic analysis for distributed biogas generation, animal feed production, and carbon credits trading. Environ Technol Innov 32:103321. https://doi.org/10.1016/J.ETI.2023.103321

Lora Grando R, de Souza Antune AM, da Fonseca FV et al (2017) Technology overview of biogas production in anaerobic digestion plants: a European evaluation of research and development. Renew Sust Energ Rev 80:44–53. https://doi.org/10.1016/j.rser.2017.05.079

Luna-Avelar KD, Barrena R, Font X et al (2021) A preliminary assessment of anaerobic co-digestion potential of mango and microalgal residue biomass using a design of experiments approach: effect of thermal, physical and biological pretreatments. Food Bioprod Process 128:143–152. https://doi.org/10.1016/J.FBP.2021.04.015

M'Arimi MM, Mecha CA, Kiprop AK, Ramkat R (2020) Recent trends in applications of advanced oxidation processes (AOPs) in bioenergy production: review. Renew Sust Energ Rev 121:109669. https://doi.org/10.1016/J.RSER.2019.109669

Maharaj BC, Mattei MR, Frunzo L et al (2018) ADM1 based mathematical model of trace element precipitation/dissolution in anaerobic digestion processes. Bioresour Technol 267:666–676. https://doi.org/10.1016/J.BIORTECH.2018.06.099

Markou G, Ilkiv B, Brulé M et al (2022) Methane production through anaerobic digestion of residual microalgal biomass after the extraction of valuable compounds. Biomass Convers Biorefinery 12:419–426. https://doi.org/10.1007/S13399-020-00703-3/FIGURES/4

Mata-Alvarez J, Macé S, Llabrés P (2000) Anaerobic digestion of organic solid wastes. An overview of research achievements and perspectives. Bioresour Technol 74:3–16. https://doi.org/10.1016/S0960-8524(00)00023-7

Mendez L, Mahdy A, Timmers RA et al (2013) Enhancing methane production of Chlorella vulgaris via thermochemical pretreatments. Bioresour Technol 149:136–141. https://doi.org/10.1016/J.BIORTECH.2013.08.136

Moestedt J, Nilsson Påledal S, Schnürer A (2013) The effect of substrate and operational parameters on the abundance of sulphate-reducing bacteria in industrial anaerobic biogas digesters. Bioresour Technol 132:327–332. https://doi.org/10.1016/J.BIORTECH.2013.01.043

Montebello AM, Mora M, López LR et al (2014) Aerobic desulfurization of biogas by acidic biotrickling filtration in a randomly packed reactor. J Hazard Mater 280:200–208. https://doi.org/10.1016/J.JHAZMAT.2014.07.075

Morais MG, Santos TD, Moraes L et al (2022) Exopolysaccharides from microalgae: production in a biorefinery framework and potential applications. Bioresour Technol Rep 18:2589–2603. https://doi.org/10.1016/j.biteb.2022.101006

Mussgnug JH, Klassen V, Schlüter A, Kruse O (2010) Microalgae as substrates for fermentative biogas production in a combined biorefinery concept. J Biotechnol 150:51–56. https://doi.org/10.1016/J.JBIOTEC.2010.07.030

Nagarajan D, Chang JS, Lee DJ (2020) Pretreatment of microalgal biomass for efficient biohydrogen production—recent insights and future perspectives. Bioresour Technol 302:122871. https://doi.org/10.1016/J.BIORTECH.2020.122871

Neves VT de C, Sales EA, Perelo LW (2016) Influence of lipid extraction methods as pre-treatment of microalgal biomass for biogas production. Renewable and Sustainable Energy Reviews 59:160–165. https://doi.org/10.1016/j.rser.2015.12.303

Olsson J, Feng XM, Ascue J et al (2014) Co-digestion of cultivated microalgae and sewage sludge from municipal waste water treatment. Bioresour Technol 171:203–210. https://doi.org/10.1016/j.biortech.2014.08.069

Ometto F, Quiroga G, Pšenička P et al (2014) Impacts of microalgae pre-treatments for improved anaerobic digestion: thermal treatment, thermal hydrolysis, ultrasound and enzymatic hydrolysis. Water Res 65:350–361. https://doi.org/10.1016/J.WATRES.2014.07.040

Passos F, Astals S, Ferrer I (2014) Anaerobic digestion of microalgal biomass after ultrasound pretreatment. Waste Manag 34:2098–2103. https://doi.org/10.1016/J.WASMAN.2014.06.004

Penaud V, Delgenès JP, Moletta R (1999) Thermo-chemical pretreatment of a microbial biomass: influence of sodium hydroxide addition on solubilization and anaerobic biodegradability. Enzym Microb Technol 25:258–263. https://doi.org/10.1016/S0141-0229(99)00037-X

Quinn JC, Hanif A, Sharvelle S, Bradley TH (2014) Microalgae to biofuels: life cycle impacts of methane production of anaerobically digested lipid extracted algae. Bioresour Technol 171:37–43. https://doi.org/10.1016/J.BIORTECH.2014.08.037

Ramos-Suárez JL, Carreras N (2014) Use of microalgae residues for biogas production. Chem Eng J 242:86–95. https://doi.org/10.1016/j.cej.2013.12.053

Richardson JW, Johnson MD, Zhang X et al (2014) A financial assessment of two alternative cultivation systems and their contributions to algae biofuel economic viability. Algal Res 4:96–104. https://doi.org/10.1016/J.ALGAL.2013.12.003

Rodriguez C, Alaswad A, Mooney J et al (2015) Pre-treatment techniques used for anaerobic digestion of algae. Fuel Process Technol 138:765–779. https://doi.org/10.1016/J.FUPROC.2015.06.027

Ruiling G, Shikun C, Zifu L (2017) Research progress of siloxane removal from biogas. Int J Agric Biol Eng 10:30–39. https://doi.org/10.3965/j.ijabe.20171001.3043

Sakhuja D, Ghai H, Rathour RK et al (2021) Cost-effective production of biocatalysts using inexpensive plant biomass: a review. 3 Biotech 11. https://doi.org/10.1007/S13205-021-02847-Z

Santos-Ballardo DUDUDU, Font-Segura X, Ferrer ASAS et al (2015) Valorisation of biodiesel production wastes: anaerobic digestion of residual Tetraselmis suecica biomass and co-digestion with glycerol. Waste Manag Res 33:250–257. https://doi.org/10.1177/0734242X15572182

Scholz MJ, Weiss TL, Jinkerson RE et al (2014) Ultrastructure and composition of the Nannochloropsis gaditana cell wall. Eukaryot Cell 13:1450–1464. https://doi.org/10.1128/EC.00183-14/SUPPL_FILE/ZEK999094360SO1.PDF

Sialve B, Bernet N, Bernard O (2009) Anaerobic digestion of microalgae as a necessary step to make microalgal biodiesel sustainable. Biotechnol Adv 27:409–416. https://doi.org/10.1016/J.BIOTECHADV.2009.03.001

Singh B, Bauddh K, Bux F (2015) Algae & environmental sustainability. Springer

Solé-Bundó M, Salvadó H, Passos F et al (2018) Strategies to optimize microalgae conversion to biogas: co-digestion, pretreatment and hydraulic retention time. Molecules 23(23):2096. https://doi.org/10.3390/MOLECULES23092096

Su X, Xu J, Liang B et al (2016) Catalytic carbon dioxide hydrogenation to methane: a review of recent studies. J Energy Chem 25:553–565. https://doi.org/10.1016/J.JECHEM.2016.03.009

Tarek M, Santos JS, Márquez V et al (2024) A critical review towards the causes of the iron-based catalysts deactivation mechanisms in the selective oxidation of hydrogen sulfide to elemental sulfur from biogas. J Energy Chem 90:388–411. https://doi.org/10.1016/J.JECHEM.2023.10.048

Thapa A, Jo H, Han U, Cho SK (2023) Ex-situ biomethanation for CO_2 valorization: state of the art, recent advances, challenges, and future prospective. Biotechnol Adv 68:108218. https://doi.org/10.1016/J.BIOTECHADV.2023.108218

Vali SA, Moral-Vico J, Font X, Sánchez A (2023) Adsorptive removal of siloxanes from biogas: recent advances in catalyst reusability and water content effect. Biomass Convers Biorefinery 1:3. https://doi.org/10.1007/s13399-023-04478-1

Vasistha S, Balakrishnan D, Manivannan A, Rai MP (2023) Microalgae on distillery wastewater treatment for improved biodiesel production and cellulose nanofiber synthesis: a sustainable biorefinery approach. Chemosphere 315:137666. https://doi.org/10.1016/J.CHEMOSPHERE.2022.137666

Vassalle L, Passos F, Rosa-Machado AT et al (2022) The use of solar pre-treatment as a strategy to improve the anaerobic biodegradability of microalgal biomass in co-digestion with sewage. Chemosphere 286:131929. https://doi.org/10.1016/j.chemosphere.2021.131929

Wang H, Zhou Q (2023) Dominant factors analyses and challenges of anaerobic digestion under cold environments. J Environ Manag 348:119378. https://doi.org/10.1016/J.JENVMAN.2023.119378

Wang SK, Stiles AR, Guo C, Liu CZ (2015) Harvesting microalgae by magnetic separation: a review. Algal Res 9:178–185. https://doi.org/10.1016/J.ALGAL.2015.03.005

Wang M, Lee E, Dilbeck MP et al (2017) Thermal pretreatment of microalgae for biomethane production: experimental studies, kinetics and energy analysis. J Chem Technol Biotechnol 92:399–407. https://doi.org/10.1002/JCTB.5018

Wu N, Moreira CM, Zhang Y et al (2019) Techno-economic analysis of biogas production from microalgae through anaerobic digestion. Anaerobic Digestion. https://doi.org/10.5772/INTECHOPEN.86090

Xia A, Cheng J, Song W et al (2015) Fermentative hydrogen production using algal biomass as feedstock. Renew Sust Energ Rev 51:209–230. https://doi.org/10.1016/J.RSER.2015.05.076

Xiao C, Liao Q, Fu Q et al (2019) A solar-driven continuous hydrothermal pretreatment system for biomethane production from microalgae biomass. Appl Energy 236:1011–1018. https://doi.org/10.1016/J.APENERGY.2018.12.014

Xiao C, Fu Q, Liao Q et al (2020) Life cycle and economic assessments of biogas production from microalgae biomass with hydrothermal pretreatment via anaerobic digestion. Renew Energy 151:70–78. https://doi.org/10.1016/J.RENENE.2019.10.145

Xu P, Li J, Qian J et al (2023) Recent advances in CO_2 fixation by microalgae and its potential contribution to carbon neutrality. Chemosphere 319:137987. https://doi.org/10.1016/J.CHEMOSPHERE.2023.137987

Yamada T, Sakaguchi K (1982) Comparative studies on chlorella cell walls: induction of protoplast formation. Arch Microbiol 132:10–13. https://doi.org/10.1007/BF00690809/METRICS

Yang G, Wang J (2019) Ultrasound combined with dilute acid pretreatment of grass for improvement of fermentative hydrogen production. Bioresour Technol 275:10–18. https://doi.org/10.1016/J.BIORTECH.2018.12.013

Zabed HM, Akter S, Yun J et al (2019) Recent advances in biological pretreatment of microalgae and lignocellulosic biomass for biofuel production. Renew Sust Energ Rev 105:105–128. https://doi.org/10.1016/J.RSER.2019.01.048

Zamalloa C, Vulsteke E, Albrecht J, Verstraete W (2011) The techno-economic potential of renewable energy through the anaerobic digestion of microalgae. Bioresour Technol 102:1149–1158. https://doi.org/10.1016/J.BIORTECH.2010.09.017

Zhang Y, Caldwell GS, Zealand AM, Sallis PJ (2018) Anaerobic co-digestion of microalgae Chlorella vulgaris and potato processing waste: effect of mixing ratio, waste type and substrate to inoculum ratio. Biochem Eng J. https://doi.org/10.1016/j.bej.2018.12.021

Zhao B, Ma J, Zhao Q, Laurens L, Jarvis E, Chen S, Frear C (2014) Efficient anaerobic digestion of whole microalgae and lipid-extracted microalgae residues for methane energy production. Bioresource Technology 161:423–430. https://doi.org/10.1016/j.biortech.2014.03.079

Microalgae Biomass as a Feedstock for Bioethanol

8

Andréia Anschau, Débora Bohrer de Lima, Evelyn Pereira Ribeiro Garcia, Miriam Domingues Guimarães, Sofia de Souza Oliveira, and Paula Fernandes Montanher

Abstract

Microalgae biomass is a promising feedstock with several benefits over conventional sources for manufacturing bioethanol. Given the depletion of fossil fuel stocks and environmental concerns, this third-generation biofuel is an appealing substitute because of its high productivity and low land consumption. Microalgae can produce much more biomass per unit area than terrestrial crops, making them attractive for biofuel production. However, challenges remain in optimizing production processes and economic viability. The high costs associated with harvesting and processing microalgae can limit commercial scalability, and effective fermentation processes and biomass hydrolysis are critical for maximizing ethanol output. Strategies such as synthetic biology, nanotechnology, omics technologies and genetic editing, bubble generation, photobioreactor designs, electro−/magnetic−/bio-flocculation, and artificial intelligence integration in microalgal production are being explored to overcome these constraints. These strategies have significant promise in improving the production of microalgae. While microalgae offer a sustainable alternative to bioethanol, ongoing research is essential to address economic and technical challenges, ensuring their viability as a mainstream biofuel source. So, microalgal bioethanol technologies have not advanced sufficiently, but expectations for the near future are strong.

Keywords

Microalgae · Fermentation feedstocks · Pretreatments Biofuels · Bioenergy

A. Anschau (✉) · D. B. de Lima · E. P. Ribeiro Garcia
M. D. Guimarães · S. de Souza Oliveira · P. F. Montanher
Department of Biotechnology and Bioprocess Engineering, Universidade Federal do Paraná – UTFPR, Paraná, Brazil
e-mail: andreiaanschau@utfpr.edu.br

8.1 Introduction

Microalgae biomass grows quickly, has a high carbohydrate content, lacks lignin and hemicellulose, and is also common in terrestrial biomass. So, it has become a promising feedstock for the generation of bioethanol. Due to its distinctive composition, the hydrolysis and fermentation processes may be carried out more easily, leading to an effective conversion to bioethanol. Besides, some microalgae like *Chlorella vulgaris* may accumulate up to 37% dry weight of starch, which makes them very useful for producing bioethanol (Dragone et al. 2011; Kumar et al. 2016). Furthermore, estimations of the amount of ethanol that microalgae can produce range from 46,760 to 140,290 L/ha, higher than the rates achieved with traditional feedstocks (Khan et al. 2016). Economic benefits can also be obtained from microalgae farming, particularly when wastewater is used as the growth medium.

This strategy lowers ethanol production costs but also considers wastewater management–related environmental concerns (Odjadjare et al. 2018). Utilizing nutrient-rich wastewater can promote microalgae growth, increasing biomass yields and, as a result, the possibility of increasing bioethanol yields (Khan et al. 2023). Furthermore, microalgae can store carbon dioxide during their growth phase, which confers an ecological advantage, reducing greenhouse gas emissions while generating biofuels (Ho et al. 2017; Klinthong et al. 2015).

Furthermore, microalgae must have optimal growing conditions to accumulate carbohydrates to the greatest extent possible. It has been demonstrated that variables like pH, light intensity, and nitrogen concentration significantly impact biomass productivity and carbohydrate concentrations (Bibi et al. 2021; Condor et al. 2022a, b). Microalgae, for instance, can boost their carbohydrate accumulation to about 50% of their biomass when nitrogen is scarce, mainly in glucose, which is perfect for fermentation (Ho et al. 2017). *Arthrospira platensis* presents high carbohydrate concentra-

D. U. Santos Ballardo, S. Rossi (eds.), *Microalgae as Promising Source of Commercial Bioproducts*, Developments in Applied Phycology 14, https://doi.org/10.1007/978-3-031-86433-9_8

tion, which makes it a good choice for producing bioethanol (Ma'mun et al. 2024).

On the other hand, the use of tertiary treated municipal wastewater can enhance biomass yield, as demonstrated with *Asterarcys quadricellulare*, which produced a high yield of 12.091 kg of dry biomass (Karthikeyan 2023). Moreover, the optimal conditions for hydrolysis include using 2.0% H_2SO_4 at 100 °C for 180 min, producing significant glucose concentrations (Karthikeyan 2023). Additionally, a biomass/acid ratio of 15% (w/v) is crucial for maximizing sugar concentration (Oliveira et al. 2023). Following hydrolysis, fermentation with *Saccharomyces cerevisiae* is essential, with studies showing bioethanol yields of up to 2.89% from *Spirulina* (Ma'mun et al. 2024).

Enhancing microalgae's economic viability as feedstocks for bioethanol requires these enhancements. Furthermore, a variety of biochemical conversion techniques are used to transform microalgal biomass into bioethanol, such as fermentation with different microbial strains and enzymatic hydrolysis (Parakh et al. 2023; Wang and Yin 2018).

Even though microalgae offers a viable path toward manufacturing bioethanol, difficulties still exist in refining the growing and processing techniques to maximize yields and sustainability.

8.2 Feedstocks Used for Ethanol Production

Bioethanol production is classified into generations, each differentiated by the raw material used and the technology employed in production, as shown in Fig. 8.1.

Food crops rich in sugars and starches, such as maize, wheat, and sugarcane, are used in the production of first-generation ethanol where three main stages are required: starch liquefaction, in which enzymes such as alpha (α)-amylase transform the starch into simpler sugars; saccharification, in which these sugars are broken down by enzymes such as glucoamylase to form glucose; and fermentation, in which yeasts such as *Saccharomyces cerevisiae* convert the glucose into bioethanol. Even though it is a widely used renewable fuel, its manufacture presents significant difficulties, as competition with food production can influence food security (Cavelius et al. 2023; Jain and Kumar 2024).

The second generation is produced from lignocellulosic waste and materials that are not edible, such as wheat straw, sugar cane bagasse, rice husks, wood, and other agricultural, forestry, and industrial waste; this generation of ethanol avoids competition with food production and reduces the waste of resources, as it uses materials that would otherwise be discarded. The physical and chemical pretreatment of biomass to break down lignin, thus removing a barrier to the accessibility of the sugars present in cellulose and hemicellulose, is an additional step in the manufacturing process. The biomass is hydrolyzed with enzymes after pretreatment, transforming the cellulose and hemicellulose polymers into simple sugars, which are then fermented into bioethanol. High processing costs mean that second-generation ethanol faces economic difficulties despite its environmental benefits (Cavelius et al. 2023; Jain and Kumar 2024; Guimarães et al. 2023; Igwebuike et al. 2024).

Derived from microalgae, macroalgae, and cyanobacteria, third-generation ethanol represents a significant development over the bioethanol of past generations. Algae are a good source of bioethanol as they are very effective in transforming carbon dioxide into biomass loaded with carbs. While microalgae, including *Chlorella* and *Spirulina*, may be grown to cumulate significant quantities of carbohydrates under specific situations of nutritional stress, macroalgae,

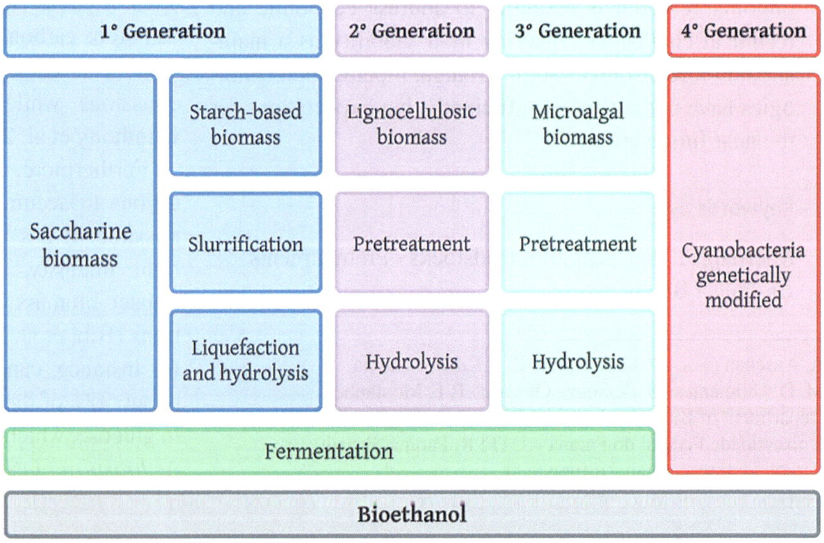

Fig. 8.1 Generations of ethanol production. (Adapted from Silva and Bertucco 2019)

such as brown and red algae, are rich in polysaccharides such as alginates, laminarin, and mannitol. Maximizing carbohydrate content and biomass production depends on the choice of the algae species and the growing environment (Müller et al. 2023). The algae are grown in bioreactors, then harvested and dehydrated, and the cellular components, such as lipids and carbohydrates, are extracted from the algae. These carbohydrates are then broken down to use in the fermentation process for bioethanol production.

Furthermore, both closed photobioreactors and open systems can be used to grow algae, and each has its advantages and disadvantages. The use of photobioreactors gives better control over cultivation conditions, which can optimize the process and produce a greater quantity of biomass. Open systems have a lower yield but are cheaper.

Moreover, one of the difficulties in growing algae is the extraction process due to the size of their cells, and flocculation, centrifugation, and filtration are some of the techniques used to harvest algae. First, the cell walls must be broken down to liberate the internal molecules, mostly carbohydrates, thus transforming algal biomass into bioethanol. Complex polysaccharides, proteins, and lipids make up the cell walls of microalgae, so cell rupture calls for either physical, chemical, or biological means. Acid or enzymatic hydrolysis may convert these polysaccharides into fermentable sugars, hence known as saccharification. However, effective hydrolysis of algae depends on best practices to reduce the development of byproducts that can stop later fermentation.

On the other hand, yeasts, including *Saccharomyces cerevisiae*, are widely employed for manufacturing first-generation bioethanol and may help ferment algae-derived sugars to generate bioethanol. Algae have a variety of sophisticated sugars, such as mannitol and alginate, and traditional yeasts cannot easily break them down. Genetically modified yeasts and new strains are being studied and created to resolve this difficulty and increase fermentation efficiency and bioethanol generation. There are also studies aimed at investigating whether it is feasible to replace fresh water with salt water to reduce the water footprint of the fermentation process (Müller et al. 2023). One of the main benefits of algae as a raw material is its fast growth rate and capacity to be grown in surroundings free of competition for terrestrial agriculture, including wastewater and marine waters, lowering fresh water usage and improving energy efficiency.

Furthermore, microalgae can be genetically altered to increase carbohydrate production, thus improving the efficiency of bioethanol generation. Still, significant obstacles, such as high prices and technical complexity related to algae cultivation and processing, stand in the way of large-scale acceptance of third-generation bioethanol. Making third-generation ethanol a reasonable and affordable substitute for sustainable biofuels in the future depends on overcoming these challenges through continuous research and development (Cavelius et al. 2023; Jain and Kumar 2024).

Besides, the bioethanol from the fourth generation is based on modern biotechnological techniques or genetically modified organisms to improve the efficiency of converting biomass into bioethanol, marking a significant advance in biofuel production. Unlike past generations, which used agricultural waste or raw food ingredients, this generation studies organisms designed to maximize sugar conversion and survive unfavorable conditions, supporting a more sustainable biofuel. To reduce the carbon footprint and increase efficiency, the technologies seek to incorporate carbon capture methods and better use of waste (Ashokkumar et al. 2024; Shokravi et al. 2022).

Research on bioethanol synthesis from many biomass sources has been extensively conducted, and notable output differences have been shown based on the employed material. Such lignocellulosic biomass as sugar cane straw and maize stover offer a chance to boost bioethanol output without appreciably increasing the area under cultivation. Usually regarded as agricultural waste, these biomasses may be turned into bioethanol by crop byproducts.

Microalgae stand out among the new options as promising third-generation bioethanol sources since they can be grown in aquatic environments, including brackish water and wastewater, without conflicting with food agriculture and have a very fast growth rate. They can also gather plenty of sugars, which fermentation techniques allow to be turned into bioethanol, and have great potential to surpass conventional crops in production (Kouhgardi et al. 2023; Müller et al. 2023).

Table 8.1 shows bioethanol productivity from various biomass sources, including traditional crops such as corn and sugar cane, as well as microalgae, which represent a more advanced and promising solution, especially in terms of productivity and sustainability.

8.3 Microalgae as Potential Feedstocks for Bioethanol Production

Microalgae are emerging as promising sources of biomass for bioethanol production due to the inherent advantages of their biological characteristics and cultivation. Among the benefits, the growth speed of microalgae and the ability to accumulate large concentrations of carbohydrates, essential for bioethanol production, stand out. Unlike traditional biomass crops, such as sugarcane and corn, microalgae do not compete directly with food production and can be cultivated in diverse environments, including wastewater, maximizing sustainability (Kusmiyati et al. 2023). This cultivation flexibility allows the use of nonarable land and the reduction of

Table 8.1 Bioethanol productivity from biomass sources

Biomass sources	Productivity (L bioethanol/ha.year)	Reference
Corn	3.450–4.4600	BNDES/CGEE (2008)
Beet	5.000–10.000	BNDES/CGEE (2008)
Sugarcane	5.400–10.800	BNDES/CGEE (2008)
Sweet sorghum	3.050–4.070	Mussatto et al. (2010)
Corn straw	1.050–1.400	Mussatto et al. (2010)
Cassava	3.310	Mussatto et al. (2010)
Wheat	2.590	Mussatto et al. (2010)
Maize	3.460–4.20	Mussatto et al. (2010)
Lignocellulosic biomass (sugarcane straw)	~10.000	Santos et al. (2012, 2014)
Microalgae low carbohydrate content	7.093–21.279	Acie et al. (2012)
Microalgae medium carbohydrate content	12.413–37.286	Acién et al. (2012)
Microalgae high carbohydrate content	17.733–53.199	Acién et al. (2012)

Source: Own authorship (2024)

freshwater consumption, contributing to the conservation of natural resources.

Several microalgae species have stood out for their efficiency in bioethanol production (Table 8.2) due to their high carbohydrate content and ability to adapt to different cultivation conditions. The species selected for bioethanol production are chosen based on their growth characteristics, chemical composition, and resistance to adverse conditions.

The bioethanol production process from microalgae involves several stages: harvesting, carbohydrate extraction, and fermentation. Efficient harvesting of microalgae is crucial as it directly influences the economic viability of the process, with techniques such as flocculation, centrifugation, and filtration being used to optimize this stage. After harvesting, carbohydrate extraction from microalgal cells is performed using acidic or enzymatic hydrolysis, which are standard methods to convert complex biomolecules into fermentable sugars. The subsequent fermentation of these sugars by specific microorganisms, such as yeasts, produces bioethanol (Chaudhary et al. 2014) (Fig. 8.2).

Recent advances in biotechnology techniques have improved the efficiency of each of these stages, making bioethanol production from microalgae more economically viable. Moreover, studies indicate that genetic engineering can increase carbohydrate content in microalgal cells and develop microalgal strains more resistant to adverse environmental conditions. Manipulating the metabolic pathways of microalgae can enhance the efficiency of converting sunlight and CO_2 into carbohydrate-rich biomass (Chandrasekhar et al. 2023).

In addition to technical advantages, cultivating microalgae for bioethanol production also offers significant environmental benefits, as they can capture large amounts of CO_2 from the atmosphere during photosynthesis, contributing to climate change mitigation. Bioethanol production from microalgae also generates valuable byproducts, such as proteins and lipids, which can be used in other industries, such as food and biofuels.

Integrating bioethanol production systems with microalgae biorefineries can maximize the use of all components of algal biomass, promoting a circular economy approach. Ongoing research in this area aims to optimize cultivation conditions, such as lighting, temperature, and nutrient composition, as well as improve conversion processes to reduce costs and increase productivity.

Therefore, using microalgae for bioethanol production represents a sustainable and efficient alternative to traditional methods. With significant potential to contribute to the transition to a low-carbon economy and reduce dependence on fossil fuels, microalgae position themselves as key components in the future of renewable energy and environmental sustainability. Continued research and technological development will be crucial to overcoming current challenges and fully realizing the potential of microalgae in bioethanol production.

8.4 Cultivation Conditions for Carbohydrate Accumulation in Microalgae

Bioethanol viable production from microalgae is directly related to the ability of these microorganisms to accumulate fermentable sugars (carbohydrates). Optimizing cultivation conditions is fundamental to ensure the production of large quantities of these carbohydrates and enable the economic viability of the bioprocess. Despite significant progress in this field, the ideal conditions for improving carbohydrate accumulation in microalgal biomass remain partially unknown (Chandrasekhar et al. 2023; Maity and Mallick 2022).

It is possible to cultivate microalgae under four different conditions, which are enlisted as follows: photoautotrophic, photoheterotrophic, chemoheterotrophic, and mixotrophic (Table 8.3).

Table 8.2 Bioethanol production from various microalgae strains

Microalgae	Pretreatment parameters of microalgal biomass	Alcoholic fermentation conditions and microorganism	Characteristics	Yield ($g.g^{-1}$)	Reference
Scenedesmus dimorphis	Organic solvent (2:1), 24 h, 750 rpm, enzymatic hydrolysis	150 rpm, 34 °C, *S. cerevisiae*	High growth rate, good carbohydrate content, adaptability to different environmental conditions	*0.266*	Chng et al. (2017)
Scenedesmus acuminatus	2 N H_2SO_4, autoclaving, pH 5.5	200 rpm, 30 °C, 80 h, *S. cerevisiae*	High biomass productivity, high carbohydrate content, resistance to adverse environments	*0.120*	Sanchez Rizza et al. (2017)
Chlorella vulgaris	200 $g.L^{-1}$, H_2SO_4 at 120° for 20 min	200 rpm, 30 °C, *S. cerevisiae*	High carbohydrate content, rapid growth, versatility in cultivation	*0.214*	Condor et al. (2022b)
Tetraselmis sp.	Chemoenzymatic hydrolysis and pH 5, 60 °C	150 rpm, 30 °C, 48 h, *S. cerevisiae*	High lipid and carbohydrate productivity, adaptability, resilience	*0.314*	Reyimu and Özçimen (2017)
Tetraselmis sp.	0.75% sodium hydroxide for 10 min			*0.073*	
Leptolyngbia sp.	1.5 N H_2SO_4, 0.8 bar, 3 h	150 rpm, 30 °C, 4 h, *S. cerevisiae*	Nitrogen fixation capacity, high carbohydrate content, adaptability	*0.113*	Chandra et al. (2020)
Leptolyngbia valderiana	1:5 biomass to liquid, H_2SO_4 and $MgSO_4$ at 121 °C for 20 min	200 rpm, 80 h, 30 °C, *S. cerevisiae*	Nitrogen fixation efficiency, high carbohydrate content, environmental resistance	*0.160*	Jafari Olia et al. (2022)
Chlorococcum sp.	H_2SO_4, pH 7, 160 °C	200 rpm, 48 h, 30 °C, *Saccharomyces bayanus*	High carbohydrate content, rapid growth, versatility	*0.520*	Tsolcha et al. (2021)
Chlorococcum sp.	Supercritical CO_2	200 rpm, 60 h, 30 °C, *S. bayanus*		*0.350*	Maity and Mallick (2022)
Desmodesmus sp.	10% solid loading/ homogeneous catalysts, 120 °C, 30 min, H_2SO_4	120 rpm, 28 °C, 30 h, *S. cerevisiae*	High biomass productivity, high carbohydrate content, environmental resilience	0.240	Harun et al. (2010)

Adapted from Khan et al. (2023)

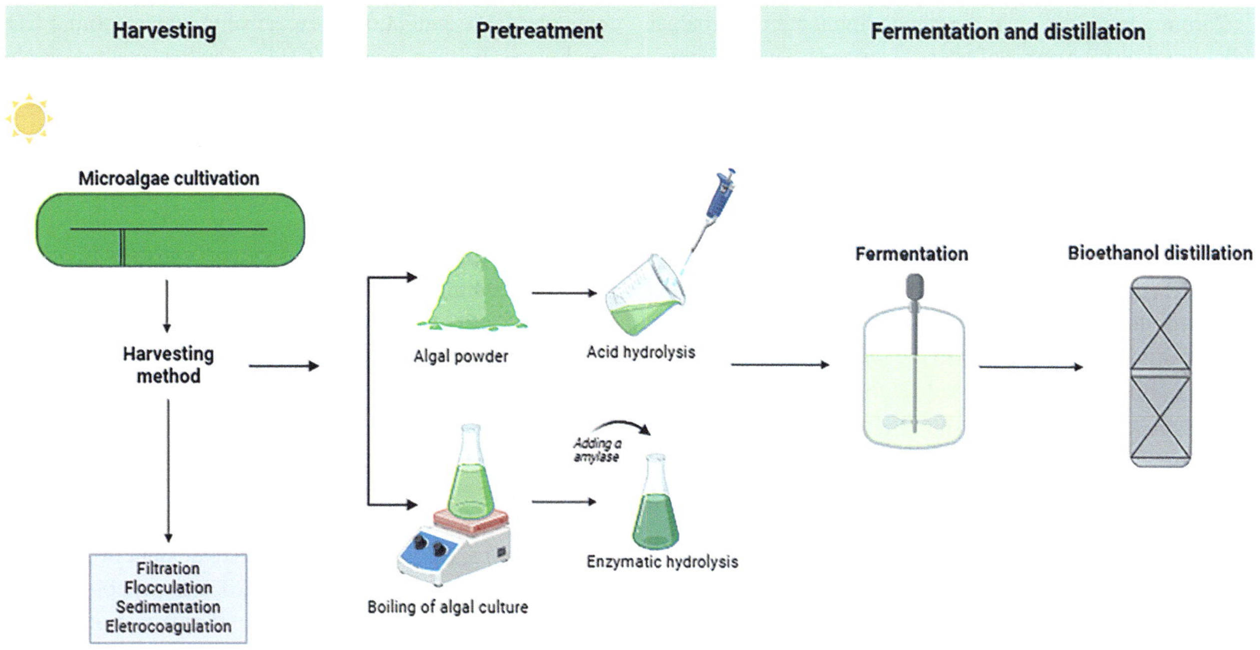

Fig. 8.2 Bioethanol production scheme from microalgae. (Source: Own authorship 2024)

Under chemoheterotrophic conditions, microalgae can develop without light and use organic carbon as a source of energy and carbon. On the other hand, mixotrophic microalgae obtain energy from light and organic carbon, as well as

Table 8.3 Nutritional pathways according to energy and carbon source

Growing conditions	Source of energy	Carbon source
Photoautotrophic	Light	CO_2
Photoheterotrophic	Light	Organic carbon
Chemoheterotrophic	Organic carbon	Organic carbon
Mixotrophic	Light and organic carbon	CO_2 and organic carbon

Source: Adapted from Abreu et al. (2022)

from CO_2 and organic carbon. Under photoautotrophic and photoheterotrophic conditions, the energy is obtained from photosynthesis (light). However, the carbon source in photoheterotrophic conditions is organic carbon, and in photoautotrophic conditions, it is CO_2 (Abreu et al. 2022; Bora et al. 2024).

The bioreactor choice for microalgae production should align with the selected cultivation conditions. Pond systems, raceway ponds, and photobioreactors (tubular and airlift) can be used for mixotrophic, photoheterotrophic, and photoautotrophic cultivation. However, it is important to select bioreactors that prevent light from entering the culture for chemoheterotrophic conditions. Currently, most large-scale microalgae cultivation systems are open tanks due to their low cost and ease of scaling. These systems can be installed in areas with low agricultural productivity, avoiding competition for land, have lower energy requirements, and are easier to maintain and clean (Smachetti et al. 2018; Maity and Mallick 2022).

Carbohydrates are an important component in microalgae cells, serving as a reserve of energy (especially in the form of starch) and a component of the cellular structure, being found in the inner layer (glycoproteins, hemicellulose, and cellulose) and external (pectin, alginates, and polysaccharides) to the cell wall. Under normal cultivation conditions, carbohydrates represent between 20% and 40% of the biochemical composition of microalgal cells, varying interspecifically (i.e., different species naturally have higher carbohydrate concentrations as shown in Table 8.4) or intraspecifically (i.e., it varies within the same species depending on the cultivation method used or the cell growth phase) (Pôjo et al. 2021). This variation highlights the potential of specific species to produce third-generation bioethanol.

During the exponential growth phase, microalgae have a high nutrient demand, prioritizing protein and lipid synthesis, so carbohydrate accumulation is not significant. However, in the stationary phase, sugars that are no longer needed for immediate growth accumulate, leading to increased carbohydrate storage (Phwan et al. 2019). Microalgal species exhibit significant variability in carbohydrate accumulation (Table 8.4) depending on the cultivation condition. The microalgae *Tetraselmis suecica* can produce 29.17 ± 0.04%, 69.59 ± 0.01%, and 71.52 ± 0.03% carbohydrates under autotrophic, chemoheterotrophic, and photoheterotrophic conditions, respectively (Kassim et al. 2021). Another example is the microalgae *Chlorella pyrenoidosa*, which produced approximately 20% carbohydrates under chemoheterotrophic and photoheterotrophic conditions and 30% carbohydrates under mixotrophic conditions (Han et al. 2022). Table 8.4 presents information on cultivation conditions, biomass concentration, biomass productivity, and carbohydrate concentration from these and other studies that analyzed different microalgae.

Other factors also significantly influence maximizing carbohydrate accumulation in microalgae for bioethanol production, including light intensity, pH, temperature, and nutrient limitation. For example, *Chlorella vulgaris* showed a four-times increase in carbohydrate content under nutrient-limited conditions (Oraby et al. 2023). *Chlorella kessleri* Bh-2 and *Scenedesmus* sp. k-7 showed carbohydrate contents of 74.5% and 71.0%, respectively, under optimal CO_2 conditions (Bauenova et al. 2024). Another example is the microalga *Ettlia texensis* that demonstrated a 1.3-fold increase in carbohydrate accumulation when cultivated in wastewater-based media (Samadhiya et al. 2023).

Furthermore, nitrogen limitation in the culture medium is a critical factor for carbohydrate accumulation, although it also decreases cell growth rate. Strategies such as two-stage cultivation, with nitrogen supplementation in the first stage to increase growth rate and subsequent cultivation without nitrogen for carbohydrate accumulation, have been proposed as solutions (Bastos 2018; Dragone et al. 2011).

Other strategies that have been used to maximize carbohydrate accumulation in microalgal biomass are the application of response surface methodology (RSM) and artificial neural networks (RNA) that have been effective in optimizing conditions, reaching a carbohydrate content of 59.5% in *Chlorella vulgaris* (Condor et al. 2022a). Although these findings are promising to improve carbohydrate accumulation, it is essential to consider variability in microalgal species and environmental factors that can influence growth and productivity.

Table 8.4 Carbohydrate production by different microalgae in relation to cultivation conditions

Microalgae	Cultivation conditions	Biomass concentration (g/L)	Biomass productivity (g/L.d)	Carbohydrates (% dry weight)	Reference
Desmodesmus sp.	Mixotrophic	0.7 ± 0.0	–	53.5 ± 14.2	Sanchez Rizza et al. (2017)
Chlorella sp.		0.7 ± 0,3	–	57.8 ± 16.6	
Chlorella pyrenoidosa	Chemoheterotrophic	0.15	0.020	≈ 20	Han et al. (2022)
	Mixotrophic	1.15	0.164	30	
	Photoautotrophic	0.10	0.014	≈ 20	
Neochloris oleoabundans	Chemoheterotrophic	14.2	–	54	Morales-Sánchez et al. (2013)
Chlorella vulgaris	Photoautotrophic	5.66	0.026 ± 0.078	54.0 ± 4.5	Chen et al. (2016)
	Mixotrophic	5.12 ± 0.29	1.002 ± 0.058	51.3 ± 4.2	
Tetraselmis suecica	Photoautotrophic	0.349 ± 0.024	0.044 ± 0.003	29.17 ± 0.04	Kassim et al. (2021)
	Chemoheterotrophic	0.635 ± 0.061	0.106 ± 0.021	69.59 ± 0.01	
	Photoheterotrophic	0.669 ± 0.01	0.111 ± 0.001	71.52 ± 0.03	
Scenedesmus almeriensis	Photoautotrophic	–	–	14.5	Hernández et al. (2015)
Nannochloropsis gaditana B-3	Mixotrophic	–	–	11.2	

Source: Own authorship (2024)

8.5 Technologies of Microalgal Carbohydrates to Bioethanol

Among the sources of bioethanol, microalgal carbohydrates are regarded as important feedstocks for their high carbohydrate content and efficient conversion to fermentable sugars. Several technologies have been developed to improve the efficacy in extraction and bioconversion processes that could render microalgae as potential alternative feedstocks compared with conventional sources employed currently for polysaccharide-to-bioethanol conversion. Typically, the strategies for the production process of microalgal carbohydrates to bioethanol are schematically shown as biomass cultivation, extraction of carbohydrates from microalgae (saccharification), hydrolysis, and fermentation.

Biomass productivity and carbohydrate accumulation in microalgae are affected by several environmental conditions, such as light intensity, nutrient availability, and carbon dioxide concentration, that can be optimized for the production of target molecules from microalgae species with an entirely different regulatory mechanism, even if conserved between them certain ones like transcription factors homologs to control lipid synthesis genes expression. For example, it has been shown that in some species, such as *Scenedesmus obliquus,* carbon accumulation can be increased upon nitrogen depletion (>50% dry biomass), which is primarily stored as glucose (Cho et al. 2020; Ho et al. 2017). After harvesting microalgal biomass, the following step is extracting carbohydrates, which different pretreatments can do. These practices are geared toward converting large, complex carbohydrates into smaller, fermentable sugars. Using pretreatment strategies, acid and enzymatic hydrolysis, and microwave-assisted extraction can enhance its efficiency (Phwan et al. 2019; Yirgu et al. 2021). Dilute acid pretreatment has been reported to enhance the fermentable sugar yields from microalgal biomass (Harun et al. 2014; Phwan et al. 2019). Studies have shown that the use of enzymatic hydrolysis after diluted acid pretreatment on *Chlorella vulgaris* and *Nannochloropsis oculata* biomasses increases the release of sugars (Agwa et al. 2017; Fetyan et al. 2022). The released sugars are fermented by microorganisms such as the yeasts *Saccharomyces cerevisiae* or *Pichia pastoris* to generate bioethanol. The efficiency of the fermentation processes depends on the type of sugars released, the pretreatment/hydrolysis technology used, and the cultivation conditions (El-Dalatony et al. 2017; Khan et al. 2016). Ethanol yield can reach up to 6 g/L from defatted biomass pretreated with diluted acid–coupled enzyme hydrolysis (Fetyan et al. 2022). In addition, it has been found that the fermentation microalgae sugars can present higher productivity than conventional lignocellulosic feedstocks and make them competitive alternatives for the bioethanol industry (Cho et al. 2020; Khan et al. 2016).

While microalgal carbohydrates hold promise for bioethanol production, there are still many hurdles to overcome to scale up these technologies so they can be used commercially. To be economically viable, microalgal bioethanol must overcome issues like low carbohydrate productivity, high pretreatment costs, and efficient harvesting/processing methods (Meng et al. 2021; Mirsiaghi and Reardon 2015). Current research is directed toward the refined functioning of these processes and integrated biorefinery approaches utilizing the whole potential of microalgal biomass to produce various biofuels and bioproducts (Chia et al. 2018; Kumar et al. 2021). Technologies have evolved in converting microalgal carbohydrates to bioethanol, providing a sustainable and economically viable way for biofuel synthesis. Further development will require even greater innovation in growing, extracting, and fermenting microalgae as energy feedstocks.

8.6 Methods of Harvesting Microalgae Biomass

In industrial applications, the efficient harvest of microalgae biomass is crucial. However, the small size of the cells and the low microalgae density in the cultivation medium are significant challenges for this task. In addition, techniques involving high chemical additives or chemical shear can cause degradation, so the chosen harvesting method must maintain the integrity of the bioactive content of interest. Centrifugation, filtration, flocculation, and sedimentation (Fig. 8.3) are traditional techniques widely used in collecting microalgae for industrial production, each with advantages and disadvantages. However, new methodologies are being developed to reduce costs and increase efficiency. Bioflocculation, for example, employs biological mechanisms to aid in the sedimentation process of the cells, offering a promising and sustainable alternative to traditional harvesting techniques (Liu et al. 2023; Pereira et al. 2024; Zhu et al. 2024).

8.6.1 Physical Methods

8.6.1.1 Centrifugation

Centrifugation is widely used in the industrial harvesting of microalgae due to its ability to process large volumes and its non-specificity for different species, rapidly concentrating algae from diluted suspensions. Industrial centrifuges can achieve recovery efficiencies of over 90–95%. However, the method is energy-intensive, with consumption ranging from 0.80 to 20 kWh/m^3, which limits its application to high-value products. Centrifugation can damage cells due to shear forces, potentially reducing the quality of the final product. Despite these drawbacks, the process's scalability is well-established, with industrial systems available that can be optimized to reduce costs. Techniques that combine precoagulation/flocculation to reduce the processed volume have demonstrated significant energy savings (Liu et al. 2023; Zhu et al. 2024).

8.6.1.2 Filtration

In filtration technology, a porous membrane separates solids from liquids, allowing the liquid medium to pass through while retaining microalgae cells. Methods such as microfiltration and ultrafiltration employ membranes with different pore sizes to concentrate biomass, separating microalgae cells from water in diluted cultures. Although filtration is scalable and allows for operation in large units, it faces challenges like frequent membrane cleaning and high operational costs, primarily due to energy consumption and membrane replacement. For larger volumes, centrifugation becomes more advantageous than filtration. Technological innovations, such as using stainless-steel membranes and vibration systems, aim to increase efficiency and reduce fouling, but obstacles remain in making filtration economically viable in large-scale processes (Liu et al. 2023; Zhu et al. 2024).

8.6.1.3 Gravity Sedimentation

Sedimentation is a simple and low-cost method used in industrial production for harvesting microalgae, especially in processes where the final product has low value, such as bio-

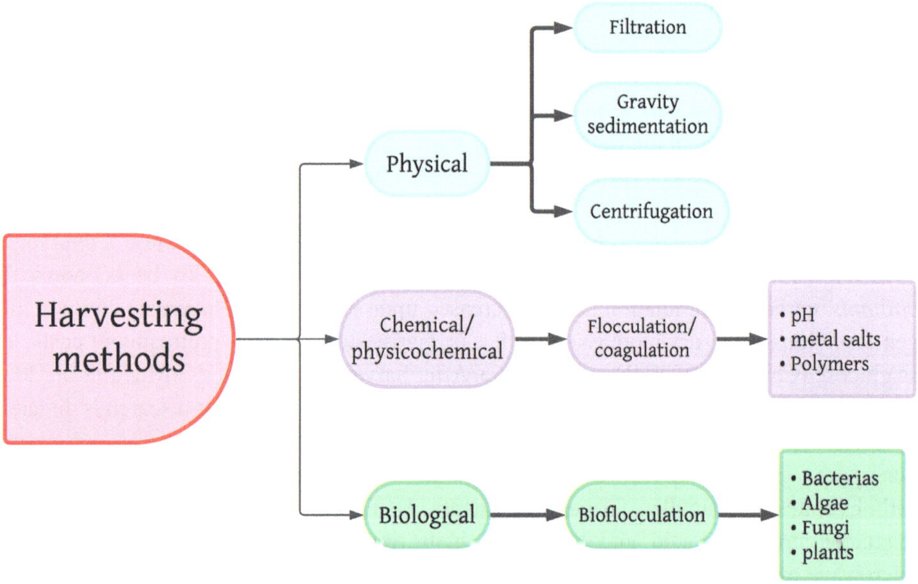

Fig. 8.3 Microalgal biomass harvest methods. (Source: Adapted from Liu et al. 2023)

fuels. This method relies on gravity to separate microalgae cells from water using sedimentation ponds or simple tanks. Sedimentation is energy-efficient and does not require chemicals or mechanical intervention, making it an environmentally friendly approach. However, sedimentation has several limitations, including low efficiency and long residence times required for separation, particularly for microalgae with small size. Moreover, a preliminary coagulation/flocculation step is often applied to enhance efficiency, which facilitates cell aggregation and speeds up the sedimentation process. Despite these improvements, sedimentation alone is still considered inefficient for large-scale commercial algae production (Kusmiyati et al. 2023; Zhu et al. 2024).

8.6.2 Chemical/Physicochemical Methods

8.6.2.1 Flocculation/Coagulation

Flocculation is widely used in industrial production to harvest microalgae, effectively forming cellular aggregates that sediment to gravity. The process involves the addition of flocculating agents, such as multivalent metal salts (iron chloride ($FeCl_3$) and aluminum sulfate ($Al_2(SO_4)_3$)), and cationic polymers (chitosan and polyacrylamides), which neutralize the negative surface load of microalgae, allowing the formation of quickly established flakes. In addition, flocculation can be performed in different ways, including bioflocculation, automatic flocculation, inorganic and organic flocculation, and combined processes. Each method has specific advantages and disadvantages, depending on the conditions of the medium and the type of microalgae. Flocculation faces challenges in large-scale production, even though it is considered very economical. Cell concentration, the type of flocculant used, pH, the environment's ionic force, and the microalgae cultivation conditions are some of the factors that impact the effectiveness of the process. Another negative factor is sludge generation and possible biomass contamination by metal that limits the application of the method in specific contexts, such as biofuel production (Liu et al. 2023; Zhu et al. 2024).

8.6.3 Biological Methods

8.6.3.1 Bioflocculation

Considered an innovative and sustainable approach, the use of biocoagulants and bioflocculants in microalgae industrial-scale harvesting has the potential to replace conventional chemical coagulants. These agents, which can be derived from plants, bacteria, or fungi, exhibit chemical coagulant efficacy without compromising the quality of the biomass harvested. Biocoagulants derived from plants, such as *Moringa*, combined with chitosan, showed high efficacy in biomass recovery, with efficiencies of up to 95.76%, without compromising biomass quality. Similarly, bacterial and fungal bioflocculation stands out for its efficiency and cost-effectiveness. For example, *Bacillus Licheniformis* bacteria and *Aspergillus niger* fungi reach harvest efficiencies above 90% using natural interactions that help in flocculation without the need for chemical agents. Although the results of a laboratory scale are promising, the transition to large-scale production requires further research to optimize operational parameters and reduce the costs of this biotechnology (Liu et al. 2023; Pereira et al. 2024; Zhu et al. 2024).

8.7 Pretreatment of Microalgae Biomass

The pretreatment of microalgal biomass is a crucial step in bioethanol production, as it aims to prepare the microalgal cells for efficient carbohydrate extraction and subsequent fermentation. The success of this stage is fundamental to the economic viability and overall efficiency of the process. Pretreatment of microalgal biomass involves various techniques (Fig. 8.4), each with advantages and challenges. Among the most common approaches are:

- Acid hydrolysis: Effective in breaking down the cell walls of microalgae and releasing the carbohydrates within. This process uses strong acids, such as hydrochloric acid, under controlled temperature and time conditions. Acid hydrolysis can be applied to different biomasses, but it may generate undesirable byproducts and require additional neutralization and purification steps (Chaudhary et al. 2014; Kusmiyati et al. 2023).
- Enzymatic hydrolysis: Utilizes specific enzymes, such as cellulase, pectinase, protease, and amylase, to degrade cell walls and convert complex carbohydrates into fermentable sugars. This technique is more selective and less aggressive than acid hydrolysis. However, it can be more expensive due to the cost of enzymes and the need for optimized conditions for enzymatic activity (Kusmiyati et al. 2023).
- Physical methods: Milling, ultrasound, and electroporation are physical techniques used to disrupt microalgae cell walls and facilitate the release of intracellular compounds. These methods can be combined with chemical or enzymatic processes to improve the overall pretreatment efficiency. Milling is effective in increasing the surface area of the biomass, while ultrasound and electroporation use acoustic waves and electric fields to promote cell membrane permeability (Kusmiyati et al. 2023).
- Combined methods: The combination of physical, chemical, and enzymatic methods has shown promise in optimizing the pretreatment of microalgal biomass. For

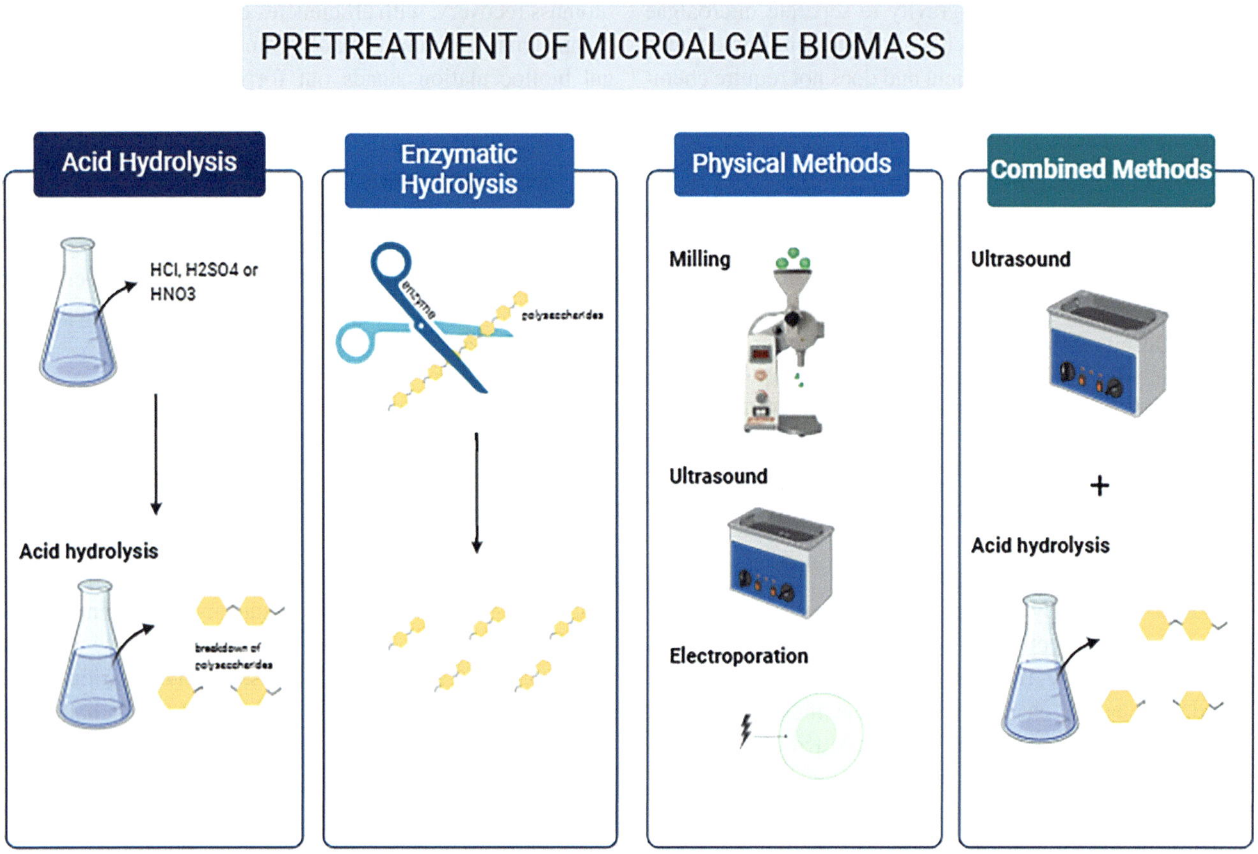

Fig. 8.4 Pretreatment of microalgae biomass for bioethanol production. (Source: Own authorship 2024)

example, combining acid hydrolysis followed by enzymatic hydrolysis can increase the efficiency of fermentable sugar release. This approach can also help minimize the use of chemical reagents and reduce the generation of undesirable byproducts (Chandrasekhar et al. 2023).

Furthermore, effective pretreatment of microalgal biomass not only enhances the conversion of carbohydrates into bioethanol but also contributes to the sustainability of the process by reducing the need for reagents and minimizing waste generation. It is a vital step in efficient bioethanol production, and choosing the appropriate technique can significantly affect economic viability. Traditional approaches, such as acid and enzymatic hydrolysis, have been widely used and show specific advantages but also face challenges related to cost and byproduct generation. Acid hydrolysis effectively breaks down cell walls but can result in undesirable byproducts and requires additional neutralization processes.

On the other hand, enzymatic hydrolysis offers a more selective and less aggressive alternative, although its cost can be a limiting factor. Physical methods, such as milling, ultrasound, and electroporation, have proven effective in disrupting cell walls and increasing carbohydrate extraction efficiency. Combining these techniques with chemical or enzymatic processes can offer more robust and economical solutions, maximizing the recovery of fermentable sugars and minimizing waste generation. Studies have shown that different pretreatment conditions can significantly impact ethanol yield (Table 8.5). For example, *Scenedesmus dimorphis* subjected to an organic solvent pretreatment (2:1), 24 h, 750 rpm, followed by enzymatic hydrolysis, resulted in an ethanol yield of 0.266 g/g. In another study, *Scenedesmus acuminatus* treated with 2 N H_2SO_4, autoclaving, pH 5.5, showed a 0.120-g/g yield. Combined techniques, such as chemoenzymatic hydrolysis, have also shown promising results. *Tetraselmis* sp. subjected to chemoenzymatic hydrolysis at pH 5 and 60 °C, achieved a yield of 0.314 g/g of ethanol. In contrast, using sodium hydroxide pretreatment at 0.75% for 10 min in *Tetraselmis* sp. resulted in a lower yield of 0.073 g/g.

Recent advances in biotechnology and genetic engineering have driven improvements in pretreatment processes, such as the genetic modification of microalgae to alter their cell walls and increase resistance to adverse conditions. This is a promising area that could reduce costs and improve process efficiency. Additionally, new technological approaches, such as heterogeneous catalysts, are emerging as innovative

Table 8.5 Pretreatment type of microalgae biomass

Type of pretreatment	Solvent used	Condition	Raw material	Advantage	Disadvantage	Reference
Supercritical carbon dioxide	CO_2 and ethanol	40 °C, 35 MPa, 30 min	*Schizochytrium limacinum*	Ease to recovery, nontoxic, nonflammable reagents	High process cost due to required infrastructure and operator	Tang et al. (2011)
	CO_2, ethanol, and methanol	35 MPa, 50 °C	*Nannochloropsis oculata*			Liau et al. (2010)
Ammonia fiber explosion	Ammonia 33 wt%	60–120 °C, 3 MPa, 30 min	*Thalassiosira weissflogii*	Lack of inhibitor production, environmentally friendly	High capital and operational cost	Bals et al. (2011)
	Ammonia	90–100 °C	*Chlorella vulgaris*			
Ultrasonication	Hexane	90 °C, 0.1 MPa	*Chlorella and Scenedesmus* sp.	Easy disruption of cell structure without changing the chemical structure	High power consumption and difficulty in upscaling	Teymouri et al. (2005)
		40 kHz, output power: 2.2 kW, 45 °C	*Chlorella vulgaris and Uronema belkae*			Sathish and Sims (2012)
		40 kHz, 15–60 min	*Chlamydomonas mexicana*			Hwang et al. (2016)
Acid pretreatment	1% (v/v) H_2SO_4	30 min, 140 °C	*Chlorococcum humicola*	Low temperature process	Corrosiveness and toxicity of reagents and production of reagent product	Eldalatony et al. (2016)
	H_2SO_4 4 M	120 °C, 40 min	*Chlorella vulgaris*			Harun et al. (2010)
						Mendez et al. (2013)
Alkaline pretreatment	0.3 M KOH, NaOH, Na_2CO_3, and NH_4OH	90 °C for 75 min	*Chlorella* spp. *and Tetraselmis suecica*	Lower reaction temperature, efficient removal of the cell wall	High cost due to the required downstream processing	Kassim and Bhattacharya (2016)
	NaOH concentrations (from 0 to 21 g/L)	121 °C and 10 bar, 30 min	*Pavlova* cf sp., *Tetraselmis* sp., and *Thalassiosira weissflogii*			Bohutskyi et al. (2014)
Enzymatic	Endogalactouronase 800 U/g	pH 6, 50 °C, 60 min	*Scenedesmus obliquus*	High specificity, lower environmental impact	High cost of enzymes and long reaction time	Ometto et al. (2014)
	Esterase 3600 U/g					
	Protease 90 U/g					

Adapted from Kusmiyati et al. (2023)

alternatives to enhance pretreatment efficiency further. These catalysts overcome some of the deficiencies of other techniques and improve substrate and reagent affinity, offering good thermal stability and high water stability, making the process efficient in biomass hydrolysis (Guv et al. 2012; Tan 2019).

In summary, the integration and optimization of various pretreatment techniques, combined with advances in genetic engineering and biotechnology, are essential for the economic and environmental viability of biomethanol production from microalgae.

8.8 Saccharification of Microalgal Biomass

Saccharification of algae biomass is carried out to produce third-generation ethanol, transforming the carbohydrates found in algae cells into fermentable sugars and then into ethanol. Pretreatment processes are usually required to break down the cell walls and release the carbohydrate polymers during algae saccharification. Standard techniques include acid, alkaline, and enzymatic treatments, which help cellulose and hemicellulose be broken down into simple sugars (Abou El-Souod et al. 2021).

In acid hydrolysis, acids such as sulfuric acid are used at different dosages and temperatures to break down the cell walls of algae and thus release the sugars. The choice of acid used will depend on the cost and efficiency in breaking down lignocellulose. Many factors, such as the concentration of the acid, the temperature of the process, and the duration of the reaction, influence the effectiveness of acid hydrolysis. The choice between using the acid in dilute or concentrated forms depends on the balance between efficiency in breaking down the sugar molecules and minimizing the generation of inhibitory byproducts, such as furfural and hydroxymethylfurfural, which can interact negatively with subsequent fermentation. Acid in modest quantities and at controlled temperatures allows for the effective conversion of polysaccharides into fermentable sugars without affecting the viability of the fermentation process (Abou El-Souod et al. 2021; Kusmiyati et al. 2023).

Alkaline hydrolysis breaks the chemical bonds in algae cell walls using alkaline solutions, including sodium hydroxide (NaOH) and potassium hydroxide (KOH), thus releasing polysaccharides such as cellulose and hemicellulose. The solubility of cell wall components, including lignin, represents a significant obstacle to effective saccharification and is one of the main advantages of alkaline hydrolysis. Generally, at high temperatures, the procedure increases the efficiency of cell destruction, thus facilitating access to carbohydrates for the following stages of fermentation. The maximum efficiency of the process depends on optimizing these conditions. However, compared to acid hydrolysis, alkaline treatment can produce byproducts that need additional treatment to neutralize the solution before fermentation, which can increase operating costs and be less effective in terms of producing fermentable sugars. Therefore, alkaline hydrolysis must be carefully managed between the need to preserve the economic and environmental sustainability of the process and the efficiency of sugar release (Abou El-Souod et al. 2021; Hamouda et al. 2018; Kusmiyati et al. 2023).

Besides using enzymes that hydrolyze carbohydrate polymers into fermentable monosaccharides, enzymatic hydrolysis is considered the most promising in terms of the environment and efficiency. Operating under milder circumstances and without the need for corrosive reagents helps this approach reduce processing expenses and environmental effects.

Enzymes, such as cellulases and amylases, break down the polysaccharides found in the cell walls of microalgae into simple sugars, such as glucose, so that yeasts can easily ferment them. Compared to chemical techniques, the main benefits of enzymatic hydrolysis are selectivity and less formation of inhibitory byproducts that could affect fermentation performance. While amylases break down starch in microalgal biomass, transforming these polymers into reducing sugars, enzymes such as cellulase act directly on cellulose. Factors such as enzyme concentration, temperature, and pH can affect the effectiveness of the enzymatic hydrolysis process.

Even so, enzymatic hydrolysis has difficulties, including the high cost of enzymes and the need to maximize process parameters for many types of microalgal biomass. Mainly because of its lower environmental impact and greater selectivity in the biomass conversion process, enzymatic hydrolysis remains a promising and sustainable method for generating bioethanol despite these difficulties (Abou El-Souod et al. 2021; Hamouda et al. 2018; Kusmiyati et al. 2023).

Table 8.6 shows the comparison between the different hydrolysis methods, the sugar concentration, and the bioethanol yield obtained by each method.

8.9 Fermentation Process for Bioethanol Production

An original approach with great potential to change how biofuels are generated is the fermentation of algae to generate ethanol. When basic sugars are released from algae biomass through hydrolysis, they become accessible for fermentation. This stage is fundamental because it is here that the sugars are transformed into bioethanol by the activity of microorganisms such as yeasts, *Saccharomyces cerevisiae* being the most frequently used. Yeasts break down glucose under anaerobic conditions, without oxygen, during fermentation to generate carbon dioxide and bioethanol as the main results.

Many elements determine the effectiveness of this fermentation process. First, bioethanol production is highly influenced by the concentration of the accessible sugar. While very high concentrations can reduce microbial activity and affect fermentation, a low initial glucose concentration can lead to suboptimal ethanol generation. In addition, the vital characteristics that must be strictly regulated to maximize yeast activity are the temperature and pH settings. Generally, a pH range of 4–6 and temperatures between 28 and 35 °C are suitable parameters to guarantee maximum bioethanol generation.

In addition to *Saccharomyces cerevisiae*, other microorganisms, including *Zymomonas mobilis* and other species of bacteria, can also be employed based on the specific characteristics of the algae substrate used (Yin et al. 2020).

The efficiency and economic viability of the fermentation techniques used to generate ethanol from algae define their basic nature. The two most widely used techniques are simultaneous saccharification and fermentation (SSF) and separate hydrolysis and fermentation (SHF).

In the SHF approach, hydrolysis and fermentation take place in different stages. First, the algae biomass is hydrolyzed, transforming the polysaccharides into simple sugars such as glucose, and microorganisms ferment these sugars to generate bioethanol. The main advantage of this approach is the possibility of precisely controlling the parameters of each stage, making it possible to maximize both hydrolysis and fermentation independently, improving the process's overall efficiency (Kusmiyati et al. 2023).

In the SSF process, hydrolysis and fermentation occur simultaneously in the same reactor; using this method, polysaccharides are broken down into simpler forms, while the rapid fermentation of these simple sugars produces bioetha-

Table 8.6 Different hydrolysis parameters for microalgae biomass in releasing sugars for ethanol production

Microalgae	Hydrolysis parameter	Sugar/glucose concentration	Bioethanol yield	Reference
Chlorella sorokiniana	5 M NaOH, 90 °C, 30 min, 60 rpm	0.008 g/L	–	Hernández et al. (2015)
Nannochloropsis gaditana	5 M NaOH, 90 °C, 30 min, 60 rpm	0.014 g/L	–	Hernández et al. (2015)
Scenedesmus almeriensis	5 M NaOH, 90 °C, 30 min, 60 rpm	0.015 g/L	–	Hernández et al. (2015)
Chlorella vulgaris	H_2SO_4 (0.1–1.5% v/v), 121 °C, 20 min	7–24 g/L	7.6–11.7 g/L	Ho et al. (2013)
Chlamydomonas reinhardtii	12 N H_2SO4, 121 °C, 120 min, 1–11 hydrolysis cycles	0.25–2 g/L	3.6 g/L	Scholz et al. (2012)
Gelidium amansii	Biomass ratio 15% (solid/liquid), 2% (w/w) H_2SO_4,121 °C, 60–240 min	2–4.5 g/L	11.8–27.6 g/L	Park et al. (2012)
Kappaphycus alvarezii	0.2 M H_2SO_4, 130 °C, 15 min	0–0.89 g/L	0.25–0.369 g/L	Dyah et al. (2012)
Chlorella vulgaris	Endoglucanase, β-glucosidase, amylase, 200 rpm, 45 °C, 2.880 min	4–11 g/L	3.55–4.3 g/L	Ho et al. (2013)
Ulva fasciata Delile	Enzymatic cellulase 1, 2, and 5% (v/v), 45 °C, 360–2.160 min	82.2–219 g/L	35.1–93.1 g/L	Trivedi et al. (2015)
Gracilaria verrucosa	Cellulase + β-glucosidase, 50 °C, 0–2.880 min	0.7–8.7 g/L sugar per g cellulase	14–9.4 g/L	Kumar et al. (2013)

Table 8.7 Microalgal biomass subjected to different types of pretreatments applied in the SSF and SHF processes

Algae	Pretreatment	Process type	Fermenting organism and process time	Bioethanol yield (%)	Reference
Chlorella vulgaris	3% H_2SO_4, 110 °C, 105 min	SHF	*Escherichia coli* SJL2526, 24 h	40.0	Lee et al. (2011)
Chlorococcum humicola	12 N H_2SO_4	SHF	*Saccharomyces cerevisiae*, 48 h	44.0	Scholz et al. (2012)
Gracilaria verrucosa	Enzymatic hydrolysis with cellulose and β-glucosidase, 30 °C	SHF	*Saccharomyces cerevisiae*, 48 h	43.0	Kumar et al. (2013)
Gelidium corneum	0.5–1% $H_2C_2O_4$, 121 °C, 30 and 60 min	SHF	*Saccharomyces cerevisiae*, 96 h	58.0	Ho et al. (2012)
Sargassum spp.	3.4–4.6% H_2SO_4, 115 °C, 90 min and enzymatic hydrolysis with cellulose and β-glucosidase	SHF	*Saccharomyces cerevisiae*, 48 h	65.0	Borines et al. (2013)
Chlamydomonas reinhardtii UTEX 90	α-amylase 90 °C, 30 min and amyloglucosidase 55 °C, 30 min	SSF	*Saccharomyces cerevisiae* S288C, 40 h	23.5	Choi et al. (2010)
Laminaria japonica	0.1 N HCl, 121 °C, 15 min and enzymatic hydrolysis with celluclast	SSF	*Escherichia coli* KO11, 72 h	16.1	Kim et al. (2011)

nol. SSF offers the advantage of reducing production costs by eliminating the need for different reactors for each stage, thus limiting the danger of contamination, as the process takes place in a single closed environment. This method can, however, be less effective in terms of bioethanol yield since the ideal conditions for hydrolysis enzyme activity may not coincide with the ideal conditions for fermentation by the microorganisms, thus reducing the efficiency of converting sugars into bioethanol (Kusmiyati et al. 2023).

Table 8.7 shows the use of different algae species, pretreatment methods, types of processes, duration of the fermentation process, and the resulting bioethanol yields (such as simultaneous saccharification and fermentation (SSF) and separate hydrolysis and fermentation (SHF)).

The ideal technique depends on the attributes of the process, the algae being converted, and the production objectives. SSF combines enzymatic hydrolysis and fermentation in a single step, reducing processing time but imposing strict requirements on process parameters. SHF splits hydrolysis and fermentation into separate steps, allowing independent optimization but increasing complexity. Fermentation generates secondary products such as carbon dioxide and water, which require purification. Distillation, which vaporizes and recondenses contaminants and water, remains the preferred purification method due to its ability to separate ethanol from impurities effectively.

Although algae fermentation to produce ethanol is very promising, many fundamental challenges remain. The effectiveness of the process depends not only on the resilience of the microbial strains used in fermentation and the effectiveness of hydrolysis but also on economic concerns such as the cost of inputs and the feasibility of mass production. Scaling up the process to industrial levels requires more research and development, mainly to reduce prices and improve efficiency, thus allowing the manufacture of algal ethanol to be a competitive and valuable alternative to fossil fuels.

8.10 Recent Advances, Challenges, and Future Directions for Microalgae-Based Bioethanol

Microalgal biomass's high photosynthetic efficiency and quick growth rate make it a promising resource for bioethanol synthesis. Microalgae offer a versatile and sustainable substitute for biofuel production since they can be grown in various conditions, such as nonarable soils and wastewater (Pandit et al. 2024).

Enhancing microalgae growth and harvesting operations is essential for raising biomass production and generating bioethanol. This is now possible thanks to recent research in several ways. Developing more effective growing methods, especially by optimizing light and nutritional conditions, is a significant area of advancement. Advances in photobioreactor design have considerably boosted photosynthetic efficiency and biomass output, allowing for more precise control over light intensity and wavelength (Singh 2024; Shitanaka et al. 2024). Additionally, it has been demonstrated that employing nutrient-rich wastewater as a growing medium is a sustainable method for both the environment and the economy that has two advantages: it reduces the cost of generating microalgae and helps with wastewater treatment (Pandit et al. 2024; Udayan et al. 2022; Chia et al. 2018).

Genetic engineering has improved the sector by enabling the development of microalgae strains with greater capacity for producing bioethanol, mainly through CRISPR/Cas9 technology.

Specific genetic changes have increased microalgae's tolerance to environmental stressors and the amount of fermentable sugars they store (Dhokane et al. 2023; Zhang et al. 2019). These modifications are necessary to promote production scalability and maximize the overall effectiveness of the bioethanol manufacturing process.

The pretreatment of microalgal biomass, an essential stage in the effective synthesis of bioethanol, has also seen notable advancements. Novel pretreatment strategies that combine chemical and biological techniques have been developed to more effectively break down the stiff cell walls of microalgae, increasing the accessibility of fermentable sugars (Khan et al. 2023). Moreover, enzymatic hydrolysis has gained traction as a more energy-efficient and environmentally friendly substitute for conventional chemical processes (Vasić et al. 2021).

However, there are still several obstacles to commercializing bioethanol derived from microalgae. Apart from the exorbitant expenses linked to extensive farming and gathering, the requirement for regulated environmental circumstances, combined with the energy requirements of conventional harvesting techniques like centrifugation, increases the process's running costs. The high energy requirements of the dehydration and drying processes—necessary to prepare microalgal biomass for bioethanol synthesis—further compound this economic difficulty (Shitanaka et al. 2024; Udayan et al. 2022).

Another difficulty lies in the selection of appropriate microalgae strains. Given the diverse species available, each with intrinsic growth requirements and biochemical profiles, identifying strains that are resilient to environmental fluctuations and capable of high bioethanol yields remains a complex task (Jabłońska-Trypuć et al. 2023; Khan et al. 2023). Because of this, the variety of microalgae species offers both a benefit and a drawback. Although some species have the potential to be highly prolific, large-scale cultivation of these species may necessitate considerable production technique improvement.

Concerns about environmental sustainability also exist, mainly because of the energy requirements and carbon impact of processing and growing microalgae. Although microalgae may aid in carbon fixation, the energy expenditures associated with growing and processing biomass must be considered when weighing the total environmental advantages (El-Moustaqim et al. 2024). The difficulty in guaranteeing high biomass productivity while keeping operating costs low exacerbates this problem (Udayan et al. 2022).

Future studies must concentrate on the creation of more affordable and environmentally friendly methods of cultivation in order to overcome these obstacles. In addition to reducing cultivation costs, using wastewater for microalgae growth has the added benefit of bioremediation, which addresses environmental concerns about pollution caused by organic overloading (Li et al. 2023; Udayan et al. 2022). Bioethanol production can be made more sustainable overall by combining microalgae growth with wastewater treatment techniques to form a symbiotic system that simultaneously removes nutrients and produces biomass (Dhokane et al. 2023; Zhang et al. 2019).

Enhancing the efficiency of bioethanol production also requires advancements in metabolism and genetic engineering. Researchers can improve total yields by increasing the accumulation of bioethanol precursors by genetically altering microalgae strains to promote metabolic pathways (Dhokane et al. 2023; Zhang et al. 2019). Furthermore, synthetic biology techniques may enable the development of entirely new metabolic pathways to optimize bioethanol synthesis in various environmental circumstances.

Numerous important problems make large-scale microalgae biomass culture for bioethanol synthesis commercially unfeasible. These obstacles must be removed in order to increase the viability of this exciting new biofuel source. These consist of financial, operational, and technological barriers. High production costs remain the primary barrier because of the complexity of cultivation techniques and the need for state-of-the-art technologies (Dębowski et al. 2023; Wang et al. 2024). In addition, insufficient funding and legal support for innovative projects further complicate the economic landscape (Dębowski et al. 2023).

The scalability of laboratory methods to industrial applications is fraught with difficulties, including species selection and the optimization of growth conditions such as light, temperature, and nutrient availability (Dębowski et al. 2023; Novoveská et al. 2023). Contamination risks and the need for effective harvesting strategies also pose significant hurdles (Novoveská et al. 2023).

A shortage of experienced personnel in phycology limits the operational efficiency of large-scale systems (Novoveská et al. 2023). Integrating waste as feedstock and optimizing cultivation techniques are essential for improving yield and cost-effectiveness (Dębowski et al. 2023). Notwithstanding these obstacles, developments in genetic engineering and process optimization provide possibilities to improve microalgae's economic feasibility for the manufacture of bioethanol (Dębowski et al. 2023; Wang et al. 2024). Nevertheless, these challenges must be addressed for the industry to reach its full potential.

The economic feasibility can be enhanced by integrating bioethanol from microalgae production with other biorefinery procedures. In addition to producing bioethanol, coproducing value-added products (such as bioplastics, pigments, or medications) can significantly boost the total profitability of microalgae production. Microalgae-based biofuels are now more competitive with traditional fossil fuels, thanks to the biorefinery approach, which optimizes resource use and diversifies revenue streams (Shitanaka et al. 2024).

To fully utilize microalgae as renewable energy sources, multidisciplinary cooperation and constant innovation are needed to develop more economical and environmentally friendly production methods. In conclusion, there is a lot of promise for using microalgae as feedstocks to produce bioethanol.

8.11 Conclusions

Microalgae's lack of lignin makes the processes involved in producing bioethanol simpler and more effective than those involving lignocellulosic biomass, which frequently necessitates intensive pretreatment. Microalgae are now more valuable as feedstocks for bioethanol due to recent improvements in extraction technologies, which have increased the efficiency of extracting important molecules from them.

The generation of bioethanol from microalgae biomass became a feasible and sustainable alternative due to its fast growth, high carbohydrate content, and capacity to cultivate using wastewater, putting it in a favorable position relative to other traditional biofuel feedstocks. However, to fully realize the potential of microalgae in the bioethanol sector, more research is still needed to improve conversion methods and growing conditions.

References

Abou El-Souod GW, Morsy EM, Hassan LHS, El-Sheekh MM (2021) Efficient saccharification of the microalga *Chlorella vulgaris* and its conversion into ethanol by fermentation. Iran J Sci Technol Trans A Sci 45(3):767–774. https://doi.org/10.1007/s40995-021-01097-1

Abreu AP, Morais RC, Teixeira JA, Nunes J (2022) A comparison between microalgal autotrophic growth and metabolite accumulation with heterotrophic, mixotrophic and photoheterotrophic cultivation modes. Renew Sust Energ Rev 159:112247. https://doi.org/10.1016/j.rser.2022.112247

Acién FG, Fernández JM, Magán JJ, Molina E (2012) Production cost of a real microalgae production plant and strategies to reduce it. Biotechnol Adv 30(6):1344–1353. https://doi.org/10.1016/j.biotechadv.2012.02.005

Agwa OK, Nwosu IG, Abu GO (2017) Bioethanol production from *Chlorella vulgaris* biomass cultivated with plantain (*Musa paradisiaca*) peels extract. Adv Biosci Biotechnol 08(12):478–490. https://doi.org/10.4236/abb.2017.812035

Ashokkumar V, Chandramughi VP, Kumar G, Ngamcharussrivichai C, Piechota G, Igliński B, Kothari R, Chen WH (2024) Advancements in lignocellulosic biomass: a critical appraisal of fourth-generation biofuels and value-added bioproduct. Fuel 365(February):130751. https://doi.org/10.1016/j.fuel.2023.130751

Bals B, Wedding C, Balan V, Sendich E, Dale B (2011) Evaluating the impact of ammonia fiber expansion (AFEX) pre-treatment conditions on the cost of ethanol production. Bioresour Technol 102(2):1277–1283. https://doi.org/10.1016/j.biortech.2010.08.058

Bastos RG (2018) Biofuels from microalgae: bioethanol. In: Green energy and technology, Vol. 0, Issue 9783319690926. Springer Verlag, pp 229–246. https://doi.org/10.1007/978-3-319-69093-3_11

Bauenova MO, Sadvakasova AK, Kossalbayev BD, Yilmaz G, Huang Z, Wang J, Balouch H, Zaletova DE, Lyaguta MA, Alharby HF, Allakhverdiev SI (2024) Optimising microalgae-derived butanol yield. Int J Hydrog Energy 49:593–601. https://doi.org/10.1016/j.ijhydene.2023.11.065

Bibi F, Yasmin H, Jamal A, AL-Harbi MS, Ahmad M, Zafar M, Ahmad B, Samra BN, Ahmed AF, Ali MI (2021) Deciphering role of technical bioprocess parameters for bioethanol production using microalgae. Saudi J Biol Sci 28(12):7595–7606. https://doi.org/10.1016/j.sjbs.2021.10.011

BNDES/CGEE (2008) Ethanol from sugarcane – energy for sustainable development. In: BNDES Brazilian Development Bank and CGEE Center of Strategic Studies (coedited with ECLAC and FAO) (Ed.), Ministério do Desenvolvimento, Indústria e Comércio Exterior

Bohutskyi P, Betenbaugh MJ, Bouwer EJ (2014) The effects of alternative pre-treatment strategies on anaerobic digestion and methane production from different algal strains. Bioresour Technol 155:366–372. https://doi.org/10.1016/j.biortech.2013.12.095

Bora A, Thondi Rajan AS, Ponnuchamy K, Muthusamy G, Alagarsamy A (2024) Microalgae to bioenergy production: recent advances, influencing parameters, utilization of wastewater – a critical review. Sci Total Environ 946:174230. https://doi.org/10.1016/j.scitotenv.2024.174230

Borines MG, de Leon RL, Cuello JL (2013) Bioresource Technology Bioethanol production from the macroalgae *Sargassum spp.* Bioresour Technol 138:22–29. https://doi.org/10.1016/j.biortech.2013.03.108

Cavelius P, Engelhart-Straub S, Mehlmer N, Lercher J, Awad D, Brück T (2023) The potential of biofuels from first to fourth generation. PLoS Biol 21(3):1–20. https://doi.org/10.1371/JOURNAL.PBIO.3002063

Chandra N, Shukla P, Mallick N (2020) Role of cultural variables in augmenting carbohydrate accumulation in the green microalga Scenedesmus acuminatus for bioethanol production. Biocatal Agric Biotechnol 26:101632. https://doi.org/10.1016/j.bcab.2020.101632

Chandrasekhar T, Varaprasad D, Gnaneswari P, Swapna B, Riazunnisa K, Anu Prasanna V, Korivi M, Wee YJ, Lebaka VR (2023) Algae: the reservoir of bioethanol. Fermentation 9(8):712. https://doi.org/10.3390/fermentation9080712

Chaudhary L, Pradhan P, Soni N, Singh P, Tiwari A (2014) Algae as a feedstock for bioethanol production: new entrance in biofuel world. Int J ChemTech Res 6(2):1381–1389. https://www.researchgate.net/publication/267752859

Chen CY, Chang HY, Chang JS (2016) Producing carbohydrate-rich microalgal biomass grown under mixotrophic conditions as feedstock for biohydrogen production. Int J Hydrog Energy 41(7):4413–4420. https://doi.org/10.1016/j.ijhydene.2015.05.163

Chia SR, Chew KW, Show PL, Yap YJ, Ong HC, Ling TC, Chang JS (2018) Analysis of economic and environmental aspects of microalgae biorefinery for biofuels production: a review. Biotechnol J 13(6). Wiley-VCH Verlag. https://doi.org/10.1002/biot.201700618

Chng LM, Lee KT, Chan DJC (2017) Synergistic effect of pre-treatment and fermentation process on carbohydrate-rich Scenedesmus dimorphus for bioethanol production. Energy Convers Manag 141:410–419. https://doi.org/10.1016/j.enconman.2016.10.026

Cho JM, Oh YK, Park WK, Chang YK (2020) Effects of nitrogen supplementation status on CO_2 biofixation and biofuel production of the promising microalga *chlorella sp.* ABC-001. J Microbiol Biotechnol 30(8):1235–1243. https://doi.org/10.4014/jmb.2005.05039

Choi SP, Nguyen MT, Sim SJ (2010) Bioresource technology enzymatic pre-treatment of *Chlamydomonas reinhardtii* biomass for ethanol production. Bioresour Technol 101(14):5330–5336. https://doi.org/10.1016/j.biortech.2010.02.026

Condor BE, de Luna MDG, Abarca RRM, Chang YH, Leong YK, Chen CY, Chen PT, Lee DJ, Chang JS (2022a) Optimization and modeling of carbohydrate production in microalgae for use as feedstock in bioethanol fermentation. Int J Energy Res 46(13):19300–19312. https://doi.org/10.1002/er.7709

Condor BE, de Luna MDG, Chang YH, Chen JH, Leong YK, Chen PT, Chen CY, Lee DJ, Chang JS (2022b) Bioethanol production from microalgae biomass at high-solids loadings. Bioresour Technol 363:128002. https://doi.org/10.1016/j.biortech.2022.128002

Dębowski M, Świca I, Kazimierowicz J, Zieliński M (2023) Large scale microalgae biofuel technology—development perspectives in light of the barriers and limitations. Energies 16(1). https://doi.org/10.3390/en16010081

Dhokane D, Shaikh A, Yadav A, Giri N, Bandyopadhyay A, Dasgupta S, Bhadra B (2023) CRISPR-based bioengineering in microalgae for production of industrially important biomolecules. Front Bioeng Biotechnol (11). Frontiers Media SA. https://doi.org/10.3389/fbioe.2023.1267826

Dragone G, Fernandes BD, Abreu AP, Vicente AA, Teixeira JA (2011) Nutrient limitation as a strategy for increasing starch accumulation in microalgae. Appl Energy 88(10):3331–3335. https://doi.org/10.1016/j.apenergy.2011.03.012

Dyah M, Meinita N, Kang J, Jeong G, Koo HM, Park SM (2012) Bioethanol production from the acid hydrolysate of the carrageenophyte Kappaphycus alvarezii (cottonii). 857–862. https://doi.org/10.1007/s10811-011-9705-0

Eldalatony MM, Kabra AN, Hwang JH, Govindwar SP, Kim KH, Kim H, Jeon BH (2016) Pre-treatment of microalgal biomass for enhanced recovery/extraction of reducing sugars and proteins. Bioprocess Biosyst Eng 39(1):95–103. https://doi.org/10.1007/s00449-015-1493-5

El-Dalatony MM, Salama ES, Kurade MB, Hassan SHA, Oh SE, Kim S, Jeon BH (2017) Utilization of microalgal biofractions for bioethanol, higher alcohols, and biodiesel production: a review. Energies 10(12). https://doi.org/10.3390/en10122110

El-Moustaqim K, Sbai SE, Mabrouki J, Hmouni D (2024) Critical review of green microalgal biomass development through CO_2 sequestration. In: Euro-Mediterranean Journal for Environmental Integration. Springer Nature. https://doi.org/10.1007/s41207-024-00548-6

Fetyan NAH, El-Sayed AEKB, Ibrahim FM, Attia YA, Sadik MW (2022) Bioethanol production from defatted biomass of Nannochloropsis oculata microalgae grown under mixotrophic conditions. Environ Sci Pollut Res 29(2):2588–2597. https://doi.org/10.1007/s11356-021-15758-6

Guv F, Fang Z, Charles Xu C, Smith RL (2012) Solid acid mediated hydrolysis of biomass for producing biofuels. Prog Energy Combust Sci:672–690

Guimarães CEC, Neto FS, Bizerra VC, Nascimento JGA, Valério RBR, Junior PGS, Braz AKS, Melo RLF, Serpa JF, Lima RKC, Guimarães AP, Souza MCM, Lopes AAS, Rios MAS, Desai AS, Bilal M, Smułek W, Jesionowski T, Santos JCS (2023) Sustainable bioethanol production from first- and second-generation sugar-based feedstocks: Advanced bibliometric analysis. Bioresource Technology Reports, 23, 101543. https://doi.org/10.1016/j.biteb.2023.101543

Hamouda RA, Sherif SA, Ghareeb MM (2018) Bioethanol production by various hydrolysis and fermentation processes with micro and macro green algae. Waste Biomass Valoriz 9(9):1495–1501. https://doi.org/10.1007/s12649-017-9936-7

Han T, Han X, Ye X, Xi Y, Zhang Y, Guan H (2022) Applying mixotrophy strategy to enhance biomass production and nutrient recovery of *Chlorella pyrenoidosa* from biogas slurry: an assessment of the mixotrophic synergistic effect. Bioresour Technol 366:128185. https://doi.org/10.1016/j.biortech.2022.128185

Harun R, Danquah MK, Forde GM (2010) Microalgal biomass as a fermentation feedstock for bioethanol production. J Chem Technol Biotechnol 85:199–203

Harun R, Danquah MK, Thiruvenkadam S (2014) Particulate size of microalgal biomass affects hydrolysate properties and bioethanol concentration. Biomed Res Int 2014:1. https://doi.org/10.1155/2014/435631

Hernández D, Riaño B, Coca M, García-González MC (2015) Saccharification of carbohydrates in microalgal biomass by physical, chemical and enzymatic pre-treatments as a previous step for bioethanol production. Chem Eng J 262:939–945. https://doi.org/10.1016/j.cej.2014.10.049

Ho M, Woo Y, Han C, Bum Y (2012) Bioresource Technology Simultaneous production of bio-ethanol and bleached pulp from red algae. Bioresour Technol 126:198–201. https://doi.org/10.1016/j.biortech.2012.08.102

Ho S, Huang S, Chen C, Hasunuma T, Kondo A (2013) Bioethanol production using carbohydrate-rich microalgae biomass as feedstock. Bioresour Technol 135:191–198. https://doi.org/10.1016/j.biortech.2012.10.015

Ho SH, Di Chen Y, Chang CY, Lai YY, Chen CY, Kondo A, Ren NQ, Chang JS (2017) Feasibility of CO_2 mitigation and carbohydrate production by microalga Scenedesmus obliquus CNW-N used for bioethanol fermentation under outdoor conditions: effects of seasonal changes. Biotechnol Biofuels 10(1):27. https://doi.org/10.1186/s13068-017-0712-5

Igwebuike CM, Awad S, Andrès Y (2024) Renewable Energy Potential: Second-Generation Biomass as Feedstock for Bioethanol Production. Molecules, 29, 1619. DOI: https://doi.org/10.3390/molecules29071619

Hwang JH, Kabra AN, Ji MK, Choi J, El-Dalatony MM, Jeon BH (2016) Enhancement of continuous fermentative bioethanol production using combined treatment of mixed microalgal biomass. Algal Res 17:14–20. https://doi.org/10.1016/j.algal.2016.03.029

Jabłońska-Trypuć A, Wołejko E, Ernazarovna MD, Głowacka A, Sokołowska G, Wydro U (2023) Using algae for biofuel production: a review. Energies 16(4). MDPI. https://doi.org/10.3390/en16041758

Jafari Olia MS, Azin M, Moazami N (2022) Application of a statistical design to evaluate bioethanol production from *Chlorella S4* biomass after acid – thermal pre-treatment. Renew Energy 182:60–68. https://doi.org/10.1016/j.renene.2021.10.019

Jain S, Kumar S (2024) A comprehensive review of bioethanol production from diverse feedstocks: current advancements and economic perspectives. Energy 296(February):131130. https://doi.org/10.1016/j.energy.2024.131130

Karthikeyan S (2023) Large scale cultivation and pre-treatment optimization of freshwater microalgae biomass for bioethanol production by yeast fermentation. Nat Environ Pollut Technol 22(2):1035–1040. https://doi.org/10.46488/NEPT.2023.v22i02.051

Kassim MA, Bhattacharya S (2016) Dilute alkaline pre-treatment for reducing sugar production from Tetraselmis suecica and *Chlorella sp.* biomass. Process Biochem 51(11):1757–1766. https://doi.org/10.1016/j.procbio.2015.11.027

Kassim MA, Ramli SH, Meng TK (2021) Analysis of microalgal growth kinetic model and carbohydrate biosynthesis cultivated using agro-industrial waste residuals as carbon source. Prep Biochem Biotechnol 52(5):514–524. https://doi.org/10.1080/10826068.2021.1969576

Khan MI, Lee MG, Seo HJ, Shin JH, Shin TS, Yoon YH, Kim MY, Choi JI, Kim JD (2016) Enhancing the feasibility of microcystis aeruginosa as a feedstock for bioethanol production under the influence of various factors. Biomed Res Int 2016:1. https://doi.org/10.1155/2016/4540826

Khan S, Das P, Abdul Quadir M, Thaher MI, Mahata C, Sayadi S, Al-Jabri H (2023) Microalgal feedstock for biofuel production: recent advances, challenges, and future perspective. Fermentation 9(3). MDPI. https://doi.org/10.3390/fermentation9030281

Kim N, Li H, Jung K, Nam H, Cheon P (2011) Bioresource technology ethanol production from marine algal hydrolysates using *Escherichia coli KO11*. Bioresour Technol 102(16):7466–7469. https://doi.org/10.1016/j.biortech.2011.04.071

Klinthong W, Yang YH, Huang CH, Tan CS (2015) A review: microalgae and their applications in CO_2 capture and renewable energy. In: Aerosol and air quality research, Vol. 15, Issue 2. AAGR Aerosol and Air Quality Research, pp 712–742. https://doi.org/10.4209/aaqr.2014.11.0299

Kouhgardi E, Zendehboudi S, Mohammadzadeh O, Lohi A, Chatzis I (2023) Current status and future prospects of biofuel production from brown algae in North America: progress and challenges. Renew Sust Energ Rev 172(May 2021):113012. https://doi.org/10.1016/j.rser.2022.113012

Kumar S, Gupta R, Kumar G, Sahoo D, Chander R (2013) Bioresource technology bioethanol production from *Gracilaria verrucosa*, a red alga, in a biorefinery approach. Bioresour Technol 135:150–156. https://doi.org/10.1016/j.biortech.2012.10.120

Kumar VB, Pulidindi IN, Kinel-Tahan Y, Yehoshua Y, Gedanken A (2016) Evaluation of the Potential of *Chlorella vulgaris* for Bioethanol Production. Energy Fuels 30(4):3161–3166. https://doi.org/10.1021/acs.energyfuels.6b00253

Kumar AN, Yoon JJ, Kumar G, Kim SH (2021) Biotechnological valorization of algal biomass: an overview. In: Systems microbiology and biomanufacturing, Vol. 1, Issue 2. Springer, pp 131–141. https://doi.org/10.1007/s43393-020-00012-w

Kusmiyati K, Hadiyanto H, Fudholi A (2023) Treatment updates of microalgae biomass for bioethanol production: a comparative study. J Clean Prod 383:135236. https://doi.org/10.1016/j.jclepro.2022.135236

Lee S, Oh Y, Kim D (2011) Converting carbohydrates extracted from marine algae into ethanol using various ethanolic Escherichia coli strains. Appl Biochem Biotechnol 164:878–888. https://doi.org/10.1007/s12010-011-9181-7

Li L, Gao K, Yang M, Zheng Q, Zhang M, Deng X (2023) Challenges and potential solutions of microalgae-based systems for wastewater treatment and resource recovery. Front Bioeng Biotechnol 11. Frontiers Media S.A. https://doi.org/10.3389/fbioe.2023.1210228

Liau BC, Shen CT, Liang FP, Hong SE, Hsu SL, Jong TT, Chang CMJ (2010) Supercritical fluids extraction and anti-solvent purification of carotenoids from microalgae and associated bioactivity. J Supercrit Fluids 55(1):169–175. https://doi.org/10.1016/j.supflu.2010.07.002

Liu Z, Hao N, Hou Y, Wang Q, Liu Q, Yan S, Chen F, Zhao L (2023) Technologies for harvesting the microalgae for industrial applications: current trends and perspectives. Bioresour Technol 387:129631. https://doi.org/10.1016/J.BIORTECH.2023.129631

Ma'mun S, Prasetio MW, Anugrah AR (2024) Bioethanol production from Spirulina (*Arthrospira platensis*) by microwave-assisted acid hydrolysis pre-treatment. Jurnal Bahan Alam Terbarukan 13(1):1–9. https://doi.org/10.15294/jbat.v13i1.49344

Maity S, Mallick N (2022) Bioprospecting marine microalgae and cyanobacteria as alternative feedstocks for bioethanol production. Sustain Chem Pharm 29:100798. https://doi.org/10.1016/j.scp.2022.100798

Mendez L, Mahdy A, Timmers RA, Ballesteros M, González-Fernández C (2013) Enhancing methane production of *Chlorella vulgaris* via thermochemical pre-treatments. Bioresour Technol 149:136–141. https://doi.org/10.1016/j.biortech.2013.08.136

Meng TK, Harun R, Kamaludin R, Kassim MA (2021) Microalgal biorefinery: challenge and strategy in bioprocessing of microalgae carbohydrate for fine chemicals and biofuel. J Appl Biol Biotechnol 9(6). https://doi.org/10.7324/JABB.2021.9601

Mirsiaghi M, Reardon KF (2015) Conversion of lipid-extracted nannochloropsis salina biomass into fermentable sugars. Algal Res 8:145–152. https://doi.org/10.1016/j.algal.2015.01.013

Morales-Sánchez D, Tinoco-Valencia R, Kyndt J, Martinez A (2013) Heterotrophic growth of *Neochloris oleoabundans* using glucose as a carbon source. Biotechnol Biofuels 6(1):1–13. https://doi.org/10.1186/1754-6834-6-100

Müller C, Scapini T, Rempel A, Abaide ER, Camargo AF, Nazari MT, Tadioto V, Bonatto C, Tres MV, Zabot GL, Colla LM, Treichel H, Alves SL (2023) Challenges and opportunities for third-generation ethanol production: a critical review. Eng Microbiology 3(1). https://doi.org/10.1016/j.engmic.2022.100056

Mussatto SI, Dragone G, Guimarães PMR, Silva JPA, Carneiro LM, Roberto IC, Vicente A, Domingues L, Teixeira JA (2010) Technological trends, global market, and challenges of bioethanol production. Biotechnol Adv 28(6):817–830. https://doi.org/10.1016/j.biotechadv.2010.07.001

Novoveská L, Nielsen SL, Eroldoğan OT, Haznedaroglu BZ, Rinkevich B, Fazi S, Robbens J, Vasquez M, Einarsson H (2023) Overview and challenges of large-scale cultivation of photosynthetic microalgae and cyanobacteria. Marine Drugs 21(8). Multidisciplinary Digital Publishing Institute (MDPI). https://doi.org/10.3390/md21080445

Odjadjare EC, Mutanda T, Chen YF, Olaniran AO (2018) Evaluation of pre-chlorinated wastewater effluent for microalgal cultivation and biodiesel production. Water (Switzerland) 10(8). https://doi.org/10.3390/w10080977

Oliveira J, Pardilhó S, Dias JM, Pires JCM (2023) Microalgae to bioenergy: optimization of *Aurantiochytrium* sp. Saccharification. Biology 12(7). https://doi.org/10.3390/biology12070935

Ometto F, Quiroga G, Pšenička P, Whitton R, Jefferson B, Villa R (2014) Impacts of microalgae pre-treatments for improved anaerobic digestion: thermal treatment, thermal hydrolysis, ultrasound and enzymatic hydrolysis. Water Res 65:350–361. https://doi.org/10.1016/j.watres.2014.07.040

Oraby S, Hegazy M, Labeeb H, Mahdy A (2023) Biochemical responses of chlorella vulgaris and Scenedesmus sp. under nutrient limitation. Sinai J Appl Sci 12(1). https://doi.org/10.21608/sinjas.2023.190718.1186

Pandit C, Pandit S, Kuhad RC, Ray S, Mishra SK, Mathuriya AS, Prasad R (2024) Microalgal bioethanol production for sustainable development: current status and future prospects. Indian J Microbiol. https://doi.org/10.1007/s12088-024-01364-w

Parakh SK, Tian Z, Wong JZE, Tong YW (2023) From microalgae to bioenergy: recent advances in biochemical conversion processes. Fermentation 9(6). MDPI. https://doi.org/10.3390/fermentation9060529

Park J, Hong J, Chul H, Geun S, Kim S, Yoon J, Jin Y (2012) Bioresource technology use of *Gelidium amansii* as a promising resource for bioethanol: a practical approach for continuous dilute-acid hydrolysis and fermentation. Bioresour Technol 108:83–88. https://doi.org/10.1016/j.biortech.2011.12.065

Pereira ASA d P, Silva TA d, Magalhães IB, Ferreira J, Braga MQ, Lorentz JF, Assemany PP, Couto E d A d, Calijuri ML (2024) Biocompounds from wastewater-grown microalgae: a review of emerging cultivation and harvesting technologies. Sci Total Environ 920:170918. https://doi.org/10.1016/j.scitotenv.2024.170918

Phwan CK, Chew KW, Sebayang AH, Ong HC, Ling TC, Malek MA, Ho YC, Show PL (2019) Effects of acids pre-treatment on the microbial fermentation process for bioethanol production from microalgae. Biotechnol Biofuels 12(1):191. https://doi.org/10.1186/s13068-019-1533-5

Pôjo V, Tavares T, Malcata FX (2021) Processing methodologies of wet microalga biomass toward oil separation: an overview. Molecules 26(3):641. https://doi.org/10.3390/molecules26030641

Reyimu Z, Özçimen D (2017) Batch cultivation of marine microalgae Nannochloropsis oculata and Tetraselmis suecica in treated municipal wastewater toward bioethanol production. J Clean Prod 150:40–46. https://doi.org/10.1016/j.jclepro.2017.02.189

Samadhiya K, Ghosh A, Bala K (2023) Scaling up of native species for a sustainable microalgal biorefinery targeting different microalgal products. Algal Res 75:103246. https://doi.org/10.1016/j.algal.2023.103246

Sanchez Rizza L, Sanz Smachetti ME, Do Nascimento M, Salerno GL, Curatti L (2017) Bioprospecting for native microalgae as an alternative source of sugars for the production of bioethanol. Algal Res 22:140–147. https://doi.org/10.1016/j.algal.2016.12.021

Santos FA, Queiróz JH d, Colodette JL, Fernandes SA, Guimarães VM, Rezende ST (2012) Potencial da palha de cana-de-açúcar para produção de etanol. Química Nova 35(5):1004–1010. https://doi.org/10.1590/s0100-40422012000500025

Santos FA, De Queiroz JH, Colodette JL, Manfredi M, Queiroz MELR, Caldas CS, Soares FEF (2014) Otimização do pré-tratamento hidrotérmico da palha de cana-de-açúcar visando à produção de etanol celulósico. Quim Nova 37(1):56–62. https://doi.org/10.1590/S0100-40422014000100011

Sathish A, Sims RC (2012) Biodiesel from mixed culture algae via a wet lipid extraction procedure. Bioresour Technol 118:643–647. https://doi.org/10.1016/j.biortech.2012.05.118

Scholz MJ, Riley MR, Cuello JL (2012) Acid hydrolysis and fermentation of microalgal starches to ethanol by the yeast Saccharomyces cerevisiae. Biomass Bioenergy 48:59–65. https://doi.org/10.1016/j.biombioe.2012.10.026

Shitanaka T, Fujioka H, Khan M, Kaur M, Du ZY, Khanal SK (2024) Recent advances in microalgal production, harvesting, prediction, optimization, and control strategies. Biores Technol 391. Elsevier Ltd:129924. https://doi.org/10.1016/j.biortech.2023.129924

Shokravi H, Heidarrezaei M, Shokravi Z, Ong HC, Lau WJ, Din MFM, Ismail AF (2022) Fourth generation biofuel from genetically modified algal biomass for bioeconomic development. J Biotechnol 360(September):23–36. https://doi.org/10.1016/j.jbiotec.2022.10.010

Silva CEF, Bertucco A (2019) Bioethanol from microalgal biomass: promising approach in biorefinery. Braz Arch Biol Technol 62

Singh AK (2024) Recent advancements in microalgae-based biofuel production. In: Shah M, Deka D (eds) Emerging sustainable technologies for biofuel production. Environmental science and engineering. Springer, Cham. https://doi.org/10.1007/978-3-031-52167-6_14

Smachetti MES, Rizza LS, Coronel CD, Do Nascimento M, Curatti L (2018) Microalgal biomass as an alternative source of sugars for the production of bioethanol. Princ Appl Ferment Technol 351–386. https://doi.org/10.1002/9781119460381.ch16

Tan ISLMKFHCYLSLKT (2019) Advances of macro algae biomass for the third generation of bio-ethanol production. Chin J Chem Eng:502–517

Tang S, Qin C, Wang H, Li S, Tian S (2011) Study on supercritical extraction of lipids and enrichment of DHA from oil-rich microalgae. J Supercrit Fluids 57(1):44–49. https://doi.org/10.1016/j.supflu.2011.01.010

Teymouri F, Laureano-Perez L, Alizadeh H, Dale BE (2005) Optimization of the ammonia fiber explosion (AFEX) treatment parameters for enzymatic hydrolysis of corn stover. Bioresour Technol *96*(18 SPEC. ISS):2014–2018. https://doi.org/10.1016/j.biortech.2005.01.016

Trivedi N, Gupta V, Reddy CRK, Jha B (2015) Bioresource technology enzymatic hydrolysis and production of bioethanol from common macrophytic green alga *Ulva fasciata Delile*. Bioresour Technol 150(2013):106–112. https://doi.org/10.1016/j.biortech.2013.09.103

Tsolcha ON, Patrinou V, Economou CN, Dourou M, Aggelis G, Tekerlekopoulou AG (2021) Utilization of biomass derived from cyanobacteria-based agro-industrial wastewater treatment and raisin residue extract for bioethanol production. Water (Switzerland) 13(4). https://doi.org/10.3390/w13040486

Udayan A, Sirohi R, Sreekumar N, Sang BI, Sim SJ (2022) Mass cultivation and harvesting of microalgal biomass: current trends and future perspectives. Bioresour Technol 344. Elsevier Ltd:126406. https://doi.org/10.1016/j.biortech.2021.126406

Vasić K, Knez Ž, Leitgeb M (2021) Bioethanol production by enzymatic hydrolysis from different lignocellulosic sources. Molecules 26(3). MDPI AG. https://doi.org/10.3390/molecules26030753

Wang J, Yin Y (2018) Fermentative hydrogen production using pretreated microalgal biomass as feedstock. Microb Cell Factories 17(1):22. BioMed Central Ltd. https://doi.org/10.1186/s12934-018-0871-5

Wang M, Ye X, Bi H, Shen Z (2024) Microalgae biofuels: illuminating the path to a sustainable future amidst challenges and opportunities. Biotechnol Biofuels Bioprod 17(1):10. BioMed Central Ltd. https://doi.org/10.1186/s13068-024-02461-0

Yin Z, Zhu L, Li S, Hu T, Chu R, Mo F, Hu D, Liu C, Li B (2020) A comprehensive review on cultivation and harvesting of microalgae for biodiesel production: environmental pollution control and future directions. Bioresour Technol 301(January):122804. https://doi.org/10.1016/j.biortech.2020.122804

Yirgu Z, Leta S, Hussen A, Khan MM, Aragaw T (2021) Optimization of microwave-assisted carbohydrate extraction from indigenous

Scenedesmus sp. grown in brewery effluent using response surface methodology. Heliyon 7(5):e07115. https://doi.org/10.1016/j.heliyon.2021.e07115

Zhang YT, Jiang JY, Shi TQ, Sun XM, Zhao QY, Huang H, Ren LJ (2019) Application of the CRISPR/Cas system for genome editing in microalgae. Appl Microbiol Biotechnol 103(8):3239–3248. Springer Science and Business Media Deutschland GmbH. https://doi.org/10.1007/s00253-019-09726-x

Zhu J, Wakisaka M, Omura T, Yang Z, Yin Y, Fang W (2024) Advances in industrial harvesting techniques for edible microalgae: recent insights into sustainable, efficient methods and future directions. J Clean Prod 436:140626. https://doi.org/10.1016/J.JCLEPRO.2024.140626

Biohydrogen from Microalgae: Future Perspectives and Constraints

9

Rodrigo T. Patino-Diaz, Blondy Canto-Canche, Hugo Virgilio Perales-Vela, Giuseppe Torzillo, Ana Margarita Silva-Benavides, Elda Espana-Gamboa, Liliana Alzate-Gaviria, and Ruby Alejandra Valdez-Ojeda

Abstract

The worldwide energy crisis has encouraged an international commitment to the most sustainable fuels, like biohydrogen, which is visualized as a strong candidate to meet energy needs. Biohydrogen production from microalgal activity has been addressed in this chapter, in which summarized information is included concerning (1) biophotolysis and the corresponding sulfur limitation as the strategy to inhibit enzymatic sensibility to oxygen presence, (2) dark fermentation (DF) strategies to improve hydrogen yield and the different alternatives of pretreatment (physical, chemical, and biological) to apply in microalgal biomass and favoring hydrogen production, (3) photo-fermentation by microalga per se or cocultivated with bacteria or using microalgal biomass under fermentative conditions, (4) integration of different fermentation processes, (5) improved electrochemical systems to get hydrogen from algal biomass and the implementation of nanomaterials in the different photobiological production processes, and (6) the recent advances in genetic engineering in microalgae. The challenge in all different biohydrogen production strategies through microalgae is still the operational complications with high-volume reactors.

Keywords

Microalgae-based hydrogen · Biophotolysis · Dark fermentation · Photo-fermentation · Nanotechnology · Genetic engineering

9.1 Introduction

Climate change is a global crisis mainly derived from the massive use of fossil fuels in the past century, with consequent carbon dioxide (CO_2) emissions and its accumulation in the atmosphere (IPCC 2023). Indeed, greenhouse gases, such as CO_2, produce dangerous planetary warming, which may cause catastrophic effects on the ecosystems. Alternative energy sources are being explored, but a bottleneck is certainly the intermittency of the systems, like photovoltaic or wind farms. Electricity is a vector that may be used to store and manage energy when a gap is found between the availability of the energy source and the requirement for its use (Bompard et al. 2020). However, in the past decades, molecular hydrogen (H_2) has been proposed as an alternative energy vector (Abdin et al. 2020), and in the context of biofuels, biohydrogen is an attractive product that can be obtained from various microorganisms.

The biological production of gaseous H_2 occurs in all three domains of life: eubacteria, archaea, and eukarya. In bacteria, hydrogen production occurs through the

R. T. Patino-Diaz
The Center for Research and Advanced Studies of the National Polytechnic Institute, Cinvestav – Unidad Merida, Merida, Yucatan, Mexico

B. Canto-Canche
Biotechnology Division, The Yucatan Scientific Research Center, Merida, Yucatan, Mexico

H. V. Perales-Vela
Laboratorio de Bioquimica, Unidad de Morfologia y Funcion, Facultad de Estudios Superiores Iztacala, Universidad Nacional Autonoma de Mexico, Tlalnepantla de Baz, Estado de Mexico, Mexico

G. Torzillo
Institute of BioEconomy, IBE, CNR, Florence, Italy

Centro de Investigacion en Ciencias Del Mar y Limnología, Universidad de Costa Rica, San Pedro, San Jose, Costa Rica

A. M. Silva-Benavides
Centro de Investigacion en Ciencias Del Mar y Limnología, Universidad de Costa Rica, San Pedro, San Jose, Costa Rica

E. Espana-Gamboa · L. Alzate-Gaviria · R. A. Valdez-Ojeda (✉)
Renewable Energy Division, The Yucatan Scientific Research Center, Merida, Yucatan, Mexico
e-mail: dubi@cicy.mx

D. U. Santos Ballardo, S. Rossi (eds.), *Microalgae as Promising Source of Commercial Bioproducts*, Developments in Applied Phycology 14, https://doi.org/10.1007/978-3-031-86433-9_9

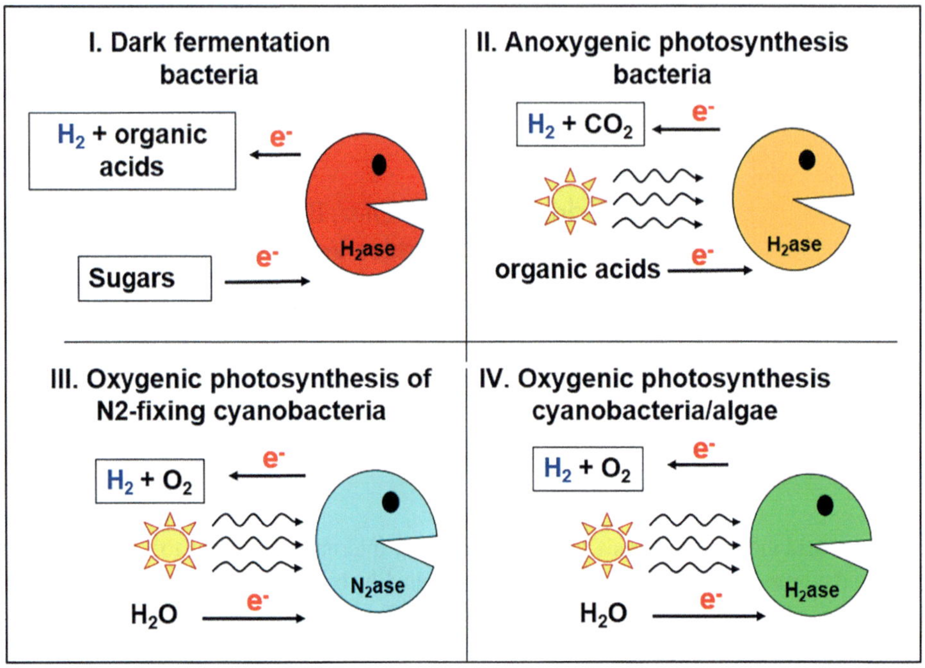

Fig. 9.1 Metabolic pathways for the biological production of hydrogen (H_2)

fermentation of substrates, primarily sugars. Fermentation does not need light, forming organic acids and hydrogen as products (Fig. 9.1., scheme I) (Redding et al. 2022).

In photo-fermentation, which occurs within the domains of archaea and eubacteria, hydrogen is produced through anoxygenic photosynthesis. The source of electrons is oxidizable substrates, such as organic acids, with a lower redox potential than water. These substrates can be oxidized by photosystems containing bacteriochlorophyll. Electrons are driven by light through electron carriers to hydrogenase, ultimately producing H_2 (Fig. 9.1., Scheme II).

Biological photoproduction of hydrogen exclusively occurs within the domains of eukaryotes and eubacteria, particularly among the group of cyanobacteria. This phenomenon is unique to oxygenic photosynthetic organisms (Fig. 9.1., schemes III and IV).

This chapter will present a revision of the advances in the state of the art to produce biohydrogen from microalgae. Initially, the oxygenic photosynthesis process is summarized, known as biophotolysis in cyanobacteria and microalgae metabolisms. This well-known phenomenon has been extensively studied and reported in the literature. Since the process produces oxygen and the enzymes involved are highly inhibited by its presence, a strategy to maintain hydrogen production is sulfur starvation with a change in the culture broth, as explained later. Dark fermentation with bacteria, but using algae as substrates, is also an interesting way to produce biohydrogen without light and leverage biomass waste from other bioprocesses with algae, always around the biorefinery concept. In order to take advantage of this biomass for fermentation, an important process is its pretreatment to hydrolyze the cells and liberate the nutrients; a review of the different biomass pretreatments is included in this work. Hydrogen yield production can be improved if fermentation is not only dark but light is added to the culture, as explained later in this chapter, in what is called photo-fermentation. Moreover, hydrogen production is also raised when coupling both dark and photo-fermentation, and some examples are also shown. It is also interesting to show additional alternatives for biohydrogen generation, such as microbial electrolysis cells (MECs), where electrochemical systems are coupled to microorganisms, the use of metal nanoparticles to improve hydrogen production during fermentation, or changes in metabolic capacities by genetic manipulation of microorganisms in order to optimize hydrogen generation.

9.2 Biophotolysis

Biological photoproduction of hydrogen exclusively occurs within the domains of eukaryotes and eubacteria, particularly among the group of cyanobacteria. This phenomenon, known as biophotolysis, is unique to oxygenic photosynthetic organisms (Fig. 9.1., schemes III and IV). It was recognized in the cyanobacterium *Anabaena* sp. more than 100 years ago by Jackson and Ellms in 1890, and the photoproduction of H_2 gas by microalgae was discovered by Gaffron and Rubin in 1942. Today, hydrogen production has been extensively studied in microalgae such as *Chlamydomonas reinhardtii*, *Scenedesmus obliquus*, *Chlorella vulgaris*, and many other green microalgae and cyanobacteria (Ghirardi 2015; Touloupakis and Torzillo 2019).

Biophotolysis (Fig. 9.2) can be carried out by nitrogen-fixing cyanobacteria using the enzyme nitrogenase (EC 1.18.6.1) and by cyanobacteria and microalgae using the enzyme hydrogenase (EC 1.12.99.6). This last enzyme can catalyze the formation and breakdown of H_2, the simplest molecule. Hydrogenases belong to the group of metalloproteins, which incorporate in their active site a redox element composed of a combination of atoms of nickel (Ni) and iron (Fe) or only iron (Fe). These metalloproteins are generally sensitive to oxygen due to their electronegativity, but they are also sensitive to high temperatures, carbon monoxide (CO), cyanide (CN), and other anions such as atrazine (N^{3-}) (Kovács et al. 2006; Redding et al. 2022).

A double function of hydrogenase in the cells of algae and cyanobacteria has been proposed: (1) obtaining chemical energy by oxidizing H_2 and (2) regulating redox balance when there is an excess of electrons moving through the transport chain and the Calvin–Benson cycle is not active (Melis et al. 2000; Redding et al. 2022).

In a comparative analysis of algae and cyanobacteria hydrogenases, Redding et al. (2022) reported that the hydrogenase of *Chlamydomonas reinhardtii*, characterized by an active site with Fe-Fe, exhibited an activity of 150 µmol/mg Chl·h. However, their hydrogenases are inhibited irreversibly, and thus, the synthesis of a new enzyme is required. This contrasts with the hydrogenase of *Synechocystis* sp. PCC 6803 showed an activity of 640 µmol/mg Chl·h and contained an active site composed of Ni-Fe; this enzyme can reactivate its activity by converting bound oxygen into water.

Biophotolysis occurs in photosynthetically active radiation (PAR, 400–700 nm) by both direct and indirect routes (Fig. 9.1). Direct production or biophotolysis has only been observed in green algae and cyanobacteria (Díaz-Cruz and Rodriguez-Torres 2023).

To describe both processes, it is necessary to indicate that the electromagnetic energy from the sun and the electrons provided by the photolysis of water are necessary elements for the photoproduction of H_2 in oxygenic photosynthetic algae. The protein nanomachinery needed to produce it resides in the thylakoidal membranes of microalgae and cyanobacteria, composed of the water oxidizer, photosystem II (P680), and the proton reducer, photosystem I (P700) (Redding et al. 2022).

The photo production of hydrogen in microalgae results from two pathways that provide electrons. One pathway involves the direct transport of electrons from the photolysis of water, while the other pathway is indirect, involving the contribution of electrons through the oxidation of starch and the entry of electrons via the NADH dehydrogenase complex type 1 (NDH-1), which incorporates electrons through plastoquinone (Fig. 9.3) (Laughlin et al. 2020). The general reaction of this process is:

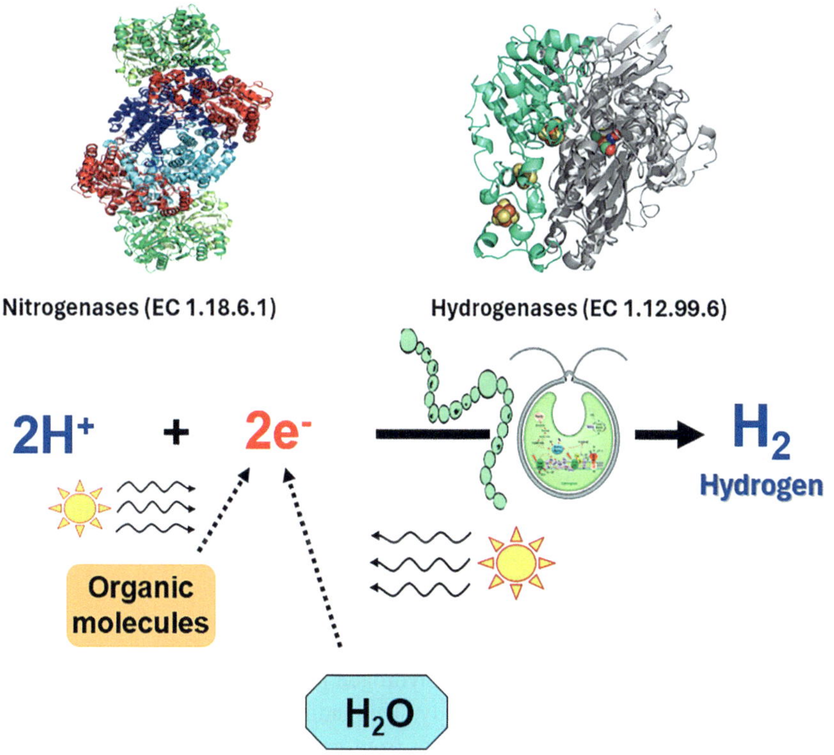

Fig. 9.2 Biophotolysis process for the production of H_2. (Protein structures obtained from: Protein Data Bank (PDB) https://www.rcsb.org/structure/8OIE; https://www.rcsb.org/6GM5)

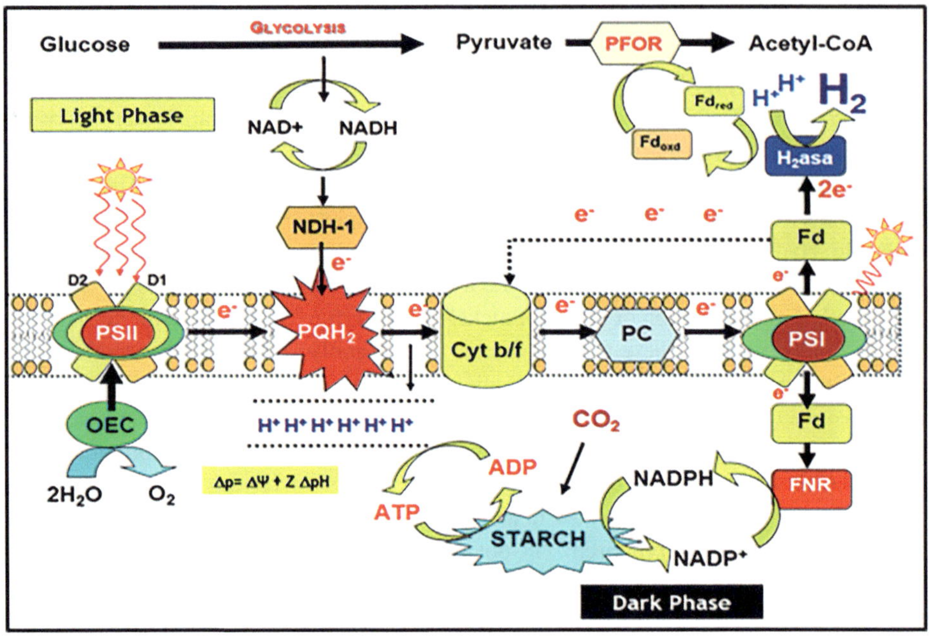

Fig. 9.3 Direct and indirect production of hydrogen (H_2) by biophotolysis. *PSII* photosystem II, *OEC* oxygen evolution complex, *PQH2* reduced plastoquinone, *Cytb/f* cytochrome b/f, *PC* plastocyanin, *PSI* photosystem I, *Fd* ferredoxin, *FNR* ferredoxin oxide reductase, *NDH-1* dehydrogenase complex I, *PFOR* pyruvate oxide/reductase

$$2H_2O(light) - O_2 \rightarrow 4H^+ + 4e^- \rightarrow \text{Hydrogenase} \rightarrow H_2 \quad (9.1)$$

The direct production of hydrogen, or biophotolysis, involves obtaining electrons through the oxidation of water in photosystem II. These electrons are then transported through the electron transport chain to photosystem I, which, in turn, reduces ferredoxin (Fd). Ferredoxin donates its electrons to hydrogenase, leading to the reversible production of diatomic hydrogen (H_2). This process is independent of the dark phase since electrons are diverted toward hydrogenase, and it does not produce NADPH, which is necessary for CO_2 fixation. However, electron transport establishes the electrochemical gradient necessary for ATP synthesis (Fig. 9.3).

In the direct photobiological production of hydrogen, the electron flux from the photolysis of water is necessary to produce hydrogen. Both photosystems II and I participate in reducing two protons ($2H^+$) to produce H_2. The electrons are extracted by water oxidation in photosystem II (PSII) with a redox potential change of $E_0 = +820$ mV. These electrons are eventually carried through electronic transporters to photosystem I (PSI), which guides them to the end of its acceptor side, ferredoxin (Fd), with a redox power change of $E_0 = -450$ mV. Ferredoxin then donates these electrons to hydrogenase, which has a redox potential change of $E_0 = -420$ mV and produces hydrogen.

Electrons coming from the photolysis of water are necessary for the direct production of hydrogen. However, they are not the only electrons participating in hydrogen production; others are added to this flow indirectly. This indirect route receives oxidation-reducing equivalents from carbohydrate stores such as starch through glycolysis, forming pyruvate. Reduced coenzymes can donate their electrons directly through the chloroplast dehydrogenase NDH-1, which reduces the quinone pool or cytochrome b_6/f. Pyruvic acid can also be decarboxylated, oxidized, and converted to acetyl-coenzyme A by pyruvate ferredoxin oxide reductase (PFOR), which directly reduces ferredoxin. Ferredoxin, in turn, reduces hydrogenase (H_2ase) to produce hydrogen (H_2) (Faraloni and Torzillo 2023; Wang et al. 2022).

The photobiological production of hydrogen using green microalgae has two significant limitations. The first is the low photosynthetic conversion of electromagnetic energy into chemical energy for hydrogen production. The second, and perhaps the most limiting, is the sensitivity of the enzyme hydrogenase to oxygen, which is manifested in enzymatic inactivation and a reduction of the expression of genes encoding hydrogenases (Faraloni and Torzillo 2023; Redding et al. 2022). Once the flow of electrons to hydrogenase ceases, the electrons are redirected toward ferredoxin oxide reductase to reduce $NADP^+$ to NADPH, a coenzyme necessary for the Calvin–Benson cycle (Fig. 9.3).

One of the most extensively studied algae species known for hydrogen production is *Chlamydomonas reinhardtii* (Ghirardi et al. 2000), which has demonstrated higher hydrogen production compared to *Scenedesmus obliquus* and *Chlorococcum littorale*, with values of 200, 150, and 55 nmoles H_2/mg Chla·h, respectively. Its easy cultivation and genetic manipulability have enabled researchers to study it under various experimental conditions to induce maximum hydrogen production (Faraloni and Torzillo 2023). Additionally, other genera of green algae, including *Scenedesmus obliquus*, *Chlorococcum littorale*, and

Chlorella fusca, have also been reported to produce hydrogen (Díaz-Cruz and Rodriguez-Torres 2023). Interestingly, researchers found no hydrogenase activity in *Chlorella vulgaris* and *Dunaliella salina* under anaerobic conditions, indicating the absence of hydrogenase in these organisms (Winkler et al. 2002).

9.3 Sulfur Starvation

The direct pathway through electrons from the photolysis of water that occurs in photosystem II is essential for H_2 production. However, this reaction inevitably produces oxygen as a by-product, which hampers the enzymatic activity of dehydrogenase, because oxygen in the cell inhibits the expression of genes coding for this enzyme (Ghirardi 2015; Redding et al. 2022). A potential solution to this challenge is the application of inert gases to culture systems, creating artificial anaerobiosis (Reeves and Greenbaum 1985). Nevertheless, this approach is operationally expensive, especially when dealing with reactors of significant culture volume (Faraloni and Torzillo 2023). To overcome this limitation, a biotechnological strategy involves the removal of sulfur from the culture medium (Melis et al. 2000), where the production of O_2 and H_2 are temporarily separated in two phases, resembling both direct and indirect routes (Redding et al. 2022).

The first phase is aerobic, during which the algae are cultivated in a complete medium under conditions that promote maximum photosynthetic growth. In this stage, cells utilize water as an electron donor for CO_2 fixation, enabling the doubling of the cells. Besides, cells naturally accumulate carbon reserves, especially carbohydrates like starch, which play a crucial role in hydrogen production (Fig. 9.4). In the second phase, the sulfur is removed from the culture medium. As a result, cells in the division process lack essential elements for proper protein synthesis, leading to a gradual impairment of photosystem II. This occurs because the biosynthesis of D1, a key protein located within the photosystem II, is impaired due to the inability of the chloroplast to synthesize sulfurous amino acids: cysteine and methionine (Wykoff et al. 1998). The D1 protein is associated with the oxygen-releasing complex, the reaction center, and the binding site of quinone B (QB) (Fig. 9.4).

During this phase, the cells continue electronic transport from water to $NADP^+$ but in significantly reduced amounts. Consequently, the oxygen concentration decreases by consumption during respiration, resulting in an anaerobic environment. This condition inhibits CO_2 fixation in the Calvin–Benson cycle, causing the electron transport route to shift toward hydrogenase, producing H_2. Simultaneously, this shift blocks the $NADP^+$ reduction pathway through the enzyme ferredoxin oxide reductase (FNR) (Fig. 9.4).

On the other hand, anaerobiosis also leads to the influx of reducing equivalents into the photosynthetic chain in NADH or NADPH, derived from the oxidation of the carbon reserves, primarily starch. These electrons contribute to what is known as the indirect route for H_2 production (Redding et al. 2022). Other electron transport complexes in the photosynthetic chain, such as cytochrome b6/f and photosystem I, remain unaffected compared to photosystem II. The Calvin-Benson cycle activity is inhibited and redirects electrons from ferredoxin to hydrogenase; hydrogenase can produce hydrogen in the absence of oxygen (Melis et al. 2000).

The sensitivity of photosystem II and one of its subunits, the D1 protein, has been another strategy to hasten the separation between the aerobic and anaerobic phases, leading to enhanced hydrogen production. Several authors have utilized D1 protein mutants to amplify the impact of sulfur removal from the culture medium on the activity of the oxygen-releasing complex. This approach aims to boost hydrogen synthesis through hydrogenase (Makarova et al. 2007; Torzillo et al. 2009). However, implementing this strategy in high-volume reactors has been challenging due to operational complications, and the search for strains and strategies enabling the establishment of high-volume reactors is ongoing (Ghirardi 2015; Faraloni and Torzillo 2023).

9.4 Dark Fermentation

Beyond hydrogen production by microalgae, dark fermentation is a strategy to produce hydrogen from bacteria, where microalgae may be the carbon source. Compared with other bioprocesses that produce H_2, dark fermentation is recognized as the one with higher production rates and lower costs (Aslam et al. 2023; Dincer and Acar 2015). Microalgae, like any cell, are composed of carbohydrates, lipids, and proteins, nutrients that may be leveraged by fermentative bacteria (Lacroux et al. 2023). This method may be helpful in treating microalgal biomass, for instance, after hydrogen production by photo-fermentation and photolysis processes. This approach can contribute to completing a biorefinery scheme around biohydrogen production with microalgae. Evidently, dark fermentation is highly dependent on the metabolism of the specific microorganism, temperature, nutrients, pH, or even the type of reactor.

Microorganisms producing hydrogen through dark fermentation could be either obligated or facultative anaerobes (Aslam et al. 2023; Bhatia et al. 2023; Lacroux et al. 2023). Table 9.1 has a list of different microorganisms that have been reported to produce hydrogen by using microalgae as feedstocks. Moreover, the main metabolic pathways for these processes are acetogenesis (Eq. 9.1) and butyric acid production (Eq. 9.2):

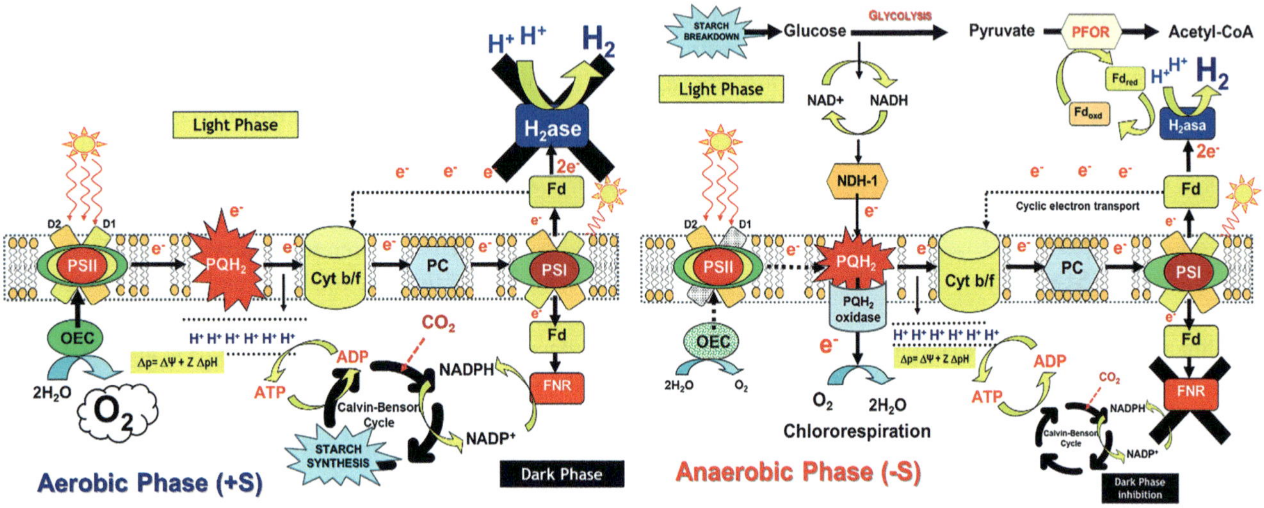

Fig. 9.4 Electronic transport in the aerobic (+S) and anaerobic (−S) phases. In the aerobic phase, the culture medium is complete, and electrons flow from water (producing oxygen) via photosystem II (PSII)–photosystem I (PSI) to FNR, producing NADPH. In the anaerobic phase, sulfur is removed from the culture medium, and electrons flow from water (producing oxygen) via photosystem II (PSII)–photosystem I (PSI) to the hydrogenase (H_2ase), producing hydrogen (H_2)

$$C_6H_{12}O_6 + 2H_2O \rightarrow 2CH_3COOH + 2CO_2 + 4H_2 \quad (9.2)$$

$$C_6H_{12}O_6 + 2H_2O \rightarrow 2CH_3CH_2CH_2COOH + 2CO_2 + 2H_2 \quad (9.3)$$

In both pathways, glucose oxidation in an aqueous medium is not totally completed and could theoretically liberate six hydrogen molecules per glucose molecule. Instead, partial oxidation is derived from synthesizing carboxylic acids with carbon dioxide and hydrogen as by-products. While the carbon dioxide may be coupled to the microalgae growth process in which photosynthesis is essential, gas separation should be implemented to recuperate pure hydrogen (Salakkam et al. 2021). On the other hand, carboxylic acids should also be directed to any commercialization following the concept of the biorefinery already mentioned.

About the substrate obtained from microalgal cultures, it should be important to consider a pretreatment to liberate carbohydrates from the algae cells and hydrolyze them into monosaccharides. Proteins and lipids can also be hydrolyzed before fermentation to produce hydrogen. However, the individual processes have low yields; they add to the total yield in the overall dark fermentation (Salakkam et al. 2021). However, it is important to consider how to leverage those cell components that may remain as waste from this process; a proposal in this sense is to use the residuals from the pretreatment as feedstocks in a second fermentation process, this time through methanogenic microorganims, in order to have a total coproduction of hydrogen and methane (Ding et al. 2020; Wieczorek et al. 2014).

Physical, chemical, biological, or combined methods are used for the pretreatment of microalgae biomass. Although these methods are explained in the next section, they are briefly described here. Among the physical methods, homogenization, grinding, autoclaving, and sonication have been compared to obtain fermentable saccharides (Kumar et al. 2016; Singh et al. 2022); other tested methods are hydrothermal (Cheng et al. 2019) and microwave (Yin et al. 2019) treatments, as well as gamma irradiation (Chen et al. 2022). The chemical treatments involve both acid (Giang et al. 2019; Sivagurunathan et al. 2017) or basic hydrolysis (de Oliveira et al. 2022; Yang et al. 2010; Yin and Wang 2019c), with the disadvantage of producing furfural, a substance that diminishes the hydrogen yield during dark fermentation. Finally, the biological methods are related to the use of purified enzymes (Nayak et al. 2014) or a cocktail or consortium of enzymes (Abdel-Kader et al. 2022; Bhujbal et al. 2022; Botta et al. 2020; Thongbunrod and Chaiprasert 2021). It is important to consider a pretreatment or a combination of them, not only based on the final hydrogen yield but also on materials, technical devices, or energy costs.

Once the substrate from pretreated microalgae is obtained, dark fermentation should be optimized, defining conditions for the specific process. In order to know these conditions, a combination matrix with different operation variables may be useful to identify how the process may produce an attractive hydrogen yield, considering production costs. Among these conditions to test during dark fermentation are the inoculum/substrate ratio (Taşkan et al. 2022), pH (Kumar et al. 2013; Liu et al. 2012; Qiu et al. 2017), temperature

Table 9.1 Microorganisms involved in dark fermentation to produce hydrogen by using microalgae as feedstocks

Fermentative microorganisms	Microalgae	Reference
Bacillus firmus/Rhodobacter sphaeroides	*Spirulina platensis*	Pandey et al. (2021)
Brevibacillus invocatus	*Oscillatoria angustissmia* and *Synchocsyts* sp.	Gad El-Rab et al. (2018)
Clostridium acetobutylicum	*Arthrospira platensis* and *Chlorococcum* sp.	Chernova and Kiseleva (2017)
Clostridium acetobutylicum	*Nannochloropsis* sp.	Efremenko et al. (2012)
Clostridium butyricum	*Scenedesmus obliquus*	Batista et al. (2018); Ferreira et al. (2013)
Clostridium butyricum	*Chlorella vulgaris*	Liu et al. (2012)
Enterobacter aerogenes	*Scenedesmus obliquus*	Batista et al. (2018)
Enterobacter aerogenes	*Nannocloropsis* sp.	Nobre et al. (2013)
Enterobacter cloacae	*Chlorella sorokiniana*	Kumar et al. (2013)
Rhodobacter sp.	*Chlorella fusca*	Abdel-Kader et al. (2022)
Rhodospirillum sp.	*Chlorella* sp.	Liu et al. (2020)
Thermococcus eurythermalis	*Dunaliella primolecta*	Chen et al. (2020)
Thermotoga neopolitana	*Chlamydomonas reinhardtii*	Nguyen et al. (2010)
Mixed consortia (isolated or natural cultures)	*Anabaena* sp., *chlorella* sp., *Nannochloropsis* sp., *Scenedesmus* sp.	Bhatia et al. (2023); Nagarajan et al. (2017)

Adaptation from Aslam et al. (2023), Bhatia et al. (2023), Nagarajan et al. (2017)

(Gadow et al. 2012), and retention time (Martins et al. 2022). Furthermore, other strategies have been proposed to improve hydrogen yield: detoxification of the pretreated microalgae—mainly when producing furfurals (Gonzalez-Fernandez et al. 2015; Kumar et al. 2016; Park et al. 2011), use of nanocatalysts (Srivastava et al. 2020), co-digestion of microalgae biomass with other substrates (Salakkam et al. 2021), co-fermentation (Ding et al. 2016; Jehlee et al. 2019; Xia et al. 2016; Xu et al. 2019), combination of dark fermentation with photo-fermentation (Cheng et al. 2012; Zheng et al. 2010), or, as already mentioned, coproduction of methane.

One of the future challenges is the number of research that remains to be developed in the following years in order to find more insights to improve the dark fermentation processes using microalgae as feedstocks, scaling from laboratory assays to pilot plants, and then commercialize, maintaining the concept of biorefinery while combining with other processes and avoiding residuals. It is also important to consider a sustainable approach, in which raw materials, energy, and residuals should be correctly balanced, and environmental and social impacts should not be negatives when the overall life cycle is proposed. Among specific challenges, the microalgae biomass production, harvesting, and pretreatment should be improved, as well as the specific dark fermentation processes.

9.5 Pretreatment of Algal Biomass to Biohydrogen Production

In order to obtain effective hydrolysis of algal biomass and subsequently produce biohydrogen, it is necessary to apply a pretreatment, which has the aim of breaking the cell wall of algal biomass and obtaining simple compounds from the complex mixture, such as fermentable sugars, one of the major prerequisites for microbial fermentation. According to what was found in the literature, pretreatment increases H_2 production by up to 50–70% (Goswami et al. 2021; Sirohi et al. 2021). As mentioned before, pretreatments are mainly classified as physical, chemical, and biological. None of these methods is universally applicable because the characteristics of microalgal biomass are diverse, and pretreatment conditions vary from strain to strain (Bhatia et al. 2023).

9.5.1 Physical Pretreatments

It has been reported that physical pretreatments improve hydrogen production to a certain extent, increasing it from 51% to 262% of the hydrogen amount compared to untreated biomass (Bhatia et al. 2023). These methods change the microalgal biomass's size, surface area, polymerization degree, or crystallinity index. They include direct steps for cellular breakdown, which leave a less environmental impact but are high-energy and power-intensive techniques. Therefore, they have a high production cost (Sirohi et al. 2021). The more employed physical pretreatments are sonication, ultrasonication, milling, autoclaving (hydrothermal), and microwave.

9.5.1.1 Sonication or Ultrasonication

Sonication or ultrasonication applies ultrasound waves at a frequency range from 10 kHz to 20 MHz (Karishma et al. 2022; Sirohi et al. 2021). This method changes the electrical energy into mechanical energy, and microbubbles are formed. Shock waves create a high pressure during the cavitation process (bubble implosion) for the cell wall breakdown, facilitating the intracellular cell breakdown. The breakdown frequency depends on the nature of the microalgae biomass, such as density, shape, size, and cell suspension, among others. Ultrasonic pretreatment is considered a powerful method for degradability; however, it is energy-intensive and energy-demanding, which restricts its application. Kant Bhatia et al. (2023) and Karishma et al. (2022) mentioned that, at high frequencies, the formation of free radicals causes oxidative damage in the biomass, altering the chemical composition and decreasing the biohydrogen yield.

9.5.1.2 Milling

This technique applies shear forces on the microalgae biomass to break the cell wall for releasing cellular content. The milling has several benefits, including improved surface area by lowering particle size, easy operation, and the ability to be conducted efficiently on a large scale (Priya et al. 2023). The materials of used beads include quartz, ceramic, glass, and steel, and several factors like biomass concentration, bead properties (shape, size, density, and composition), structure and integrity of the cell wall, chamber design, agitation speed, and exposure time are required to optimize the cell breakdown (Sirohi et al. 2021). The main disadvantage of this process is the enormous amount of energy required to operate the grinding machines and mills and the high costs, which are the key drawbacks that limit its large-scale implementation (Priya et al. 2023). However, it is reported that milling-assisted pretreatment for *S. obliquus* was effective in enhancing the production of biohydrogen from 32.7 to 116.3 mL H_2 g^{-1} VS (El-Dalatony et al. 2020).

9.5.1.3 Autoclaving (Hydrothermal)

This pretreatment employs high temperatures to disintegrate the cellular structure, improve biomass solubilization, and enhance the availability of organic compounds and sugars for biohydrogen production (El-Dalatony et al. 2020; Priya et al. 2023; Sirohi et al. 2021). Moreover, temperature and time are the primary controlling factors influencing the degree of hydrolysis (Priya et al. 2023). Besides, Kant Bhatia et al. (2023) mentioned that the optimum temperature and time for algal biomass pretreatment for hydrothermal processes are 135–140 °C and 15–20 min, respectively. However, this time and temperature may differ according to the microalgal strain (Sirohi et al. 2021). In addition, it is reported that when the pretreatment temperature is below 135 °C and the time is less than 15 min, most recalcitrant components, such as cellulose, are not hydrolyzed efficiently to smaller sugar molecules.

On the other hand, an increase in temperature beyond the threshold value (above 140 °C for 20 min) affects the pretreatment efficiency. It can lead to the formation of toxic by-products such as 5-hydroxymethylfurfural furfural (5-HMF), one of the major intermediate products in the Maillard reaction between reducing sugars and proteins, that can inhibit the activity of fermentative bacteria involved in biohydrogen production. In addition, inhibitory organic acids such as levulinic and formic acids generated from 5-HMF accumulate in the fermentation medium, reducing the biocatalyst activity and inhibiting biohydrogen production (Bhatia et al. 2023; Priya et al. 2023).

9.5.1.4 Microwave

This treatment uses electromagnetic microwaves in the frequency range from 0.3 to 300 GHz. Microwaves interact with polar and dielectric molecules of the cell suspension, which consequently increases the pressure and heat to break the cell wall (Karishma et al. 2022; Sirohi et al. 2021). The microwave method is a widely used pretreatment technique owing to its efficiency and ease of use; it is also a selective process and has a rapid pretreatment time, and the reaction rate can be accelerated and controlled during this treatment process. In algae biomass, a wavelength of 2450 MHz microwave is enforced to disrupt the cell wall. Priya et al. (2023) mentioned that microwave irradiation at 160 °C for 30 min enhanced biohydrogen production from *Laminaria japonica* by 1.9-fold. Also, biohydrogen production increases when the algae biomass is pretreated with microwave and an oxidizing agent such as H_2O_2 and iron oxide nanoparticles. However, overheating, nonuniform heating, higher cost of maintenance, and the involvement of high energy and operational costs limit its application in large-scale processes (Karishma et al. 2022; Sirohi et al. 2021).

9.5.2 Chemical Pretreatment

The chemical pretreatment consists of the use of acids, alkalis, surfactants, organic solvents, and alcohols for hydrolysis of the cell wall of algal biomass (El-Dalatony et al. 2020; Priya et al. 2023; Sirohi et al. 2021). This pretreatment method has several advantages over mechanical methods: it does not require an additional energy source, it is a quick process, and it separates other biomolecules or pigments; these characteristics make it one of the easiest and least expensive methods (Goswami et al. 2021; Sirohi et al. 2021). Moreover, it is reported that chemical pretreatment using the acid/alkaline method combines saccharification and liquefaction in a single step, which improves the hydrogen yield on dark fermentation. However, it has some disadvantages, such as corrosion in bioreactors and issues with recycling and draining solvents, and the development of specific inhibitors for hydrogen production, such as furfurals, are reported in this kind of treatment (Goswami et al. 2021; Priya et al. 2023).

9.5.2.1 Acid Treatment

Acid-based hydrolytic pretreatment stimulates the degradation of cell wall polymers, leading to cell breakdown of algal biomass, and improves hydrogen production through dark fermentation. The main dilute acids used are sulfuric acid (H_2SO_4), nitric acid (HNO_3), phosphoric acid (H_3PO_4), and hydrochloric acid (HCl). Different studies have mentioned that H_2SO_4 speeds up the sugar recovery rate and hydrogen output over other dilute acids; however, the efficiency of pretreatment depends on the type of algal biomass, biomass loading, treatment temperature, acid concentration, and hydrolysis time (Bhatia et al. 2023; Karishma et al. 2022; Priya et al. 2023; Sirohi et al. 2021). Furthermore, Kant Bhatia et al. (2023) reported that 1079 mL/L of hydrogen was produced when the biomass of *Chlorella* sp. was pretreated with 0.75% v/v H_2SO_4, 160 °C, 30 min, and 40 g biomass/L, which was higher than the not-pretreated biomass.

9.5.2.2 Alkali Treatment

Alkali pretreatment is preferred for breaking the cell wall components and releasing the biochemicals suitable for converting different products from microalgae. These processes create saponification and solvation reactions, forming pores in the cell wall and releasing intracellular components from the cell wall. It also decreases the size and crystallinity of starch and cellulose. Monovalent alkalis lead to better solubilization when compared with divalent alkalis since divalent alkalis ionize only partially in aqueous media, and the main alkaline compounds used are calcium hydroxide [$Ca(OH)_2$], sodium hydroxide (NaOH), potassium hydroxide (KOH), and ammonium hydroxide (NH_4OH) (Bhatia et al. 2023; Karishma et al. 2022; Priya et al. 2023; Sirohi et al. 2021).

9.5.3 Biological Pretreatment

Biological pretreatment consists of using microorganisms containing various types of biological enzymes or a cocktail of enzymes with the potential to degrade the polymeric compounds on the microalgal cell wall and convert these compounds into simple fermentable feedstocks for hydrogen production. This pretreatment prevents the production of furfurals, and the main enzymes used in the pretreatment of algal biomass include cellulose, hemicellulose, pectinase, protease, amylase, lysozyme, and papain, among others. Biological pretreatment is considered an effective and environmentally friendly process with low energy requirements compared to chemical and mechanical pretreatments. Besides, this pretreatment process can be categorized into enzymatic pretreatments and microbial pretreatments (bacterial, fungal, and microbial consortia) (Bhatia et al. 2023; Goswami et al. 2021; Karishma et al. 2022; Priya et al. 2023; Sirohi et al. 2021).

9.5.3.1 Enzymatic Pretreatments

Enzymes are biocatalysts having highly specific proteins, so the degradation of compounds is selective and reduces the risk of generating compounds that interfere with hydrogen production, such as furfurals. The enzymes can be used in a pure form or a cocktail for cell wall degradation on microalgal biomass. Sirohi et al. (2021) mentioned that enzyme cocktails are more efficient than single enzymes for microalgae biopolymer cell breakdown. Moreover, specific enzyme requirements depend on the strain type, as it differs in cell wall structure and stored polysaccharide content; however, cellulose is the main compound in the cell wall on microalgal biomass, and the cellulolytic enzymes break the algal biomass into compounds with five or six carbon atoms, such as glucose, xylose, arabinose, or mannose, which can be utilized by microorganisms in the fermentation process for hydrogen production (Priya et al. 2023). Kant Bhatia et al. (2023) reported cyanobacterial biomass with enzymatic treatment that produced 1600 mL/L of hydrogen during dark fermentation, whereas untreated algal biomass could not produce any hydrogen. However, although enzymatic pretreatment has several benefits, it also has limitations like the high cost of hydrolytic enzymes, which reduce its use when compared to other available methods; in addition, it is a slow process, requiring proper time and optimum conditions for its effective work (El-Dalatony et al. 2020; Goswami et al. 2021).

9.5.3.2 Microbial Pretreatments

Microalgal pretreatment with hydrolytic microorganisms, such as fungi and bacteria, is a promising green process for the solubilization of polymers due to its fast growth, shorter time, and high metabolic activities (Sirohi et al. 2021). In this process, the microorganisms produce different kinds of enzymes, such as xylanases, cellulases, and laccases, that can hydrolyze the cell wall in algal biomass and release single compounds able to be transformed into biohydrogen. The main difference between enzymatic pretreatments is that, in this process, the microorganism can produce enzymes in continuous mode. Hence, it does not affect the problem of enzyme deactivation. This characteristic makes microbial pretreatment more cost-effective than pure enzyme purchasing (Sirohi et al. 2021).

As can be seen, there are different methods of pretreatment of microalgal biomass for hydrogen production. Selecting the most appropriate one will depend on the characteristics of the microalgal biomass and the cost-benefit.

9.6 Photo-Fermentation

As a strategy for hydrogen production, photo-fermentation offers the advantages of biophotolysis by utilizing light energy, as well as dark fermentation through the degradation of organic material. Besides, researchers have reported successful H_2 production from photo-fermentation using *Chlamydomonas*, *Chlorella*, and *Scenedesmus*, among others (Putatunda et al. 2023). One of the first studies about microalgae photo-fermentation for hydrogen production was with *Chlamydomonas reinhardtii*. This species demonstrated, as reported by René Gfeller and Martín Gibbs (1984), higher hydrogen production in light than in the dark, with a difference of 6.49 mol H_2 per mol of glucose, additional to other fermentation products such as formate, ethanol, and glycerol, although with minor production from H_2O and starch, endogenously synthesized. Interestingly, a decrease in starch breakdown due to a light presence, reverted by the proton ionophore carbonyl cyanide p-trifluoromethoxy-phenylhydrazone (FCCP) (Gfeller and Gibbs 1984), indicating that light provokes an energy exchange that probably affects the enzymes involved in glycolysis or pyruvate oxidation.

$$CH_3COOH + 2H_2O \rightarrow 4H_2 + 2CO_2 \qquad (9.4)$$

To maximize hydrogen production yield, other strategies based on microalgae cocultivation with bacteria have been implemented, which is advantageous because bacteria consume O_2 by respiration at low light intensity, creating an anoxic environment for microalgae hydrogenase enzyme activity. This results in higher H_2 production compared to microalgae pure cultures; in fact, studies suggest that it can obtain a four-times higher yield of H_2 in comparison (Li et al. 2013). However, this depends on the added carbon source, bacteria species, and light intensities. To this respect, it has been demonstrated that acetic acid in the medium reduces the photosynthetic activity of *Chlamydomonas* for O_2 generation and CO_2 fixation and improves mitochondrial respiration (Asada and Miyake 1999); therefore, it is important to choose the carbon source. Concerning light intensities, low intensity is preferred in low cell density cultures with acetate-containing media because it promotes the consumption of O_2 from headspaces to reach anaerobic condition after 24 h, the appropriate conditions for H_2 production (Jurado-Oller et al. 2015). The coculturing of *Chlamydomonas* with bacteria species such as *Escherichia colli*, *Pseudomonas putida*, and *Pseudomonas stutzeri* results in the assimilation of acetic acid and consumes O_2 in the headspace, which improves H_2 photoproduction, although with evidently diminished *Chlamydomonas* growth (Fakhimi and Tavakoli 2019).

Certainly, bacteria and microalgae are biologically convenient for H_2 production because neither bacteria pure culture can produce hydrogen nor microalgae alone can produce high H_2 yields. However, it is important to find the optimal conditions for each microalgae–bacteria pairing to reach high yields of H_2.

Another alternative is using microalgae biomass (rich in carbohydrates, lipids, and proteins) associated with bacteria. This kind of organic material is advantageous in economic terms due to its multiplication in a short time and fast-growing carbohydrates that anaerobic organisms can easily digest (Karishma et al. 2022). Microalgae lack complex structural components such as lignin and hemicellulose, making hydrolysis easier. Different pretreatments, such as physical (microwave and ultrasonication), biological (using enzymes), chemical (acid or base reactants), or combined methods, have been evaluated to improve the microbial accessibility and solubility of algal biomass (Karishma et al. 2022).

The most common fermentative microorganisms include *Clostridium* sp., *Klebsiella* sp., *Enterobacter* sp., and *Citrobacter* sp. (Shobana et al. 2017). Moreover, purple non-sulfur bacteria (PNSB) form another group that exhibits metabolic versatility and can grow in different conditions, including photo-fermentation. This occurs through diverse metabolic pathways, depending on the PNSB species and the carbon source. Under photo-fermentation conditions, PNSB use a single photosystem that resembles photosystem II, incapable of oxidizing water, and thus, no O_2 is produced. PNSB oxidize the organic compounds (such as microalgae biomass) through different oxidative pathways to obtain carbon and electrons (Hädicke et al. 2011). Then, electrons are energized by the photosystem using light and channeled through an H^+-pumping electron transfer chain.

Meanwhile, protons can make ATP by ATP-synthase; the electrons can be donated to $NADP^+$ to generate NADPH for biosynthesis through reverse electron transfer. Parallelly, electrons are repeatedly energized and cycled through the electron transfer chain, which allows the continuous maintenance of proton motive force and ATP pools. This process, called cyclic phosphorylation, is advantageous as cycling a few electrons can generate usable energy for cell maintenance and repair (McKinlay 2014). However, the excess of reactants must be oxidized to maintain its homeostasis through CO_2 fixation via the Calvin-Benson cycle or through H_2 production by the hydrogenase activity (at the expense of 4 mol ATP per mole H_2 released) or by the reverse direction of the uptake hydrogenase (without consumption of ATP). The photoheterotrophic conditions have the advantage that the high energy cost of hydrogen production by enzymatic activity could be covered by the ATP generated by photosynthesis (Hädicke et al. 2011).

On the other hand, nitrogenase enzyme presence has been documented in some PNSB species, bacteria, and cyanobacteria. It is better known for producing NH_3 from N_2, a process known as nitrogen fixation. However, as it was referred before, H_2 is an obligate product to alleviate reductant excess. In fact, under photoheterotrophic conditions, the main excretion route of H_2 is via nitrogenase or hydrogenase (Hädicke et al. 2011). Also, nitrogenase is conformed by over 20 accessory genes that, for its proper assembly and the individual subunits of the active enzyme, must associate and dissociate eight times in one catalytic cycle to convert one mole N_2 into two mol NH_3, expending 16 ATPs in the process (McKinlay 2014).

Nitrogenase is an enzymatic complex consisting of two proteins named according to their metal composition, MoFe protein (dinitrogenase) and Fe protein (dinitrogenase reductase), which catalyze the reduction of dinitrogen to ammonia (Hädicke et al. 2011). Besides, nitrogen fixation represents significant demands on cellular energy supplies, that is, a large amount of ATP and reducing power. Therefore, the synthesis and activity of nitrogenase are strictly regulated at the transcriptional level and posttranslational level in response to environmental stimuli, such as the absence or presence of fixed nitrogen, ammonium, and/or oxygen (Akköse et al. 2009). The switch-off/switch-on mode of nitrogenase in *Rhodobacter capsulatus* (Yakunin and Hallenbeck 1998) modulates its in vivo activity in a reversible way. A similar mechanism is in *Rhodobacter sphaeroides* O.U.001, NH_4^+ concentration or aerobic conditions modified the gene expression of nitrogenase enzymes (*nif* genes) (Akköse et al. 2009). Nitrogenase activity is proportional to the amount of added NH^+_4, similar to a full recovery of the initial rate of nitrogenase activity (Yakunin and Hallenbeck 1998). Nitrogenase produces hydrogen gas while catalyzing the reduction of dinitrogen to ammonia; however, in the absence of N_2, nitrogenases behave like hydrogenases, producing H_2 as the sole product.

$$\begin{aligned}\text{Mo}-\text{nitrogenase}: \text{N}_2 + 8\text{H}^+ + 8e^- + 16\text{ATP} \\ \rightarrow 2\text{NH}_3 + \text{H}_2 + 16\text{ADP} + 16\text{Pi} \end{aligned} \tag{9.5}$$

$$\begin{aligned}\text{Mo}-\text{nitrogenase in the absence of N}_2: \\ 8\text{H}^+ + 8e^- + 16\text{ATP} \rightarrow 4\text{H}_2 + 16\text{ADP} \end{aligned} \tag{9.6}$$

Furthermore, there are more differences among hydrogenases: their turnover rate is an order of magnitude lower than that of Ni-hydrogenase and three orders of magnitude lower than that of Fe-hydrogenase. Its enzymatic activity requires ATP (4 mol ATP per mole H_2 released), which is easily obtained by cyclic photophosphorylation, whereas hydrogenase requires only one electron donor. All those differences allow nitrogenases to produce higher levels of H_2 without slowing the reaction and eventually running in reverse, as in hydrogenase. As it is exposed further, sensitivity to O_2 is high in both enzymes, although nitrogenase is also inhibited in the presence of NH_4^+.

Future Challenge Strategies to couple dark and photo-fermentation are still visualized as the preferred methods to reach maximum yield in H_2 production by fermentative procedures (Bhatia et al. 2023). However, it is also necessary to consider combined pretreatment methods to provide more nutrimental accessibility to bacteria. It is worth mentioning that the challenges for stable hydrogen production by fermentative methods are associated with a metabolic shift of hydrogen-producing bacteria that causes changes in hydrogen partial pressure, which must be maintained low (Osman et al. 2023). On the other hand, numerous authors have estimated the price of hydrogen from biological sources to be, on average, USD 2 to 3 per kg (Lee, 2016). This positively highlights the potential of photo-fermentation in addition to dark fermentation, as hydrogen yields and prices may vary depending on the feedstock and operating conditions, such as the fermenter and illumination. Thus, Bhatia et al. (2023) identify algal biomass as a definitive resource for providing a continuous supply of hydrogen at a lower cost, while also contributing to the reduction of global warming.

9.7 Co-fermentation

The strategy based on fermentation to improve H_2 yield production is coupling dark fermentation with photo-fermentation by utilizing organic biomass. Dark fermentation products (volatile fatty acids) are used as substrates in photo-fermentation. Regarding this point, some aspects have to be highlighted. First, different pretreatments can be evaluated to select the one through which more valuable products are obtained to feed in a photo-fermenter. Second, dark fermentation products must be treated to reduce the presence of inhibitors such as ammonia and then used for the photo-fermentation process. In most of the cases of study, coupling of both fermentative processes rendered a superior yield compared to the single fermentative method owing to making use of the energy losses in the residues of dark fermentation. The hydrogen yield increased by 4.8 times by using combined dark fermentation and photo-fermentation than only dark fermentation, indicating this strategy is an efficient hydrogen-producing process using microalgal biomass of *Nannochloropsis oceanica* (Xia et al. 2013b). In this way, a two-stage reactor or hybrid bioreactor is utilized sequentially for both fermentation methods.

Different organic biomasses could be used, with limited use of microalgal biomass, because only species such as *Chlorella pyrenoidosa* have been reported (Xia et al. 2013a) even though algal biomass utilization can provide a continuous supply of hydrogen at a lower price, with reduced global warming (Bhatia et al. 2023).

On the other hand, another strategy is coupling both fermentative methods to methane production, such as in *Nannochloropsis oceanica* (Xia et al. 2013b); in this way, the total energy yield of hydrogen is 1.7 times higher than that obtained through the two-stage (dark fermentation and photo-fermentation) process, and methane is 1.3 times higher from microalgae biomass than through single-stage methanogenesis. In this way, the total energy yield of hydrogen is 1.7 times higher than that obtained through the two-stage process (dark fermentation and photo-fermentation), while methane yield is 1.3 times higher from microalgae biomass compared to the single-stage methanogenesis process. (Cheng et al. 2012).

9.8 Biohydrogen Production Using a Microbial Electrolysis Cell (MEC)

Microbial electrolysis cells (MECs), known as catalyzed electrolysis cells or electrofermentation, are bioelectrochemical systems (BESs) that produce hydrogen (H_2) from a wide variety of substrates. The configuration of an MEC is similar to a microbial fuel cell (MFC), which has two electrodes, anode and cathode, that can be placed in a single chamber or separated into two individual chambers. In a two-chamber configuration, a proton exchange membrane (PEM) is commonly used to separate them, and the organic waste to be consumed is placed in the anodic chamber. In contrast, the cathodic chamber can be filled with different solutions, such as phosphate buffer, bicarbonate buffer, and saline solutions, among others (Osman et al. 2023).

The operating principle of MEC is the same in both configurations, with a single-chamber or a two-chamber system. Electroactive bacteria or exoelectrogens are microorganisms capable of oxidizing organic matter, donating electrons to the anode, and generating protons (H^+). Both H^+ and electrons move to the cathode; however, the latter do so through an external circuit. At the cathode, in the absence of oxygen and by applying an electrochemical voltage to raise the cathodic potential, it is feasible to directly create H_2 by the reduction of H^+ ions with electrons produced by microorganisms in the anode (Nandikes et al. 2023; Nguyen et al. 2020; Osman et al. 2023).

The advantage of an MEC is that H_2 production requires a lower additional voltage (0.4–0.8 V) than water electrolysis (1.8–3.5 V). The mechanism of organic oxidation and evolution of H_2 in the MEC is as follows (He et al. 2022):

$$\textbf{Anode}: \text{organic compounds} + H_2O \rightarrow CO_2 + e^- + H^+ \quad (9.7)$$

$$\textbf{Cathode}: 2H^+ + 2e^- \rightarrow H_2 \quad (9.8)$$

According to the review by He et al. in 2022, it is difficult to maintain high levels of H_2 production for long periods of operation in MEC systems, regardless of the substrate used, which limits their practical application. Therefore, research must be strengthened in three main topics:

1. Inhibition of methanogenic archaea and methane-producing bacteria (CH_4) is necessary, as they often grow in the MEC, to prevent the conversion of produced hydrogen into methane.
2. Synthesis of noble metal-free catalysts for the optimization of electrode materials and reactor configuration.
3. Exploration of strategies to enhance electron transfer between microorganisms and electrode materials.

In recent research, the combination of dark fermentation and MEC processes has been proposed to obtain high yields of H_2 production. Recent research has proposed the combination of dark fermentation and MEC processes to achieve high yields of hydrogen production. In dark fermentation, incomplete oxidation of organic matter occurs, resulting in by-products such as volatile fatty acids (VFAs), with a yield of 4 mol H_2 per mol of glucose. During this coupling, the recalcitrant polymers or complex molecules are metabolized into simpler molecules in dark fermentation. Then, the exoelectrogens in the MEC continue oxidizing these organic compounds to generate H_2.

Nguyen et al., in 2020, implemented for the first time a hybrid with dark fermentation (DF) and the MEC system for the simultaneous production of biohydrogen in a single reactor, called sDFMEC (Fig. 9.5). In this system, macroalgae *Saccharina japonica* was used as substrate. The biomass was transformed by dark fermentation into H_2, VFAs, and other molecules in the liquid phase. Meanwhile, the produced VFAs were simultaneously oxidized by the exoelectrogens in the biofilm of the anode, generating an additional amount of H_2 at the cathode. In this sDFMEC system, the total H_2 production was 438.7 ± 13.3 mL/g-TS (applying 0.8 V), while in a single DF process, 54.6 ± 0.8 mL/g-TS of H_2 was produced, and in a two-stage system of DF and MEC, 403.5 ± 7.9 mL/g-TS of H_2 was reached, demonstrating that the sDFMEC system presents a better performance in H_2 production.

Concerning the catalysts, platinum is ideal due to its electrochemical performance for hydrogen production, but it is not feasible to scale it due to the high costs (Nandikes et al. 2023). For this reason, Yun et al. (2021) proposed a stainless steel mesh (SSM) as a promising material for the cathode due to its low cost, high electrical conductivity, high mechanical strength, and lack of toxicity on microorganisms. This SSM cathode was modified in its surface through anodiza-

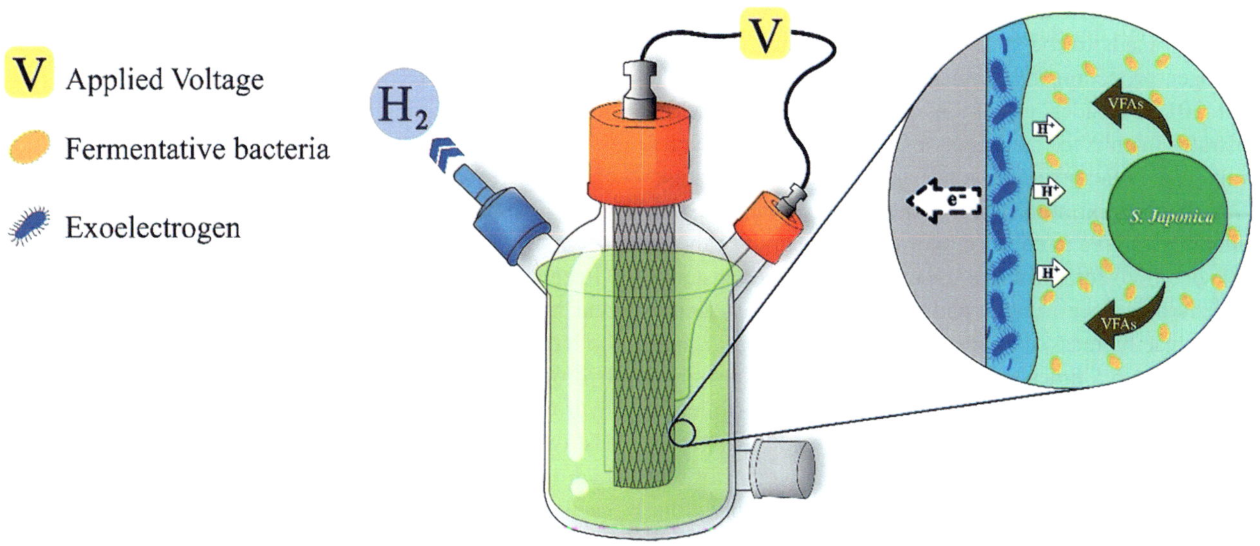

Fig. 9.5 sDFMEC system for *S. Japonica* degradation. (Modified from Nguyen et al. 2020)

tion to increase the specific surface area and provide surface pores; indeed, it was called surface-modified stainless steel mesh (smSSM). Moreover, when using the sDFMEC system with *Saccharina japonica* as substrate, a 150% higher H_2 production was obtained with an smSSM cathode compared to the SSM cathode. Moreover, the electrochemical performance with the smSMM was 67–75% similar to a Pt/C cathode. Additionally, the smSSM cathode performance was stable for 12 cycles over 60 days.

Despite the benefits of the sDFMEC system, some bottlenecks affect H_2 production, such as methane formation and low energy efficiency. For this reason, Phan et al. (2023) proposed to add chloroform ($CHCl_3$) in low concentrations (<100 ppm) as an inhibitor of methanogenic and acetogenic microorganisms and apply an intermittent potential to evaluate three different species of macroalgae: *Gracilaria tenuistipitata* (*Rhodophyta*), *Sargassum polycystum* (*Phaeophyta*), *Ulva lactuca* (*Chlorophyta*). From those, *S. polycystum* showed the highest H_2 yield with 472.4 ± 9.2 mL-H_2/g VS. However, intermittent application of potential and the addition of chloroform can help to 691.0 ± 18.1 mL-H_2/g VS.

On the other hand, a working group in China implemented hydrogen production in an MEC to increase methane production. This was done through a three-chamber reactor (anode-anaerobic digestion-cathode) using blue algae from Lake Taihu as substrates for the anaerobic digestion. In this system, the production of biohydrogen was carried out using the ammoniacal nitrogen migrated to the cathode during the anaerobic digestion of the algae. The improvement of in situ biogas production using the hydrogen collected at the cathode allowed for the achievement of 73.67% of ammonium removal and the production of 202.87 mL of hydrogen when a potential of 0.4 V was applied. The highest methane production (286.62 mL) and ammonium removal (75.73%) were when 0.25 bar of H_2 was initially injected in situ into the anaerobic digestion chamber (Yun et al. 2021).

In conclusion, the production of H_2 through MECs using algae as substrates presents a promising future, but further research needs to be carried out on other operational parameters such as substrate concentration, temperature, fermentative bacteria/exoelectrogenic ratio, biogas collection and storage, electrode/electrolyte ratio, and the search for more environmentally friendly inhibitors for the implementation of these systems on a large scale.

9.9 Metal Nanoparticles

Nanoparticles (NPs) of different metals such as iron, nickel, copper, gold, silver, and titanium have been used to improve hydrogen production; mainly, Fe and Ni NPs exhibit a significant role in hydrogen production by dark and photo-fermentation (Taherdanak et al. 2015). Because NPs provide better dispersion that supplies a wider surface, they influence the metabolic activity of the microorganisms involved (Yin and Wang 2019a).

Zero-valent iron (Fe^0) NPs are especially good because they are reductive materials that, after their addition, react with oxidants in the broth, diminishing the oxidation-reduction potential (ORP) (Feng et al. 2014). This is because NPs act as oxygen scavengers (Srivastava et al. 2020), promoting, in this way, a suitable environment for hydrogen generation. Furthermore, Fe^0 NPs corrode after being added, forming ferrous (Fe^{+2}) and ferric (Fe^{+3}) ions. Fe^{+2} stimulates electron transfer, which, in turn, promotes the expression of functional genes for hydrogenase and dehydrogenase, essential enzymes for hydrogen production (Zhao et al., 2017). This also leads to the presence of proteins involved in the hydrogen-producing metabolism, such as ferredoxins (Wu et al. 2017). Besides, nickel ions (Ni^{+2}) have also been demonstrated to enhance the activity of dehydrogenase enzymes, as they are cofactors in the active site of NiFe-hydrogenase.

In this context, numerous studies have demonstrated the suitability of NPs to enhance hydrogen production, either in dark or photo-fermentation (Li et al. 2022). For instance, using fermentative bacteria and different substrates such as macroalgae biomass, hydrogen production was significantly enhanced by 6.5 times compared with the control test under dark fermentation (Yin and Wang 2019a). Meanwhile, for biophotolysis, silica-based NPs have been used to scatter the supplied and incident light in a tubular photobioreactor with Chlamydomonas reinhardtii under low-sulfur conditions. The hydrogen production was 3121.5 ± 178.9 ml, and the chlorophyll content increased by 23% (Giannelli and Torzillo 2012). It is important to remark that this work is the only report found in the literature.

On the other hand, Fe^0 NPs with enhanced hydrogenase activity in other dark fermentation processes may be adopted for the biophotolysis process of hydrogen production (Shanmugam et al. 2020). Although Fe^0 NPs have proven effective in enhancing hydrogen production, their dosing must be done cautiously, as it depends on the substrate source and operating conditions. A high dose typically promotes hydrogen consumption in almost every case (Zhen et al., 2015) and may even alleviate the inhibitory effect of high hydrogen pressure (Feng et al. 2014). Evidently, in dark fermentation, the presence of Fe^0 NPs changes the microbial distribution. Thus, it affects the metabolic pathways, obtaining different products from fermentation and, therefore, energy conversion (Yin and Wang 2019a).

The use of NPs in catalyzing various reactions during the hydrogen production process has been successful. Thus, research needs to be focused on developing highly efficient, environmentally sustainable, and cost-effective catalysts (Pathy et al. 2022). The synthesis of NPs by green routes may reduce the economic cost of the overall process (Srivastava et al. 2020). In this context, bacteria, fungi, and yeast are utilized to produce nanomaterials by green synthesis (Michael et al. 2022). In this way, nanoparticles with low toxicity, low energy consumption, and high catalytic activity can be obtained (Pathy et al. 2022).

Dark fermentation applications use the Fe^0 NPs to pretreat lignocellulosic biomass and promote the total energy recovery from this biomass. Fe_3O_4 NPs protonated under acidic conditions were used in a semi-continuum system of hydrogen production for the efficient harvesting of *Chlamydomonas vulgaris* biomass and restart the biophotolysis process for hydrogen production (Wang et al. 2016). Another advantage is the recovery of magnetic particles via detachment, to be reused at least five times with high harvesting efficiency (Savvidou et al. 2021).

9.10 Genetic Engineering for Biohydrogen Production in Algae

Green algae produce H_2 under fermentative dark conditions and by photosynthetic water oxidation (photohydrolysis), both driven by hydrogenases. Three different main classes of hydrogenases are known: [NiFe]-, [FeFe]-, and [Fe]-hydrogenases; the first ones occur in methanogens, and the last two classes are widely distributed. In algae, the [FeFe]-hydrogenases are the most efficient hydrogen conversion catalysts but are essentially oxygen-sensitive. Low concentration of O_2 affects the [4Fe-4S] cluster present in the active site of these hydrogenases and inactivates them. Therefore, the enzymes need to be resynthesized after they are used as a catalyst (Schumann et al. 2023). To overcome the oxygen sensitivity of the process, the production of hydrogen was attempted under anoxygenic photosynthesis, but this option has some troubles, such as the high cost and low abundance of the electron donors, preventing sustainable, large-scale evolution of hydrogen through this way. Then, the oxygenic photosynthesis that produces H_2 from water remains the best option (Redding et al. 2022).

Two distinct hydrogenase-encoding genes, *hydA1* and *hydA2*, have been identified in the model green alga, *Chlamydomonas reinhardtii*. Both genes are encoded in the nucleus, but these proteins are up to 90% of all chloroplast proteins. The study of HydA1 and HydA2 hydrogenases in *C. reinhardtii* under anoxic conditions showed a predominance of HydA1 hydrogenase (75% of the activity), the reason why most efforts of genetic engineering on algae's hydrogenases have focused on HydA1 (El-Dalatony et al. 2020).

In a pioneer work, Reifschneider-Wegner et al. (2014) modified the *hydA1* gene in *C. reinhardtii*, eliminating the sequence corresponding to the chloroplast peptide and the introns and optimizing the codon use, that is, the relatively AT-rich composition of the chloroplast genome compared to the GC-rich nuclear genome in this alga. The enzyme was successfully expressed in the chloroplast, demonstrating that hydrogenase may be synthesized in situ. Besides, oxygen is deleterious not only for hydrogenase activity and stability but also for its transcription and maturation (Srirangan et al. 2011). Interestingly, the expression and translation of the *hydA1* gene in the chloroplast overcame the native regulation, and the gene was expressed under both anaerobic and aerobic conditions (Reifschneider-Wegner et al. 2014).

Recently, Yang et al. (2019) modified the amino acid residues A105I, V265W, G113I, and V273I in the HydA hydrogenase protein of *Chlorella* sp.; these amino acids were part of the gas tunnel in HydA, and the purpose was to prevent the O_2 accession to the enzyme active site. Transgenic lines with single or double mutations produced up to sevenfold more H_2 than the control in the presence of 5% O_2 in vitro. Also, some mutants could produce in vivo up to 30-fold more H_2 than the control for several days in the presence of O_2 (Yang et al. 2019), supporting this strategy to continue the efforts in this research line.

Furthermore, the heterologous expression of the *hydA1* gene is the co-expression of the genes *hydE, hydF, and hydG*, which are involved in the assembly and maturation of functional [FeFe]-hydrogenase (Posewitz et al. 2004). For example, the expression of *hydA1* in the bacterium *Escherichia coli*, a host lacking these genes, produced HydA1 polypeptide but not hydrogen (Srirangan et al. 2011). On the contrary, when *Clostridium acetobutylicum*'s *hydA1* was co-expressed with *C. reinhardtii*'s *hydEGF* cluster in *E. coli*, the hydrogenase was assembled correctly and produced hydrogen (Nagarajan et al. 2016). These maturation enzymes (*HydEFG*) have been exclusively found in algal organisms (Schumann et al. 2023; Wegelius et al. 2021).

To date, only a few hydrogenases have been characterized in detail. Recently, the characterization of [FeFe]-hydrogenases from *Clostridium beijerinckii* and *Desulfovibrio* sp. uncovered oxygen-stable enzymes (Morra 2022), and they are interesting alternatives for genetic engineering in coming times.

[FeFe]-hydrogenases accept electrons from ferredoxin (Fdx), which is reduced by Photosystem I (PSI), a multisubunit protein complex located in the thylakoid membranes. Chien et al. (2012) conducted the engineering of *C. vulgaris* [FeFe] hydrogenase for coupling HydA to PSI to reduce competition for electrons from PSI with ferredoxin-$NADP^+$ reductase (FNR); this strategy resulted in up to a tenfold increase in H_2 production. A similar strategy was followed later for *C. reinhardtii* by fusing native Fdx1 and HydA, resulting in up to a fourfold increase in H_2 production; in addition, an enhanced O_2 tolerance was observed. However, up to 85% of the chimeric Fdx1-HydA1 protein remained in the inactive apoprotein form (Weiner et al. 2018). Improving hydrogenase assembly and maturation must solve this bottleneck.

In the case of fermentative dark conditions, starch is oxidized in a fermentative process, with H_2 as a by-product. While HydA1 is dedicated to hydrogen production, HydA2 appears directed toward H_2 oxidation. The fusion of the stromal PSI subunit PsaC to *C. reinhardtii* HydA2 resulted in an increment of 60% of hydrogenase activity in anoxic dark conditions, leading to fermentative H_2 production. The PSI-HydA2 construct was sevenfold less than PSI in the control strain; this reduction leads to decreased PSII, reducing oxygen production and O_2 production. As a result, the inactivation of the hydrogenase decreases, and H_2 production increases (Hambourger et al. 2008).

Furthermore, the integration of genomics, transcriptomics, proteomics, and metabolomics on *C. reinhardtii* allowed other critical genes involved in H_2 production to be identified, such as sulfate permease (Sulp), truncated light-harvesting antenna 1 (Tla1), starchless isoamylase (Sta7–10), and pyruvate formate lyase (PFL1). Mutation of the two former genes has a negative effect on H_2 production, while mutation of the two latter genes has a positive effect on H_2 production (Xu et al. 2019).

The close relationship between photosynthesis and hydrogen production is evident. Many efforts are focused on engineering photosynthesis-involved proteins, such as components of light harvesting, truncation of light-harvesting complexes, and rewiring electron channeling, among others. A comprehensive review of all strategies followed so far to improve H_2 production through genetic engineering on photosynthesis was recently published by Schumann et al. (2023).

Further challenges for genetic engineering to improve H_2 production include the evaluation of known oxygen-resistant hydrogenases and the development of novel oxygen-resistant ones. It is necessary to identify chassis algal organisms (algae that house and support the foreign genes and allow them to function) to develop synthetic biology.

C. reinhardtii is an attractive model as a potential chassis organism since its genome has been largely studied, and this alga is suitable for knock-in and knock-out by CRISPR-cas9 (Kelterborn et al. 2022; Nievergelt et al. 2023), and many genetic building blocks from *C. reinhardtii* are currently available (Schroda and Remacle 2022). The expression of chimeric HydA-HydEFG proteins in a host with alterations in PSI and PSII systems that channel electrons to the [FeFe]-hydrogenase must significantly increase the H_2 production. In addition, the heterologous expression of algal genes in any of the known chassis organisms (Chi et al. 2019) should not be ruled out either. Moreover, the expression in algae of oxygen-resistant bacterial HydA genes from *C. beijerinckii* and *Desulfovibrio* sp. (Morra 2022) should be included in the strategic route planning toward sustainable alternatives to H_2 production.

9.11 Conclusion

For more than one century, hydrogen production by photoautotrophic microorganisms has attracted the attention of researchers, in combination with the interest in revealing metabolic mechanisms and genetic bases of cyanobacteria and microalgae. This hydrogen production has been extended and complemented with other non-photosynthetic microorganisms and novel techniques, such as microbial cells, nanoparticles, and genetic engineering. Facing an urgent energy transition, alternative energy sources, including renewable sources, and biofuels are included as options to conventional fossil fuels. In particular, hydrogen as an energy vector and biohydrogen from microalgae are being explored. Here, important efforts have been shown in this direction, but getting practical and sustainable applicative processes remains a challenge.

References

Abdel-Kader HAA, Abdel-Basset R, Danial AW (2022) Yeast and enzymatic hydrolysis in converting Chlorella biomass into hydrogen gas by Rhodobacter sp. and Rhodopseudomonas palustris. Int J Hydrog Energy 47(3):1516–1528. https://doi.org/10.1016/j.ijhydene.2021.10.126

Abdin Z, Zafaranloo A, Rafiee A, Mérida W, Lipiński W, Khalilpour KR (2020) Hydrogen as an energy vector. Renew Sust Energ Rev 120:109620. https://doi.org/10.1016/j.rser.2019.109620

Akköse S, Gündüz U, Yücel M, Eroglu I (2009) Effects of ammonium ion, acetate and aerobic conditions on hydrogen production and expression levels of nitrogenase genes in Rhodobacter sphaeroides O.U.001. Int J Hydrog Energy 34(21):8818–8827. https://doi.org/10.1016/j.ijhydene.2009.08.040

Asada Y, Miyake J (1999) Photobiological hydrogen production. J Biosci Bioeng 88(1):1–6. https://doi.org/10.1016/S1389-1723(99)80166-2

Aslam A, Bahadar A, Liaquat R, Muddasar M (2023) Recent advances in biological hydrogen production from algal biomass: a comprehensive review. Fuel 350:128816. https://doi.org/10.1016/j.fuel.2023.128816

Batista AP, Gouveia L, Marques PASS (2018) Fermentative hydrogen production from microalgal biomass by a single strain of bacterium Enterobacter aerogenes – effect of operational conditions and fermentation kinetics. Renew Energbiore 119:203–209. https://doi.org/10.1016/j.renene.2017.12.017

Bhatia SK, Rajesh Banu J, Singh V, Kumar G, Yang Y-H (2023) Algal biomass to biohydrogen: pretreatment, influencing factors, and conversion strategies. Bioresour Technol 368:128332. https://doi.org/10.1016/j.biortech.2022.128332

Bhujbal SK, Ghosh P, Vijay VK et al (2022) Biotechnological potential of rumen microbiota for sustainable bioconversion of lignocellulosic waste to biofuels and value-added products. Sci Total Environ 814:152773. https://doi.org/10.1016/j.scitotenv.2021.152773

Bompard E, Botterud A, Corgnati S et al (2020) An electricity triangle for energy transition: application to Italy. Appl Energy 277:115525. https://doi.org/10.1016/j.apenergy.2020.115525

Botta LS, Delforno TP, Rabelo CABS, Silva EL, Varesche MBA (2020) Microbial community analyses by high-throughput sequencing of rumen microorganisms fermenting office paper in mesophilic and thermophilic lysimeters. Process Saf Environ 136:182–193. https://doi.org/10.1016/j.psep.2020.01.030

Chen S, Qu D, Xiao X, Miao X (2020) Biohydrogen production with lipid-extracted Dunaliella biomass and a new strain of hyper-thermophilic archaeon Thermococcus eurythermalis A501. Int J Hydrog Energy 45(23):12721–12730. https://doi.org/10.1016/j.ijhydene.2020.03.010

Chen Y, Yin Y, Wang J (2022) Biohydrogen production using macroalgal biomass of Laminaria japonica pretreated by gamma irradiation as substrate. Fuel 309:122179. https://doi.org/10.1016/j.fuel.2021.122179

Cheng J, Xia A, Liu Y, Lin R, Zhou J, Cen K (2012) Combination of dark- and photo-fermentation to improve hydrogen production from Arthrospira platensis wet biomass with ammonium removal by zeolite. Int J Hydrog Energy 37(18):13330–13337. https://doi.org/10.1016/j.ijhydene.2012.06.071

Cheng J, Yue L, Ding L et al (2019) Improving fermentative hydrogen and methane production from an algal bloom through hydrothermal/steam acid pretreatment. Int J Hydrog Energy 44(12):5812–5820. https://doi.org/10.1016/j.ijhydene.2019.01.046

Chernova NI, Kiseleva SV (2017) Microalgae biofuels: induction of lipid synthesis for biodiesel production and biomass residues into hydrogen conversion. Int J Hydrog Energy 42(5):2861–2867. https://doi.org/10.1016/j.ijhydene.2016.05.302

Chi H, Wang X, Shao Y et al (2019) Engineering and modification of microbial chassis for systems and synthetic biology. Synth Syst Biotechnol 4(1):25–33. https://doi.org/10.1016/j.synbio.2018.12.001

Chien L-F, Kuo T-T, Liu B-H, Lin H-D, Feng T-Y, Huang C-C (2012) Solar-to-bioH2 production enhanced by homologous overexpression of hydrogenase in green alga *Chlorella* sp. DT. Int J Hydrog Energy 37(23):17738–17748. https://doi.org/10.1016/j.ijhydene.2012.09.068

de Oliveira MC, Bassin ID, Cammarota MC (2022) Microalgae and cyanobacteria biomass pretreatment methods: a comparative analysis of chemical and thermochemical pretreatment methods aimed at methane production. Fermentation 8(10):497

Díaz-Cruz EB, Rodriguez-Torres MP (2023) Biotechnology in hydrogen generation. In: Rodriguez-Torres MP, Martinez-Alonso C (eds) Biotechnology in the generation of biofuels. Springer Nature, Singapore, pp 111–133

Dincer I, Acar C (2015) Review and evaluation of hydrogen production methods for better sustainability. Int J Hydrog Energy 40(34):11094–11111. https://doi.org/10.1016/j.ijhydene.2014.12.035

Ding L, Cheng J, Xia A, Jacob A, Voelklein M, Murphy JD (2016) Co-generation of biohydrogen and biomethane through two-stage batch co-fermentation of macro- and micro-algal biomass. Bioresour Technol 218:224–231. https://doi.org/10.1016/j.biortech.2016.06.092

Ding L, Cheng J, Lin R, Deng C, Zhou J, Murphy JD (2020) Improving biohydrogen and biomethane co-production via two-stage dark fermentation and anaerobic digestion of the pretreated seaweed Laminaria digitata. J Clean Prod 251:119666. https://doi.org/10.1016/j.jclepro.2019.119666

Efremenko EN, Nikolskaya AB, Lyagin IV et al (2012) Production of biofuels from pretreated microalgae biomass by anaerobic fermentation with immobilized Clostridium acetobutylicum cells. Bioresour Technol 114:342–348. https://doi.org/10.1016/j.biortech.2012.03.049

El-Dalatony MM, Zheng Y, Ji M-K, Li X, Salama E-S (2020) Metabolic pathways for microalgal biohydrogen production: current progress and future prospectives. Bioresour Technol 318:124253. https://doi.org/10.1016/j.biortech.2020.124253

Fakhimi N, Tavakoli O (2019) Improving hydrogen production using co-cultivation of bacteria with Chlamydomonas reinhardtii microalga. Mater Sci Energ Technol 2(1):1–7. https://doi.org/10.1016/j.mset.2018.09.003

Faraloni C, Torzillo G (2023) Interplay between photobiological hydrogen production by microalgae and bioeconomy. In: Oncel SS (ed) A sustainable green future: perspectives on energy, economy, industry, cities and environment. Springer International Publishing, Cham, pp 541–559

Feng Y, Zhang Y, Quan X, Chen S (2014) Enhanced anaerobic digestion of waste activated sludge digestion by the addition of zero valent iron. Water Res 52:242–250. https://doi.org/10.1016/j.watres.2013.10.072

Ferreira AF, Ortigueira J, Alves L, Gouveia L, Moura P, Silva C (2013) Biohydrogen production from microalgal biomass: energy requirement, CO2 emissions and scale-up scenarios. Bioresour Technol 144:156–164. https://doi.org/10.1016/j.biortech.2013.06.079

Gad El-Rab SMF, Hifney AF, Abdel-Basset R (2018) Costless and huge hydrogen yield by manipulation of iron concentrations in the new bacterial strain Brevibacillus invocatus SAR grown on algal biomass. Int J Hydrog Energy 43(41):18896–18907. https://doi.org/10.1016/j.ijhydene.2018.08.116

Gadow SI, Li Y-Y, Liu Y (2012) Effect of temperature on continuous hydrogen production of cellulose. Int J Hydrog Energy 37(20):15465–15472. https://doi.org/10.1016/j.ijhydene.2012.04.128

Gfeller RP, Gibbs M (1984) Fermentative metabolism of Chlamydomonas reinhardtii: I. Analysis of fermentative products from starch in dark and light. Plant Physiol 75(1):212–218. https://doi.org/10.1104/pp.75.1.212

Ghirardi ML (2015) Implementation of photobiological H2 production: the O2 sensitivity of hydrogenases. Photosynth Res 125(3):383–393. https://doi.org/10.1007/s11120-015-0158-1

Ghirardi ML, Zhang L, Lee JW et al (2000) Microalgae: a green source of renewable H(2). Trends Biotechnol 18(12):506–511. https://doi.org/10.1016/s0167-7799(00)01511-0

Giang TT, Lunprom S, Liao Q, Reungsang A, Salakkam A (2019) Improvement of hydrogen production from chlorella sp. biomass by acid-thermal pretreatment. PeerJ 7:e6637. https://doi.org/10.7717/peerj.6637

Giannelli L, Torzillo G (2012) Hydrogen production with the microalga Chlamydomonas reinhardtii grown in a compact tubular photobioreactor immersed in a scattering light nanoparticle suspension. Int J Hydrog Energy 37(22):16951–16961. https://doi.org/10.1016/j.ijhydene.2012.08.103

Gonzalez-Fernandez C, Sialve B, Molinuevo-Salces B (2015) Anaerobic digestion of microalgal biomass: challenges, opportunities and research needs. Bioresour Technol 198:896–906. https://doi.org/10.1016/j.biortech.2015.09.095

Goswami RK, Mehariya S, Obulisamy PK, Verma P (2021) Advanced microalgae-based renewable biohydrogen production systems: a review. Bioresour Technol 320:124301. https://doi.org/10.1016/j.biortech.2020.124301

Hädicke O, Grammel H, Klamt S (2011) Metabolic network modeling of redox balancing and biohydrogen production in purple non-sulfur bacteria. BMC Syst Biol 5(1):150. https://doi.org/10.1186/1752-0509-5-150

Hambourger M, Gervaldo M, Svedruzic D et al (2008) [FeFe]-hydrogenase-catalyzed H2 production in a photoelectrochemical biofuel cell. J Am Chem Soc 130(6):2015–2022. https://doi.org/10.1021/ja077691k

He K, Li W, Tang L, Li W, Lv S, Xing D (2022) Suppressing methane production to boost high-purity hydrogen production in microbial electrolysis cells. Envir Sci Tech 56(17):11931–11951. https://doi.org/10.1021/acs.est.2c02371

IPCC (2023) Summary for policymakers. In: Lee H, Romero J (eds) Climate change 2023: synthesis report contribution of working groups I, II and III to the sixth assessment report of the intergovernmental panel on climate change. IPCC, Geneva, pp 1–34

Jehlee A, Rodjaroen S, Waewsak J, Reungsang A, O-Thong S (2019) Improvement of biohythane production from *Chlorella* sp. TISTR 8411 biomass by co-digestion with organic wastes in a two-stage fermentation. Int J Hydrog Energy 44(32):17238–17247. https://doi.org/10.1016/j.ijhydene.2019.03.026

Jurado-Oller JL, Dubini A, Galván A, Fernández E, González-Ballester D (2015) Low oxygen levels contribute to improve photohydrogen production in mixotrophic non-stressed Chlamydomonas cultures. Biotechnol Biofuels 8(1):149. https://doi.org/10.1186/s13068-015-0341-9

Karishma S, Saravanan A, Senthil Kumar P, Rangasamy G (2022) Sustainable production of biohydrogen from algae biomass: critical review on pretreatment methods, mechanism and challenges. Bioresour Technol 366:128187. https://doi.org/10.1016/j.biortech.2022.128187

Kelterborn S, Boehning F, Sizova I, Baidukova O, Evers H, Hegemann P (2022) Gene editing in green alga *Chlamydomonas reinhardtii* via CRISPR-Cas9 ribonucleoproteins. Methods Mol Biol 2379:45–65. https://doi.org/10.1007/978-1-0716-1791-5_3

Kovács KL, Maróti G, Rákhely G (2006) A novel approach for biohydrogen production. Int J Hydrog Energy 31(11):1460–1468. https://doi.org/10.1016/j.ijhydene.2006.06.011

Kumar K, Roy S, Das D (2013) Continuous mode of carbon dioxide sequestration by *C. sorokiniana* and subsequent use of its biomass for hydrogen production by *E. cloacae* IIT-BT 08. Bioresour Technol 145:116–122. https://doi.org/10.1016/j.biortech.2013.01.137

Kumar G, Sivagurunathan P, Thi NBD et al (2016) Evaluation of different pretreatments on organic matter solubilization and hydrogen fermentation of mixed microalgae consortia. Int J Hydrog Energy 41(46):21628–21640. https://doi.org/10.1016/j.ijhydene.2016.05.195

Lacroux J, Llamas M, Dauptain K et al (2023) Dark fermentation and microalgae cultivation coupled systems: outlook and challenges. Sci Total Environ 865:161136. https://doi.org/10.1016/j.scitotenv.2022.161136

Laughlin TG, Savage DF, Davies KM (2020) Recent advances on the structure and function of NDH-1: the complex I of oxygenic photosynthesis. Biochim Biophys Acta Bioenerg 1861(11):148254. https://doi.org/10.1016/j.bbabio.2020.148254

Lee D-H (2016) Cost-benefit analysis, LCOE and evaluation of financial feasibility of full commercialization of biohydrogen. Int J Hydrog Energy 41(7):4347–4357. https://doi.org/10.1016/j.ijhydene.2015.09.071

Li X, Huang S, Yu J, Wang Q, Wu S (2013) Improvement of hydrogen production of *Chlamydomonas reinhardtii* by co-cultivation with isolated bacteria. Int J Hydrog Energy 38(25):10779–10787. https://doi.org/10.1016/j.ijhydene.2013.02.102

Li S, Li F, Zhu X, Liao Q, Chang J-S, Ho S-H (2022) Biohydrogen production from microalgae for environmental sustainability. Chemosphere 291:132717. https://doi.org/10.1016/j.chemosphere.2021.132717

Liu C-H, Chang C-Y, Cheng C-L, Lee D-J, Chang J-S (2012) Fermentative hydrogen production by *Clostridium butyricum* CGS5 using carbohydrate-rich microalgal biomass as feedstock. Int J Hydrog Energy 37(20):15458–15464. https://doi.org/10.1016/j.ijhydene.2012.04.076

Liu H, Zhang Z, Zhang H et al (2020) Evaluation of hydrogen yield potential from *Chlorella* by photo-fermentation under diverse substrate concentration and enzyme loading. Bioresour Technol 303:122956. https://doi.org/10.1016/j.biortech.2020.122956

Makarova VV, Kosourov S, Krendeleva TE et al (2007) Photoproduction of hydrogen by sulfur-deprived *C. reinhardtii* mutants with impaired Photosystem II photochemical activity. Photosyn Res 94(1):79–89. https://doi.org/10.1007/s11120-007-9219-4

Martins I, Surra E, Ventura M, Lapa N (2022) BioH_2 from dark fermentation of OFMSW: effect of the hydraulic retention time and organic loading rate. Appl Sci 12(9):4240

McKinlay JB (2014) Systems biology of photobiological hydrogen production by purple non-sulfur bacteria. In: Zannoni D, De Philippis R (eds) Microbial BioEnergy: hydrogen production. Springer, Dordrecht, pp 155–176

Melis A, Zhang L, Forestier M, Ghirardi ML, Seibert M (2000) Sustained photobiological hydrogen gas production upon reversible inactivation of oxygen evolution in the green alga

Chlamydomonas reinhardtii 1. Plant Physiol 122(1):127–136. https://doi.org/10.1104/pp.122.1.127

Michael A, Singh A, Roy A, Islam MR (2022) Fungal- and algal-derived synthesis of various nanoparticles and their applications. Bioinorg Chem Appl 2022:3142674. https://doi.org/10.1155/2022/3142674

Morra S (2022) Fantastic [FeFe]-hydrogenases and where to find them. Front Microbiol 13. https://doi.org/10.3389/fmicb.2022.853626

Nagarajan D, Lee DJ, Kondo A, Chang JS (2016) Recent insights into biohydrogen production by microalgae – from biophotolysis to dark fermentation. Bioresour Technol 227:373. https://doi.org/10.1016/j.biortech.2016.12.104

Nagarajan D, Lee D-J, Kondo A, Chang J-S (2017) Recent insights into biohydrogen production by microalgae – from biophotolysis to dark fermentation. Bioresour Technol 227:373–387. https://doi.org/10.1016/j.biortech.2016.12.104

Nandikes G, Pathak P, Singh L (2023) Cathodic catalysts for microbial electrolysis cell to produce biohydrogen transition metal-based electrocatalysts: applications in green hydrogen production and storage, ACS Symposium Series. American Chemical Society, pp 21–42

Nayak BK, Roy S, Das D (2014) Biohydrogen production from algal biomass (*Anabaena* sp. PCC 7120) cultivated in airlift photobioreactor. Int J Hydrog Energy 39(14):7553–7560. https://doi.org/10.1016/j.ijhydene.2013.07.120

Nguyen T-AD, Kim K-R, Nguyen M-T, Kim MS, Kim D, Sim SJ (2010) Enhancement of fermentative hydrogen production from green algal biomass of *Thermotoga neapolitana* by various pretreatment methods. Int J Hydrog Energy 35(23):13035–13040. https://doi.org/10.1016/j.ijhydene.2010.04.062

Nguyen PKT, Das G, Kim J, Yoon HH (2020) Hydrogen production from macroalgae by simultaneous dark fermentation and microbial electrolysis cell. Bioresour Technol 315:123795. https://doi.org/10.1016/j.biortech.2020.123795

Nievergelt AP, Diener DR, Bogdanova A, Brown T, Pigino G (2023) Efficient precision editing of endogenous *Chlamydomonas reinhardtii* genes with CRISPR-Cas. Cell Rep Methods 3(8):100562. https://doi.org/10.1016/j.crmeth.2023.100562

Nobre BP, Villalobos F, Barragán BE et al (2013) A biorefinery from *Nannochloropsis* sp. microalga – extraction of oils and pigments. Production of biohydrogen from the leftover biomass. Bioresour Technol 135:128–136. https://doi.org/10.1016/j.biortech.2012.11.084

Osman AI, Deka TJ, Baruah DC, Rooney DW (2023) Critical challenges in biohydrogen production processes from the organic feedstocks. Biomass Convers Biorefinery 13(10):8383–8401. https://doi.org/10.1007/s13399-020-00965-x

Pandey A, Sinha P, Pandey A (2021) Hydrogen production by sequential dark and photofermentation using wet biomass hydrolysate of Spirulina platensis: response surface methodological approach. Int J Hydrog Energy 46(10):7137–7146. https://doi.org/10.1016/j.ijhydene.2020.11.205

Park J-H, Yoon J-J, Park H-D, Kim YJ, Lim DJ, Kim S-H (2011) Feasibility of biohydrogen production from *Gelidium amansii*. Int J Hydrog Energy 36(21):13997–14003. https://doi.org/10.1016/j.ijhydene.2011.04.003

Pathy A, Nageshwari K, Ramaraj R, Pragas Maniam G, Govindan N, Balasubramanian P (2022) Biohydrogen production using algae: potentiality, economics and challenges. Bioresour Technol 360:127514. https://doi.org/10.1016/j.biortech.2022.127514

Phan TP, Nguyen TL, Nguyen PKT (2023) Evaluation and enhancement of hydrogen production from Vietnamese macroalgae in the single-stage of dark fermentation and microbial electrolysis cell. Biomass and Bioenergy 175:106885.https://doi.org/10.1016/j.biombioe.2023.106885

Posewitz MC, King PW, Smolinski SL, Zhang L, Seibert M, Ghirardi ML (2004) Discovery of two novel radical S-adenosylmethionine proteins required for the assembly of an active [Fe] hydrogenase. J Biol Chem 279(24):25711–25720. https://doi.org/10.1074/jbc.M403206200

Priya A, Naseem S, Pandey D et al (2023) Innovative strategies in algal biomass pretreatment for biohydrogen production. Bioresour Technol 369:128446. https://doi.org/10.1016/j.biortech.2022.128446

Putatunda C, Behl M, Solanki P et al (2023) Current challenges and future technology in photofermentation-driven biohydrogen production by utilizing algae and bacteria. Int J Hydrog Energy 48(55):21088–21109. https://doi.org/10.1016/j.ijhydene.2022.10.042

Qiu C, Yuan P, Sun L, Wang S, Lo S, Zhang D (2017) Effect of fermentation temperature on hydrogen production from xylose and the succession of hydrogen-producing microflora. J Chem Technol Biotechnol 92(8):1990–1997. https://doi.org/10.1002/jctb.5190

Redding KE, Appel J, Boehm M et al (2022) Advances and challenges in photosynthetic hydrogen production. Trends Biotechnol 40(11):1313–1325. https://doi.org/10.1016/j.tibtech.2022.04.007

Reeves M, Greenbaum E (1985) Long-term endurance and selection studies in hydrogen and oxygen photoproduction by *Chlamydomonas reinhardtii*. Enzym Microb Technol 7(4):169–174. https://doi.org/10.1016/0141-0229(85)90059-6

Reifschneider-Wegner K, Kanygin A, Redding KE (2014) Expression of the [FeFe] hydrogenase in the chloroplast of Chlamydomonas reinhardtii. Int J Hydrog Energy 39(8):3657–3665. https://doi.org/10.1016/j.ijhydene.2013.12.157

Salakkam A, Sittijunda S, Mamimin C, Phanduang O, Reungsang A (2021) Valorization of microalgal biomass for biohydrogen generation: a review. Bioresour Technol 322:124533. https://doi.org/10.1016/j.biortech.2020.124533

Savvidou MG, Dardavila MM, Georgiopoulou I et al (2021) Optimization of microalga Chlorella vulgaris magnetic harvesting. Nano 11(6):1614

Schroda M, Remacle C (2022) Molecular advancements establishing Chlamydomonas as a host for biotechnological exploitation. Front Plant Sci 13:911483. https://doi.org/10.3389/fpls.2022.911483

Schumann C, Fernández Méndez J, Berggren G, Lindblad P (2023) Novel concepts and engineering strategies for heterologous expression of efficient hydrogenases in photosynthetic microorganisms. Front Microbiol 14. https://doi.org/10.3389/fmicb.2023.1179607

Shanmugam S, Hari A, Pandey A, Mathimani T, Felix L, Pugazhendhi A (2020) Comprehensive review on the application of inorganic and organic nanoparticles for enhancing biohydrogen production. Fuel 270:117453. https://doi.org/10.1016/j.fuel.2020.117453

Shobana S, Saratale GD, Pugazhendhi A et al (2017) Fermentative hydrogen production from mixed and pure microalgae biomass: key challenges and possible opportunities. Int J Hydrog Energy 42(42):26440–26453. https://doi.org/10.1016/j.ijhydene.2017.07.050

Singh H, Rout S, Das D (2022) Dark fermentative biohydrogen production using pretreated Scenedesmus obliquus biomass under an integrated paradigm of biorefinery. Int J Hydrog Energy 47(1):102–116. https://doi.org/10.1016/j.ijhydene.2021.10.018

Sirohi R, Ummalyma SB, Sagar NA et al (2021) Strategies and advances in the pretreatment of microalgal biomass. J Biotechnol 341:63–75. https://doi.org/10.1016/j.jbiotec.2021.09.010

Sivagurunathan P, Kumar G, Kobayashi T, Xu K, Kim S-H (2017) Effects of various dilute acid pretreatments on the biochemical hydrogen production potential of marine macroalgal biomass. Int J Hydrog Energy 42(45):27600–27606. https://doi.org/10.1016/j.ijhydene.2017.05.106

Srirangan K, Pyne ME, Perry Chou C (2011) Biochemical and genetic engineering strategies to enhance hydrogen production in photosynthetic algae and cyanobacteria. Bioresour Technol 102(18):8589–8604. https://doi.org/10.1016/j.biortech.2011.03.087

Srivastava N, Srivastava M, Mishra PK et al (2020) Advances in nanomaterials induced biohydrogen production using waste biomass. Bioresour Technol 307:123094. https://doi.org/10.1016/j.biortech.2020.123094

Taherdanak M, Zilouei H, Karimi K (2015) Investigating the effects of iron and nickel nanoparticles on dark hydrogen fermentation from starch using central composite design. Int J Hydrog Energy 40(38):12956–12963. https://doi.org/10.1016/j.ijhydene.2015.08.004

Taşkan B, Köroğlu EO, Taşkan E (2022) Cladophora sp. as a sustainable feedstock for dark fermentative biohydrogen production. Int J Hydrog Energy 47(34):15410–15418. https://doi.org/10.1016/j.ijhydene.2022.03.024

Thongbunrod N, Chaiprasert P (2021) Efficacy and metagenomic analysis of the stabilized anaerobic lignocellulolytic microbial consortium from bubalus bubalis rumen with rice straw enrichment for methane production. Bioenergy Res 14(3):870–890. https://doi.org/10.1007/s12155-020-10167-y

Torzillo G, Scoma A, Faraloni C, Ena A, Johanningmeier U (2009) Increased hydrogen photoproduction by means of a sulfur-deprived Chlamydomonas reinhardtii D1 protein mutant. Int J Hydrog Energy 34(10):4529–4536. https://doi.org/10.1016/j.ijhydene.2008.07.093

Touloupakis E, Torzillo G (2019) Chapter 14 – Photobiological hydrogen production. In: Calise F, D'Accadia MD, Santarelli M, Lanzini A, Ferrero D (eds) Solar hydrogen production. Academic, pp 511–525

Wang T, Yang W-L, Hong Y, Hou Y-L (2016) Magnetic nanoparticles grafted with amino-riched dendrimer as magnetic flocculant for efficient harvesting of oleaginous microalgae. J Chem Eng 297:304–314. https://doi.org/10.1016/j.cej.2016.03.038

Wang Y, Chen X, Spengler K et al (2022) Pyruvate: ferredoxin oxidoreductase and low abundant ferredoxins support aerobic photomixotrophic growth in cyanobacteria. eLife 11:e71339. https://doi.org/10.7554/eLife.71339

Wegelius A, Land H, Berggren G, Lindblad P (2021) Semisynthetic [FeFe]-hydrogenase with stable expression and H2 production capacity in a photosynthetic microbe. Cell Rep Phys Sci 2(3):100376. https://doi.org/10.1016/j.xcrp.2021.100376

Weiner I, Shahar N, Feldman Y et al (2018) Overcoming the expression barrier of the ferredoxin-hydrogenase chimera in *Chlamydomonas reinhardtii* supports a linear increment in photosynthetic hydrogen output. Algal Res 33:310–315. https://doi.org/10.1016/j.algal.2018.06.011

Wieczorek N, Kucuker MA, Kuchta K (2014) Fermentative hydrogen and methane production from microalgal biomass (Chlorella vulgaris) in a two-stage combined process. Appl Energ 132:108–117. https://doi.org/10.1016/j.apenergy.2014.07.003

Winkler M, Hemschemeier A, Gotor C, Melis A, Happe T (2002) [Fe]-hydrogenases in green algae: photo-fermentation and hydrogen evolution under sulfur deprivation. Int J Hydrog Energy 27(11):1431–1439. https://doi.org/10.1016/S0360-3199(02)00095-2

Wu H, Wang C, Chen P et al (2017) Effects of pH and ferrous iron on the co-production of butanol and hydrogen by *Clostridium beijerinckii* IB4. Int J Hydrog Energy 42(10):6547–6555. https://doi.org/10.1016/j.ijhydene.2017.02.094

Wykoff DD, Davies JP, Melis A, Grossman AR (1998) The regulation of photosynthetic electron transport during nutrient deprivation in *Chlamydomonas reinhardtii*. Plant Physiol 117(1):129–139. https://doi.org/10.1104/pp.117.1.129

Xia A, Cheng J, Ding L et al (2013a) Improvement of the energy conversion efficiency of *Chlorella pyrenoidosa* biomass by a three-stage process comprising dark fermentation, photofermentation, and methanogenesis. Bioresour Technol 146:436–443. https://doi.org/10.1016/j.biortech.2013.07.077

Xia A, Cheng J, Lin R, Lu H, Zhou J, Cen K (2013b) Comparison in dark hydrogen fermentation followed by photo hydrogen fermentation and methanogenesis between protein and carbohydrate compositions in *Nannochloropsis oceanica* biomass. Bioresour Technol 138:204–213. https://doi.org/10.1016/j.biortech.2013.03.171

Xia A, Jacob A, Tabassum MR, Herrmann C, Murphy JD (2016) Production of hydrogen, ethanol and volatile fatty acids through co-fermentation of macro- and micro-algae. Bioresour Technol 205:118–125. https://doi.org/10.1016/j.biortech.2016.01.025

Xu L, Fan J, Wang Q (2019) Omics application of bio-hydrogen production through green alga *Chlamydomonas reinhardtii*. Front Bioeng Biotechnol 7. https://doi.org/10.3389/fbioe.2019.00201

Yakunin AF, Hallenbeck PC (1998) Short-term regulation of nitrogenase activity by NH4+ in *Rhodobacter capsulatus*: multiple *in vivo* nitrogenase responses to NH4+ addition. J Bacteriol 180(23):6392–6395. https://doi.org/10.1128/jb.180.23.6392-6395.1998

Yang Z, Guo R, Xu X, Fan X, Li X (2010) Enhanced hydrogen production from lipid-extracted microalgal biomass residues through pretreatment. Int J Hydrog Energy 35(18):9618–9623. https://doi.org/10.1016/j.ijhydene.2010.07.017

Yang D-W, Syn J-W, Hsieh C-H, Huang C-C, Chien L-F (2019) Genetically engineered hydrogenases promote biophotocatalysis-mediated H2 production in the green alga Chlorella sp. DT. Int J Hydrog Energy 44(5):2533–2545. https://doi.org/10.1016/j.ijhydene.2018.11.088

Yin Y, Wang J (2019a) Enhanced biohydrogen production from macroalgae by zero-valent iron nanoparticles: insights into microbial and metabolites distribution. Bioresour Technol 282:110–117. https://doi.org/10.1016/j.biortech.2019.02.128

Yin Y, Wang J (2019b) Hydrogen production and energy recovery from macroalgae Saccharina japonica by different pretreatment methods. Renew Energ 141:1–8. https://doi.org/10.1016/j.renene.2019.03.139

Yin Y, Hu J, Wang J (2019) Fermentative hydrogen production from macroalgae Laminaria japonica pretreated by microwave irradiation. Int J Hydrog Energy 44(21):10398–10406. https://doi.org/10.1016/j.ijhydene.2019.03.034

Yun WH, Yoon YS, Yoon HH, Nguyen PKT, Hur J (2021) Hydrogen production from macroalgae by simultaneous dark fermentation and microbial electrolysis cell with surface-modified stainless steel mesh cathode. Int J Hydrog Energy 46(79):39136–39145. https://doi.org/10.1016/j.ijhydene.2021.09.168

Zhao X, Xing D, Qi N, Zhao Y, Hu X, Ren N (2017) Deeply mechanism analysis of hydrogen production enhancement of *Ethanoligenens harbinense* by Fe^{2+} and Mg^{2+}: monitoring at growth and transcription levels. Int J Hydrog Energy 42(31):19695–19700. https://doi.org/10.1016/j.ijhydene.2017.06.038

Zhen G, Lu X, Li Y-Y, Liu Y, Zhao Y (2015) Influence of zero valent scrap iron (ZVSI) supply on methane production from waste activated sludge. J Chem Eng 263:461–470. https://doi.org/10.1016/j.cej.2014.11.003

Zheng GH, Wang L, Kang ZH (2010) Feasibility of biohydrogen production from tofu wastewater with glutamine auxotrophic mutant of *Rhodobacter sphaeroides*. Renew Energ 35(12):2910–2913. https://doi.org/10.1016/j.renene.2010.04.030

Part III

Microalgae as Source of Edible Products

Properties, Extraction, Characterization, and Market Opportunities of Edible Oil from a Residual Microalgae Biomass

10

Nidia Araiza Lizarde and Claudia Amezcua Vega

Abstract

Oils extracted from microalgae have been studied for use in generating clean energy. However, they have been studied for human consumption in the food industry due to their nutritional properties. This chapter shows the importance of the physicochemical properties of oils, the most used extraction methods, the importance of cultivation systems for oil use, and the market potential of these resources.

Keywords

Oil · Microalgal · Edible · Extraction · Lipid · Linolenic acid

10.1 Introduction

Nowadays, society has sought new alternatives to consuming edible oil with antioxidant properties, pleasant flavor, low cost, and health benefits, as with oil extracted from microalgae. Different microalgae species have been analyzed to determine which ones contain properties that allow them to be used for human consumption under strict safety regulations and provide benefits to improve health. Some genera of microalgae have been the most studied and offer such properties: *Chlorella, Scenedesmus, Arthrospira, Dunaliella, Nannochloropsis, and Tetraselmis.*

In this chapter, different microalgae species have been analyzed to determine which ones are viable for human consumption, in addition to the oil extraction methods, physicochemical characteristics, and market opportunities as edible oil sources.

N. Araiza Lizarde (✉) · C. Amezcua Vega
Ingeniería en Biotecnología Universidad Politécnica de Sinaloa, Mazatlán, Sinaloa, México
e-mail: naraiza@upsin.edu.mx

10.2 Microalgae with Properties for Edible Oil Extraction

The production of food supplements with microalgae is significant due to their high protein content, approximately 50%, as reported in *Chlorella, Scenedesmus, and Arthrospira*, among others. Moreover, this protein shows a greater nutritional value than the one found in several cereals and legumes usually consumed, although not higher than that of animal sources (Shahidi et al. 2017).

The consumption of microalgae for humans is ruled by strict alimentary security control, as well as by commercial factors, market demand, and specifications in their preparation. The three genera that stand out in the aspects previously mentioned are *Chlorella, Spirulina,* and *Dunaliella*, which are commercialized as food supplements (Mendoza et al. 2011).

Some species have been commercialized due to their health benefits. *Chlorella* is used as a supplement for kidney failure and intestinal *Lactobacillus* growth. Furthermore, it favors the decrease in blood cholesterol concentrations. On the other hand, *Dunaliella* is commercialized due to its beta (β)-carotene content (Brennan and Owende 2010). Other authors have demonstrated that a diet with *Chlorella* biomass increases the antioxidant level in smokers, which confirms that it may act as a supplement that favors health (Lee et al. 2008).

Norzagaray-Valenzuela et al. (2017) performed in vitro research demonstrating that the residual biomass extract of the three microalgal species: *Dunaliella tertiolecta, Tetraselmis suecica,* and *Nannochloropsis* sp. has high antioxidant and antiaging potential of the residual extracts; likewise, two peptides (tripeptide and heptapeptide) were also

D. U. Santos Ballardo, S. Rossi (eds.), *Microalgae as Promising Source of Commercial Bioproducts*, Developments in Applied Phycology 14, https://doi.org/10.1007/978-3-031-86433-9_10

isolated and purified from the species *Nannochloropsis oculata*, showing antihypertensive activity by inhibiting the angiotensin-converting enzyme (ACE) I to angiotensin II, which is responsible for contracting blood vessels and is the first therapeutic line against hypertension (Samarakoon et al. 2013).

Furthermore, the high content of polysaccharides and oligosaccharides in the form of soluble and insoluble fibers of some species is a property attributed to their prebiotic activity, which promotes the adequate balance of intestinal microbiota. Besides, biochemical and physiological effects improve the consumer's metabolism (Raposo et al. 2016).

Usually, lipids are found in the cellular membrane and produced from the smooth endoplasmic reticulum. Furthermore, microalgae accumulate lipids in oil granules in peroxisomes and spherosomes. Some species contain 40% of their weight, but it may vary from 5% to 77% from one species to another. *Spirulina* has a great application in the food industry and as a vitamin and protein supplement in aquaculture. In Africa, it is used as a food supplement; in Asia, it is used as a protein supplement and healthy food (FAO 2008). Also, *Spirulina* has a high protein content (65–70%); essential and nonessential amino acids; minerals such as potassium, calcium, magnesium, manganese, selenium, iron, and phosphorus; vitamins such as pyridoxine (B6), biotin, pantothenic, folic, inositol, niacin or nicotinic acids, riboflavin (B12), thiamine (B1), tocopherol (E), and cyanocobalamin (B12); natural complex sugars; carotenoids; enzymes; and essential fatty acids (Álvarez and Bague 2011). Furthermore, *Spirulina* has many polyunsaturated fatty acids (1.5–2.0% of 5–6% of total lipids). It is rich in gamma (γ)-linolenic acid (GLA) (36% total polyunsaturated fatty acids) and also provides linoleic acid (36% of the total), besides stearidonic, eicosapentaenoic (EPA), and other acids such as docosahexaenoic (DHA) and arachidonic (AA) acids (FAO 2008). Some fatty acids are precursors of prostaglandins, which have a beneficial effect on arthritis, obesity, alcoholism, neuropsychiatry, and inflammatory disorders (Álvarez and Bague 2011).

On the other hand, *Chlorella* sp. and *Dunaliella* show higher lipid content than other species. Moreover, the mixotrophic cultures of *Chlorella* have shown the highest cellular growth rates. On the other hand, *Chlorella* shows lower saturated fatty acids than *Dunaliella salina*, placing it in a lower-density oil, higher boiling, and better calorific properties that can be used in the food intustry in the food industry (Praveenkumar et al. 2014).

Besides, Tejeda-Benítez et al. (2015a, b) analyzed the lipid content of different species of microalgae isolated in the laboratory of the Centro Internacional Náutico Fluvial y Portuarios SENA in Cartagena, where the most significant lipid percentage was found in *Chlorella* sp. *with* 40.23%.

Also, Liang and Sarkany (2009) showed that *Chlorella vulgaris* contains between 33% and 38% lipids. On the other hand, Praveenkumar et al. (2012) obtained 31.2% lipid content and 32% (Zhao et al. 2014).

Paggi Matos et al. (2016) analyzed six microalgal species to evaluate the physiochemical and bromatological characteristics of their oils, as well as their application in the food industry, where they found that *Nannochloropsis gaditana* has a high content of saturated fatty acids, mainly palmitic acid (5.1 g/100 g), while *Spirulina platensis* and *C. vulgaris* are rich in gamma-linolenic acid (GLA) (1.9 g/100 g) and alpha-linolenic acid (ALA) (2.8 g/100 g), respectively. *Porphyridium cruentum* differs from other algae because it contains a significant amount of arachidonic acid (AA) (3.7 g/100 g). *Nannochloropsis oculata* and *Phaeodactyum tricornutum* are also sources of long-chain polyunsaturated fatty acids (ω3 LC-PUFA) and are composed mainly of EPA and DHA.

10.3 Oil Extraction from Microalgal Biomass

10.3.1 Soxhlet Extractor

The AOAC MD international method 960.39, USA (AOAC 2001) is usually used to assess oil content from organic materials (including microalgae). Place 10 g of microalgae biomass in a previously weighed thimble and place it in the Soxhlet extractor for n-hexane extraction for 16 h; the solvent is then vacuum distilled in a rotary evaporator (Buchi, ModelR-205, USA). The thimble is placed at 103 2 °C in the oven for 24 h to remove the residual solvent, cooled in a desiccator, and weighed on an analytical balance (Sartorius AX124, Goettingen, Germany). The oil could be stored in amber bottles for physicochemical analysis (Araiza et al. 2017).

10.3.2 Bligh and Dyer's Method

A mixture of methanol, chloroform, and water is used in the Bligh and Dyer technique in a ratio of 2:1:1.8. Solvents and biomass are mixed with their given proportions. Then, the same percentage of chloroform is added. Therefore, the ratio of methanol, chloroform, and water is 2:1:1, and the ratio of solvent in tissues is [(3 + 1):1] (Araiza et al. 2017). For the wet route, the ratio should be considered 2:1:1.8, and for the dry route, [(3 + 1):1]; it is important to investigate if the water content is lower than the biomass. Finally, fractional distillation is applied to isolate or release lipids (Iverson et al. 2001).

10.3.3 Slow and Fast Pyrolysis

Pyrolysis is a thermal process that involves heating biomass to high temperatures, typically between 450 and 550 °C or higher, in an oxygen-free environment. The primary product of pyrolysis is pyrolyzed oil or bio-oil, with biochar and gaseous products as by-products, which can be directly used as fuels. An experimental study on algal bio-oil demonstrated that various algal species, including *Polytrichum commune, Sphagnum palustre, Dicranum scoparium, Thuidium tamariscinum, Drepanocladua revolvens, Chlorella protothecoides, and Cladophora fracta*, yielded bio-oil in the range of 34.3–53.3% at 775 °C with a heating rate of 10 K/s via fast pyrolysis. Another experimental study indicated that *Spirulina* sp. yielded 32.42% of bio-oil. (Hossain et al. 2020).

10.3.4 Solvent-Free Extraction

Aqueous enzymatic extraction (AEE) is a proven technique that is utilized intensely to obtain lipids from different oil seeds. This procedure mainly relies on oilseed squashing, the insolubility of oil in water, and the thickness distinction between water and oil. Enzymes perform a significant role in disordering lipoproteins and lipopolysaccharides to speed up the lipid release. Organic solvent consumption can be wiped out effectively by this strategy, and this study does not contribute to generating noxious waste (Xue et al. 2018).

10.3.5 Hydrothermal Liquefaction (HTL)

A stainless steel reactor with a capacity of 500 ml (Trident Labortek, Mumbai, India) equipped with an electric heating unit could be used. Internal temperature monitoring uses a type K thermocouple with a sensor coupled to one end of the reactor. A pressure-measuring manometer is loaded with water and microalgae according to the desired proportion. After the reactor and its headspace are closed, it is purged with nitrogen gas to remove air and create an inert environment. After purging the nitrogen, the valve is closed, and the reactor is pressurized until an initial pressure of 10 psi. HTL runs are performed in different temperatures of 200–320 °C for a residence time of 30 min. After the reaction, the reactor must be cooled (Panga et al. 2018).

10.3.6 Supercritical Fluid Extraction

Supercritical fluid extraction (SFE) is a green, innovative technology. It has drawn attention worldwide for its high selectivity and utilization of elements with the characteristics of fluids and gases when subjected to high pressure and temperature (Zhang et al. 2015).

10.3.7 Folch Method

The Folch method suggests various organic solvents or combinations of different solvents to extract lipids from a complex mixture of organic compounds selectively. The Folch method (Folch et al. 1957) employs chloroform-methanol (2:1 by volume) to extract lipids from endogenous cells. The homogenized cells are equilibrated with a quarter volume of saline and mixed well to separate them into two layers, and the lipids are deposited in the upper phase. This method is one of the oldest in lipid extraction and constitutes the basis for developing future extraction procedures with improvements. The method is still used for spectrophotometric estimation of algal lipids. However, it is less sensitive than newer procedures (Ranjith et al. 2015).

10.3.8 Bead Milling

Bead milling is a mechanical method for cell destruction, where direct cell damage is caused by high-speed spinning of the biomass suspension with fine beads. In bead mills, cells are broken by the impact of the grinding beads against the cells. Shaker vessels and stirred beads are the two common types of bead mills. Therefore, this method can be used exclusively on a laboratory scale. Undoubtedly, the combined effect of shaking, collision, and crushing of the beads produces a more efficient alteration process. Likewise, it is unnecessary to dehydrate the algae suspension, unlike the press method, which contributes to the reduction of processing costs. Different beads are used for different cell types. The optimal bead diameter for microalgae cells is 0.5 mm, and the optimal bead loading volume fraction is approximately 0.5 ml. Beads made of zirconium-silica, zirconium oxide, or titanium carbide can improve the extraction efficiency of microalgal cells, presumably due to their higher hardness and density (Ranjith et al. 2015).

10.3.9 Microwave Extraction

Microwave (MW) is a promising technology for microalgal cell disruption in scientific and industrial areas. It is commonly used as a pretreatment method when applied to raw biomass materials or as an assisted extraction method when biomass is already resuspended in an appropriate solvent. Both approaches have already been partially tested on microalgae. This approach, however, requires an extra step for lipid extraction, which can increase processing duration and solvent requirements. Hence, the cost is increased as well.

Microwave-assisted extraction saves time and solvent by applying the microwave directly to the system of microalgae and extraction solvent, although lipid yields are still too low in the case of microalgae. Therefore, the microwave processes should be optimized to guarantee good performance in lipid extraction from microalgae (Shang et al. 2022).

10.3.10 Cell Membrane Electroporation

Electroporation applications are specifically developed to extract lipids and other intracellular metabolites from microalgae. It is based on a phenomenon in which the cell wall is exposed to a high electric field intensity in a given period. During exposure to the electric field, the cell membrane temporarily destabilized. The increase in the intensity of the electric field will cause the cell membrane to undergo reversible or irreversible electroporation. Meanwhile, irreversible electroporation is when the cell membrane is permanently damaged and unable to recover its original shape (Joannes et al. 2024).

10.3.11 Ultrasound Extraction

It consists of exposing microalgae to acoustic waves. These waves are produced by transmitting electrical current to a mechanical system. After ultrasound waves in a liquid medium, millions of microscopic bubbles are generated that expand and collapse (cavitation), causing the rupture of the cell wall. Applying low-frequency ultrasound for a continuous exposure of 15 min causes high algal cell destruction, even greater than applying high-frequency waves (Sanchez et al. 2020).

10.3.12 Autoclaved

The principle of extraction by autoclaved is similar to extraction by subcritical water, which is favorable since it avoids the drying stage of the microalgae biomass, preventing the lipids present in the algae from degrading (Salazar 2012).

According to Jooanes et al. (2024), this pretreatment technique uses conditions of 300 °C and 100 Megapascal (MPa) with times of between 5 and 6 min with wet biomass; the technique can be compared with extraction with supercritical fluids.

10.3.13 Osmotic Shock

The cells are initially balanced in a high osmotic pressure medium; then, by osmosis, water suddenly enters the cell, increases the internal pressure, and causes rupture; it is a relatively easy method to use (Gonzáles et al. 2009).

10.4 Physicochemical Properties of Microalgal Edible Oils

Environmental and nutritional factors are important in microalgal biomass growth rate and strongly influence the fatty acid profile. Some freshwater microalgae show fatty acids, such as myristic acid (C14:0), palmitic acid (C16:0), oleic acid (C18:1), linoleic acid (C18:2), and ɣ-linoleic acid C18:3, wich improve the digestion and absorption of nutrients in the human gastrointestinal system (Teixeira and Morales 2006). On the other hand, it is well-reported that nutrient deficiency, such as nitrogen and saline stress, enhances the accumulation of C18:1.

Tejeda-Benítez et al. (2015a) analyzed *Chlorella* sp. fatty acid profiles and found 34.9% of oleic acid, 20.56% of linoleic acid, and 39.33% of linolenic acid. On the other hand, Praveenkumar et al. (2014) observed 21.92% of linolenic acid; Zhao et al. (2014) found 21.93% of oleic acid and 21.9% of linoleic acid. Besides, *Dunaliella salina* reported 24.71% oleic acid, 18.77% linoleic acid, and 50.65% linolenic acid.

On the other hand, other important physicochemical parameters in the quality of the oil are density and the peroxide index. Tejeda-Benítez et al. (2015a, b) characterized the oils obtained from microalgae *D. Salina* and *Chlorella* sp., finding a density (g/cm^3) of 0.898 and 0.893. Additionally, a refraction index of 1.492 and 1.746 and an iodine index of 164.35 and 126.79.

Furthermore, in human physiology and development, polyunsaturated acids are vital in cardiovascular illness. These fatty acids are usually found in fish because, in the food chain, they consume microalgae at larvae stages, which also show a high content of unsaturated fatty acids (Brennan and Owende 2010; Mendoza et al. 2011).

On the other hand, microalgae are important sources of docosahexaenoic acid (DHA), demonstrating that mice and hamsters fed with *Chlorella* had decreased levels of triglycerides, total cholesterol, and low-density lipoprotein (LDL) cholesterol, which in turn helps, prevent arteriosclerosis.

Moreover, meeting the demand for plant-based edible oils has been challenging. Furthermore, consumers demand quality oil to make nutritional supplements and cosmetic products that improve their quality of life (Xue et al. 2018).

Table 10.1 shows the physicochemical specifications that should be complied with for using plant-based edible oils

Table 10.1 Physicochemical specifications for using edible oil

	NMX-F-808-SCFI-2018		Authors	Parameter
Specifications	Minimum	Maximum		
Free fatty acids (such as oleic) in %*		0.05	Vásquez-Suárez et al. (2007) Ejim and Kamen (2013)	3.12 0.96
Humidity %		0.05	Tejeda-Benítez et al. (2015a, b) (*D. Salina*) Ejim and Kamen (2013)	0.043 0.85
Color (Lovibond scale)		30A–3.5R	Ejim and Kamen (2013)	Greenish-yellow
Peroxide index in meq/Kg*		1.0	Tejeda-Benítez et al. (2015a, b) (*D. Salina*)	3.963
Insoluble impurities in %		0.02		
Trans fatty acids in %		3.5		
Density (g/cm^3)			Tejeda-Benítez et al. (2015a, b) (*D. Salina*) Ejim and Kamen (2013)	0.898 0.892 g/cm^3

A = Yellow
R = Red

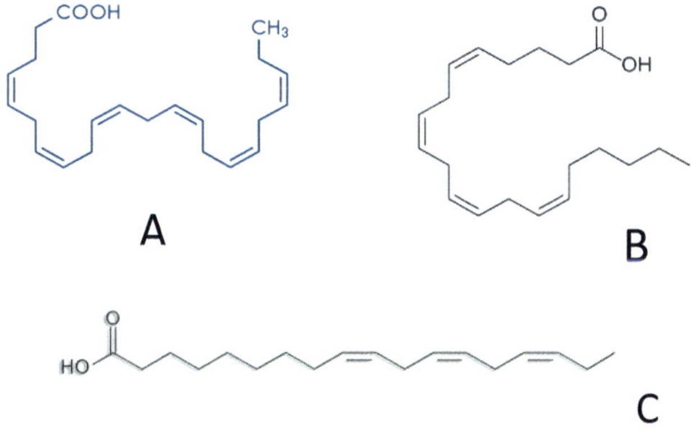

Fig. 10.1 Structures of unsaturated fatty acids: (**a**) docosahexaenoic acid (DHA), (**b**) eicosapentaenoic acid (EPA), and (**c**) arachidonic acid (AA), obtained from microalgae

according to the Mexican Norm NMX-F-808-SCFI-2018, besides the physicochemical characteristics that other authors have pointed out in their research.

Currently, there are no specific international standards to evaluate the quality of microalgae oil as a consumer food. However, there are standards for edible oils extracted from other species of oilseeds, such as safflower and olive, among others. On the other hand, olive oil is considered one of the best oils for the food industry. Based on the above, according to the Official Journal of the European Union, in its regulation EU 2022/2104 (established on July 29, 2022), it mentions a peroxide index ≤ 20.0 (mEq O2/kg) and a content of free fatty acids (% oleic) of 55.00–85.00, to the values reported by other authors (Table 10.1) referring to the oil obtained from microalgae, this material is a viable alternative to be used in the food industry.

10.5 Market Opportunities of Microalgal Oil

One of the microalgal oil applications that stands out is undoubtedly biofuel (Sundaram et al. 2023; Khan et al. 2018). However, it has been incorporated into diverse food, pharmaceutical, and cosmetic products in recent years, mainly due to its health benefits. The growing consciousness of consumers about healthy food and its advantages boosts market growth in Europe and America. The microalgal oil market is estimated to have an annual increase of 4.7% and sales that may exceed USD 3.1 billion by 2032 (FMI 2023).

Microalgal oil is a mix of saturated and unsaturated fatty acids ranging from C_{12} to C_{24} and C_{16} to C_{18}, respectively. The total oil content varies from 20% to 50% (dry weight). Bellou et al. (2014) reported that the amount of oil and the presence and position of double links depend on the microalga species and culture conditions used.

The oils with high commercial importance are fatty acids: arachidonic acid (ARA), eicosapentaenoic acid (EPA), and docosahexaenoic acid (DHA) (Fig. 10.1), known as omega 6 and 3, respectively, which have demonstrated to be fundamental for the formation and functionality of the central nervous system, retina, and cellular membranes (Valenzuela et al. 2013). Additionally, these fatty acids reduce blood cholesterol levels and promote healthy skin in humans, improving intestinal function and resistance to diseases and increasing animal reproductive function (Camacho et al. 2019). The multiple benefits associated with omega 3 allowed for a rise in the EMR (Aclaight Enterprise 2023) Latin American market of USD 275 million in 2021 with an expected annual growth rate of 7.3% for 2023–2028.

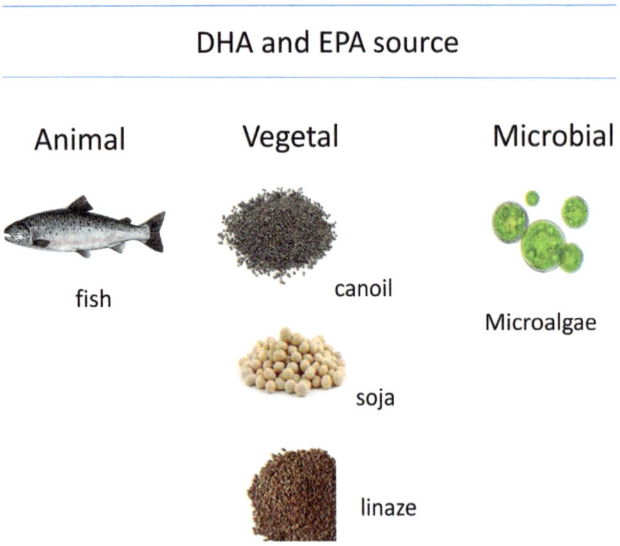

Fig. 10.2 Various sources of docosahexaenoic acid (DHA), eicosapentaenoic acid (EPA) and alpha-linoleic acid

The sources for obtaining omega 3 are not exclusively microalgae; it can also be obtained from different plants or animals (Fig. 10.2). However, advantages have been detected (Robles et al. 1998) in using microalgae to obtain omega 3 compared to the other two origins, such as:

1. Sustainable production.
2. No overexploitation of fishery resources.
3. Low oil oxidation, which reduces "bad fish odor."
4. Selective culture conditions, which decrease contamination risk.
5. Lower contamination by heavy metals.
6. No presence of cholesterol.
7. Microalgae culture does not compromise food production.

In addition to the properties of microalgae oil for human consumption, there are also advantages in obtaining the oil according to the following characteristics:

- Microalgae are capable of reproducing throughout the year.
- Microalgae grow in aqueous medium but need less water than terrestrial cultures.
- Microalgae may be cultured in brackish water, reducing the use of freshwater sources.
- Design of microalgal strains to improve yield by molecular biology tools.
- Renewable resource.
- Study diversity.

Detected disadvantages:

- Commercial production at a large scale may still be considered low (Valenzuela et al. 2015).
- Lack of regulation and norms.
- Lack of financing and risk/venture capital available.
- Lack of better cooperation between the industry and academia.

10.6 Conclusion

Microalgae have nutritional physicochemical and nutritional properties that make them potential for the food industry. Different extraction methods can be used depending on the physicochemical characteristics to obtain better performance. In addition, advances are being made in their cultivation systems. In the future, microalgal oil will present a significant use in the market.

References

Álvarez N, Bague S (2011) Los alimentos funcionales: una oportunidad para una mejor salud. Ed. Madrid, p 208
AOAC (Association Society of Official Agricultural Chemists) (2001) Official methods of analysis of AOAC, 16th edn. CDROM
Araiza L, Alcaraz L, Angulo M, Reynoso G, Cruz H, Calderón C (2017) Physicochemical composition of seed oil of wild jojoba populations in Northwestern Mexico. J Food Nutr Res 5(6):443–450
Bellou S, Baeshen M, Elazzazy M et al (2014) Microalgal lipids biochemestry and biotechnological perspectives. Biotechnol Adv 32:1476–1493
Brennan L, Owende P (2010) Biofuels from microalgae-a review of technologies for production, processing, and extractions of biofuels and co-products. Renew Sust Energ Rev 14:557–577
Camacho F, Macedo A, Malcata F (2019) Potential industrial applications and commercialization of microalgae in the functional food and feed industries: a short review. Mar Drugs 17:312
Diario Oficial de la Unión Europea (2022) Reglamento delegado (UE) 2022/2104 de la comisión de 29 de julio de 2022. Normas de comercialización del aceite de oliva. https://eur-lex.europa.eu/legal-ntent/ESTXT/PDF/?uri=CELEX:32022R
Ejim IF, Kamen FL (2013) Physiochemical characterization of algae oil from microalgae of Nike Lake Enugu, Department of Chemical Engineering, Institute of Management and Technology Enugu State. Department of Chemical Engineering, Federal University of Technology, Owerri, Imo State. J Eng Appl Sci 5:181–187
EMR Aclaight Enterprise (2023) Perspectiva del Mercado Latinoamericano de Omega 3. https://www.informesdeexpertos.com/informes/mercado-latinoamericano-de-omega3
FAO (2008) A review on culture, production and use of Espirulina as food for humans and feeds for domestic animals and fish. In: ftp://ftp.fao.org/docrep/fao/011/i0424e/i0424e00.pdf. Consulted: Agosto 2022
Future Market Insights (FMI) (2023) Algae oil market outlook 2022–2032. https://www.futuremarketinsights.com/reports/algae-oil-market

Folch J, Lees M, Sloane Stanley GH (1957) A simple method for the isolation and purification of total lipids from animal tissues. Journal of Biological Chemistry, 226, 497–509

Gonzáles A, Vatchesla K, Guzmán A (2009) Desarrollo de métodos de extracción de aceites en la cadena de producción de biodiesel a partir de microalgas. Rev Prospect 7(2):53–60

Hossain N, Chowdhuryb T, Chowdhuryc H, Ahmedd A, Nizamuddine S, Griffina G, Meurah T, Young-Kwon P (2020) Edible bio-oil production from microalgae. Chapter 4. Elsevier, Inc, pp 14–15. https://doi.org/10.1016/B978-0-12-821218-9.00004-9

Iverson S, Lang L, Cooper M (2001) Comparison of the Bligh and Dyer and Folch methods for total lipid determination in a broad range of marine tissue. Lipids 36:1283–1287

Joannes C, Sipaut S, Dayou J, Yasir S, Mansa R (2015) Review paper on cell membrane electroporation of microalgae using electric field treatment method for microalgae lipid extraction. IOP Conf Ser: Mater Sci Eng 78:012034. https://doi.org/10.1088/1757-899X/78/1/012034

Khan M, Shin J, Kim J (2018) The promising future of microalgae: current status, challenges, and optimization of a sustainable and renewable industry for biofuels, feed, and other products. Microb Cell Factories 17(1):36

Lee E, Park J, Choi J, Huh B et al (2008) A randomized study to establish the effects of Spirulina in type 2 diabetes mellitus patients. Nutr Res Pract 2(4):295–300. https://doi.org/10.4162/nrp.2008.2.4.295

Liang N, Sarkany C (2009) Biomass and lipid productivities of Chlorella vulgaris under autotrophic, heterotrophic and mixotrophic growth conditions. Biotechnol Lett 31:1043–1049

Mendoza H, de la Jara A, Portillo E (2011) Planta piloto de cultivo de microalgas: Desarrollo potencial de nuevas actividades económicas asociadas a la biotecnología en Canarias. Instituto Tecnológico de Canarias, Santa Cruz de Tenerife, p 60

NMX-F-808-SCFI-2018. Alimentos – aceite vegetal comestible – especificaciones. Dirección general de normas. SINEC – 20181031114039808

Norzagaray-Valenzuela C, Valdez-Ortiz A, Shelton L et al (2017) Residual biomasses and protein hydrolysates of three green microalgae species exhibit antioxidant and anti-aging activity. J Appl Phycol 29(1):189–198. https://doi.org/10.1007/s10811-016-0938-9

Paggi Matos A, Feller R, Siegel M, Vladimir de Olivera J et al (2016) Chemical characterization of six microalgae with potential utility for food application. J Am Oil Chem Soc. https://doi.org/10.1007/s11746-016-2849-y

Panga K, Saranga V, Verma K et al (2018) Bio oil production from microalgae via hydrothermal liquefaction technology under subcritical water conditions. J Microbiol Methods 153:108–117

Praveenkumar K, Shameera G, Mahalakshmi M et al (2012) Influence of nutrient deprivations on lipid accumulation in a dominant indigenous microalga Chlorella *sp.*, BUM11008: evaluation for biodiesel production. Biomass Bioenergy 37:60–66

Praveenkumar B, Choi E, Lee K et al (2014) Improved biomass and lipid production in a mixotrophic culture of *Chlorella sp.* KR-1 with addition of coal-fired flue-gas. Bioresour Technol 171:500–505

Ranjith R, Hanumatha P, Arumugam M (2015, 2014) Lipid extraction methods from microalgae: a comprehensive review. Sec Bioenergy Biofuels 2. https://doi.org/10.3389/fenrg.2014.00061

Raposo F, Morais R, Morais A (2016) Emergent sources of prebiotics: seaweeds and microalgae. Mar Drugs 14(2):1–27. https://doi.org/10.3390/md14020027. PMID: 26828501

Robles A, Molina E, Giménez A (1998) Downstream processing of algal polyunsaturated fatty acids. Biotechnol Adv 16:517

Salazar E (2012) Evaluación de métodos de extracción de aceites de microalgas para producción de biodiesel. Editorial Universidad de Piura, Perú

Samarakoon K, O-Nam K, Ko J (2013) Purification and identification of novel angiotensin-I converting enzyme (ACE) inhibitory peptides from cultured marine microalgae (*Nannochloropsis oculata*) protein hydrolysate. J Appl Phycol 25(5):1595–1606. https://doi.org/10.1007/s10811-013-9994-6

Sanchez J, Loaña J, Agualongo M, Espinoza K (2020) Cultivation techniques and methods of extraction of fatty acids based on microalgae for the benefit of humanity. Agroind Sci 10(3):319–328. https://doi.org/10.17268/agroind.sci.2020.03.15

Shahidi F, Rajapakse N, Kim S (2017) Marine nutraceuticals and functional foods (Chap. 15). Taylor and Francis, Boca Ratón, pp 367–403

Shang Y, Rehman H, Mehmood K, Xu A, Iftikhar Y, Wang Y, Sharma R (2022) The nexuses between social media marketing activities and consumers' engagement behaviour: A two-wave time-lagged study. Frontiers in Psychology, 13, Article 811282

Sundaram T, Rajendran S, Gnanasekaran L et al (2023) Bioengineering strategies of microalgae biomass for biofuel production: recent advancement and insight. Bioengineered 14(1):2252228

Teixeira CM, Morales E (2006) Microalga como matéria-prima para a producao de biodiesel. Annals of I Congress of the Brazilian Network of Biodiesel Technology, Brasilia, p 2006

Tejeda-Benítez, Henao-Argumedo D, Alvear-Alayón M (2015a) Caracterización y perfil lipídico de aceites de microalgas. Rev Fac Ing (Fac. Ing.) 24:39

Tejeda-Benítez L, Henao-Argumedo D, Alvear-Alayón M et al (2015b) Caracterización y perfil lipídico de aceites de microalgas Characterization and lipid profile of oil from microalgae Caracterización de perfil lipídico de aceites de microalgas

Valenzuela B, Morales P, Sanhueza C et al (2013) Docosahexanoic acid (DHA), an essential fatty acid at the brain. Rev Chil Nutr 40:383–390

Valenzuela B, Sanhueza C, Valenzuela B (2015) Microalgae: a renewable source for obtaining omega-3 long-chain fatty acids for human and animal nutrition. Rev Chil Nutr 42(3):306–310

Vásquez-Suárez A, Guevara M, Salazar G (2007) Crecimiento y composición bioquímica de cuatro especies de Dunaliella para ser utilizadas en Acuicultura. Bol Cent Investig Biol Univ Zulia 41:181–194

Xue Z, Wan F, Yu W, Liu J, Zhang Z, Kou X (2018) Edible oil production from microalgae: a review. Eur J Lipid Sci Technol 120:170042. https://doi.org/10.1002/ejlt.201700428

Zhang Y, Li Y, Zhang X, Tan T (2015) Biodiesel production by direct transesterification of microalgal biomass with co-solvent. Bioresour Technol 196:712–715. https://doi.org/10.1016/j.biortech.2015.07.052

Zhao X, Yu J, Li X et al (2014) Enhancing lipid productivity by co-cultivation of Chlorella sp. U4341 and *Monoraphidium sp.* FXY-10. J Biosci Bioeng 118:72–77

Pigments from Microalgae: Chlorophyll, Carotenoids, and Phycobiliproteins as Source of Natural Colorants for Food

11

Angel Valdez-Ortiz, Karelia Arizbeth Meza-Ayala, Jose Manuel García-Padilla, Rafael Valdez-Ortiz, and Lourdes Janeth Germán-Báez

Abstract

The color of food is a determining factor for consumer decisions; therefore, the industry has resorted to large-scale synthetic pigment production to meet their growing demand. However, artificial colorants have been frequently associated with health problems ranging from behavioral alterations to allergic reactions and even cancer promotion. For this reason, and due to current trends in population demands, the use of colorants has been revolutionized toward natural alternatives, increasing this trend in various sectors of the food industry. Natural colorants are generally considered harmless compounds; consequently, their use limitations are lower than those affecting synthetic colorants. Microalgae offers several technological advantages for the production and extraction of natural pigments. These advantages include their ability to produce a wide range of pigments, the high concentrations of pigments generated within their cells, their rapid growth rate, and their adaptability to various environments.

Keywords

Natural pigments · Microalgae · Phycoerythrin · Phycocyanin · Allophycocyanin · Carotenoids · Chlorophyll

A. Valdez-Ortiz · K. A. Meza-Ayala
J. M. García-Padilla · L. J. Germán-Báez (✉)
Laboratorio de Biotecnología e Ingeniería Genética, Facultad de Ciencias Químico-Biológicas, Universidad Autónoma de Sinaloa, Culiacán, Sinaloa, Mexico
e-mail: ljgerman@uas.edu.mx

R. Valdez-Ortiz
Service of Nephrology, General Hospital of Mexico, Dr. Eduardo Liceaga, Mexico City, Mexico

11.1 Introduction

Dyes or pigments are molecules characterized by their ability to absorb or emit light in the visible range (400–700 nm). These structures and production systems have been intensely studied in recent years due to their increasing demand in various areas of the food, pharmaceutical, and cosmetic industries, among others (Rao et al. 2017).

In particular, in the food industry, these additives are used to fulfill different functions, such as restoring the original color of the food when it has been destroyed by processing or storage conditions, correcting natural variations in color intensity, maintaining its uniformity, intensifying the food's color when it is weak and uneven and give to the food a color that identifies it and provides an attractive appearance. Thus, adding them to foods, dyes, or pigments helps improve their acceptability, as they provide a defined, uniform, and consistent color for the same product even across different production batches and times of the year. The pigments used in foods can be classified into two groups: those that exist in nature or are isolated from natural sources and those that are chemically synthesized (Salem et al. 2009).

Synthetic pigments have been widely used in the industry over the years; they are those created by humans through chemical synthesis processes and do not exist naturally. The use of these compounds quickly advanced, as they were easy to produce, less expensive, and did not impart flavor to foods. Additionally, they were superior in color properties, requiring only small amounts. Manufacturers of these pigments offered more than 80 artificial colors; however, many were never tested for toxicity and adverse effects, leading to negative side effects on health and the environment (Downham and Collins 2000). It is now known that dyes like tartrazine, which ranges from red to yellow, can cause allergies alone and in combination with other pigments. In this sense, the Food and Drug Administration (FDA) in the United States, the European Food Safety Authority (EFSA) in Europe, and many other organizations worldwide have restricted the use

D. U. Santos Ballardo, S. Rossi (eds.), *Microalgae as Promising Source of Commercial Bioproducts*, Developments in Applied Phycology 14, https://doi.org/10.1007/978-3-031-86433-9_11

of synthetic colorants in foods, confectionery, and beverages due to their confirmed or suspected association with cancer development, allergic reactions, hyperactivity disorders, allergies, attention deficits, among other issues (Manikprabhu and Lingappa 2015; Martelli et al. 2014; Kobylewski and Jacobson 2010).

Natural pigments are extracted from renewable sources, generally from plants, insects, fungi, algae, and diverse microorganisms such as microalgae and cyanobacteria. These pigments are usually obtained using water or organic solvents such as different types of alcohols, being the most commonly obtained carotenoids, chlorophylls, flavonoids, betalains, tannins, quinones, xanthones, and others (Mulders et al. 2014; Aberoumand 2011). These offer uniform color, low risk of microbial contamination, and relatively low production costs. However, they have a lower staining capacity compared to chemically synthesized dyes, as they are more sensitive to light, temperature, oxygen, and pH. Despite this, the demand for natural pigments has been increasing, mainly due to the toxic and carcinogenic effects found in synthetic pigments. It has also been demonstrated that some natural colorants, including carotenoids and phycobiliproteins, have preventive effects against certain diseases such as diabetes, obesity, and cardiovascular diseases (De Mejia et al. 2020; Manikprabhu and Lingappa 2015; Salem et al. 2009).

Until now, the most well-known natural colorants obtained from various plants are green, derived from chlorophyll and obtained from various vegetables and grasses; yellow, derived from carotenoids, lutein, and curcumin, obtained from palm oil, marigold, and turmeric; orange, such as bixin and capsanthin, obtained from annatto and Aji (Capsicum); red, such as lycopene obtained from tomato; purples, derived from anthocyanins and proanthocyanidins, obtained from red cabbage, grapes, blackberries, cherries, and strawberries; and pink, such as betanin, obtained from beetroot. There are many others obtained from bacteria, yeasts, and fungi, as well as from insects, such as carminic acid, a red colorant obtained from cochineal, as well as those obtained from microalgae and cyanobacteria (Vera-López Portillo and Martínez Jiménez 2021; De Mejia et al. 2020).

Microalgae are a source of various beneficial compounds with health benefits that can enhance the nutritional value of dietary supplements. Beyond being a good protein source, species like *Arthrospira*, *Chlorella*, and *Nannochloropsis* also significantly produce pigments, carbohydrates, and lipids. Nowadays, the generally recognized as safe (GRAS) designation has been granted to *Arthrospira*, *Chlorella*, *Dunaliella*, *Haematococcus*, *Schizochytrium*, *Porphyridium cruentum*, and *Crypthecodinium cohnii*. Thus, the products derived from microalgae, such as the blue pigment phycocyanin from *Arthrospira*, DHA from *C. cohnii*, and β-carotene from *Dunaliella*, are available on the market (Udayan et al. 2021).

Microalgae are considered a natural source of food colorants due to their rich content of photosynthetic pigments, making them widely used as food coloring agents. These pigments not only provide color but also offer neuroprotective, antioxidant, and hepatoprotective benefits. As a result, they are utilized in the nutraceutical, pharmaceutical, cosmetic, and aquaculture industries. In Brazil, chlorophyll extracted from *Spirulina* is used as a natural colorant in the food industry. In poultry feed, *Chlorella vulgaris* consumption enhances the pigmentation of egg yolks. *Dunaliella* contains high levels of beta-carotene, which is more easily absorbed by the body than synthetic pigments. Powder made from *Dunaliella* biomass is commonly used as a food and feed additive. Additionally, the consumption of microalgae rich in astaxanthin, like *Haematococcus pluvialis,* by fish enhances the red color of their flesh and provides high levels of vitamin E due to its antioxidant properties (Udayan et al. 2017, 2021; De Mejia et al. 2020).

11.2 Pigments from Microalgae

Microalgae offers several advantages as a source of natural pigments for food, such as a wide range of pigments, rapid growth compared to higher plants, pigment concentrations that can exceed those found in higher plants, and ease of cultivation (Mulders et al. 2014). Among the pigments most produced by microalgae are chlorophylls *a*, *b* and *c*, β-carotene, astaxanthin, phycocyanin, xanthophylls, and phycoerythrin (PE), which are frequently used in pharmacology, food, textile industry, and cosmetics, due to their proven preventive effects against diseases such as diabetes, obesity, and cardiovascular diseases (De Mejia et al. 2020).

The primary role of these pigments in microalgae is to capture energy at several wavelengths within the visible light spectrum, enhancing the efficiency of the photosynthesis process; this is why they are commonly called photosynthetic pigments and are classified into three categories: chlorophylls, phycobiliproteins, and carotenoids; additionally, they are responsible for giving microalgae their characteristic color, resulting in a wide variety of microalgae, including red, brown, green, and blue-green types (Table 11.1) (Vera-López Portillo and Martínez Jiménez 2021; Mulders et al. 2014).

In this regard, the primary function of pigments in photosynthetic organisms is to enhance the capture of photons and their translocation toward the photosystems. However, pigments also serve as photoprotectors and reservoirs of carbon and nitrogen. Light-capturing pigments can be divided into two categories: primary and secondary. Among the primary pigments, only chlorophyll *a* is present, as it is the pigment responsible for splitting the water molecule, consequently releasing the proton and capturing the electron, which will

Table 11.1 Main pigments from microalgae and their applications

Pigment	Microalgae	Application	Compound	Absorption (λmax, nm)	Pigment color
Chlorophylls	Green microalgae: *Dunaliella tertiolecta, Dunaliella salina, Rhodomonas* sp., *Chlorella vulgaris, Scenedesmus obliquus, Microcystis aeruginosa, Gloeothece membranacea*	Food dye and antioxidant	Chlorophyll a	642–653	Dark green Dark green
			Chlorophyll b	660–665	Green Green
Carotenoids	*Dunaliella salina, Dunaliella tertiolecta, Phaeodactylum tricornutum, Asterarcys quadricellulare, Scenedesmus* sp.	Vitamin A precursor, antioxidant properties, prevents macular degeneration	β-Carotene	455	Orange Orange
	Chlorella protothecoides, Chlorella minutissima, Scenedesmus obliquus, Scenedesmus sp., *Coccomyxa acidophila, Chlamydomonas* sp.	Prevents age-related macular degeneration, food additive, reduces the risk of cataracts and cardiovascular diseases	Lutein	445	Golden yellow Golden yellow
	Haematococcus pluvialis, Chlorella sp., *Chlorococcum* sp.	Protect from UV rays, antioxidants, food coloring, improve immune system functions	Astaxanthin	482	Red Red
Phycobiliproteins	*Porphyridium cruentum, Porphyridium marinum, Porphyridium purpureum, Pseudanabaena* sp.	Fluorescent markers, dyes used in the food and cosmetic industries	Phycoerythrin	500–580	Red Red
	Arthrospira platensis, Arthrospira máxima, Galdieria sulphuraria	Fluorescent marker, hepatoprotective activity, antioxidant, anti-inflammatory, and neuroprotective	Phycocyanin	620	Blue Blue
	Arthrospira platensis, Arthrospira máxima, Galdieria sulphuraria	Biochemical techniques such as a fluorescent protein probe, especially for flow cytometry. Antioxidant and anti-enterovirus properties	Allophycocyanin A	650	Light blue Light blue
			Allophycocyanin B	671	Cyan Cyan

Adapted from Vera-López Portillo and Martínez Jiménez (2021), Conceição et al. (2020), Mulders et al. (2014), Su et al. (2009)

be used in photosynthesis for the subsequent formation of energy in the form of ATP and the formation of NADPH, which is used to reduce CO_2 into sugars. The secondary pigments include other chlorophylls, phycobiliproteins, and carotenes. On the other hand, pigments are considered to have a photoprotective function in preventing photooxidation resulting from photosynthesis. They are mainly produced when there is an excess of light intensity or long exposure times, as well as in the presence of external oxidizing agents.

Photoprotective pigments are carotenoids such as β-carotene, astaxanthin, lutein, zeaxanthin, violaxanthin, and diadinoxanthin. Understanding the role of pigments in microalgae is crucial, as it facilitates the development of strategies to enhance the production and accumulation of these pigments. These strategies generally involve manipulating the conditions of microalgal cultivation by adjusting factors such as light intensity, exposure times or photoperiods, type of lighting, composition of the culture medium, and exposure to reactive oxygen species, among others (Mulders et al. 2014; Borowitzka 2013; Aberoumand 2011).

11.2.1 Chlorophyll

Chlorophyll is the natural pigment that gives plants and algae their dark green color. It was isolated in 1817 by the French chemists Pelletier and Caventou, who were the first to separate it and named it chlorophyll, derived from the Greek adjective *chlorós* and the noun *phýllon*, meaning "green leaf"

(Gopi et al. 2014). This pigment is located in the thylakoids of chloroplasts and is responsible for capturing and storing light energy, which initiates the process of photosynthesis. During this process, the energy is transformed into chemical energy capable of converting carbon dioxide and water into carbohydrates, thereby promoting the release of oxygen into the environment. It absorbs light most effectively in the blue part of the electromagnetic spectrum, followed by the red part. In contrast, chlorophyll poorly absorbs light in the green and near-green regions, which is why chlorophyll-containing tissues appear green (Vera-López Portillo et al. 2021; Gopi et al. 2014).

In nature, there are different types of chlorophylls, such as those of type *a* and *b*, the two primary photoreceptor pigments of higher plants. Chlorophyll *a* is the most widely distributed; it is always present in oxygen-producing photosynthetic cells; in addition, it is found in all green plants together with chlorophyll *b* and in brown algae, diatoms, and dinoflagellates with chlorophyll *c*. Many plants and algae contain only chlorophyll; an example is cyanophyte algae. On the other hand, chlorophyll *b* is found in green algae and higher plants, while chlorophylls *c* and *d* are only observed in certain algae (Hosikian et al. 2010; Rasmussen and Morrissey 2007).

11.2.1.1 Structure and Classification

As previously mentioned, there are four types of chlorophyll found in nature: *a*, *b*, *c*, and *d*. Among these, chlorophylls *a* and *b* are the most abundant. Structurally, chlorophylls are composed of a porphyrin ring consisting of four pyrrole rings connected by methine bridges (--CH=). At the center of this porphyrin ring, there is a magnesium ion. Additionally, chlorophylls have a long hydrophobic phytol tail, which helps anchor the molecule in the thylakoid membrane of chloroplasts. The specific types of side groups attached to the porphyrin ring differ among the various types of chlorophyll, contributing to their distinct absorption properties. Chlorophylls are considered part of the tetrapyrrole compounds, including phycobiliproteins (accessory pigments in blue and red algae). The tetrapyrrole structure is the basic framework of porphyrins, which include not only chlorophylls but also hemoglobins and cytochromes (Fig. 11.1) (Vera-López Portillo et al. 2021; Alarcón-Bonilla et al. 2020; Gopi et al. 2014; Scheer et al. 2004; Humphrey 1980).

The chromophoric characteristic of chlorophyll is due to its conjugated double bonds formed by the linkage of pyrrole rings through methine groups. The ring system of chlorophyll has distinct features, such as a central magnesium metal atom. In its third pyrrole ring, there is an additional isocyclic structure, the condensed cyclopentanone ring, which contains a carboxylic group esterified with methanol. The fourth ring is esterified at its propionic acid group with an alcohol called phytol, which consists of a 20-carbon chain with a double bond. This phytol "tail" gives pure chlorophyll a waxy nature, explaining its insolubility in water and good solubility in organic solvents. Consequently, chlorophyll is an amphipathic compound with a hydrophobic phytol tail and a hydrophilic porphyrin core (Alarcón-Bonilla et al. 2020; Hosikian et al. 2010; Delgado-Vargas and Paredes-López 2002).

The differences in the substituents on the porphyrin ring and the phytol tail give rise to the various types of chlorophyll; a large number of chlorophyll structures have been elucidated, all sharing the same basic structure with variations in one or more of the positions 3, 7, 8, 17, 18, and 20 (Delgado-Vargas and Paredes-López 2002).

Among the chlorophylls, *a* and *b* stand out due to their abundance in photosynthetic organisms. These chlorophylls have identical structures, except that chlorophyll *a* has a methyl group at the C-3 position of ring II, while chlorophyll *b* has a formyl group at the same position; these variations result in each type of chlorophyll having a characteristic maximum absorbance at specific wavelengths of visible light due to their conjugated double-bond systems and, which accounts for the blue-green color of chlorophyll *a* which has a maximum absorbance in the range of 433 to 666 nm and the brilliant-green color of chlorophyll *b* which has a maximum absorbance in the range of 462 to 650 nm. Both chlorophylls exhibit maximum absorption in the red and blue-violet regions, in the range of 600 to 670 nm and 400 to 470 nm respectively, and show intense red fluorescence in solution. The chlorophyll *c* exhibits a yellow-green color, and the chlorophyll *d* exhibits a brilliant forest-green color (Vera-López Portillo et al. 2021; Mulders et al. 2014; Hosikian et al. 2010).

Pure chlorophyll *a* absorbs light energy at a wavelength of 663 nm (red-orange light). However, a complete chloroplast containing all the plant pigments absorbs more red light at a wavelength of 680 nm. This is due to other molecules interacting with the chlorophyll, altering its absorption spectrum. In other words, chlorophyll is associated with other molecules and can absorb more energy up to a wavelength of 700 nm. Although all chlorophylls can absorb light energy, only the *a* type can transfer that energy to acceptor molecules; the other chlorophylls (b, c, and d) capture the energy and transmit it to chlorophylls *a*. The same process occurs with carotenoids and phycobiliproteins, which capture light energy and pass it on to neighboring chlorophyll molecules (Koller et al. 2014; Mulders et al. 2014; Hosikian et al. 2010).

11.2.1.2 Biological Properties of Chlorophyll

Few clinical studies demonstrate the long-term benefits of chlorophyll or supplements containing it; however, some animal studies have suggested that it may reduce inflammation, aid in wound healing, and fight infections by inhibiting

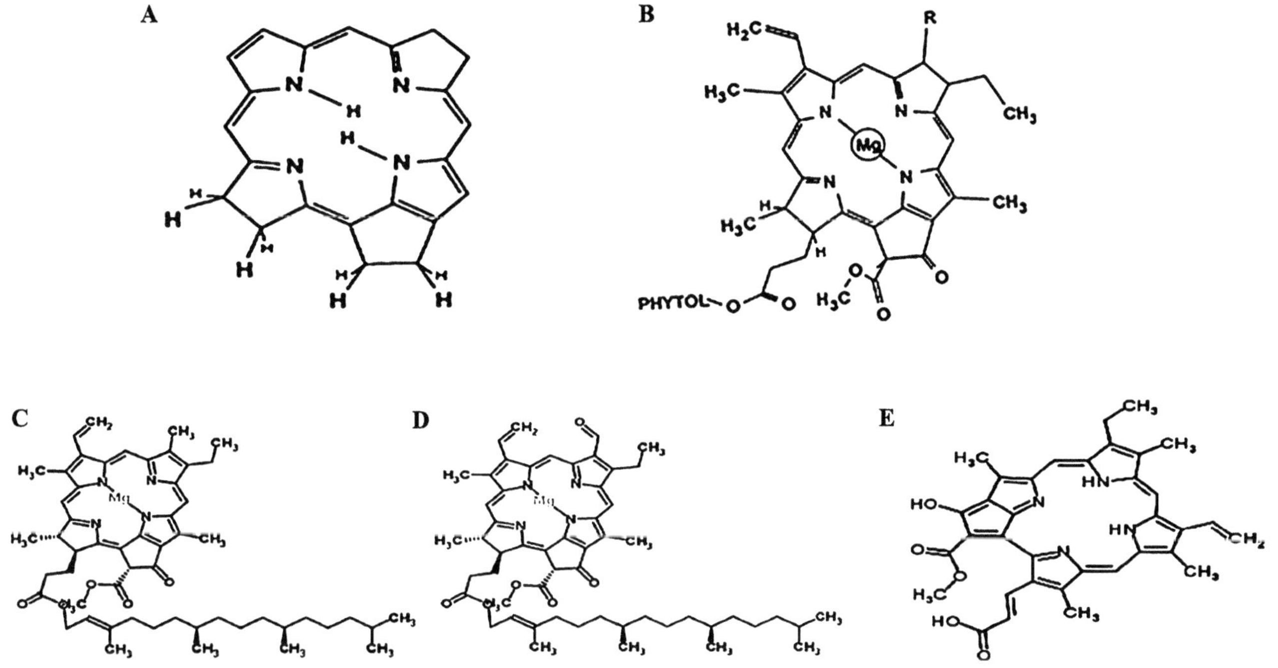

Fig. 11.1 Structure of chlorophylls. (**a**) Structure of porphyrin; (**b**) Porphyrin bound to its hydrophobic side chain, phytol, which together form chlorophyll; (**c**) Structure of chlorophyll a; (**d**) Structure of chlorophyll b; (**e**) Structure of chlorophyll c. (Adapted from Gopi et al. (2014); Scheer et al. (2004); Humphrey (1980))

the growth of certain bacteria (Subramoniam et al. 2011). Besides, Mancebo-Dorvigny et al. (2011) demonstrated the wound-healing activity of chlorophyll by formulating a cream containing 5% chlorophyll concentrate and other bioactive compounds (carotenes and vitamins). This cream was applied daily to wounds inflicted on mice and evaluated throughout the tissue regeneration period. The study confirmed the healing effect through a reduction in wound area compared to the control group. Additionally, the study observed that chlorophyll eliminated wound odors after just a few applications (Goldberg 1943). Chlorophyll's non-toxic nature, antibacterial properties, and deodorizing function have made it a key compound in treating oral sepsis (Goldberg 1943; Gispert-Abreu et al. 2001).

Another notable medicinal property is its promotion of healthy digestion by maintaining intestinal flora, preventing constipation through stimulation of peristalsis, and alleviating discomfort caused by gastritis due to its ability to heal damaged intestinal tissues. Chlorophyll also aids in colon cleansing, preventing the proliferation of colonocytes (Balder et al. 2006).

It has also been shown that chlorophyll can block the formation of certain cancer-causing compounds, including heterocyclic amines from cooked meat, polycyclic hydrocarbons from tobacco smoke, and aflatoxins from moldy peanuts and cereals. For instance, a study by Vibeke (1994) evaluated chlorophyll's inhibitory capacity against aflatoxin B1, a potential liver carcinogen, in rainbow trout fry. The fish were acclimated in several tanks for two weeks, then fed a diet rich in chlorophyll and its derivatives (2% of their body weight) alongside doses of aflatoxin B1. After 14 days, 15 fish were randomly selected from each tank for histopathological examination to identify the number and morphology of tumors. The results showed fewer tumors in the fish fed a chlorophyll-rich diet thanks to chlorophyll's capacity to prevent mutagenesis by binding to harmful substances.

Studies have identified chlorophyll as a substance that reduces carcinogenesis caused by genetic mutations. Evidence of this is the study by Carter et al. (2004), which cultured human colon cancer cells in vitro and treated them with chlorophyll and its derivatives. The treated cells showed sensitivity to the treatment, evidenced by decreased growth. Studies have shown that the higher the chlorophyll content in the chosen treatment or diet, the greater the protection against carcinogenesis.

Another property of chlorophyll is its ability to scavenge free radicals, making it sought after in the market as an antioxidant agent, helping to slow aging and cellular oxidation (Hsu et al. 2013; Lanfer-Marquez et al. 2005).

Overall, the positive data from in vitro and in vivo studies have led to growing interest in the potential use of chlorophylls as preventive agents for populations at high risk of chronic diseases (Gopi et al. 2014).

11.2.1.3 Uses of Chlorophyll

Given its numerous nutraceutical benefits, chlorophyll is increasingly attracting attention as a product to fulfill the market demand for food colorants and additives that provide extra health advantages (Koller et al. 2014). Chlorophyll and its derivatives are produced worldwide and used as supplements and colorants in food products, including pre-cooked pasta and noodles, absinthe, cheeses and dairy products, fats and oils, canned goods, sweets, ice creams, jams, wines, beverages, and infant food formulations. They are also widely used in producing pharmaceutical and cosmetic products due to their action as natural dyes and biological properties (Hosikian et al. 2010).

However, chlorophyll's low water solubility and instability during industrial production processes significantly limit its direct use in commercial product manufacturing. Consequently, it is necessary to obtain molecular derivatives of chlorophyll, which can be used as direct ingredients in product formulations. Among the chlorophyll derivatives with the most significant market impact are metallochlorophyllins. Unlike chlorophyll, these compounds are water-soluble, which has increased their commercial importance in recent years because they can be used in a broader range of formulations. Metallochlorophyllins are more stable molecules that can withstand abrupt pH changes and high temperatures, conditions often encountered during industrial product manufacturing (Selig et al. 2020; Ferruzzi and Schwartz 2005).

The metallochlorophyllins are salts derived from the pigment chlorophyll. During the conversion process of chlorophyll to metallochlorophyllins, sodium salts are added, the phytol chain is lost, and the central magnesium atom is replaced by other metal ions such as copper, magnesium, zinc, iron, and tin, forming chlorophyllins like sodium and magnesium chlorophyllin (CSM), sodium and iron chlorophyllin (CSI), sodium and tin chlorophyllin (CSE), sodium and zinc chlorophyllin (CSZ), and copper and sodium chlorophyllin (CCS), with the latter two being the most common. The addition reaction varies depending on the metal to be introduced. An acidic solution is required to prepare copper, cobalt, iron, nickel, silver, zinc, manganese, or gold complexes. Meanwhile, under weakly alkaline solutions, mercury, tin, potassium, lead, or thallium complexes are formed. Strongly alkaline conditions are required to form lithium and magnesium complexes (Rani and Banu 2017; Kephart 1955).

Research on the biological properties of metallochlorophyllins and their constituent compounds has demonstrated their high commercial potential in the cosmetic and pharmaceutical industries, as they possess antimicrobial, antimutagenic, antioxidant, and anti-aging properties. Additionally, they act as an internal deodorant to reduce odors associated with incontinence, colostomies, similar procedures, and general body odor. They are also available as a topical preparation, practical for both the treatment and control of wound odor, lesions, and other skin conditions, particularly radiation burns (Rani and Banu 2017; McCook et al. 2015; Ferruzzi and Schwartz 2005; Lanfer-Márquez et al. 2005).

11.2.1.4 Main Sources of Chlorophyll

Chlorophyll is a bioactive compound that can be obtained from both plants and some aquatic organisms. Traditionally, commercially extracted chlorophyll comes from bamboo leaves and green vegetables such as chard, spinach, purslane, watercress, broccoli, lettuce, and wheat and alfalfa sprouts. However, aquatic microorganisms like microalgae are an alternative source of chlorophyll that has recently gained importance in the food, cosmetic, and pharmaceutical industries. This is due to their high capacity to accumulate large amounts of this pigment and to the multiple applications of the byproducts from the extraction process (Li et al. 2016; Torres et al. 2014).

Among the microalgae species most commonly used for chlorophyll extraction are those from the genus *Arthrospira* (*Spirulina*), *Chlorella*, and *Dunaliella*. The latter has attracted particular interest because it consists of marine microalgae that do not require freshwater for growth and do not compete for arable land or essential food resources. This makes chlorophyll production through *Dunaliella* more economical and sustainable (Pitchford 2002).

11.2.1.5 Methods for Chlorophyll Extraction

The traditional method for obtaining chlorophyll involves extraction with organic solvents; this process consists of the solvent penetrating through the cell membrane, dissolving the lipids and lipoproteins of the chloroplast membranes (Simon and Helliwell 1998; Jeffrey et al. 1997). It has been shown that the effectiveness of organic solvent extraction is influenced by various parameters such as the duration and number of stages in the process, the filtration method used before analysis, and the nature of the solvents employed. Additionally, scientific evidence suggests that performing a cell disruption before extraction, whether by grinding, sonication, or ultrasound, significantly improves yield (Macías-Sánchez et al. 2009; Simon and Helliwell 1998; Jeffrey et al. 1997). One limitation of organic solvent extraction compared to other processes is that it does not quantify the amount of product obtained. However, spectrophotometry can be used for quantification, measuring sample absorbances at specific wavelengths and using simultaneous equations formulated to calculate present concentrations (Jeffrey and Humphrey 1975).

Another extraction method used is chromatographic techniques, which fractionate chlorophyll into different types or derived products and allow for its quantification. The three types of chromatography widely used for extracting and fractionating chlorophylls are paper chromatography, thin-layer chromatography (TLC), and high-performance liquid chromatography (HPLC). Due to technological advancements, the chromatographic method most commonly used today is HPLC, which is superior to TLC as it requires fewer samples for analysis, is faster, and has an automatic detection system. Additionally, it is more precise and has a higher degree of sensitivity, although it requires expensive equipment for its development (Hosikian et al. 2010; Mortensen and Geppel 2007).

It is known that the formation of degradation products significantly affects the yield. In the case of chlorophyll, degradation products form when its molecules are exposed to excessive light, oxygen/air, high temperatures, and elevated pH, commonly encountered during extraction processes (Cubas et al. 2008; Jeffrey et al. 1997).

11.2.2 Carotenoids

Carotenoids are a varied group of isoprenoid pigments produced by all photosynthetic organisms and some non-photosynthetic bacteria and fungi. They are characterized by their C40 polyene chain structure, enabling them to absorb visible light and thus play critical roles in capturing light and providing photoprotection within the photosynthetic system. Carotenoids are structurally categorized into carotenes, which are hydrocarbons, and xanthophylls, which include oxygen (Britton et al. 2004). The synthesis of carotenoids starts with the formation of isopentenyl pyrophosphate (IPP) and dimethylallyl pyrophosphate (DMAPP), which are converted into geranylgeranyl pyrophosphate (GGPP). Phytoene synthase catalyzes the condensation of two GGPP molecules to produce phytoene, a colorless carotene. Phytoene undergoes desaturation and isomerization reactions to form lycopene, a key intermediate for synthesizing various carotenoids. Lycopene is then cyclized to create β-carotene and α-carotene, and further hydroxylation and epoxidation reactions transform these two carotenes into a wide variety of xanthophylls (Cuttriss et al. 2011).

11.2.2.1 Structure and Types of Carotenoids

As mentioned earlier, the carotenoids produced in microalgae belong to the groups of carotenes and xanthophylls.

Carotenes

Among the carotenoids in the carotenes group, the microalgae can produce three types: α and β carotene, and echinenone, which are described below.

α-Carotene is a carotenoid with a molecular structure comprising 40 carbon atoms arranged in a C40 polyene chain. The molecular formula of α-carotene is $C_{40}H_{56}$. Its structural configuration includes a series of conjugated double bonds, imparting color and antioxidant properties (National Center for Biotechnology Information 2024a). The antioxidant properties of α-carotene derive from its ability to quench singlet oxygen and neutralize free radicals. The conjugated double bonds in its structure enable it to stabilize and neutralize reactive oxygen species, thereby protecting cells from oxidative stress and mitigating the risk of chronic diseases. Additionally, α-carotene is a precursor to vitamin A, contributing to its health benefits, including support for immune function and vision (Tanumihardjo 2013).

β-Carotene has a molecular structure characterized by a carbon skeleton comprising 40 carbon atoms arranged in a long hydrocarbon chain, known as a C40 polyene chain. The molecular formula of β-carotene is $C_{40}H_{56}$ as well. At each end of this chain are two β-ionone rings, which are cyclohexene rings that are both cyclized. This cyclization is a key feature that distinguishes β-carotene from α-carotene (National Center for Biotechnology Information 2024b). β-Carotene biological activity derives from its capacity to neutralize free radicals by donating electrons or hydrogen atoms, thereby reducing oxidative stress. Besides its capacity to neutralize free radicals, β-Carotene can also quench singlet oxygen, a highly reactive form of oxygen whose neutralization helps prevent oxidative damage to proteins, DNA, and lipids molecules (Paiva and Russell 1999).

Although it is not a carotene, echinenone is a xanthophyll derived from β-carotene and is structurally similar to it, since it also has a C40 polyene chain arranged in a linear hydrocarbon chain. However, unlike β-carotene, echinenone has a keto group (C=O) at the C4 position of one of its β-ionone rings, which makes echinenone acquire different properties such as increased electron withdrawing capacity, which can stabilize free radicals more efficiently. Furthermore, the keto group's presence can influence how echinenone interacts with light and dissipates the absorbed energy. Also, the presence of the polar keto group can increase the solubility and absorption of echinenone, hence improving its effectiveness as a dietary antioxidant (National Center for Biotechnology Information 2024c; Kusama et al. 2015).

Xanthophylls

On the other hand, microalgae can produce a wide variety of carotenoids belonging to the xanthophylls group, which can be found next.

Alloxanthin is a carotenoid whose composition has the particularity of containing oxygen in it, which is a characteristic of the xanthophyll category. Alloxanthin structure is characterized by a C40 carbon skeleton besides an acetylenic bond, which is a carbon-carbon triple bond, and two hydroxyl groups (–OH) located at the 3 and 3′ positions of the terminal rings of its carbon skeleton and its chemical formula is $C_{40}H_{54}O_2$. Moreover, multiple conjugated double bonds are contained along the carbon skeleton, contributing to its light absorption properties and color (National Center for Biotechnology Information 2024d). Alloxanthin's biological activity lies in its oxidative stress protective properties and free radical scavenging. These properties come primarily from the alloxanthin capacity to neutralize free radicals, promoting the protection of cellular components. Furthermore, alloxanthin is involved in the light-harvesting process in diverse photosynthetic organisms, which helps convert light into chemical energy (Kabir et al. 2022; West et al. 2016).

Antheraxanthin is a xanthophyll with the chemical formula $C_{40}H_{54}O_3$. It acts as an intermediate in the biosynthesis of zeaxanthin from violaxanthin as a part of the xanthophyll cycle in plants and algae. Structurally, it consists of a C40 backbone with conjugated double bonds as well as hydroxyl groups in the terminal position at each end of the molecule, specifically on carbons 3 and 3′, which contributes to its polar character in comparison to other carotenoids as β-carotene. Moreover, antheraxanthin possesses an epoxide group located between carbons 5 and 6. This epoxide group allows the conversion of antheraxanthin into different xanthophylls, such as zeaxanthin or violaxanthin (National Center for Biotechnology Information 2024e; Yamamoto et al. 2006). As part of the biological function antheraxanthin has, photoprotection plays a crucial role when it comes to protecting plants and microalgae from damage caused by light excess thanks to the conversion of antheraxanthin into zeaxanthin dissipating this way the heat. Moreover, antheraxanthin can help neutralize reactive oxygen species thanks to conjugated double bonds that might absorb the excess energy and donate electrons, thereby decreasing oxidative stress (Shimode et al. 2018; Gilmore et al. 1994).

Astaxanthin is a carotenoid pigment with the chemical formula $C_{40}H_{52}O_4$ characterized by a polyene chain of conjugated double bonds, contributing to its intense red color. Besides the polyene chain, astaxanthin is composed of two β-ionone rings at each end of the polyene chain, contributing to its stability and solubility in fats. Moreover, it possesses two hydroxyl and one carbonyl group attached to each β-ionone ring, providing the molecule with antioxidant activity (National Center for Biotechnology Information 2024f). Thanks to all its structural features, astaxanthin is known to have antioxidant and anti-inflammatory properties, adding to its role in eye health and skin protection. Although astaxanthin is a carotenoid, it cannot be converted into retinol due to its molecular structure, which differs significantly from provitamin A. Moreover, astaxanthin can improve eye health because the molecule can cross the blood-retinal barrier and provide antioxidant and anti-inflammatory effects (Guerin et al. 2003).

Canthaxanthin is an orange-red xanthophyll with a molecular formula of $C_{40}H_{52}O_2$ and a chemical structure characterized by its C40 polyene chain. It possesses keto groups at the 4 and 4′ positions on the β-ionone rings at each end of the molecule (National Center for Biotechnology Information 2024g). Canthaxanthin has several biological properties; it acts as a powerful antioxidant due to its capacity to neutralize free radicals and singlet oxygen, thus protecting cells from oxidative stress. Also, it has been demonstrated that canthaxanthin possesses anti-inflammatory properties due to reduced production of pro-inflammatory cytokines. It has been reported that canthaxanthin has a photoprotective effect stronger than β-carotene but weaker than astaxanthin (Camera et al. 2009; Chan et al. 2009; Abe et al. 2007;).

Diadinoxanthin is a xanthophyll pigment whose chemical formula is $C_{40}H_{56}O_4$. It comprises 40 carbon atoms arranged in a chain with conjugated double bonds that allow the molecule to absorb light. It has epoxy and hydroxyl groups that contribute to its photoprotective and polarity properties, respectively (National Center for Biotechnology Information 2024h).

Diatoxanthin is derived from diadinoxanthin along the xanthophyll cycle. Its molecular formula is $C_{40}H_{56}O_4$, and its structure consists of a polyene chain with conjugated double bonds, which confers the molecule with the ability to absorb light and enhance its role in photoprotection. Also, it possesses two cyclohexane rings, each with hydroxyl groups that help stabilize the molecule, thus enhancing its antioxidant properties. The characteristic that distinguishes diatoxanthin from diadinoxanthin is the presence of acetylenic bonds within its structure and its function is related to the energy dissipation and absorption of light, also, they are implied in the enhanced reactivity that allows diatoxanthin to be able to neutralize reactive oxygen species (National Center for Biotechnology Information 2024i; Pistelli et al. 2021; Kuczynska and Jemiola-Rzeminska 2016).

On the other hand, fucoxanthin is a brown/orange xanthophyll whose chemical formula is $C_{42}H_{58}O_6$. Its composition comprises a C40 polyene chain with conjugated double bonds. It also has a cyclohexane ring with a carbonyl group attached to it at one end of the molecule. On the other end of the molecule, a furan ring is attached to the polyene chain and contributes to the characteristic color properties of fucoxanthin. Moreover, the molecule exhibits epoxide groups, one fused to the cyclohexane ring and the other fused to the polyene chain, and exhibits hydroxyl groups, which augment the polar character of the molecule and contribute to its biological activity (National Center for Biotechnology Information 2024j). It is known for its antiproliferative, anti-inflammatory, and antioxidant activity since it can act as an electron donor and neutralizer agent to decrease oxidative stress. Moreover, it is known that fucoxanthin can act as a photoprotective agent against damage induced by UV-B light (Oliyaei et al. 2021; Tavares et al. 2020; Wang et al. 2012).

Besides, lutein is a yellowish-green/brown carotenoid with the chemical formula $C_{40}H_{56}O_2$ composed of a C40 polyene chain fused to two β-ionone rings, one at each end of the chain. The β-ionone ring composition contributes to its color and biological activity. Along the C40 polyene chain, it is possible to find conjugated double bonds that contribute to its light absorption properties, as well as hydroxyl groups attached to the β-ionone rings, which define the polar character of the molecule (National Center for Biotechnology Information 2024k). Lutein is known for its photoprotective activity against UV light. It has a more potent antioxidant activity than α-tocopherol. It is a pigment with potential anticancer effects. Also, it improves eye health since it protects against the damage caused by the blue light oxidative effects (Jahns and Holzwarth 2012; Tanaka et al. 2012; Ouchi et al. 2010).

Myxoxanthophyll is a yellowish xanthophyll with the chemical formula $C_{46}H_{66}O_7$. It is composed of a polyene chain, two cyclohexane rings, and two hydroxyl groups fused to the cyclohexane groups (National Center for Biotechnology Information 2024l). It is known that this pigment plays a role in photoprotection and is associated with neutralizing the effects of UV-B light (Ehling-Schulz et al. 1997).

Furthermore, neoxanthin is a light-yellow xanthophyll with the chemical formula $C_{40}H_{56}O_4$, and its composition is defined by a C40 polyene chain composed of conjugated double bonds; it also contains an epoxide group and several methyl groups associated with the polyene chain (National Center for Biotechnology Information 2024m). Neoxanthin has an interesting biological function since it can induce apoptosis and act as an antiproliferative agent in different cancer cell lines (Kotake-Nara et al. 2005; Asai et al. 2004; Kotake-Nara et al. 2001). Furthermore, it has been reported to play an important role in photoprotection due to its ability to dissipate excess energy. It has also been reported that it can prevent oxidative damage in the DNA (Giossi et al. 2020; Şahin et al. 2020).

Violaxanthin is pale yellow/orange xanthophyll with the chemical formula $C_{40}H_{56}O_4$. It has a chemical structure composed of a C40 polyene chain involved in its light absorption ability; it also possesses two epoxide groups involved in its transformation into zeaxanthin and antheraxanthin along the xanthophyll cycle. Moreover, it has two hydroxyl groups fused to each end of the molecule (National Center for Biotechnology Information 2024n).

Violaxanthin's biological activity is related to its anti-inflammatory effect, its antioxidant activity, and its anticancer potential since it has been demonstrated that this carotenoid can induce apoptosis in cancer cells (Soontornchaiboon et al. 2012; Pasquet et al. 2011; Jahns et al. 2009).

Zeaxanthin is a yellow xanthophyll with the chemical formula $C_{40}H_{56}O_2$. It has a molecular composition composed of a C40 polyene chain arranged in a conjugated double-bond configuration. As a specific trait, zeaxanthin possesses two hydroxyl groups on each molecule's end, increasing its polar character (National Center for Biotechnology Information 2024o). As part of its biological properties, zeaxanthin exhibits a strong antioxidant power due to its ability to neutralize free radicals, thus protecting cells from oxidative stress; all this is thanks to a chain of isoprene residues containing conjugated double bonds (Murillo et al. 2019). Also, it is a known anti-inflammatory agent since it helps to reduce the presence of oxidized low-density lipoprotein (LDL), thereby reducing the risk of age-related inflammatory diseases such as atherosclerosis and age-related macular degeneration, which usually is associated with blindness (Firdous et al. 2015; Wang et al. 2007).

11.2.2.2 Microalgae Sources of Carotenoids

The food industry primarily sources carotenoids from plants and microalgae. Common plant sources include carrots, sweet potatoes, tomatoes, and leafy greens, abundant in carotenoids such as β-carotene, lycopene, and lutein. These compounds are extracted for use as natural colorants and dietary supplements because of their antioxidant properties and health benefits, which include enhancing vision and reducing the risk of chronic diseases as a consequence of oxidative stress (Victoria-Campos et al. 2022).

Microalgae have become increasingly important in carotenoid production for the food industry due to their ability to produce large quantities of these pigments. They can be cultivated in controlled environments to achieve high yields and have the unique capability of synthesizing carotenoids not typically found in terrestrial plants. For instance, microalgae like *Dunaliella salina*, *H. pluvialis*, and *Spirulina platensis* are known producers of β-carotene, astaxanthin, and zeaxanthin, respectively. Besides them, other microalgae species from the genus *Dunaliella*, *Chlorella*, *Coelastrum*, *Desmodesmus*, and some more have been confirmed as carotenoid producers (Galasso et al. 2017).

The main strains of microalgae to produce carotenes are mentioned below: α-Carotene has been reported to be produced by microalgae species such as *C. vulgaris* BIN, *D. salina*, and *Heterochlorella luteoviridis* (Fernandes et al. 2021; Şahin et al. 2019; Diprat et al. 2017). β-Carotene can be extracted from several microalgae species, such as *D. salina*, which is one of the most traditional sources since this microalga can produce high β-Carotene quantities under high salinity, high light intensity, and nutrient deficiency conditions (Pourkarimi et al. 2020). Besides *D. salina*, β-Carotene can also be extracted from, *Arthrospira platensis*, *Chlamydomonas reinhardtii*, *Chlorella sorokiniana*, *Dunaliella tertiolecta*, *Phaeodactylum tricornutum* (Wang et al. 2023a, b; Rengel et al. 2022; Diprat et al. 2017).

Moreover, one of the notable sources of echinenone is the microalga *H. pluvialis* under mixotrophic conditions or in environments with specific stress conditions, which enhances echinenone production (Orosa et al. 2001). Besides *H. pluvialis*, other microalgae have been reported to produce echinenone, such as *Phormidium autumnale*, *Synechocystis lividus*, and *Synechocystis pevalekii* (Paliwal et al. 2016; Rodrigues et al. 2014).

Regarding xanthophylls, the main microalgal strains that produce them are *D. salina*, *Nannochloropsis salina*, and *C. sorokinia* for alloxanthin (Safafar et al. 2015). *Besides, C. vulgaris* and *H. pluvialis* are recognized for their xanthophyll content, specifically antheraxanthin; besides them, microalgae like *Chlorella variablis*, *Monoraphidium minutum*, and *Scenedesmus bijuga* are confirmed to be antheraxanthin producers (León-Vaz et al. 2023; Fernandes et al. 2021; Paliwal et al. 2016).

On the other hand, astaxanthin is commonly extracted from several microalgae species; a great example of this is *H. pluvialis*, which is the primary source of astaxanthin production due to its high content and biosynthesis efficiency (Oslan et al. 2021). Besides *H. pluvialis*, microalgae like *Desmodesmus subspicatus*, *D. salina*, *Grasiella emersonni* CCNM 1001, *M. minutum*, and *S. dimorphus* are known to be able to produce astaxanthin (Achour et al. 2022; Ahmad et al. 2021; Paliwal et al. 2016). There are several microalgae considered sustainable and efficient canthaxanthin sources; one of them is *Chlorella zofingensis* since this microalga accumulates canthaxanthin under traditional growth conditions or stress conditions such as high light intensity or nutrient deprivation (Pelah et al. 2004). Other species known as canthaxanthin producers are *Acutodesmus dimorphus*, *C. variablis*, *D. subspicatus*, *and Nannochloropsis* sp. G1–5 and *Nannochloropsis granulata* (Villanova et al. 2022; Kim et al. 2021; Paliwal et al. 2016).

On the other hand, diadinoxanthin is found particularly in diatoms where it is involved in the xanthophyll cycle, protecting the microalgae from light-produced stress due to its conversion into diatoxanthin which dissipates the excess of energy as heat (Lavaud et al. 2002). Also, diadinoxanthin could be extracted and purified from microalgae like *P. tricornutum* through a process that involves four key steps: Algae cultivation, solvent extraction, ODS open-column chromatography, and ethanol precipitation, yielding high purity levels and recovery rates of approximately 55% (Zhuang et al. 2023). Diatoxanthin can be found naturally in diatoms like *P. tricornutum* and *Nitzschia* sp. (Wang et al. 2023a, b; Pennington et al. 1988). Fucoxanthin is a photosynthetic pigment in microalgae like *P. tricornutum* and *Nitzschia Closterium* (Ma et al. 2023; Zhuang et al. 2023). Since microalgae can produce 3–6 times more lutein per hectare compared to their primary competitors (the marigold flowers), these microorganisms are considered a sustainable platform for lutein production. Examples of lutein microalgae producers are species like *C. sorokiniana*, *D. salina*, *D. tertiolecta*, *H. pluvialis* and *P. tricornutum* (León-Vaz et al. 2023; Wang et al. 2023a, b; Diprat et al. 2017; Safafar et al. 2015). Microalgae like *A. platensis*, *S. lividus*, and *S. pevalekii* are considered myxoxanthophyll producers (Wang et al. 2023a, b; Paliwal et al. 2016).

Moreover, neoxanthin is produced by microalgae like *C. reinhardtii*, *Tetraselmis* sp. CTP4, and *Chlorella pyrenoidosa* (Şahin et al. 2020; Schüler et al. 2020; Inbaraj et al. 2006). Violaxanthin can be obtained from microalgae such as *Chlorella zofingiensis*, *Chlorella ellipsoidea*, and *D. tertiolecta* (Soontornchaiboon et al. 2012; Pasquet et al. 2011; Del Campo et al. 2004). Microalgae like *C. sorokiniana*, *C. reinhardtii*, and *C. vulgaris* are known for their ability to produce zeaxanthin as part of the xanthophyll cycle (Cordero et al. 2011).

11.2.2.3 Applications of Carotenoids in Food

One of the main applications of carotenoids in food is as a natural colorant due to the increasing demand for natural products since they offer an attractive alternative to synthetic dyes. Furthermore, the antioxidant properties of carotenoids contribute to their role in protecting cells from oxidative stress, which is associated with the incidence of chronic diseases such as cancer and cardiovascular disease.

α-Carotene is used in the food industry due to its powerful antioxidant properties added to its provitamin A character, thus enhancing the nutritional profile of food products. It is often used as a colorant in beverages and dairy products due to its orange color. On the other hand, β-Carotene is extensively used in the food industry because of its dual role as a provitamin A and its colorant attractiveness. β-Carotene can be found in products such as margarine, bakery, and cheese, improving their nutritional value and adding a vibrant yellow color to them, making these products more appealing to consumers. Besides, echinenone is gaining popularity due to its antioxidant properties, which place this carotenoid as one of the future candidates to combat oxidative stress (Meléndez-Martínez et al. 2021; Zakynthinos and Varzakas 2016).

Alloxanthin might not be as popular as other carotenoids; nevertheless, due to its antioxidant properties, it is a xanthophyll with potential application in developing methods to preserve freshness and extend the shelf life of dietary products. Antheraxanthin can neutralize free radicals, thus maintaining the nutritional quality and sensory attributes of food products. This added to its vibrant yellow color, makes this pigment an appealing alternative to be applied in a wide range of products. Whereas alloxanthin and antheraxanthin are not the top-used xanthophyll to this date, astaxanthin stands out among carotenoids due to its antioxidant power. These antioxidant properties offer important health benefits when they are used in food products, such as anti-inflammatory effects and cardiovascular disease prevention. Moreover, astaxanthin enhances the organoleptic characteristics of beverages and other dietary products, thus improving consumer appreciation. Canthaxanthin is a red-orange xanthophyll extensively used in the food industry due to its colorful nature and antioxidant properties. Its primary use is as a colorant in products such as sausages and fish. However, it also adds healthy characteristics to food products, such as those derived from its role as an antioxidant agent and immune response booster (Meléndez-Martínez et al. 2021).

Diadinoxanthin and diatoxanthin are antioxidant agents that help preserve food, providing a yellow-orange color to the products. Fucoxanthin is a brownish xanthophyll with powerful antioxidant activity, making this pigment an appealing candidate for use as an additive in food products. Nowadays, it is used as a natural colorant, providing products with a brown to olive-green hue that enhances public acceptance. Lutein is a yellow carotenoid used in industry due to its vibrant yellow color and its attractive health benefits, particularly in eye health, since it acts as a filter for harmful blue light, thus preventing age-related macular degeneration. It is commonly used in products such as fortified foods and beverages to enhance their nutraceutical profile (Meléndez-Martínez et al. 2021; Galasso et al. 2017). Neoxanthin is a yellow-orange xanthophyll whose main application in food industries is related to its antioxidant nature. Since its antioxidant properties help prevent food products from oxidative damage, it is used to extend the aliments' shelf life. Additionally, it is known that neoxanthin possesses anti-inflammatory and anticancer properties, which makes it a desirable additive in functional foods (Galasso et al. 2017).

Violaxanthin is a pigment with an orange color used mainly as a natural colorant to enhance the appearance of food products such as beverages, sauces, and different processed foods. Additionally, it possesses a strong antioxidant property, which helps provide aliments with longer shelf life. Finally, zeaxanthin is well-known for being present in aliments like paprika and corn, where it is commonly used to enhance color and nutritional profile due to its vibrant yellow-orange color and its photoprotective activity, which prolongs the shelf life of aliments (Meléndez-Martínez et al. 2021).

11.2.2.4 Biological Properties of Carotenoids

Carotenoids are a group of diverse molecules responsible for the colors observed in different microalgae; nevertheless, beyond their role in providing coloration, these molecules are indispensable for diverse biological functions such as those related to photosynthesis and all the processes involved in capturing the light energy, providing photoprotection, prevention of damage caused by harmful reactive oxygen species and free radicals. Furthermore, some carotenoids have been found to enhance immune response, improve skin and eye health, and help in the anticancer mechanisms. This section summarizes the diverse functional and health benefits these natural pigments provide in Table 11.2.

11.2.3 Phycobiliproteins

Phycobiliproteins, also called biliproteins, are a family of protein pigments that provide a wide range of bright colors and are biological macromolecules with linear tetrapyrrole prosthetic groups called phycobilins, which are covalently linked to the apoprotein chain, which is composed of two subunits (α and β), by specific bonds in cysteine residues; they are water-soluble and form antenna complexes called phycobilisomes (PBS), these last ones are spherical and approximately 600 nm in diameter, with a mass of around 2500 kDa, they are attached to the external face of the thylakoid membrane and act as antennas that capture light by channeling the excitation energy. Biliproteins constitute almost 80% of the mass of the phycobilisome, and it has been suggested that they play a role in the assembly and stabilization of phycobilisomes. Additionally, they are believed to help optimize the absorbance characteristics and energy transfer, facilitating a unidirectional flow of energy from the peripheral phycobiliproteins to the reaction centers through the phycobilisome core. The phycobiliproteins are organized in the phycobilisomes in order of their maximum absorption, finding phycoerythrins at the periphery, phycocyanins in the center, and allophycocyanins very close to the reaction centers in the thylakoid membrane (Fig. 11.2) (Sonani et al. 2016; Pandey et al. 2013; Ruiz-Ortega 2012). Therefore, the energy transfer route within the cell is as depicted in Fig. 11.2.

In addition to its role as an antenna complex in light harvesting, the phycobilisomes are also crucial for photoprotection mechanisms under high light conditions. They may serve as a nutrient source for microalgae during nitrogen and phosphorus deficiency (Sonani et al. 2016).

11.2.3.1 Structure and Classification

As previously mentioned, the phycobiliproteins are composed by two groups of chemical molecules, the first group is a linear tetrapyrrole prosthetic group, called phycobilins, which are covalently linked to the second group, the apoprotein chain, which is composed of two subunits (α and β). Phycobilins or bilins are composed of linear tetrapyrroles containing four pyrrolic rings, A, B, C, and D, connected by carbon. Each A and D ring contains an oxygen atom, while the middle rings, B and C, are associated with propionic acid residues. In contrast to heme groups and chlorophylls, bilins do not contain metal atoms and have a molecular weight of approximately 600 kDa (Stadnichuk and Tropin 2017). To date, four types of bilin chromophores have been reported, including phycocyanobilin (PCB, blue), phycoerythrobilin

Table 11.2 Functional and health benefits provided by carotenoids

Carotenoid	Functional and health benefits	References
α-Carotene	Provitamin A activity Anticarcinogenic activity Antioxidant activity	Cooperstone et al. (2017) Murakoshi et al. (1992) Paiva and Russell (1999)
β-Carotene	Provitamin A activity Antioxidant activity Photoprotective activity Anticarcinogenic activity	Mein et al. (2008) Paiva and Russell (1999) Moshell and Bjornson (1977) Nishino et al. (2002)
Echinenone	Photoprotective activity Antioxidant activity	Kusama et al. (2015)
Alloxanthin	Anti-inflammatory activity Photoprotective activity	Konishi et al. (2008) West et al. (2016)
Antheraxanthin	Photoprotective activity Xanthophyll cycle pigment Antioxidant activity	Niyogi et al. (1997) Yamamoto et al. (2006) Shimode et al. (2018)
Astaxanthin	Anti-inflammatory activity Antidiabetic activity Antioxidant activity Photoprotective activity Anticarcinogenic activity	Guerin et al. (2003)
Canthaxanthin	Antioxidant activity Photoprotective activity Anti-inflammatory activity	Abe et al. (2007) Camera et al. (2009) Chan et al. (2009)
Diadinoxanthin	Photoprotective activity	Lavaud et al. (2002)
Diatoxanthin	Photoprotective activity Anticarcinogenic activity Antioxidant activity	Pistelli et al. (2021) Kuczynska and Jemiola-Rzeminska (2016)
Fucoxanthin	Antioxidant activity Anticarcinogenic activity Anti-inflammatory activity Antidiabetic activity Photoprotective activity	Wang et al. (2012) Oliyaei et al. (2021) Tavares et al. (2020)
Lutein	Photoprotective activity Antioxidant activity	Jahns and Holzwarth (2012) Miller et al. (1996)
Myxoxanthophyll	Photoprotective activity	Ehling-Schulz et al. (1997)
Neoxanthin	Antiproliferative agent Photoprotective activity Antioxidant activity	Asai et al. (2004) Giossi et al. (2020) Şahin et al. (2020)
Violaxanthin	Photoprotective activity Antiproliferative activity Anti-inflammatory activity	Jahns et al. (2009) Pasquet et al. (2011) Soontornchaiboon et al. (2012)
Zeaxanthin	Antioxidant Anti-inflammatory	Murillo et al. (2019) W. Wang et al. (2007)

(PEB, red), phycoviolobillin (PVB, purple), and phycourobilin (PUB, yellow). They are structurally distinguished only by minor differences in the groups attached to the pyrrole rings (Fig. 11.3) (Mulders et al. 2014).

Biliproteins are mainly composed of hexameric and trimeric disks of bilins. In this sense, they can be divided into four main groups: phycoerythrins, which are red biliproteins whose chromophores are phycoerythrobilin and phycourobilin, have two or three bilins in the α subunit and three in the β subunit, presenting absorption maxima between 540 and 570 nm; phycocyanins are blue proteins whose chromophore is phycocyanobilin, they also have phycoerythrobin and their absorption maxima are between 610 and 620 nm, allophycocyanins are blue-green biliproteins whose chromophore is phycocyanobilin found in the α subunit and another chromophore in the β subunit, their absorption maximum is between 650 and 655 nm; finally there are phycoerythrocyanins, whose absorption maxima are between 570 and 595 nm and range from purple to deep blue. The variations in absorption lengths among the different phycobiliproteins arise from minor differences in the proteins and the structures of the bilin groups they contain (Vera-López Portillo and Martínez-Jiménez 2021; Sonani et al. 2016; Ruiz-Ortega 2012). These chromophores confer high extinction coefficients to phycobiliproteins in a wide area of the electromagnetic spectrum, which, together with their high quantum yield, provide excellent fluorescent properties (phycoerythrin $\lambda_{F\ max}$ = 575–590 nm, phycocyanin $\lambda_{F\ max}$ = 645–653 nm, allo-

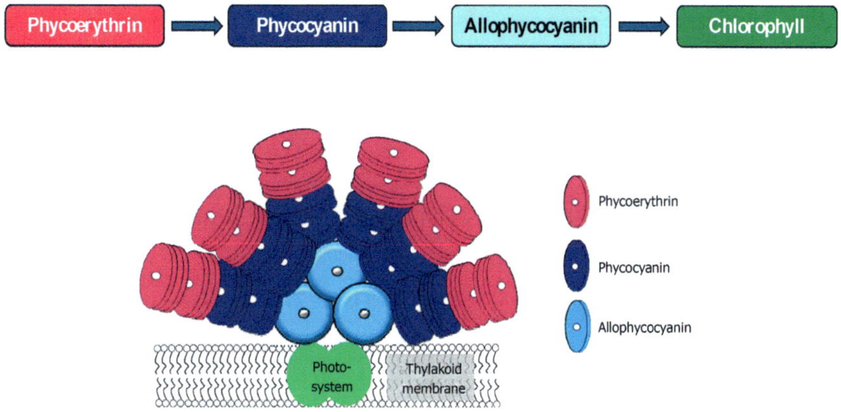

Fig. 11.2 Phycobilisome located on the external side of the thylakoid membrane. Phycobilisomes of cyanobacteria, algae, and microalgae are formed by phycoerythrin and, phycocyanin hexamers, and allophycocyanin trimers, which are represented by pink, blue, and light blue circles, respectively. (Adapted from Sonani et al. (2016))

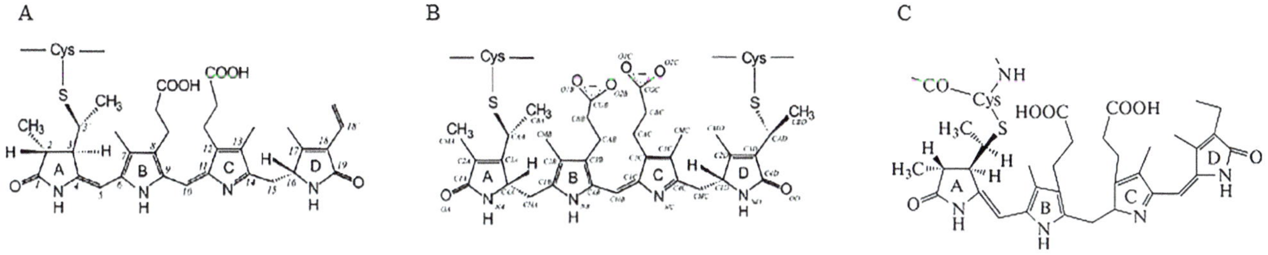

Fig. 11.3 Structure of the main bilin chromophores. (**a**) Cysteine-phycourobilin (Cys-PUB); (**b**) Cysteine-phycoerythrobilin (Cys-PEB); (**c**) Cysteine-phycocyanobilin (Cys-PCB). (Adapted from Stadnichuk and Tropin (2017) and Bacarizo-Roa (2012))

phycocyanin $\lambda_{F\ max}$ = 657–660 nm) (Fig. 11.4). The spectroscopic properties of phycobiliproteins depend on their structure, the number of bilins they contain, the protein environment surrounding these chromophores, and the steric relationships between them (Sonani et al. 2016). Because they absorb light across a spectrum from green to red-orange, phycobiliproteins allow the microorganisms that contain them to carry out photosynthesis in deep waters, where only green light can reach (Sonani et al. 2016; De Marsac 2005).

Phycoerythrin

Phycoerythrin (PE) is the most abundant phycobiliproteins found in red algae, microalgae, and some cyanobacteria. It is a brilliant red-pink color because it contains two bilins, chromophores, phycourobilin (PUB), and phycoerythrobilin (PEB). Structurally, PE forms very stable hexameric complexes in the shape of discs 12 nm in diameter and 6 nm in height, with each hexamer containing 34 bilins (Schoenleber et al. 1984; Glazer 1994). The absorption spectrum of this protein shows peaks at 480 nm and 570 nm. It can be classified into three main types, based on its absorption spectrum and origin: β-Phycoerythrin (BPE), commonly found in microalgae, with peaks at 545 and 565 nm, and a trough at 499 nm; R-Phycoerythrin, abundant in red algae, with peaks at 499 and 565 nm, and a trough at 545 nm; and finally, C-Phycoerythrin isolated from cyanobacteria, which has a single peak at 565 nm. The molecular weights of PE subunits vary depending on the organism from which they are derived (Leney et al. 2017; Galland-Irmouli et al. 2000).

The molecular structure of β-Phycoerythrin has three subunits in the apoprotein region; the α and β subunits with molecular weights of 17.82 kDa and 18.55 kDa, respectively, while a third subunit, γ, has a molecular weight of 30 to 33 kDa. The α subunit contains two phycoerythrobilins, which are covalently bonded to cysteine residues 82 and 139; the β subunit contains three phycoerythrobilins, linked to Cys82, Cys158, and the third linked to both Cys50 and Cys61 (Fig. 11.5). The γ subunit is estimated to contain four bilin subunits, two phycoerythrobilins, and two phycourobilins (Leney et al. 2017).

The αβ monomers can subsequently form triangular-shaped aggregates called trimers (heterohexamers) and hexamers (heterododecamers), which serve as the building blocks of phycobilisomes, the antenna pigments anchored to thylakoid membranes. Physiologically, β-Phycoerythrin is considered a macromolecular complex composed of three hexamers [(αβ)2] and a γ polypeptide. Inter- and intramolecular interactions play a role within the complex and are crucial for β-Phycoerythrin's function in light transmission. The residual electron density in crystallographic data has suggested that the γ subunit is located in the center of the [(αβ)2]3 hexamers. However, due to rotation, structural data for the γ subunit has not been obtained, and its structure has only been inferred (Leney et al. 2017; Camara-Artigas et al. 2012).

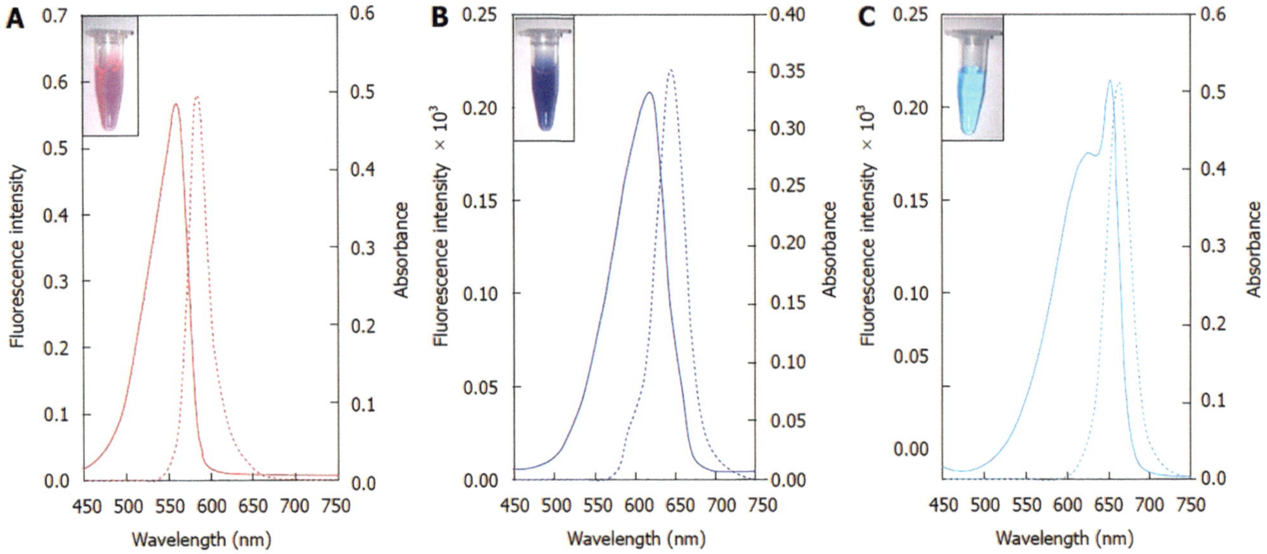

Fig. 11.4 Fluorescence emission (dotted line) and UV-vis absorption (solid line) spectra, along with the appearance (inset) of (**a**) phycoerythrin, (**b**) phycocyanin, and (**c**) allophycocyanin. (Taken from Sonani et al. 2016)

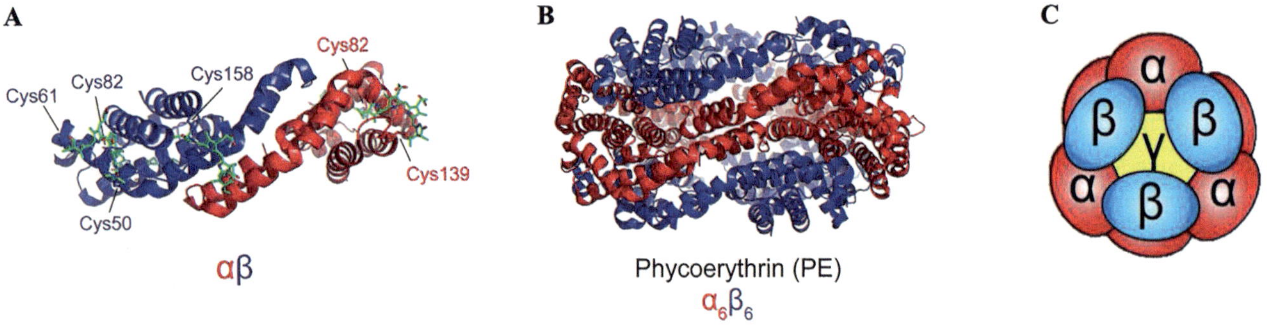

Fig. 11.5 Model of the β-Phycoerythrin protein complex. The α and β subunits are shown in red and blue, respectively. (**a**) αβ dimer of β-Phycoerythrin, shows the bilin molecules. (**b**) β-Phycoerythrin model (α6β6). (**c**) A representation of the β-phycoerythrin complex at physiological pH shows that BPE primarily consists of a highly fluorescent α6β6γ structure, with the γ subunit positioned in the center of the complex. (Adapted from Leney et al. (2017))

Regarding its current applications, β-Phycoerythrin (BPE) is employed as an indicator in quantitative assays due to its high absorption coefficient and exceptional fluorescence characteristics. It is recognized as the brightest natural fluorophore, being ten times more luminous and intense than the best organic dyes. BPE is positioned in a spectral region that generally differs from the emission region of basic organic dyes used as fluorescent indicators. This makes it a promising candidate for use as a biosensor. Regarding its use in antibody production, BPE demonstrated superior quality for immunoassays (Manirafasha 2016).

Phycocyanin and Allophycocyanin

The phycocyanins include both phycocyanin and allophycocyanin; they are the most abundant phycobiliprotein found in cyanobacteria, especially in those of the genus *Arthrospira* (*Spirulina*), where they can represent up to 20% of the protein fraction. They are bright blue and are used as a dye in food and cosmetic products (Grover et al. 2021). Phycocyanin is composed of two polypeptide chains, α and β, each with a molecular weight between 17 and 18 kDa, held together by non-covalent interactions. It contains three linear tetrapyrrolic prosthetic groups (chromophore or bilin) known as phycocyanobilin (PCB), one of which is attached to cysteine 84 of the α chain and the other two to cysteines 82 and 153 of the β chain, respectively; these binding positions represent highly conserved sites in this phycobiliprotein; phycocyanin also exists in red algae in its R type and cyanobacteria in its C type, with absorption maxima of 620 nm as phycocyanin and 650 nm as allophycocyanin (González-Fosados et al. 2021; Llopiz et al. 2016; Marín-Prida et al. 2015).

Phycocyanin exists as monomeric units that tend to form multimeric structures, typically trimers or hexamers and sometimes oligomers, with molecular weights ranging between 44 and 260 kDa. Factors such as temperature, pH, light, pressure, heavy metal cations, and denaturants influence the binding and structure of phycocyanin, which in turn affects its biological activity and coloration (González-Fosados et al. 2021; Ferraro et al. 2020).

11.2.3.2 Main Sources of Phycobiliproteins

Phycobiliproteins can be found in high quantities in algae, microalgae, and cyanobacteria of the *Rhodophyceae, Cyanophyceae, Cryptophycea,* and some *Pirrophyceae* classes. Sometimes, depending on the source of the phycobiliprotein, the prefix "C" (C-phycocyanin, C-phycoerythrin) is added for those extracted from cyanophytes (blue-green algae), or the prefix "R" (R-phycocyanin, R-phycoerythrin) for those obtained from Rhodophytes (red algae or microalgae) (Stadnichuk and Tropin 2017).

For C-phycocyanin, the highest production rates are observed from cyanobacteria, especially of the genus *Arthrospira (Spirulina)*; they are one of the few photosynthetic microorganisms that have been produced on a large scale and globally commercialized as food and a source of various high-value products. This microalga is symbiotic, multicellular, and filamentous, with a blue-green color and a notable ability to fix atmospheric nitrogen. It is characterized by multicellular cylindrical trichomes arranged in a left-handed open helix along its entire length. The surface of the *Spirulina* body is smooth and lacks of a coating, which facilitates its digestion through simple enzymatic systems, and its primary photosynthetic pigment is phycocyanin, which is blue (González-Fosada et al. 2021; De Mejia et al. 2020). *Spirulina* has been used as a food source for over 1000 years due to its high protein content, which accounts for 55% to 70% of the total dry weight (Prabakaran et al. 2020). Phycocyanin can constitute more than 60% of the water-soluble proteins in *Arthrospira* (*Spirulina*) *platensis*, or 20% of the dry cell weight (De Mejia et al. 2020).

On the other hand, the red microalgae, particularly *P. cruentum*, have shown to be efficient in the production of phycoerythrin, as this protein represents approximately 1.66% of the dry weight of the microalga due to phycoerythrin being an important pigment for light absorption in red microalgae (Borowitzka 2013; Basaca-Loya 2009). They are predominantly marine organisms found in both tropical and cold waters; they are the only photosynthetic eukaryotic organisms in the evolution of algae that contain phycobilisomes in their thylakoid membranes. Red microalgae, such as *P. cruentum* or *Rhodella reticulata*, can easily grow in large quantities in pure cultures in the laboratory (Borowitzka 2013; Ruiz-Ortega 2012).

The principal strain used for obtention of phycobiliproteins are *A. platensis, Arthrospira maxima, Synechococcus, Aphanizomenon gracile, Nostoc calcicole, Galdieria sulphuraria,* and *Phorphyridium aerugineum* for phycocyanin; and the red algae *Corallina elongata,* the cyanobacterium *Lyngbya arboricola, Phormidium valerianum,* and *Spirulina fusiformis*; and the red microalgae *P. cruentum, Rhodosorus marinus, Rhodomonas salina,* and *Porphyridium purpureum* for phycoerythrin (Wang et al. 2023a, b; Stadnichuk and Tropin 2017; Sonani et al. 2016).

11.2.3.3 Uses of Phycobiliproteins

Phycobiliproteins are closely linked to natural colors: phycoerythrins exhibit shades of pink and red; phycocyanins and phycoerythrocyanins vary from purple to deep blue; and allophycocyanins have a distinctive blue-green hue. These macromolecules, therefore, have significant potential as natural colorants, offering a viable alternative to synthetic dyes and contributing to the expansion and diversification of the natural dye market (Ruiz-Ortega 2012). Its excellent water solubility and a negative charge at physiological pH facilitate its use as a coloring agent in the food industry, especially in beverages and sweets; they are also used in many other food products such as fermented milk products, ice creams, soft drinks, desserts, sweet cake decoration, milkshakes, maintaining the color for at least 1 month at room temperature. They exhibit antioxidant activity, act as color stabilizers, improve rheological properties, and extend the shelf life of products. Due to their non-toxic and anticancer properties, phycobiliproteins have also been used as nutraceutical agents. Additionally, phycobiliproteins are used in cosmetology and in the production of lipsticks and eyeliners in countries such as Japan, Thailand, and China (Sonani et al. 2016; Pandey et al. 2013; Sekar and Chandramohan 2007).

For manufacturers of food and cosmetic products, the quest to find a vibrant, stable source of natural blue colorants has proved to be elusive. Nowadays, Phycocyanin production and stability from Spirulina have been well documented and reviewed. For this reason, the FDA's 2013 certification added to the portfolio of natural colorants the phycocyanin obtained from *Spirulina* (*A. platensis*), considering it a new natural blue colorant, and granted it generally recognized as safe (GRAS) status. Nowadays, phycocyanins from Spirulina provide the only approved natural blue colorant in the United States, Europe, and Asia (De Mejia et al. 2020). Phycocyanins maintain stability only within a pH range of 5.0–7.5 at 25 °C. Due to this limitation, their initial use was restricted to coloring candies and chewing gum. However, recent advancements in food processing have expanded their application to include dairy products, soft drinks, and cosmetics (Pandey et al. 2013). While their incorporation into beverages has been the primary use, ongoing issues with stability at low pH levels continue to hinder broader use (De Mejia et al. 2020). In a study, phycocyanin was found to be a more adaptable blue dye than gardenia and indigo, offering a vibrant blue hue in jelly gums and coated soft candies, despite being less stable under heat and light conditions.

Phycocyanin is a food coloring and cosmetic pigment (Pandey et al. 2013).

Additionally, some candies for cake decoration (that fluoresce under UV light), like transparent lollipops made from sugar solution and dry sugar-drop candies, as well as soft drinks and fluorescence alcoholic beverages, have been prepared using phycoerythrin; these products were tested to exploit not only the coloring properties of this pigment but also its yellow fluorescence (Sonani et al. 2016).

Phycobiliproteins are also used in immunological assays, offering advantages over fluorescein, the most commonly used dye as a biochemical and biological marker. However, it has some drawbacks, such as reducing the water solubility of the proteins it labels and having a tendency to bind nonspecifically to other substances. Additionally, fluorescein's fluorescence is endogenously produced by specific natural components in biological fluids and tissues, leading to a lack of specificity in the fluorescent signal. In contrast, phycobiliproteins exhibit high quantum yield fluorescence, producing a distinct orange fluorescence that is easy to differentiate. Their emission is stable across a wide range of temperatures and pH levels, they are highly water-soluble, and they can be stored for months due to their stability. These properties make phycobiliproteins excellent fluorescent markers, applicable in various study techniques, diagnostic procedures, and biomedical research, including immunofluorescence, flow cytometry, and fluorometric microplate assays, among others (Juin et al. 2014; Ruiz-Ortega 2012; Eriksen 2008).

11.2.3.4 Biological Properties of Phycobiliproteins

As previously mentioned, phycobiliproteins are characterized by having a tetrapyrrolic ring covalently attached to their structure. This pigment is partly responsible for the properties associated with these phycobiliproteins, primarily including hepatoprotective, anti-inflammatory, and antioxidant activities (Romay et al. 2003; Bhat and Madyastha 2000). It has been demonstrated that they possess therapeutic value due to their protective effect, anticancer activity, and potential for treating diseases such as Alzheimer's and Parkinson's (Pandey et al. 2013; Rimbau et al. 2001). In addition, phycobiliproteins exhibit potential pharmacological activities in neuroprotection and demonstrate strong inhibitory effects on the proliferation of cancer cells in the lung, liver, and breast. These properties suggest significant application potential in treating injuries and sequelae associated with COVID-19. Consequently, the demand in these areas for phycobiliproteins could be markedly increased in the future (Wang et al. 2023a, b).

Various authors have demonstrated the functionality of phycoerythrin. Sonani et al. (2014) reported its antioxidant and anti-aging activities and its potential use in Alzheimer's treatment from the protein isolated from *Lyngbya* sp. In their study, different doses of phycoerythrin were tested in the diets of both wild-type and transgenic *Caenorhabditis elegans* cells to evaluate its anti-aging potential. They found that the survival rate of *C. elegans* increased compared to those under normal conditions, and the treatment also moderated the decline in anti-aging functions associated with the locomotion of N_2 nematodes. Additionally, phycoerythrin treatment increased stress tolerance in adults from 22% to 41.6% under thermal stress conditions and 40.1% to 63.1% under oxidative stress conditions. This treatment also moderated the activity expression of amyloid-beta peptide (Aβ1–42), associated with muscular paralysis in the transgenic tissues of *C. elegans* CL4176 (a model for Alzheimer's disease).

In addition, various reports have demonstrated that β-phycoerythrin exhibits immunosuppressive and antihypertensive activities (Pandey et al. 2013; Sekar and Chandramohan 2007; Glazer 1994). Other researchers, such as Bei et al. (2002), conducted experiments with mouse S180 tumor cells and SMC 7721 liver carcinoma cells using photodynamic therapy with argon irradiation. They found that the α, β, and γ subunits of β-phycoerythrin may play a role in enhancing the selectivity of this type of cancer therapy. Najdenski et al. (2013) reported that the phycoerythrin from *P. cruentum* acts as an antibacterial agent with a concentration of 0.29 μg/mL against *S. aureus*. Finally, phycoerythrins are currently being used to develop bifunctional biological nanoparticles, with potential applications in biomedical and photochemical fields (Bekasova 2024).

In the specific case of phycocyanin, various studies have focused on its research, demonstrating that this biliprotein has additional functions beyond its recognized photosynthetic role. It is suggested that it may have functions such as antioxidant, anti-inflammatory, antiviral, hepatoprotective, cholesterol-lowering effects, and radical scavenging properties. There is scientific evidence of its antiproliferative and antimetastatic roles against different cancer cell lines. Additionally, it has been shown to improve the immune system (De Mejia et al. 2020; Jiang et al. 2017; Udayan et al. 2017).

As the FDA has classified *Spirulina* products as generally recognized as safe (GRAS) for human consumption, in humans, research has shifted mainly from examining phycocyanins as a standalone compound to exploring the overall health benefits of *Spirulina*, which includes a range of important nutrients beyond just phycocyanins. In this sense, several studies have shown that *Spirulina* can significantly reduce body weight and body mass index in obese individuals, as well as lower serum triglycerides and improve insulin sensitivity in people with diabetes, even at a dose of 1.0 g/day, which is considered a low dose (Yousefi et al. 2019).

11.2.3.5 Strategies for Phycobiliproteins Extraction

The process of extracting phycobiliproteins from different species of algae and microalgae has been widely researched due to their significant economic value. The success of phycobiliprotein recovery largely depends on the extraction method, which generally involves selecting a suitable biomass source, followed by cell disruption and protein release. However, there is a lack of studies targeting the extraction and purification of these proteins on medium and large scales. As a result, the potential of phycobiliproteins remains underutilized, primarily due to their limited availability and the absence of efficient and systematic methods for their isolation and purification. Typically, obtaining these compounds from microalgae involves two main stages. The first stage includes the pre-treatment of the biomass to generate a protein-rich extract, while the second stage focuses on refining the purification process, often utilizing chromatographic techniques (Morales et al. 2010; Patil et al. 2008).

The extraction of phycobiliproteins is generally carried out through a process that involves cell disruption, extraction in water or a specific buffer solution, and centrifugation to separate the filtrate. This filtrate can then be partially purified or sterilized through microfiltration and freeze-dried. The purification of this type of biliprotein requires multiple costly stages and processes due to the time involved, the solvents used, and the labor-intensive steps, which ultimately reduce extraction yields and result in a high production cost (Juin et al. 2014). For example, purification of the supernatant by ammonium sulfate precipitation and ion-exchange chromatography produces nearly pure phycobiliproteins in the form of α and β subunits (Rowan 1989). However, the chemical similarities between different types of phycobiliproteins make it relatively difficult to isolate using standard methods such as electrophoresis and ion-exchange chromatography. The stages and sub-stages of these methodologies are generally not scalable; additionally, the high number of required unit operations increases production costs and the time needed to optimize purification (Ruiz-Ortega 2012).

The commercial value of phycobiliproteins is determined by their purity, which is defined by the ratio of A_{545nm}/A_{280nm}. A purity of 0.7 is considered food grade, 3.9 is considered reagent grade, and greater than four is considered analytical grade, indicating high purity (Ruiz-Ortega 2012; Morales et al. 2010). Traditionally, the purity issue has been addressed through successive separations of phycobiliproteins from associated cellular pigments and other cellular proteins. This process is facilitated by the high-water solubility of phycobiliproteins, allowing for their separation from chlorophylls and carotenoids through high-speed centrifugation of the homogenate following cell disruption (Rowan 1989).

The growing demand and importance of phycobiliproteins have driven ongoing research and the search for new methodologies to extract this microalgal product more efficiently. These new methods aim to reduce production costs and increase purity for potential use as a food colorant and other previously described applications. This represents an unresolved issue in the systematic production of phycobiliproteins at industrial levels; without it, the utilization of this protein with significant potential in the food, pharmaceutical, and cosmetic industries will be virtually nonexistent.

11.3 Strategies for Obtaining Pigments from Microalgae

The vibrant color, non-toxic characteristics, and antioxidant properties of natural pigments from microalgae offer significant potential for their use across different industries. Numerous researchers are working on enhancing these pigments' production and purification processes to satisfy the growing market demand. In this sense, the strategies for obtaining them include biomass accumulation, recovery, and processing, as well as extraction processes (Wang et al. 2023a, b; Sonani et al. 2016).

The first step in obtaining pigments from microalgae is to identify and select the microalgal strain, as the type and content of pigments vary depending on the strain. Once the microalgal strain is selected, the cultivation method, type of bioreactor, and cultivation parameters such as light conditions, pH, and culture medium can be established. The quantity and characteristics of the biomass obtained depend not only on the strain but also on its growth characteristics, which are linked to the cultivation method and conditions (Wang et al. 2023a, b; Vera-López Portillo and Martínez Jiménez 2021).

The choice of bioreactor is an important factor in microalgal cultivation. There are two types: open and closed bioreactors. Open bioreactors, such as natural lakes and raceway-type ponds, are more suitable for large-scale cultivation, with the latter being the most commonly used. Closed bioreactors, or photobioreactors, are typically used at the laboratory scale. They are generally smaller but have the advantage of being customizable to meet the specific growth requirements of the microalgal strain. These include various types, such as tubular or flat, inclined or spiral, and biofilm bioreactors (Wang et al. 2023a, b; Vera-López Portillo and Martínez Jiménez 2021; Sonani et al. 2016).

Since pigments in microalgae are primarily used as energy harvesters for photosynthesis, light has one of the most significant effects on their production. Changes in light intensity, wavelength, and exposure time result in variations in pigment accumulation. Chlorophyll and phycobiliproteins are the pigments most affected by low light exposure, as short light exposure times or low light intensity lead to reduced energy capture, and consequently, the cells increase the number of antenna pigments to enhance the capture of the energy reaching the photosystems. In contrast, carotenoids, while also serving as antenna complexes, are mainly synthesized in response to the presence of reactive oxygen species or overexcitation of chlorophyll. In this sense, the use of unfavorable growth conditions (often called "stress conditions") leads to a significant increase in the concentration of secondary pigments (Vera-López Portillo and Martínez Jiménez 2021; Begum et al. 2016; Mulders et al. 2014).

It is still unknown why the use of certain carbon or nitrogen sources promotes the accumulation of specific pigments; however, it has been observed that some microalgae can accumulate phycobiliproteins as a nitrogen reservoir in cultures where the carbon source is low, and there is an excess of nitrogen. Similarly, it has been reported that the pigment content can vary depending on the carbon source used, such as glucose, acetate, and malonate, among others, or depending on the oxidation state of nitrogen (nitrates, nitrites, ammonium, and urea; carbon and nitrogen sources) (Manirafasha et al. 2018; Casal et al. 2011).

11.4 Conclusions

Microalgae represent an emerging source of natural pigments that can be utilized in the food, pharmaceutical, and cosmetic industries. These pigments provide the desired color and offer beneficial biological effects for the consumer's health. As described here, the range of colors and shades offered by microalgal pigments is extensive, making them a viable replacement for synthetic pigments, which have been shown to have harmful effects on consumers. However, developing and implementing strategies to produce these pigments in more significant quantities is necessary to meet market demand. In this regard, various techniques can be applied for both the production of microalgal biomass and the standardization of a simple and scalable method for extracting the desired pigment.

References

Abe K, Hattori H, Hirano M (2007) Accumulation and antioxidant activity of secondary carotenoids in the aerial microalga *Coelastrella striolata* var. multistriata. Food Chem 100(2):656–661. https://doi.org/10.1016/j.foodchem.2005.10.026

Aberoumand A (2011) A review article on edible pigments properties and sources as natural biocolorants and foodstuff and food industry. WJDFS 6(1):71–78. https://www.idosi.org/wjdfs/wjdfs6(1)/11.pdf

Achour HY, Llamero CB, Saadi SA, Zitouni A, Señoráns FJ (2022) Ultrasound assisted extraction of carotenoids from green and orange biomasses of Dunaliella Salina DunaDZ1. The Annals of the University Dunarea De Jos of Galati Fascicle VI. Food Technol 46(2):22–34. https://doi.org/10.35219/foodtechnology.2022.2.02

Ahmad N, Mounsef JR, Lteif R (2021) Pigment production by Scenedesmus dimorphus using different low-cost and alternative culture media. J Chem Technol Biotechnol 97(1):287–294. https://doi.org/10.1002/jctb.6940

Alarcón-Bonilla J, Arriaga-Montoya SI, Chávez-Ávalos F, Jiménez-Zúñiga MI, Rodríguez-Cuevas E, Fuentes-Hernández J (2020) Evaluación de la actividad citotóxica de derivados de la clorofila microalgal en líneas celulares. PÄDI Boletín Científico De Ciencias Básicas E Ingenierías Del ICBI 8(Especial):28–32. https://doi.org/10.29057/icbi.v8iespecial.6374

Asai A, Terasaki M, Nagao A (2004) An Epoxide–Furanoid rearrangement of spinach neoxanthin occurs in the gastrointestinal tract of mice and in vitro: formation and cytostatic activity of neochrome stereoisomers. J Nutr 134(9):2237–2243. https://doi.org/10.1093/jn/134.9.2237

Bacarizo-Roa JL (2012) Cambios conformacionales de la B-ficoeritrina procedente de Porphyridium cruentum en función del pH. Tesis de Maestría. Universidad de Almería

Balder HF, Vogel J, Jansen MCJF, Weijenberg MP, Van Den Brandt PA, Westenbrink S, Van Der Meer R, Goldbohm RA (2006) HEME and chlorophyll intake and risk of colorectal cancer in The Netherlands cohort study. Cancer Epidemiol Biomarkers Prev 15(4):717–725. https://doi.org/10.1158/1055-9965.epi-05-0772

Básaca-Loya G, Valdez M, Enríquez-Guevara E, Gutierrez-Millán L, Burboa M (2009) Extraction and purification of B-phycoerythrin from the red microalga Rhodosorus marinus. Cienc Mar 35(4):359–368. https://doi.org/10.7773/cm.v35i4.1614

Begum H, Yusoff FMD, Banerjee S, Khatoon H, Shariff M (2016) Availability and utilization of pigments from microalgae. Crit Rev Food Sci Nutr 56:2209–2222. https://doi.org/10.1080/10408398.2013.764841

Bei H, Guang-Ce W, Chen-Kui Z, Zhen-Gang L (2002) The experimental research of R-Phycoerythrin subunits on cancer treatment: a new photosensitizer in PDT. Cancer Biother Radiopharm 17(1):35–42. https://doi.org/10.1089/10849780252824055

Bekasova O (2024) Properties and potential applications of bioconjugates of R-phycoerythrin with Ag° or CdS nanoparticle synthesized in its tunnel cavity: a review. Int J Biol Macromol 255:128181. https://doi.org/10.1016/j.ijbiomac.2023.128181

Bhat VB, Madyastha KM (2000) C-Phycocyanin: a potent peroxyl radical Scavenger in vivo and in vitro. Biochem Biophys Res Commun 275(1):20–25. https://doi.org/10.1006/bbrc.2000.3270

Borowitzka MA (2013) High-value products from microalgae—their development and commercialisation. J Appl Phycol 25(3):743–756. https://doi.org/10.1007/s10811-013-9983-9

Britton G, Liaaen-Jensen S, Pfander H (2004) Carotenoids. In Birkhäuser Basel eBooks. https://doi.org/10.1007/978-3-0348-7836-4

Camara-Artigas A, Bacarizo J, Andujar-Sanchez M, Ortiz-Salmeron E, Mesa-Valle C, Cuadri C, Martin-Garcia JM, Martinez-Rodriguez S, Mazzuca-Sobczuk T, Ibañez MJ, Allen JP (2012) pH-dependent structural conformations of B-phycoerythrin from Porphyridium cruentum. FEBS J 279(19):3680–3691. https://doi.org/10.1111/j.1742-4658.2012.08730.x

Camera E, Mastrofrancesco A, Fabbri C, Daubrawa F, Picardo M, Sies H, Stahl W (2009) Astaxanthin, canthaxanthin and β-carotene differently affect UVA-induced oxidative damage and expression of oxidative stress-responsive enzymes. Exp Dermatol 18(3):222–231. https://doi.org/10.1111/j.1600-0625.2008.00790.x

Carter O, Bailey GS, Dashwood RH (2004) The dietary phytochemical chlorophyllin alters E-cadherin and B-catenin expression in human colon cancer cells. J Nutr 134(12):3441S–3444S. https://doi.org/10.1093/jn/134.12.3441s

Casal C, Cuaresma M, Vega JM, Vilchez C (2011) Enhanced productivity of a lutein-enriched novel acidophile microalga grown on urea. Mar Drugs 9:29–42. https://doi.org/10.3390/md9010029

Chan K, Mong M, Yin M (2009) Antioxidative and anti-inflammatory neuroprotective effects of astaxanthin and canthaxanthin in nerve growth factor differentiated PC12 cells. J Food Sci 74(7):H225. https://doi.org/10.1111/j.1750-3841.2009.01274.x

Conceição D, Lopes RG, Derner RB, Cella H, Carmo APBD, D'Oca MGM, Petersen R, Passos MF, Vargas JVC, Galli-Terasawa LV, Kava V (2020) The effect of light intensity on the production and accumulation of pigments and fatty acids in Phaeodactylum tricornutum. J Appl Phycol 32(2):1017–1025. https://doi.org/10.1007/s10811-019-02001-6

Cooperstone JL, Goetz HJ, Riedl KM, Harrison EH, Schwartz SJ, Kopec R (2017) Relative contribution of α-carotene to postprandial vitamin A concentrations in healthy humans after carrot consumption. Am J Clin Nutr 106(1):59–66. https://doi.org/10.3945/ajcn.116.15082

Cordero BF, Obraztsova I, Couso I, Leon R, Vargas MA, Rodriguez H (2011) Enhancement of lutein production in Chlorella sorokiniana (Chorophyta) by improvement of culture conditions and random mutagenesis. Mar Drugs 9(9):1607–1624. https://doi.org/10.3390/md9091607

Cubas C, Lobo MG, González M (2008) Optimization of the extraction of chlorophylls in green beans (Phaseolus vulgaris L.) by N,N-dimethylformamide using response surface methodology. J Food Compos Anal 21(2):125–133. https://doi.org/10.1016/j.jfca.2007.07.007

Cuttriss AJ, Cazzonelli CI, Wurtzel ET, Pogson BJ (2011) Carotenoids. In Advances in botanical research, pp 1–36. https://doi.org/10.1016/b978-0-12-386479-6.00005-6

De Marsac NT (2005) Phycobiliproteins and phycobilisomes: the early observations. Adv Photosynth Resp:443–451. https://doi.org/10.1007/1-4020-3324-9_43

De Mejia EG, Zhang Q, Penta K, Eroglu A, Lila MA (2020) The colors of health: chemistry, bioactivity, and market demand for colorful foods and natural food sources of colorants. Annu Rev Food Sci Technol 11(1):145–182. https://doi.org/10.1146/annurev-food-032519-051729

Del Campo JA, Rodríguez H, Moreno J, Vargas M, Rivas J, Guerrero MG (2004) Accumulation of astaxanthin and lutein in chlorella zofingiensis (Chlorophyta). Appl Microbiol Biotechnol 64(6):848–854. https://doi.org/10.1007/s00253-003-1510-5

Delgado-Vargas F, Paredes-López O (2002) Natural colorants for food and nutraceutical uses. CRC Press, Boca Raton, pp 221–231

Diprat AB, Menegol T, Boelter JF, Zmozinski A, Vale MGR, Rodrigues E, Rech R (2017) Chemical composition of microalgae Heterochlorella luteoviridis and Dunaliella tertiolecta with emphasis on carotenoids. J Sci Food Agric 97(10):3463–3468. https://doi.org/10.1002/jsfa.8159

Downham A, Collins P (2000) Colouring our foods in the last and next millennium. Int J Food Sci Technol 35(1):5–22. https://doi.org/10.1046/j.1365-2621.2000.00373.x

Ehling-Schulz M, Bilger W, Scherer S (1997) UV-B-induced synthesis of photoprotective pigments and extracellular polysaccharides in the terrestrial cyanobacterium Nostoc commune. J Bacteriol 179(6):1940–1945. https://doi.org/10.1128/jb.179.6.1940-1945.1997

Eriksen NT (2008) Production of phycocyanin—a pigment with applications in biology, biotechnology, foods and medicine. Appl Microbiol Biotechnol 80(1):1–14. https://doi.org/10.1007/s00253-008-1542-y

Fernandes AS, Nascimento TC, Pinheiro PN, Vendruscolo RG, Wagner R, De Rosso VV, Jacob-Lopes E, Zepka LQ (2021) Bioaccessibility of microalgae-based carotenoids and their association with the lipid matrix. Food Res Int 148:110596. https://doi.org/10.1016/j.foodres.2021.110596

Ferraro G, Imbimbo P, Marseglia A, Illiano A, Fontanarosa C, Amoresano A, Olivieri G, Pollio A, Monti DM, Merlino A (2020) A thermophilic C-phycocyanin with unprecedented biophysical and biochemical properties. Int J Biol Macromol 150:38–51. https://doi.org/10.1016/j.ijbiomac.2020.02.045

Ferruzzi MG, Schwartz SJ (2005) Thermal degradation of commercial grade sodium copper chlorophyllin. J Agric Food Chem 53(18):7098–7102. https://doi.org/10.1021/jf051010s

Firdous AP, Kuttan G, Kuttan R (2015) Anti-inflammatory potential of carotenoidmeso-zeaxanthin and its mode of action. Pharm Biol 53(7):961–967. https://doi.org/10.3109/13880209.2014.950673

Galasso C, Corinaldesi C, Sansone C (2017) Carotenoids from marine organisms: biological functions and industrial applications. Antioxidants 6(4):96. https://doi.org/10.3390/antiox6040096

Galland-Irmouli AV, Pons L, Luçon M, Villaume C, Mrabet NT, Guéant JL, Fleurence J (2000) One-step purification of R-phycoerythrin from the red macroalga Palmaria palmata using preparative polyacrylamide gel electrophoresis. J Chromatogr B Biomed Appl 739(1):117–123. https://doi.org/10.1016/s0378-4347(99)00433-8

Gilmore AM, Mohanty N, Yamamoto HY (1994) Epoxidation of zeaxanthin and antheraxanthin reverses non-photochemical quenching of photosystem II chlorophyll a fluorescence in the presence of trans-thylakoid ΔpH. FEBS Lett 350(2–3):271–274. https://doi.org/10.1016/0014-5793(94)00784-5

Giossi C, Cartaxana P, Cruz S (2020) Photoprotective role of neoxanthin in plants and algae. Molecules 25(20):4617. https://doi.org/10.3390/molecules25204617

Gispert-Abreu E, Cantillo-Estrada E, Rivero-López A, Oramas-Rodríguez B, Barreto-Banderas E (2001) Evaluación de una crema dental con extracto de cuproclorofila. Rev Cuba Estomatol 38(3):149–154

Glazer AN (1994) Phycobiliproteins—a family of valuable, widely used fluorophores. J Appl Phycol 6(2):105–112. https://doi.org/10.1007/bf02186064

Goldberg SL (1943) The use of water soluble chlorophyll in oral sepsis. Am J Surg 62(1):117–123. https://doi.org/10.1016/s0002-9610(43)90301-0

González-Fosados LBG, González-Chávez BK, Garrido-Hernández A, Carrillo-Sancen G, Cerón-Montes GI, Martínez-Valdez FJ (2021) Caracterización de ficocianina por espectrofotometría de masas. PÄDI Boletín Científico De Ciencias Básicas E Ingenierías Del ICBI 9(Especial2):149–156. https://doi.org/10.29057/icbi.v9iespecial2.7985

Gopi S, Varma K, George R (2014) A short review on the medicinal properties of chlorophyll juice. Asian J Pharm Technol Innov 2(9):89–94. https://www.asianpharmtech.com/articles/a-short-review-on-the-medicinalproperties-of-chlorophyll-juice.pdf

Grover P, Bhatnagar A, Kumari N, Bhatt AN, Nishad DK, Purkayastha J (2021) C-Phycocyanin-a novel protein from Spirulina platensis-in vivo toxicity, antioxidant and immunomodulatory studies. Saudi J Biol Sci 28:1853–1859. https://doi.org/10.1016/j.sjbs.2020.12.037

Guerin M, Huntley ME, Olaizola M (2003) Haematococcus astaxanthin: applications for human health and nutrition. Trends Biotechnol 21(5):210–216. https://doi.org/10.1016/s0167-7799(03)00078-7

Hosikian A, Lim S, Halim R, Danquah MK (2010) Chlorophyll extraction from microalgae: a review on the process engineering aspects. Int J Chem Eng 2010:1–11. https://doi.org/10.1155/2010/391632

Hsu C-Y, Chao P-Y, Hu S-P, Yang C-M (2013) The antioxidant and free radical scavenging activities of chlorophylls and pheophytins. Food Nutr Sci 04(08):1–8. https://doi.org/10.4236/fns.2013.48a001

Humphrey AM (1980) Chlorophyll. Food Chem 5(1):57–67. https://doi.org/10.1016/0308-8146(80)90064-3

Inbaraj BS, Chien JT, Chen BH (2006) Improved high performance liquid chromatographic method for determination of carotenoids in the microalga Chlorella pyrenoidosa. J Chromatogr A 1102(1–2):193–199. https://doi.org/10.1016/j.chroma.2005.10.055

Jahns P, Holzwarth AR (2012) The role of the xanthophyll cycle and of lutein in photoprotection of photosystem II. Biochimica et

Biophysica Acta (BBA)—Bioenergetics 1817(1):182–193. https://doi.org/10.1016/j.bbabio.2011.04.012

Jahns P, Latowski D, Strzalka K (2009) Mechanism and regulation of the violaxanthin cycle: the role of antenna proteins and membrane lipids. Biochimica et Biophysica Acta (BBA)—Bioenergetics 1787(1):3–14. https://doi.org/10.1016/j.bbabio.2008.09.013

Jeffrey SW, Humphrey GF (1975) New spectrophotometric equations for determining chlorophylls a, b, c1 y c2 in higher plants, algae and natural phytoplankton. Biochem Physiol Pflanz 167(2):191–194

Jeffrey SW, Mantoura RFC, y Wright SW. (1997) Phytoplankton pigments in oceanography: guidelines to modern methods. UNESCO, Paris

Jiang L, Wang Y, Yin Q, Liu G, Liu H, Huang Y, Li B (2017) Phycocyanin: a potential drug for cancer treatment. J Cancer 8(17):3416–3429. https://doi.org/10.7150/jca.21058

Juin C, Chérouvrier J-R, Thiery V, Gagez A-L, Bérard J-B, Nicolas J, Kaas R, Cadoret J-P, Picot L (2014) Microwave-assisted extraction of phycobiliproteins from porphyridium purpureum. Appl Biochem Biotechnol 175(1):1–15. https://doi.org/10.1007/s12010-014-1250-2

Kabir MT, Rahman MH, Shah M, Jamiruddin MR, Basak D, Al-Harrasi A, Bhatia S, Ashraf GM, Najda A, El-Kott AF, Mohamed HR, Al-Malky HS, Germoush MO, Altyar AE, Alwafai EB, Ghaboura N, Abdel-Daim MM (2022) Therapeutic promise of carotenoids as antioxidants and anti-inflammatory agents in neurodegenerative disorders. Biomed Pharmacother 146:112610. https://doi.org/10.1016/j.biopha.2021.112610

Kephart JC (1955) Chlorophyll derivatives—their chemistry? Commercial preparation and uses. Econ Bot 9(1):3–38. https://doi.org/10.1007/bf02984956

Kim SY, Kwon YM, Kim KW, Kim JYH (2021) Exploring the potential of nannochloropsis sp. extract for cosmeceutical applications. Marine Drug 19(12):690. https://doi.org/10.3390/md19120690

Kobylewski S, Jacobson MF (2010) Food dyes: a rainbow of risks. Center for Science in the Public Interest

Koller M, Muhr A, Braunegg G (2014) Microalgae as versatile cellular factories for valued products. Algal Res 6:52–63. https://doi.org/10.1016/j.algal.2014.09.002

Konishi I, Hosokawa M, Sashima T, Maoka T, Miyashita K (2008) Suppressive effects of alloxanthin and diatoxanthin from halocynthia roretzi on LPS-induced expression of pro-inflammatory genes in RAW264.7 cells. J Oleo Sci 57(3):181–189. https://doi.org/10.5650/jos.57.181

Kotake-Nara E, Miyashita K, Nagao A, Kushiro M, Zhang H, Sugawara T (2001) Carotenoids affect proliferation of human prostate cancer cells. J Nutr 131(12):3303–3306. https://doi.org/10.1093/jn/131.12.3303

Kotake-Nara E, Asai A, Nagao A (2005) Neoxanthin and fucoxanthin induce apoptosis in PC-3 human prostate cancer cells. Cancer Lett 220(1):75–84. https://doi.org/10.1016/j.canlet.2004.07.048

Kuczynska P, Jemiola-Rzeminska M (2016) Isolation and purification of all-trans diadinoxanthin and all-trans diatoxanthin from diatom Phaeodactylum tricornutum. J Appl Phycol 29(1):79–87. https://doi.org/10.1007/s10811-016-0961-x

Kusama Y, Inoue S, Jimbo H, Takaichi S, Sonoike K, Hihara Y, Nishiyama Y (2015) Zeaxanthin and echinenone protect the repair of photosystem II from inhibition by singlet oxygen in Synechocystis sp. PCC 6803. Plant Cell Physiol 56(5):906–916. https://doi.org/10.1093/pcp/pcv018

Lanfer-Marquez UM, Barros RMC, Sinnecker P (2005) Antioxidant activity of chlorophylls and their derivatives. Food Res Int 38(8–9):885–891. https://doi.org/10.1016/j.foodres.2005.02.012

Lavaud J, Rousseau B, Van Gorkom HJ, Etienne A (2002) Influence of the diadinoxanthin pool size on photoprotection in the marine planktonic diatom phaeodactylum tricornutum. Plant Physiol 129(3):1398–1406. https://doi.org/10.1104/pp.002014

Leney AC, Tschanz A, Heck AJR (2017) Connecting color with assembly in the fluorescent B-phycoerythrin protein complex. FEBS J 285(1):178–187. https://doi.org/10.1111/febs.14331

León-Vaz A, León R, Vigara J, Funk C (2023) Exploring Nordic microalgae as a potential novel source of antioxidant and bioactive compounds. New Biotechnol 73:1–8. https://doi.org/10.1016/j.nbt.2022.12.001

Li T, Xu J, Wu H, Wang G, Dai S, Fan J, He H, Xiang W (2016) A saponification method for chlorophyll removal from microalgae biomass as oil feedstock. Mar Drugs 14(9):162. https://doi.org/10.3390/md14090162

Llopiz A, Milián A, Díaz-Domínguez G (2016) Purificación y caracterización de la ficocianobilina. Bionatura 1(4):202–207. https://doi.org/10.21931/rb/2016.01.04.8

Ma R, You Y, Liu X, Ho S, Xie Y, Chen J (2023) Highly efficient co-production of fucoxanthin and eicosapentaenoic acid by heterotrophic cultivation of a newly isolated microalga Nitzschia sp. FZU62. Algal Research 71:103046. https://doi.org/10.1016/j.algal.2023.103046

Macías-Sánchez MD, Mantell C, Rodríguez M, De La Ossa EM, Lubián LM, Montero O (2009) Comparison of supercritical fluid and ultrasound-assisted extraction of carotenoids and chlorophyll a from Dunaliella salina. Talanta 77(3):948–952. https://doi.org/10.1016/j.talanta.2008.07.032

Mancebo-Dorvigny B, Sánchez-Perera LM, Díaz-Aguirre S, Bulnes-Goucohea C, Ivis-Regalado C, Escobar-Medina A, Cordero-Machado E (2011) Efecto cicatrizante de la pasta de clorofila-caroteno de Pinus caribaea var. caribaea sobre heridas abiertas asépticas. Revista Cubana de Plantas Medicinales 16(1):24–33

Manikprabhu D, Li W-J (2015) Antimicrobial agents from actinomycetes: chemistry and applications. Antimicrobials synthetic and natural compounds. In: CRC Press eBooks 8(1113):99–116. https://doi.org/10.1201/b19224-8

Manirafasha E, Ndikubwimana T, Zeng X, Lu Y, Jing K (2016) Phycobiliprotein: potential microalgae derived pharmaceutical and biological reagent. Biochem Eng J 109:282–296. https://doi.org/10.1016/j.bej.2016.01.025

Manirafasha E, Murwanashyaka T, Ndikubwimana T, Rashid Ahmed N, Liu J, Lu Y, Zeng X, Ling X, Jing K (2018) Enhancement of cell growth and phycocyanin production in Arthrospira (Spirulina) platensis by metabolic stress and nitrate fed-batch. Bioresour Technol 255:293–301. https://doi.org/10.1016/j.biortech.2017.12.068

Marín-Prida J, Llópiz-Arzuaga A, Pavón N (2015) Aplicaciones de la C-ficocianina: métodos de obtención y propiedades farmacológicas. Revista de ciencias farmacéuticas y alimentarias 1:29–43. Recuperado a partir de https://revistas.uh.cu/rcfa/article/view/5185

Martelli G, Folli C, Visai L, Daglia M, Ferrari D (2014) Thermal stability improvement of blue colorant C-Phycocyanin from Spirulina platensis for food industry applications. Process Biochem 49(1):154–159. https://doi.org/10.1016/j.procbio.2013.10.008

McCook J, Dorogi P, Vasily D, Cefalo D (2015) In vitro inhibition of hyaluronidase by sodium copper chlorophyllin complex and chlorophyllin analogs. Clin Cosmet Investig Dermatol 443. https://doi.org/10.2147/ccid.s86863

Mein JR, Lian F, Wang X (2008) Biological activity of lycopene metabolites: implications for cancer prevention. Nutr Rev 66(12):667–683. https://doi.org/10.1111/j.1753-4887.2008.00120.x

Meléndez-Martínez AJ, Mandić AI, Bantis F, Böhm V, Borge GIA, Brnčić M, Bysted A, Cano MP, Dias MG, Elgersma A, Fikselová M, García-Alonso J, Giuffrida D, Gonçalves VSS, Hornero-Méndez D, Kljak K, Lavelli V, Manganaris GA, Mapelli-Brahm P, O'Brien N (2021) A comprehensive review on carotenoids in foods and feeds: status quo, applications, patents, and research needs. Crit Rev Food Sci Nutr 62(8):1999–2049. https://doi.org/10.1080/10408398.2020.1867959

Miller NJ, Sampson J, Candeias LP, Bramley PM, Rice-Evans CA (1996) Antioxidant activities of carotenes and xanthophylls. FEBS Lett 384(3):240–242. https://doi.org/10.1016/0014-5793(96)00323-7
Morales CC, De Medeiros Burkert JF, Kalil SJ (2010) C-phycocyanin extraction process for large-scale use. J Food Biochem 34(1):133–148
Mortensen A, Geppel A (2007) HPLC–MS analysis of the green food colorant sodium copper chlorophyllin. Innov Food Sci Emerg Technol 8:419–425
Moshell AN, Bjornson L (1977) Photoprotection in erythropoietic protophyria: mechanism of photoprotection by beta carotene. J Invest Dermatol 68(3):157–160. https://doi.org/10.1111/1523-1747.ep12492479
Mulders KJM, Lamers PP, Martens DE, Wijffels RH (2014) Phototrophic pigment production with microalgae: biological constraints and opportunities. J Phycol 50(2):229–242. https://doi.org/10.1111/jpy.12173
Murakoshi M, Nishino H, Satomi Y, Takayasu J, Hasegawa T, Tokuda H, Iwashima A, Okuzumi J, Okabe H, Kitano H, Iwasaki R (1992) Potent preventive action of alpha-carotene against carcinogenesis: spontaneous liver carcinogenesis and promoting stage of lung and skin carcinogenesis in mice are suppressed more effectively by alpha-carotene than by beta-carotene. PubMed 52(23):6583–6587. https://pubmed.ncbi.nlm.nih.gov/1423303
Murillo AG, Hu S, Fernandez ML (2019) Zeaxanthin: metabolism, properties, and antioxidant protection of eyes, heart, liver, and skin. Antioxidants 8(9):390. https://doi.org/10.3390/antiox8090390
Najdenski HM, Gigova LG, Iliev II, Pilarski PS, Lukavský J, Tsvetkova IV, Ninova MS, Kussovski VK (2013) Antibacterial and antifungal activities of selected microalgae and cyanobacteria. Int J Food Sci Technol 48(7):1533–1540. https://doi.org/10.1111/ijfs.12122
National Center for Biotechnology Information (2024a) PubChem Compound Summary for CID 6419725, alpha-Carotene. Retrieved August 2, 2024 from https://pubchem.ncbi.nlm.nih.gov/compound/alpha-Carotene
National Center for Biotechnology Information (2024b) PubChem Compound Summary for CID 5280489, Beta-Carotene. Retrieved August 2, 2024 from https://pubchem.ncbi.nlm.nih.gov/compound/Beta-Carotene
National Center for Biotechnology Information (2024c) PubChem Compound Summary for CID 5281236, Echinenone. Retrieved August 2, 2024 from https://pubchem.ncbi.nlm.nih.gov/compound/Echinenone
National Center for Biotechnology Information (2024d) PubChem Compound Summary for CID 6443740, Alloxanthin. Retrieved August 2, 2024 from https://pubchem.ncbi.nlm.nih.gov/compound/Alloxanthin
National Center for Biotechnology Information (2024e) PubChem Compound Summary for CID 5281223, Antheraxanthin. Retrieved August 2, 2024 from https://pubchem.ncbi.nlm.nih.gov/compound/Antheraxanthin
National Center for Biotechnology Information (2024f) PubChem Compound Summary for CID 5281224, Astaxanthin. Retrieved August 2, 2024 from https://pubchem.ncbi.nlm.nih.gov/compound/Astaxanthin
National Center for Biotechnology Information (2024g) PubChem Compound Summary for CID 5281227, Canthaxanthin. Retrieved August 2, 2024 from https://pubchem.ncbi.nlm.nih.gov/compound/Canthaxanthin
National Center for Biotechnology Information (2024h) PubChem Compound Summary for CID 6449888, Diadinoxanthin. Retrieved August 2, 2024 from https://pubchem.ncbi.nlm.nih.gov/compound/Diadinoxanthin
National Center for Biotechnology Information (2024i). PubChem Compound Summary for CID 6449888, Diadinoxanthin. Retrieved August 2, 2024 from https://pubchem.ncbi.nlm.nih.gov/compound/Diadinoxanthin
National Center for Biotechnology Information (2024j) PubChem Compound Summary for CID 5281239, Fucoxanthin. Retrieved August 2, 2024 from https://pubchem.ncbi.nlm.nih.gov/compound/Fucoxanthin
National Center for Biotechnology Information (2024k) PubChem Compound Summary for CID 5281243, Lutein. Retrieved August 2, 2024 from https://pubchem.ncbi.nlm.nih.gov/compound/Lutein
National Center for Biotechnology Information (2024l) PubChem Compound Summary for CID 102601544, CID 102601544. Retrieved August 2, 2024 from https://pubchem.ncbi.nlm.nih.gov/compound/Myxoxanthophyll
National Center for Biotechnology Information (2024m) PubChem Compound Summary for CID 5282217, Neoxanthin. Retrieved August 2, 2024 from https://pubchem.ncbi.nlm.nih.gov/compound/Neoxanthin
National Center for Biotechnology Information (2024n). PubChem Compound Summary for CID 448438, Violaxanthin. Retrieved August 2, 2024 from https://pubchem.ncbi.nlm.nih.gov/compound/Violaxanthin
National Center for Biotechnology Information (2024o). PubChem Compound Summary for CID 5280899, Zeaxanthin. Retrieved August 2, 2024 from https://pubchem.ncbi.nlm.nih.gov/compound/Zeaxanthin
Nishino H, Murakoshi M, Ii T, Takemura M, Kuchide M, Kanazawa M, Mou XY, Wada S, Masuda M, Ohsaka Y, Yogosawa S, Satomi Y, Jinno K (2002) Carotenoids in cancer chemoprevention. Cancer Metastasis Rev 21(3/4):257–264. https://doi.org/10.1023/a:1021206826750
Niyogi KK, Björkman O, Grossman AR (1997) The roles of specific xanthophylls in photoprotection. Proc Natl Acad Sci USA 94(25):14162–14167. https://doi.org/10.1073/pnas.94.25.14162
Oliyaei N, Moosavi-Nasab M, Tamaddon AM, Tanideh N (2021) Antidiabetic effect of fucoxanthin extracted from Sargassumangustifolium on streptozotocin-nicotinamide-induced type 2 diabetic mice. Food Sci Nutr 9(7):3521–3529. https://doi.org/10.1002/fsn3.2301
Orosa M, Franqueira D, Cid A, Abalde J (2001) Carotenoid accumulation in haematococcus pluvialis in mixotrophic growth. Biotechnol Lett 23(5):373–378. https://doi.org/10.1023/a:1005624005229
Oslan SNH, Tan JS, Oslan SN, Matanjun P, Mokhtar RM, Shapawi R, Huda N (2021) Haematococcus pluvialis as a potential source of Astaxanthin with diverse applications in industrial sectors: current research and future directions. Molecules 26(21):6470. https://doi.org/10.3390/molecules26216470
Ouchi A, Aizawa K, Iwasaki Y, Inakuma T, Terao J, Nagaoka S, Mukai K (2010) Kinetic study of the quenching reaction of singlet oxygen by carotenoids and food extracts in solution. Development of a singlet oxygen absorption capacity (SOAC) assay method. J Agric Food Chem 58(18):9967–9978. https://doi.org/10.1021/jf101947a
Paiva SA, Russell RM (1999) B-carotene and other carotenoids as antioxidants. J Am Coll Nutr 18(5):426–433. https://doi.org/10.1080/07315724.1999.10718880
Paliwal C, Ghosh T, George B, Pancha I, Maurya R, Chokshi K, Ghosh A, Mishra S (2016) Microalgal carotenoids: potential nutraceutical compounds with chemotaxonomic importance. Algal Res 15:24–31. https://doi.org/10.1016/j.algal.2016.01.017
Pandey VD, Pandey A, Sharma V (2013) Biotechnological applications of cyanobacterial phycobiliproteins. Int J Curr Microbiol App Sci 2(9):89–97. https://www.ijcmas.com/vol-2-9/V.D.%20Pandey,%20et%20al.pdf
Pasquet V, Morisset P, Ihammouine S, Chepied A, Aumailley L, Bérard J, Serive B, Kaas R, Lanneluc I, Thiéry V, Lafferriere M, Piot J, Patrice T, Cadoret J, Picot L (2011) Antiproliferative activity of violaxanthin isolated from bioguided fractionation of Dunaliella ter-

tiolecta extracts. Mar Drugs 9(5):819–831. https://doi.org/10.3390/md9050819

Patil V, Tran KQ, Giselrod HR (2008) Towards sustainable production of biofuels from microalgae. Int J Mol Sci 9(7):1188–1195

Pelah D, Sintov A, Cohen E (2004) The effect of salt stress on the production of Canthaxanthin and Astaxanthin by chlorella zofingiensis grown under limited light intensity. World J Microbiol Biotechnol 20(5):483–486. https://doi.org/10.1023/b:wibi.0000040398.93103.21

Pennington F, Guillard RR, Liaaen-Jensen S (1988) Carotenoid distribution patterns in bacillariophyceae (diatoms). Biochem Syst Ecol 16(7–8):589–592. https://doi.org/10.1016/0305-1978(88)90067-1

Pistelli L, Sansone C, Smerilli A, Festa M, Noonan DM, Albini A, Brunet C (2021) MMP-9 and IL-1B as targets for diatoxanthin and related microalgal pigments: potential chemopreventive and photoprotective agents. Mar Drugs 19(7):354. https://doi.org/10.3390/md19070354

Pitchford P (2002) Healing with whole foods: Asian traditions and modern nutrition. North Atlantic Books 3:227–234. (ISBN: 1-55643-471-5)

Pourkarimi S, Hallajisani A, Alizadehdakhel A, Nouralishahi A, Golzary A (2020) Factors affecting production of beta-carotene from Dunaliella salina microalgae. Biocatal Agric Biotechnol 29:101771. https://doi.org/10.1016/j.bcab.2020.101771

Prabakaran G, Sampathkumar P, Kavisri M, Moovendhan M (2020) Extraction and characterization of phycocyanin from Spirulina platensis and evaluation of its anticancer, antidiabetic and antiinflammatory effect. Int J Biol Macromol 153:256–263. https://doi.org/10.1016/j.ijbiomac.2020.03.009

Rani GR, Banu N (2017) Antibacterial activity of sodium copper Chlorophyllin (SCC) from Leucas aspera L. Res J Pharm Technol 10(3):792. https://doi.org/10.5958/0974-360x.2017.00149.4

Rao MPN, Xiao M, Li W-J (2017) Fungal and bacterial pigments: secondary metabolites with wide applications. Front Microbiol 8. https://doi.org/10.3389/fmicb.2017.01113

Rasmussen RS, Morrissey MT (2007) Marine biotechnology for production of food ingredients. Adv Food Nutr Res:237–292. https://doi.org/10.1016/s1043-4526(06)52005-4

Rengel R, Giraldez I, Díaz MJ, García T, Vigara J, León R (2022) Simultaneous production of carotenoids and chemical building blocks precursors from chlorophyta microalgae. Bioresour Technol 351:127035. https://doi.org/10.1016/j.biortech.2022.127035

Rimbau V, Camins A, Pubill D, Sureda F, Romay C, González R, Jiménez A, Escubedo E, Camarasa J, Pallàs M (2001) C-Phycocyanin protects cerebellar granule cells from low potassium/serum deprivation-induced apoptosis. Naunyn Schmiedeberg's Arch Pharmacol 364(2):96–104. https://doi.org/10.1007/s002100100437

Rodrigues DB, Flores ÉM, Barin JS, Mercadante AZ, Jacob-Lopes E, Zepka LQ (2014) Production of carotenoids from microalgae cultivated using agroindustrial wastes. Food Res Int 65:144–148. https://doi.org/10.1016/j.foodres.2014.06.037

Romay C, Gonzalez R, Ledon N, Remirez D, Rimbau V (2003) C-Phycocyanin: a Biliprotein with antioxidant, anti-inflammatory and neuroprotective effects. Curr Protein Pept Sci 4(3):207–216. https://doi.org/10.2174/1389203033487216

Rowan KS (1989) The biliproteins. Photosynthetic pigments of algae. Cambridge University Press, New York, pp 166–210

Ruiz-Ortega E (2012) Obtención y purificación de ficoeritrina mediante cromatografía de adsorción en lecho expandido. Escalado del proceso, caracterización fisicoquímica y aplicación como colorante natural. Tesis doctoral Escuela politécnica superior de linares Universidad de Jaén. https://hdl.handle.net/10953/354

Safafar H, Van Wagenen J, Møller P, Jacobsen C (2015) Carotenoids, phenolic compounds and tocopherols contribute to the antioxidative properties of some microalgae species grown on industrial wastewater. Mar Drugs 13(12):7339–7356. https://doi.org/10.3390/md13127069

Şahin S, Nasir NTBM, Erken İ, Çakmak ZE, Çakmak T (2019) Antioxidant composite films with chitosan and carotenoid extract from Chlorella vulgaris: optimization of ultrasonic-assisted extraction of carotenoids and surface characterization of chitosan films. Mater Rese Exp 6(9):095404. https://doi.org/10.1088/2053-1591/ab2def

Şahin S, Aybastıer Ö, Dawbaa S, Karkar B, Çakmak T (2020) Study of the ability of lutein and neoxanthin as standards and in the extract of Chlamydomonas reinhardtii to prevent oxidatively induced DNA Base damage using ultrasensitive GC–MS/MS analysis. Chromatographia 83(8):919–926. https://doi.org/10.1007/s10337-020-03918-8

Salem MA, Al-Ghonemiy AF, Zaki AB (2009) Photocatalytic degradation of Allura red and Quinoline yellow with polyaniline/TiO2 nanocomposite. Appl Catal B 91(1–2):59–66. https://doi.org/10.1016/j.apcatb.2009.05.027

Scheer H, William JL, Lane MD (2004) Chlorophylls and carotenoids. Encyclopedia of biological chemistry. Elsevier, pp 430–437

Schoenleber RW, Lundell DJ, Glazer AN, Rapoport H (1984) Bilin attachment sites in the alpha and beta subunits of B-phycoerythrin. Structural studies on 118 the singly linked phycoerythrobilins. J Biol Chem 259(9):5485–5489. PMID: 6715355

Schüler LM, Gangadhar KN, Duarte P, Placines C, Molina-Márquez AM, Léon-Bañares R, Sousa VS, Varela J, Barreira L (2020) Improvement of carotenoid extraction from a recently isolated, robust microalga, Tetraselmis sp. CTP4 (chlorophyta). Bioprocess Biosyst Eng 43(5):785–796. https://doi.org/10.1007/s00449-019-02273-9

Sekar S, Chandramohan M (2007) Phycobiliproteins as a commodity: trends in applied research, patents and commercialization. J Appl Phycol 20(2):113–136. https://doi.org/10.1007/s10811-007-9188-1

Selig MJ, Gamaleldin S, Celli GB, Marchuk MA, Smilgies D-M, Abbaspourrad A (2020) The stabilization of food grade copper-chlorophyllin in low pH solutions through association with anionic polysaccharides. Food Hydrocoll 98:105255. https://doi.org/10.1016/j.foodhyd.2019.105255

Shimode S, Miyata K, Araki M, Shindo K (2018) Antioxidant activities of the antheraxanthin-related carotenoids, antheraxanthin, 9-cis-antheraxanthin, and mutatoxanthins. J Oleo Sci 67(8):977–981. https://doi.org/10.5650/jos.ess18060

Simon D, Helliwell S (1998) Extraction and quantification of chlorophyll a from freshwater green algae. Water Res 32(7):2220–2223. https://doi.org/10.1016/s0043-1354(97)00452-1

Sonani RR, Singh NK, Awasthi A, Prasad B, Kumar J, Madamwar D (2014) Phycoerythrin extends life span and health span of Caenorhabditis elegans. Age 36(5):9717. https://doi.org/10.1007/s11357-014-9717-1

Sonani RR, Rastogi RP, Patel R, Madamwar D (2016) Recent advances in production, purification and applications of phycobiliproteins. World J Biol Chem 7(1):100. https://doi.org/10.4331/wjbc.v7.i1.100

Soontornchaiboon W, Joo SS, Kim SM (2012) Anti-inflammatory effects of violaxanthin isolated from microalga *Chlorella ellipsoidea* in RAW 264.7 macrophages. Biol Pharm Bull 35(7):1137–1144. https://doi.org/10.1248/bpb.b12-00187

Stadnichuk IN, Tropin IV (2017) Phycobiliproteins: structure, functions and biotechnological applications. Appl Biochem Microbiol 53(1):1–10. https://doi.org/10.1134/s0003683817010185

Su H-N, Xie B-B, Chen X-L, Wang J-X, Zhang X-Y, Zhou B-C, Zhang Y-Z (2009) Efficient separation and purification of allophycocyanin from spirulina (Arthrospira) platensis. J Appl Phycol 22(1):65–70. https://doi.org/10.1007/s10811-009-9427-8

Subramoniam A, Asha VV, Nair SA, Sasidharan SP, Sureshkumar PK, Rajendran KN, Karunagaran D, Ramalingam K (2011) Chlorophyll revisited: anti-inflammatory activities of chlorophyll A and inhibition of expression of TNF-A gene by the same. Inflammation 35(3):959–966. https://doi.org/10.1007/s10753-011-9399-0

Tanaka T, Shnimizu M, Moriwaki H (2012) Cancer chemoprevention by carotenoids. Molecules 17(3):3202–3242. https://doi.org/10.3390/molecules17033202

Tanumihardjo S (2013) Carotenoids: health effects. In Elsevier eBooks, pp 292–297. https://doi.org/10.1016/b978-0-12-375083-9.00045-3

Tavares RSN, Kawakami CM, De Castro Pereira K, Amaral GTD, Benevenuto CG, Maria-Engler SS, Colepicolo P, Debonsi HM, Gaspar LR (2020) Fucoxanthin for topical administration, a phototoxic vs. photoprotective potential in a tiered strategy assessed by in vitro methods. Antioxidants 9(4):328. https://doi.org/10.3390/antiox9040328

Torres PB, Chow F, Furlan CM, Mandelli F, Mercadante A, Santos DYACD (2014) Standardization of a protocol to extract and analyze chlorophyll a and carotenoids in Gracilaria tenuistipitata Var. Liui Zhang and Xia (Rhodophyta). Brazil J Oceanograph 62(1):57–63. https://doi.org/10.1590/s1679-87592014068106201

Udayan A, Arumugam M, Pandey A (2017) Nutraceuticals from algae and cyanobacteria. Algal Green Chem:65–89. https://doi.org/10.1016/B978-0-444-63784-0.00004-7

Udayan A, Pandey AK, Sharma P, Sreekumar N, Kumar S (2021) Emerging industrial applications of microalgae: challenges and future perspectives. Syst Microbiol Biomanuf 1(4):411–431. https://doi.org/10.1007/s43393-021-00038-8

Vera-López Portillo F, Martínez Jiménez A (2021) Pigmentos en microalgas: funciones, aplicaciones y técnicas de sobreproducción. Revista De La Sociedad Mexicana De Biotecnología Y Bioingeniería AC 25(5):35–51. https://smbb.mx/wp-content/uploads/2021/12/Vera-Lopez-Portillo-y-Martinez-Jimenez-2021.pdf

Vibeke B (1994) Chlorophyllin anticarcinogenesis in the rainbow trout model. Tesis de Doctorado Doctorado en Toxicología Universidad de Oregon:35–38

Victoria-Campos CI, Ornelas-Paz J, Ruiz-Cruz S, De Jesús Ornelas-Paz J, Cervantes-Paz B, Rios-Velasco C, Pérez-Martínez JD, Gardea-Béjar AA, Yahia EM, Ibarra-Junquera V (2022) Dietary sources, bioavailability and health effects of carotenoids. BIOtecnia 25(1):156–168. https://doi.org/10.18633/biotecnia.v25i1.1809

Villanova V, Galasso C, Vitale GA, Della Sala G, Engelbrektsson J, Strömberg N, Shaikh KM, Andersson MX, Esposito FP, Ekendahl S, De Pascale D, Spetea C (2022) Mixotrophy in a local strain of nannochloropsis granulata for renewable high-value biomass production on the west coast of Sweden. Mar Drugs 20(7):424. https://doi.org/10.3390/md20070424

Wang W, Connor SL, Johnson EJ, Klein ML, Hughes S, Connor WE (2007) Effect of dietary lutein and zeaxanthin on plasma carotenoids and their transport in lipoproteins in age-related macular degeneration. Am J Clin Nutr 85(3):762–769. https://doi.org/10.1093/ajcn/85.3.762

Wang T, Jónsdóttir R, Liu H, Gu L, Kristinsson HG, Raghavan S, Ólafsdóttir G (2012) Antioxidant capacities of phlorotannins extracted from the brown algae Fucus vesiculosus. J Agric Food Chem 60(23):5874–5883. https://doi.org/10.1021/jf3003653

Wang M, Morón-Ortiz Á, Zhou J, Benítez-González A, Mapelli-Brahm P, Meléndez-Martínez AJ, Barba FJ (2023a) Effects of pressurized liquid extraction with dimethyl sulfoxide on the recovery of carotenoids and other dietary valuable compounds from the microalgae spirulina, chlorella and phaeodactylum tricornutum. Food Chem 405:134885. https://doi.org/10.1016/j.foodchem.2022.134885

Wang J, Qin S, Lin J, Wang Q, Li W, Gao Y (2023b) Phycobiliproteins from microalgae: research progress in sustainable production and extraction processes. Biotechnol Biofuels Bioprod 16(1):170. https://doi.org/10.1186/s13068-023-02387-z

West R, Keşan G, Trsková E, Sobotka R, Kaňa R, Fuciman M, Polívka T (2016) Spectroscopic properties of the triple bond carotenoid alloxanthin. Chem Phys Lett 653:167–172. https://doi.org/10.1016/j.cplett.2016.04.085

Yamamoto HY, Bugos RC, Hieber AD (2006) Biochemistry and molecular biology of the xanthophyll cycle. In Kluwer Academic Publishers eBooks, pp 293–303. https://doi.org/10.1007/0-306-48209-6_16

Yousefi R, Saidpour A, Mottaghi A (2019) The effects of Spirulina supplementation on metabolic syndrome components, its liver manifestation and related inflammatory markers: a systematic review. Complement Ther Med 42:137–144. https://doi.org/10.1016/j.ctim.2018.11.013

Zakynthinos G, Varzakas T (2016) Carotenoids: from plants to food industry. Curr Res Nutr Food Sci J 4(1):38–51. https://doi.org/10.12944/crnfsj.4.special-issue1.04

Zhuang G, Ye Y, Zhao J, Zhou C, Zhu J, Li Y, Zhang J, Yan X (2023) Valorization of *Phaeodactylum tricornutum* for integrated preparation of diadinoxanthin and fucoxanthin. Bioresour Technol 385:129412. https://doi.org/10.1016/j.biortech.2023.129412

Microalgae as Source of Alimentary Ingredients: Valorization of Microalgal Protein, Lipids, and Vitamins from the Perspective of Food Applications

12

Ana Luisa Félix Sámano, Janitzio Xiomara Korina Perales Sánchez, Jennifer Vianey Félix Medina, and Roberto Gutiérrez Dorado

Abstract

Modern lifestyles in developed countries have incited the consumption of high-calorie foods, resulting in chronic diseases considered a public health issue, which affects the well-being of the general population (malnourishment, obesity, hypertension, diabetes, and others). To prevent these consequences, it is required to follow a healthy diet based on vitamins, antioxidants, polyunsaturated fatty acids (FA), and other bioactive compounds able to improve human health. Microalgae biomass contains proteins similar to common protein sources and offers many carbohydrates, lipids, and vitamins. Ingredients, supplements, and additives from microalgae are the principal products obtained by its processing with identifiable benefits for human health. Nonetheless, the use of microalgae-based food has not prospered despite its nutritional advantages, mainly due to the production cost, which increases are reflected in the final product's price. The principal problem that has hindered the success of microalgae products is the absence of incentives and deficient information on their health benefits. This chapter aims to provide hard data on the most important components that possess microalgae and how they can boost food nutrimental value while being used as a source of nutrients and nutraceutical compounds.

Keywords

Microalgae-based food · Nutritional compounds · Nutraceuticals · Functional food

12.1 Introduction

12.1.1 Nutritional Value

In recent decades, microalgae have been noticed as an alternative source of nutrients (polysaccharides, proteins, lipids, vitamins, and minerals), fiber, and bioactive compounds. The benefits of this compound can be obtained using the whole biomass or extracted from it (Barros de Medeiros et al. 2022). The content of each compound depends on the type of microalgae and the growth conditions, such as nutrient requirements, light exposition, temperature, and stage phase (Barkia et al. 2019). Consequently, the microalgae can be manipulated during the culture to increase the interest compound accumulation (Barros de Medeiros et al. 2022). Therefore, the microalgae chemical composition is variable, with reported values between 9–77%, 6–54%, and 4–74% for proteins, carbohydrates, and lipids, respectively (Bernaerts et al. 2019).

As mentioned, consumers recognize the advantages of microalgae chemical compounds and the inherent increases, and they have been considered a natural alternative for food.

A. L. Félix Sámano
Food Science and Technology Graduate Program Faculty of Chemical and Biological Sciences (FCQB) of the Autonomous University of Sinaloa (UAS), University City, Culiacan, Sinaloa, Mexico

J. X. K. Perales Sánchez · R. Gutiérrez Dorado (✉)
Food Science and Technology Graduate Program Faculty of Chemical and Biological Sciences (FCQB) of the Autonomous University of Sinaloa (UAS), University City, Culiacan, Sinaloa, Mexico

Biotechnology Graduate Integral Program, Faculty of Chemical and Biological Sciences (FCQB) of the Autonomous University of Sinaloa (UAS), University City, Culiacan, Sinaloa, Mexico
e-mail: rgutierrez@uas.edu.mx

J. V. Félix Medina
CONAHCYT-Agroindustrial Engineering, Polytechnic University of Sea and the Sierra, Elota, Sinaloa, Mexico

National Technological of Mexico (TecNM), Technological Institute of Culiacan (ITC), Culiacan, Sinaloa, Mexico

D. U. Santos Ballardo, S. Rossi (eds.), *Microalgae as Promising Source of Commercial Bioproducts*, Developments in Applied Phycology 14, https://doi.org/10.1007/978-3-031-86433-9_12

Lafarga et al. (2021) evaluated the preference for incorporating microalgae in food like bread, proving that microalgae can be a reliable source of nutrients and that consumers accept products that contain it. Nowadays, consumers prefer to include non-animal alternative sources for nutrition in their diet due to the multiple benefits they can provide (Boukid and Castellari 2021; Geada et al. 2021).

At this moment, it is well known that microalgae can biosynthesis substances with multiple applications (e.g., food products, supplements, cosmetics, biofuels, wastewater treatment, animal feeding, fertilizers, and others) (Gohara-Beirigo et al. 2022; Ruzik 2023).

Due to the high content of beneficial compounds that microalgae present, beneficial health properties have been identified, such as the protective potential against various health conditions (oxidation, cancer, inflammation, hypertension, diabetes, coagulation, dyslipidemia, cardiovascular and immunological diseases, viral and microbial infection, among others) (Elaya Perumal and Sundararaj 2020; Nova et al. 2020).

Microalgae harvesting with food attention has been used principally in Asiatic countries for hundreds of years. Microalgae commercial production originated around 60 years ago. In the decade of sixties, Japan and Taiwan cultivated *Chlorella* to obtain its biomass; besides, in the 1980s, Australia, China, Israel, Thailand, and the United States of America, produced biomass mainly from *Spirulina*, *Dunaliella salina*, and *Haematococcus pluvialis* (Barkia et al. 2019). China produces 60–70% of the total microalgae biomass worldwide, with 35.000 tons/ year of biomass production (Barros de Medeiros et al. 2022).

Whole microalgal biomass products are considered "superfoods" due to their nutrient content, mainly elevated protein content, and high bioactive content (Pulz et al. 2013). The most commercially produced microalgae species are *Aphanizomenon*, *Chlorella*, *Dunaliella*, and *Spirulina* (Ruzik 2023). For example, *Spirulina* is a reliable source of excellent quality protein, omega 6 (essential fatty acid (FA) of the ω-6 series), and phycocyanins. Also, this microalga has properties versus viral infections, cancer, oxidative stress, and inflammation. However, the species that have dominated the functional food market in the world are the microalgae *Chlorella* and *Spirulina*, owing their popularity to their high content of nutritional and nutraceutical compounds of interest (Sathasivam et al. 2019). The aqueous extracts of green algae of the genus *Chlorella* are considered a "growth factor" in plants due to their immunomodulatory effect, and they are composed of nucleic acids, amino acids, vitamins, minerals, polysaccharides, glycoproteins, and β-glucans (Barkia et al. 2019). Likewise, Chlorella extracts also have beneficial properties for human health, such as lowering cholesterol, antioxidant capacity, and effect against bacteria and tumors (Reyna-Martinez et al. 2018). *Chlorella strains* are supplements that can prevent or diminish the effects of chronic ailments like Alzheimer's, hypertension, cancer, etc. (Caporgno and Alexander 2018). *Spirulina*, known colloquially as a "superfood," is an exceptional source of vitamin A and complex B vitamins, mainly B1, B2, and B12, including polyunsaturated fatty acids and pigments with antioxidant properties. As it is common knowledge, the bioactive compounds found in *Spirulina* must be provided by external sources for the human body to absorb (Sathasivam et al. 2019). Albeit *Chlorella* and *Spirulina* are the most popular microalgae within the food industry, other varieties have been studied to ascertain their viability as a source of nutrients, always considering factors such as safety regulations and market demand (Abd El-Razik and Mohamed 2013).

The use of algal products as ingredients to enhance the nutritional value of prepared food has triggered the need for extensive R&D by the food industry to identify new species or enhance the already used by selecting targeted secondary metabolites toward genetic engineering, always looking for a balance between cost and the benefits given by such modifications (Barkia et al. 2019).

12.2 Proteins

The proteins have a fundamental role in microalgae's cellular structure and metabolism. In a plant's membrane and light-harvesting complex, proteins are integral components, representing different catalytic enzymes related to photosynthesis (Barkia et al. 2019).

On the other hand, microalgae are like bacteria concerning chemical composition. Due to this, microalgal biomass is considered a non-animal good protein source containing all essential amino acids in higher quantities than recommended by the WHO/FAO (Koyande et al. 2019; Kratzer and Murkovic 2021). Essential amino acids, such as Thr, Leu, Val, Ile, Lys, Met, and His, must be obtained from external sources because the human body cannot produce them (Bleakley and Hayes 2017). Besides, they present very high values of protein, higher values than soybean, corn, and wheat; when the microalgae proteins are quantitatively and qualitatively compared with proteins from conventional sources, they have better properties, usually with good amino acid profile, so it is feasible to use them alternatively to meal, soy and the egg (Gohara-Beirigo et al. 2022; Ideny et al. 2022).

Furthermore, microalgae are a source of non-essential amino acids (Pro, Arg, Asp, Ser, Gly, Cys, Tyr, and Glu), which provide health benefits (Table 12.1) (Kusmayadi et al. 2021). These elements benefit the immune system, antioxidant responses, and cell conditions (Mahata et al. 2022). Concerning the protein concentration in microalgae species, they present a high quantity of this chemical component; the

Table 12.1 Protein and amino acids content (g/100 g of dry matter and g/100 g of protein, respectively) of microalgae and conventional sources

Source	Egg	Soybean	*Aphanizomenon flos-aquae*	*Arthrospira platensis*	*Chlorella vulgaris*	*Dunaliella tertiolecta*	*Dunaliella salina*	*Haematococcus pluvialis*	*Nannochloropsis* sp.	*Spirulina* sp.
Essential amino acids										
Histidine	2.4	2.6	0.81	2.2	2.0	2.1	1.73	0.31	2.6	1.0
Isoleucine	6.6	5.3	2.34	6.7	3.8	4.8	4.09	4.32	4.7	3.6
Leucine	8.8	7.7	4.00	9.8	8.8	8.4	9.58	3.64	9.4	5.5
Lysine	5.3	6.4	2.23	4.8	8.4	6.0	5.99	2.68	6.8	3.0
Methionine	3.2	1.3	1.09	2.5	2.2	1.4	2.79	0.65	2.3	1.4
Phenylalanine	5.8	5.0	2.16	5.3	5.0	6.7	6.98	1.4	5.5	2.8
Threonine	5.0	4.0	2.38	6.2	4.8	4.7	5.16	5.47	4.8	3.3
Tryptophan	1.7	1.4	0.51	0.3	2.1	1.5	0.18	–	–	1.0
Valine	7.2	5.3	2.43	7.1	5.5	6.2	7.23	2.45	6.0	4.5
Non-essential acids										
Tyrosine	4.2	3.7	2.40	5.3	3.4	5.0	4.86	2.22	3.9	3.0
Alanine	–	5.0	3.70	9.5	7.9	8.0	10.99	5.6	6.8	4.7
Arginine	6.2	7.4	3.24	9.5	6.4	6.8	8.16	10.26	6.0	4.4
Asparagine	11.0	1.3	4.67	7.3	9.0	9.0	9.56	5.01	9.1	6.0
Glutamic	12.6	19.0	5.48	10.3	11.6	10.3	12.41	10.41	13.8	9.2
Glycine	4.2	4.5	2.44	5.7	5.8	6.0	8.71	6.61	5.2	3.2
Proline	4.2	5.3	2.00	4.2	4.8	6.5	5.23	1.24	8.3	2.7
Serine	6.9	5.8	2.75	5.1	4.1	5.2	4.81	3.43	4.2	3.3
Cystine	2.3	1.9	0.06	0.9	1.4	0.44	1.63	0.25	0.1	0.7
Protein content (% of dry matter)	47.0	36.0	62.0	63.0	51.0	20.0	43.0	48.0	29.0	62.24

Adapted by Brown (1991), Christaki et al. (2011), Liestianty et al. (2019), Koyande et al. (2019), Zhu et al. (2019), Sandgruber et al. (2021), and Wang et al. (2021)

protein content ranges from 40% to 60% on a dry weight basis (Barros de Medeiros et al. 2022; Mahata et al. 2022), and its content can be modified by different parameters (light, pH, temperature, and media composition) (Tsvetanova and Yankov 2022).

Proteins of microalgae have an important role in the synthesis of enzymes, antibodies, and cytokines (Tsvetanova and Yankov 2022); they also promote health due to ingested was correlated with antioxidant, antihypertensive, anticancer, anti-atherosclerosis, immunosuppressive and hepatoprotective properties, in this sense this material is a good candidate for functional products formulations (Bleakley and Hayes 2017; Barros de Medeiros et al. 2022). However, the presence of chlorophyll and other non-protein components in microalgae significantly affects the color and flavor of products based on them, limiting their application (Becker 2007). Likewise, in some species of microalgae, mainly green microalgae, such as *Chlorella* and *Tetraselmis*, the extraction of intracellular proteins can be difficult due to the rigidity of their cell wall, resulting in low digestibility for humans and, in consequence, applications of the whole microalgal biomass is limited (Becker 2007; Barkia et al. 2019). In addition, microalgae are a viable alternative to common proteins of vegetal origin consumed by vegans and vegetarians because microalgae proteins are of high quality and present all the essential amino acids in an adequate quantity. In contrast, most plant proteins do not have a balanced amino acid profile, which represents a serious health problem for this type of consumer (Mahata et al. 2022).

12.3 Lipids

Microalgae can synthesize polar lipids, neutral lipids, glycolipids, phytosterols, and waxes (Santos-Sánchez et al. 2016). Polar lipids include phospholipids and galactolipids. About neutral lipids, also known as reserve/structural lipids, contain polyunsaturated and free fatty acids, acylglycerols, and carotenoids (Barkia et al. 2019).

The phospholipids in microalgae are phosphatidylserine (PS), phosphatidylcholine (PC), phosphatidylinositol (PI), and phosphatidylethanolamine (PE) (Mahata et al. 2022). The most common long-chain polyunsaturated fatty acids (PUFA) found in microalgae are α-linoleic acid (ALA), γ-linolenic acid (GLA), arachidonic acid (AA), eicosapentaenoic acid (EPA), docosahexaenoic acid (DHA), and arachidonic acid (ARA). The fatty acid profile depends on the type of microalgae; however, the most abundant are C14, C16, and C18 saturated and unsaturated fatty acids and some longer carbon-chain lengths (Barros de Medeiros et al. 2022).

The lipid content in microalgae ranges from 1% to 70% of the dry biomass (Table 12.2). In general, microalgae are mainly formed by polar lipids, representing 10–50% of the total lipids contained in microalgae (Barkia et al. 2019). The considerable difference in the type and content of lipids is related to different factors, including the microalgae species and cultivation conditions such as light exposure, temperature, culture medium, and even harvest efficiency (Barkia et al. 2019; Tsvetanova and Yankov 2022). Furthermore, under stress conditions, the accumulation of intracellular lipids is higher (especially in the absence of nitrogen or phosphorus), increasing the radical and antioxidant levels, and, consequently, oxidative stress (Kratzer and Murkovic 2021; Hu et al. 2019).

Due to this potential elevated lipid content, microalgae can be an interesting source of edible and non-edible oils, especially promising sources of PUFA, mostly omega-6 family (ω6) such as γ-linolenic acid (GLA) and arachidonic acid (AA), and omega-3 family (ω3) as eicosapentaenoic acid (EPA) and docosahexaenoic acid (DHA) (Table 12.3)

Table 12.2 Lipid profile of principal commercial microalga

Lipids (% of dry matter)	Microalgae	Reference
3.0	*Aphanizomenon flos-aquae*	Christaki et al. (2011) and Ideny et al. (2022)
4.0–9.0	*Arthrospira platensis*	Ideny et al. (2022)
14.0–22.0	*Chlorella vulgaris*	Christaki et al. (2011) and Ideny et al. (2022)
11.04–14.70	*Crypthecodinium cohnii*	Taborda et al. (2020)
6.0	*Dunaliella salina*	Ideny et al. (2022)
14.36	*Dunaliella* sp.	Kent et al. (2015)
11.64	*Dunaliella tertiolecta*	da Silva Gorgônio and Aranda (2013)
15.0	*Haematococcus pluvialis*	Christaki et al. (2011) and Ideny et al. (2022)
21.78	*Nannochloropsis* sp.	Kent et al. (2015)
9.0–14.0	*Porphyridium cruentum*	Ideny et al. (2022)
15.07	*Scenedesmus* sp.	Kent et al. (2015)
17.83–50.35	*Schizochytrium* sp.	Mahata et al. (2022)
21.83	*Spirulina* sp.	Bhakar et al. (2013)

Table 12.3 Fatty acid composition reported in different microalgae

Fatty acid	Microalgae *Aphanizomenon flos-aquae*	*Arthrospira platensis*	*Chlorella vulgaris*	*Dunaliella* sp.	*Haematococcus pluvialis*	*Nannochloropsis* sp.	*Scenedesmus* sp.	*Spirulina* sp.
ALA (C18:3 n-3)	8.80	0.01–0.04	6.24–12.99	21.18	3.26	0.23	21.22	9.0
EPA (C20:5 n-3)	0.06	0.006–0.09	0.04–0.05	ND	0.01	36.76	0.11	–
DHA (C22:6 n-3)	≈0	0.005–0.02	≈0	ND	ND	ND	ND	–
LA (C18:2 n-6)	2.17	6.35–14.89	21.96–31.43	8.43	5.12	4.11	6.19	9.5
GLA (C18:3 n-6)	≈0	4.89–9.15	0.02	2.53	0.08	0.82	0.49	16.0
ARA (C20:4 n-6)	0.02	0.01–0.37	0.0119	ND	0.002	7.45	ND	15.0
PUFA (%FA)	25.21	33.58–44.49	43.52–41.88	71.67	29.11	44.64	51.14	60.96
Ω6/Ω3 Ratio	0.0025	1.55–9.56	0.01–0.05	0.43	0.01	0.35	0.22	4.5
Total Ω-3 FA	8.87	0.05–0.15	6.3–13.04	33.05	3.27	36.99	34.21	9.0
Total Ω-6 FA	2.20	14.00–24.66	22.06–31.53	2.53	5.22	8.27	0.49	40.5
Total FA	43.91	43.21–55.81	86.90–83.80	66.32	29.16	114.57	82.08	81.20

Adapted by Kent et al. (2015), Liestianty et al. (2019), and Sandgruber et al. (2021)

Values expressed: fatty acid content (FA), total FA content (mg/g dry weight), and polyunsaturated FAs (PUFA) content [percentage (%) of total FAs]

ALA α-Linolenic acid, *EPA* Eicosapentaenoic acid, *DHA* Docosahexaenoic acid, *LA* Linoleic acid, *GLA* γ-Linolenic acid, *ARA* Arachidonic acid, *ND* Not detected

(Andrade et al. 2018). The EPA content in microalgae varied from 7% to 34% (Ideny et al. 2022).

On the other hand, the human body cannot produce PUFAs, which are essential and related to physiological functions in human organisms (Tsvetanova and Yankov 2022). It is well known that PUFA activity in humans and animals has a significant role in maintaining cholesterol and triglycerides, reducing the probability of suffering certain illnesses mainly related to the cardiovascular system (Mozaffarian and Wu 2011).

Moreover, polyunsaturated fatty acids can inhibit cancerous tissue growth (Qian and Xu 2014), balance metabolic disorders (Li et al. 2019), prevent neurodegenerative disorders (Kaewsutas et al. 2017), and could enhance the central nervous system development in infants (Guesnet and Alessandri 2011).

Due to the importance and benefits of microalgal lipids, they have received attention for their extraction and commercialization (Ideny et al. 2022). For example, PUFAs like DHA and EPA improve mental health by helping neurons to create new connections within them, having an impact on the processes that take place in the temporal lobe, such as memory, and in the first stages of childhood prevent the appearance of cognitive disorders, also related with the prevention of inflammation and cardiovascular disease (Mahata et al. 2022). Besides, DHA and EPA are commonly used in the supplementation of infant formula (Barkia et al. 2019; Barros de Medeiros et al. 2022). Furthermore, PUFAs are considered vegetarian natural sources of safe edible oils with high purity and quality and great sensory qualities compared to animal-based oils (Barros de Medeiros et al. 2022).

The principal microalgae characterized due to the high content of PUFAs are *Arthrospira*, *Chlorella*, *Haematococcus*, *Dunaliella*, *Crypthecodinium cohnii*, *Schizochytrium*, and *Porphyridium cruentum* (Gohara-Beirigo et al. 2022).

12.4 Vitamins

The human and animal body's health depends on certain micronutrients like vitamins and minerals; usually, the organic requirements of these compounds are low in quantities but are essential for different processes and functions in the organisms (Del Mondo et al. 2020).

The main functions of these micronutrients in biochemical pathways are hormones, antioxidants, and enzymatic cofactors. Likewise, they have numerous cellular functions, such as immunity stimulators, growth, development, and repair of cellular damage. Also, a deficiency in vitamins can cause vitaminosis, resulting in beriberi, rickets, and night blindness (Gohara-Beirigo et al. 2022).

Moreover, vitamins are classified according to their solubility in water-soluble or fat-soluble. In group one (water-soluble vitamins), find C and the B group (B1, B2, B3, B5, B6, B7, B9, and B12, also named thiamin, riboflavin, niacin = nicotinic acid, pantothenic acid, pyridoxine, biotin, folic acid, and cobalamin, respectively) vitamins.

Photosynthetic organisms can synthesize almost all the vitamins; however, some vitamins B and K can be ingested across the diet because these are produced principally by bacteria (Del Mondo et al. 2010).

Vitamin A (retinol with antioxidant potential) and B complex vitamins are vital for living organisms (Camacho et al. 2019). The vitamins from microalgae are easily absorbed because they accumulate naturally, like plant models. In contrast, vitamin absorption is more complicated in other foods, such as artificial ones.

Besides, the accumulation or synthesis of vitamins in microalgae is inconstant, and it can be determined by varied factors, such as strain, light exposure, type and composition of culture medium, and growth phase (Ideny et al. 2022).

Microalgae are a reliable source of vitamins not found in higher plants, such as vitamins B12, K, and D (Table 12.4) (Del Mondo et al. 2020). In addition, they contain good levels of provitamin A, vitamin B9, and vitamin E; the last one has excellent health benefits (prevention of eye, skin, heart, and cancer diseases). It also plays an important role in male fertility and functions as a natural antioxidant, protecting the membrane lipids from oxidative stress (Mahata et al. 2022). Furthermore, vitamin B12 is found mainly in animal sources, so vegan and vegetarian diets can observe deficiencies in this vitamin. Therefore, for humans, microalgae can represent a significant alternative source of vitamin B12, especially for vegan and vegetarian consumers (Koyande et al. 2019). In summary, microalgae represent an important source of vitamins, making them a potential supplement or food additive (Tsvetanova and Yankov 2022).

12.5 Nutraceutical Compounds

Marine microalgae have been known for their nutritional value since millennial cultures. However, they are considered substances with nutraceutical properties as they provide different bioactive compounds with many applications, such as food supplements for medical and nutritional purposes (Matos et al. 2017).

Among the bioactive compounds naturally found in microalgae it is possible to observe the presence of vitamins (B1, B3, B7, B12, C, and E), proteins (essential amino acids, phycobiliproteins, hormone-like bioactive peptides, and enzymes), polyunsaturated fatty acids (EPA, DHA, linolenic acid), minerals, pigments (lutein, astaxanthin, β-carotene, chlorophylls, among others), fiber, phenolic compounds, and phytosterols (Islam et al. 2017).

The amount and type of nutraceutical compounds depend on the microalgae species, and the biosynthesis of bioactive compounds in microalgae could be enhanced by manipulating culture conditions such as temperature, nutrients, light intensity, and exposure, which affects growth and yield (Mahata et al. 2022; Mahendran et al. 2021). Furthermore, due to their nutritional and nutraceutical potential, microalgae have become one of the most promising and innovative sources of new functional products. Likewise, microalgae can be used as functional ingredients to improve food's nutritional and nutraceutical value, affecting human health and quality of life (Matos et al. 2017).

12.6 Microalgae-Based Food

During the last decade, microalgae have received attention principally in biofuel production. However, these microorganisms could nourish providing several essential bioactive compounds like amino acids, essential lipids, vitamins, minerals, carotenoids, and other high molecules (Mahata et al. 2022). The global market for functional foods from natural and organic components has increased recently. This behavior results from consumers' interest in acquiring products to satisfy specific demands in accordance with health deficiencies or even preferences. Some requirements are needed depending on a specific age group, for example, infants that need specific nutrients to develop organs and systems. Moreover, these days, life expectancy has become an important issue for older adults because, at this age, bodies experiment less efficiently and are necessary for nutrient supplementation (Gohara-Beirigo et al. 2022). In this sense, microalgae provide different natural nutrients that can address these demands. Besides, they are a better source of nutritional antioxidants attributable to their higher capacity than conventional plant-derived sources (Koyande et al. 2019).

Furthermore, microalgae biomass, its isolated interest elements, and bioactivity components are well known and present vast applications, such as innate pigments and nutritional and nutraceutical components that could be incorporated into matrix foods and as an innovative ingredient,

Table 12.4 Content of vitamins in microalgae

Microalgae	Vitamins (μg/g DW) A (Retinol)	B1 (Thiamine)	B2 (Riboflavin)	B3 (Niacina)	B5 (Pantothenic acid)	B6 (Pyridoxine)	B7 (Biotin)	B9 (Folic acid)	B12 (Cobalamin)	C (Ascorbic acid)	D (Calciferol)	E (Tocopherol)	K1 (Phylloquinone)
Anabaena	280	5.8	55	8	88	7	0.18	15	1.55	2000	n.r.	4000	200
Aphanizomenon	4200	40	6	130	8	13	n.r	1.0	0.006	n.r.	n.r.	100–140	n.r.
Arthrospira	650	10–24	33–45	0.13–149	13	9.6	n.r.	0.27–4.8	0.50–6.6	100	0.004	110–2500	13–50
Chaetoceros	520–970	655	12	25	n.r.	4	n.r.	n.r.	8	120–1879	n.r.	890–1630	n.r.
Chlorella	0.02–650	18–23	20–68	0.15–250	21–190	1.9–25	0.5–1.1	3.1–34	0.08–2.5	100–150	0.004	10–2000	0.73–66
Dunaliella	10–630	9–29	9–31	10	95–13	2.2–4	0.9	0.4–53.7	0.04–0.7	160–2200	n.r.	120–190	0.1
Euglena	30	n.r.	55.71	n.r.	n.r.	14.71	0.22	n.r.	n.r.	1820	n.r.	280–1200	n.r.
Haematococcus	540	4.7	17	66	14	3.6	n.r.	2.9	0.001	n.r	n.r.	270–880	n.r.
Isochrysis	10–270	14–462	14–30	2690	9.1	1.8–183	1	3	0.6–89	120–4450	5	60–120	8
Nannochloropsis	50–80	70	22–25	0.12	n.r.	3.6	1.1	17–22	0.3–1.7	5240	0.35–1	20–4720	0.17
Pavlova	10–260	36–290	6–50	955	n.r.	4–8.4	1.9	23	1.7–1162	840–1300	0.35–0.39	140–350	6.5
Porphyridium	750	n.r.	n.r.	n.r.	n.r.	n.r.	n.r.	5.39	n.r.	n.r.	n.r.	n.r.	n.r.
Skeletonema	140	710	37	511	n.r.	134	n.r.	n.r.	117	60–6700	11	110	5.5
Spirulina	5800	48	55	0.15	2	8	0.55	3600	n.r.	n.r.	n.r.	410	25.5
Stichococcus	60	29	25	n.r.	n.r.	17	1.3	24	1.95	2500	0.35	130–440	n.r.
Tetraselmis	4–4280	1.8–627	19–42	1410	37.7	2.8–155	0.8–1.3	3–20	1.95–9	190–3000	0.35–14	40–6320	28–201

Adapted by Fabregas and Herrero (1990), Bishop and Zubeck (2012), Liestianty et al. (2019), Pereira et al. (2019), Del Mondo et al. (2020), Sandgruber et al. (2021), and Fithriani and Melanie (2022)

n.r. Not reported

sometimes enhancing physicochemical and nutritional w in the development products (Barros de Medeiros et al. 2022).

Consequently, even though microalgae products have different applications, around 75% of the total production is created for human food, animal feed, and nutraceutical compounds. Respect for its nutraceutical applications is expected to increase because of the imminent demand for healthy products (Rahman 2020; Barros de Medeiros et al. 2022). In fact, between 2015 and 2019, around 13,090 innovative nourishment products were formulated with biomass microalgae or its isolated components; these developed products are divided into 79% for solid nourishment and 21% for drinks (De Oliveira and Bragotto 2022).

On the other hand, nowadays, the microalgae market is focused on a few countries such as the USA, Asia, Japan, Australia, China, and Oceania; nonetheless, during previous years, Europe has been paying attention to the development of microalgae products and is expected to gain a remarkable position in the worldwide top. Moreover, the commercial production of microalgae biomass has been dominated by the following species that include *Arthrospira* sp., *Chlorella* sp., *Dunaliella salina*, *Spirulina* spp., *Haematococcus* spp., and *Scenedesmus* (Barros de Medeiros et al. 2022; Ideny et al. 2022).

The use of other microalgal species can be determined by the nourishment requirements, consumer necessities, commercial opportunities, production capacities, and cost of cultivation (Abd El-Razik and Mohamed 2013). Furthermore, the commercial production of microalgae presents disadvantages like low production capacities, high production costs, a predominant color in green microalgae, and fishy taste and aroma (Lafarga 2019).

Innocuousness during food processing is the main goal for national departments dedicated to human health protection, for example, The American Food and Drug Administration (FDA) and the European Food Safety Authority (EFSA); these agencies constrict the application of artificial color due to effects of intolerance symptoms and toxicity. Consequently, natural ingredients have more preference in food industries (Vaz et al. 2016). Microalgae can provide natural pigments such as chlorophylls (green), carotenoids (yellow/orange/red), anthocyanin (red/blue), and betanin (red) (De Oliveira and Bragotto 2022), utilized as an added ingredient to increase nutritional value (García-Vaquero et al. 2021).

Microalgae commercial development products are food supplements, additives, and ingredients, nevertheless mainly sold as nutritional supplements in tablet, capsule, or powder form, and some products are recommended as superfoods, rich in protein and omega-3 (Lafarga 2019). Studies have found that microalgae biomass can improve rheological aspects in products, appropriate sensory acceptance, enhance color, and provide health benefits to consumers (Abd El-Razik and Mohamed 2013). Nevertheless, the type and proportion of microalgae need to be explored to understand the changes related to the physical and rheological properties of the final nourishment (Agustini et al. 2016). The acceptability of microalgae products depends on the type of diet of the target population, mainly those who need a personalized diet or people who choose to be healthier (Lafarga 2019).

Recently, the tendency in food formulation has been to integrate microalgae biomass or its derived compound in the food matrix (Table 12.5). Food products from microalgae are divided into two major classes: complete microalgae incorporation in the food matrix and isolated incorporation of microalgae compounds. The industrial production of microalgae-based foods is being commercialized worldwide, including baked products, beverages, and snacks (Lafarga 2019).

Microalgae have been incorporated in baked products to increment their protein or PUFA content. However, the limiting application is their color, which can also alter the taste, odor, and flavor of baked products; an alternative solution to this problem could be to extract the microalgae-derived bioactive instead of the microalgae biomass (Delwiche 2012; Lafarga 2019).

Furthermore, functional beverages are divided into lacteous, vegetable, or fruit-based beverages and stimulant or power drinks (Corbo et al. 2014). The market for beverages is the functional foods category that is increasing quickly due to easy distribution. Moreover, functional beverages are principally formulated for infants and the elderly. In the case of infants and children, the main beverages produced are juices, milk, and other drinks enriched in essential fatty acids like DHA or EPA. Beverages for the elderly population are focused on protein content due to high deficiency at this age (Singh et al. 2020; Araújo and Peteiro 2021). Increasing protein content over recommended dietary intake has improved complications related to the cardiovascular system, wound healing, bone tissue, and nervous system (Wolfe et al. 2008). The elaboration of juices, shakes, smoothies, and other beverages, along with adding microalgae, presents advantages due to the attractive colors desired in these products, which could be interesting to consumers (Lafarga 2019). During 2018, plant-based products thrived. As a result, the food industry has focused on developing products that are free from preservatives, reduced in carbohydrates, and enhanced with functional compounds (Mintel 2019).

For example, soups containing microalgae have been formulated, demonstrating improved viscosity, antioxidant properties, and bioactive compounds (Lafarga 2019).

Changes in lifestyle in most countries, mainly developed ones, make people travel long distances to their working locations, and the lack of time to prepare complete meals has made consumers choose quick foods to satisfy their physiological need to eat. As a result of the changes in society's

Table 12.5 Addition of microalgae-based food products

Type of product	Microalgae-based food	Microalgae	Incorporated amount	Improvement	Reference
Supplements	Energy gel	*Spirulina* sp.	Microalgae biomass 0.35%	Provides protein and carbohydrates, besides providing antioxidant activity due to these microalgae having carotenoid, phycocyanin, and phenolic compounds. Sensory acceptance	Moreira et al. (2018)
	Supplement and shake	*Spirulina platensis*	750 mg/100 g	Increased the lipid value, decreased carbohydrate value	Freitas et al. (2019)
	Shake	*Spirulina* sp.	750 mg/100 g	Improve protein concentration. Sensory acceptance	Santos et al. (2016)
	Tablets	*Dunaliella salina*	555 mg powder microalgae-Effective dose 100 mg/kg	Enhanced antifibrotic effect in rats	El-Baz et al. (2020)
Beverages	Infant formulae	*Schizochytrium* sp.	Oil from microalgae 63 mg/100 mL	Oils rich in DHA requirements for infants and young children, and approved for safe consumption	EFSA NDA Panel (2021)
	Smoothies	*Chlorella*, *Spirulina*	Additional microalgae biomass of 2.2%	Incremented the vitamin B12 value throughout the expiration date with respect to microbiological and sensory quality	Castillejo et al. (2018)
	Yogurt	*Spirulina platensis*	Incorporation of 7% microalgae biomass	Increment in protein, chlorophyll, carotenoid, and β-carotene content. Also, an increase in probiotic count due to time reduction in fermentation, as well as acceptability to consumers	Patel et al. (2019)
	Chocolate milk	*Spirulina platensis*	Microencapsulated microalgae 5% and 8.75%	Increase in protein, antioxidant activity, and phenolic compounds content, besides reduction in total lipids. Maintains consistency parameters and consumer acceptance	De Oliveira et al. (2021)
Baked products	Bread	*Dunaliella salina*	Incorporation of power microalgae 10%	Increment in protein and essential amino acids; also improve organoleptic quality and rheological parameters	Tertychnaya et al. (2020)
	Wheat bread	*Tetraselmis chuii*	Addition of 12%	Increase of essential amino acids and protein. Improved dough structure and physical properties of bread. Positive impact on the color and aroma profile of bread	Qazi et al. (2021)
	Butter cookies	*Arthrospira* (*Spirulina*), *Chlorella vulgaris*, *Tetraselmis suecica*, and *Phaeodactylum tricornutum*	Biomass incorporation 2–6%	Improvement of the phenolic compounds, antioxidant activity, digestibility, and rheological properties, as well as presented an attractive color	Batista et al. (2017)
	3D printed cookies	*Arthrospira platensis* *Chlorella vulgaris*	Addition of microalgae biomass 0.5–2%	Improved printability and more stable and bake-resistant 3D structures due to protein supplementation across the addition of microalgae biomass	Uribe-Wandurraga et al. (2020)
	Gluten-free bread	*Chlorella sorokiniana*	Addition of 2.5 or 5.0 g/100 g microalgae powder	Improvement of proteins, carotenoids, and polyunsaturated fatty acids. Incorporating microalgae has no impact on rheological properties but impacts the bread color. Besides it is acceptable by consumers	Diprat et al. (2020)
Snacks	Snacks	*Spirulina* sp.	Addition biomass 2.6%	Improved protein, carotenoids, and mineral content. Sensorial acceptance	Lucas et al. (2018)
	Corn extrudates	*Spirulina*	Incorporation of microalgae powder 8%	Improved chemical composition; also, they have a higher content of carotenoids and chlorophylls. However, expansion and malleability were reduced in extrudates with vast *Spirulina* concentration.	Tańska et al. (2017)

(continued)

Table 12.5 (continued)

Type of product	Microalgae-based food	Microalgae	Incorporated amount	Improvement	Reference
Other products	Pasta	*Chlorella sorokiniana*	Additional biomass of 5%	Enriched content of protein, PUFA, chlorophyll, and carotenoids	Bazarnova et al. (2021)
	Pasta	*Dunaliella salina*	Addition of algae biomass 1% with semolina	Increased contents of protein, PUFA, and ash; also, an increment in phytochemicals was observed. Good sensory acceptance	El-Baz et al. (2017)
	Natural colorant in beverages	*Porphyridium cruentum*	Isolate B-phycoerythrin concentration used 0.75 g/L	Presented an enormous potential for being used as a coloring additive, being comparable to commercial additives	Carmona et al. (2022)
	Ice cream	*Spirulina platensis*	Powder microalgae biomass 1.2%	Change positive protein, total solid, lipid, sugar, boiling temperature, and consumers' acceptance	Agustini et al. (2016)
	Soft cheese	*Spirulina platensis*	Powder microalgae biomass 1%	Change positive protein, water, lipid, β-carotene content, and texture	Agustini et al. (2016)
	Functional crispy chocolate bar	*Scenedesmus obliquus*	Incorporation encapsulated microalgae 5%	Potential source of PUFA and amino acids, and oxidative stability	Hlaing et al. (2020)
	Soup	*Chlorella* sp. *Spirulina* sp. *Tetraselmis* sp.	Incorporation of microalgae biomass concentration from 0.5 and 1.0% (w/v)	Improved phenolic acids content and antioxidant capacity. Increased bioaccessibility of polyphenols. Sensorially accepted	Lafarga et al. (2019)

diets, the snack industry has thrived during the last decade, addressing their marketing efforts, particularly to children, as they represent a remarkable vehicle for improving consumer health (Félix-Medina et al. 2020). The development of formulation of snack products that include microalgae biomass increases protein, lipid, and mineral content; however, these increases are related to the quantity of biomass added to matrix food. Besides, flavor, textures, and taste obtained suggest a high sensorial acceptance (Lucas et al. 2018). The principal disadvantages of developing snacks added with microalgae were intense color and crispness (Lafarga 2019).

Due to economic uncertainty derived from COVID and different variants since early 2020, pasta consumption has increased due to healthy, sensory attributes, low cost, easy to cook, and better diet quality. The improvement of pasta formulation has been studied to enhance its nutritional and performance attributes. Regarding nutritious and performance attributes, incorporating microalgae into pasta products enhanced the protein and the EPA and DHA presence and showed good consumer acceptance (Lafarga 2019).

Despite the great potential presented by microalgae in the food industry, it still presents several challenges, including increased production efficiency, cost-effectiveness, and legislative and regulatory concerns. Also, the intense color, aroma, and unpleasant flavors limited the inclusion of microalgae biomass in food products. The off-flavors are related to volatile organic compounds (VOCs) such as hydrocarbons, phenolic and sulfur components, ketones, lactones, aldehydes, alcohols, esters, and short and medium-chain free fatty acids. Different technological processes allow removing, isolating, or neutralizing off-flavor components (Gohara-Beirigo et al. 2022).

12.7 Conclusion and Future Prospects

Microalgae-based products provide great nourishing and nutraceutical potential as they contain proteins (essential amino acids), vital elements (polyunsaturated fatty acids, vitamins, minerals, pigments, fiber), and bioactive compounds that could have a positive effect on health by providing antioxidant activity and improving the prognosis of different degenerative diseases such as cancer, diabetes, hepatic, and hypertension. Furthermore, microalgae-derived biomass is frequently used to obtain biofuels and to incorporate them into animal food.

Despite the nutritional characteristics already discovered in microalgae, only a few percent (approximately 5%) of the existing species have been studied to define their human and animal applications. Several species of microalgae that exist in nature possess great potential in health and nutrition and have yet to be explored, which means that there is an excellent opportunity to find other options that can impact the quality of life, health, and life expectancy.

Recently, microalgae-based food products have thrived. Nonetheless, functional foods by microalgae cannot be employed efficiently yet, mainly because of the high production costs (related to harvesting and extraction methods for compounds). Consequently, massive microalgae production must be improved to guarantee economic viability and sustainability in microalgae production systems. Affordable and eco-friendly supplies and procedures could promote a sustainable system to obtain nutraceutical compounds from microalgae, which are important to the biotech industry. The versatility and relevance of microalgae characteristics could positively modify industries in areas such as energy, personal care, health, and food in the subsequent years.

References

Abd El-Razik MM, Mohamed AG (2013) Utilization of acid casein curd enriched with *Chlorella vulgaris* biomass as substitute of egg in mayonnaise production. World Appl Sci J 26:917–925

Agustini WT, Ma'ruf FW, Widayat W, Suzery M, Hadiyanto H, Benjakul S (2016) Application of *Spirulina platensis* on ice cream and soft cheese with respect to their nutritional and sensory perspectives. J Teknol 78(4–2):245–251

Andrade LM, Andrade CJ, Dias M, Nascimento CAO, Mendes MA (2018) Chlorella and Spirulina Microalgae as Sources of Functional Foods, Nutraceuticals, and Food Supplements; an Overview. MOJFPT 6(1):45-58. https://doi.org/10.15406/mojfpt.2018.06.00144

Araújo R, Peteiro C (2021) Algae as food and food supplements in Europe. EUR 30779 EN, Publications Office of the European Union, Luxembourg. https://doi.org/10.2760/049515.JRC125913

Barkia I, Saari N, Manning SR (2019) Microalgae for high-value products towards human health and nutrition. Mar Drugs 17(5):304

Batista AP, Niccolai A, Fradinho P, Fragoso S, Bursic I, Rodolfi L, Natascia B, Tredici MR, Sousa I, Raymundo A (2017) Microalgae biomass as an alternative ingredient in cookies: Sensory, physical and chemical properties, antioxidant activity and in vitro digestibility. Algal Res 26:161-171. http://dx.doi.org/10.1016/j.algal.2017.07.017

Barros de Medeiros VP, da Costa WKA, da Silva RT, Pimentel TC, Magnani M (2022) Microalgae as source of functional ingredients in new-generation foods: challenges, technological effects, biological activity, and regulatory issues. Crit Rev Food Sci Nutr 62(18):4929–4950

Bazarnova J, Nilova L, Trukhina E, Bernavskaya M, Smyatskaya Y, Aktar T (2021) Use of microalgae biomass for fortification of food products from grain. Foods 10(12):3018

Becker EW (2007) Micro-algae as a source of protein. Biotechnol Adv 25(2):207–210

Bernaerts TM, Gheysen L, Foubert I, Hendrickx ME, Van Loey AM (2019) The potential of microalgae and their biopolymers as structuring ingredients in food: a review. Biotechnol Adv 37(8):107419

Bhakar RN, Kumar R, Pabbi S (2013) Total lipids and fatty acid profile of different Spirulina strains as affected by salinity and incubation time. Vegetos 26(2s):148–154

Bishop WM, Zubeck HM (2012) Evaluation of microalgae for use as nutraceuticals and nutritional supplements. J Nutr Food Sci 2(5):1–6

Bleakley S, Hayes M (2017) Algal proteins: extraction, application, and challenges concerning production. Foods 6(5):33

Boukid F, Castellari M (2021) Food and beverages containing algae and derived ingredients launched in the market from 2015 to 2019: a front-of-pack labeling perspective with a special focus on Spain. Foods 10:1

Brown MR (1991) The amino-acid and sugar composition of 16 species of microalgae used in mariculture. J Exp Mar Bio Ecol 145(1):79–99

Camacho F, Macedo A, Malcata F (2019) Potential industrial applications and commercialization of microalgae in the functional food and feed industries: A short review. Mar Drugs 17(6):31

Caporgno MP, Alexander M (2018) Trends in microalgae incorporation into innovative food products with potential health benefits. Front Nutr 5:58

Carmona R, Murillo MC, Lafarga T, Bermejo R (2022) Assessment of the potential of microalgae-derived phycoerythrin as a natural colorant in beverages. J Appl Phycol 34(6):3025–3034

Castillejo N, Martínez-Hernández GB, Goffi V, Gómez PA, Aguayo E, Artés F, Artés-Hernández F (2018) Natural vitamin B12 and fucose supplementation of green smoothies with edible algae and related quality changes during their shelf life. J Sci Food Agric 98(6):2411–2421

Christaki E, Florou-Paneri P, Bonos E (2011) Microalgae: a novel ingredient in nutrition. Int J Food Sci Nutr 62(8):794–799

Corbo MR, Bevilacqua A, Petruzzi L, Casanova FP, Sinigaglia M (2014) Functional beverages: the emerging side of functional foods: commercial trends, research, and health implications. Compr Rev Food Sci Food Saf 13:1192–1206

da Silva Gorgônio CM, Aranda DAG (2013) Morphological and chemical aspects of *Chlorella pyrenoidosa*, *Dunaliella tertiolecta*, *Isochrysis galbana* and *Tetraselmis gracilis* microalgae. Nat Sci 5(7):783–791

De Oliveira APF, Bragotto APA (2022) Microalgae-based products: food and public health. J Future Foods 6:100157

De Oliveira TTB, dos Reis IM, de Souza MB, da Silva Bispo E, Maciel LF, Druzian JI, Lordelo PPGT, de Oliveira AC, dos Santos EBM, Abreu MBG, Lima VD, de Santana LRR (2021) Microencapsulation of *Spirulina sp.* LEB-18 and its incorporation in chocolate milk: Properties and functional potential. LWT-Food Sci Technol 148:111674

Del Mondo A, Smerilli A, Sané E, Sansone C, Brunet C (2020) Challenging microalgal vitamins for human health. Microbl Cell Fact 19:1-23. https://doi.org/10.1186/s12934-020-01459-1

Del Mondo G, Stell JG, Claramunt C, Thibaud R (2010) A graph model for spatio-temporal evolution. J Univ Comput Sci 16(11):1452–1477

Delwiche JF (2012) You eat with your eyes first. Physiol Behav 107:502–504

Diprat AB, Thys RCS, Rodríguez E, Rech R (2020) *Chlorella sorokiniana*: a new alternative source of carotenoids and proteins for gluten-free bread. LWT-Food Sci Technol 134:109974

EFSA NDA Panel (EFSA Panel on Nutrition, Novel Foods and Food Allergens), Turck D, Castenmiller J, De Henauw S, Hirsch-Ernst KI, Kearney J, Maciuk A, Mangelsdorf I, McArdle HJ, Naska A, Pelaez C, Pentieva K, Siani A, Thies F, Tsabouri S, Vinceti M, Cubadda F, Frenzel T, Heinonen M, Marchelli R, Neuhauser-Berthold M, Poulsen M, Prieto Maradona M, Schlatter JR, van Loveren H, Turla E, Knutsen HK (2021) Scientific Opinion on the safety of oil from *Schizochytrium limacinum* (strain FCC-3204) for use in infant and follow-on formula as a novel food pursuant to Regulation (EU) 2015/2283. EFSA J 19(1):19, e06344. https://doi.org/10.2903/j.efsa.2021.6344

Elaya Perumal U, Sundararaj R (2020) Algae: a potential source to prevent and cure the novel coronavirus–a review. Int J Emerg Technol 11(2):479–483

El-Baz FK, Abdo SM, Hussein AM (2017) Microalgae *Dunaliella salina* for use as food supplement to improve pasta quality. Int J Pharm Sci Rev Res 46(2):45–51

El-Baz FK, Ali SI, Basha M, Kassem AA, Shamma RN, Elgohary R, Salama A (2020) Design and evaluation of bioenhanced oral tablets of *Dunaliella salina* microalgae for treatment of liver fibrosis. J Drug Deliv Sci Technol:101845

Fabregas J, Herrero C (1990) Vitamin content of four marine microalgae. Potential use as a source of vitamins in nutrition. J Ind Microbiol Biotechnol 5(4):259–263

Félix-Medina JV, Montes-Ávila J, Reyes-Moreno C, Perales-Sánchez JXK, Gómez-Favela MA, Aguilar-Palazuelos E, Gutiérrez-Dorado R (2020) Second-generation snacks with high nutritional and antioxidant value produced by an optimized extrusion process from corn/common bean flours mixtures. LWT-Food Sci Technol 124:109172

Freitas BCB, Santos TD, Moreira JB, Zanfonato K, Morais MG, Costa JAV (2019) Novel foods: a meal replacement shake and a high-calorie food supplemented with Spirulina biomass. Int Food Res J26:59–65

Fithriani D, Melanie S (2022) Vitamin and mineral content of microalgae Phorpyridium and Chlorella and development prospects for food raw materials in Indonesia. IOP Conf Ser: Earth Environ Sci 1034(1):012043. https://doi.org/10.1088/1755-1315/1034/1/012043

García-Vaquero M, Brunton N, Lafarga T (2021) Microalgae as a source of pigments for food applications. In: Lafarga T, Acién G (eds) Cultured microalgae for the food industry: current and potential applications. Academic Press, London, pp 177–198

Geada P, Moreira C, Silva M, Nunes R, Madureira L, Rocha CMR, Pereira RN, Vicente AA, Teixeira JA (2021) Algal proteins: production strategies and nutritional and functional properties. Bioresour Technol 332:125

Gohara-Beirigo AK, Matsudo MC, Cezare-Gomes EA, de Carvalho JCM, Danesi EDG (2022) Microalgae trends toward functional staple food incorporation: Sustainable alternative for human health improvement. Trends Food Sci Technol 125:185–199

Guesnet P, Alessandri JM (2011) Docosahexaenoic acid (DHA) and the developing central nervous system (CNS)–implications for dietary recommendations. Biochim 93(1):7-12. https://doi.org/10.1016/j.biochi.2010.05.005

Hlaing SAA, Sadiq MB, Anal AK (2020) Enhanced yield of *Scenedesmus obliquus* biomacromolecules through medium optimization and development of microalgae based functional chocolate. J Food Sci Technol 57(3):1090–1099

Hu Y, Gong M, Feng S, Xu CC, Bassi A (2019) A review of recent developments of pre-treatment technologies and hydrothermal liquefaction of microalgae for bio-crude oil production. Renew Sust Energ Rev 101:476–492

Ideny JN, Eya JC, Nwankwegu AS, Nwoba EG (2022) Aquaculture sustainability through alternative dietary ingredients: microalgal value-added products. Eng Microbiol:100049

Islam MN, Alsenani F, Schenk PM (2017) Microalgae as a sustainable source of nutraceuticals. In: Gupta VK, Treichel H, Shapaval VO, de Oliveira LA, Tuohy MG (eds) Microbial functional foods and nutraceuticals. Wiley, Chichester, pp 1–19

Kaewsutas M, Sarikaphuti A, Nararatwanchai T, Sittiprapaporn P, Patchanee P (2017) Electroencephalographic study of microalgae DHA omega-3 egg consumption on cognitive function. J Funct Foods 29:46–52

Kent M, Welladsen HM, Mangott A, Li Y (2015) Nutritional evaluation of Australian microalgae as potential human health supplements. PLoS One 10(2):e0118985

Koyande AK, Chew KW, Rambabu K, Tao Y, Chu DT, Show PL (2019) Microalgae: a potential alternative to health supplementation for humans. Food Sci Human Wellness 8(1):16–24

Kratzer R, Murkovic M (2021) Food ingredients and nutraceuticals from microalgae: main product classes and biotechnological production. Foods 10(7):1626

Kusmayadi A, Leong YK, Yen HW, Huang CY, Chang JS (2021) Microalgae as sustainable food and feed sources for animals and humans–biotechnological and environmental aspects. Chemosphere 271:129800

Lafarga T (2019) Effect of microalgal biomass incorporation into foods: Nutritional and sensorial attributes of the end products. Algal Res 41:101566

Lafarga T, Acién-Fernández FG, Castellari M, Villaró S, Bobo G, Aguiló-Aguayo I (2019) Effect of microalgae incorporation on the physicochemical, nutritional, and sensorial properties of an innovative broccoli soup. LWT-Food Sci Technol 111:167–174

Lafarga T, Rodríguez-Bermúdez R, Morillas-España A (2021) Consumer knowledge and attitudes toward microalgae as food: the case of Spain. Algal Res 54:102174. https://doi.org/10.1016/J.ALGAL.2020.102174

Li TT, Tong AJ, Liu YY, Huang ZR, Wan XZ, Pan YY, Jia RB, Liu B, Chen XH, Zhao C (2019) Polyunsaturated fatty acids from microalgae *Spirulina platensis* modulates lipid metabolism disorders and gut microbiota in high-fat diet rats. Food Chem Toxicol 131:110558

Liestianty D, Rodianawati I, Arfah RA, Assa A, Patimah, Sundari, Muliadi (2019) Nutritional analysis of *Spirulina sp* to promote as superfood candidate. In: IOP conference series: materials science and engineering, vol 509. IOP Publishing Ltd, p 012031. https://doi.org/10.1088/1757-899X/509/1/012031

Lucas BF, de Morais MG, Santos TD, Costa JAV (2018) Spirulina for snack enrichment: Nutritional, physical and sensory evaluations. LWT-Food Sci Technol 90:270–276

Mahata C, Das P, Khan S, Thaher MI, Abdul M, Annamalai SN, Al Jabri H (2022) The potential of marine microalgae for the production of food, feed, and fuel (3F). Fermentation 8(7):316

Mahendran MS, Dhanapal ACTA, Wong LS, Kasivelu G, Djearamane S (2021) Microalgae as a potential source of bioactive food compounds. Curr Res Nutr Food Sci 9(3):917

Matos J, Cardoso C, Bandarra NM, Afonso C (2017) Microalgae as healthy ingredients for functional food: a review. Food Funct 8(8):2672-2685. https://doi.org/10.1039/C7FO00409E

Mintel (2019) Functional health ingredients to watch, pp 1–31. www.mintel.com

Moreira JB, Carvalho LF, Santos TD, Oliveira MS, Costa JAV (2018) Microalgal biotechnology for development of energy gel in sports. Int Food Res J 25:5

Mozaffarian D, Wu HY (2011) Omega-3 fatty acids and cardiovascular disease: effects on risk factors, molecular pathways, and clinical events. J Am Coll Cardiol 58(20):2047–2067

Nova P, Pimenta-Martins A, Silva JL, Silva AM, Gomes AM, Freitas AC (2020) Health benefits and bioavailability of marine resources components that contribute to health – what's new? Crit Rev Food Sci Nutr 60(21):3680–3692

Patel P, Jethani H, Radha C, Vijayendra SVN, Mudliar SN, Sarada R, Chauhan VS (2019) Development of a carotenoid enriched probiotic yogurt from fresh biomass of Spirulina and its characterization. J Food Sci Technol 56:3721–3731

Pereira J, Simões M, Silva JL (2019) Microalgal assimilation of vitamin B12 toward the production of a superfood. J Food Biochem 43(8):e12911. https://doi.org/10.1111/jfbc.12911

Pulz O, Broneske J, Waldeck P (2013) IGV GmbH experience report, industrial production of microalgae under controlled conditions: innovative prospects. In: Richmond A, Hu Q (eds) Handbook of microalgal culture: applied phycology and biotechnology. Wiley, Oxford, UK, pp 445–460

Qazi WM, Ballance S, Uhlen AK, Kousoulaki K, Haugen JE, Rieder A (2021) Protein enrichment of wheat bread with the marine green microalgae *Tetraselmis chuii*–Impact on dough rheology and bread quality. LWT-Food Sci Technol 143:111115

Qian S, Xu Y (2014) Anticancer activities of ω-6 polyunsaturated fatty acids. Biom J 37(3):112–119. https://doi.org/10.4103/2319-4170.131378

Rahman KM (2020) Food and high value products from microalgae: market opportunities and challenges. In: Alam M, Xu JL, Wang Z (eds) Microalgae biotechnology for food, health, and high value products. Springer Nature Singapore Pte Ltd, Singapore, pp 3–27. https://doi.org/10.1007/978-981-15-0169-2_1

Reyna-Martinez R, Gomez-Flores R, López-Chuken U, Quintanilla-Licea R, Caballero-Hernandez D, Rodríguez-Padilla C, Tamez-

Guerra P (2018) Antitumor activity of *Chlorella sorokiniana* and *Scenedesmus sp.* microalgae native of Nuevo León State, México. PeerJ 6:e4358. https://doi.org/10.7717/peerj.4358
Ruzik L (2023) Microalgae with active biological metal-nanoparticles as a novel food. Biosynthesis, characterization and bioavailability investigation–review. Trends Food Sci Technol 139:104127. https://doi.org/10.1016/j.tifs.2023.104127
Sandgruber F, Gielsdorf A, Baur AC, Schenz B, Müller SM, Schwerdtle T, Dawczynski C (2021) Variability in macro-and micronutrients of 15 commercially available microalgae powders. Mar Drugs 19(6):310
Santos TD, Freitas BCB, Moreira JB, Zanfonato K, Costa JAV (2016) Development of powdered food with the addition of Spirulina for food supplementation of the elderly population. Innovative Food Sci Emerg Technol 37:216–220
Santos-Sánchez NF, Valadez-Blanco R, Hernández-Carlos B, Torres-Ariño A, Guadarrama-Mendoza PC, Salas-Coronado R (2016) Lipids rich in ω-3 polyunsaturated fatty acids from microalgae. Appl Microbiol Biotechnol 100:8667–8684
Sathasivam R, Radhakrishnan R, Abeer Hashem EF, Abd_Allah (2019) Microalgae metabolites: a rich source for food and medicine. Saudi J Biol Sci 26(4):709–722
Singh H, Singh J, Singh SK, Singh N, Paul S, Sohal HS, Gupta U, Jain SK (2020) Vitamin E TPGS based palatable, oxidatively and physically stable emulsion of microalgae DHA oil for infants, children and food fortification. J Dispers Sci Technol 41(11):1674–1689
Taborda T, Moniz P, Reis A, da Silva TL. (2020) Evaluating low-cost substrates for Crypthecodinium cohnii lipids and DHA production, by flow cytometry. J App Phycol 33:263-274. https://doi.org/10.1007/s10811-020-02304-z
Tańska M, Konopka I, Ruszkowska M (2017) Sensory, physico-chemical and water sorption properties of corn extrudates enriched with Spirulina. Plant Foods Hum Nutr 72:250–257
Tertychnaya TN, Manzhesov VI, Andrianov EA, Yakovleva SF (2020) New aspects of application of microalgae Dunaliella Salina in the formula of enriched bread. IOP Conf Ser: Earth and Environ Sci 422(1): 012021. https://doi.org/10.1088/1755-1315/422/1/012021
Tsvetanova F, Yankov D (2022) Bioactive compounds from red microalgae with therapeutic and nutritional value. Microorganisms 10(11):2290
Uribe-Wandurraga ZN, Igual M, Reino-Moyón J, García-Segovia P, Martínez-Monzón J (2020) Effect of microalgae (*Arthrospira platensis* and *Chlorella vulgaris*) addition on 3D printed cookies. Food Biophys 16:27–39
Vaz BS, Moreira JB, Morais MG, Costa JAV (2016) Microalgae as a new source of bioactive compounds in food supplements. COFS 7:73–77. https://doi.org/10.1016/j.cofs.2015.12.006
Wang Y, Tibbetts SM, McGinn PJ (2021) Microalgae as sources of high-quality protein for human food and protein supplements. Foods 10(12):3002
Wolfe RR, Miller SL, Miller KB (2008) Optimal protein intake in the elderly. Clin Nutr 27:675–684
Zhu Y, Zhao X, Zhang X, Liu H (2019) Extraction, structural and functional properties of *Haematococcus pluvialis* protein after pigment removal. Int J Biol Macromol 140:1073–1108

Part IV

Health Products Developed from Microalgal Biomass

13 Microalgal Compounds with Therapeutic Properties with Potential Uses in Medicine and Human Health

Xóchitl Ariadna Ruiz-Armenta, Xiomara Patricia Perea-Domínguez, Feliznando Isidro Cárdenas-Torres, and Mario Armando Gómez-Favela

Abstract

Marine organisms are considered an excellent source of bioactive compounds; within these options, microalgae have been of significant interest in recent years. Microalgae are rich in nutrients and biologically active substances such as proteins, polysaccharides, lipids, polyunsaturated fatty acids (PUFAs), vitamins, pigments, enzymes, etc. These bioactive substances are demanded in different areas such as pharmacology, cosmetics, medicine, the chemical industry, and agriculture, among others, because they can present various effects such as antioxidant, immunostimulant, neuroprotective, antitumor, antihypertensive, anti-viral, anti-bacterial, and regenerative, among others. This chapter refers to the bioactive compounds produced by microalgae, such as polyphenols, carotenoids, fatty acids, terpenoids, polysaccharides, and peptides, as well as the biological activities attributed to them, such as antioxidants, anti-inflammatory, anticarcinogenic, and antimicrobial, among others.

Keywords

Microalgae · Bioactive compounds · Polyphenols · Carotenoids · Peptides · Polysaccharides

13.1 Introduction

Marine organisms are considered an excellent source of bioactive compounds; among them, microalgae have had significant interest recently (Le-Goff et al. 2019; Alves-Ferraz et al. 2022). Microalgae comprise a diverse group of microorganisms with 72,500 cataloged species (Jacob-Lopes et al. 2019), estimated to comprise between 3000 and 1'000,000 species between red, green, and brown grouped into 12 different phyla (Alves-Ferraz et al. 2022); classifying into two categories: prokaryotes called blue-green algae (cyanobacteria) and eukaryotes, which include green algae and diatoms (Khavari et al. 2021). Microalgae are photosynthetic, unicellular, and microscopic organisms that can be prokaryotes or eukaryotes that have plastids linked to a double membrane that is rich in pigments where chlorophyll and some other pigments, mainly carotenoids (Cichónski and Chrzanowski 2022). They are microscopic photosynthetic organisms with the ability to adapt to extremely adverse conditions, making them good candidates for drug discovery due to the presence of certain compounds developed for their defense and survival (El-Hack et al. 2019). The use of microalgae as raw material for industrial production has been defended by the ability to grow on non-productive land for agriculture and

X. A. Ruiz-Armenta
Programa de Posgrado en Ciencia y Tecnología de Alimentos, Facultad de Ciencias Químico Biológicas, Universidad Autónoma de Sinaloa, Culiacan, Sinaloa, Mexico
e-mail: xochitlruiz@uas.edu.mx

X. P. Perea-Domínguez
Doctorado en Sustentabilidad, Universidad Autónoma de Occidente, Unidad Regional Guasave, Guasave, Sinaloa, Mexico
e-mail: xiomara.perea@uadeo.mx

F. I. Cárdenas-Torres
Posgrado de Ciencias de la Nutrición y Alimentos Medicinales, Facultad de Ciencias de la Nutrición y Gastronomía, Universidad Autónoma de Sinaloa, Culiacan, Sinaloa, Mexico
e-mail: feliznando@uas.edu.mx

M. A. Gómez-Favela (✉)
Programa de Posgrado en Ciencia y Tecnología de Alimentos, Facultad de Ciencias Químico Biológicas, Universidad Autónoma de Sinaloa, Culiacan, Sinaloa, Mexico

Posgrado de Ciencias de la Nutrición y Alimentos Medicinales, Facultad de Ciencias de la Nutrición y Gastronomía, Universidad Autónoma de Sinaloa, Culiacan, Sinaloa, Mexico

D. U. Santos Ballardo, S. Rossi (eds.), *Microalgae as Promising Source of Commercial Bioproducts*, Developments in Applied Phycology 14, https://doi.org/10.1007/978-3-031-86433-9_13

the non-use of drinking water, in addition to having greater productivity in surface area and photosynthetic effectiveness compared to terrestrial crops (Kumar et al. 2022). Besides, microalgae are rich in nutrients and biologically active substances such as proteins, polysaccharides, lipids, polyunsaturated fatty acids (PUFAs), vitamins, pigments, enzymes, etc. These bioactive substances are demanded in different areas such as pharmacology, cosmetics, medicine, chemical industry, agriculture, etc., because they can present various effects such as antioxidant, immunostimulant, neuroprotective, antitumor, antihypertensive, anti-viral, anti-bacterial, regenerative, among others (Dolganyuk et al. 2020).

Furthermore, the proximal composition of microalgae varies depending on the species, as well as the conditions in which they were grown, finding values of protein, lipids, and carbohydrates of 6–71%, 1.9–40%, and 4–64%, respectively (Koyande et al. 2019). *Besides, Spirulina* and *Chlorella* are two microalgae species used as food or dietary supplements due to their high nutritional value since *Spirulina* has high proteins, vitamins, minerals, and γ-linolenic acid contents. It has been reported that this microalga has therapeutic effects in arthritis, diabetes, cardiovascular diseases, and cancer. For its part, *Chlorella* has high contents of proteins, vitamins, carotenoids, and β-glucans (Khavari et al. 2021).

The total protein content in the biomass of microalgae depends on the type of microalgae, as well as the amino acid content, but the biological content is not lower than that of animal protein, varying between 30% and 70% of the dry weight; finding that in diatom microalgae the primary amino acids present are serine, alanine, arginine, leucine, cysteine and valine (Dolganyuk et al. 2020). In the case of lipids, microalgae can produce several types of these compounds, including triacylglycerides, phospholipids, glycolipids, or phytosterols containing fatty acids from C12 to C14, frequently with C16 and C18 mono- or polyunsaturated acids. The lipid content in microalgae varies from 20% to 50% of dry weight. These lipids can be used to store energy and as a structural component of the cell membrane or for metabolic processes such as signal transduction, transcriptional and translational control, intracellular interactions, and vesicle secretion and transfer. Many microalgae contain polyunsaturated fatty acids (PUFAs) such as arachidonic acid, eicosapentaenoic acid, and docosahexaenoic acid, as well as palmitic acid as the main saturated fatty acid (Dolganyuk et al. 2020). On the other hand, microalgae are rich in vitamins C, E, K, thiamine, pyridoxine, riboflavin, nicotinic acid, biotin, and tocopherol, in addition to vitamin A (Jacob-Lopes et al. 2019); as well as minerals such as iron, iodine, potassium, calcium, and manganese, which makes them a primary source of food in Asian countries, especially China, Korea and Japan (Rizwan et al. 2018).

It has been reported that microalgae present secondary metabolites such as pigments (mainly chlorophyll and carotenoids), phenolic compounds, sterols, and terpenes, which, together with proteins (amino acids and peptides), lipids (PUFAs), and carbohydrates (polysaccharides) present biological activities beneficial to health against some conditions of interest (El-Hack et al. 2019; Jacob-Lopes et al. 2019). Next, some of these bioactive compounds (polyphenols, carotenoids, polyunsaturated fatty acids, sterols, terpenoids, polysaccharides, and peptides) present in microalgae will be discussed, as well as the biological activities that have been studied in them and the attributed medicinal potential.

13.2 Biological Activities and Medicine Potential of Microalgal Bioactive Compounds

13.2.1 Polyphenols

Polyphenols are secondary metabolites generated by various plant species to protect themselves from predators, ultraviolet rays, and pathogens. Its structure consists of one or more aromatic rings and hydroxyl groups. They can exist as glycosides or aglycones (in a matrix form or free compounds). More than 8000 polyphenol structures, polymerized or monomeric, are known (Alara et al. 2021). Various studies have reported that polyphenols present biological activities of interest, such as anti-bacterial, anti-fungal, anti-viral, anti-inflammatory, anti-cancer, anti-proliferative, and antioxidant effects (Kumar et al. 2020).

Diverse investigations report the presence of polyphenols in different varieties of microalgae. The presence of around 20 phenolic compounds has been detected in *Spirulina platensis*, where succinic acid, quinic acid, 3,4-hydroxubenzoic acid, catechin, citric acid, vanillic acid, and gallic acid stand out as the most abundant (Seghiri et al. 2019). Likewise, coumaric acid, gallic acid, and salicylic acid were detected in *Nannochloris* sp. and *Picochlorum* sp. (Pereira et al. 2015). Haoujar et al. (2019) reported the presence of protocatechuic acid, caffeic acid, quercetin, catechin, and kaempferol in *Phaedactylium tricornutum, Nannochloropsis gaditana, Nannochloris sp*, and *Tetraselmis suecica*. Mallikarjun-Gouda et al. (2015b) reported contents of 21.0 mg GAE/100 g of biomass and 27.4 mg rutin equivalents (RE)/100 g of biomass for total phenolic compounds and flavonoids, respectively, in *S. platensis*. Moreover, Seghiri et al. (2019) reported for this same variety, concentrations of 0.287 mg GAE/g of dry extract and 0.166 mg of quercetin equivalents (QE)/g of dry extract, for total phenolic compounds and total flavonoids, respectively. Also, Maadane et al. (2015) found CFT values of 8.2 to 32.0 mg of gallic acid/mg of extract when evaluating different varieties of microalgae (*Nannochlorpsis gaditana, Dunaliella* sp., *Dunaliella salina, P. tricornutum, Isochrysis* sp., *Navicula* sp., *Chaetoceros* sp., *Chlorella* sp., *and Tetraselmis* sp.), with *N. gaditana* being the variety that presented the highest polyphenolic content. Likewise, Pereira et al. (2015) reported values of 114 mg GAE/g in *Picochlorum* sp. and 5.8 mg GAE/g in *Nannochloris* sp. Besides, Arguelles (2018) found values of 606.91 µg gallic acid equivalents (GAE)/mL in the total phenolic compounds of *Scenedesmus quadricauda*. According to Haoujar et al. (2019), *P. tricornotum* presented a total phenolic compound content of 39.34 mg GAE/g, while *N. gaditana* presented 5.18 mg quercetin equivalents (QE)/g in total flavonoids. On the other hand, *Galdieria sulphuraria* (312 mg GAE/g extract) and *Crypthecodinium cohnii* (117 mg GAE/g extract) presented the highest content of phenolic compounds among 11 varieties of microalgae (Gürlek et al. 2020). Furthermore, Andriopoulos et al. (2022) evaluated aqueous and methanolic extracts of five microalgae (*Chlorella minutissima, D. salina, Isochrysis galbana, Nannochloropsis oculata*, and *Tisochrysis lutea*). In the aqueous extracts, the values of phenolic compounds varied from 2.81 to 8.78 mg gallic acid equivalents (GAE)/g of biomass, with *C. minutissima and D. salina* being the microalgae that presented the lowest and highest values, respectively. In the methanolic extracts, the varieties *T. lutea and C. minutissima* were the ones that presented the lowest value (1.25 mg GAE/g of biomass) and the highest (9.04 mg GAE/g of biomass).

On the other hand, the demands for natural antioxidant compounds grow daily. The aim is to reduce oxidative stress and the deterioration of food products, drugs, and/or cosmetics. Oxidative stress is an imbalance between reactive oxygen species (ROS) compared to antioxidants (Hayes et al. 2020).

Also, polyphenols are considered antioxidant agents because they can donate a hydrogen atom and/or an electron to oxidizing agents such as free radicals and can cause a break in the oxidation chain. The amount and position of their hydroxyl groups provide the antioxidant effects of these compounds (Albuquerque et al. 2021).

The antioxidant potential of various species of microalgae has been reported. Maadane et al. (2015) studied the DPPH antioxidant potential of ethanolic, hydroaqueous (ethanol/water) and aqueous extracts of nine microalgae (*N. gaditana, Dunaliella* sp., *D. salina, P. tricornutum, Isochrysis* sp., *Navicula* sp., *Chaeotoceros* sp., *Chlorella* sp., *and Tetraselmis* sp.). The IC_{50} values for inhibiting the DPPH radical ranged between 247 and 464 µg/mL, where *Tetraselmis* sp. was the variety that presented the highest IC_{50} value. In *S. platensis*, IC_{50} values of DPPH inhibition, reducing power, and hydroxyl radical inhibition were reported at 109.6, 81.66, and 112.70 µg/mL, respectively (Mallikarjun-Gouda et al. 2015a) and EC_{50} values = 23 µg/mL for the inhibition of the DPPH radical (Seghiri et al. 2019). Moreover, for *Desmochlorus* sp., inhibition of the DPPH radical (61%) has been reported, as well as the greatest chelation of Fe^{+2} (81%) and Cu^{+2} (61%) ions at a concentration of 10 mg/mL (Pereira et al. 2015). Besides, in *S. quadricauda*, an inhibition of the DPPH radical of 15.25% and an absorbance of 0.767 for cupper reduction antioxidant capacity were reported.

Furthermore, the antioxidant activity increased as the concentration of phenolic compounds in the extract increased (Arguelles 2018). Haoujar et al. (2019) evaluated the antioxidant activity of methanolic extracts of four microalgae: *P. tricornutum (Bacillariophyceae), N. gaditana (Eustigmatophyceae), Nannochloris sp (Trebouxiophyceae), and T. suecica (Chlorodendrophycea)*. These authors found that *Nannochloris* sp. presented the best IC_{50} value (356 µg/mL), *T. suecica* showed the best IC_{50} value (2.81 µg/mL) in iron ion chelation, and *P. triconitum* had the best value with 23.98 equivalents of ascorbic acid/ mL for FRAP.

On the other hand, for *G. sulfuraria and C. cohnii* varieties, DPPH radical inhibition values greater than 90% were reported (Gürlek et al. 2020). *C. minutissima* presented high values of antioxidant activity by FRAP in aqueous extract (0.84 mg of Trolox equivalents (TE)/g of biomass) and methanolic extract (9.04 mg TE/g of biomass). Likewise, the same variety of microalgae presented high values of antioxidant activity by DPPH (3.74 mg TE/g biomass) and ion chelation (ICA, 19.61 mg EDTA/g biomass). Likewise, *D. salina* presented the highest antioxidant activity using the DPPH method (1.32 mg TE/g biomass) and the microalgae *I. galbana* in ICA (16.74 mg EDTA/g biomass) in the aqueous extract (Andriopoulos et al. 2022).

On the other hand, different microalgae species have antimicrobial activity due to their content of bioactive compounds, where the content of phenolic compounds stands out. Due to this antimicrobial property, microalgae could be used pharmacologically to combat pathogenic microorganisms (Mallikarjun-Gouda et al. 2015a; Maadane et al. 2017). Such is the case of *S. platensis*, which has been studied and proven to have high antimicrobial activity for Gram-positive bacteria such as *Staphylococcus aureus and Bacillus cereus* and moderate activity for Gram-negative bacteria *Escherichia coli and Y. enterocolitica*. It is worth mentioning that Gram-negative bacteria are more resistant due to the lipopolysaccharides of the outer membranes (Mallikarjun-Gouda et al. 2015a). Furthermore, it has been reported that *S. platensis* presented anti-fungal inhibition on *C. versicolor* with a minimum inhibitory concentration (MIC) of 0.001 mg/mL; however, it did not present inhibition on *Aspergillus niger* (Seghiri et al. 2019). Moreover, phenolic and flavonoid compounds in *S. platensis* allow this microalga to have high antimicrobial activity, mainly in Gram-positive bacteria (Mallikarjun-Gouda et al. 2015b). Another species that has shown antimicrobial activity is *S. quadricauda*. The methanolic extracts of this microalga presented high activity for *S. aureus and Pseudomonas aeruginosa*, with a minimum inhibitory concentration (MIC) of 250 µg/mL. While for *Listeria monocytogenes and Methicillin-resistant S. aureus*, the inhibition was moderate (MIC = 1000 µg/mL). However, these species did not present antimicrobial activity against *Aeromonas hydrophila, E. coli, Enterobacter aerogenes, and Salmonella typhimurium* (Arguelles 2018). Maadane et al. (2017) made ethanolic extracts of nine microalgae (*N. gaditana, D. salina, Dunaliella* sp., *P. tricornutum, Isochrysis* sp., *Navicula* sp., *Chaeotoceros* sp., *Chlorella* sp., *and Tetraselmis* sp.) and evaluated the activity antimicrobial activity of these extracts on two Gram-negative bacteria (*E. coli* ATCC-8739, *P. aeruginosa* ATCC-9027), one Gram-positive bacteria (*S. aureus* ATCC-6538), one yeast (*Candida albicans* ATCC-10231), and a mold (*A. niger* ATCC-16404). The authors reported that microalgae *Tetraselmis* sp. strongly inhibited *E. coli* (MIC = 2.6 µg of extract/mL), *S. aureus* (MIC = 3 µg of extract/mL), and *P. aeruginosa* (MIC = 2.9 µg of extract/mL). While *D. salina* strongly inhibited the growth of *P. aeruginosa* (MIC = 2.6 µg of extract/mL). On the other hand, microalgae extract of *N. gaditana and P. tricornutum* achieved a medium inhibition of the yeast *C. albicans*, with MIC values of 4.0 and 4.7 µg of extract/mL, respectively. No extract of the microalgae studied had an inhibitory effect on *A. niger* at the concentration used (5 µg of extract/mL).

S. platensis species can be used as a nutritional supplement or functional food for patients with type 2 diabetes mellitus because it can suppress the absorption of carbohydrates in the intestine and reduce the postprandial increase in blood glucose. Mallikarjun-Gouda et al. (2015b) observed a significant decrease in blood glucose content in fed rats and an IC_{50} = 37.5 µg/mL in the inhibition of intestinal α-glucosidase in methanol/butanol extracts of *S. platensis*.

Another property of microalgae that has been reported is antihypertensive activity. Hypertension falls within the group of cardiovascular diseases with a high incidence in the world. Among the phenolic compounds that can prevent and/or suppress hypertension are gallic acid, ferulic acid, chlorogenic acid, and caffeic acid (Yu et al. 2022). Some of these phenolic compounds have been found in different species of microalgae, such as gallic acid (Pereira et al. 2015; Seghiri et al. 2019), and the polyphenol content in microalgae has been found to play an important role in preventing various activities like hypertensive. Villalpando et al. (2020) discussed that *Spirulina platensis* microalga could have antihypertensive properties since it can improve the vascular function of the arteries by increasing the release/bioavailability/function of NO. These authors reported that this microalga species increased NO production and decreased hydrogen peroxide production in spontaneously hypertensive rats (SHR).

Furthermore, it increased the vasodilatory responses of ACh, SNP and pindadicil and the expression of p-Akt and HO-1 proteins. Likewise, Mallikarjun-Gouda et al. (2015a, b) found that *S. platensis* presented a high inhibition of low-density lipoprotein (LDL) oxidation (LDLox) and that as the concentration of the methanol/butanol extract increased, the inhibition also increased. The most significant inhibition (90%) of LDLox occurred after 4 h at a concentration of 250 µg/mL. Furthermore, values of IC_{50} = 274 µg/mL have been reported in the inhibition of angiotensin-converting enzyme (ACE), which plays a crucial role in regulating blood pressure. The authors attributed the antihypertensive activity to the content of phenolic compounds present in the microalgae.

According to various studies, microalgae could potentially isolate pharmacologically active metabolites and be candidates for future therapeutic use since beneficial effects against metabolic disorders have been recorded (Samarakoon et al. 2013; Pereira et al. 2015).

Moreover, some microalgae species, such as *Galdieria sulphuraria and Crypthecodinium cohnii*, present antiproliferative activity. In a study, these two microalgae species presented IC_{50} values of 118.9–607.1 µg/mL. However, the cytotoxic effect of the extracts against HepG2 was not strong (Gürlek et al. 2020). On the other hand, microalgal species with antitumor activity and enzyme inhibitory potential have been reported. Pereira et al. (2015) analyzed the cytotoxicity of three cell lines: human hepatocarcinoma (HepG2), cervical carcinoma (HeLa) and non-tumoral murine stromal (S17) and evaluated the antitumor activity of methanolic extracts of four microalgae: *Nannochloris* sp. (SBL1 and SBL4), *Picochlorum* sp. (SBL2) and *Desmochlorus* sp. (SBL3). Of the varieties studied, SBL2 and SBL4 significantly reduced the viability of the HepG2 and HeLa tumor lines at a concentration of 125 µg/mL and presented low toxicity in the non-tumor control lines. However, the SBL4 extract presented a higher selectivity index, indicating that the compounds present can act selectively on cancer cells. Additionally, these authors evaluated the inhibition of acetylcholinesterase (AChE), butyrylcholinesterase (BuChE), and tyrosinase. Inhibition of BuChE could be related to preventing the loss of cholinergic neurons and controlling the onset of Alzheimer's. The SBL2 extract presented the highest inhibition of BuChE (69.3%) at a concentration of 10 mg/mL, while the inhibition of ACh E was low and null. The enzyme tyrosinase is involved in the formation of neuromelanin in the human brain, and its inhibition helps prevent neurodegeneration associated with Parkinson's by preventing the formation of catecholamine quinone derivatives due to the oxidative activity of the enzyme. All microalgae presented a moderate tyrosinase inhibition effect except for SBL4, which did not present any inhibition.

On the other hand, other studies have found that microalgae have anti-inflammatory activity. For example, methanolic extracts of four microalgae (*Phaeodactylim tricornutum Bohlin, Chlorella ovalis Butcher, Amphidinium carterae, and Nannchloropsis oculata (Droop)*) were evaluated to estimate their anti-inflammatory activity through the production of nitric oxide (NO) in macrophage cell lines (RAW 264.7). Results showed that the highest activity occurred in the microalgae *N. oculata* at a concentration of 6.25 μg/mL (methanol/hexane and methanol/chloroform extracts). Also, the anti-cancer activity of the different species of microalgae was evaluated by inhibiting the growth and proliferation of three cell lines: HL-60 (a human promyelocytic leukemia cell line), A549 (a human lung carcinoma cell line) and B16F10 (a mouse melanoma cell line). The results obtained showed significant suppression of the growth of HL-60 cells at concentrations of 50 and 25 μg/mL in *C. ovalis* (methanol/ethyl acetate extract) and *A. carterae* (methanol chloroform extract), respectively (Samarakoon et al. 2013). Another activity reported in microalgae is antiparasitic. Pereira et al. (2015) evaluated the antileishmanial activity of methanolic extracts of four microalgae: *Nannochloris* sp. (SBL1 and SBL4), *Picochlorum* sp. (SBL2) and *Desmochlorus* sp. (SBL3). The results showed that only the SBL1 extract reduced the viability of *Leishmania infantum* promastigotes by up to 62% at a concentration of 250 μg/mL, showing a high potential to combat this disease.

Among the phenolic compounds that could be related to the different therapeutic activities of microalgae is catechin. Catechin is a flavonoid that has great potential to protect against the development of different types of cancer, inhibits enzymes, and interrupts cancer cell growth processes (Cardoso et al. 2020).

13.2.2 Carotenoids

Carotenoids are tetraterpene pigments with various colors, such as yellow, orange, red, and purple. These pigments are widely distributed in nature. The general structure of these compounds is composed of a polyene chain with nine conjugated double bonds and a terminal group at both of the chain ends (Maoka 2020). All photosynthetic organisms synthesize these isoprenoid metabolites, including plants, algae, microalgae, and cyanobacteria, as they are essential for photosynthesis and photoprotection. Also, they are considered light harvesters and structural components in photosystems (Sun et al. 2022).

Microalgae can synthesize large amounts of mixtures of carotenoids such as astaxanthin, β-carotene, canthaxanthin, and echinenone under unfavorable or stressful conditions. Some microalgae species such as *Spirulina platensis, Dunaliella salina, Nanochloropsis* sp., *Chlorella* sp., *Chlorococcum* sp., *Coelastrella striolata, Botryococcus braunii,* Haematococcus pluvialis, and *Scenedesmus* sp. are known for the production of β-carotene, lutein, canthaxanthin, astaxanthin and fucoxanthin (Singh et al. 2019; Ibrahim et al. 2021). Zeaxanthin, fucoxanthin, and β-carotin have been found in *P. tricornutum* (Neumann et al. 2018). While echinenone, diatoxanthin, and antheraxanthin have been found in *Scenedesmus obliquus* (Ibrahim et al. 2021).

Carotenoids are health promoters and can trap toxic free radicals, preventing and/or reducing damage to living cells. Due to their antiradical capacity, they are used to extend the shelf life of some pharmaceutical products (Srivastava 2021). In addition, various studies have reported its protective effect against inflammatory, chronic degenerative, metabolic, and age-related diseases (Di Lena et al. 2019; Casagrande-do Nascimento et al. 2020). Regarding this, *Chlorella vulgaris* can inhibit albumin denaturation, proteinase, hypotonicity-induced hemolysis, and lipoxygenase (Prabakaran et al. 2018). *Phaeodactylum tricornutum* presents an anti-inflammatory effect exerted by inhibiting the activation of nuclear factor-κB (NF-κB) and depends on the MAP kinase pathway (Neumann et al. 2018). Likewise, *Euglena gracilis* also presented anti-inflammatory activity in a neuroinflammation in vitro model. Moreover, Piovan et al. (2021) reported that *E. gracilis* has a total carotenoid content of 17.8 mg/g of extract, with diadinoxanthin being the most abundant carotenoid in this microalga. These authors found that a carotenoid concentration of 10 μg/mL significantly blocks the release of nitric oxide (NO) due to lipopolysaccharide (LPS) stimulation in microglial cells and that a concentration of 100 μg/mL could inhibit it. In addition, the inhibition of the expression of the iNOS protein, which is the enzyme responsible for the production of large amounts of NO, has been observed. In addition to this, the carotenoids of this microalga can reduce the release of pro-inflammatory cytokines (TNF-α and IL-1β) in cells stimulated with LPS, starting at a concentration of 25 μg/mL. Besides, the carotenoid extracts were able to completely reverse the downregulation of Gas6 induced by LPS, suggesting that the effect on phagocytosis could be part of the anti-inflammatory mechanism of the extract. Also, the activation of the NF-κB transcription factor by LPS stimulation was suppressed by the carotenoid extract of *E. gracilis*, suggesting that the anti-inflammatory effect of the carotenoids in the extract could be through the mechanism of suppression or non-activation of the NF-κB factor. Finally, the carotenoid extract increased the levels of Nrf2 and HO-1 related to antioxidant pathways.

On the other hand, different species of microalgae (*Haematococcus pluvialis, Porphyridium cruentum, Nannochloropsis oceanica, and Tisochrysis lutea*) were studied by Gallego et al. (2022). These authors evaluated the neuroprotective potential of extracts rich in carotenoids by estimating antioxidant, anti-inflammatory, and anticholinergic activities; the cytotoxicity in different cell lines (HK-2, THP-1, and SH-SYS5) was also estimated. All the microalgae used presented a moderate inhibition of acetylcholinesterase (AChE, IC_{50} = 47.17–89.21 μg/mL) and butyrylcholinesterase (BuChE, IC_{50} = 101.24–231.15 μg/mL), in addition to high antioxidant (IC_{50} = 10.0–18.74 μg/mL) and anti-inflammatory ((IC_{50} = 28.45–73.30 μg/mL) activities. Besides, *H. pluvialis, P. cruentum, and T. lutea* presented a partial anti-inflammatory effect, while *N. oceanica* presented significant inhibition in the secretion of the TNF-α factor, being the most promising microalga to obtain a neuroprotective effect.

In another study, Ibrahim et al. (2021) reported that the microalgae *S. platensis and D. salina* presented a significantly high inhibition in the proliferation of cell lines, such as HepG2 cells (both with 100%), MCF7 (70.5% and 69.7%, respectively) and HCT116 (91% and 80.7%, respectively), while *Scenedesmus obliquus* presented an inhibition in the proliferation of HepG2 cells of 60.2%. In addition, the authors reported 69% and 39% cholinesterase inhibition for *Dunaliella salina and Scenedesmus obliquus*, respectively, at a concentration of 100 μg/mL; while *Spirulina platensis* was the opposite, showing the most significant inhibition under the lowest concentration used (1 μg/mL) reaching a value of 31%. According to the authors, the presence of β-carotene as the most significant component in *D. salina* could have given its anti-cancer activity. Furthermore, anticholinesterase activity (an enzyme found in the microalgae studied) could help treat Alzheimer's due to the inhibition of acetylcholinesterase. The authors concluded that due to the properties of the microalgae studied, *D. salina and S. platensis* could be used to support anti-cancer treatments. At the same time, *D. salina and S. obliquus* could be considered in the treatments of neurodegenerative diseases such as Alzheimer's.

On the other hand, *Scenedesmus obliquus* was studied by Casagrande-do Nascimento et al. (2020). These authors investigated the effects of carotenoid extracts from this microalga on cellular peroxidation, activity of the endogenous antioxidant system, and serum lipid profile in eutrophic mice. The main carotenoid compounds found in this species were all-trans-β-carotene (29.4%) and all-trans-lutein (28.2%), and the antioxidant activity in vitro was evaluated by the Lipophilic Oxygen Radical Absorbance Capacity (ORAC-L) methodology, obtaining 1779.9 μmol Trolox equivalents (TE)/g. The authors observed no differences in the growth and health of the mice after administering the carotenoid extract. Also, the carotenoid concentration of *S. obliquus* used was not toxic. However, even though there were no changes in most of the components (TC, HDL, LDL, TGO, and TGP) of the serum lipid profile, a difference was observed in the triglycerides (TAG) content in the group of mice that were supplied with a high concentration (2.5 mg/kg body weight) increasing by 60% compared to the control. This increase could be related to lipid peroxidation in the liver, negatively affecting physiological regulators such as lipoprotein lipase (LPL) and hepatic lipase (HL) synthesized in that tissue. Furthermore, lipid peroxidation was decreased in the heart, liver, and kidney for the group supplied with a low dose of carotenoids (0.25 mg/kg body weight), while a high dose (2.5 mg/kg body weight) only increased lipid peroxidation in the liver compared to the control group. According to the authors, this result indicates that carotenoids can change from antioxidants to pro-oxidants in high concentrations. Also, there was an increase in the enzymes of the endogenous antioxidant system, such as GPx (heart and kidney) and GR (heart), indicating that the carotenoids could increase the detoxifying effectiveness by stimulating the activity of this endogenous system. Finally, the authors concluded that according to the results, a low dose of carotenoids from *S. obliquus* could be exploited as an alternative against tissue oxidative stress due to the demonstrated capacity to improve endogenous antioxidant activity and the reduction of lipid peroxidation. However, more studies are still needed to establish the in vivo mechanisms of action.

13.2.3 Polyunsaturated Fatty Acids

Fatty acids can be classified into saturated and unsaturated fatty acids depending on double bonds in the chemical structure. Unsaturated fatty acids can be monounsaturated (MUFAs) and polyunsaturated fatty acids (PUFAs), depending on the number of double bonds (Lee et al. 2016; Kapoor et al. 2021). Besides, PUFAs are structural components of phospholipids of the cell membranes and are important for maintaining cellular function and homeostasis. PUFAs can influence cellular function via effects on membrane properties and acting as a precursor pool for lipid mediators (Burdge and Calder 2015; Dyall et al. 2022). PUFAs, including omega 6 (ω-6) and omega 3 (ω-3) fatty acids, are essential nutrients to regulate human health and development (Lee et al. 2016; Sun et al. 2019).

According to the number of carbons, lipids from microalgae can be considered as long-chain polyunsaturated fatty acids (LC-PUFAs), usually with 14–20 carbons, which have been used for the production of biodiesel by the transesterification process. Moreover, LC-PUFAs have been related to health care and heart disease prevention and modulation of the nervous and immune systems (Sun et al. 2018; Ghiffary et al. 2019).

LC-PUFAs such as eicosapentaenoic acid (EPA) and docosahexaenoic acid (DHA) are also found as major fatty acids in some microalgae (Yang et al. 2020). Furthermore, microalgae are important producers of omega-3 and omega-6 fatty acids, which are generated during the photosynthesis process. The main PUFA biosynthesis pathways in microalgae include the desaturase/elongase pathway and the polyketide synthase (PKS) pathway (Sun et al. 2019; Khoo et al. 2023). Therefore, considering that microalgae naturally produce PUFAs, optimizing the production of these biomolecules requires few genetic modifications in the existing biosynthesis pathways. However, many species of microalgae are widely accepted for human consumption, being considered "generally regarded as safe" (GRAS) (Ghiffary et al. 2019), which represents an advantage for the use of their nutrients and potential health benefits.

Furthermore, omega-3 fatty acids have multiple health-promoting properties associated with antioxidant activity, anti-inflammation, and immunological regulation. Recently, oral administration of microalgal oil (with a high content of DHA) to mice was evaluated to determine its anti-inflammatory effects. Research showed that DHA-rich microalgal oil may be a promising natural agent for preventing monosodium urate-induced inflammation and potentially acute gout. The anti-inflammatory mechanism may be due to decreased oxidative stress in organisms (Zhang et al. 2019). Additionally, lipids from microalgae have been identified as possible mediators in inflammatory bowel diseases. As mentioned, PUFAs (and also derivative oxylipins) play a key role in the resolution of inflammation, which has been demonstrated by the administration of lyophilized biomass of microalgae in a murine recurrent colitis model, establishing the potential use of microalgae as nutraceuticals in the treatment of this disease (Ávila-Román et al. 2016).

On the other hand, Novichkova et al. (2020) demonstrated that the omega-6 LC-PUFA and dihomo-γ-linolenic acid (DGLA) from microalga *Lobosphaera incsa* P127 modulate the inflammatory response. These results concord with previous studies, which have demonstrated in different models that DGLA could attenuate atherosclerosis, a disease whose complications are responsible for one in three global deaths (Gallagher et al. 2019). Besides, it has been shown in animal models that supplementation with omega-3-rich oil from *Isochrysis* sp. biomass increases the concentration of DHA in the brain and improves the lipid profile; likewise, it improves the antioxidant status of the organism without toxic effects on its organs. In addition, in this research, the expression of pro-inflammatory markers decreased while anti-inflammatory markers were upregulated. Due to their results, the authors concluded that *Isochrysis* sp. is a safe source of ω-3 fatty acids and can be used as a dietary supplement (Balakrishnan et al. 2019).

According to Amaro et al. (2013), PUFAs from microalgae, especially DHA, exhibit cancer chemopreventive features at different stages of carcinogenesis, especially in gastric cancer. Recently, the anti-proliferative effects of fatty acid extracts rich in PUFAs (from *Chlorella* sp.) via cytotoxic action against MCF-7 breast cancer cell line and A549 lung cancer cell line have been demonstrated. Besides, modulation effects on stress oxidative indicators (catalase, glutathione, and lipidic peroxidation) have been observed. Hence, this suggests that *Chlorella* sp. may be a suitable source of fatty acids rich in PUFAs with anti-cancer potentials (Vilakazi et al. 2021).

Yang et al. (2020) evaluated the lipid content of different species of microalgae, finding that microalgal lipids containing abundant PUFAs, such as EPA and DHA, could be related to beneficial health effects and can be used as nutraceutical lipids with great commercial potential. Some functional foods have been developed incorporating microalgae biomass or microalgae lipid extracts in formulations. Matos et al. (2021) elaborated on a functional yogurt incorporating *Isochrysis galbana* in the formulation, increasing the content of ω-3 LC-PUFAs and the ω3/ω6 ratio. The authors suggest that this functional yogurt can be considered a suitable source of essential fatty acids in the human diet; however, more bioavailability assays are needed.

13.2.4 Sterols

Microalgal sterols are the primary source of steroidal materials for many marine organisms. Sterols are located in the cell membrane and can be classified into phytosterols and zoosterols. In addition, they can occur in microalgae in two forms: free and conjugated. Its primary function is to give rigidity and permeability of the cell membrane. Among different microalgal sterols, phytosterols are valorized due to their potential applications in human foods and their beneficial health effects (Stonik and Stonik 2018; Fagundes and Mesadri 2023). An important variety and amounts of sterols have been reported in different classes of microalgae, being of particular interest the production of phytosterols; besides, the type and quantity of sterols in microalgae vary according to the species and the culture conditions (Volkman 2016); the predominant phytosterol from microalgae has Δ5 double bond and has no methyl groups at C4, and most of them have 27–29 carbon atoms. The profile of phytosterols depends on different factors such as strain, light, temperature, and other aspects (Luo et al. 2015).

Currently, microalgae are considered an important alternative for the sustainable production of phytosterols, mainly because these microorganisms have more efficient growth rates and better nutrient use than terrestrial plants; in addition, they can be grown in water not suitable for human consumption and can produce a wide variety of sterols and sterol precursors (Randhir et al. 2020). Different efforts in scientific research have been aimed at improving the production of phytosterols in green microalgae; such is the case of the work carried out by Potijun et al. (2021); they use *Chlamydomonas reinhardtii* mutant strains and studied the effect of terbinafine resistance on the sterol content, finding that the mutant strains showed overaccumulation of the predominant sterols and other bioactive compounds, this being a novel alternative to improve lipid production in microalgae.

On the other hand, phytosterols have attracted significant attention due to their bioactive potential, highlighting their ability to reduce blood cholesterol and their antioxidant, anti-inflammatory, and anti-cancer activities, as well as their anti-atherogenicity and cardiovascular protection (Randhir et al. 2020; Ampofo and Abbey 2022). Microalgal phytosterols as campesterol and β-sitosterol can reduce the cholesterol levels in the blood through different mechanisms: the down-regulation of hepatic HMG-CoA reductase and the stimulation of the LDL-C receptors (Le-Goff et al. 2019). Fractions with a high content of phytosterols obtained from *Dunaliella tertiolecta* and *Chlorella sorokiniana* have shown potential anti-inflammatory and anti-proliferative effects in vitro (Ciliberti et al. 2017; Ciliberti et al. 2019). Also, stigmasterol has been reported as a compound with potent antitumor capacity against different types of cancer (Zhang et al. 2022). Furthermore, according to Kim et al. (2014), stigmasterol from marine microalgae *Navicula incerta* has a strong apoptosis inductive effect in the liver cancer cell line (HepG2) and can be an option to support the development of anti-cancer drugs. The bioactive effects and biochemical mechanisms of phytosterols from different plant sources are widely known. However, the evaluation of the biological activity of phytosterols from different microalgae is still scarce; in this sense, scientific research is required to lead to the sustainable use of these compounds with potential beneficial effects on human health.

13.2.5 Terpenoids

Terpenoids are one of the largest and most important groups of chemical compounds in nature; their importance is due to their diversity of biological and pharmacological activities and food applications. Terpenoids are constructed by units of isoprene and play an important role in different cellular processes such as photosynthesis and respiration; besides, terpenoids are membrane components (sterols) and antioxidants (carotenoids) (Huang et al. 2021). Microalgal terpenoids comprise mono-, di-, tri-, tetra-, mero-, and sesquiterpenoids. Tetraterpenoid, mostly carotenoid, is a widely studied microalgal terpenoid (Sathasivam and Ki 2018) and will be described below. Besides these tetraterpenoids, other terpenoids from microalgae have been poorly studied compared to seaweed.

Many plant terpenoids have been used for medicinal purposes and have been a valuable source of medical discoveries. Many more have shown medicinal properties, but the testing processes are painstaking. The precise number of terpenoids in nature with medicinal properties is not known for sure, but it is likely >105 (Bergman et al. 2019).

Hepatocellular carcinoma cell (HepG2) inhibitory activity has been reported for two stereoisomers of the monoterpenoid lactora (loliolide and epiloliolide) present in the crude extract of *Tisochrysis lutea*, a marine microalga that is being used in aquaculture as food for oyster and shrimp larvae (Gangadhar et al. 2020).

On the other hand, geraniol is a high-value monoterpenoid that has applications as a fragrance and insect repellent and is a key intermediate in the production of pharmaceuticals. In the work carried out by George et al. (2020), they engineered the model diatom *Phaeodactylum tricornutum* to produce geraniol using synthetic biology, laying the foundations to elevate *Phaeodactylum tricornutum* to a chassis organism containing the genetic elements necessary to function as a producer of the compound of interest at a biotechnological level.

Furthermore, diterpenoids stand out for their broad biological activity and application in producing pharmaceuticals and nutraceuticals. Synthetic biology and metabolic engineering can redirect metabolic flux to efficiently synthesize heterologous diterpenoids in the green microalga *Chlamydomonas reinhardtii*. High and sustainable production of the diterpenoid sclareol has been demonstrated, which is recognized for its bioactivity, including antitumor, anti-inflammation and anti-pathogenic microbes, even anti-diabetes and hypertension. For this reason, microalgae are considered powerful green factories for the industrial biotechnological production of valuable terpenoids (Einhaus et al. 2022; Zhou et al. 2022).

Phytol is an alcohol isoprenoid classified as a diterpene and is involved in the synthesis of vitamin E and K (Gutbrod et al. 2021; Sadukha et al. 2023). Phytol has been marketed as a fragrance with potential applications in the pharmaceutical and biotechnology industries because it has been reported to have antioxidant, anxiolytic, anti-inflammatory, cytotoxic, and antimicrobial effects, among others (Islam et al. 2018). *Dunaliella* sp. CCNM 1161 has been identified as a strain that produces high amounts of phytol and lutein, being a promising strain for the simultaneous production of both metabolites to achieve a microalgae biorefinery model or applications in the multi-product nutraceutical industry (Sadukha et al. 2023).

Regarding triterpenoids, there are few studies that characterize and evaluate the bioactive potential of these compounds from microalgae. Vieira et al. (2021) evaluated the triterpenoid profile of four species of microalgae: *Porphyridium purpureum, Chlorella vulgaris, Arthorspira platensis,* and *Nannochloropsis oculata* and reported the presence of at least 12 compounds such as alphitolic, corosolic, oleonolic, and ursolic acids, among others. It was reported that the antioxidant activity was positively correlated with the triterpenoid content of the extracts.

On the other hand, squalene is a terpenoid produced by plants and animals, to which important biological activity has been attributed in human organisms, highlighting its anti-inflammatory, detoxifying, and antioxidant effects. Squalene is obtained mainly from sharks, but recently, microalgae strains have been identified as sustainable sources of this compound. Microalgae Thraustochytrids are squalene producers, and they are recognized as commercial strains for the production of nutraceuticals (Yarkent and Oncel 2022). Studies evaluating the production of squalene in a bioreactor scale are scarce; this is an area of opportunity for conducting scientific research. Noteworthy is the work of Patel et al. (2020), who optimized the production of squalene and DHA using *Schizochytrium limacinum* SR21, reporting that this strain is suitable for simultaneously producing both metabolites for nutraceutical purposes. Moreover, according to Potijun et al. (2021), resistance to terbinafine in green microalgae such as *Chlamydomonas* can lead to an overaccumulation of squalene, representing a sustainable alternative to produce this compound.

On the other hand, it has been shown that treatment with the phytohormone methyl jasmonate induces the synthesis of metabolic intermediates of the biochemical pathway for triterpenoid production in the green microalgae *Chlamydomonas reinhardtii*, being one of the first attempts to elucidate the regulation of the metabolism of these compounds in microalgae and laying the foundations for future metabolic engineering work that allows the larger scale production of bioactive compounds of commercial interest (Commault et al. 2019). As mentioned above, metabolic engineering and synthetic biology constitute powerful tools to increase the synthesis of metabolites of interest in microalgae, strengthening their potential as a source of commercial bioproducts.

13.2.6 Polysaccharides

Carbohydrates from microalgae can be divided into three groups according to their function: (1) cell wall structure polymers, (2) storage polysaccharides (intracellular), and (3) extracellular polymeric substances or EPS (Tunsi et al. 2022). In microalgae, polysaccharides are mainly found as structural or energy storage polymers that carry out various metabolic processes and EPS. The composition of the polysaccharides depends on various factors such as species, strain, and growth conditions, with galactose, xylose, and glucose being the main components of the microalgae polysaccharides; furthermore, the presence of carboxyl and sulfate groups, as well as acidic compounds, give the polysaccharides an anionic nature and consequently various biological activities. EPS from microalgae have gained importance mainly due to their bioactive properties (Moreira et al. 2022). There are reports in the literature that polysaccharides derived from microalgae present some biological activity such as antioxidant, anti-inflammatory, antiproliferative, and antimicrobial, among others (Moreira et al. 2022). There are various methodologies to evaluate the antioxidant activity of compounds, among which are those that evaluate the capacity to capture or eliminate radicals such as DPPH, ABTS, hydroxyl, superoxide, and FRAP, which are relatively simple and quick to evaluate. The DPPH and ABTS radical scavenging capacity evaluate the ability to eliminate synthetic free radicals. In contrast, the superoxide radical and hydroxyl radical scavenging capacity estimate the scavenging capacity of natural free radicals, and FRAP measures the reducing capacity of the polysaccharide. Hafsa et al. (2017) evaluated the antioxidant activity, by inhibition of the DPPH radical, of polysaccharides extracted from *Isochrysis galbana* and *Nannochloropsis oculata*, reporting that the effect of the polysaccharides showed a concentration-dependent manner, presenting maximum inhibitions of 41.45% and 59.07% at a concentration of 10 μg/mL. For their part, Trabelsi et al. (2016) found values of 8.08 mg ascorbic acid equivalents (AAE)/g of sample, 43.2% inhibition, IC_{50} = 0.33 mg/mL, IC_50 = 0.87 mg/mL and 15.2% inhibition for the antioxidant activity of polysaccharides extracted from *Grasiella* sp. evaluated by the methodologies of total antioxidant capacity (TAC), DPPH radical inhibition, iron chelation capacity, hydroxyl radical inhibition and superoxide radical inhibition, respectively. Moreover, Sun et al. (2014) evaluated the antioxidant activity of three polysaccharides isolated from *I. galbana* (IPSI-A, IPSI-B, and IPSII), reporting that the IPSII polysaccharide presented greater inhibition of the superoxide radical (53%), hydroxyl radical (63.6%), and hydrogen peroxide (48.7%) than IPSI-A (33.8%, 34.3%, and 11.4%, respectively) and IPSI-B (32.3%, 37.5%, and 10.2%, respectively), as well as a greater reducing power of said polysaccharide attributing that the high antioxidant activity was due to the high presence of uronic acid and sulfates in the IPSII polysaccharide, since it has been reported that the presence of uronic acid confers a negative charge to the polysaccharide, making it have a greater radical trapping affinity. Similarly, Chen et al. (2019a) evaluated the antioxidant activity of polysaccharides extracted from *Disctyosphaerium* sp. reporting values of 84.84%, 86.48%, 27.03%, 9.03%, and 348.56 μmol $FeSO_4$/g for the inhibition of DPPH radical, hydroxyl radical, ABTS radical, superoxide radical and FRAP, respectively, at a concentration of 10 mg/mL, suggesting that radical trapping activity is related to the concentration and properties of the polysaccharide evaluated. Also, Kashif et al. (2018) reported DPPH values of 19.85 μmol/g min and 60.33 μmol/g min, ABTS of 65.67 μmol/g min and 18.73 μmol/g min, and FRAP of 0.64 μmol/g min and 0.16 μmol/g min, for polysaccharides extracted from two strains of the microalgae *Tetraselmis* sp.

designated T1 (KCTC 12236 BP) and T2 (KCTC 12432 BP), respectively.

Besides, the antioxidant activity of polysaccharides extracted from the species *Spirulina platensis*, *Scenedesmus obliquus* and *Dunaniella salina* was reported by Singab et al. (2018) who evaluated their capacity to inhibit the DPPH radical, finding values of 206.12%, 308.16% and 308.16% at a concentration of 100 μg/mL, respectively.

Another biological activity evaluated in polysaccharides extracted from microalgae is their ability to inhibit the proliferation of cells of various types of cancer such as HepG2 (liver cancer), HeLa (cervical cancer), MCF7 (breast cancer), Caco-2, and HCT16 (colon cancer). Hafsa et al. (2017) evaluated the anti-proliferative activity of polysaccharides from *I. galbana* and *N. oculata*, reporting that both polysaccharides completely inhibited the proliferation of HeLa cells in a dose-dependent manner.

Another study related to the inhibition of the proliferation of HepG2 cells due to the effect of microalgae polysaccharides was reported by Trabelsi et al. (2016) using the species *Graciella* sp. Also, Chen et al. (2019b) use the species *Tribonema* sp., while Singab et al. (2018) use the species *S. platensis*, *S. obliquus*, and *D. salina*. These recent results showed that the inhibition of proliferation was concentration-dependent, reaching an inhibition of 66.8% at a dose of 250 μg/mL and increasing the rate of cell apoptosis by increasing the concentration of the polysaccharide, reaching a value of 38.65% but without influencing the cell cycle of HepG2, while the third indicated inhibition of 43.5%, 50.4%, and 35% in the proliferation of HepG2 for the polysaccharides of *S. platensis*, *S. obliquus*, and, *D. salina*, respectively.

Regarding the inhibition of proliferation in colon cancer cells, Trabelsi et al. (2016) found a value of IC_{50} = 0.3 mg/mL in the inhibition of Caco-2 cells by polysaccharides from the microalga *Graciella* sp., while Singab et al. (2018) reported that polysaccharides obtained from the species *S. platensis*, *S. obliquus*, and *D. salina* presented an inhibition of 31.4%, 17.6% and 64.2% in the proliferation of HCT16 cells, respectively. Likewise, the inhibitory effect of polysaccharides from *S. platensis*, *S. obliquus*, and *D. salina* on the proliferation of MCF7 cells was evaluated by Singab et al. (2018), who reported inhibition of 30.7%, 30.7%, and 35.1%, respectively. These results suggest that microalgae polysaccharides have a high potential to be an alternative in the treatment of various types of cancer; however, studies in vivo models are necessary to elucidate better the mechanism of action of polysaccharides on cancerous cells.

On the other hand, various biological activities such as antimicrobial, anticholinesterase, anti-viral, anti-inflammatory, and immunostimulant/immunomodulatory have been reported for polysaccharides derived from different microalgae species. Hafsa et al. (2017) evaluated the effect of polysaccharides from *I. galbana* and *N. oculata* on Gram-positive bacteria (*E. faecalis, S. aureus*), Gram-negative bacteria (*E. coli, P. aeruginosa*), and three species of *Candida* (*C. glabrata, C. parapsilosis,* and *C. Krusei*) finding that both polysaccharides presented positive antimicrobial activity, being the anti-fungal effect more substantial than the anti-bacterial. For their part, Kashif et al. (2018) evaluated the anti-fungal activity against *C. albicans* and *P. italicum* due to the effect of polysaccharides extracted from two lines of the microalgae *Tetraselmis* sp. (T1 and T2). They reported that the polysaccharides extracted from T1 presented more significant inhibition than polysaccharides extracted from T2, inhibiting 80% of the growth of *C. albicans* and 70% of the growth of *P. italicum*. Also, the anti-bacterial and anti-fungal activity of polysaccharides from *Chlamidomonas* sp. was evaluated by Sun et al. (2017) who reported that these polysaccharides presented high activity against *P. aureaginosa* with an inhibition zone diameter of 16.43 mm, weak activity against *E. coli*, *S. albus* and *S. aureus*, and null activity against *B. subtilis* and *S. typhimurium*; Likewise, the anti-fungal activity was weak against *S. cereviseae* and null against *R. chinensis* and *A. niger*. They indicated that these activities vary according to the type of microalgae and the type of polysaccharide extract used.

On the other hand, the anticholinesterase activity was evaluated by Hafsa et al. (2017) in polysaccharides from *I. galbana* and *N. oculata*, finding that the polysaccharides from both microalgae presented good activity, being the most important for *N. oculata* compounds, suggesting that both polysaccharides could be used as a natural treatment for Alzheimer's disease, this is because inhibiting the cholinesterase enzyme prevents the hydrolysis of two neurotransmitters such as acetylcholine and butyrylcholine, reducing the risk of suffering from said neurodegenerative disease.

Also, the anticholinesterase activity for polysaccharides of *S. platensis*, *S. obliquus*, and *D. salina* was evaluated by Singab et al. (2018), finding that this effect was dependent on the concentration for S. obliquus polysaccharides, reaching an inhibition of 46.69% at a concentration of 100 μg/mL. However, for *S. platensis* it was the opposite, presenting more significant inhibition at a lower concentration that was 31.20% at one μg/mL; while the *D. salina* polysaccharide did not present a significant increase or decrease with concentration, maintaining inhibition of 40.15%.

On the other hand, viruses are the smallest and most common entities on Earth, they are generally considered non-living because they are made up of a core of genetic material surrounded by a protective shell called a capsid. Viruses have been an evolutionary success due to massive genetic variation, variety in transmission methods, good replication mechanisms, and survival capacity (Carbone et al. 2021). There are some reports on the anti-viral activity of polysaccharides extracted from some species of microalgae, such as the work of Singab et al. (2018) who evaluated the effect of polysaccharides from *S. platensis*, *S. obliquus*, and *D. salina* against HCV genotype 4a replicon, coxsackievirus B4, rotavirus and herpes simplex virus 1, finding that the polysaccharides of *S. platensis* reduced replication by 50–87.6% considering this as a high anti-viral activity, while the polysaccharides of the other two species presented moderate activity with a reduction 10–40%. Also, the immunomodulatory and immunostimulant activity was reported by Qui and Kim (2017) and Chen et al. (2019b), respectively. The first carried out the extraction of polysaccharides from *Chlorella ellipsoidea* and separated them, obtaining two fractions (F1 and F2), which were evaluated for their immunomodulatory activity using a murine macrophage cell line RAW 264.7. They reported that the production of nitric oxide (NO) due to the effect of the polysaccharide and the fractions was in a concentration-dependent manner, where F1 presented the lowest production of NO, suggesting that its production was due to the stimulation of the expression of iNOS by the polysaccharides.

Also, the expression of cytokines such as IL-1β, IL-6, IL-10, and IL-12, of which IL-6 and IL-1β are known to be pro-inflammatory cytokines, stimulating the immune system to repair cells and tissues damaged, protecting the host's survival. However, their excessive production triggers a cascade of multiple reactions. Hence, the expression of anti-inflammatory cytokines such as IL-10 and IL-12 is necessary to help regulate or prevent damage due to excessive activation of macrophages. Furthermore, different authors evaluated the activation of NF-κB and MAPK, which regulate the expression of pro-inflammatory molecules in immune responses, reporting that the polysaccharides stimulated the expression of markers such as Iκ-Bα, p-JNK, p-ERK, and p-p38 in RAW 264.7 cells by activation of these two pathways, concluding that the results indicated that polysaccharides have an immunomodulatory effect. For their part, Chen et al. (2019b) indicated that the polysaccharides of *Tribonema* sp. could increase the viability of RAW 264.7 macrophages and stimulate cytokines such as IL-6, IL-10, and TNF-α, since they reported an increase in these cytokines by increasing the concentration of polysaccharide, indicating an immunostimulating activity in macrophages, which play an important role in defending against infections, increasing the stability of the body and controlling the immune system. The inflammation process is a physiological phenomenon of the body, and it is the immune system's first response when an injury, infection, or stress occurs. It is a natural protective response through the secretion of NO and pro-inflammatory cytokines, which, when prolonged, becomes chronic, causing various conditions such as fever, asthma, atherosclerosis, neurodegenerative diseases, and cancer. The treatment of inflammation is based on anti-inflammatory medications that have side effects, limiting their prolonged use. Natural alternatives are required, with polysaccharides being suitable (Hou et al. 2020). Levy-Ontman et al. (2017) carried out the extraction of polysaccharides from the microalgae *Porphyridium* sp. They evaluated its anti-inflammatory effect in human coronary artery endothelial cells (HCAEC) stimulated by TNF-α. They reported that, in the viability test, the TNF-α factor decreased the level of cell survival to almost 43% compared to the control and that the cells treated with the polysaccharide almost completely repaired cell viability. They also reported that these polysaccharides decreased the expression of ICAM-1, VCAM-1, and P-65 in a concentration-dependent manner, while the expression of IκB was the opposite. Both P-65 and IκB are related to the activation of the NF-κB factor, playing an important role in the inflammation process. The authors evaluated the intracellular levels of reactive oxygen species (ROS), finding that polysaccharides drastically decreased the levels of ROS induced by TNF-α stimulation in HCAECs. They indicated that the mechanism of the anti-inflammatory action is given by the inactivation of the NF-κB factor since the results they obtained demonstrated that these polysaccharides significantly reduced the expression of cytokines in cells stimulated with TNF-α, as well as the decrease in levels of ROS, indicating that the anti-inflammatory activity of the polysaccharide is probably related to its antioxidant activity. Finally, they concluded that since the polysaccharide does not present toxicity, it could be a candidate as a therapeutic additive and a potential healthy ingredient to preserve endothelial function, but in vivo studies are needed to establish the mechanisms of action more conclusively.

13.2.7 Peptides

Proteins are macronutrients that, in addition to fulfilling their nutritional functions, can perform other types of functions in the body. During digestion, enzymes hydrolyze proteins and produce various bioactive peptides with different functional activities (Mada et al. 2020; Bhandari et al. 2020). Bioactive peptides are amino acids joined by covalent bonds known as amide or peptide bonds. It has been reported that these present pharmacological activities and are classified according to their mode of action as antioxidants, antimicrobials, antihypertensives, antithrombotics, opioids, and immunomodulators, among others (Sánchez and Vázquez 2017).

Some authors (Aiello et al. 2019; Pereira et al. 2019; Akbarbaglu et al. 2022; Moaveni et al. 2022) reported that peptides from different species of microalgae present antioxidant activity, where the hydrolysis time to obtain peptides significantly influences the antioxidant activity of protein hydrolysates. Antioxidant activity increases as hydrolysis time is longer. Besides, inhibition values from 18% to 42%

have been reported for DPPH and ABTS radicals for *Spirulina* sp. LEB 18 establishes the most significant inhibition at 90 min of digestion. Other factors that can influence the antioxidant activity of protein hydrolysates are pH and temperature. pH conditions can affect the ionization potential and electron transfer capacity of peptides, while high temperatures affect the secondary structure of the peptides, causing them to lose their biological activity. Likewise, it has been reported that high concentrations of salt (8%) can help the functional groups of the peptides to be more exposed and increase the capture of radicals (Pereira et al. 2019). Moreover, Akbarbaglu et al. (2022) found that hydrolysates of *Spirulina platensis* with alcalase present high hydroxyl radical inhibition (76%), nitric oxide inhibition (NO, 28%), and iron chelation (73.5%). Likewise, Barriga-Montalvo et al. (2019) found that the protein hydrolysates of *Arthrospira maxima* with subtilisin A, presented a high affinity for the chelation of the iron ion (IC_{50} = 6.98 µg/mL) and a high anti-inflammatory activity through the inhibition of type hyaluronidase IV with an IC_{50} = 1.63 mg/mL. Also, Tejano et al. (2019) reported studying hydrolysates of the microalga *Chlorella sorokiniana*, and they reported that hydrolysates with pepsin presented inhibition of the DPPH radical (48.86%) and reducing power (0.2101), while hydrolyzed with thermolysin (CSHTh) presented a low inhibition (27.12%) of this radical.

On the other hand, in protein hydrolysates of *Schizochytrium limacinum* through simulated gastrointestinal digestion using pepsin (stomach stage), trypsin and chymotrypsin (intestinal stage), it was observed that the values of DPPH, FRAP and ABTS increased after being digested in the stomach phase by action of pepsin and after digestion with trypsin and chymotrypsin, the antioxidant values increased again. Some amino acids are hidden within the tertiary structure of proteins, so they cannot access free radicals and oxidizing agents, limiting their potential. However, after hydrolysis, these amino acids become more available to interact with free radicals (Moaveni et al. 2022).

Unfortunately, in some cases, enzymatic hydrolysis may not increase antioxidant activity if specific peptides with this activity are not released or are inactivated by digestion. Such is the case of Derbel et al. (2023), who made protein hydrolysates from *Rhodomonas* sp. using alcalase and pepsin, finding that the protein concentrate presented greater iron-reducing power and more significant inhibition of the DPPH radical (74%) than the peptic hydrolysates (approximately 56.5% pepsin and 47.5% alcalase).

The antioxidant potential found in protein hydrolysates from various microalgae has generated great interest and various investigations have been carried out focused on evaluating the nutraceutical/functional potential, such as antihypertensive, anti-inflammatory, and anti-bacterial activity, among others, using different enzymes for protein hydrolysis. Alcalase, trypsin, pancreatin, and pepsin are some enzymes used to obtain protein hydrolysates from *Spirulina platensis*. The hydrolysates from pepsin (SP) and trypsin (ST) present a high antioxidant and inhibitory activity of ACE enzymes and dipeptidyl peptidase IV (DPP-IV) with IC_{50} values that vary from 0.1 to 3.4 mg/mL, both enzymes play a crucial role in the suffering of hypertension. Hence, their inhibition indicates an antihypertensive effect (Aiello et al. 2019).

On the other hand, Tejano et al. (2019) obtained protein hydrolysates from *Chlorella sorokiniana* using pepsin, thermolysin, and bromelain, finding that the hydrolysates of this microalgae had ACE inhibitory activity, with hydrolysis with pepsin presenting the greatest inhibition (24.29%). Moreover, Barkia et al. (2018) performed protein hydrolysates from six microalgae using enzymes such as trypsin, flavourzyme, papain, and pepsin. The biological activities of these protein hydrolysates depended on the degree of hydrolysis, with papain being the enzyme that presented the highest degree of hydrolysis. *Nitzschia* sp. presented the lowest (61%) inhibition of ACE, while *Bellerochae malleus* presented the highest inhibition of this enzyme (84.35%), a value close to that reported for casein. The inhibition of ACE could be due to the activity of small peptides released during enzymatic hydrolysis and because papain has a broad enzymatic specificity, which is why it releases a more significant number of small peptides.

Also, the protein extracts of the *B. malleus* species were used for antihypertensive evaluation in vivo by administration via intraperitoneal (75 and 100 mg/kg body weight) and orally (200 and 400 mg/kg body weight) after 5 days of treatment. The group of rats fed with the *B. malleus* extract intraperitoneally presented a significant decrease in systolic blood pressure (SBP) compared to the control group. Furthermore, SBP values were similar to those produced by the drug captopril, while the group administered orally only the concentration of 400 mg/kg of body weight presented a significant decrease in SBP compared to the control group.

On the other hand, Tejano et al. (2019) obtained protein hydrolysates from *Chlorella sorokiniana* using pepsin, thermolysin, and bromelain, finding that only the hydrolyzed with pepsin presented an inhibitory effect against the growth of *Staphylococcus aureus* (Gram + bacteria) and *Escherichia coli* (Gram − bacteria), the inhibition being greater for *S. aureus*. In addition, hydrolysates of *Spirulina platensis* with alcalase also present anti-bacterial activity against *E. coli and B. cereus*, probably due to the presence of hydrophobic amino acids and the composition of amino acids in the peptides and hydrolysates, as well as positively charged amino acids which inhibit the bacterial growth by penetrating and reacting with the cell membrane of bacteria (Akbarbaglu et al. 2022). Moreover, Sadeghi et al. (2018) obtained protein hydrolysates and fractions (<3, 3–10, >10 kDa) from

Spirulina platensis using trypsin and chymotrypsin sequentially, which were evaluated for their antimicrobial activity (*E. coli and S. aureus*), and anti-cancer activity in human colon adenocarcinoma cell line. The authors reported that the peptidic fraction <3 kDa presented the greatest inhibition of *E. coli and S. aureus* with 15.2% and 19.6%, respectively, with an MIC value of 625 μg/mL. Also, Barriga-Montalvo et al. (2019) found that protein hydrolysates from *Arthrospira maxima* with pepsin presented high antimicrobial activity against *B. subtilis, S. aureus*, and *S. typhi* with IC_{50} > 1 mg/mL. However, the use of pepsin and subtilisin A in sequence to obtain hydrolysates allowed the inhibition of *E. coli* (IC_{50} = 0.79 mg/mL) and increased the anti-inflammatory activity, inhibiting the enzyme hyaluronidase type IV (IC_{50} = 0.92 mg/mL). mL) and the enzyme collagenase (IC_{50} = 32.5 μg/mL). According to the authors, the peptides present in the hydrolyzed with both enzymes contained cysteine and methionine or aromatic side chains with histidine and tyrosine, facilitating hydrogen atom donation. In addition, these peptides can help maintain skin elasticity, and by inhibiting the hydrolysis of hyaluronic acid, they increase the proliferation, regeneration, and repair of skin tissue. Likewise, peptides can prevent collagen breakdown, delaying the process of formation of pre-collagen fibers and the loss of elasticity, flexibility, and resistance of the skin. Therefore, this microalga has a high potential for application in the pharmacological industry.

Furthermore, Lee et al. (2015) found that the PPY1 peptide derived from the microalgae *Pyropia yezoensis* has anti-inflammatory activity. This peptide reduced more than 65% of NO production at a concentration of 1000 ng/mL and inhibited more than 50% of the level of intracellular reactive oxygen species. It also suppressed the expression of inflammatory enzymes such as iNOS and COX-2, the expression of the markers TNF-α and IL-1β, and the reduction of the levels of p-ERK1/2, p-JNK, and p-p38 by suppressing the phosphorylation of ERK, p38, and JNK. Besides, Pei et al. (2022) evaluated the mechanism of the nonapeptide EMFGTSSET (ETT) on epithelial cell inflammation and apoptosis induced by LDL oxidation in atherosclerosis using human umbilical vein endothelial cells (HUVEC). The authors reported that the ETT peptide significantly reduced the expression of LOX-1 by inhibiting the oxidation of LDL, also decreasing the production of reactive oxygen species (ROS), and significantly regulating the expression of inflammatory factors such as IL-6, IL-1β, and TNF-α. Furthermore, ETT decreased the expression of ICAM-1 and VCAM-1 protein, which are related to atherosclerosis, indirectly protecting cells and preventing the appearance of this disease. Likewise, ETT also decreased the expression levels of the NF-κB pathway, preventing the phosphorylation of IκBα and p65 as well as their entry into the cell nucleus for the production of pro-inflammatory proteins, while the expression levels of Nrf2 and HO-1 increased, which are related to the regulation of antioxidant enzymes and oxidative stress. Similarly, the authors reported that ETT inhibited in a dose-dependent manner the phosphorylation of JNK, ERK, and p38 involved in the MAPK pathway that participates in cell growth, apoptosis, and proliferation. While, there was a decrease in the expression of Bax, c-c-3, and c-c-9 proteins and an increase in the expression of Bcl-2 and p-Akt, proteins involved in the apoptosis process. The ETT peptide presented a protective effect against inflammation induced by LDL oxidation, oxidative stress, and apoptosis in HUVEC, asserting that this peptide can treat atherosclerosis; however, animal studies and testing in other routes are needed.

On the other hand, through *Spirulina maximum* protein digestion with pepsin, trypsin, and chymotrypsin, two peptides, LDAVNR (P1) and MMLDF (P2), were identified, which inhibited histamine and IL-8 in a dose-dependent manner. Histamine is the central molecule mediating the acute inflammation process causing pain, acute rhinitis, pruritus, and vasodilation, so inhibiting its release can contribute to relieving the inflammatory response. Furthermore, the presence of histamine promotes the production of inflammatory cytokines such as IL-8 in endothelial cells; inhibiting the release of histamine inhibits IL-8 and, therefore, the inflammatory process. The inhibitory effects of peptides P1 and P2 were mediated by the suppression of intracellular production of ROS in cells and could be used as ingredients in functional foods and/or pharmaceuticals against inflammatory diseases (Vo et al. 2013). Also, Sadeghi et al. (2018) evaluated the anti-cancer activity in the human colon adenocarcinoma cell line of hydrolysates and protein fractions (<3, 3–10, >10 kDa) of *Spirulina platensis* using sequentially trypsin and chymotrypsin enzymes. The authors reported that the peptic fraction <3 kDa presented the greatest inhibition of *E. coli and S. aureus* with 15.2% and 19.6%, respectively, with an MIC value of 625 μg/mL. When evaluating the anti-cancer activity of the hydrolyzed and the fractions, they found that the <3 kDa fraction presented the most significant growth inhibition of the human colon adenocarcinoma cell line at a concentration of 0.01 μg/mL, allowing only 36% cell viability and yielding an IC_{50} value of 1.04 μg/mL. They concluded that the results suggest that the peptides could potentially be used to treat colon cancer. However, more studies are needed to evaluate their behavior after preclinical and clinical stages.

It is important to keep in mind that, in some cases, enzymatic hydrolysis may not increase the antioxidant activity of protein hydrolysates. For example, if specific peptides with this activity are not released or are inactivated by digestion. Such is the case of Derbel et al. (2023), who made protein hydrolysates from *Rhodomonas* sp. using alcalase and pepsin enzymes, finding that the protein concentrate presented greater iron-reducing power and more significant inhibition of the DPPH radical (74%) than the peptic hydrolysates (approximately 56.5% pepsin and 47.5% alcalde).

13.3 Conclusions

The bioactive compounds in various microalgae species, such as polyphenols, carotenoids, polyunsaturated fatty acids, sterols, terpenoids, polysaccharides, and peptides, have been evaluated. The potential in their various biological activities, such as antioxidant, anti-inflammatory, antihypertensive, antimicrobial, antiparasitic, and anti-proliferative, among others, has been proved by different investigations. Due to this, microalgae and its bioactive compounds show potential to be used to develop new and safe therapeutic agents and could be used in the treatment of various diseases such as oxidative stress, inflammation, arthritis, Parkinson's, Alzheimer's, and cancer, among others. It is important to note that many of these studies were carried out at laboratory levels, and it is important to prove this in animal and human models to ensure the functionality of these potentials.

References

Aiello G, Li Y, Boschin G et al (2019) Chemical and biological characterization of spirulina protein hydrolysates: focus on ACE and DPP-IV activities modulation. J Funct Foods 63:103592. https://doi.org/10.1016/j.jff.2019.103592

Akbarbaglu Z, Ayaseh A, Ghanbarzadeh B et al (2022) Techno-functional, biological and structural properties of Spirulina platensis peptides from different proteases. Algal Res 66:102755. https://doi.org/10.1016/j.algal.2022.102755

Alara OR, Abdurahman NH, Ukaegbu CI (2021) Extraction of phenolic compounds: a review. Curr Res Food Sci 4:200–214. https://doi.org/10.1016/j.crfs.2021.03.011

Albuquerque BR, Heleno SA, Oliveira MBP et al (2021) Phenolic compounds: current industrial applications, limitations and future challenges. Food Funct 12(1):14–29. https://doi.org/10.1039/D0FO02324H

Alves-Ferraz CA, Grougnet R, Nicolau E et al (2022) Carotenoids from marine microalgae as antimelanoma agents. Mar Drugs 20:618. https://doi.org/10.3390/md20100618

Amaro HM, Barros R, Guedes AC et al (2013) Microalgal compounds modulate carcinogenesis in the gastrointestinal tract. Trends Biotechnol 31(2):92–98. https://doi.org/10.1016/j.tibtech.2012.11.004

Ampofo J, Abbey L (2022) Microalgae: bioactive composition, health benefits, safety and prospects as potential high-value ingredients for the functional food industry. Food Secur 11(1744). https://doi.org/10.3390/foods11121744

Andriopoulos V, Gkioni MD, Koutra E et al (2022) Total phenolic content, biomass composition, and antioxidant activity of selected marine microalgal species with potential as aquaculture feed. Antioxidants 11:1320. https://doi.org/10.3390/antiox11071320

Arguelles ED (2018) Proximate analysis, antibacterial activity, total phenolic content and antioxidant capacity of a green microalga *Scenedesmus quadricauda* (Turpin) Brébisson. Asian J Microbiol Biotechnol Environ Sci 20(1):150–158. https://www.researchgate.net/publication/324490066

Ávila-Román J, Talero E, Rodríguez-Luna A et al (2016) Anti-inflammatory effects of an oxylipin-containing lyophilised biomass from a microalga in a murine recurrent colitis model. Br J Nutr 116(12):2044–2052. https://doi.org/10.1017/S0007114516004189

Balakrishnan J, Dhavamani S, Sadasivam S et al (2019) Omega-3-rich Isochrysis sp. biomass enhances brain docosahexaenoic acid levels and improves serum lipid profile and antioxidant status in Wistar rats. J Sci Food Agric 99(13):6066–6075. https://doi.org/10.1002/jsfa.9884

Bhandari D, Rafiq S, Gat Y et al (2020) A review on bioactive peptides: physiological functions, bioavailability and safety. Int J Pept Res Ther 26:139–150. https://doi.org/10.1007/s10989-019-09823-5

Barkia I, Al-Haj L, Hamid AA et al (2018) Indigenous marine diatoms as novel sources of bioactive peptides with antihypertensive and antioxidant properties. Int J Food Sci Technol 54(5):1514–1522. https://doi.org/10.1111/ijfs.14006

Barriga-Montalvo GE, Thomaz-Soccol V, Vandenberghe LPS et al (2019) Arthrospira maxima OF15 biomass cultivation at laboratory and pilot scale from sugarcane vinasse for potential biological new peptides production. Bioresour Technol 273:103–113. https://doi.org/10.1016/j.biortech.2018.10.081

Bergman ME, Davis B, Phillips MA (2019) Medically useful plant terpenoids: biosynthesis, occurrence, and mechanism of action. Molecules 24(21):3961. https://doi.org/10.3390/molecules24213961

Burdge GC, Calder PC (2015) Introduction to fatty acids and lipids. World Rev Nutr Diet 112:1–16. https://doi.org/10.1159/000365423

Carbone DA, Pellone P, Lubritto C et al (2021) Evaluation of microalgae anti-viral activity and their bioactive compounds. Antibiotics 10:746. https://doi.org/10.3390/antibiotics10060746

Cardoso RR, Neto RO, dos Santos D'Almeida CT et al (2020) Kombuchas from green and black teas have different phenolic profile, which impacts their antioxidant capacities, anti-bacterial and anti-proliferative activities. Food Res Int 128:108782. https://doi.org/10.1016/j.foodres.2019.108782

Casagrande-do Nascimento T, Betim-Cazarin CB, Maróstica-Jr MR et al (2020) Microalgae carotenoids intake: influence on cholesterol levels, lipid peroxidation and antioxidant enzymes. Food Res Int 128:108770. https://doi.org/10.1016/j.foodres.2019.108770

Chen C, Zhao Z, Ma S et al (2019a) Optimization of ultrasonic-assisted extraction, refinement and characterization of water-soluble polysaccharide from Dictyosphaerium sp. and evaluation of antioxidant activity in vitro. J Food Meas Charact 14:963–977. https://doi.org/10.1007/s11694-019-00346-7

Chen X, Song L, Wang H et al (2019b) Partial characterization, the immune modulation and anti-cancer activities of sulfated polysaccharides from filamentous Microalgae Tribonema sp. Molecules 24:322. https://doi.org/10.3390/molecules24020322

Cichónski J, Chrzanowski G (2022) Microalgae as a source of valuable phenolic compounds and carotenoids. Molecules 27:8852. https://doi.org/10.3390/molecules27248852

Ciliberti MG, Albenzio M, Francavilla M et al (2019) Extracts from Microalga Chlorella sorokiniana exert an anti-proliferative effect and modulate cytokines in sheep peripheral blood mononuclear cells. Animals 9(2):45. https://doi.org/10.3390/ani9020045

Ciliberti MG, Francavilla M, Intini S et al (2017) Phytosterols from Dunaliella tertiolecta reduce cell proliferation in sheep fed flaxseed during postpartum. Mar Drugs 15(7):216. https://doi.org/10.3390/md15070216

Commault A, Fabris M, Kuzhiumparambil U et al (2019) Methyl jasmonate treatment affects the regulation of the 2-C-methyl-D-erythritol 4-phosphate pathway and early steps of the triterpenoid biosynthesis in Chlamydomonas reinhardtii. Algal Res 39(101462):101462. https://doi.org/10.1016/j.algal.2019.101462

Derbel H, Elleuch J, Mahfoudh W et al (2023) In vitro antioxidant and anti-inflammatory activities of bioactive proteins and peptides

from Rhodomonas sp. Appl Sci 13:3202. https://doi.org/10.3390/app13053202
Di Lena G, Casini I, Lucarini M et al (2019) Carotenoid profiling of five microalgae species from large-scale production. Food Res Int 120:810–818. https://doi.org/10.1016/j.foodres.2018.11.043
Dolganyuk V, Belova D, Babich O et al (2020) Microalgae: a promising source of valuable bioproducts. Biomol Ther 10:1153. https://doi.org/10.3390/biom10081153
Dyall SC, Balas L, Bazan NG et al (2022) Polyunsaturated fatty acids and fatty acid-derived lipid mediators: recent advances in the understanding of their biosynthesis, structures, and functions. Prog Lipid Res 86(101165):101165. https://doi.org/10.1016/j.plipres.2022.101165
Einhaus A, Steube J, Freudenberg RA et al (2022) Engineering a powerful green cell factory for robust photoautotrophic diterpenoid production. Metab Eng 73:82–90. https://doi.org/10.1016/j.ymben.2022.06.002
El-Hack MEA, Abdelnour S, Alagawany M et al (2019) Microalgae in modern cancer therapy: current knowledge. Biomed Pharmacother 111:42–50. https://doi.org/10.1016/j.biopha.2018.12.069
Fagundes MB, Mesadri J (2023) Microalgal sterols. In: Jacob-Lopes E, Queiroz MI, Manzoni Maroneze M, Queiroz Zepka L (eds) Handbook of food and feed from microalgae. Academic press, Massachusetts, pp 171–179. https://doi.org/10.1016/B978-0-323-99196-4.00006-1
Gallagher H, Williams JO, Ferekidis N et al (2019) Dihomo-γ-linolenic acid inhibits several key cellular processes associated with atherosclerosis. Biochim Biophys Acta Mol basis Dis 1865(9):2538–2550. https://doi.org/10.1016/j.bbadis.2019.06.011
Gallego R, Valdés A, Suarez-Montenegro ZJ et al (2022) Anti-inflammatory and neuroprotective evaluation of diverse microalgae extracts enriched in carotenoids. Algal Res 67:102830. https://doi.org/10.1016/j.algal.2022.102830
Gangadhar KN, Rodrigues MJ, Pereira H et al (2020) Anti-hepatocellular carcinoma (HepG2) activities of monoterpene hydroxy lactones isolated from the marine microalga Tisochrysis lutea. Mar Drugs 18(567). https://doi.org/10.3390/md18110567
George J, Kahlke T, Abbriano RM et al (2020) Metabolic engineering strategies in diatoms reveal unique phenotypes and genetic configurations with implications for algal genetics and synthetic biology. Front Bioeng Biotechnol 8(513). https://doi.org/10.3389/fbioe.2020.00513
Ghiffary MR, Kim HU, Chang YK (2019) Metabolic engineering strategies for the enhanced microalgal production of long-chain polyunsaturated fatty acids (LC-PUFAs). Biotechnol J 14(6):e1900043. https://doi.org/10.1002/biot.201900043
Gürlek C, Yarkent Ç, Köse A et al (2020) Screening of antioxidant and cytotoxic activities of several microalgal extracts with pharmaceutical potential. Health Technol 10:111–117. https://doi.org/10.1007/s12553-019-00388-3
Gutbrod P, Yang W, Grujicic GV et al (2021) Phytol derived from chlorophyll hydrolysis in plants is metabolized via phytenal. JBC 296:100530. https://doi.org/10.1016/j.jbc.2021.100530
Hafsa MB, Ismail MB, Garrab M et al (2017) Antimicrobial, antioxidant, cytotoxic and anticholinesterase activities of water-soluble polysaccharides extracted from microalgae Isochrysis galbana and Nannochloropsis oculate. J Serb Chem Soc 82(5):509–522. https://doi.org/10.2298/JSC161016036B
Haoujar I, Cacciola F, Abrini J et al (2019) The contribution of carotenoids, phenolic compounds, and flavonoids to the antioxidative properties of marine Microalgae isolated from Mediterranean Morocco. Molecules 24:4037. https://doi.org/10.3390/molecules24224037
Hayes JD, Dinkova-Kostova AT, Tew KD (2020) Oxidative stress in cancer. Cancer Cell 38(2):167–197. https://doi.org/10.1016/j.ccell.2020.06.001
Hou C, Chen L, Yang L et al (2020) An insight in to anti-inflammatory effects of natural polysaccharides. Int J Biol Macromol 153:248–255. https://doi.org/10.1016/j.ijbiomac.2020.02.315
Huang PW, Wang LR, Geng SS et al (2021) Strategies for enhancing terpenoids accumulation in microalgae. Appl Microbiol Biotechnol 105(12):4919–4930. https://doi.org/10.1007/s00253-021-11368-x
Ibrahim NA, Matloub AA, El-Sayed AEKB et al (2021) Phytochemical and biological studies of carotenoids in some microalgae (Dunaliella salina, Scenedesmus obliquus and Spirulina platensis). Egypt J Chem 64(5):2343–2353. https://doi.org/10.21608/EJCHEM.2021.60663.3306
Islam MT, Ali ES, Uddin SJ et al (2018) Phytol: a review of biomedical activities. FCT 121:82–94. https://doi.org/10.1016/j.fct.2018.08.032
Jacob-Lopes E, Maroneze MM, Deprá MC et al (2019) Bioactive food compounds from microalgae: an innovative framework on industrial biorefineries. Curr Opin Food Sci 25:1–7. https://doi.org/10.1016/j.cofs.2018.12.003
Kapoor B, Kapoor D, Gautam S, Singh R et al (2021) Dietary polyunsaturated fatty acids (PUFAs): uses and potential health benefits. Curr Nutr Rep 10(3):232–242. https://doi.org/10.1007/s13668-021-00363-3
Kashif SA, Hwang YJ, Park JK (2018) Potent biomedical applications of isolated polysaccharides from marine microalgae Tetraselmis species. Bioprocess Biosyst Eng 41:1611–1620. https://doi.org/10.1007/s00449-018-1987-z
Khavari F, Saidijam M, Taheri M et al (2021) Microalgae: therapeutic potentials and applications. Mol Biol Rep 48:4757–4765. https://doi.org/10.1007/s11033-021-06422-w
Khoo KS, Ahmad I, Chew KW et al (2023) Enhanced microalgal lipid production for biofuel using different strategies including genetic modification of microalgae: a review. Prog Energy Combust Sci 96:101071. https://doi.org/10.1016/j.pecs.2023.101071
Kim YS, Li XF, Kang KH et al (2014) Stigmasterol isolated from marine microalgae Navicula incerta induces apoptosis in human hepatoma HepG2 cells. BMB Rep 47(8):433–438. https://doi.org/10.5483/bmbrep.2014.47.8.153
Koyande AK, Chew KW, Rambabu K et al (2019) Microalgae: a potential alternative to health supplementation for humans. Food Sci Human Wellness 8:16–24. https://doi.org/10.1016/j.fshw.2019.03.001
Kumar S, Abedin MM, Singh AK et al (2020) Role of phenolic compounds in plant-defensive mechanisms. In: Lone R, Shuab R, Kamili A (eds) Plant phenolics in sustainable agriculture. Springer, Singapore. https://doi.org/10.1007/978-981-15-4890-1_22
Kumar R, Hegde AS, Sharma K et al (2022) Microalgae as a sustainable source of edible proteins and bioactive peptides – current trends and future prospects. Food Res Int 157:111338. https://doi.org/10.1016/j.foodres.2022.111338
Le-Goff M, Le-Ferrec E, Mayer C et al (2019) Microalgal carotenoids and phytosterols regulate biochemical mechanisms involved in human health and disease prevention. Biochimie 167:106–118. https://doi.org/10.1016/j.biochi.2019.09.012
Lee HA, Kim IH, Nam TJ (2015) Bioactive peptide from Pyropia yezoensis and its anti-inflammatory activities. Int J Mol Med 36:1701–1706. https://doi.org/10.3892/ijmm.2015.2386
Lee JM, Lee H, Kang S et al (2016) Fatty acid desaturases, polyunsaturated fatty acid regulation, and biotechnological advances. Nutrients 8(1):23. https://doi.org/10.3390/nu8010023
Levy-Ontman O, Huleihel M, Hamias R et al (2017) Ananti-inflammatory effect of red microalga polysaccharides in coronary artery endothelial cells. Atherosclerosis 264:11–18. https://doi.org/10.1016/j.atherosclerosis.2017.07.017
Luo X, Su P, Zhang W (2015) Advances in microalgae-derived phytosterols for functional food and pharmaceutical applications. Mar Drugs 13(7):4231–4254. https://doi.org/10.3390/md13074231

Maadane A, Merghoub N, El-Mernissi N et al (2017) Antimicrobial activity of marine microalgae isolated from Moroccan coastlines. J Microbiol Biotechnol Food Sci 6(6):1257–1260. https://doi.org/10.15414/jmbfs.2017.6.6.1257-1260
Maadane A, Merghoub N, Ainane T et al (2015) Antioxidant activity of some Moroccan marine microalgae: Pufa profiles, carotenoids and phenolic content. J Biotechnol 215:13–19. https://doi.org/10.1016/j.jbiotec.2015.06.400
Mada SB, Ugwu CP, Abarshi MM (2020) Health promoting effects of food-derived bioactive peptides: a review. Int J Pept Res Ther 26:831–848. https://doi.org/10.1007/s10989-019-09890-8
Mallikarjun-Gouda KG, Udaya-Sankar K, Sarada R et al (2015a) Supercritical CO2 extraction of functional compounds from Spirulina and their biological activity. J Food Sci Technol 52:3627–3633. https://doi.org/10.1007/s13197-014-1426-3
Mallikarjun-Gouda KG, Kavitha MD, Sarada R (2015b) Antihyperglycemic, antioxidant and antimicrobial activities of the butanol extract from Spirulina platensis. J Food Biochem 39:594–602. https://doi.org/10.1111/jfbc.12164
Maoka T (2020) Carotenoids as natural functional pigments. J Nat Med 74(1):1–16. https://doi.org/10.1007/s11418-019-01364-x
Matos J, Afonso C, Cardoso C et al (2021) Yogurt enriched with Isochrysis galbana: an innovative functional food. Food Secur 10(7):1458. https://doi.org/10.3390/foods10071458
Moaveni S, Salami M, Khodadadi M et al (2022) Investigation of S.limacinum microalgae digestibility and production of antioxidant bioactive peptides. LWT- Food Sci Technol 154:112468. https://doi.org/10.1016/j.lwt.2021.112468
Moreira JB, da Silva-Vaz B, Barcelos-Cardias B et al (2022) Microalgae polysaccharides: an alternative source for food production and sustainable agriculture. Polysaccharides 3:441–457. https://doi.org/10.3390/polysaccharides3020027
Neumann U, Louis S, Guille A et al (2018) Anti-inflammatory effects of Phaeodactylum tricornutum extracts on human blood mononuclear cells and murine macrophages. J Appl Phycol 30:2837–2846. https://doi.org/10.1007/s10811-017-1352-7
Novichkova E, Chumin K, Eretz-Kdosha N et al (2020) DGLA from the microalga lobosphaera incsa p127 modulates inflammatory response, inhibits iNOS expression and alleviates NO secretion in RAW264.7 Murine Macrophages. Nutrients 12(9):2892. https://doi.org/10.3390/nu12092892
Patel A, Liefeldt S, Rova U et al (2020) Co-production of DHA and squalene by thraustochytrid from forest biomass. Sci Rep 10:1992. https://doi.org/10.1038/s41598-020-58728-7
Pei Y, Liu Y, Cai S et al (2022) A novel peptide isolated from microalgae Isochrysis zhanjiangensis exhibits anti-apoptosis and anti-inflammation in Ox-LDL induced HUVEC to improve atherosclerosis. Plant Foods Hum Nutr 77:181–189. https://doi.org/10.1007/s11130-022-00965-4
Pereira AM, Lisboa CR, Santos TD et al (2019) Bioactive stability of microalgal protein hydrolysates under food processing and storage conditions. J Food Sci Technol 56:4543–4551. https://doi.org/10.1007/s13197-019-03915-2
Pereira H, Custódio L, Rodrigues MJ et al (2015) Biological activities and chemical composition of methanolic extracts of selected autochthonous microalgae strains from the Red Sea. Mar Drugs 13:3531–3549. https://doi.org/10.3390/md13063531
Piovan A, Filippini R, Corbioli G et al (2021) Carotenoid extract derived from Euglena gracilis overcomes lipopolysaccharide-induced neuroinflammation in microglia: role of NF-κB and Nrf2 signaling pathways. Mol Neurobiol 58:3515–3528. https://doi.org/10.1007/s12035-021-02353-6
Potijun S, Jaingam S, Sanevas N et al (2021) Green microalgae strain improvement for the production of sterols and squalene. Plan Theory 10(8):1673. https://doi.org/10.3390/plants10081673
Prabakaran G, Moovendhan M, Amurugam A et al (2018) Evaluation of chemical composition and in vitro antiinflammatory effect of marine microalgae Chlorella vulgaris. Waste Biomass Valor 10:3263–3270. https://doi.org/10.1007/s12649-018-0370-2
Qui J, Kim SM (2017) Characterization and immunomodulatory activities of polysaccharides extracted from green alga Chlorella ellipsoidea. Int J Biol Macromol 95:106–114. https://doi.org/10.1016/j.ijbiomac.2016.11.039
Randhir A, Laird D, Maker G et al (2020) Microalgae: a potential sustainable commercial source of sterols. Algal Res 46(101772):101772. https://doi.org/10.1016/j.algal.2019.101772
Rizwan M, Mujtaba G, Memon SA et al (2018) Exploring the potential of microalgae for new biotechnology applications and beyond: a review. Renew Sust Energ Rev 92:394–404. https://doi.org/10.1016/j.rser.2018.04.034
Sadeghi S, Jalili H, Ranaei-Siadat SO et al (2018) Anticancer and antibacterial properties in peptide fractions from hydrolyzed spirulina protein. J Agric Sci Technol 20:673–683. http://jast.modares.ac.ir/article-23-19830-en.html
Sadukha S, Bhayani A, Padariya H et al (2023) Performance evaluation of microalgal strains for concurrent production of high-value bio-actives lutein and phytol: a step forward towards the multiproduct paradigm. Biocatal Agric Biotechnol 50:102737. https://doi.org/10.1016/j.bcab.2023.102737
Sathasivam R, Ki JS (2018) A Review of the Biological Activities of Microalgal Carotenoids and Their Potential Use in Healthcare and Cosmetic Industries. Mar Drugs, 16 (1): 6. https://doi.org/10.3390/md16010026
Samarakoon KW, Ko JY, Rahman-Shah MM et al (2013) In vitro studies of anti-inflammatory and anti-cancer activities of organic solvent extracts from cultured marine microalgae. Algae 28(1):111–119. https://doi.org/10.4490/algae.2013.28.1.111
Sánchez A, Vázquez A (2017) Bioactive peptides: a review. Food Qual Saf 1(1):29–46. https://doi.org/10.1093/fqsafe/fyx006
Seghiri R, Kharbach M, Essamri A (2019) Functional composition, nutritional properties, and biological activities of Moroccan Spirulina Microalga. J Food Qual 707219:11. https://doi.org/10.1155/2019/3707219
Singab ANB, Ibrahim NA, El-Sayed AEKB et al (2018) Anti-viral, cytotoxic, antioxidant and anticholinesterase activities of polysaccharides isolated from microalgae Spirulina platensis, Scenedesmus obliquus, and Dunaliella salina. Arch Pharm Sci Ain Shams Univ 2(2):121–137. https://doi.org/10.21608/aps.2018.18740
Singh DP, Khattar JS, Rajput A et al (2019) High production of carotenoids by the green microalga Asterarcys quadricellulare PUMCC 5.1. 1 under optimized culture conditions. PLoS One 14(9):e0221930. https://doi.org/10.1371/journal.pone.0221930
Srivastava R (2021) Physicochemical, antioxidant properties of carotenoids and its optoelectronic and interaction studies with chlorophyll pigments. Sci Rep 11:18365. https://doi.org/10.1038/s41598-021-97747-w
Stonik VA, Stonik IV (2018) Sterol and sphingoid glycoconjugates from microalgae. Mar Drugs 16(12):514. https://doi.org/10.3390/md16120514
Sun J, Zhao J, Fu D et al (2017) Extraction, optimization and antimicrobial activity of IWSP from Oleaginous Microalgae Chlamydomonas sp. YB-204. Food Sci Technol Res 23(6):819–826. https://doi.org/10.3136/fstr.23.819
Sun T, Rao S, Zhou X et al (2022) Plant carotenoids: recent advances and future perspectives. Mol Hortic 2(1):3. https://doi.org/10.1186/s43897-022-00023-2
Sun XM, Ren LJ, Zhao QY et al (2018) Microalgae for the production of lipid and carotenoids: a review with focus on stress regulation and adaptation. Biotechnol Biofuels 11(272):272. https://doi.org/10.1186/s13068-018-1275-9

Sun XM, Ren LJ, Zhao QY et al (2019) Enhancement of lipid accumulation in microalgae by metabolic engineering. Biochim Biophys Acta Mol Cell Biol Lipids 1864(4):552–566. https://doi.org/10.1016/j.bbalip.2018.10.004

Sun Y, Wang H, Guo G et al (2014) The isolation and antioxidant activity of polysaccharides from the marine microalgae Isochrysis galbana. Carbohydr Polym 113:22–31. https://doi.org/10.1016/j.carbpol.2014.06.058

Tejano LA, Peralta JP, Yap EES et al (2019) Bioactivities of enzymatic protein hydrolysates derived from Chlorella sorokiniana. Food Sci Nut 7:2381–2390. https://doi.org/10.1002/fsn3.1097

Trabelsi L, Chaieb O, Mnari A et al (2016) Partial characterization and antioxidant and anti-proliferative activities of the aqueous extracellular polysaccharides from the thermophilic microalgae Graesiella sp. BMC Complement Altern Med 16:210. https://doi.org/10.1186/s12906-016-1198-6

Tunsi L, Hentati F, Hlima HB et al (2022) Microalgae as feedstock for bioactive polysaccharides. Int J Biol Macromol 221:1238–1250. https://doi.org/10.1016/j.ijbiomac.2022.08.206

Vieira MV, Turkiewicz IP, Tkacz K et al (2021) Microalgae as a potential functional ingredient: evaluation of the phytochemical profile, antioxidant activity and in-vitro enzymatic inhibitory effect of different species. Molecules 26(7593). https://doi.org/10.3390/molecules26247593

Vilakazi H, Olasehinde TA, Olaniran AO (2021) Chemical Characterization, Antiproliferative and Antioxidant Activities of Polyunsaturated Fatty Acid-Rich Extracts from Chlorella sp. S14. Molecules, 26: 4109. https://doi.org/10.3390/molecules26144109

Villalpando DM, Verdasco-Martín CM, Plaza I et al (2020) Beneficial effects of spirulina aqueous extract on vasodilator function of arteries from hypertensive rats. Int J Vasc Med 6657077:9. https://doi.org/10.1155/2020/6657077

Vo TS, Riu B, Kim SK (2013) Purification of novel anti-inflammatory peptides from enzymatic hydrolysate of the edible microalgal Spirulina maxima. J Funct Foods 5:1336–1346. https://doi.org/10.1016/j.jff.2013.05.001

Volkman JK (2016) Sterols in Microalgae. In: Borowitzka M, Beardall J, Raven J (eds) The physiology of microalgae. Developments in Applied Phycology (6). Springer, Cham, Edinburgh. https://doi.org/10.1007/978-3-319-24945-2_19

Yang Y, Du L, Hosokawa et al (2020) Total lipids content, lipid class and fatty acid composition of ten species of microalgae. J Oleo Sci 69(10):1181–1189. https://doi.org/10.5650/jos.ess20140

Yarkent Ç, Oncel SS (2022) Recent progress in microalgal squalene production and its cosmetic application. Biotechnol Bioprocess Eng 27(3):295–305. https://doi.org/10.1007/s12257-021-0355-z

Yu M, Kim HJ, Heo H et al (2022) Comparison of the antihypertensive activity of phenolic acids. Molecules 27(19):6185. https://doi.org/10.3390/molecules27196185

Zhang X, Wang J, Zhu L et al (2022) Advances in stigmasterol on its antitumor effect and mechanism of action. Front Oncologia 12:1101289. https://doi.org/10.3389/fonc.2022.1101289

Zhang Y, Liu L, Sun D et al (2019) DHA protects against monosodium urate-induced inflammation through modulation of oxidative stress. Food Funct 10(7):4010–4021. https://doi.org/10.1039/c9fo00573k

Zhou J, Xie X, Tang H et al (2022) The bioactivities of sclareol: a mini review. Front Pharmacol 13(1014105). https://doi.org/10.3389/fphar.2022.1014105

Microalgae as Source of Antioxidants: Potential Role in Nutraceutical Products for Human Health

14

Nayely Leyva-López, Laura Aracely Contreras-Angulo, Manuel de Jesús Bernal-Millán, Luis Angel Cabanillas-Bojórquez, Melissa Garcia-Carrasco, Erick Paul Gutiérrez-Grijalva, and José Basilio Heredia

Abstract

Microalgae are present in aquatic surroundings. These unicellular and photosynthetic microorganisms are attracting the attention of researchers and industries in different areas, such as pharmaceutical and nutraceutical, due to their content of antioxidant and bioactive molecules, such as polyunsaturated fatty acids (PUFAs), carotenoids (pigments), polysaccharides, vitamins, and phenolic compounds. These antioxidants have shown the ability to reduce or inhibit oxidative stress, which has been strongly associated with the appearance of several chronic diseases, such as diabetes, autoimmune disorders, cardiovascular ailments, and cancer, among several others. In addition to *Chlorella* sp., *Dunaliella* sp., and *Spirulina* sp., which are the most studied microalgae species, other species are worth exploring as sources of antioxidant and bioactive substances. This chapter aims to identify microalgae species that can be used as sources of bioactive compounds with antioxidant and nutraceutical properties.

N. Leyva-López · L. A. Cabanillas-Bojórquez
Postdoc SECIHTI-Centro de Investigación en Alimentación y Desarrollo, Culiacan, Sinaloa, Mexico

Centro de Investigación en Alimentación y Desarrollo, Culiacan, Sinaloa, Mexico

L. A. Contreras-Angulo · M. d. J. Bernal-Millán
M. Garcia-Carrasco · J. B. Heredia (✉)
Centro de Investigación en Alimentación y Desarrollo, Culiacan, Sinaloa, Mexico
e-mail: jbheredia@ciad.mx

E. P. Gutiérrez-Grijalva
Programa Investigadoras e investigadores por México SECIHTI-Centro de Investigación en Alimentación y Desarrollo, Culiacan, Sinaloa, Mexico

Keywords

Haematococcus pluvialis · *Nannochloropsis* spp. · *Porphyridium cruentum* · Bioactive compounds · Anticancer agents

14.1 Introduction

Microalgae are those microorganisms that contain chlorophyll *a* and other photosynthetic pigments, capable of oxygenic photosynthesis to transform CO_2 and sunlight into biomass (Chisti 2007). In this context, cyanobacteria, or blue-green algae with a prokaryotic cellular structure, have raditionally been considered within microalgae. Microalgae can be found in freshwater or saline environments (Luna 2007).

Many microalgae species can substantially accumulate antioxidant molecules, namely polyunsaturated fatty acids (PUFAs), carotenoids, polysaccharides, vitamins, and phenolic compounds. All these components of the microalgae have potential in the development of nutraceutical products. The definition of the word *nutraceuticals* can be recently found in the Encyclopedia of Human Nutrition, in which is stated that nutraceuticals "are products manufactured from both food and non-food sources that provide health benefits in support of structure or function of the body, prevention or mitigation of diseases, delay of the aging process, and promotion of wellbeing" (Chen and Mah 2023).

In recent years, microalgae have been the focus of attention as biofactories in obtaining molecules of interest in the food, pharmaceutical, and nutraceutical industries. The main advantages of microalgae usage as a source of antioxidant and bioactive substances over other sources (i.e. plants) are that (1) they grow faster than plants and show the highest yield of biomass, (2) depending on the microalgae species, they can be selected to produce a specific group of bioactive

D. U. Santos Ballardo, S. Rossi (eds.), *Microalgae as Promising Source of Commercial Bioproducts*, Developments in Applied Phycology 14, https://doi.org/10.1007/978-3-031-86433-9_14

molecules, (3) the cultivation of microalgae does not need the use of pesticides and reduces soil erosion, and (4) they do not require organic substances for energy (Abd El Baky and El-Baroty 2013). With all these characteristics, identifying microalgae species with the potential to be used as sources of bioactive compounds with antioxidant and nutraceutical properties is scientifically and commercially relevant.

14.2 Antioxidants Present in Microalgae

It is well known that an antioxidant is a substance capable of reacting with free radicals to prevent the oxidation of biologically important macromolecules ((DNA) (Deoxyribonucleic acid), lipids, proteins) and protect the organism from oxidative damage.

Free radicals are molecules and reactive species with an unpaired electron in their last orbital. They need stability, which is achieved by receiving electrons from other molecules. The most common reactive species are anion superoxide (O_2•⁻), hydroxyl radical (OH•), peroxyl (RO_2•), and alkoxyl (RO•) radicals and oxides of nitrogen (NO•, NO_2•). Other reactive species that are not in radical form are hydrogen peroxide (H_2O_2), hypochlorous acid (HOCl), ozone (O_3), and singlet oxygen ($^1O_2$) (Lushchak 2014). Organisms like humans have an endogenous antioxidant system that protects against reactive species. This system is composed of the antioxidant enzymes superoxide dismutase (SOD), catalase (CAT), and glutathione peroxidase (GPX), as well as glutathione S-transferase (GST) and NAD(P)H quinone oxidoreductase 1(NQO1) (Ighodaro and Akinloye 2018). When high levels of reactive species and a diminished endogenous antioxidant defense are present, it is said that the organism has reached the state of oxidative stress.

Since oxidative stress is related to several illnesses, an antioxidant might help improve health. Microalgae have proved to promote the antioxidant defense systems against oxidative stress and to be an excellent source of antioxidants; therefore, they are a relevant alternative to obtain molecules with nutraceutical potential. The presence of antioxidant and bioactive compounds studied in microalgae is mentioned next.

14.2.1 Polyunsaturated Fatty Acids (PUFAs)

The PUFAs have more than a double bond in their molecular structure. These are the main structural components of the cell wall and are classified into four families: the omega-3 family, generated from α-linolenic acid; the omega-6 family, produced from linoleic acid; the omega-7 family, originated from palmitoleic acid; and the omega-9 family, generated from oleic acid (Kothri et al. 2020). The consumption of PUFAs (especially omega-3 and omega-6 fatty acids) is related to important health benefits. Among the PUFAs' main functions are the regulation of blood glucose and pressure levels, improvement of metabolic rates, and non-alcoholic fatty liver treatment; they modulate brain development and cognition, in addition to playing an important role in inflammation, cancer, osteoarthritis, autoimmune disorders, obesity, and diverse chronic diseases (Lee et al. 2016; Santin et al. 2021; Kapoor et al. 2021).

The omega-3 PUFAs (EPA: eicosapentaenoic acid and DHA: docosahexaenoic acid) are essential for humans due to their cardioprotective effects, presenting benefits on cardiac hemodynamic factors. When inserted into cellular membranes, the PUFAs act as antioxidants and regulate the antioxidant signaling pathways. The DHA reduces oxidative stress and the oxidase cytochrome c activity in the mitochondria (Oppedisano et al. 2020). Also, the omega 3 PUFAs can reduce lipoperoxidation levels and increase SOD2 levels in the liver of diabetic rats with a high-fat diet. In addition, the supplementation with omega-3 PUFAs increases the expression of heme oxygenase-1 (HO-1) through the activation of Nrf-2, proving that EPA and DHA protect the adipocytes against the production of reactive oxygen species (ROS) induced by H_2O_2. Furthermore, it has been demonstrated that supplementation with EPA and DHA reduces lipoperoxide concentrations in the plasma of patients with mild cognitive deterioration (Giordano and Visioli 2014).

The most important PUFA producers worldwide are microalgae (Han et al. 2019). According to parameters like biomass amount, oil content, and PUFA percentage, the strains, or species with the most significant potential for PUFA extraction can be selected (Lu et al. 2021). Microalgae have been isolated from cold environments or areas with high altitude, for example, *Phaeodactylum tricornutum,* obtaining a biomass yield of 60 mg/L/day and an EPA percentage: 7.33% of total fatty acids (TFA) and DHA: 0.58% of TFA (Steinrücken et al. 2017). Also, Gour et al. (2016) extracted PUFAs from *Scenedesmus dimorphus* and *Ankistrodesmus fusiformis,* obtaining 39.03% and 40.24% of total fatty acids (TFA). Other reported species are *Botryococcus braunii*, *Chlorella* sp., *Chlorococcum humicola,* and *Chlamydomonas* sp., with biomass yields that vary between 158.9 and 249.2 mg/L/day and PUFAs percentage between 11.49% and 55.52% of TFA (Hena et al. 2015).

However, microalgae can synthesize PUFAs in different growing conditions. It has been reported that factors like salinity, glucose, pH, and nitrogen increase the content of PUFAs. Accumulation of high quantities of DHA and EPA has been observed in *Nannochloropsis*, *Crypthecodinium cohnii*, *Phaeodactylum*, *Schizochytrium,* and *Thraustochytrium* strains (Ramesh Kumar et al. 2019). Besides, *C. cohnii* has been recognized as a DHA producer since it can reach up to 65% of PUFAs in its lipid profile

(Barta et al. 2021). Moreover, *Schizochytrium* sp. has been proposed as an omega-3 source since it contains 37–47% DHA and 3–10% of EPA. Moreover, other species like *B. braunii* and *Isochrysis galbana* have been reported to have the potential to produce DHA and EPA, respectively (Santos-Sánchez et al. 2016). Other microalgae where EPA has been detected are *Florideophyceae*, *Cryptophyceae*, *Chlorophyceae*, *Chrysophyseae*, and *Bacillariophyceae* (Li et al. 2019). Ryckebosch et al. (2014) evaluated 11 different microalgae, reporting *Isochrysis T-iso* and *Pavlova lutheri* with the highest DHA content (46 and 40.9 mg/g oil, respectively), while *Nannochloropsis gaditana* and *Nannochloropsis oculate* presented high EPA content (175 and 193 mg/g oil, respectively).

14.2.2 Carotenoids

Carotenoids are an important natural pigment group with health-promoting properties used commercially as animal feed supplements, colorants, and cosmetic and pharmaceutic ingredients (Xie et al. 2022). Carotenoids comprise a wide number of molecules (over 600 members) with long carbon chains (40 carbon atoms) that are soluble in organic solvents such as methanol, chloroform, hexane, and acetone, among others (Jaeschke et al. 2016). Carotenoids are divided into two types by their chemical structure: carotenes, namely to the group of hydrocarbon molecules, and xanthophylls, namely to the group of oxygenated carotenoids (Ambati et al. 2019; Safafar et al. 2015). Carotenoids are important in photosynthetic organisms due to energy transference and protection against UV-light in microorganisms and plants during light harvesting (Diprat et al. 2017; Paliwal et al. 2016).

Microalgae are an important source of carotenoids synthesized during specific growth stages on different stress strategies and cultivation protocols (Tzima et al. 2023; Xie et al. 2022). In this sense, it has been reported that high light and nutrient deficient have the most effective stresses to enhance the carotenoid production of microalgae; however, in recent years, the effect of temperature, the presence of reactive oxygen species (Bazarnova et al. 2022), as well as the impact of plant hormone has been studied (Canelli et al. 2022; Gui et al. 2022; Rodrigues et al. 2014; Van den Berg et al. 2019; Xie et al. 2022). On the other hand, different carotenoid extraction techniques on microalgae have been studied. In this sense, solvent extraction is the most reported; nevertheless, green extraction techniques such as high pressure, ionic liquid extraction, electric pulse field, pressurized liquid extraction, supercritical CO_2 extraction, and ultrasound-assisted extraction have been on the rise (Achour et al. 2022; Jaeschke et al. 2016; Morcelli et al. 2021; Paliwal et al. 2022; Sánchez-Camargo et al. 2018; Sanzo et al. 2018; Tzima et al. 2023; Wang et al. 2023).

Furthermore, it has been reported the extraction of the carotenes α-carotene, β-carotene, and echinenone, as well as xanthophylls such as antheraxanthin, astaxanthin, canthaxanthin, diadinoxanthin, fucoxanthin, lutein, myxoxanthophyll, neoxanthin, peridin, vaucheriaxanthin, violaxanthin, and zeaxanthin from different microalgae species (Table 14.1) (Diprat et al. 2017; Gui et al. 2022; Paliwal et al. 2022; Reyes et al. 2016; Rodrigues et al. 2014; Safafar et al. 2015; Wang et al. 2023). The most reported microalgae for carotenoid extraction are *Chlorella*, *Dunaliella*, *Haematococcus*, *Nannochloropsis*, and *Scenedesmus* species (León-Vaz et al. 2023; Paliwal et al. 2016; Safafar et al. 2015). Likewise, solvent extraction is the most studied technique; however, supercritical CO_2 extraction (SFE- CO_2) is the technique with the higher carotenoid content obtained from microalgae (Liau et al. 2010; Molino et al. 2018; Morcelli et al. 2021; Şahin et al. 2019; Sanzo et al. 2018).

14.2.3 Polysaccharides

Polysaccharides are polymeric carbohydrate molecules with complex structures obtained by polymerizing long chains of monosaccharides (xylose, galactose, fructose, rhamnose, and mannose) (Gaignard et al. 2019). These polysaccharides can occur mainly as structural polymers, part of the cell membrane, or energy storage polymers in different metabolic processes. In addition, the exopolysaccharides that are found provide protection and cellular interactions (Moreira et al. 2022; Costa et al. 2021).

The carbohydrates in polysaccharides are mainly synthesized in the chloroplasts of eukaryotic cells during the Calvin cycle in the dark phase. The Calvin cycle presents three main stages for carbohydrate production (fixation, reduction, and regeneration). Firstly, CO_2 is added to ribulose-1,5-bisphosphate, a reaction catalyzed by ribulose-1,5-bisphosphate carboxylase oxygenase (RuBisCo), forming two phosphoglycerate molecules. Phosphoglycerate is converted to two molecules of glyceraldehyde-3-phosphate. One of the molecules continues in the Calvin cycle, while the other is used as a substrate to form sucrose. The glyceraldehyde-3-phosphate is not immediately used as an energy source and is converted to starch in the chloroplast. The biosynthesis and sulfation of polysaccharides in eukaryotic cells occur in the Golgi complex, whose primary role is the synthesis of exopolysaccharides (Raines 2003; Leegood 1990; Hügler and Sievert 2010). The bioactivity of these compounds depends mainly on the type of monosaccharide, molecular size, and degree of sulfation of the molecule (Costa et al. 2021; Garcia-Perez et al. 2023).

Table 14.1 Carotenoids obtained from microalgae by different extraction methods

Microalgae	Extraction method	Carotenoid content	References
Acutodesmus dimorphus	Solvent extraction	Antheraxanthin (0.02 ± 0.01 mg/g), astaxanthin (0.07 ± 0.01), β-carotene (0.32 ± 0.05), canthaxanthin (0.06 ± 0.02), lutein (2.00 ± 1.42), neoxanthin (4.13 ± 2.45), violaxanthin (0.02 ± 0.01), zeaxanthin (0.54 ± 0.26)	(Paliwal et al. 2016)
Arthrospira platensis	Pressurized liquid extraction	β-carotene (29.11 mg/L), myxoxanthophyll (1.54), zeaxanthin (24.93)	(Wang et al. 2023)
Bracteacoccus pseudominor	Solvent extraction	β-carotene (0.17 ± 0.06 mg/g), lutein (1.16 ± 0.42), neoxanthin (0.23 ± 0.05), violaxanthin (0.09 ± 0.04), zeaxanthin (0.38 ± 0.21)	(Paliwal et al. 2016)
Chlamydomonas reinhardtii	Solvent extraction	β-carotene (1.9 ± 0.2 mg/g), lutein (2.9 ± 0.3), neoxanthin (0.6 ± 0.09), violaxanthin (1.3 ± 0.1)	(Rengel et al. 2022)
Chlorella miniata	Solvent extraction	β-carotene (14.20 mg/mL), canthaxanthin (62.30), lutein (42.30)	(Ova et al. 2022)
Chlorella saccharophila	Solvent extraction	Antheraxanthin (0.16 ± 0.02 mg/g), lutein (2.64 ± 0.16), neoxanthin (0.53 ± 0.04), violaxanthin (0.60 ± 0.08)	(León-Vaz et al. 2023)
Chlorella sorokiniana	Solvent extraction	Alloxanthin (94.06 ± 0.8 μg/g), astaxanthin (48.42 ± 0.5), β-carotene (1039 ± 17), fucoxanthin (104.8 ± 5.5), neoxanthin (48.29 ± 1.2), lutein (3220 ± 54), peridin (16.70 ± 0.1), zeaxanthin (151.1 ± 1.7)	(Safafar et al. 2015)
		α-carotene (63.68 ± 0.4 μg/g), β-carotene (157.92 ± 0.72), lutein (873.43 ± 7.15), neoxanthin (146.80 ± 0.51), violaxanthin (55.29 ± 0.31), zeaxanthin (41.68 ± 0.24)	(Fernandes et al. 2021)
		α and β-carotene (0.756 ± 0.052 mg/g), lutein (2.79 ± 0.05), violaxanthin (0.22 ± 0.02), zeaxanthin (0.30 ± 0.01)	(Morcelli et al. 2021)
	SFE- CO_2	α and β-carotene (0.05–0.60 mg/g), lutein (0.01–0.59), violaxanthin (0.00–0.06), zeaxanthin (0.00–0.06)	(Morcelli et al. 2021)
Chlorella variablis	Solvent extraction	Antheraxanthin (0.07 ± 0.02 mg/g), β-carotene (0.17 ± 0.04), canthaxanthin (0.03 ± 0.01), lutein (1.13 ± 1.02), neoxanthin (0.30 ± 0.19), violaxanthin (0.08 ± 0.03), zeaxanthin (0.36 ± 0.12)	(Paliwal et al. 2016)
Chlorella vulgaris BIN	Ultrasound-assisted solvent extraction	β-carotene (0.52 ± 0.05 mg/mL), fucoxanthin (1.4 ± 0.1), lutein (5.4 ± 0.5)	(Bazarnova et al. 2022)
	Pulse electric field and high-pressure homogenization	β-carotene (0.62 ± 0.01 mg/g), lutein (1.15 ± 0.01), zeaxanthin (0.07 ± 0.00)	(Canelli et al. 2022)
	Solvent extraction	Antheraxanthin (0.22 ± 0.07 mg/g), lutein (2.62 ± 0.27), neoxanthin (0.64 ± 0.08), violaxanthin (0.66 ± 0.06)	(León-Vaz et al. 2023)
	SFE- CO_2	α-carotene (0.28 ± 0.04 mg/g), β-carotene (0.26 ± 0.02), lutein (4.84 ± 0.78)	(Şahin et al. 2019)
Chlorella zofingiensis	Solvent extraction	Astaxanthin (0.14–6.02 mg/g), canthaxanthin (0.8–8.5)	(Pelah et al. 2004)
Coelastrum astroideum	Solvent extraction	Antheraxanthin (0.13 ± 0.01 mg/g), lutein (5.65 ± 0.69), neoxanthin (1.82 ± 0.08), violaxanthin (1.44 ± 0.08)	(León-Vaz et al. 2023)
Coelastrum microporum	Solvent extraction	Antheraxanthin (0.06 ± 0.00 mg/g), lutein (3.73 ± 0.18), neoxanthin (2.97 ± 0.17), violaxanthin (1.44 ± 0.28)	(León-Vaz et al. 2023)
Desmodesmus opoliensis	Solvent extraction	Antheraxanthin (0.13 ± 0.03 mg/g), lutein (8.67 ± 0.29), neoxanthin (2.01 ± 0.16), violaxanthin (2.07 ± 0.15)	(León-Vaz et al. 2023)
Desmodesmus subspicatus	Solvent extraction	Antheraxanthin (0.07 ± 0.01 mg/g), astaxanthin (0.37 ± 0.06), canthaxanthin (0.04 ± 0.02), β-carotene (0.26 ± 0.07), lutein (1.98 ± 1.39), neoxanthin (2.73 ± 1.23), zeaxanthin (0.23 ± 0.09)	(Paliwal et al. 2016)
Dunaliella bardawil	Solvent extraction	α-carotene (0.36 ± 0.02 μg/mL), β-carotene (0.15 ± 0.01), lutein (0.97 ± 0.02), zeaxanthin (0.28 ± 0.01)	(Xie et al. 2022)

(continued)

Table 14.1 (continued)

Microalgae	Extraction method	Carotenoid content	References
Dunaliella salina	Solvent extraction	α-carotene (76.01 ± 0.1 μg/g), alloxanthin (55.73 ± 0.5), antheraxanthin (344.4 ± 4.6), β-carotene (743.5 ± 44.3), diadinoxanthin (43.44 ± 1.0), lutein (3014 ± 76), neoxanthin (103.2 ± 12), violaxanthin (83.01 ± 11.7), zeaxanthin (195.8 ± 6.7)	(Safafar et al. 2015)
		Astaxanthin (0.76 ± 0.05 μg/mg), β-carotene (79.82 ± 7.02), fucoxanthin (0.24 ± 0.01), lutein (1.93 ± 0.02)	(Gui et al. 2022)
	Ultrasound-assisted extraction	Astaxanthin (40.5–131.43 mg/g), α-carotene (39.59–84.21), β-carotene (38.53–131.83), lutein (60.82–393.19), zeaxanthin (37.5–75.20)	(Achour et al. 2022)
Dunaliella tertiolecta	Solvent extraction	α-Carotene (0.04 ± 0.01 mg/g), β-carotene (0.83 ± 0.22), lutein (1.79 ± 0.21), violaxanthin (0.67 ± 0.16), zeaxanthin (0.08 ± 0.02)	(Diprat et al. 2017)
Grasiella emersonni CCNM 1001	Solvent extraction	Astaxanthin (0.03 ± 0.01 mg/g), canthaxanthin (0.06 ± 0.02), β-carotene (0.1 ± 0.03), lutein (0.89 ± 0.24), neoxanthin (0.13 ± 0.05), violaxanthin (0.18 ± 0.01), zeaxanthin (0.34 ± 0.11)	(Paliwal et al. 2016)
Haematococcus pluvialis	SFE- CO_2	Astaxanthin (18.4 mg/g), lutein (7.15)	(Molino et al. 2018)
		Astaxanthin (19.72 mg/g), lutein (4.03)	(Sanzo et al. 2018)
	Solvent extraction	Astaxanthin (5.01 mg/g)	(Ding et al. 2019)
		Astaxanthin (2.65 ± 0.22 μg/mg), β-carotene (4.12 ± 0.08), lutein (1.49 ± 0.08)	(Gui et al. 2022)
		Astaxanthin (0.29 ± 0.04 mg/g), antheraxanthin (0.26 ± 0.01), lutein (4.34 ± 0.16), neoxanthin (0.92 ± 0.04), violaxanthin (0.92 ± 0.07)	(León-Vaz et al. 2023)
I. galbana	Solvent extraction	Astaxanthin (9.54 ± 0.07 μg/mg), fucoxanthin (20.46 ± 0.12), lutein (1.15 ± 0.02)	(Gui et al. 2022)
Heterochlorella luteoviridis	Solvent extraction	α-Carotene (0.47 ± 0.03 mg/g), β-carotene (0.79 ± 0.05), lutein (1.30 ± 0.19), violaxanthin (0.79 ± 0.27), zeaxanthin (0.14 ± 0.05)	(Diprat et al. 2017)
	Moderate electric fields coupled with solvent extraction	β-carotene (185 ± 2 μg/g), lutein (856 ± 56), zeaxanthin (244 ± 20)	(Jaeschke et al. 2016)
Monoraphidium minutum	Solvent extraction	α-carotene (0.17 ± 0.03 mg/g), antheraxanthin (0.16 ± 0.06), astaxanthin (0.27 ± 0.04), β-carotene (0.53 ± 0.13), canthaxanthin (0.15 ± 0.06), lutein (2.20 ± 1.11), neoxanthin (0.59 ± 0.05), violaxanthin (0.43 ± 0.11), zeaxanthin (1.11 ± 0.65)	(Paliwal et al. 2016)
Myrmecia incisa	Solvent extraction	β-carotene (15.21 ± 0.06 μg/mg), fucoxanthin (0.73 ± 0.01), lutein (3.97 ± 0.25)	(Gui et al. 2022)
Nannochloropsis sp. G1–5	Solvent extraction	Astaxanthin (0.78 ± 0.02 mg/g), β-carotene (5.28 ± 0.21), canthaxanthin (1.93 ± 0.04), vaucheriaxanthin (0.82 ± 0.02), violaxanthin (1.81 ± 0.04), zeaxanthin (0.13 ± 0.00)	(Kim et al. 2021)
N. gaditana	SFE- CO_2	Violaxanthin (0.18 mg/g)	(Sánchez-Camargo et al. 2018)
Nannochloropsis granulate	Solvent extraction	β-carotene (0.07 ± 0.04 mg/g), canthaxanthin (0.02 ± 0.00), lutein (0.05 ± 0.02), violaxanthin (0.29 ± 0.03), zeaxanthin (0.03 ± 0.01)	(Villanova et al. 2022)
Nannochloropsis limnetica	Solvent extraction	β-carotene (284.5 ± 2 μg/g), fucoxanthin (183.2 ± 9.8), neoxanthin (423.4 ± 28), violaxanthin (1228 ± 61), zeaxanthin (136.8 ± 1.0)	(Safafar et al. 2015)
Nannochloropsis oculate	SFE- CO_2	Zeaxanthin (13.7 mg/g)	(Liau et al. 2010)
Nannochloropsis salina	Solvent extraction	α-carotene (84.16 ± 3.2 μg/g), alloxanthin (130.5 ± 2.9), β-carotene (2223 ± 88), fucoxanthin (13.05 ± 0.1), neoxanthin (53.45 ± 3.2), violaxanthin (1679 ± 83), zeaxanthin (584.9 ± 3.6)	(Safafar et al. 2015)
Neochloris oleoabundans	Absorbent-assisted SFE- CO_2	Astaxanthin (1.00 ± 0.11 mg/g), canthaxanthin (3.53 ± 0.31), lutein (1.68 ± 0.14)	(Reyes et al. 2016)
Odontella aurita	Solvent extraction	Fucoxanthin (19.6 ± 0.01 mg/g)	(Xia et al. 2013)

(continued)

Table 14.1 (continued)

Microalgae	Extraction method	Carotenoid content	References
P. tricornutom	Solvent extraction	β-carotene (348.7 ± 4.4 μg/g), diadinoxanthin (2166 ± 68), fucoxanthin (264.5 ± 29)	(Safafar et al. 2015)
		β-carotene (2.17 mg/mL), zeaxanthin (12.47)	(Ova et al. 2022)
	Pressurized liquid extraction	Diatoxanthin (2.28 mg/L), fucoxanthin (33.12), lutein (0.82), β-carotene (2.95),	(Wang et al. 2023)
Phormidium autumnale	Ultrasound-assisted solvent extraction	β-carotene (70.22 μg/ g), echinenone (35.57), lutein (21.92), zeaxanthin (26.25)	(Rodrigues et al. 2014)
Scenedesmus almeriansis	SFE- CO_2	Lutein (2.97 mg/g)	(Mehariya et al. 2019)
Scenedesmus bijuga	Solvent extraction	α-carotene (51.72 ± 0.47 μg/g), antheraxanthin (28.89 ± 0.69), β-carotene (236.26 ± 3.77), lutein (974.45 ± 5.52), neoxanthin (149.73 ± 1.17), violaxanthin (79.86 ± 0.87), zeaxanthin (13.20 ± 0.31)	(Fernandes et al. 2021)
S. dimorphus	Solvent extraction	Astaxanthin (3.6 ± 0.2 mg/g), violaxanthin (4.3 ± 0.1), zeaxanthin (4.1 ± 0.3)	(Ahmad et al. 2022)
Scenedesmus obliquus	Solvent extraction	Antheraxanthin (0.31 ± 0.05 mg/g), lutein (3.38 ± 0.11), neoxanthin (1.18 ± 0.03), violaxanthin (0.90 ± 0.06)	(León-Vaz et al. 2023)
Spirulina sp	Solvent extraction	β-carotene (15.64 ± 1.03 μg/mg), fucoxanthin (0.34 ± 0.01), lutein (1.25 ± 0.04)	(Gui et al. 2022)
Synechocystis lividus	Solvent extraction	β-carotene (0.14 ± 0.04 mg/g), canthaxanthin (0.03 ± 0.02), echinenone (0.11 ± 0.04), myxoxanthophyll (0.02 ± 0.01), zeaxanthin (0.10 ± 0.05)	(Paliwal et al. 2016)
Synechocystis pevalekii	Solvent extraction	β-carotene (0.37 ± 0.10 mg/g), canthaxanthin (0.18 ± 0.03), echinenone (0.55 ± 0.26), myxoxanthophyll (0.29 ± 0.08), zeaxanthin (0.65 ± 0.21)	(Paliwal et al. 2016)

Microalgae are photosynthetic microorganisms that have the property of creating a large amount of biomass from domestic and industrial wastewater and thus transforming it into byproducts with great commercial value. Within these byproducts are the polysaccharides to which they have been attributed antioxidant, antimicrobial, antiviral, anti-inflammatory, antibacterial, antiproliferative, antitumor, anti-adhesive, and anticoagulant properties (Moreira et al. 2022; Costa et al. 2021; Sansone et al. 2020; Ahirwar et al. 2022; Han et al. 2021; Montero-Lobato et al. 2018; Stengel and Connan 2015). It has been reported that the microalgae species *Spirulina platensis*, *S. obliquus,* and *D. salina* showed a total polysaccharide content of 7%, 12%, and 9% w/w, respectively (Singab et al. 2018). Other microalgae species, such as *Picochlorum* sp., *C. sorokiniana,* and *Chlorella* sp., presented a content of exopolysaccharides of 67, 95, and 103 mg/g dry mass, respectively. It was also reported that these species exerted DPPH and ABTS-radical scavenging activities (Mousavian et al. 2022). The data presented probe that polysaccharide content is strongly related to the microalgae species. Obtaining polysaccharides is determined by the growth conditions of the microalgae (temperature, light intensity, amount of nitrogen, and salinity in the medium), so the obtaining process can be complex. For instance, it has been stated that the red microalgae *Porphyridium cruentum* can grow in several salinity concentrations. At an optimum salinity level of 25 g/L, *P. cruentum* can produce 55 mg/L of polysaccharides per day (Decamp et al. 2023).

14.2.4 Vitamins

Microalgae are marine species with a high biological potential due to their chemical composition. Among the compounds present are vitamins, which are micronutrients important in human health since their deficiency produces diseases such as rickets, beriberi, scurvy, anemia, and osteoporosis, among others (Koyande et al. 2019). Besides, vitamins have precursor properties of enzyme cofactors and are considered antioxidants (Levasseur et al. 2020). Vitamins synthesized by microalgae include vitamin A, vitamin C, vitamin E, vitamin K, thiamine (vitamin B_1), riboflavin (vitamin B_2), niacin (vitamin B_3), pantothenic acid (vitamin B_5), Vitamin B_6, biotin (vitamin B_7), folic acid (vitamin B_9), choline (vitamin B_{11}), and vitamin B_{12} (Table 14.2). The presence of these bioactive compounds in microalgae depends on the species's growth phase and culture conditions (Galasso et al. 2019; Levasseur et al. 2020).

Furthermore, vitamin A, a micronutrient of a fat-soluble nature, exists in two main forms: as vitamin A (retinol is converted to retinal, form active of vitamin A) obtained from animals, and the other form is as provitamin A (β-carotene is converted to vitamin A in the intestine by the enzyme oxygenase), and the main source of these compounds are plants. Both forms of the vitamin are characterized by having a beta-ionone ring and an isoprenoid side chain in their structure (Ferdous and Yusof 2021; Khadim and Al Fartusie 2021; Yakıncı and Süntar 2022). This vitamin is widely recognized

Table 14.2 Vitamins content in different microalgae species

Microalgae specie	Vitamins	References
A. platensis	A, B_1, B_2, B_3, B_5, B_6, B_9, B_{11}, C, E, K	(Masuda and Chitundu 2019; Bortolini et al. 2022; Hashem et al. 2021; Abd El-Hack et al. 2019)
Arthrospira fusiformis	B_1, B_2, B_3, B_6, B_{12}, E, K	(Michael et al. 2019)
Arthrospira maxima	E	(Santiago-Morales et al. 2018)
C. vulgaris	A, C, E	(Igual et al. 2022; Tarento et al. 2018; Goiris et al. 2015)
Anabaena cylindrica	B_{12}, K	(Tarento et al. 2018; Goiris et al. 2015)
Desmodesmus asymmetricus	K	
D. salina	A, K	
N. oculata	K	
P. tricornutum	C, E, K	
D. tertiolecta	B_2, B_3, B_9, B_{12}, C, E	(Abd El Baky and El-Baroty 2013)
Chlorella stigmatiphora	C, E, B_3, B_5, B_7	
Parietochloris incise	C	(Rizwan et al. 2018)
Cylindrospermum sp.	B_{12}	
Euglena gracilis	B_7, C, E	(Priyadarshani and Rath 2012)
Prototheca moriformis	C	
Nannochloropsis oceanica	D	(Ljubic et al. 2020)
Cytoseira fucales	C	(Hashem et al. 2021)
Turbinaria decurrens	C	
Tetraselmis suecica	B_1, B_3, C, E	(Islam et al. 2017; Goiris et al. 2015)
Prorocentrum cordatum	E	(Häubner et al. 2014)

for its beneficial effect on vision. Also, it is important for the growth and development of bone and teeth, as in cell division, reproduction, and carrying on the immune system (Khadim and Al Fartusie 2021). In microalgae, β-carotene, α-carotene, fucoxanthin, violaxanthin, and neoxanthin have activity as the provitamin A; the species *Dunaliella*, *Spirulina*, *Chlorella,* and *Haematococcus* contain these metabolites, therefore could be source of provitamin A (Anusree et al. 2023).

Vitamin C is a micronutrient of great importance and is considered essential for humans since it is impossible to synthesize it; also, its deficiency mainly causes scurvy disease. It has a high antioxidant potential due to its ability to donate electrons and contribute to immune defense; likewise, it is involved in collagen synthesis (Carr and Maggini 2017; Grosso et al. 2013). Moreover, vitamin C content was evaluated in eight species of microalgae and macroalgae (*Dictyota flabellata*, *Hypnea spinella*, *Gracilaria textorii*, *Gracilaria vermicullophyla*, *Caulerpa sertularioides*, *Codium simulans*, *Codium amplivesiculatum*, and *Ulva lactuca*), finding a range of 1.24–6.08 mg/g, and *C. amplivesiculatum* showed the highest value (Tenorio-Rodríguez et al. 2013). Besides, Del Mondo et al. (2020) reported a range of vitamin C between 0.06 and 18.79 mg/g in different microalgae species.

On the other hand, vitamin K is a nutrient produced by microalgae from different species. This vitamin has been related to some biochemical processes such as blood coagulation, modulation of inflammation, neuroprotection, atherosclerosis, and bone health (Tarento et al. 2018). Vitamin K is reduced in the cell to its metabolic form, phylloquinone (vitamin K_1), by VKOR enzyme (vitamin K epoxide reductase). This vitamin is a cofactor in forming and activating blood coagulation proteins (Halder et al. 2019). It also acts as an enzyme cofactor for a vitamin K-dependent (VKD) carboxylase, which converts glutamate residues into γ-carboxyglutamic acid (Gla) residues and presents an important function during the blood-clotting process (Azuma et al. 2014; Vermeer 2012). According to Tarento et al. (2018), microalgae species *A. cylindrica* contains a higher concentration of vitamin K (200 μg/g) than several vegetable sources. The microalgae *Sargassum miticum* and *A. cylindrica* have been reported to present a high vitamin K content with values of 750 and 200.25 μg/g, respectively (Simes et al. 2020).

On the other hand, vitamin E (tocopherol) is a fat-soluble molecule that includes eight isoforms (α-, β-, γ-, and δ-tocopherol, and α-, β-, γ-, and δ-tocotrienol). The deficiency of vitamin E causes neurological disorders. Vitamin E is an important antioxidant and protects against lipid peroxyl radicals damage to plasma and red blood cells. Two mechanisms achieve this action: (i) formation of tocopherol quinone and (ii) charge transfer (Tenorio-Rodríguez et al. 2013; Coulombier et al. 2021; Chaturvedi et al. 2021). Santiago-

Morales et al. (2018) reported an amount of 2.5 mg α-tocopherol/g biomass of *A. maxima* (*Spirulina maxima*), while *Nanochloropsis* sp. and *Tetraselmis suecica* presented the same concentration of 1.7 mg α-tocopherol/g biomass. Also, *Chaetoceros muelleri* showed the lowest content (0.89 mg α-tocopherol/g biomass). The isoform of vitamin E γ-tocopherol was highest in *T. suecica* (17.0 mg/g biomass), while *C. muelleri*, *A. maxima*, and *Nannochloropsis* sp. showed 10.5, 1.4, and 0.4 mg γ-tocopherol/g biomass, respectively. The results of antioxidant capacity expressed as μmol Trolox equivalents/g showed a high correlation with tocopherol content, especially in *T. suecica* and *C. muelleri*.

Moreover, the vitamin E from microalgae species, *Nodularia spumigena*, *P. tricornutum*, *Skeletonema costatum*, *D. tertiolecta*, *P. cordatum*, and *Rhodomonas salina* were studied by Häubner et al. (2014), under abiotic factors, finding concentrations of approximately 0.01 mg/g up to 0.8 mg/g depending on the species.

Another lipid-soluble molecule is vitamin D; its deficiency could cause osteoporosis, osteomalacia, and rickets. This vitamin is considered an antioxidant and anti-inflammatory agent because it can inhibit or reduce reactive oxygen species, induce the synthesis of pro-inflammatory cytokines, and increase IkappaBalpha mRNA stability (Chaturvedi et al. 2021). Vitamin D is present in sea edible organisms such as microalgae; its production could be due to the UV radiation action, and it is found as vitamin D_3 (cholecalciferol) in microalgae. Some microalgae species showed an increase in vitamin D_3 after their exposure to UV radiation, such as *N. oceanica* (285 ng/g), *D. salina* (145 ± 11 ng/g), and *N. limnetica* (2700 ± 198 ng/g) (Del Mondo et al. 2020; Ljubic et al. 2021).

Vitamins B have several hydroxyl groups in their molecular structure, which allows them to be water-soluble. Some B-group vitamins, such as B_1, B_7, and B_{12}, play an important role in marine ecology. Furthermore, in human health, these are also very important in preventing and controlling diseases such as beriberi, anemia, and pellagra, among others (Sañudo-Wilhelmy et al. 2014).

Graziani et al. (2013) reported values of vitamin B_2 and vitamin B_3 of 30 ± 2 and 32 ± 3 mg/kg, respectively, in microalgae *Galdieria sulphuraria* grown under heterotrophic conditions. Besides, in *A. fusiformis* cultivated using the NPK 10-20-20 fertilizer, the content of vitamins B_1, B_3, B_6, and B_{12} was 41.62, 49.57, 183.05, 3.32, and 4.73 mg/100 g, respectively; while cultivated with a Zarroulk medium, the vitamin concentrations were lowest, except in B_1 (44.59 mg/100 g) and B_6 (9.21 mg/100 g) (Michael et al. 2019). In the same microalga species, Masuda and Chitundu (2019) found the vitamins B_1, B_2, B_3, B_6, and B_9 at concentrations of 0.048, 0.039, 0.39, 0.009, and 0.73 mg/g, respectively. Likewise, vitamin B_9 was highest in *Picochlorum* sp. (6470 μg/100 g), followed by *C. vulgaris* (3460 μg/100 g) (Woortman et al. 2020).

14.2.5 Phenolic Compounds

Phenolic compounds are secondary metabolites ubiquitously present in the plant kingdom. These compounds are characterized by a phenol group attached to a benzene ring. Phenolic compounds are classified as weak acids due to the lability of the phenolic hydroxyl, which significantly influences their stability in extraction procedures, storage, and once consumed. Phenolic compounds can be classified depending on the number and distribution of their OH- radicals. Most authors agree that phenolics can be classified as simple phenolics, phenolic acids, cinnamic acids, coumarins, chalcones, flavans, flavones, flavanones, flavanonols, anthocyanidins, and anthocyanins (Vermerris and Nicholson 2008). Phenolic compounds have attracted the attention of researchers worldwide since these molecules are associated with preventing non-communicable diseases like cardiovascular illnesses, cancer, diabetes, and Parkinson's disease, among many others.

Although microalgae are a rich source of bioactive compounds like carotenoids, PUFAs, and polysaccharides, individual phenolic compounds (Table 14.3) are not among the most abundant; however, the mixtures of these compounds obtained from the different microalgae species can be used as ingredients in the development of dietary supplements and food development (Kiuru et al. 2014; Ghosh et al. 2022).

14.3 Nutraceutical Potential of Microalgae Antioxidants

There are several antioxidant and bioactive substances found in microalgae species. These molecules have probed the characteristic of exerting activity against numerous diseases, such as metabolic disorders, chronic inflammation, cancer, and viral infections, among others. Next, the nutraceutical potential of antioxidant and bioactive compounds found in microalgae is mentioned.

14.3.1 Nutraceutical Potential of PUFAs

The polyunsaturated fatty acids of microalgae have been studied for their potential as nutraceuticals in different pathologies. For example, Go et al. (2016) evaluated the anti-obesity effect of a mixture of PUFAs, including EPA and DHA, extracted from the marine microalgae *Thraustochytriidae* sp. in mice with obesity induced by a high-fat diet. They reported a weight reduction of 10% at 9 weeks and an inhibitory effect on the accumulation of lipids in the liver. Also, Ran et al. (2022) studied the oil effect of microalgae *Schizochytrium* sp. (rich in DHA) in obese mice, discovering that the microalgal oil inhibited the expression of genes related to lipogenesis and induced the expres-

Table 14.3 Common phenolic compounds obtained from microalgae species

Microalgae species	Extraction	Phenolic compounds	References
Pediastrum boryanum	Methanol 40%	Gallic acid, protocatechuic acid, chlorogenic acid, hydroxybenzoic acid, vanillic acid	(Corrêa da Silva et al. 2022)
C. reinhardtii	Methanol 100%	Chlorogenic acid, gallic acid, coumaric acid	(Faraloni et al. 2021)
G. sulphuraria	Acetone	4 aminobenzoic acid, salicylic acid, quercetin, ellagic acid, catechin	(Zimermann et al. 2020)
Tetraselmis sp.	Ethanol 55.4% – ultrasound-assisted extraction	Epigallocatechin gallate	(Gam et al. 2020)
Chlorella sp. PR-1	Methanol 99.9%	2,4-bis (1,1-dimethylethyl)-phenol	(Gupta et al. 2021)
Chlorella sp., *Spirulina* sp., *Dunaliella* sp., and *Isochrysis* sp.	Ethanol 70%	Gallic acid, protocatechuic acid, caftaric acid, p-hydroxybenzoic acid, catechin, chlorogenic acid, caffeic acid, diosmin, kaempferol	(Li et al. 2022)
Chloromonas cf. *reticulata*	Methanol	Gallic acid, catechin, vanillic acid, epicatechin, syringic acid	(Santiago-Díaz et al. 2022)
Pseudopediastrum boryanum	Methanol	Gallic acid, protocatechuic acid, catechin, vanillic acid, epicatechin, syringic acid, rutin	
Chloroidium saccharophilum	Methanol	Gallic acid, protocatechuic acid, catechin, vanillic acid, epicatechin, syringic acid, rutin, gentisic acid, coumaric acid, ferulic acid	
Chlorella sp.	Subcritical water extraction	p-coumaric acid, ferulic acid, caffeic acid	(Zakaria et al. 2020)
Spirulina sp.	Pulsed electric fields (PEF)	Cirsilineol, quercetin-3-O- galactoside, ellagic acid, p-anisaldehyde, kaempferol	(Zhou et al. 2022)
	Pressurized liquid extraction (PLE)	Ellagic acid, p-anisaldehyde, quercetin-3-O-glucuronide	
	Combined PEF + PLE	Cirsilineol, quercetin-3-O-galactoside, quercetin-3-O-glucuronide, ellagic acid, p-anisaldehyde, cinnamic acid, kaempferol	

sion of genes related to lipolysis; it also had modulatory effects on the intestinal microbiota of mice. Moreover, Yu et al. (2017) also evaluated the oil of *Schizochytrium* sp., which prevents abdominal fat accumulation. Furthermore, Mayer et al. (2022) assessed the effects of *Diacronema lutheri* rich in omega-3 LC-PUFA against metabolic disorders in Wistar rats subjected to a high-fat and fructose diet, showing that *D. lutheri* decreases the levels of plasmatic triacylglycerols (−75%), cholesterol (−84%), and leptin (−68%), while the anti-inflammatory cytokines IL-4 and IL-10 increased (+74% and +58%, respectively) in rats with cardiac insufficiency.

Furthermore, Neumann et al. (2018) studied the bioavailability and safety consumption for mice of *C. vulgaris*, *N. oceanica*, and *P. tricornutum* microalgae, finding high levels of EPA in the liver after feeding them with *N. oceanica*; in addition, no species of microalgae presented cardiac, renal, or hepatic toxicity induced by any of the microalgae diets. On the other hand, Wan et al. (2019) evaluated in mice the anti-diabetic activity of *Chlorella pyrenoidosa* and *S. platensis* extracts (rich in PUFAs), observing an improvement in the tolerance to glucose and alteration in the intestinal microbiota composition. Additionally, Qin et al. (2023) evaluated the effect of DHA in mice with type I diabetes induced by streptozotocin, demonstrating that the intake of DHA significantly improved the levels of serum lipids and cardiac interstitial fibrosis. In another study, Balakrishnan et al. (2019) proved that the supplementation with *Isochrysis* sp. biomass (an alternative source of PUFAs) improves DHA status in the brain and liver. It also increases serum cholesterol high-density lipoprotein (HDL) levels and decreases oxidative stress markers with a reduction in creatinine and urea levels in a murine model. Also, Morgese et al. (2016) reported an improvement in short-term memory in rats treated with *C. sorokiniana* extract, attributing this function to omega-3 and -6 PUFAs content.

On the other hand, studies with humans have investigated the effect of PUFAs on coronary diseases. Sawada et al. (2016) evaluated the treatment with EPA for 6 h in patients with coronary artery disease with impaired glucose metabolism, reporting that EPA corrected the hyperglycemia, postprandial hypertriglyceridemia, and insulin secretion capacity as well as improved recovery from concomitant endothelial dysfunction. Besides, Chang et al. (2023), reported a reduction of somatic symptoms and fatigue in patients with major depressive disorder due to cardiovascular diseases when ingesting omega-3 PUFAs for 12 weeks. Hammer et al. (2019) conducted a similar study in which they informed that supplementation with omega-3 PUFAs in patients with peripheral arterial disease improved flow-mediated vasodilation. Moreover, the PUFAs' effect on the appearance of cerebral vasospasm was also reported (Nakagawa et al. 2017).

In general, PUFAs have shown potential against various diseases through clinical and animal studies.

14.3.2 Nutraceutical Potential of Carotenoids

Carotenoids are natural compounds found in microalgae with potential health benefits against different non-concomitance diseases such as cancer, cardiovascular diseases, and diabetes, among others (Ambati et al. 2019; Kim et al. 2021; Martínez et al. 2022). These potential activities of carotenoids have been linked to the antioxidant activity, regulation of gene transcription, provitamin A activity, and induction of defense enzymes (Ambati et al. 2019; Fernandes et al. 2021; Huy et al. 2022; León-Vaz et al. 2023). In this sense, Fernandes et al. (2021) reported that microalgae biomass from *Scenedesmus bijuga* and *C. sorokiniana* have a high bioaccessibility of carotenoids in an *in vitro* model, which could affect health. Likewise, Xia et al. (2013) found that fucoxanthin from *O. aurita* microalgae exhibited antioxidant properties on 1,1-dihpenyl-2-picrylhydrazyl (DPPH) radical and 2,2′-Azino-bis (3-ethylbenzthiazoline-6-sulfonic acid (ABTS) radicals. Besides, Safafar et al. (2015) reported that microalgae (*Phaeodactylum* sp., *Nannochloropsis* sp., *Chlorella* sp., *Dunaliella* sp., and *Desmodesmus* sp.) carotenoids have antioxidant capacity against (DPPH) radical scavenging activity, ferrous reduction power (FRAP), and ABTS-radical scavenging capacity activity *in vitro* tests. Recently, Wang et al. (2023) found that carotenoids (including β-carotene, α-carotene, lutein, and zeaxanthin) extracted from *Chlorella* microalgae strain have been linked with the high antioxidant capacity studied on ORAC and ABTS tests. On the other hand, microalgae have been used as ingredients of functional foods, fish feed oil supplements, and fish pigmentation due to their high content of lipophilic compounds such as lipids and carotenoids (Fernandes et al. 2021; Hulatt et al. 2017; Huy et al. 2022). Also, microalgae have been used for fish pigmentation and nutraceutical enrichment (Huy et al. 2022).

Microalgae such as *Chlamydomonas pulsatilla*, *Chlamydomonas klinobasis*, *Chlamydomonas platystigma*, *Chlamydomonas malina*, *Tetraselmis chui*, *Koliella antarctica*, and *N. gaditana* have been studied for potential application on salmon feed (Hulatt et al. 2017). Likewise, microalgae are a promised source of carotenoids with potential application in cosmetic, nutraceutical, and pharmaceutic industries because it has been reported that astaxanthin, β-carotene, zeaxanthin, lutein, canthaxanthin, neoxanthin, and violaxanthin reduced oxidative stress (*in vitro* model), which has related to the prevention of aging, protection of skin damage by UV-light, and anti-inflammatory process, among other activities; therefore, these carotenoids have been widely studied as promised sources for industry (Borowitzka 2013; Huy et al. 2022; Koller et al. 2014). In this sense, Gheysen et al. (2019) reported that a higher carotenoid content by *Nannochloropsis* biomass reduced the lipid oxidation levels in a food model. Also, Kim et al. (2021) found that ethanolic extract from *Nannochloropsis* showed potential skin protection due to its antioxidant, anti-inflammatory, and anti-melanogenic activities, UV protection, aging prevention, and skin humectation. Besides, Martínez et al. (2022) reported that a carotenoid (crocoxanthin) extracted from *D. tertiolecta* showed antiproliferative activity against A2058 and HCC827 cells. Similarly, *N. granulata* microalgae extracts with lipophilic compounds such as β-carotene, canthaxanthin, lutein, violaxanthin, zeaxanthin, and polyunsaturated fatty acids showed antiproliferative activity against human prostate cell line PC3 (Villanova et al. 2022). Therefore, a positive correlation between carotenoids extracted from microalgae and the potential human effect can be stated. In addition, some microalgae strains, such as *Chlorococcum* sp. and *Scenedesmus* sp., under cold stress conditions could enhance the bioactive compounds production, which could be used for carotenoid production with potential effects on humans (León-Vaz et al. 2023).

14.3.3 Nutraceutical Potential of Microalgal Polysaccharides

Different microalgae species are used to obtain sulfated polysaccharides, and their bioactivity may vary. Singab et al. (2018) determined the antioxidant, antiviral, cytotoxic, and anti-cholinesterase activity of polysaccharides extracted from three different species of microalgae *D. salina*, *S. platensis*, and *S. obliquus*, finding that *S. obliquus*, presented a more significant number of polysaccharides compared to the other two microalgae. Nevertheless, the polysaccharides present in *S. platensis* showed promising antiviral activity by reducing the replication between 50 and 88% of Hepatitis C virus, Coxsackievirus, Rotavirus, and Herpes simplex at non-toxic concentrations. Regarding the polysaccharides extracted from *D. salina*, they presented high cytotoxicity in HTC 116 colon cancer cells. The polysaccharides of *D. salina and S. obliquus* demonstrated a high antioxidant capacity against the DPPH radical at 100 μg/mL. Finally, the polysaccharides extracted from *S. obliquus* showed an increased inhibition of anti-cholinesterase activity. This assures us that depending on the species, the polysaccharides

Table 14.4 Species of microalgae and cyanobacteria, their nutraceutical effect, extraction method, and yield

Microalgae or cyanobacteria	Nutraceutical effect	Yield (%)/extraction method	Results	References
C. sorokiniana *Chlorella* sp. *Picochorum* sp.	Antioxidant, anticoagulant	14–19/using heated acid	Antioxidant capacity against DPPH and ABTS radicals and anticoagulant activity	(Mousavian et al. 2022)
Navicula sp.	Antioxidant	3.4–4.4/gravity sedimentation method	Antioxidant capacity against DPPH and ABTS radicals	(Fimbres-Olivarria et al. 2018)
Oscillatoria simplicissima and *acutissima*	Antioxidant, cytotoxic	Not reported	Antioxidant capacity against DPPH radical. *O. simplicissima* induced around 50% of viability in lung cancer cells (A-549) at 100 μg/mL	(Elkomy and Ismail 2021)
Chaetoceros muelleri	Antioxidant, cytotoxic, hypoglicemiant (male Winstar rats)	2.2/gravity sedimentation method	Antioxidant capacity against DPPH radical, reduction in the glycemic index (50% lower than the control)	(Miranda-Arizmendi et al. 2022)
I. galbana, Nannochloropsis sp., *Tetraselmis* sp., *and P. tricornutum*	Antioxidant, antiinflamatory	0.40–21/precipitation method	Antioxidant capacity against ABTS radical	(Pagels et al. 2022)
S. platensis	Antioxidant	13.5/ultrasound-assisted extraction and ethanol/hot water extraction	Antioxidant capacity against DPPH and OH radicals	(Kurd and Samavati 2015)
Chlorophyte Graesiella sp.	Antioxidant, antiproliferative	65/centrifugation method	Antioxidant capacity against DPPH, OH^- and H_2O_2, antiproliferative activity against HepG2 and Caco-2	(Gongi et al. 2021; Trabelsi et al. 2016)
Chlorella sp.	Antioxidant	3.57–7.38/hot water extraction method	Antioxidant capacity against DPPH, OH^- and H_2O_2	(Song et al. 2018)
Tetraselmis sp.	Antioxidant, antifungal (*Candida albicans* and *Penicillium italicum*)	~35/basic extraction	Antioxidant capacity against DPPH and ABTS radicals, tyrosinase inhibitory activity was 86%, inhibited 80% of both fungus	(Amna Kashif et al. 2018)
C. pyrenoidosa	Antioxidant (*Caenorhabditis elegans)*	Not reported	Improved antioxidant enzyme activities, reduced the accumulation of ROS and MDA	(Wan et al. 2021)
C. zofingiensis and C. vulgaris	Antioxidant, antitumor	10–18/centrifugation method	Antioxidant capacity against DPPH and OH^- radicals, reduction on the viability of HTC8	(Zhang et al. 2019)
Nostoc sphaeroides	Antioxidant, antitumor (*Caenorhabditis elegans)*	~30/ethanol precipitation	Reduction of ROS levels, inhibition of lipid peroxidation, increase in the activity of superoxide dismutase and catalase, reduction of the proliferation and induction of apoptosis of lung adenocarcinoma A549 cells via caspase-3 activation	(Li et al. 2018)

synthesized by microalgae will present different nutraceutical potential (Table 14.4).

Furthermore, in most of the *in vitro* studies on healthy cell lines, it was observed that cell viability was greater than 90%, in some exceeding 100%, which tells us that the antioxidant capacity of the polysaccharides extracted from microalgae can help reduce the damage of free radicals present in cells, which has an impact on the reduction of diseases caused by oxidative stress, such as heart disease, cancer, among many others (Miranda-Arizmendi et al. 2022). Some of the published works handle different mechanisms of the action exerted by these polysaccharides, among which the regulation of factors or enzymes regulating oxidative stress, such as superoxide dismutase and catalase (Zhong et al. 2019), the overexpression of genes responsible for the activation of different pathways such as AMPK, TOR, and JNK, which is given by the overexpression of Daf-16 Mrna (Wan et al. 2021), regarding anti-inflammatory mechanisms, a decrease in NF-κB and TNF-α was found, which induces oxidative stress (Levy-Ontman et al. 2017). Therefore, these compounds have potential activity against different diseases related to oxidative stress, so they are a good prospect for use as a nutraceutical.

14.3.4 Nutraceutical Potential of Vitamins from Microalgae

Although various studies report a high content of vitamins in microalgae, and these molecules have been related to important antioxidant potential, there is scarce information aimed at using microalgae as a source of these molecules for therapeutic and nutraceutical purposes.

Vitamin C and E from microalgae can act as scavenging agents since they react with carcinogens and reduce their harmful potential (Abd El Baky and El-Baroty 2013). In this sense, microalgae *Spirulina* sp. presents a metal-protective role, attributed mainly to its contents of vitamin E and vitamin, among other compounds (Hoseini et al. 2013). Likewise, *C. pyrenoidosa* is a source of folic acid, vitamin B_{12}, and iron, important compounds in controlling anemia and hypertensive disorders common in human pregnancy. Therefore, Nakano et al. (2010) performed a study with pregnant women who were administrated daily for 6 weeks from the 12th week of gestation. The results showed a reduction in the number of anemic subjects; additionally, proteinuria and edema (signs of hypertension) were lower in pregnant women with supplementation with *C. pyrenoidosa*. The possible mechanism might be due to the protection vitamins exert on endothelial cells against oxidative stress. Besides, Armaini et al. (2018) conducted a study to reduce anemia with *S. dimorphus*. The experiment was performed using mice with aplastic anemia (induced with chloramphenicol); the *S. dimorphus* doses administrated were 2.5, 5.0, and 7.5 g/50 mL for 28 days. The results showed increased erythrocytes and hemoglobin levels; likewise, a weight gain in mice at 5 g/50 mL doses was reported. These effects were attributed to the content of folic acid, vitamin B_{12}, and iron in the microalga, concluding that *S. dimorphus* biomass has a positive effect against anemia.

14.3.5 Nutraceutical Potential of Microalgal Phenolic Compounds

Regular intake of foods rich in phenolic compounds has been associated with a lower incidence of the onset of non-communicable diseases. This has attracted worldwide attention due to the burden of non-communicable diseases (NCDs), as they are the leading cause of death (World Health Organization 2023). Numerous reports exist on the potential use of phenolic compounds to improve overall health.

Phenolic compounds have been associated with anti-inflammatory, anticancer, anti-diabetic, and antimicrobial. Thus, the valorization strategies evaluated by researchers and technologists point to using microalgae as a source of nutraceuticals and ingredients for functional foods (Cichoński and Chrzanowski 2022).

The antioxidant, anti-inflammatory, anticancer, and anti-diabetic potential of phenolic compounds has been widely reported. However, in the particular case of those obtained from microalgae, research has focused on optimizing the extraction methods of bioactive compounds to enhance the yield and promote further studies using microalgae, using both fresh material and also as part of the valorization process of their biomass (Almendinger et al. 2021; Bulut et al. 2019; Corrêa da Silva et al. 2022; Gam et al. 2020; León-Vaz et al. 2023; Zakaria et al. 2020). Nonetheless, as far as our literature research has been concerned, we have found no *in vivo* reports. Literature is limited to *in vitro* assays using oxidant or free radical generator probe molecules.

Finally, even though there are reports on the many health benefits of phenolic compounds, it has been reported that phenolics can exert synergistic effects with each other; thus, the mixture of microalgae extracts could have different attributes or mechanisms of action. Further research is needed.

14.4 Conclusions

Microalgae are nowadays considered biofactories for the obtention of antioxidant and bioactive compounds used in developing food, pharmaceutical, and nutraceutical products. Several *in vitro* studies assessing the antioxidant capacity and nutraceutical effects of microalgae components are available; nevertheless, cellular and *in vivo* studies are still scarce. Because of this, exploring the effects of microalgae antioxidants on human health is very promising.

Several microalgae species are proposed for further examination as sources of nutraceutical and antioxidant molecules. For instance, *Scenedesmus dimorphus* presents important content of PUFAs, carotenoids, and vitamins. The species *Haematococcus pluvialis* might be a good source of carotenoids and vitamins, mainly, while *Porphyridium cruentum* shows potential to be a source of polysaccharides. These might be excellent candidates for the development of nutraceutical products.

References

Abd El-Hack ME, Abdelnour S, Alagawany M, Abdo M, Sakr MA, Khafaga AF, Mahgoub SA, Elnesr SS, Gebriel MG (2019) Microalgae in modern cancer therapy: current knowledge. Biomed Pharmacother 111:42–50. https://doi.org/10.1016/j.biopha.2018.12.069

Abd El Baky HH, El-Baroty G (2013) Healthy benefit of microalgal bioactive substances. J Aquat Sci 1(1):11–23. https://doi.org/10.12691/jas-1-1-3

Achour HY, Llamero CB, Saadi SA, Zitouni A, Senorans FJ (2022) Ultrasound assisted extraction of carotenoids from green and orange biomasses of *Dunaliella Salina* DUNADZ1. An Univ Dunărea de Jos Galaţi Fasc VI Food Technol 46(2):22–34. https://doi.org/10.35219/foodtechnology.2022.2.02

Ahirwar A, Kesharwani K, Deka R, Muthukumar S, Khan MJ, Rai A, Vinayak V, Varjani S, Joshi KB, Morjaria S (2022) Microalgal drugs: a promising therapeutic reserve for the future. J Biotechnol 349:32–46. https://doi.org/10.1016/j.jbiotec.2022.03.012

Ahmad N, Mounsef JR, Lteif R (2022) Pigment production by *Scenedesmus dimorphus* using different low-cost and alternative culture media. J Chem Technol Biotechnol 97(1):287–294. https://doi.org/10.1002/jctb.6940

Almendinger M, Saalfrank F, Rohn S, Kurth E, Springer M, Pleissner D (2021) Characterization of selected microalgae and cyanobacteria as sources of compounds with antioxidant capacity. Algal Res 53:8. https://doi.org/10.1016/j.algal.2020.102168

Ambati RR, Gogisetty D, Aswathanarayana RG, Ravi S, Bikkina PN, Bo L, Yuepeng S (2019) Industrial potential of carotenoid pigments from microalgae: current trends and future prospects. Crit Rev Food Sci Nutr 59(12):1880–1902. https://doi.org/10.1080/10408398.2018.1432561

Amna Kashif S, Hwang YJ, Park JK (2018) Potent biomedical applications of isolated polysaccharides from marine microalgae Tetraselmis species. Bioprocess Biosyst Eng 41(11):1611–1620. https://doi.org/10.1007/s00449-018-1987-z

Anusree MK, Manasa Leela K, Sreehari M, Raj S, Sreenikethanam A, Bajhaiya AK (2023) Chapter 14 – Marine microalgae: an emerging source of pharmaceuticals and bioactive compounds. In: Meena SN, Nandre V, Kodam K, Meena RS (eds) New horizons in natural compound research. Academic Press, pp 251–265. https://doi.org/10.1016/B978-0-443-15232-0.00025-4

Armaini A, Salim M, Pribadi P (2018) Induction effect of microalgae *Scenedesmus dimorphus* against hematology on mice (*Mus musculus*) suffering anemia diseases. Asian J Pharm Clin Res 11(7):348–352. https://doi.org/10.22159/ajpcr.2018.v11i7.24825

Azuma K, Ouchi Y, Inoue S (2014) Vitamin K: novel molecular mechanisms of action and its roles in osteoporosis. Geriatr Gerontol Int 14(1):1–7. https://doi.org/10.1111/ggi.12060

Balakrishnan J, Dhavamani S, Sadasivam SG, Arumugam M, Vellaikumar S, Ramalingam J, Shanmugam K (2019) Omega-3-rich *Isochrysis* sp. biomass enhances brain docosahexaenoic acid levels and improves serum lipid profile and antioxidant status in Wistar rats. J Sci Food Agric 99(13):6066–6075. https://doi.org/10.1002/jsfa.9884

Barta DG, Coman V, Vodnar DC (2021) Microalgae as sources of omega-3 polyunsaturated fatty acids: biotechnological aspects. Algal Res 58:102410. https://doi.org/10.1016/j.algal.2021.102410

Bazarnova J, Smyatskaya Y, Shlykova A, Balabaev A, Đurović S (2022) Obtaining fat-soluble pigments-carotenoids from the biomass of *Chlorella* microalgae. Appl Sci 12(7). https://doi.org/10.3390/app12073246

Borowitzka MA (2013) High-value products from microalgae—their development and commercialisation. J Appl Phycol 25(3):743–756. https://doi.org/10.1007/s10811-013-9983-9

Bortolini DG, Maciel GM, Fernandes IAA, Pedro AC, Rubio FTV, Branco IG, Haminiuk CWI (2022) Functional properties of bioactive compounds from *Spirulina* spp.: current status and future trends. Food Chem Mol Sci 5:100134. https://doi.org/10.1016/j.fochms.2022.100134

Bulut O, Akin D, Sonmez C, Oktem A, Yucel M, Oktem HA (2019) Phenolic compounds, carotenoids, and antioxidant capacities of a thermo-tolerant *Scenedesmus* sp. (Chlorophyta) extracted with different solvents. J Appl Phycol 31(3):1675–1683. https://doi.org/10.1007/s10811-018-1726-5

Canelli G, Tevere S, Jaquenod L, Dionisi F, Rohfritsch Z, Bolten CJ, Neutsch L, Mathys A (2022) A novel strategy to simultaneously enhance bioaccessible lipids and antioxidants in hetero/mixotrophic *Chlorella vulgaris* as functional ingredient. Bioresour Technol 347:126744. https://doi.org/10.1016/j.biortech.2022.126744

Carr AC, Maggini S (2017) Vitamin C and immune function. Nutrients 9(11):1211. https://doi.org/10.3390/nu9111211

Chang JPC, Chang SS, Chen HT, Chien YC, Yang HT, Huang SY, Tseng PT, Chang CH, Galecki P, Su KP (2023) Omega-3 polyunsaturated fatty acids (n-3 PUFAs), somatic and fatigue symptoms in cardiovascular diseases comorbid major depressive disorder (MDD): a randomized controlled trial. Brain Behav Immun 112:125–131. https://doi.org/10.1016/j.bbi.2023.06.008

Chaturvedi S, Chaudhary R, Tiwari S (2021) Contribution of crop biofortification in mitigating vitamin deficiency globally. In: Kumar Upadhyay S (ed) Genome engineering for crop improvement, pp 112–130. https://doi.org/10.1002/9781119672425.ch7

Chen O, Mah E (2023) Nutraceuticals: health effects and clinical applications. In: Caballero B (ed) Encyclopedia of human nutrition, 4th edn. Academic Press, Oxford, pp 501–512. https://doi.org/10.1016/B978-0-12-821848-8.00043-3

Chisti Y (2007) Biodiesel from microalgae. Biotechnol Adv 25(3):294–306. https://doi.org/10.1016/j.biotechadv.2007.02.001

Cichoński J, Chrzanowski G (2022) Microalgae as a source of valuable phenolic compounds and carotenoids. Molecules 27(24). https://doi.org/10.3390/molecules27248852

Corrêa da Silva MG, Pires Ferreira S, Dora CL, Hort MA, Giroldo D, Prates DF, Radmann EM, Bemvenuti RH, Costa JAV, Badiale-Furlong E, Muccillo-Baisch AL (2022) Phenolic compounds and antioxidant capacity of *Pediastrum boryanum* (Chlorococcales) biomass. Int J Environ Health Res 32(1):168–180. https://doi.org/10.1080/09603123.2020.1744113

Costa JA, Lucas BF, Alvarenga AG, Moreira JB, de Morais MG (2021) Microalgae polysaccharides: an overview of production, characterization, and potential applications. Polysaccharides 2(4):759–772. https://doi.org/10.3390/polysaccharides2040046

Coulombier N, Jauffrais T, Lebouvier N (2021) Antioxidant compounds from microalgae: a review. Mar Drugs 19(10):549. https://doi.org/10.3390/md19100549

Decamp A, Martineau E, Grizeau D, Pruvost J, Gonçalves O (2023) Effects of the salinity on the biosynthesis of the polysaccharides of the marine microalgae *Porphyridium cruentum*. Algal Res 71:103089. https://doi.org/10.1016/j.algal.2023.103089

Del Mondo A, Smerilli A, Sané E, Sansone C, Brunet C (2020) Challenging microalgal vitamins for human health. Microb Cell Factories 19(1):201. https://doi.org/10.1186/s12934-020-01459-1

Ding W, Cui J, Zhao Y, Han B, Li T, Zhao P, Xu J-W, Yu X (2019) Enhancing *Haematococcus pluvialis* biomass and γ-aminobutyric acid accumulation by two-step cultivation and salt supplementation. Bioresour Technol 285:121334. https://doi.org/10.1016/j.biortech.2019.121334

Diprat AB, Menegol T, Boelter JF, Zmozinski A, Rodrigues Vale MG, Rodrigues E, Rech R (2017) Chemical composition of microalgae *Heterochlorella luteoviridis* and *Dunaliella tertiolecta* with emphasis on carotenoids. J Sci Food Agric 97(10):3463–3468. https://doi.org/10.1002/jsfa.8159

Elkomy RG, Ismail MM (2021) Crude sulfated polysaccharides extracted from marine cyanobacterium Oscillatoria simplicissima with evaluation antioxidant and cytotoxic activities. Iran J Microbiol 13(2008–3289 (Print)):7. https://doi.org/10.18502/ijm.v13i4.6983

Faraloni C, Di Lorenzo T, Bonetti A (2021) Impact of light stress on the synthesis of both antioxidants polyphenols and carotenoids, as fast photoprotective response in *Chlamydomonas reinhardtii*: new prospective for biotechnological potential of this microalga. Symmetry 13(11). https://doi.org/10.3390/sym13112220

Ferdous UT, Yusof ZNB (2021) Medicinal prospects of antioxidants from algal sources in cancer therapy. Front Pharmacol 12. https://doi.org/10.3389/fphar.2021.593116

Fernandes AS, Nascimento TC, Pinheiro PN, Vendruscolo RG, Wagner R, de Rosso VV, Jacob-Lopes E, Zepka LQ (2021) Bioaccessibility of microalgae-based carotenoids and their association with the lipid matrix. Food Res Int 148:110596. https://doi.org/10.1016/j.foodres.2021.110596

Fimbres-Olivarria D, Carvajal-Millan E, Lopez-Elias JA, Martinez-Robinson KG, Miranda-Baeza A, Martinez-Cordova LR, Enriquez-Ocaña F, Valdez-Holguin JE (2018) Chemical characterization and antioxidant activity of sulfated polysaccharides from *Navicula* sp. Food Hydrocoll 75:229–236. https://doi.org/10.1016/j.foodhyd.2017.08.002

Gaignard C, Laroche C, Pierre G, Dubessay P, Delattre C, Gardarin C, Gourvil P, Probert I, Dubuffet A, Michaud P (2019) Screening of marine microalgae: investigation of new exopolysaccharide producers. Algal Res 44:101711. https://doi.org/10.1016/j.algal.2019.101711

Galasso C, Gentile A, Orefice I, Ianora A, Bruno A, Noonan DM, Sansone C, Albini A, Brunet C (2019) Microalgal derivatives as potential nutraceutical and food supplements for human health: a focus on cancer prevention and interception. Nutrients 11(6). https://doi.org/10.3390/nu11061226

Gam D, Kim SY, Kim JW (2020) Optimization of ultrasound-assisted extraction condition for phenolic compounds, antioxidant activity, and epigallocatechin gallate in lipid-extracted microalgae. Molecules 25(3):17. https://doi.org/10.3390/molecules25030454

Garcia-Perez P, Cassani L, Garcia-Oliveira P, Xiao J, Simal-Gandara J, Prieto MA, Lucini L (2023) Algal nutraceuticals: a perspective on metabolic diversity, current food applications, and prospects in the field of metabolomics. Food Chem 409:135295. https://doi.org/10.1016/j.foodchem.2022.135295

Gheysen L, Lagae N, Devaere J, Goiris K, Goos P, Bernaerts T, Van Loey A, De Cooman L, Foubert I (2019) Impact of *Nannochloropsis* sp. dosage form on the oxidative stability of n-3 LC-PUFA enriched tomato purees. Food Chem 279:389–400. https://doi.org/10.1016/j.foodchem.2018.12.026

Ghosh S, Sarkar T, Pati S, Kari ZA, Edinur HA, Chakraborty R (2022) Novel bioactive compounds from marine sources as a tool for functional food development. Front Mar Sci 9. https://doi.org/10.3389/fmars.2022.832957

Giordano E, Visioli F (2014) Long-chain omega 3 fatty acids: molecular bases of potential antioxidant actions. Prostaglandins Leukot Essent Fatty Acids 90(1):1–4. https://doi.org/10.1016/j.plefa.2013.11.002

Go RE, Hwang KA, Park GT, Lee HM, Lee GA, Kim CW, Jeon SY, Seo JW, Hong WK, Choi KC (2016) Effects of microalgal polyunsaturated fatty acid oil on body weight and lipid accumulation in the liver of C57BL/6 mice fed a high fat diet. J Biomed Res 30(3):234–242. https://doi.org/10.7555/JBR.30.2016K0004

Goiris K, Van Colen W, Wilches I, León-Tamariz F, De Cooman L, Muylaert K (2015) Impact of nutrient stress on antioxidant production in three species of microalgae. Algal Res 7:51–57. https://doi.org/10.1016/j.algal.2014.12.002

Gongi W, Cordeiro N, Pinchetti JLG, Sadok S, Ben Ouada H (2021) Extracellular polymeric substances with high radical scavenging ability produced in outdoor cultivation of the thermotolerant chlorophyte *Graesiella* sp. J Appl Phycol 33(1):357–369. https://doi.org/10.1007/s10811-020-02303-0

Gour RS, Chawla A, Singh H, Chauhan RS, Kant A (2016) Characterization and screening of native *Scenedesmus* sp. isolates suitable for biofuel feedstock. PLoS One 11(5):e0155321. https://doi.org/10.1371/journal.pone.0155321

Graziani G, Schiavo S, Nicolai MA, Buono S, Fogliano V, Pinto G, Pollio A (2013) Microalgae as human food: chemical and nutritional characteristics of the thermo-acidophilic microalga *Galdieria sulphuraria*. Food Funct 4(1):144–152. https://doi.org/10.1039/c2fo30198a

Grosso G, Bei R, Mistretta A, Marventano S, Calabrese G, Masuelli L, Giganti MG, Modesti A, Galvano F, Gazzolo D (2013) Effects of vitamin C on health: a review of evidence. Front Biosci 18(3):1017–1029. https://doi.org/10.2741/4160

Gui L, Xu L, Liu Z-y, Zhou Z-g, Sun Z (2022) Carotenoid-rich microalgae promote growth and health conditions of *Artemia nauplii*. Aquaculture 546:737289. https://doi.org/10.1016/j.aquaculture.2021.737289

Gupta P, Banerjee A, Castillo A, Bandopadhyay R (2021) Novel phenolic compound from southern ocean microalgae *Chlorella* sp. PR-1 and its antibacterial activity. Gayana Bot 78(1):29–37. https://doi.org/10.4067/S0717-66432021000100029

Halder M, Petsophonsakul P, Akbulut AC, Pavlic A, Bohan F, Anderson E, Maresz K, Kramann R, Schurgers L (2019) Vitamin K: double bonds beyond coagulation insights into differences between vitamin K1 and K2 in health and disease. Int J Mol Sci 20(4):896. https://doi.org/10.3390/ijms20040896

Hammer A, Moertl D, Schlager O, Matschuck M, Seidinger D, Koppensteiner R, Steiner S (2019) Effects of n-3 PUFA on endothelial function in patients with peripheral arterial disease: a randomised, placebo-controlled, double-blind trial. Br J Nutr 122(6):698–706. https://doi.org/10.1017/S0007114519001582

Han P, Li J, Zhong H, Xie J, Zhang P, Lu Q, Li J, Xu P, Chen P, Leng L, Zhou W (2021) Anti-oxidation properties and therapeutic potentials of spirulina. Algal Res 55:102240. https://doi.org/10.1016/j.algal.2021.102240

Han P, Lu Q, Fan L, Zhou W (2019) A review on the use of microalgae for sustainable aquaculture. Appl Sci (Switzerland) 9(11). https://doi.org/10.3390/app9112377

Hashem SM, Abd El-Lahot M, Helal AM, Massoud MI (2021) Evaluation of the phytochemicals and nutritional characteristics of some microalgae grown in Egypt as healthy food supplements. Egypt J Food Sci 49(1):173–185. https://doi.org/10.21608/ejfs.2021.66232.1100

Häubner N, Sylvander P, Vuori K, Snoeijs P (2014) Abiotic stress modifies the synthesis of alpha-tocopherol and beta-carotene in phytoplankton species. J Phycol 50(4):753–759. https://doi.org/10.1111/jpy.12198

Hena S, Abida N, Tabassum S (2015) Screening of facultative strains of high lipid producing microalgae for treating surfactant mediated municipal wastewater. RSC Adv 5(120):98805–98813. https://doi.org/10.1039/c5ra20019a

Hoseini SM, Khosravi-Darani K, Mozafari MR (2013) Nutritional and medical applications of spirulina microalgae. Mini-Rev Med Chem 13(8):1231–1237. https://doi.org/10.2174/1389557511313080009

Hügler M, Sievert SM (2010) Beyond the Calvin cycle: autotrophic carbon fixation in the ocean. Annu Rev Mar Sci 3(1):261–289. https://doi.org/10.1146/annurev-marine-120709-142712

Hulatt CJ, Berecz O, Egeland ES, Wijffels RH, Kiron V (2017) Polar snow algae as a valuable source of lipids? Bioresour Technol 235:338–347. https://doi.org/10.1016/j.biortech.2017.03.130

Huy M, Kristin Vatland A, Kumar G (2022) Nutraceutical productions from microalgal derived compounds via circular bioeconomy perspective. Bioresour Technol 347:126575. https://doi.org/10.1016/j.biortech.2021.126575

Ighodaro O, Akinloye O (2018) First line defence antioxidants-superoxide dismutase (SOD), catalase (CAT) and glutathione peroxidase (GPX): their fundamental role in the entire antioxidant defence grid. Alex J Med 54(4):287–293

Igual M, Uribe-Wandurraga ZN, García-Segovia P, Martínez-Monzó J (2022) Microalgae-enriched breadsticks: analysis for vitamin C, carotenoids, and chlorophyll a. Food Sci Technol Int 28(1):26–31. https://doi.org/10.1177/1082013221990252

Islam MN, Alsenani F, Schenk PM (2017) Microalgae as a sustainable source of nutraceuticals. In: Microbial functional foods and nutraceuticals, pp 1–19. https://doi.org/10.1002/9781119048961.ch1

Jaeschke DP, Menegol T, Rech R, Mercali GD, Marczak LDF (2016) Carotenoid and lipid extraction from *Heterochlorella luteoviridis* using moderate electric field and ethanol. Process Biochem 51(10):1636–1643. https://doi.org/10.1016/j.procbio.2016.07.016

Kapoor B, Kapoor D, Gautam S, Singh R, Bhardwaj S (2021) Dietary polyunsaturated fatty acids (PUFAs): uses and potential health benefits. Curr Nutr Rep 10(3):232–242. https://doi.org/10.1007/s13668-021-00363-3

Khadim R, Al-Fartusie F (2021) Antioxidant vitamins and their effect on immune system. J Phys Conf Ser 1853:012065. https://doi.org/10.1088/1742-6596/1853/1/012065

Kim SY, Kwon YM, Kim KW, Kim JYH (2021) Exploring the potential of *Nannochloropsis* sp. extract for cosmeceutical applications. Mar Drugs 19(12). https://doi.org/10.3390/md19120690

Kiuru P, D'Auria MV, Muller CD, Tammela P, Vuorela H, Yli-Kauhaluoma J (2014) Exploring marine resources for bioactive compounds. Planta Med 80(14):1234–1246. https://doi.org/10.1055/s-0034-1383001

Koller M, Muhr A, Braunegg G (2014) Microalgae as versatile cellular factories for valued products. Algal Res 6:52–63. https://doi.org/10.1016/j.algal.2014.09.002

Kothri M, Mavrommati M, Elazzazy AM, Baeshen MN, Moussa TAA, Aggelis G (2020) Microbial sources of polyunsaturated fatty acids (PUFAs) and the prospect of organic residues and wastes as growth media for PUFA-producing microorganisms. FEMS Microbiol Lett 367(5). https://doi.org/10.1093/femsle/fnaa028

Koyande AK, Chew KW, Rambabu K, Tao Y, Chu D-T, Show P-L (2019) Microalgae: a potential alternative to health supplementation for humans. Food Sci Human Wellness 8(1):16–24. https://doi.org/10.1016/j.fshw.2019.03.001

Kurd F, Samavati V (2015) Water soluble polysaccharides from *Spirulina platensis*: extraction and *in vitro* anticancer activity. Int J Biol Macromol 74:498–506. https://doi.org/10.1016/j.ijbiomac.2015.01.005

Lee JM, Lee H, Kang SB, Park WJ (2016) Fatty acid desaturases, polyunsaturated fatty acid regulation, and biotechnological advances. Nutrients 8(1). https://doi.org/10.3390/nu8010023

Leegood RC (1990) 2 - Enzymes of the Calvin cycle. In: Lea PJ (ed) Methods in plant biochemistry, vol 3. Academic Press, pp 15–37. https://doi.org/10.1016/B978-0-12-461013-2.50009-5

León-Vaz A, León R, Vigara J, Funk C (2023) Exploring Nordic microalgae as a potential novel source of antioxidant and bioactive compounds. New Biotechnol 73:1–8. https://doi.org/10.1016/j.nbt.2022.12.001

Levasseur W, Perré P, Pozzobon V (2020) A review of high value-added molecules production by microalgae in light of the classification. Biotechnol Adv 41:107545. https://doi.org/10.1016/j.biotechadv.2020.107545

Levy-Ontman O, Huleihel M, Hamias R, Wolak T, Paran E (2017) An anti-inflammatory effect of red microalga polysaccharides in coronary artery endothelial cells. Atherosclerosis 264:11–18. https://doi.org/10.1016/j.atherosclerosis.2017.07.017

Li H, Su L, Chen S, Zhao L, Wang H, Ding F, Chen H, Shi R, Wang Y, Huang Z (2018) Physicochemical characterization and functional analysis of the polysaccharide from the edible microalga *Nostoc sphaeroides*. Molecules 23(2). https://doi.org/10.3390/molecules23020508

Li K, Duan X, Zhou L, Hill DRA, Martin GJO, Suleria HAR (2022) Bioaccessibility and bioactivities of phenolic compounds from microalgae during *in vitro* digestion and colonic fermentation. Food Funct 14(2):899–910. https://doi.org/10.1039/d2fo02980d

Li X, Liu J, Chen G, Zhang J, Wang C, Liu B (2019) Extraction and purification of eicosapentaenoic acid and docosahexaenoic acid from microalgae: a critical review. Algal Res 43:101619. https://doi.org/10.1016/j.algal.2019.101619

Liau B-C, Shen C-T, Liang F-P, Hong S-E, Hsu S-L, Jong T-T, Chang C-MJ (2010) Supercritical fluids extraction and anti-solvent purification of carotenoids from microalgae and associated bioactivity. J Supercrit Fluids 55(1):169–175. https://doi.org/10.1016/j.supflu.2010.07.002

Ljubic A, Jacobsen C, Holdt SL, Jakobsen J (2020) Microalgae *Nannochloropsis oceanica* as a future new natural source of vitamin D3. Food Chem 320:126627. https://doi.org/10.1016/j.foodchem.2020.126627

Ljubic A, Thulesen E, Jacobsen C, Jakobsen J (2021) UVB exposure stimulates production of vitamin D3 in selected microalgae. Algal Res 59:102472. https://doi.org/10.1016/j.algal.2021.102472

Lu Q, Li H, Xiao Y, Liu H (2021) A state-of-the-art review on the synthetic mechanisms, production technologies, and practical application of polyunsaturated fatty acids from microalgae. Algal Res 55:102281. https://doi.org/10.1016/j.algal.2021.102281

Luna LMG (2007) Microalgas: Aspectos ecológicos y biotecnológicos. Rev Cub Quim 19(2):3–20

Lushchak VI (2014) Free radicals, reactive oxygen species, oxidative stress and its classification. Chem Biol Interact 224:164–175

Martínez KA, Saide A, Crespo G, Martín J, Romano G, Reyes F, Lauritano C, Ianora A (2022) Promising antiproliferative compound from the green microalga *Dunaliella tertiolecta* against human cancer cells. Front Mar Sci 9. https://doi.org/10.3389/fmars.2022.778108

Masuda K, Chitundu M (2019) Multiple micronutrient supplementation using *Spirulina platensis* and infant growth, morbidity, and motor development: evidence from a randomized trial in Zambia. PLoS One 14(2):e0211693. https://doi.org/10.1371/journal.pone.0211693

Mayer C, Côme M, Ulmann L, Martin I, Zittelli GC, Faraloni C, Ouguerram K, Chénais B, Mimouni V (2022) The potential of the marine microalga *Diacronema lutheri* in the prevention of obesity and metabolic syndrome in high-fat-fed Wistar rats. Molecules 27(13). https://doi.org/10.3390/molecules27134246

Mehariya S, Iovine A, Di Sanzo G, Larocca V, Martino M, Leone GP, Casella P, Karatza D, Marino T, Musmarra D, Molino A (2019) Supercritical fluid extraction of lutein from *Scenedesmus almeriensis*. Molecules 24(7). https://doi.org/10.3390/molecules24071324

Michael A, Kyewalyanga MS, Lugomela CV (2019) Biomass and nutritive value of Spirulina (*Arthrospira fusiformis*) cultivated in a cost-effective medium. Ann Microbiol 69(13):1387–1395. https://doi.org/10.1007/s13213-019-01520-4

Miranda-Arizmendi V, Fimbres-Olivarria D, Miranda-Baeza A, Martínez-Robinson K, Rascón-Chu A, De Anda-Flores Y, Lizardi-Mendoza J, Mendez-Encinas MA, Brown-Bojorquez F, Canett-Romero R, Carvajal-Millan E (2022) Sulfated polysaccharides from *Chaetoceros muelleri*: macromolecular characteristics and bioactive properties. Biology 11(10). https://doi.org/10.3390/biology11101476

Molino A, Mehariya S, Iovine A, Larocca V, Di Sanzo G, Martino M, Casella P, Chianese S, Musmarra D (2018) Extraction of astaxanthin and lutein from microalga *Haematococcus pluvialis* in the red phase using CO_2 supercritical fluid extraction technology with ethanol as co-solvent. Mar Drugs 16(11). https://doi.org/10.3390/md16110432

Montero-Lobato Z, Vázquez M, Navarro F, Fuentes JL, Bermejo E, Garbayo I, Vílchez C, Cuaresma M (2018) Chemically-induced production of anti-inflammatory molecules in microalgae. Mar Drugs 16(12). https://doi.org/10.3390/md16120478

Morcelli A, Cassel E, Vargas R, Rech R, Marcílio N (2021) Supercritical fluid (CO_2+ethanol) extraction of chlorophylls and carotenoids from *Chlorella sorokiniana*: COSMO-SAC assisted prediction of properties and experimental approach. J CO2 Util 51:101649. https://doi.org/10.1016/j.jcou.2021.101649

Moreira JB, Vaz BD, Cardias BB, Cruz CG, Almeida AC, Costa JA, Morais MG (2022) Microalgae polysaccharides: an alternative source for food production and sustainable agriculture. Polysaccharides 3(2):441–457. https://doi.org/10.3390/polysaccharides3020027

Morgese MG, Mhillaj E, Francavilla M, Bove M, Morgano L, Tucci P, Trabace L, Schiavone S (2016) *Chlorella sorokiniana* extract improves short-term memory in rats. Molecules 21(10). https://doi.org/10.3390/molecules21101311

Mousavian Z, Safavi M, Azizmohseni F, Hadizadeh M, Mirdamadi S (2022) Characterization, antioxidant and anticoagulant properties of exopolysaccharide from marine microalgae. AMB Express 12(1):27. https://doi.org/10.1186/s13568-022-01365-2

Nakagawa I, Yokoyama S, Omoto K, Takeshima Y, Matsuda R, Nishimura F, Yamada S, Yokota H, Motoyama Y, Park YS, Nakase H (2017) ω-3 fatty acids ethyl esters suppress cerebral vasospasm and improve clinical outcome following aneurysmal subarachnoid hemorrhage. World Neurosurg 99:457–464. https://doi.org/10.1016/j.wneu.2016.12.018

Nakano S, Takekoshi H, Nakano M (2010) *Chlorella pyrenoidosa* supplementation reduces the risk of anemia, proteinuria and edema in pregnant women. Plant Foods Hum Nutr 65(1):25–30. https://doi.org/10.1007/s11130-009-0145-9

Neumann U, Derwenskus F, Gille A, Louis S, Schmid-Staiger U, Briviba K, Bischoff SC (2018) Bioavailability and safety of nutrients from the microalgae *Chlorella vulgaris*, *Nannochloropsis oceanica* and *Phaeodactylum tricornutum* in C57BL/6 mice. Nutrients 10(8). https://doi.org/10.3390/nu10080965

Oppedisano F, Macrì R, Gliozzi M, Musolino V, Carresi C, Maiuolo J, Bosco F, Nucera S, Zito MC, Guarnieri L, Scarano F, Nicita C, Coppoletta AR, Ruga S, Scicchitano M, Mollace R, Palma E, Mollace V (2020) The anti-inflammatory and antioxidant properties of n-3 PUFAs: their role in cardiovascular protection. Biomedicines 8(9). https://doi.org/10.3390/biomedicines8090306

Ova OD, Ata A, Ovez B (2022) Identification of photosynthetic pigments extracted from *Phaeodactylum tricornutum* as high-value bioactive compounds. Res J Biotechnol 17(12):91–99. https://doi.org/10.25303/1712rjbt91099

Pagels F, Amaro HM, Tavares TG, Amil BF, Guedes AC (2022) Potential of microalgae extracts for food and feed supplementation: a promising source of antioxidant and anti-inflammatory compounds. Life 12(11). https://doi.org/10.3390/life12111901

Paliwal C, Ghosh T, George B, Pancha I, Maurya R, Chokshi K, Ghosh A, Mishra S (2016) Microalgal carotenoids: potential nutraceutical compounds with chemotaxonomic importance. Algal Res 15:24–31. https://doi.org/10.1016/j.algal.2016.01.017

Paliwal C, Rehmanji M, Shaikh KM, Zafar SU, Jutur PP (2022) Green extraction processing of lutein from *Chlorella saccharophila* in water-based ionic liquids as a sustainable innovation in algal biorefineries. Algal Res 66:102809. https://doi.org/10.1016/j.algal.2022.102809

Pelah D, Sintov A, Cohen E (2004) The effect of salt stress on the production of canthaxanthin and astaxanthin by *Chlorella zofingiensis* grown under limited light intensity. World J Microbiol Biotechnol 20(5):483–486. https://doi.org/10.1023/B:WIBI.0000040398.93103.21

Priyadarshani I, Rath B (2012) Commercial and industrial applications of micro algae–a review. J Algal Biomass Util 3(4):89–100

Qin L, Mei Y, An C, Ning R, Zhang H (2023) Docosahexaenoic acid administration improves diabetes-induced cardiac fibrosis through enhancing fatty acid oxidation in cardiac fibroblast. J Nutr Biochem 113:109244. https://doi.org/10.1016/j.jnutbio.2022.109244

Raines CA (2003) The Calvin cycle revisited. Photosynth Res 75(1):1–10. https://doi.org/10.1023/A:1022421515027

Ramesh Kumar B, Deviram G, Mathimani T, Duc PA, Pugazhendhi A (2019) Microalgae as rich source of polyunsaturated fatty acids. Biocatal Agric Biotechnol 17:583–588. https://doi.org/10.1016/j.bcab.2019.01.017

Ran L, Yu J, Ma R, Yao Q, Wang M, Bi Y, Yu Z, Wu Y (2022) Microalgae oil from *Schizochytrium* sp. alleviates obesity and modulates gut microbiota in high-fat diet-fed mice. Food Funct 13(24):12799–12813. https://doi.org/10.1039/d2fo01772e

Rengel R, Giraldez I, Díaz MJ, García T, Vigara J, León R (2022) Simultaneous production of carotenoids and chemical building blocks precursors from chlorophyta microalgae. Bioresour Technol 351:127035. https://doi.org/10.1016/j.biortech.2022.127035

Reyes FA, Mendiola JA, Suárez-Alvarez S, Ibañez E, del Valle JM (2016) Adsorbent-assisted supercritical CO_2 extraction of carotenoids from *Neochloris oleoabundans* paste. J Supercrit Fluids 112:7–13. https://doi.org/10.1016/j.supflu.2016.02.005

Rizwan M, Mujtaba G, Memon SA, Lee K, Rashid N (2018) Exploring the potential of microalgae for new biotechnology applications and beyond: a review. Renew Sust Energ Rev 92:394–404. https://doi.org/10.1016/j.rser.2018.04.034

Rodrigues DB, Flores ÉMM, Barin JS, Mercadante AZ, Jacob-Lopes E, Zepka LQ (2014) Production of carotenoids from microalgae cultivated using agroindustrial wastes. Food Res Int 65:144–148. https://doi.org/10.1016/j.foodres.2014.06.037

Ryckebosch E, Bruneel C, Termote-Verhalle R, Goiris K, Muylaert K, Foubert I (2014) Nutritional evaluation of microalgae oils rich in omega-3 long chain polyunsaturated fatty acids as an alternative for fish oil. Food Chem 160:393–400. https://doi.org/10.1016/j.foodchem.2014.03.087

Safafar H, Van Wagenen J, Møller P, Jacobsen C (2015) Carotenoids, phenolic compounds and tocopherols contribute to the antioxidative properties of some microalgae species grown on industrial wastewater. Mar Drugs 13(12):7339–7356. https://doi.org/10.3390/md13127069

Şahin S, Tsiqah Binti Mohd Nasir N, Erken İ, Elibol Çakmak Z, Çakmak T (2019) Antioxidant composite films with chitosan and carotenoid extract from *Chlorella vulgaris*: optimization of ultrasonic-assisted extraction of carotenoids and surface characterization of chitosan films. Mater Res Express 6(9):095404. https://doi.org/10.1088/2053-1591/ab2def

Sánchez-Camargo AP, Pleite N, Mendiola JA, Cifuentes A, Herrero M, Gilbert-López B, Ibáñez E (2018) Development of green extraction processes for *Nannochloropsis gaditana* biomass valorization. Electrophoresis 39(15):1875–1883. https://doi.org/10.1002/elps.201800122

Sansone C, Brunet C, Noonan DM, Albini A (2020) Marine algal antioxidants as potential vectors for controlling viral diseases. Antioxidants 9(5). https://doi.org/10.3390/antiox9050392

Santiago-Díaz P, Rivero A, Rico M, Gómez-Pinchetti JL (2022) Characterization of novel selected microalgae for antioxidant activity and polyphenols, amino acids, and carbohydrates. Mar Drugs 20(1). https://doi.org/10.3390/md20010040

Santiago-Morales IS, Trujillo-Valle L, Márquez-Rocha FJ, Hernández JFL (2018) Tocopherols, phycocyanin and superoxide dismutase from microalgae: as potential food antioxidants. Appl Food Biotechnol 5(1):19–27. https://doi.org/10.22037/afb.v5i1.17884

Santin A, Russo MT, Ferrante MI, Balzano S, Orefice I, Sardo A (2021) Highly valuable polyunsaturated fatty acids from microalgae: strategies to improve their yields and their potential exploitation in aquaculture. Molecules 26(24). https://doi.org/10.3390/molecules26247697

Santos-Sánchez NF, Valadez-Blanco R, Hernández-Carlos B, Torres-Ariño A, Guadarrama-Mendoza PC, Salas-Coronado R (2016) Lipids rich in ω-3 polyunsaturated fatty acids from microalgae. Appl Microbiol Biotechnol 100(20):8667–8684. https://doi.org/10.1007/s00253-016-7818-8

Sanzo GD, Mehariya S, Martino M, Larocca V, Casella P, Chianese S, Musmarra D, Balducchi R, Molino A (2018) Supercritical carbon dioxide extraction of astaxanthin, lutein, and fatty acids from *Haematococcus pluvialis* microalgae. Mar Drugs 16(9). https://doi.org/10.3390/md16090334

Sañudo-Wilhelmy SA, Gómez-Consarnau L, Suffridge C, Webb EA (2014) The role of B vitamins in marine biogeochemistry. Annu Rev Mar Sci 6(1):339–367. https://doi.org/10.1146/annurev-marine-120710-100912

Sawada T, Tsubata H, Hashimoto N, Takabe M, Miyata T, Aoki K, Yamashita S, Oishi S, Osue T, Yokoi K, Tsukishiro Y, Onishi T, Shimane A, Taniguchi Y, Yasaka Y, Ohara T, Kawai H, Yokoyama M (2016) Effects of 6-month eicosapentaenoic acid treatment on postprandial hyperglycemia, hyperlipidemia, insulin secretion ability, and concomitant endothelial dysfunction among newly-diagnosed impaired glucose metabolism patients with coronary artery disease. An open label, single blinded, prospective randomized controlled trial. Cardiovasc Diabetol 15(1). https://doi.org/10.1186/s12933-016-0437-y

Simes DC, Viegas CSB, Araújo N, Marreiros C (2020) Vitamin K as a diet supplement with impact in human health: current evidence in age-related diseases. Nutrients 12(1):138. https://doi.org/10.3390/nu12010138

Singab AN, Ibrahim N, Elsayed AE-k, El-Senousy W, Aly H, Abd Elsamiae A, Matloub AA (2018) Antiviral, cytotoxic, antioxidant and anti-cholinesterase activities of polysaccharides isolated from microalgae *Spirulina platensis*, *Scenedesmus obliquus* and *Dunaliella salina*. Arch Pharm Sci Ain Shams Univ 2(2):121–137. https://doi.org/10.21608/aps.2018.18740

Song H, He M, Gu C, Wei D, Liang Y, Yan J, Wang C (2018) Extraction optimization, purification, antioxidant activity, and preliminary structural characterization of crude polysaccharide from an Arctic *chlorella* sp. Polymers 10:3. https://doi.org/10.3390/polym10030292

Steinrücken P, Erga SR, Mjøs SA, Kleivdal H, Prestegard SK (2017) Bioprospecting North Atlantic microalgae with fast growth and high polyunsaturated fatty acid (PUFA) content for microalgae-based technologies. Algal Res 26:392–401. https://doi.org/10.1016/j.algal.2017.07.030

Stengel DB, Connan S (2015) Marine algae: a source of biomass for biotechnological applications. In: Stengel DB, Connan S (eds) Natural products from marine algae: methods and protocols. Springer, New York, NY, pp 1–37. https://doi.org/10.1007/978-1-4939-2684-8_1

Tarento T, McClure D, Vasiljevski E, Schindeler A, Dehghani F, Kavanagh J (2018) Microalgae as a source of vitamin K1. Algal Res 36:77–87. https://doi.org/10.1016/j.algal.2018.10.008

Tenorio-Rodríguez P, Méndez-Rodrìguez L, Serviere-Zaragoza E, OHara T, Zenteno-Savín T (2013) Antioxidant substances and trace element content in macroalgae from a subtropical lagoon in the West Coast of the Baja California Peninsula. Vitam Trace Elem 2(108):2167–0390.1000. https://doi.org/10.4172/2167-0390.1000108

Trabelsi L, Chaieb O, Mnari A, Abid-Essafi S, Aleya L (2016) Partial characterization and antioxidant and antiproliferative activities of the aqueous extracellular polysaccharides from the thermophilic microalgae *Graesiella* sp. BMC Complement Altern Med 16(1):210. https://doi.org/10.1186/s12906-016-1198-6

Tzima S, Georgiopoulou I, Louli V, Magoulas K (2023) Recent advances in supercritical CO_2 extraction of pigments, lipids and bioactive compounds from microalgae. Molecules 28:3. https://doi.org/10.3390/molecules28031410

Van den Berg TE, Chukhutsina VU, van Amerongen H, Croce R, van Oort B (2019) Light acclimation of the colonial green alga *Botryococcus braunii* strain showa. Plant Physiol 179(3):1132–1143. https://doi.org/10.1104/pp.18.01499

Vermeer CV (2012) Vitamin K: the effect on health beyond coagulation—an overview. Food Nutr Res 56(1):5329. https://doi.org/10.3402/fnr.v56i0.5329

Vermerris W, Nicholson R (2008) Families of phenolic compounds and means of classification. In: Phenolic compound biochemistry. Springer, Dordrecht, The Netherlands, pp 1–34

Villanova V, Galasso C, Vitale GA, Della Sala G, Engelbrektsson J, Strömberg N, Shaikh KM, Andersson MX, Palma Esposito F, Ekendahl S, De Pascale D, Spetea C (2022) Mixotrophy in a local strain of *Nannochloropsis granulata* for renewable high-value biomass production on the West Coast of Sweden. Mar Drugs 20(7). https://doi.org/10.3390/md20070424

Wan X, Li X, Liu D, Gao X, Chen Y, Chen Z, Fu C, Lin L, Liu B, Zhao C (2021) Physicochemical characterization and antioxidant effects of green microalga *Chlorella pyrenoidosa* polysaccharide by regulation of microRNAs and gut microbiota in *Caenorhabditis elegans*. Int J Biol Macromol 168:152–162. https://doi.org/10.1016/j.ijbiomac.2020.12.010

Wan XZ, Li TT, Zhong RT, Chen HB, Xia X, Gao LY, Gao XX, Liu B, Zhang HY, Zhao C (2019) Anti-diabetic activity of PUFAs-rich extracts of *Chlorella pyrenoidosa* and *Spirulina platensis* in rats. Food Chem Toxicol 128:233–239. https://doi.org/10.1016/j.fct.2019.04.017

Wang M, Morón-Ortiz Á, Zhou J, Benítez-González A, Mapelli-Brahm P, Meléndez-Martínez AJ, Barba FJ (2023) Effects of pressurized liquid extraction with dimethyl sulfoxide on the recovery of carotenoids and other dietary valuable compounds from the microalgae *Spirulina*, *Chlorella* and *Phaeodactylum tricornutum*. Food Chem 405. https://doi.org/10.1016/j.foodchem.2022.134885

Woortman DV, Fuchs T, Striegel L, Fuchs M, Weber N, Brück TB, Rychlik M (2020) Microalgae a superior source of folates: quantification of folates in halophile microalgae by stable isotope dilution assay. Front Bioeng Biotechnol 7. https://doi.org/10.3389/fbioe.2019.00481

World Health Organization (2023) Non-communicable diseases. https://www.who.int/news-room/fact-sheets/detail/noncommunicable-diseases. Accessed 14 September 2023

Xia S, Wang K, Wan L, Li A, Hu Q, Zhang C (2013) Production, characterization, and antioxidant activity of fucoxanthin from the marine diatom *Odontella aurita*. Mar Drugs 11(7):2667–2681. https://doi.org/10.3390/md11072667

Xie S-R, Li Y, Chen H-H, Liang M-H, Jiang J-G (2022) A strategy to promote carotenoids production in *Dunaliella bardawil* by melatonin combined with photoinduction. Enzym Microb Technol 161:110115. https://doi.org/10.1016/j.enzmictec.2022.110115

Yakıncı ÖF, Süntar I (2022) Chapter4.21 - Vitamin A. In: Nabavi SM, Silva AS (eds) Antioxidants effects in health. Elsevier, pp 527–534. https://doi.org/10.1016/B978-0-12-819096-8.00064-1

Yu J, Ma Y, Sun J, Ran L, Li Y, Wang N, Yu T, Gao W, Jia W, Jiang R, Guo M, Bi Y, Wu Y (2017) Microalgal oil from *Schizochytrium* sp. prevents hfd-induced abdominal fat accumulation in mice. J Am Coll Nutr 36(5):347–356. https://doi.org/10.1080/07315724.2017.1302366

Zakaria SM, Kamal SMM, Harun R, Omar R, Siajam SI (2020) Characterization on phenolic acids and antioxidant activity of *Chlorella* sp. microalgae using subcritical water extraction. Sains Malays 49(4):765–774. https://doi.org/10.17576/jsm-2020-4904-05

Zhang J, Liu L, Chen F (2019) Production and characterization of exopolysaccharides from *Chlorella zofingiensis* and *Chlorella vulgaris* with anti-colorectal cancer activity. Int J Biol Macromol 134:976–983. https://doi.org/10.1016/j.ijbiomac.2019.05.117

Zhong Q, Wei B, Wang S, Ke S, Chen J, Zhang H, Wang H (2019) The antioxidant activity of polysaccharides derived from marine organisms: an overview. Mar Drugs 17(12). https://doi.org/10.3390/md17120674

Zhou J, Wang M, Berrada H, Zhu Z, Grimi N, Barba FJ (2022) Pulsed electric fields (PEF), pressurized liquid extraction (PLE) and combined PEF + PLE process evaluation: effects on Spirulina microstructure, biomolecules recovery and Triple TOF-LC-MS-MS polyphenol composition. Innovative Food Sci Emerg Technol 77:102989. https://doi.org/10.1016/j.ifset.2022.102989

Zimermann JDF, Sydney EB, Cerri ML, de Carvalho IK, Schafranski K, Sydney ACN, Vitali L, Gonçalves S, Micke GA, Soccol CR, Demiate IM (2020) Growth kinetics, phenolic compounds profile and pigments analysis of *Galdieria sulphuraria* cultivated in whey permeate in shake-flasks and stirred-tank bioreactor. J Water Process Eng 38:101598. https://doi.org/10.1016/j.jwpe.2020.101598

Part V

Cosmetic Applications from Microalgae

Microalgae as a Potential Source of Cosmeceutical Products with Anti-aging Properties

15

Lourdes Janeth Germán-Báez, Rafael Valdez-Ortiz, Karelia Arisbeth Meza-Ayala, and Angel Valdez-Ortiz

Abstract

Skin is the largest organ in the human body and acts as the first line of defense against various types of biotic and abiotic stresses coming from environmental stresses and internal factors. Thus, permanent efforts have been made to develop topical products known as cosmeceuticals that prevent skin disorders and keep it in good health. In this sense, in the last decade, numerous actions have focused on developing skin care products formulated from bioactive, non-toxic, non-allergenic, and environmentally friendly ingredients obtained from natural sources, replacing the increasingly less accepted synthetic counterparts. A natural source recognized for its richness in bioactive molecules is microalgae since it has been reported that these microorganisms widely accumulate a broad spectrum of substances, preventing adverse effects of stresses and aging. These compounds act on human skin cells in such a way that several cosmetic products such as moisturizers, whitening, skincare, serums, sunscreen, anti-aging, anti-wrinkle, and anti-cellulite agents, among others, are currently being formulated using bioactive compounds or biomass of microalgal origin.

Keywords
Bioactive compounds · Cell senescence · Skin aging · Microalgae · Cosmeceuticals

L. J. Germán-Báez · K. A. Meza-Ayala · A. Valdez-Ortiz (✉)
Laboratorio de Biotecnología e Ingeniería Genética, Facultad de Ciencias Químico-Biológicas, Universidad Autónoma de Sinaloa, Culiacán, Sinaloa, Mexico
e-mail: avaldez@uas.edu.mx

R. Valdez-Ortiz
Service of Nephrology, General Hospital of Mexico, Dr. Eduardo Liceaga, Mexico City, Mexico

15.1 Introduction

The demand for cosmetic and skincare products represents a global market with high commercial yielding profits in the order of billions of dollars, which has driven its evolution toward the development of so-called cosmeceuticals, produced by merging the cosmetic and pharmaceutical areas, which, in addition to being used for beauty purposes they provide therapeutic benefits promoting skin health and wellness (Eom and Kim 2013; Rodrigues et al. 2018; Parikh et al. 2024). Under this global context, microalgae are considered one of the main microorganisms capable of producing several biomolecules of interest in cosmeceutical industries to be used in anti-aging and skin care products, emerging as a biological model with high potential for commercial exploitation to achieve the long-awaited desire to replace the use of synthetic cosmetic products by naturally derived bioactive compounds (Puchkova et al. 2021).

Due to the development and adaptation of numerous biochemical and cellular processes, microalgae could evolve even under the most harmful stress conditions. Proof of this is its already demonstrated ability to produce numerous bioactive molecules intended to neutralize reactive oxygen species (ROS) and free radicals or inactivate potentially harmful molecules and microorganisms, displaying its high biochemical and cellular adaptive capacity. Because of the above, microalgae produce a wide variety of biomolecules attracting interest in the cosmeceutical market for their potential use in the development of high-value products in the beauty and personal care industry (El Gamal 2010; Ariede et al. 2017; Pereira 2018; Martínez-González et al. 2022). In fact, several microalgal-based products are already commercially available, and because of their acceptance and economic impact, efforts for research, development, and innovation have increased exponentially, getting more new commercial products and applications every day (Yarkent et al. 2020; Parikh et al. 2024).

D. U. Santos Ballardo, S. Rossi (eds.), *Microalgae as Promising Source of Commercial Bioproducts*, Developments in Applied Phycology 14, https://doi.org/10.1007/978-3-031-86433-9_15

15.2 Aging and Cellular Senescence

It is generally assumed that the concept of cell senescence is based on the fact that the number of divisions a cell can carry out throughout its life under normal conditions is a finite process. Hence, the accumulation of cell division events triggers a growth inhibition, similar to what occurs in response to several types of stress, causing a differential genetic regulation and modifying the molecular and metabolic environment to either protect the cell from damage or ultimately direct it toward apoptosis mediated by triggering the senescence process. These processes expand the conceived role that senescence plays in the cellular life cycle; until today, we understand it as a way to trigger the molecular mechanisms that underlie aging (DiLoreto and Murphy 2015; González-Gualda et al. 2022).

Furthermore, senescence is a cellular program elicited by stimuli such as several types of damage, stress, or aging, characterized by molecular and metabolic changes causing the accumulation of macromolecular damage and inducing harmful processes in senescent cells (Silva-Álvarez and Collado 2023). Multifactorial processes drive the fact of aging and cellular senescence, which means that they are partially genetically regulated; this may explain that under exposure to various sources of stress, a variety of DNA damages could occur, ranging from different types of potentially lethal genetic mutations to increases in the rate of telomere shortening characteristic of aging and cell senescence (d'Adda di Fagagna et al. 2003; Krishnamurthy et al. 2004; Wong et al. 2020; Silva-Álvarez and Collado 2023).

Moreover, the alteration of genetic control includes aspects such as dysregulation and consequent degradation of cellular processes. Some authors consider that this cellular scenario could correspond to a succession of programmatic events that senescent cell triggers depending on its possibilities of staying alive or is better to move toward apoptosis, or that corresponds in itself to a predetermined mechanism to incite its transition sooner or later toward aging and senescence, such as obligatory stages of their life cycle (DiLoreto and Murphy 2015; González-Gualda et al. 2021, 2022).

Once the cell begins its natural or stress-induced aging process, failures begin to accumulate in all the molecular, biochemical, and structural processes that sustain cellular health in such a way that failures that begin from the proper ordering and genetic regulation incite the lack of control in metabolic processes which in turn end up damaging the entire cellular order and structure. In the case of higher organisms, the above tends to induce a domino effect due to the damage caused to extracellular matrix interactions and signaling, affecting entire tissues such as the skin and other ones that may be vital for the health of the whole organism (Kim et al. 2004; DiLoreto and Murphy 2015; González-Gualda et al. 2022).

15.2.1 Skin Protective Activity: The Term Cosmeceuticals

Cosmeceuticals are products in the form of gel, spray, or cream, among many others, that, in addition to their cosmetic use, provide skin care and health since they are formulated with biologically active ingredients by acting topically protecting against degenerative skin conditions such as photoaging, hyperpigmentation, and wrinkles (Kadam Vaishali et al. 2013; Michalak et al. 2020). Furthermore, to meet the growing global demand for cosmeceutical skin care products, screening for new bioactive ingredients is being constantly developed, mainly from natural sources, by several corporations in the cosmetic, pharmaceutical, and biotechnology markets. Due to this, new knowledge and discoveries related to the molecular, biochemical, cellular, and physiological mechanisms that control the functioning and health of the skin are provided daily. Thus, cosmeceuticals serve as a bridge between personal care products and their pharmaceutical benefits, maintaining features such as efficacy, safety, formulation stability, novelty, and economic viability (Kadam Vaishali et al. 2013; Saini et al. 2024).

15.3 Microalgae as a Source of Cosmeceuticals

In the last decade, photosynthetic microorganisms such as microalgae have been considered a rich but insufficiently explored source of natural substances possessing biological activities that can prevent or repair skin damage associated with aging and several types of stress. Furthermore, the fact that the main species studied to date are classified as generally regarded as safe (GRAS) by the United States Food and Drug Administration (FDA) suggests that, at least in theory, most of the bioactive substances obtained from them could be less harmful and more sustainably produced than their synthetic counterparts (Berthon et al. 2017; Martínez-González et al. 2022).

Microalgae are one of the most diverse groups of organisms; estimating that more than 50,000 species are present on earth, among which only about 30,000 have been studied and analyzed, but only a modest portion of these species are being exploited and commercialized (Kholssi et al. 2023). In this sense, the large-scale biotechnological production already developed in some microalgal species opens the possibility that the massive production of microalgal biomass rich in cosmeceutical compounds becomes a reality. For this reason, the efforts to investigate and characterize new cosmeceutical substances of microalgal origin have increased exponentially (Puchkova et al. 2021).

Microalgae biomasses contain high value compounds such as sulfated polysaccharides; polyunsaturated fatty acids

(PUFAs); proteins, peptides, and mycosporine-like amino acids (MAAs); vitamins; phenolic compounds as tocopherols, polyphenols, flavonoids, tannins, and phlorotannins; pigments such as chlorophylls, carotenoids, and phycobiliproteins; and a lot of metabolites and bioactive compounds (Ryu et al. 2015; German-Báez et al. 2017; Pereira 2018; Puchkova et al. 2021; Parikh et al. 2024).

15.3.1 Natural Pigments

Some microalgae species have striking colors because they accumulate natural pigments such as chlorophylls and carotenoids (such as β-carotene and astaxanthin), showing colors ranging from yellow, orange, and red (Patel et al. 2022). Another group of typical microalgae pigments are the phycobiliproteins as β-phycoerythrin and phycocyanin, which are of various fluorescent colors, including fuchsia, purple-blue, and cyan (Dagnino-Leone et al. 2022).

Because of the biological activity already demonstrated in each one of the aforementioned microalgal pigments, these can be used to produce skincare products that go beyond coloring and beautifying the skin while imparting a protective effect, such as makeups, lipsticks, blushes, shadows, and eyeliner, allowing them to function at the same time as sunscreens, antioxidants, or anti-aging cosmetics instead of only imparting color as happens with synthetic colorants which, in addition to not being environmentally friendly, are often harmful to the skin (Berthon et al. 2017; García et al. 2017; Parikh et al. 2024; Michalak et al. 2020; Patel et al. 2022).

15.3.2 Antimicrobial Activity

The ability of microalgae's bioactive compounds to maintain the balance of skin flora because of their high antimicrobial activity has also been reported. The activity against bacteria, such as *Escherichia coli*, *Staphylococcus aureus*, and *Bacillus cereus*, and fungus, such as *Candida albicans*, *Aspergillus niger*, and *Mucor ramaniannus*, has been mainly attributed to antimicrobial peptides produced in some microalgae species, that when added to skin care products, allowing to keep the skin flora in state of balance, meanwhile providing various skin benefits as extracellular matrix homeostasis, acne management, cell renewal, skin whitening, antioxidant protection, and pro-collagen effects, all these capabilities being desirable in various cosmeceutical products (Wang et al. 2015; Alencar-Silva et al. 2018; Guzmán et al. 2019).

15.3.3 Neutralization of ROS-Induced Damage

Oxidative cellular stress, defined as an imbalance in the redox state of the cell, promotes the generation of ROS whose increase is mainly due to prolonged exposure to environmental stress factors such as ultraviolet radiation and pollution. This causes various diseases or alterations in the function and appearance of the skin leading to laxity, line and wrinkle formation, dryness, skin thinning, loosening, increased dryness, and pore growth, as a result of the degradation of elastin, elastic-fibers and collagen (Wang et al. 2015; Rodrigues et al. 2018; Saini et al. 2024).

As mentioned, one of the distinctive biochemical processes that occur during natural or stress-induced skin aging is the production of high ROS levels, which stimulates dermal fibroblasts to overexpress metalloproteinase and serine proteinases including collagenase, hyaluronidase, and elastase, mainly into the extracellular matrix responsible for providing structural support to surrounding cells, which causes the degradation of key molecules responsible for skin health such as collagen, elastin and hyaluronic acid, leading to skin damage and wrinkles (Kim et al. 2004; Masaki 2010). In this sense, it has been reported that some microalgae compounds stimulate collagen synthesis in the skin, give support to skin tissue, and increase the rebirth of the new tissue (Kumar et al. 2018; Mobin et al. 2019); other microalgae extracts react with skin's proteins to form a protective gel preventing moisture loss (Ryu et al. 2015). On the other hand, giving added value to by-products arising from microalgae application for other purposes represents a strategy that can be exploited within the so-called circular economy; in this sense, Norzagaray-Valenzuela et al. (2017) obtained protein hydrolysates with antioxidant and inhibitory activity of collagenase, hyaluronidase, and elastase enzymes, from the residual biomass of microalgal species *Dunaliella tertiolecta*, *Tetraselmis suecica*, and *Nannochloropsis* sp., generated as a by-product of oil extraction, making the production of biofuels commercially viable, and at the same time generating a product of interest in the cosmeceutical market.

15.3.4 Photoaging Protection by MAAs

Of the extrinsic stimuli that cause skin aging, cumulative exposure to ultraviolet radiation (UVA and UVB) is perhaps the most damaging external factor that threatens skin health, also known as photoaging, because sunlight radiation is capable of penetrating and damaging both the epidermal and dermal layers of the skin, UV light also degrades organic compounds such as proteins and nucleic acids, giving rise to structural changes that directly affect their biological function (Berthon et al. 2017). In this sense, microalgae synthesize a broad spectrum of compounds capable of efficient optical screening and absorption of harmful UVA and UVB radiation, among which mycosporine-like amino acids (MAAs) are typical natural UV radiation filters. Therefore, they are usually bioactive ingredients widely used in the production of skin care products exerting photoprotective, sunscreen, and anti-aging activities in human skin (Řezanka et al. 2004). The photoprotective effect of MAAs has already been widely demonstrated in research such as that reported by Suh et al. (2014), which demonstrated the biological and molecular role of UV-absorbing compounds of microalgae-derived MAAs in human cells analyzing the expression levels of genes associated with antioxidant activity, inflammation, and skin aging in response to UV irradiation. Moreover, Ryu et al. (2014) reported that porphyra-334, a MAA produced by *Porphyra yezoensis*, suppressed ROS production and the expression levels of matrix metalloproteinase in UVA irradiated human skin fibroblasts while increasing levels of extracellular matrix components such as pro-collagen and elastin; suggesting the potential application of this MAA as an anti-aging agent in cosmeceutical products because its preventive photoaging activities. Besides, the MAAs can also be used as antioxidant agents to inhibit lipid peroxidation, peroxide, superoxide, and hydroxyl ions formulation to protect against skin damage (Ibrahim et al. 2017).

15.3.5 Emollient Role of Squalene

A bioactive compound of great interest in both cosmetic and cosmeceutical industries is squalene, mainly due to its ability to impair the antistatic and emollient actions of moisturizing creams, which, added to its antioxidant capacity, allows to decrease age spots, hyperpigmentation, and is highly effective in the treatment of psoriasis or eczema, so that it is currently used even in products such as lotions, eye pencil, eye shadows, eye makeup remover and perfumes (Ryu et al. 2015; Ariede et al. 2017). Currently, the main source of squalene is shark liver oil, however, in response to the global increase in consumers demanding wildlife-friendly products, microalgae species such as *Botryococcus braunii* and *Aurantiochytrium mangrovei* can be evaluated as a good candidate for squalene production. This would also open the possibility of reducing and facilitating the production costs of this important ingredient; this is viable because of the already existence of technology for the design and implementation of highly efficient microalgae bioreactors capable of producing almost any microalgae product of commercial interest (Borowitzka 2013; Ryu et al. 2015; Ariede et al. 2017; Randhir et al. 2020).

15.3.6 Biotechnological Production of Microalgae Bioactive Compounds

In addition to the aspects already described related to the broad spectrum of bioactive compounds from the microalgal origin with applicability in the skincare market, their commercial exploitation appears as a viable challenge because of the favorable characteristics of the microalgae cultures, such as the relatively short multiplication rate and low production and maintenance costs, added to the fact that these microalgae products can be directly used as raw powder, with a little or no extra processing purification methods prior to industrial use, highlighting the feasibility of producing microalgae-derived highly-valued compounds system as an eminently viable alternative (Kumar et al. 2018; Norzagaray-Valenzuela et al. 2018; Mu et al. 2019).

15.4 Conclusions

Despite advances in the knowledge of the molecular, biochemical, and structural mechanisms associated with skin aging, there is still no proven molecular strategy capable of reversing natural or stress-induced skin aging. Faced with cosmetic and pharmacological research's challenge to transform the way we age, enhancing our overall well-being and fostering a healthier global population. Until now, it has not been possible to induce a kind of cellular reprogramming as an anti-aging strategy; the preventive use of healthcare products is practically the only real way currently available by effectively ameliorating several harmful signs of aging, mainly of the larger tissue that surrounds and protects our body, and on whose quality and good appearance the feeling of wellness and therefore happiness largely depends. This has made microalgae a highly viable model to meet global demand for new and innovative cosmeceutical products.

References

Alencar-Silva T, Braga MC, Santana GOS, Saldanha-Araujo F, Pogue R, Dias SC, Franco OL, Carvalho JL (2018) Breaking the frontiers of cosmetology with antimicrobial peptides. Biotechnol Adv 36(8):2019–2031. https://doi.org/10.1016/jbiotechadv.2018.08.005

Ariede MB, Candido TM, Jacome ALM, Velasco MVR, de Carvalho JCM, Baby AR (2017) Cosmetic attributes of algae-a review. Algal Res 25:483–487. https://doi.org/10.1016/j.algal.2017.05.019

Berthon JY, Nachat-Kappes R, Bey M, Cadoret JP, Renimel I, Filaire E (2017) Marine algae as attractive source to skin care. Free Radic Res 51(6):555–567. https://doi.org/10.1080/10715762.2017.1355550

Borowitzka MA (2013) High-value products from microalgae-their development and commercialisation. J Appl Phycol 25:743–756. https://doi.org/10.1007/s10811-013-9983-9

d'Adda di Fagagna F, Reaper PM, Clay-Farrace L, Fiegler H, Carr P, Von Zglinicki T, Saretzki G, Carter NP, Jackson SP (2003) A DNA damage checkpoint response in telomere-initiated senescence. Nature 426(6963):194–198. https://doi.org/10.1038/nature02118

Dagnino-Leone J, Pinto-Figueroa C, Latorre-Castañeda M, Donoso-Youlton A, Vallejos-Almirall A, Agurto-Muñoz A, Pavón-Pérez J, Agurto-Muñoz C (2022) Phycobiliproteins: structural aspects, functional characteristics, and biotechnological perspectives. Comput Struct Biotechnol J 20:1506–1527. https://doi.org/10.1016/j.csbj.2022.02.016

DiLoreto R, Murphy CT (2015) The cell biology of aging. Mol Biol Cell 26(25):4524–4531. https://doi.org/10.1091/mbc.E14-06-1084

El Gamal AA (2010) Biological importance of marine algae. Saudi Pharm J 18(1):1–25. https://doi.org/10.1016/j.jsps.2009.12.001

Eom SH, Kim SK (2013) Cosmeceutical applications from marine organisms. In: Farris PK (ed) Cosmeceuticals and cosmetic practice. Wiley, pp 200–208

García JL, de Vicente M, Galán B (2017) Microalgae, old sustainable food and fashion nutraceuticals. Microb Biotechnol 10(5):1017–1024. https://doi.org/10.1111/1751-7915.12800

German-Báez LJ, Valdez-Flores M, Félix-Medina J, Norzagaray-Valenzuela CD, Santos-Ballardo DU, Reyes-Moreno C, Shelton LM, Valdez-Ortiz A (2017) Chemical composition and physicochemical properties of *Phaeodactylum tricornutum* microalgal residual biomass. Food Sci Technol Int 23(8):681–689. https://doi.org/10.1177/1082013217717611

González-Gualda E, Baker AG, Fruk L, Muñoz-Espín D (2021) A guide to assessing cellular senescence *in vitro* and *in vivo*. FEBS J 288(1):56–80. https://doi.org/10.1111/febs.15570

González-Gualda E, Hui-Ling O, Macías D, Muñoz-Espín D (2022) Cellular senescence: from old to new testament. In: Serrano M, Muñoz-Espín D (eds) Cellular senescence in disease. Academic Press, pp 3–26. https://doi.org/10.1016/B978-0-12-822514-1.00016-X

Guzmán F, Wong G, Román T, Cárdenas C, Álvarez C, Schmitt P, Albericio F, Rojas V (2019) Identification of antimicrobial peptides from the microalgae *Tetraselmis suecica* (Kylin) butcher and bactericidal activity improvement. Mar Drugs 17(8):453. https://doi.org/10.3390/md17080453

Ibrahim M, Salman M, Kamal S, Rehman S, Razzaq A, Akash SH (2017) Algae-based biologically active compounds. In: Zia KM, Zuber M, Ali M (eds) Algae based polymers, blends, and composites. Elsevier, pp 155–271. https://doi.org/10.1016/B978-0-12-812360-7.00006-9

Kadam Vaishali S, Chintale Ashwini G, Deshmukh Kshitija P, Nalwad Digambar N (2013) Cosmeceuticals an emerging concept: a comprehensive review. Int J Res Pharm Chem 3(2):308–316

Kholssi R, Lougraimzi H, Moreno-Garrido I (2023) Effects of global environmental change on microalgal photosynthesis, growth and their distribution. Mar Environ Res 184:105877. https://doi.org/10.1016/j.marenvres.2023.105877

Kim Y, Uyama H, Kobayashi S (2004) Inhibition effects of (+)-catechinaldehyde polycondensates on proteinases causing proteolytic degradation of extracellular matrix. Biochem Biophys Res Commun 320(1):256–261. https://doi.org/10.1016/j.bbrc.2004.05.163

Krishnamurthy J, Torrice C, Ramsey MR, Kovalev GI, Al-Regaiey K, Su L, Sharpless NE (2004) *Ink4a/Arf* expression is a biomarker of aging. J Clin Invest 114(9):1299–1307. https://doi.org/10.1172/JCI22475

Kumar M, Enamala S, Chavali M, Donepudi J (2018) Production of biofuels from microalgae-a review on cultivation, harvesting, lipid extraction, and numerous applications of microalgae. Renew Sust Energ Rev 94:49–68. https://doi.org/10.1016/j.rser.2018.05.012

Martínez-González CA, Kim DH, Santiesteban-Romero B, Reyes-Pardo H, Villaseñor-Zepeda KR, Meléndez-Sánchez ER, Ramírez-Gamboa D, Díaz-Zamorano AL, Sosa-Hernández JE, Coronado-Apodaca KG, Gámez-Méndez AM, Hafiz MN, Parra-Saldivar R (2022) Microalgae bioactive compounds to topical applications products—a review. Molecules 27:3512. https://doi.org/10.3390/molecules27113512

Masaki H (2010) Role of antioxidants in the skin: anti-aging effects. J Dermatol Sci 58:85–90. https://doi.org/10.1016/j.jdermsci.2010.03.003

Michalak I, Dmytryk A, Chojnacka K (2020) Algae cosmetics. Encycl Mar Biotechnol 1:65–85. https://doi.org/10.1002/9781119143802.ch3

Mobin SMA, Chowdhury H, Alam F (2019) Commercially important bioproducts from microalgae and their current applications-a review. Energy Procedia 160:752–760. https://doi.org/10.1016/j.egypro.2019.02.183

Mu N, Mehar JG, Mudliar SN, Shekh AY (2019) Recent advances in microalgal bioactives for food, feed, and healthcare products: commercial potential, market space, and sustainability. Compr Rev Food Sci Food Saf 18:1882–1897

Norzagaray-Valenzuela CD, Valdez-Ortiz A, Shelton LM, Jiménez-Edeza M, Rivera-López J, Valdez-Flores MA, Germán-Báez LJ (2017) Residual biomasses and protein hydrolysates of three green microalgae species exhibit antioxidant and anti-aging activity. J Appl Phycol 29:189–198. https://doi.org/10.1007/s10811-016-0938-9

Norzagaray-Valenzuela CD, Germán-Báez LJ, Valdez-Flores MA, Hernández-Verdugo S, Shelton LM, Valdez-Ortiz A (2018) Establishment of an efficient genetic transformation method in *Dunaliella tertiolecta* mediated by *Agrobacterium tumefaciens*. J Microbiol Methods 150:9–17. https://doi.org/10.1016/j.mimet.2018.05.010.s

Parikh HS, Singh PK, Tiwari A (2024) Algal biorefinery: focus on cosmeceuticals. Syst Microbiol Biomanuf 4:1239. https://doi.org/10.1007/s43393-024-00287-3

Patel AK, Tambat VS, Chen CW, Chauhan AS, Kumar P, Vadrale AP, Huang CY, Dong CD, Singhania RR (2022) Recent advancements in astaxanthin production from microalgae: a review. Bioresour Technol 364:128030. https://doi.org/10.1016/j.biortech.2022.128030

Pereira L (2018) Seaweeds as source of bioactive substances and skin care therapy—cosmeceuticals, lgotheraphy and thalassotherapy. Cosmetics 5:68. https://doi.org/10.3390/cosmetics5040068

Puchkova TV, Khapchaeva SA, Zotov VS, Lukyanov AA, Solovchenko AE (2021) Marine and freshwater microalgae as a sustainable source of cosmeceuticals. Mar Biol J 6:67–81. https://doi.org/10.1016/10.21072/mbj.2021.06.1.06

Randhir A, Laird DW, Maker G, Trengove R, Moheimani NR (2020) Microalgae: a potential sustainable commercial source of sterols. Algal Res 46:101772. https://doi.org/10.1016/j.algal.2019.101772

Řezanka T, Temina M, Tolstikov A, Dembitsky V (2004) Natural microbial UV radiation filters—Mycosporine-like amino acids. Folia Microbiol 49:339–352. https://doi.org/10.1007/BF03354663

Rodrigues F, Cádiz-Gurrea ML, Nunes MA, Pinto D, Vinha AF, Borrás-Linares I, Oliveira MBPP, Segura-Carretero A (2018) Cosmetics. In: Galanakis GM (ed) Polyphenols: properties, recovery, and applications. Woodhead Publishing, pp 393–427. https://doi.org/10.1016/B978-0-12-813572-3.00012-9

Ryu J, Park SJ, Kim IH, Choi YH, Nam TJ (2014) Protective effect of porphyra-334 on UVA-induced photoaging in human skin fibroblasts. Int J Mol Med 34:796–803. https://doi.org/10.3892/ijmm.2014.1815

Ryu B, Himaya SWA, Kim SK (2015) Applications of microalgae-derived active ingredients as cosmeceuticals. In: Kim SK (ed) Handbook of marine microalgae. Academic Press, pp 309–316. https://doi.org/10.1016/B978-0-12-800776-1.00020-0

Saini KC, Rani A, Gupta SK, Sharma P (2024) Algae a potential source in cosmetics: current status, challenges, and economic implications. Indian J Microbiol 64:1445. https://doi.org/10.1007/s12088-024-01263-0

Silva-Álvarez SD, Collado M (2023) Cellular senescence. In: Bradshaw RA, Hart GW, Stahl PD (eds) Encyclopedia of cell biology, 2nd edn. Academic Press, pp 127–135. https://doi.org/10.1016/B978-0-12-821618-7.00219-4

Suh SS, Hwang J, Park M, Seo HH, Kim HS, Lee JH, Moh SH, Lee TK (2014) Anti-inflammation activities of mycosporine-like amino acids (MAAs) in response to UV radiation suggest potential anti-skin aging activity. Mar Drugs 12:5174–5187. https://doi.org/10.3390/md12105174

Wang HD, Chen C, Huynh P, Chang J (2015) Exploring the potential of using algae in cosmetics. Bioresour Technol 184:355–362. https://doi.org/10.1016/j.biortech.2014.12.001

Wong A, Kieu T, Robbins PD (2020) The Ercc1$^{-/\Delta}$ mouse model of accelerated senescence and aging for identification and testing of novel senotherapeutic interventions. Aging 12(24):24481–24483. https://doi.org/10.18632/aging.202321

Yarkent Ç, Gürlek C, Oncel SS (2020) Potential of microalgal compounds in trending natural cosmetics: a review. Sustain Chem Pharm 17:100304. https://doi.org/10.1016/j.scp.2020.100304

Main Prospects and Constraints for New Cosmetics Generation Based on Microalgae Bioactive Compounds

16

Renata Miliani Martinez, Mariana Santos Pinheiro, Amol D. Gholap, Maíra Bueno Ariede, João Carlos Monteiro de Carvalho, André Rolim Baby, and Fabiana Vieira Lima Solino Pessoa

Abstract

Microalgae can produce bioactives that are interesting to be incorporated in cosmetics. Commercial uses for microalgae and their products are new trends after the COVID-19 pandemic in the search for renewable ingredients. This chapter focused on cosmetic and nutricosmetic applications of the main classes of microalgae. The physical and chemical characteristics and biological potential of carotenoids, vitamins, and fatty acids derived from microalgae are analyzed. Also, the benefits and challenges of the technology involved in the harvest and extraction regarding large-scale and susta 121inable processes were discussed. Finally, the current and future perspectives on using microalgae in cosmetics are presented.

Keywords
cosmetics · microalgae · bioactive · sustainability

R. Miliani Martinez · M. B. Ariede · A. R. Baby (✉)
Department of Pharmacy, Faculty of Pharmaceutical Sciences, University of São Paulo, São Paulo, SP, Brazil
e-mail: andrerb@usp.br

M. S. Pinheiro · F. V. L. S. Pessoa
Department of Health Sciences, Federal University of Espírito Santo, São Mateus, Espírito Santo, ES, Brazil

A. D. Gholap
Department of Pharmaceutics St. John Institute of Pharmacy and Research, Palghar, Maharashtra, India

J. C. Monteiro de Carvalho
Biochemical and Pharmaceutical Technology Department, Faculty of Pharmaceutical Sciences, University of São Paulo, São Paulo, SP, Brazil

16.1 Introduction

The exponential growth of the cosmetic market brought awareness of environmental issues regarding sustainable alternatives for ingredients and cosmetics manufacturing processes. Consumer pressure in the cosmetic industry is driving a new trend of renewable products—biocosmetics (Aslam et al. 2021; Morone et al. 2020). After the COVID-19 pandemic, novel opportunities for cosmetics have arisen, especially with a focus on the circular economy. The main drives post-COVID-19 were good sustainability practices during manufacturing, upcycling cosmetic-food industries, and sustainable and biodiversity-friendly ingredients (Goyal and Jerold 2021). In this context, a generation of new cosmetics based on microalgae bioactive compounds perfectly fits consumer needs and market trends.

Microalgae can be grown in an open or enclosed environment on an industrial scale, with a high growth rate, high reproduction speed, and the possibility of being cultivated in a controlled environment (Chauton et al. 2021). Those characteristics make using microalgae and its bioactive molecules interesting and profitable for the cosmetic industry and suppliers.

This chapter covers the target microalgae for cosmetics, bioactive molecules, technologies involved in culture and harvest, and the benefits and challenges of using this source of ingredients. We present up-to-date literature regarding microalgae application on topical formulations and highlight oral intake of microalgae as supplements when cosmetic benefits were found.

16.2 Cosmetics

The definition of cosmetics is important to set the boundaries for microalgae application in this field. The most accepted definition for cosmetics is "items with mild action on the

D. U. Santos Ballardo, S. Rossi (eds.), *Microalgae as Promising Source of Commercial Bioproducts*, Developments in Applied Phycology 14, https://doi.org/10.1007/978-3-031-86433-9_16

human body for cleaning, beautifying, adding to the attractiveness, altering the appearance, or keeping or promoting the skin or hair in good condition" including "items fulfilling specific actions like skin whitening, minimizing the appearance of lines in the face and body, protecting from the sun and sun tanning" (Gagliardi and Dorato 2007). Specific rules for cosmetics are set by local regulating agencies; however, they are almost harmonized globally. Independently of definitions, cosmetics are health products. Unlike food and drugs that have a systemic effect, cosmetics are not supposed to permeate through the skin. We highlighted some cosmetic effects when microalgae ingredients and/or biomass were administered as an oral supplement. However, supplements are regulated as food under the Dietary Supplement Health and Education Act of 1994 (DSHEA), and sometimes the term "nutricosmetic" is used for products that benefit skin, hair, or nails (Dini 2023).

16.3 Microalgae for Cosmetic Applications

Algae are organisms found mostly in aquatic environments and are characterized as macroalgae (seaweeds)—when the organism is multicellular and macroscopic—and as microalgae—when the organism is unicellular and microscopic. Both organisms can produce bioactive compounds when developing photosynthesis, which can be a source for use in cosmetic industry applications. The synthesis of these compounds in microalgae may vary during their growth according to exposure to external stress, amount of light, and carbon dioxide, becoming a promising source for obtaining bioactives of interest (Ariede et al. 2017; Gauthier et al. 2020).

The main microalgae used for commercial applications are *Spirulina, Chlorella, Haematococcus, Dunaliella, Ankistrodesmus, Botryococcus, Phaeodactylum*, and *Porphyridium* (Ariede et al. 2017). The potential use of biomass and/or bioactive derived from microalgae is still under development, and new harvesting and extraction technologies may extend their use in commercial cosmetics (Cezare-Gomes et al. 2019).

16.4 Main Bioactives Derived from Microalgae

16.4.1 Carotenoids

Carotenoid is a class of bioactive used in cosmetic products with anti-aging appeal and natural lipophilic pigment (de Mendes-Silva et al. 2020; Yang et al. 2023). In general, these ingredients have a linear structure containing eight isoprenoid units. They are derived from an acyclic chain (C40H46) with single bonds and a double system distributed along its length (de Mendes-Silva et al. 2020). Carotenoids protect photosynthetic organisms, such as algae, from excessive light by eliminating reactive oxygen species, thus being potential antioxidants for topical products. Several carotenoids have already been identified and separated into two groups: carotenes (secondary carotenoids that act mainly in the photoprotection of the microalgae) and xanthophylls (primary carotenoids which are essential for the life of microalgae as they are functional and structural components of the photosynthetic apparatus). Among the carotenoids produced by microalgae, some are of commercial interest, such as astaxanthin, lutein, lycopene, and β-carotene (Coulombier et al. 2020; Pourkarimi et al. 2020).

16.4.1.1 Carotenes

Carotenes are hydrocarbon carotenoids with no substituents (or even oxygen) in their structure (Sathasivam and Ki 2018).

16.4.1.1.1 β-carotene

β-carotene is a compound that can be converted into retinol since it is an active provitamin A (Fig. 16.1). This bioactive has an antioxidant action capable of acting against singlet oxygen and interrupting the formation of free radicals, helping improve the immune system and preventing eye diseases such as cataracts (Gong and Bassi 2016).

This active is widely used as an additive in the food industry because it has a characteristic light orange color and is used as a colorant (Meléndez-Martínez et al. 2007). In the medicinal area, studies have shown that the use of β-carotene extracted from *Dunaliella salina* may have potential action against types of cancer, such as leukemia and prostate cancer, if consumed in adequate amounts (Jayappriyan et al. 2013). Finally, in the cosmetic area, β-carotene is used as a natural pigment and is a promising asset for use in photoprotective formulations due to its antioxidant action, promoting action against skin aging and damage caused by ultraviolet (UV) radiation (Arct and Mieloch 2016; Zhuang et al. 2022). Due to its antioxidant action and being a precursor of vitamin A, this compound, when penetrating the epidermis, can minimize the deleterious effects caused by UV radiation, fighting free radicals and preventing the reduction of vitamin A levels in the skin (Antille et al. 2004).

β-Carotene can be found in several species of microalgae, including *Dunaliella* sp., *Botryococcus braunii, Streptomyces platensis, Coelastrella striolata, Chlorella zofingiensis, Arthrospira* sp., *Chlamydomonas, Muriellopsis, Haematococcus* spp., and *Coccomyxa onubensis*. Of these species, *Dunaliella salina* is the one that has the highest amount of active in its dry mass, with up to 10–12% of the active, favoring its production when this microalga grows under stress conditions, such as high intensity of light and salt, and therefore is used industrially to obtain the molecule (Cezare-Gomes et al. 2019; Pourkarimi et al. 2020).

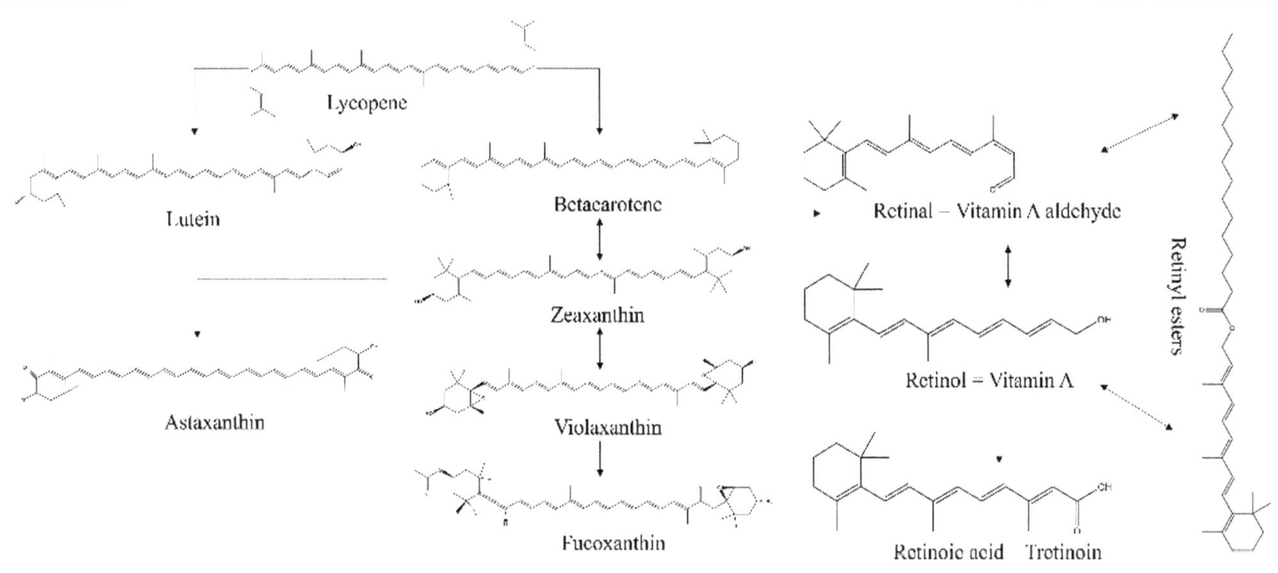

Fig. 16.1 Simplified Carotenoid's Biotransformation

16.4.1.1.2 Lycopene

Lycopene (Fig. 16.1) is a red-colored compound that is an intermediate for synthesizing other carotenoids (Meléndez-Martínez et al. 2007). In addition to being found in plant sources, it has also been identified in microalgae of the species Chlorella marina, being a biocompound used as a dye in food and in the nutraceutical area; besides, it is known to have antioxidant activity, which may help prevent cancer and reduce cardiovascular events, and also has anti-inflammatory effects (Gauthier et al. 2020; Gong and Bassi 2016; Zeng et al. 2019). Moreover, in the cosmetic area, it is an active of great interest since it can be used in anti-aging formulations, action against UV radiation damage, and helps to reduce skin wrinkles (Darvin et al. 2008; de Mendes-Silva et al. 2020).

Furthermore, the presence of lycopene in the skin of volunteers showed lower levels of wrinkle formation through non-invasive tests (resonance Raman spectroscopy); therefore, its topical use could help as an anti-aging cosmetic product (Darvin et al. 2008).

Besides, due to its antioxidant potential, lycopene is one of the carotenoids that can be used in photoprotective formulations to combat free radicals caused by UV radiation and protect the skin from erythema (Kiki 2023). The action of lycopene in a topical emulgel-type formulation on human skin was evaluated, making clear the potential of this active in improving skin hydration and elasticity, as well as in helping to reduce erythema due to its antioxidant potential (Sohail et al. 2022). To our knowledge, no studies of lycopene extraction using microalgae were found in the literature.

16.4.2 Xanthophylls

The second carotenoid family, the xanthophylls, has a distinct chemical structure from carotenes, as they contain oxygen, which donates them to antioxidant activities and cosmetic applications (Zittelli et al. 2023).

16.4.2.1 Astaxanthin

Astaxanthin is a pink/red pigment derived from β-carotene (Fig. 16.1) (Ekpe et al. 2018). The main bioactive property of astaxanthin is its powerful antioxidant capacity, admittedly superior to that of β-carotene and vitamin E (Cezare-Gomes et al. 2019). Some authors even claim that the antioxidant properties of astaxanthin are 65 times more potent than vitamin C and 50 times greater than that of vitamin E in terms of protecting cell membranes (Ekpe et al. 2018). Also, it can interact with lipid membranes in the presence of polar groups to anchor in the bilayer cell (Ekpe et al. 2018; Snell and Carberry 2022).

On the other hand, in the cosmetic area, astaxanthin shows potential in products for sun protection and anti-aging (Suganuma et al., 2010). Astaxanthin can mediate antioxidant, anti-aging, anti-inflammatory, and skin anticancer activity over UV exposure (Dutta et al. 2023). Also, it can contribute as an ingredient for sunscreen formulation by three main mechanisms: (1) the absorption of UV radiation, (2) the neutralization of the formation of free radicals formed by the incidence of UV light, and (3) the inhibition of matrix metalloproteinases, which are understood as agents that participate in aging and damaging the cutaneous tissue (Dutta et al. 2023; Ekpe et al. 2018). On the other hand, its low solubility in water and low oral instability limit its application in cosmetics and food supplementation. However, it has been circumvented by delivery systems (Lima et al. 2021; Stachowiak and Szulc 2021). Moreover, its bioactivity could be influenced by stereoisomer composition and extraction methods (Snell and Carberry 2022).

Among the different species of known microalgae, the essential source of astaxanthin is the species *Haematococcus pluvialis*, which accumulates approximately 3% of its dry biomass (Cezare-Gomes et al. 2019; Lima et al. 2021).

16.4.2.2 Fucoxanthin

It is a yellow-orange pigment that is highly unsaturated (Fig. 16.1). Thus, it is sensitive to oxidation and very reactive (Zittelli et al. 2023). Its biological function is associated with photosynthesis, helping to capture light and transfer energy (Kim et al. 2012). Numerous bioactive properties of fucoxanthin have already been described, such as antioxidant capacity, cytoprotective, anti-inflammatory, antidiabetic, anti-obesity, neuroprotective, anticancer effect, and skin protective effects (Lourenço-Lopes et al. 2021).

The most traditional natural sources of obtaining fucoxanthin are macroalgae, especially *Laminaria japonica, Eisenia bicyclis, Undaria pinnatifida*, and *Hizikia fusiformis*. However, macroalgae are consumed as food in Asia, which makes microalgae promising as producers of this. Fucoxanthin is the most abundant carotenoid produced by marine microalgae, representing more than 10% of total carotenoids (Zittelli et al. 2023). Despite this, few species have been used commercially for this purpose; among them, the most relevant is *Phaeodactylum tricornutum*. Finally, some especially *Diatom*, like *Haptophyta* and *Chrysophyte* microalgae, have been studied to synthesize fucoxanthin, obtaining ranges of 2–7% of dry biomass (Zittelli et al. 2023).

16.4.2.3 Lutein

Lutein is a yellowish color xanthophyll (Fig. 16.1). It is soluble in organic solvents and may become unstable under light, oxygen, high temperatures, and extreme pH exposure (Ochoa Becerra et al. 2020). In nature, lutein is an essential component for photon absorption and energy transfer for chlorophylls (Othman et al. 2014) and protection against oxidative damage caused by reactive oxygen species (ROS), sometimes photoinduced (Alves-Rodrigues and Shao 2004). When extracted, it is used as a food coloring in various parts of the world (Ochoa Becerra et al. 2020). It is known as a potent antioxidant through in vitro and in vivo tests, with antimutagenic and antiproliferative properties (Ribaya-Mercado and Blumberg 2004). Furthermore, it is related to have protective action on human vision for its strong absorption of visible light, acting as a photoprotective filter, and antioxidant protection on lipid peroxidation, associated with the development of cataracts (Alves-Rodrigues and Shao 2004).

Regarding the skin, lutein can improve the immune response and decrease oxidative damage induced by UV exposure (Bhosale and Bernstein 2005). However, Benefits such as improved skin elasticity, hydration, and reduction of skin roughness were reported, mainly for lutein oral intake (Shegokar and Mitri 2012; Žmitek et al. 2020). For topical applications, lutein may prevent skin damage caused by radiation when used in sunscreens, especially by absorbing UVA rays and regulating the inflammatory UV-induced cascade (Resende et al. 2022). Skincare products have recently explored the use of lutein; however, attention is required to maintain its stability in formulations and to control the yellow color of this ingredient, which may contribute to the product.

Commercially available lutein is usually obtained from plant sources such as petals of flowers of the genus Tagetes. Examples of lutein-producing microalgae are 0.5% of dry biomass for *Chlorella protothecoides* (Shi and Chen 2002), 0.7% of dry biomass for *Chlorella sorokiniana* (Cordero et al. 2011), and 0.6% of dry biomass for *Scenedesmus almeriensis* (Molino et al. 2020).

16.4.2.4 Violaxanthin

It is an orange color compound used as a dye in beverages and foods (Fig. 16.1) (Meléndez-Martínez et al. 2007; Novais et al. 2022). Its primary biological function is the photoprotection of the photosynthetic system under the excess light incidence by the reversible conversion of violaxanthin to zeaxanthin. This reaction leads to the dissipation of excess energy by heat, also called "non-photochemical quenching" (Avenson et al. 2008).

Regarding bioactive properties, anti-inflammatory effects have been described by in vitro tests, based on inhibiting the NF-κB signaling pathway, promising for treating inflammatory diseases (Soontornchaiboon et al. 2012). Violaxanthin also inhibits the proliferation of MS-7 breast cancer cells by inducing biochemical changes typical of apoptosis and seems to have a negative correlation with colon cancer cell proliferation (Milani et al. 2017).

Furthermore, in dermocosmetics, violaxanthin may be used in sunscreens as a UVB filter. The molecule showed cellular protection of human fibroblasts under elevated production of ROS (Kim et al. 2019). However, no cosmetics are currently exploring this molecule commercially.

Some examples of violaxanthin-producing microalgae are *Chlorella ellipsoidea* (Soontornchaiboon et al. 2012), *Dunaliella tertiolecta* (Pasquet et al. 2011), and *Nannochloropsis* sp. as *N. gaditana* (Ryckebosch et al. 2014).

16.4.2.5 Zeaxanthin

It is an orange/yellow (Fig. 16.1) safe food coloring (Torregrosa-Crespo et al. 2018) and antioxidant supplement

(Ashenafi et al. 2023; Bouyahya et al. 2021). Its biological function is associated with the xanthophyll cycle, in which violaxanthin is reversibly converted into zeaxanthin to dissipate excess energy from light incidence, avoiding damage to the photosynthetic apparatus (Demmig-Adams et al. 2020). Also, zeaxanthin is related to the reduction of the risk of developing age-related macular degeneration and cataracts (Jia et al.2017), and it shows a protective factor against the development of some forms of cancer when associated with lutein (Ribaya-Mercado and Blumberg 2004). Moreover, in the skin, the oral intake of the mixture of these two xanthophylls reduced lipid peroxidation and improved skin moisture (Obana et al. 2020).

In cosmetics, combining a zeaxanthin supplement and a topical serum improved skin hydration and facial lines and wrinkles (Schwartz et al. 2016). Commercially, several brands currently use zeaxanthin as an oral supplement for skincare benefits. Even if the market is still developing for topical formulation, the zeaxanthin market is forecasted to grow 8.2% throughout 2030 (Transparency Market Research 2019). The production of zeaxanthin is 4.5% of dry biomass for *Chlorella ellipsoidea*, 0.6% of dry biomass for *Dunaliella salina* (mutant), and *Porphyridium cruentum* (Cezare-Gomes et al. 2019).

16.4.3 Fat-Soluble Vitamins

A, E, and K are the fat-soluble vitamins of interest for incorporation into cosmetics. The synthesis and/or accumulation of vitamins by photosynthetic organisms is highly variable and intensely related to physiological comebacks to environmental stimuli. The above-mentioned organisms synthesize most of the vitamins, while others (some vitamins B and K) are bioaccumulated through the diet and largely produced by bacteria (Del Mondo et al. 2020).

16.4.3.1 Vitamin A

Vitamin A and its derivatives can regulate keratinization. The most used vitamin A derivatives are vitamin A alcohol (retinol), vitamin A esters (retinol palmitate and acetate), vitamin A aldehyde (retinal), and tretinoin (retinoic acid) (Fig. 16.1). Besides, retinoids (synthetic and natural analogs of vitamin A) are a class of compounds derived from vitamin A that have structural and/or functional similarities with vitamin A (Zasada and Budzisz 2019).

The great interest in producing vitamin A and its derivatives, mainly retinol, is due to the practical form of these compounds that delay the aging process. Besides, due to its lipophilicity, retinol penetrates the stratum corneum and slightly penetrates the dermis. In this way, retinol provides benefits for the skin such as stimulating the cellular activity of keratinocytes, fibroblasts, melanocytes, and Langerhans cells; promoting the profiling of keratinocytes; strengthening the epidermal protective function; reducing transepidermal water loss; protecting collagen against degradation and inhibiting the activity of metalloproteinases responsible for extracellular matrix degradation. Retinol also enhances the remodeling of reticular fibers and stimulates angiogenesis in the dermis papillary layer (Zasada and Budzisz 2019).

The main form of vitamin A production by microalgae is from the provitamin A carotenoids (e.g., β-carotene, β-cryptoxanthin, and α-carotene) (Del Mondo et al. 2020). Obtaining retinol from microalgae such as *Tetraselmis* (0.0378% of retinol equivalent) and *Chatetoceros* (0.009% of retinol equivalents) is greater than the amount obtained from edible carrots and oranges (0.00099 and 0.000027%, respectively). Another example of microalgae that produces vitamin A is *Dunaliella salina*, which produces a large amount of provitamin A (β-carotene) (Gohara-Beirigo et al. 2022).

16.4.3.2 Vitamin E

Vitamin E has four different chemical forms (alpha, beta, gamma, and sigma), with alpha-tocopherol being the most potent molecule in terms of antioxidant potential and biological activity (Esfandiari et al. 2022; Traber and Bruno 2020). Also, vitamin E is relevant in humans due to its antioxidant capacity, mainly in preventing lipid peroxidation and diseases related to oxidative stress, especially in epithelial tissues (Del Mondo et al. 2020; Lupo 2001).

Furthermore, tocopherols are widely used in cosmetic skin care products (Sivakumar et al. 2014). Also, due to its liposoluble properties, vitamin E can permeate into the stratum corneum's deeper layers through the secretions of the sebaceous glands. This ability of vitamin E provides its permanence within cell membranes, protecting them from oxidative stress, and may reduce UV-stimulated hyperpigmentation (Rattanawiwatpong et al. 2020).

In microalgae, vitamin E is an antioxidant and combats photooxidative stress. It has a complementary role in the xanthophyll cycling pigments as a photoprotector and is associated with producing polyphenols in response to abiotic stress, such as light, nutrients, or metals (Del Mondo et al. 2020).

Using microalgae to obtain vitamin E is more efficient than using food from the diet. The percentage of vitamin E in relation to the fresh weight of vegetables and microorganisms can be exemplified as follows: 0.003% for green olives, 0.002% for raw spinach, and 0.001% for blackberries and cranberries. The microalgae *Tetraselmis, Nannochloropsis*, and *Anabaena* reached 0.063, 0.040, and 0.048% of α-tocopherol, respectively (Del Mondo et al. 2020). Moreover, *Dunaliella salina* synthesizes vitamin E in a higher quantity (0.123%) than soybean oil (0.114%) (Gohara-Beirigo et al. 2022).

16.4.3.3 Vitamin K

Vitamin K, more specifically phylloquinone (vitamin K1), has shown therapeutic effects on the skin: (1) prevention of acneiform rash, which is a skin reaction of using epidermal growth factor receptor inhibitors, (2) prevention and cleaning of laser-induced bruises, and (3) contribution to wound healing. These effects might result from the antioxidant effect of the reduced form of vitamin K (vitamin K-hydroquinone) acting against ROS (Goto et al. 2020). Additionally, vitamin K can prevent and remove extravasated blood in the skin, when applied topically at least twice a day in concentrations of 1%, helping to reduce dark circles and bruises (Lupo 2001).

Topical application of vitamin K has benefits for the skin; however, cosmetics containing phylloquinone are limited in Europe due to their photosensitizing properties, such as photodegradation and phototoxicity. An alternative would be using ester derivatives of phyllohydroquinone (PKH). PKH ester derivatives overcome the light instability of vitamin K and do not produce ROS, which is related to cellular toxicity due to UVA light. In this way, topical application of PKH derivatives can be considered an approach to increase the efficacy of vitamin K and guarantee optimal delivery into the skin (Goto et al. 2020).

Microalgae biosynthesis is able to produce only the active E-isomer of phylloquinone, while chemical processes usually produce 10–20% of the inactive Z-isomer. Microalgae *Anabaena cylindrica* shows the highest production with 0.02% of dry biomass, six times higher than that found in foods such as spinach and parsley (Tarento et al. 2018).

16.4.4 Fatty Acids

Fatty acids are molecules containing a carboxyl group and a carbon chain, which generally contain 4 to 28 carbons and can be saturated or polyunsaturated (Moss et al. 1995). Microalgae have become interesting due to the composition of their lipid fraction being rich in fatty acids, including essential and polyunsaturated fatty acids (PUFAs). The main omega-3 PUFAs consist of eicosapentaenoic acid (EPA) and docosahexaenoic acid (DHA). EPA and DHA are anti-inflammatories and are important compounds for brain development, especially in early childhood, and also for improving mental health (Gohara-Beirigo et al. 2022). Other examples of PUFAs are linolenic acid (LA; omega-6) and alpha-linolenic acid (ALA; omega-3). They are present in marine fish, chia seeds (*Salvia hispanica*), flax (*Linum usitatissium*), and microalgae (*Chlorella pyrenoidosa-1; Chlorella pyrenoidosa-2; Chlorella pyrenoidosa-3; Chlorella vulgaris*) (Ötleô and Pire 2011; Saini and Keum 2018).

Currently, the primary sources of EPA and DHA are marine fish that consume seaweed. However, the number of fish has been decreasing, and sometimes, they may be contaminated, making it difficult to obtain these PUFAs. Therefore, microalgae can be used as an alternative and more sustainable source for obtaining PUFAs (Ibrahim et al. 2023). Additionally, considering the sensory aspects, products of microalgal origin do not have the strong smell characteristic of oil extracted from fish, which can be an advantage for certain consumers (Gohara-Beirigo et al. 2022). Finally, using microalgae decreases the accumulation of heavy metals in fish lipids, which occur by bioaccumulation in the food chain (Bin Lee et al. 2016).

On the other hand, the PUFAs properties make them interesting raw materials for cosmetics, such as combating inflammatory processes and ongoing oxidative stress that are associated with photoaging and skin cancer resulting from exposure to UV radiation (Huang et al. 2018), in addition to integrating formulations to mitigate skin irritations (Brunt and Burgess 2018). Besides, effects in combating UV-induced inflammatory processes were observed through oral supplementation and topical application of omega-3 and omega-6 PUFAs (Huang et al. 2018).

Some microalgae synthesize relevant amounts of EPA and DHA, such as those of the genus *Thraustochytrium* sp. which contains 0.87% of EPA/dry mass and 0.13% of DHA/dry mass (Scott et al. 2011). For comparison purposes, salmon species have an average of 0.014% of EPA + DHA mixture/wet mass, while other fish, such as eels, can contain up to 0.0084% of EPA + DHA mixture/fish wet mass (Sprague et al. 2020). In terms of dry mass comparison, a study with 12 species of Black Sea fish determined that the highest concentration of 0.0075% EPA + DHA mixture/dry mass was in *Trachurus trachurus* (Huang et al. 2018).

Furthermore, microalgae can produce alpha-linolenic acid (ALA), with a fraction of high added value obtained from the fractionation of lipids from *Chlorella sorokiniana* in the production of biodiesel (Parsaeimehr et al. 2015). Similar results were also obtained with *Chlorella pyrenoidosa* (Kokkiligadda et al. 2017) and *Desmodesmus* sp. (Sijil et al. 2019), with a productivity of up to 15.9 mg L^{-1} day^{-1} of ALA (Nagappan and Kumar Verma 2018). Besides, linolenic acid (LA) can be found in different photosynthetic microorganisms, such as *Spirulina* sp. (Ötleô and Pire 2011), *Chlorella* sp., and *Chlorococcum* sp. (Sahu et al. 2013), rarely exceeding 20% of detected fatty acids.

Table 16.1 shows specific studies regarding bioactive molecules obtained from microalgae, and Fig. 16.2 summarizes the applications of microalgae in cosmetics.

Table 16.1 Potential application of bioactives from microalgae in cosmetics

Activities in cosmetics	Bioactive	Experimental model	References
Anti-aging	Astaxanthin	In vitro assay after UVA exposure in human dermal fibroblast and evaluation of MMP-1 activity, elastase/neutral endopeptidase	Suganuma et al. (2010)
	Lutein	In vivo oral supplementation in humans.	Juturu et al. (2016), Krisnayana et al. (2021), Žmitek et al. (2020)
		In silico model of lutein as inhibitor of MMP-1	
		In vivo oral supplementation in humans using Cutometer®, Chromameter, and Visia-CR®	
	Microalgae peptides	In vitro, in cells, in animal models, and in human clinical trials	Xia et al. (2021)
	DHA	In vivo study in female HR-1 hairless mice skin	Rahman et al. (2011)
		Topical application of DHA prior to UVB irradiation	
	Astaxanthin	Human clinical study (oral and topic) evaluated skin wrinkle, age spot size elasticity, moisture content, sebum oil and skin texture and TEWL	Tominaga et al. (2012)
Anti-wrinkle		In vivo human topical cream evaluated moisture content	Lima et al. (2021)
	Lycopene	Non-invasive tests on volunteers in vivo	Darvin et al. (2008)
	Vitamin A	Human clinical study: Individuals with 80 years old or more were treated topically for 7 days with 1% retinol and its vehicle. 4 mm full-thickness punch biopsies were obtained from each individual	Varani et al. (2000)
Antioxidants	Astaxanthin	Hydrogel and lipogel of algae extract and/or oleoresin formulations were performed in vitro assay for DPPH and ABTS	Eren et al. (2019)
		In vitro antioxidative assay of liposomal formulation	Hama et al. (2012)
		UV exposure of mouse dorsal skin and pigmentation evaluation	
	Lutein	In vivo oral supplementation in humans	Antille et al. (2004)
	Vitamin E	In vitro assay with HaCaT cell line. Cells were either treated with vitamin E for 24 h prior to UVA irradiation	Delinasios et al. (2018)
	Extract of *Neochloris oleoabundans*	Hydrogels evaluated in vivo by skin superficial hydration, TEWL, and anti-inflammatory essays (laser Doppler)	Morocho-Jácome et al. (2022)
Adjuvants	Biopolymers (Carrageenans, alginates)	Gelling properties, emulsion stabilizers, and rheology modifiers	Lopez-Hortas et al. (2021)
Dyes/pigments	β-carotene	–	Zhuang et al. (2022)
	Fucoxanthin	–	Zittelli et al. (2023)
Hair growth	Me-meijicoccene/C32 botryococcene	In vitro with hair follicle dermal papilla cells (HFDPCs)	Oliva et al. (2022)
Moisturizer	Lycopene	Non-invasive tests in human skin (Corneometer®)	Sohail et al. (2022)
Sunscreen/UV filter	Astaxanthin	Protection against UV-induced inflammation by decreasing iNOs and COX2 and inhibiting apoptosis in keratinocytes	Yoshihisa et al. (2014)
	Fucoxanthin	Solid lipid nanoparticle systems were used as carrier of fucoxanthin in sunburn protection factor analysis	Lee and Nam (2020)
		Fibroblast cells were exposed to UV-B and were performed ROS measurement, cell viability (MTT), and induced DNA damage by comet assay by UVB	Heo and Jeon (2009)
	Lycopene	Non-invasive tests in human skin for skin erythema (Mexameter®)	Sohail et al. (2022)
	Lutein	In vivo oral supplementation in humans	Astner et al. (2007), Stahl and Sies (2002), Žmitek et al. (2020)
		In vivo topical application in mice reduced UVB erythema	
		In vivo oral supplementation in mice	
	Zeaxanthin	In vivo oral supplementation in mice	Astner et al. (2007), Roberts et al. (2009)
		In vitro human dermal fibroblasts protection by MMPs inhibition	
	MAAS	In vitro human fibroblasts	Peng et al. (2023), Raj et al. (2021)
	Extract of *Chlorella vulgaris*	In vitro sun protection factor	dos Santos et al. (2023)
Whitening	Fucoxanthin	In vitro tyrosinase activity and melanogenesis	Shimoda et al. (2010)
		Analysis of mice skin mRNA expression related to melanogenesis after topical and oral supplementation (after UVB irradiation)	
	Linoleic acid (liposomal)	Application of liposomal linoleic acid to UV-stimulated hyperpigmented dorsal porcine skin of brown guinea pigs	Shigeta et al. (2004)
	Zeaxanthin	In vitro tyrosinase inhibitor	Ji et al. (2021)

DPPH 1,1-Diphenyl-2-picrylhydrazyl, *ABTS* 2,2'azinobis-(3-ethylbenzothiazoline-6-sulfo- nic acid), *MTT* 3-(4,5-dimethylthiazol-2-yl)2,5-diphenyltetrazolium bromide, *ROS* reactive oxygen species, *DHA* Docosahexaenoic Acid, *MAAS* Amino acids like mycosporin, *TEWL* transepidermal water loss

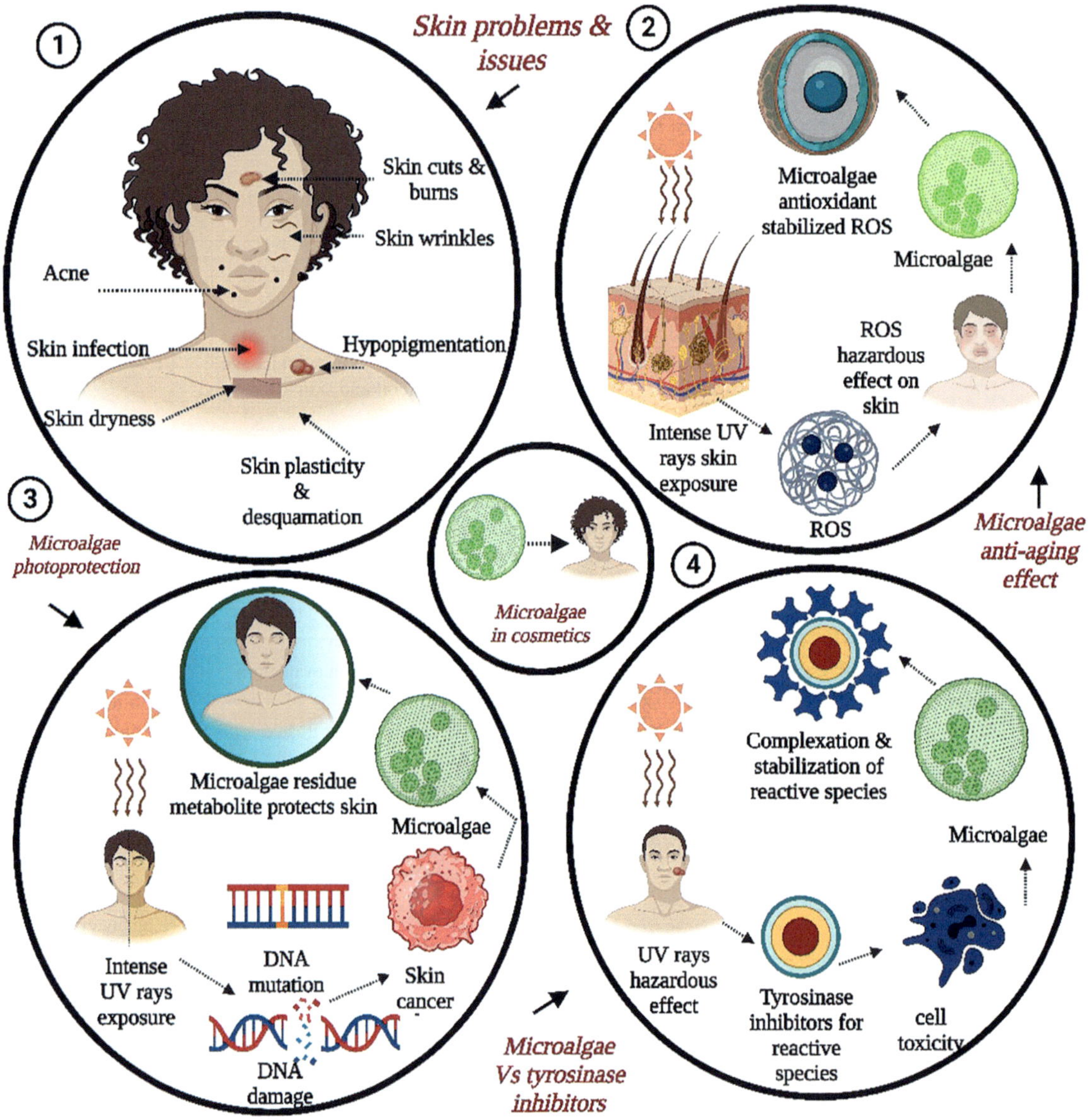

Fig. 16.2 Summary of applications of microalgae in the cosmetic industry for photoprotection, skin whitening, and anti-aging effect

16.5 Technologies and Limitations

Microalgae bioactives have been used in the cosmetic industry, but their applications are limited mainly due to their isolation process, especially for bioderived peptides. Isolating a single peptide fraction from the microalgae is quite a challenging process. These applications require a specific peptide for better interaction with the selected target rather than non-specific interactions with biological components. The second issue is associated with potential allergens obtained from unpurified mixtures of peptides, which induce contamination and toxic effects (Costa et al. 2021; Khan et al. 2018). For that reason, several tests are crucial for implementing natural source ingredients in cosmetics, including phototoxicity, carcinogenicity, genetic toxicity, and photogenotoxicity, along with the toxicokinetic for safety measurements. Especially for microalgae drive, cosmetic compounds must have cytotoxicity tests (Jain et al. 2018; Oh et al. 2015). Many chemical substances used in cosmetics show some adverse effects, including hypersensitivity and anaphylactic reactions, requiring proper health surveillance. The effective monitorization and detection of cosmetic preparation's side effects in humans are called "cosmetovigilance," a growing sector (Pereira and Pereira 2018; Wang et al. 2015). Other limitations of microalgal natural ingredients in cosmetics are stability and the organoleptic characteristics of ingredients that may affect the final product's appearance and performance.

Also, the lower productivity and low growth rate of microalgae are responsible for scale-up issues in the process.

The complexity of cultivation and processing factors, including light intensity, temperature, stress conditions, and starvation nutrients, put extra pressure on the final biomass composition and make the process expensive. The extraction, isolation, and purification process also contributed to high economic costs incurred during the same (De Luca et al.2021; Jones et al. 2019).

16.6 Conclusion and Future Perspectives

Microalgae are a feasible and sustainable source of ingredients for cosmetics that can positively interact with and improve several skin attributes. Microalgae bioactives show great potential for oral and topical applications. However, they are still little explored as a new renewable claim of cosmetics.

References

Alves-Rodrigues A, Shao A (2004) The science behind lutein. Toxicol Lett 150(1). https://doi.org/10.1016/j.toxlet.2003.10.031

Antille C, Tran C, Sorg O, Saurat JH (2004) Topical β-carotene is converted to retinyl esters in human skin ex vivo and mouse skin in vivo. Exp Dermatol 13(9):558–561. https://doi.org/10.1111/j.0906-6705.2004.00194.x

Arct J, Mieloch M (2016) β-carotene in skin care. Pol J Cosmetol 19(3):206–213. https://www.researchgate.net/publication/347513828

Ariede MB, Candido TM, Jacome ALM, Velasco MVR, de Carvalho JCM, Baby AR (2017) Cosmetic attributes of algae - a review. Algal Res 25(June):483–487. https://doi.org/10.1016/j.algal.2017.05.019

Ashenafi EL, Nyman MC, Shelley JT, Mattson NS (2023) Spectral properties and stability of selected carotenoid and chlorophyll compounds in different solvent systems. Food Chem Adv 2. https://doi.org/10.1016/j.focha.2022.100178

Aslam A, Bahadar A, Liaquat R, Saleem M, Waqas A, Zwawi M (2021) Algae as an attractive source for cosmetics to counter environmental stress. Sci Total Environ 772:144905. https://doi.org/10.1016/j.scitotenv.2020.144905

Astner S, Wu A, Chen J, Philips N, Rius-Diaz F, Parrado C, Mihm MC, Goukassian DA, Pathak MA, González S (2007) Dietary lutein/zeaxanthin partially reduces photoaging and photocarcinogenesis in chronically UVB-irradiated Skh-1 hairless mice. Skin Pharmacol Physiol 20(6). https://doi.org/10.1159/000107576

Avenson TJ, Tae KA, Zigmantas D, Niyogi KK, Li Z, Ballottari M, Bassi R, Fleming GR (2008) Zeaxanthin radical cation formation in minor light-harvesting complexes of higher plant antenna. J Biol Chem 283(6). https://doi.org/10.1074/jbc.M705645200

Bhosale P, Bernstein PS (2005) Microbial xanthophylls. Appl Microbiol Biotechnol 68(4). https://doi.org/10.1007/s00253-005-0032-8

Bouyahya A, El Omari N, Hakkur M, El Hachlafi N, Charfi S, Balahbib A, Guaouguaou FE, Rebezov M, Maksimiuk N, Shariati MA, Zengin G, El Menyiy N, Chamkhi I, Bakrim S (2021) Sources, health benefits, and biological properties of zeaxanthin. Trends Food Sci Technol 118. https://doi.org/10.1016/j.tifs.2021.10.017

Brunt EG, Burgess JG (2018) The promise of marine molecules as cosmetic active ingredients. Int J Cosmet Sci 40(1):1–15). Blackwell Publishing Ltd. https://doi.org/10.1111/ics.12435

Cezare-Gomes EA, del Mejia-da-Silva LC, Pérez-Mora LS, Matsudo MC, Ferreira-Camargo LS, Singh AK, de Carvalho JCM (2019) Potential of microalgae carotenoids for industrial application. Appl Biochem Biotechnol 188(3). https://doi.org/10.1007/s12010-018-02945-4

Chauton MS, Forbord S, Mäkinen S, Sarno A, Slizyte R, Mozuraityte R, Standal IB, Skjermo J (2021) Sustainable resource production for manufacturing bioactives from micro- and macroalgae: examples from harvesting and cultivation in the Nordic region. Physiol Plant 173(2):495–506. https://doi.org/10.1111/ppl.13391

Costa JAV, Lucas BF, Alvarenga AGP, Moreira JB, de Morais MG (2021) Microalgae polysaccharides: An overview of production, characterization, and potential applications. Polysaccharides 2(4). https://doi.org/10.3390/polysaccharides2040046

Coulombier N, Nicolau E, Le Déan L, Antheaume C, Jauffrais T, Lebouvier N (2020) Impact of light intensity on antioxidant activity of tropical microalgae. Mar Drugs 18(2). https://doi.org/10.3390/md18020122

Darvin M, Patzelt A, Gehse S, Schanzer S, Benderoth C, Sterry W, Lademann J (2008) Cutaneous concentration of lycopene correlates significantly with the roughness of the skin. Eur J Pharm Biopharm 69(3):943–947. https://doi.org/10.1016/j.ejpb.2008.01.034

de Mendes-Silva TCD, da Andrade RFS, Ootani MA, Mendes PVD, de Sá RAQC, de Silva MRF, da Souza KS, dos Correia MTS, da Silva MV, de Oliveira MBM (2020) Biotechnological potential of carotenoids produced by Extremophilic microorganisms and application prospects for the cosmetics industry. Adv Microbiol 10(08):397–410. https://doi.org/10.4236/aim.2020.108029

De Luca C, Lievore G, Bozza D, Buratti A, Cavazzini A, Ricci A, Macis M, Cabri W, Felletti S, Catani M (2021) Downstream Processing of Therapeutic Peptides by Means of Preparative Liquid Chromatography. Molecules 26. https://doi.org/10.3390/molecules26154688

Del Mondo A, Smerilli A, Sané E, Sansone C, Brunet C (2020) Challenging microalgal vitamins for human health. In: Microbial cell factories, vol 19(1). BioMed Central Ltd. https://doi.org/10.1186/s12934-020-01459-1

Delinasios GJ, Karbaschi M, Cooke MS, Young AR (2018) Vitamin E inhibits the UVAI induction of "light" and "dark" cyclobutane pyrimidine dimers, and oxidatively generated DNA damage, in keratinocytes. Sci Rep 8(1). https://doi.org/10.1038/s41598-017-18924-4

Demmig-Adams B, Stewart JJ, Adams WW, López-Pozo M, Polutchko SK (2020) Zeaxanthin, a molecule for photoprotection in many different environments. Molecules 25(24). https://doi.org/10.3390/MOLECULES25245825

Dini I (2023) The potential of algae in the nutricosmetic sector. Molecules 28(10). https://doi.org/10.3390/molecules28104032

dos Santos BB, Morocho-Jácome AL, Prieto ZA, Morowvat MH, Lima FV, Velasco MVR, de Carvalho JCM, Baby AR (2023) Algae materials for cosmetics and cosmeceuticals. In: Algae materials: applications benefitting health. https://doi.org/10.1016/B978-0-443-18816-9.00003-4

Dutta S, Kumar SPJ, Banerjee R (2023) A comprehensive review on astaxanthin sources, structure, biochemistry and applications in the cosmetic industry. In: Algal research, vol 74. Elsevier B.V. https://doi.org/10.1016/j.algal.2023.103168

Ekpe L, Inaku K, Ekpe V (2018) Antioxidant effects of astaxanthin in various diseases—a review. J Molec Pathophysiol 7(1):1. https://doi.org/10.5455/jmp.20180627120817

Eren B, Tuncay Tanrıverdi S, Aydın Köse F, Özer Ö (2019) Antioxidant properties evaluation of topical astaxanthin formulations as anti-aging products. J Cosmet Dermatol 18(1):242–250. https://doi.org/10.1111/jocd.12665

Esfandiari M, Bagheri H, Mirarab-Razi V, Mirarab Razi S, Rashidinejad A (2022) Vitamin E. In: Handbook of food bioactive

ingredients. Springer International Publishing, pp 1–23. https://doi.org/10.1007/978-3-030-81404-5_24-1
Gagliardi L, Dorato S (2007) General concepts. Current legislation on cosmetics in different countries. In: Analysis of cosmetic products. https://doi.org/10.1016/B978-044452260-3/50024-3
Gauthier MR, Senhorinho GNA, Scott JA (2020) Microalgae under environmental stress as a source of antioxidants. In: Algal research, vol 52. Elsevier B.V. https://doi.org/10.1016/j.algal.2020.102104
Gohara-Beirigo AK, Matsudo MC, Cezare-Gomes EA, de Carvalho JCM, Danesi EDG (2022) Microalgae trends toward functional staple food incorporation: sustainable alternative for human health improvement. In: Trends in food science and technology, vol 125. Elsevier Ltd., pp 185–199. https://doi.org/10.1016/j.tifs.2022.04.030
Gong M, Bassi A (2016) Carotenoids from microalgae: a review of recent developments. In: Biotechnology advances, vol 34(8). Elsevier Inc., pp 1396–1412. https://doi.org/10.1016/j.biotechadv.2016.10.005
Goto S, Setoguchi S, Nagata-Akaho N, Terada K, Watase D, Yamakawa H, Toki E, Koga M, Matsunaga K, Karube Y, Takata J (2020) Ester derivatives of phyllohydroquinone effectively deliver the active form of vitamin K1 topically, owing to their non-photosensitivity. Eur J Pharm Sci 155. https://doi.org/10.1016/j.ejps.2020.105519
Goyal N, Jerold F (2021) Biocosmetics: technological advances and future outlook. Environ Sci Pollut Res 0123456789. https://doi.org/10.1007/s11356-021-17567-3
Hama S, Takahashi K, Inai Y, Shiota K, Sakamoto R, Yamada A, Tsuchiya H, Kanamura K, Yamashita E, Kogure K (2012) Protective effects of topical application of a poorly soluble antioxidant astaxanthin liposomal formulation on ultraviolet-induced skin damage. J Pharm Sci 101(8):2909–2916. https://doi.org/10.1002/jps.23216
Heo SJ, Jeon YJ (2009) Protective effect of fucoxanthin isolated from Sargassum siliquastrum on UV-B induced cell damage. J Photochem Photobiol B Biol 95(2):101–107. https://doi.org/10.1016/j.jphotobiol.2008.11.011
Huang TH, Wang PW, Yang SC, Chou WL, Fang JY (2018) Cosmetic and therapeutic applications of fish oil's fatty acids on the skin. In: Marine Drugs, vol 16(8). MDPI AG. https://doi.org/10.3390/md16080256
Ibrahim TNBT, Feisal NAS, Kamaludin NH, Cheah WY, How V, Bhatnagar A, Ma Z, Show PL (2023) Biological active metabolites from microalgae for healthcare and pharmaceutical industries: a comprehensive review. In: Bioresource technology, vol 372, Elsevier Ltd. https://doi.org/10.1016/j.biortech.2023.128661
Jain AK, Singh D, Dubey K, Maurya R, Mittal S, Pandey AK (2018) Models and methods for in vitro toxicity. In: In Vitro toxicology. Elsevier, pp 45–65. https://doi.org/10.1016/B978-0-12-804667-8.00003-1
Jayappriyan KR, Rajkumar R, Venkatakrishnan V, Nagaraj S, Rengasamy R (2013) In vitro anticancer activity of natural β-carotene from Dunaliella salina EU5891199 in PC-3 cells. Biomed Prevent Nutr 3(2):99–105. https://doi.org/10.1016/j.bionut.2012.08.003
Jia Y-P, Sun L, Yu H-S, Liang L-P, Li W, Ding H, Song X-B, Zhang L-J (2017) The Pharmacological Effects of Lutein and Zeaxanthin on Visual Disorders and Cognition Diseases. Molecules 22. https://doi.org/10.3390/molecules22040610
Ji K, Kim Y, Kim YT (2021) A study on the tyrosinase inhibitory and antioxidant effect of microalgae extracts. Microbiol Biotechnol Lett 49(2). https://doi.org/10.48022/mbl.2012.12001
Jones AD, Boundy-Mills KL, Barla GF, Kumar S, Ubanwa B, Balan V (2019) Microbial lipid alternatives to plant lipids. In: Methods in molecular biology, vol 1995. https://doi.org/10.1007/978-1-4939-9484-7_1
Juturu V, Bowman JP, Deshpande J (2016) Overall skin tone and skin-lightening-improving effects with oral supplementation of lutein and zeaxanthin isomers: a double-blind, placebo-controlled clinical trial. Clin Cosmet Investig Dermatol 9. https://doi.org/10.2147/CCID.S115519
Khan MI, Shin JH, Kim JD (2018) The promising future of microalgae: current status, challenges, and optimization of a sustainable and renewable industry for biofuels, feed, and other products. Microb Cell Factories 17(1). https://doi.org/10.1186/s12934-018-0879-x
Kiki MJ (2023) Biopigments of microbial origin and their application in the cosmetic industry. In: Cosmetics, vol 10(2). MDPI. https://doi.org/10.3390/cosmetics10020047
Kim SM, Jung YJ, Kwon ON, Cha KH, Um BH, Chung D, Pan CH (2012) A potential commercial source of fucoxanthin extracted from the microalga Phaeodactylum tricornutum. Appl Biochem Biotechnol 166(7):1843–1855. https://doi.org/10.1007/s12010-012-9602-2
Kim HM, Jung JH, Kim JY, Heo J, Cho DH, Kim HS, An S, An IS, Bae S (2019) The protective effect of Violaxanthin from Nannochloropsis oceanica against ultraviolet B-induced damage in normal human dermal fibroblasts. Photochem Photobiol 95(2). https://doi.org/10.1111/php.13030
Kokkiligadda S, Pandey B, Ronda SR (2017) Effect of plant growth regulators on production of alpha-linolenic acid from microalgae Chlorella pyrenoidosa. Sadhana - Acad Proc Eng Sci 42(10):1821–1824. https://doi.org/10.1007/s12046-017-0723-8
Krisnayana IGB, Febyani PD, Sari IAYP, Laksmiani NPL (2021) Molecular docking of lutein as anti-photoaging agent in silico. Pharm Rep 1(1). https://doi.org/10.51511/pr.15
Lee YJ, Nam GW (2020) Sunscreen boosting effect by solid lipid nanoparticles-loaded fucoxanthin formulation. Cosmetics 7(1). https://doi.org/10.3390/cosmetics7010014
Lee JB, Kim MK, Kim BK, Kim JY, Lee KG (2016) Analysis of polychlorinated biphenyls (PCBs), heavy metals and omega-3 fatty acids in commercially available Korean functional fish oil supplements. Int J Food Sci Technol 51(10):2217–2224. https://doi.org/10.1111/ijfs.13198
Lima SGM, Freire MCLC, da Oliveira VS, Solisio C, Converti A, de Lima ÁAN (2021) Astaxanthin delivery systems for skin application: a review. In: Marine Drugs, vol 19(9). MDPI. https://doi.org/10.3390/MD19090511
Lopez-Hortas L, Florez-Fernandez N, Torres MD, Ferreira-Anta T, Casas MP, Balboa EM, Falque E, Domínguez H (2021) Applying seaweed compounds in cosmetics, cosmeceuticals and nutricosmetics. Mar Drugs 19(10):1–30. https://doi.org/10.3390/md19100552
Lourenço-Lopes C, Fraga-Corral M, Jimenez-Lopez C, Carpena M, Pereira AG, Garcia-Oliveira P, Prieto MA, Simal-Gandara J (2021) Biological action mechanisms of fucoxanthin extracted from algae for application in food and cosmetic industries. In: Trends in food science and technology, vol 117. Elsevier Ltd., pp 163–181. https://doi.org/10.1016/j.tifs.2021.03.012
Lupo MP (2001) Antioxidants and vitamins in cosmetics. Clin Dermatol 19(4):467
Meléndez-Martínez AJ, Britton G, Vicario IM, Heredia FJ (2007) Relationship between the colour and the chemical structure of carotenoid pigments. Food Chem 101(3). https://doi.org/10.1016/j.foodchem.2006.03.015
Milani A, Basirnejad M, Shahbazi S, Bolhassani A (2017) Carotenoids: biochemistry, pharmacology and treatment. Br J Pharmacol 174(11):1290–1324. https://doi.org/10.1111/bph.13625
Molino A, Mehariya S, Iovine A, Casella P, Marino T, Karatza D, Chianese S and Musmarra D (2020) Enhancing Biomass and Lutein Production From Scenedesmus almeriensis: Effect of Carbon Dioxide Concentration and Culture Medium Reuse. Front Plant Sci 11. https://doi.org/10.3389/fpls.2020.00415
Morocho-Jácome AL, Dos Santos BB, de Carvalho JCM, de Almeida TS, Rijo P, Velasco MVR, Rosado C, Baby AR (2022) Microalgae as a sustainable, natural-oriented and Vegan Dermocosmetic bioactive ingredient: the case of Neochloris oleoabundans. Cosmetics 9(1). https://doi.org/10.3390/cosmetics9010009

Morone J, Lopes G, Preto M, Vasconcelos V, Martins R (2020) Exploitation of filamentous and picoplanktonic cyanobacteria for cosmetic applications: potential to improve skin structure and preserve dermal matrix components. Mar Drugs 18(9). https://doi.org/10.3390/MD18090486

Moss G, Smith P, Tavernier D, Bos H, Boulton A, Godly E, Griinanger P, McNaught A, PanicO R, Rigaudy J, Stocker J, Jones R, March J, McBnde J, Miiller P (1995) Glossary of class names of organic compounds and reative intermediates based on structure (IUPAC Recomendations 1995). Pure Appl Chem 67:1307–1375

Nagappan S, Kumar Verma S (2018) Co-production of biodiesel and alpha-linolenic acid (omega-3 fatty acid) from microalgae, Desmodesmus sp. MCC34. Energy Sour A Recov Utiliz Environ Effec 40(24):2933–2940. https://doi.org/10.1080/15567036.2018.1514434

Novais C, Molina AK, Abreu RMV, Santo-Buelga C, Ferreira ICFR, Pereira C, Barros L (2022) Natural food colorants and preservatives: a review, a demand, and a challenge. J Agric Food Chem 70(9). https://doi.org/10.1021/acs.jafc.1c07533

Obana A, Gohto Y, Nakazawa R, Moriyama T, Gellermann W, Bernstein PS (2020) Effect of an antioxidant supplement containing high dose lutein and zeaxanthin on macular pigment and skin carotenoid levels. Sci Rep 10(1). https://doi.org/10.1038/s41598-020-66962-2

Ochoa Becerra M, Mojica Contreras L, Hsieh Lo M, Mateos Díaz J, Castillo Herrera G (2020) Lutein as a functional food ingredient: stability and bioavailability. J Funct Foods 66. https://doi.org/10.1016/j.jff.2019.103771

Oh GW, Ko SC, Heo SY, Nguyen VT, Kim G, Jang CH, Park WS, Choi IW, Qian ZJ, Jung WK (2015) A novel peptide purified from the fermented microalga Pavlova lutheri attenuates oxidative stress and melanogenesis in B16F10 melanoma cells. Process Biochem 50(8). https://doi.org/10.1016/j.procbio.2015.05.007

Oliva AK, Bejaoui M, Hirano A, Arimura T, Linh TN, Uchiage E, Nukaga S, Tominaga K, Nozaki H, Isoda H (2022) Elucidation of the potential hair growth-promoting effect of Botryococcus terribilis, its novel compound methylated-Meijicoccene, and C32 Botryococcene on cultured hair Follicle Dermal Papilla cells using DNA microarray gene expression analysis. Biomedicines 10(5). https://doi.org/10.3390/biomedicines10051186

Othman R, Mohd Zaifuddin FA, Hassan NM (2014) Carotenoid biosynthesis regulatory mechanisms in plants. J Oleo Sci 63(8). https://doi.org/10.5650/jos.ess13183

Ötleô S, Pire R (2011) Fatty acid composition of Chlorella and Spirulina microalgae species. J AOAC Int 84(6) https://academic.oup.com/jaoac/article/84/6/1708/5656907

Parsaeimehr A, Sun Z, Dou X, Chen YF (2015) Simultaneous improvement in production of microalgal biodiesel and high-value alpha-linolenic acid by a single regulator acetylcholine. Biotechnol Biofuels 8(1). https://doi.org/10.1186/s13068-015-0196-0

Pasquet V, Morisset P, Ihammouine S, Chepied A, Aumailley L, Berard JB, Serive B, Kaas R, Lanneluc I, Thiery V, Lafferriere M, Piot JM, Patrice T, Cadoret JP, Picot L (2011) Antiproliferative activity of violaxanthin isolated from bioguided fractionation of Dunaliella tertiolecta extracts. Mar Drugs 9(5). https://doi.org/10.3390/md9050819

Peng J, Guo F, Liu S, Fang H, Xu Z, Wang T (2023) Recent advances and future prospects of Mycosporine-like amino acids. Molecules 28(14):5588. https://doi.org/10.3390/molecules28145588

Pereira JX, Pereira TC (2018) Cosmetics and its health risks. Glob J Med Res:63–70. https://doi.org/10.34257/GJMRBVOL18IS2PG63

Pourkarimi S, Hallajisani A, Nouralishahi A, Alizadehdakhel A, Golzary A (2020) Factors affecting production of beta-carotene from Dunaliella salina microalgae. In: Biocatalysis and agricultural biotechnology, vol 29, Elsevier Ltd. https://doi.org/10.1016/j.bcab.2020.101771

Rahman MM, Kundu JK, Shin JW, Na HK, Surh YJ (2011) Docosahexaenoic acid inhibits UVB-induced activation of NF-κB and expression of COX-2 and NOX-4 in HR-1 hairless mouse skin by blocking MSK1 signaling. PLoS One 6(11). https://doi.org/10.1371/journal.pone.0028065

Raj S, Kuniyil AM, Sreenikethanam A, Gugulothu P, Jeyakumar RB, Bajhaiya AK (2021) Microalgae as a source of mycosporine-like amino acids (Maas); advances and future prospects. Int J Environ Res Public Health 18(23). https://doi.org/10.3390/ijerph182312402

Rattanawiwatpong P, Wanitphakdeedecha R, Bumrungpert A, Maiprasert M (2020) Anti-aging and brightening effects of a topical treatment containing vitamin C, vitamin E, and raspberry leaf cell culture extract: a split-face, randomized controlled trial. J Cosmet Dermatol 19(3):671–676. https://doi.org/10.1111/jocd.13305

Resende DISP, Jesus A, Sousa Lobo JM, Sousa E, Cruz MT, Cidade H, Almeida IF (2022) Up-to-Date Overview of the Use of Natural Ingredients in Sunscreens. Pharmaceuticals 15. https://doi.org/10.3390/ph15030372

Ribaya-Mercado JD, Blumberg JB (2004) Lutein and Zeaxanthin and their potential roles in disease prevention. J Am Coll Nutr 23. https://doi.org/10.1080/07315724.2004.10719427

Roberts RL, Green J, Lewis B (2009) Lutein and zeaxanthin in eye and skin health. Clin Dermatol 27(2):195–201. https://doi.org/10.1016/j.clindermatol.2008.01.011

Ryckebosch E, Bruneel C, Termote-Verhalle R, Muylaert K, Foubert I (2014) Influence of extraction solvent system on extractability of lipid components from different microalgae species. Algal Res 3(1). https://doi.org/10.1016/j.algal.2013.11.001

Sahu A, Pancha I, Jain D, Paliwal C, Ghosh T, Patidar S, Bhattacharya S, Mishra S (2013) Fatty acids as biomarkers of microalgae. Phytochemistry 89:53–58. https://doi.org/10.1016/j.phytochem.2013.02.001

Saini RK, Keum YS (2018) Omega-3 and omega-6 polyunsaturated fatty acids: dietary sources, metabolism, and significance — a review. In: Life sciences, vol 203. Elsevier Inc., pp 255–267. https://doi.org/10.1016/j.lfs.2018.04.049

Sathasivam R, Ki JS (2018) A review of the biological activities of microalgal carotenoids and their potential use in healthcare and cosmetic industries. In: Marine drugs, vol 16(1). MDPI AG. https://doi.org/10.3390/md16010026

Schwartz S, Frank E, Gierhart D, Simpson P, Frumento R (2016) Zeaxanthin-based dietary supplement and topical serum improve hydration and reduce wrinkle count in female subjects. J Cosmet Dermatol 15(4). https://doi.org/10.1111/jocd.12226

Scott SD, Armenta RE, Berryman KT, Norman AW (2011) Use of raw glycerol to produce oil rich in polyunsaturated fatty acids by a thraustochytrid. Enzym Microb Technol 48(3):267–272. https://doi.org/10.1016/j.enzmictec.2010.11.008

Shigeta Y, Imanaka H, Ando H, Ryu A, Oku N, Baba N, Makino T (2004) Skin whitening effect of linoleic acid is enhanced by liposomal formulations. Notes Biol Pharm Bull 27(4):591

Shegokar R, Mitri K (2012) Carotenoid Lutein: A Promising Candidate for Pharmaceutical and Nutraceutical Applications. J Diet Suppl 9. https://doi.org/10.3109/19390211.2012.708716

Shi X-M, Chen F (2002) High-Yield Production of Lutein by the Green Microalga Chlorella protothecoides in Heterotrophic Fed-Batch Culture. Biotechnol Progress 18. https://doi.org/10.1021/bp0101987

Shimoda H, Tanaka J, Shan SJ, Maoka T (2010) Anti-pigmentary activity of fucoxanthin and its influence on skin mRNA expression of melanogenic molecules. J Pharm Pharmacol 62(9):1137–1145. https://doi.org/10.1111/j.2042-7158.2010.01139.x

Sijil PV, Sarada R, Chauhan VS (2019) Enhanced accumulation of alpha-linolenic acid rich lipids in indigenous freshwater microalga Desmodesmus sp.: the effect of low-temperature on nutrient replete,

UV treated and nutrient stressed cultures. Bioresour Technol 273:404–415. https://doi.org/10.1016/j.biortech.2018.11.028

Sivakumar G, Jeong K, Lay JO (2014) Biomass and RRR-α-tocopherol production in Stichococcus bacillaris strain siva2011 in a balloon bioreactor. Microb Cell Factories 13(1). https://doi.org/10.1186/1475-2859-13-79

Snell TW, Carberry J (2022) Astaxanthin bioactivity is determined by Stereoisomer composition and extraction method. In: Nutrients, vol 14(7). MDPI. https://doi.org/10.3390/nu14071522

Sohail M, Muhammad Faran Ashraf Baig M, Akhtar N, Chen Y, Xie B, Li B (2022) Topical lycopene emulgel significantly improves biophysical parameters of human skin. Eur J Pharm Biopharm 180:281–288. https://doi.org/10.1016/j.ejpb.2022.10.016

Soontornchaiboon W, Joo SS, Kim SM (2012) Anti-inflammatory effects of violaxanthin isolated from microalga Chlorella ellipsoidea in RAW 264.7 macrophages. Biol Pharm Bull 35(7). https://doi.org/10.1248/bpb.b12-00187

Sprague M, Fawcett S, Betancor MB, Struthers W, Tocher DR (2020) Variation in the nutritional composition of farmed Atlantic salmon (Salmo salar L.) fillets with emphasis on EPA and DHA contents. J Food Compos Anal 94. https://doi.org/10.1016/j.jfca.2020.103618

Stachowiak B, Szulc P (2021) Astaxanthin for the food industry. In: Molecules, vol 26(9). MDPI AG. https://doi.org/10.3390/molecules26092666

Stahl W, Sies H (2002) Carotenoids and protection against solar UV radiation. Skin Pharmacol Appl Ski Physiol 15(5):291–296. https://doi.org/10.1159/000064532

Suganuma K, Nakajima H, Ohtsuki M, Imokawa G (2010) Astaxanthin attenuates the UVA-induced up-regulation of matrix-metalloproteinase-1 and skin fibroblast elastase in human dermal fibroblasts. J Dermatol Sci 58(2):136–142. https://doi.org/10.1016/j.jdermsci.2010.02.009

Tarento TDC, McClure DD, Vasiljevski E, Schindeler A, Dehghani F, Kavanagh JM (2018) Microalgae as a source of vitamin K1. Algal Res 36:77–87. https://doi.org/10.1016/j.algal.2018.10.008

Tominaga K, Hongo N, Karato M, Yamashita E (2012) Cosmetic benefits of astaxanthin on humans subjects*. www.actabp.pl

Torregrosa-Crespo J, Montero Z, Fuentes JL, García-Galbis MR, Garbayo I, Vílchez C, Martínez-Espinosa RM (2018) Exploring the valuable carotenoids for the large-scale production by marine microorganisms. Marine Drugs 16(6). https://doi.org/10.3390/md16060203

Traber MG, Bruno RS (2020) Vitamin E. In: Present knowledge in nutrition. Elsevier, pp 115–136. https://doi.org/10.1016/B978-0-323-66162-1.00007-X

Transparency Market Research (2019, June) *Zeaxanthin market*. https://www.Transparencymarketresearch.Com/Zeaxanthin-Market.Html

Varani J, Warner RL, Gharaee-Kermani M, Phan SH, Kang S, Chung JH, Wang ZQ, Datta SC, Fisher GJ, Voorhees JJ (2000) Vitamin A antagonizes decreased cell growth and elevated collagen- degrading matrix metalloproteinases and stimulates collagen accumulation in naturally aged human skin. J Invest Dermatol 114(3):480–486. https://doi.org/10.1046/j.1523-1747.2000.00902.x

Wang HMD, Chen CC, Huynh P, Chang JS (2015) Exploring the potential of using algae in cosmetics. Bioresour Technol 184:355–362. https://doi.org/10.1016/j.biortech.2014.12.001

Xia E, Zhu X, Gao X, Ni J, Guo H (2021) Anti-aging potential of peptides from underused marine bioresources. Marine Drugs 19(9). https://doi.org/10.3390/MD19090513

Yang N, Zhang Q, Chen J, Wu S, Chen R, Yao L, Li B, Liu X, Zhang R, Zhang Z (2023) Study on bioactive compounds of microalgae as antioxidants in a bibliometric analysis and visualization perspective. In: Frontiers in plant science, vol 14. Frontiers Media S.A. https://doi.org/10.3389/fpls.2023.1144326

Yoshihisa Y, ur Rehman M, Shimizu T (2014) Astaxanthin, a xanthophyll carotenoid, inhibits ultraviolet-induced apoptosis in keratinocytes. Exp Dermatol 23(3):178–183. https://doi.org/10.1111/exd.12347

Zasada M, Budzisz E (2019) Retinoids: active molecules influencing skin structure formation in cosmetic and dermatological treatments. In: Postepy Dermatologii i Alergologii, vol 36(4). Termedia Publishing House Ltd., pp 392–397. https://doi.org/10.5114/ada.2019.87443

Zeng J, Zhao J, Dong B, Cai X, Jiang J, Xue R, Yao F, Dong Y, Liu C (2019) Lycopene protects against pressure overload-induced cardiac hypertrophy by attenuating oxidative stress. J Nutr Biochem 66:70–78. https://doi.org/10.1016/j.jnutbio.2019.01.002

Zhuang D, He N, Khoo KS, Ng EP, Chew KW, Ling TC (2022) Application progress of bioactive compounds in microalgae on pharmaceutical and cosmetics. Chemosphere 291. https://doi.org/10.1016/j.chemosphere.2021.132932

Zittelli GC, Lauceri R, Faraloni C, Silva Benavides AM, Torzillo G (2023) Valuable pigments from microalgae: phycobiliproteins, primary carotenoids, and fucoxanthin. Photochem Photobiol Sci. https://doi.org/10.1007/s43630-023-00407-3

Žmitek K, Žmitek J, Rogl Butina M, Hristov H, Pogačnik T, Pravst I (2020) Dietary lutein supplementation protects against ultraviolet-radiation-induced erythema: results of a randomized double-blind placebo-controlled study. J Funct Foods 75. https://doi.org/10.1016/j.jff.2020.104265

Part VI

Alternative High Value Application for Microalgae Biomass

Microalgae as Biofertilizer Source in Advanced Agriculture

17

Henri Márquez-Pacheco, José Belisario Leyva-Morales, Jesús Mateo Amillano-Cisneros, María Anel Fuentes-Valencia, and Luz Adriana Ontiveros-García

Abstract

The constant studies developed in research into biofertilizers from microalgae have generated enormous knowledge of the various advantages that these presents in agriculture, improving the fertility and quality of the soil and plant growth and the environment. The constant use of biofertilizers is presented as an alternative to reducing the use of chemical fertilizers and reducing their impact on the environment. Biofertilizers could allow bioremediation and reduce the loss of arable land, allowing soils considered infertile to recover their properties economically and more sustainably. According to these considerations, microalgae, due to their functional characteristics, rapid growth, and ease of reproduction, have gained important attention worldwide as a biological source for the development of biofertilizers. This chapter provides an overview of using novel microalgae as biofertilizers and soil improvers. In addition, the current strategies that have been developed to optimize the obtaining of biofertilizers from microalgal cultures for advanced agriculture are specified.

H. Márquez-Pacheco (✉)
Maestría en Biotecnología Agropecuaria, Universidad Politécnica del Mar y la Sierra, La Cruz, Elota, Sinaloa, Mexico
e-mail: hmarquez@upmys.edu.mx

J. B. Leyva-Morales
Área Académica de Química, Instituto de Ciencias Básicas e Ingeniería, Universidad Autónoma del Estado de Hidalgo, Mineral de la Reforma, Hidalgo, Mexico

J. M. Amillano-Cisneros · M. A. Fuentes-Valencia
L. A. Ontiveros-García
Ingeniería en Agrobiotecnología, Universidad Politécnica del Mar y la Sierra, La Cruz, Elota, Sinaloa, Mexico

Keywords

Novel microalgae · Microalgae biomass · Soil improvers · Phytohormone · Biomass harvest · Biomass processing

17.1 Introduction

Due to the accelerated growth of the human population, which increased from 2.5 billion in 1950 to 8 billion people in 2022 (ONU 2023), the demand for food has been considerably high over the years. Therefore, agricultural production worldwide is constantly growing, increasing from 6,169 million tons in 2000 to 9,500 million tons by 2021 (FAO 2021). Cereal and sugarcane crops represented the main crops with a production of 32% and 22%, respectively. Besides, vegetables and oilseed crops contributed 12% each, followed by fruits (10%) and roots and tubers (9%) (FAO 2022). Agricultural activity occupies about 40% of the surface of the earth, coupled with the fact that it has a significant demand for water with 70% of global use (Brooks et al. 2019). Despite the large global production, one of agriculture's challenges is obtaining higher production yields at a lower cost and with low environmental impact (Brooks et al. 2019). Therefore, to increase agricultural production capacity, intensive practices have been implemented over the years, which include the use of chemical additives for pest control (Khan et al. 2023; Rani et al. 2021) and the use of chemical fertilizers aimed at the growth and production of agricultural resources (Liu et al. 2021; Penuelas et al. 2023).

Despite the effectiveness of the application of chemical fertilizers to increase production in the agricultural sector, the prolonged use in large quantities of these chemicals causes acidification, compaction, and loss of soil fertility (Gosal et al. 2018; Kholssi et al. 2022; Liu et al. 2021), as well as the increase in the impact of pests on crops (Krasilnicov et al. 2022). In addition, these chemical sub-

D. U. Santos Ballardo, S. Rossi (eds.), *Microalgae as Promising Source of Commercial Bioproducts*, Developments in Applied Phycology 14, https://doi.org/10.1007/978-3-031-86433-9_17

stances have caused damage to aquatic ecosystems, increasing the contamination of groundwater and surface water, causing eutrophication of water bodies (Alori et al. 2017; Alvarez et al. 2021; Osorio-Reyes et al. 2023).

Due to these problems caused by chemical fertilizers, using organisms as a source of fertilizers represents a sustainable alternative in agriculture. According to FAO (2021), biofertilizers are natural or biological origin fertilizers that provide plants with the nutrients necessary for their biological functions, in addition to increasing soil quality through the maintenance of a stable environment of microorganisms associated with plants. Moreover, massive biological components (biomass) have been used as fertilizers in agricultural activity. Since the 1940s, there have been reliable records of the use of macroalgae in agriculture (Kholssi et al. 2022). It is a fact that the application of seaweed as fertilizers has been widespread since ancient times (e.g., in the Roman Empire, China, and Japan), especially in coastal areas (Mukherjee and Patel 2020). Regarding the use of microalgae in agriculture, of the approximately 50,000 species described (Ronga et al. 2019), the first patent registered was in 1982 in Japan, where *Chlorella* extract was used for plant growth; moreover, for the year 2019, the number of accumulated patents was 132 for microalgae used in this productive activity (Murata et al. 2021).

Microalgae represent a great source of nutrients such as nitrogen (N), phosphorus (P), potassium (K), and calcium (Ca), in addition to trace elements such as copper (Cu), iron (Fe), manganese (Mn), and zinc (Zn) (Alvarez et al. 2021; Manjunath et al. 2016; Parmar et al. 2023). Also, phytohormones are a result of the metabolism of microalgae, which are bioavailable to plants (Elakbawy et al. 2022; Song et al. 2022; Wang et al. 2021). These microorganisms are producers of amino acids, vitamins, and polysaccharides that can be absorbed by plants (Chanda et al. 2019; Franco-Morgado et al. 2023; Moreira et al. 2022; Parwani et al. 2021; Prisa and Spagnuolo 2023; Santini et al. 2021; Singh et al. 2016). In addition, they provide drought resistance by keeping the crop soil moist (Parmar et al. 2023; Uysal et al. 2015). Besides, microalgae and derived extracts as biofertilizers have gained significant global attention in recent years.

17.2 Beneficial Effects of the Use of Microalgae in Crops

17.2.1 Improved Fertility and Soil Quality

As mentioned above, soil fertility is crucial for plant development and directly influences their yields. Improper agricultural practices can cause soil degradation and depletion. Nowadays, applying chemical fertilizers is a common strategy to improve soil fertility and provide essential nutrients for plant development (Chandini et al. 2019). However, overreliance on chemical fertilizers can have several adverse consequences for soil health and the environment, such as changes in soil pH, pest development, acidification, and soil crusting, which causes a decrease in soil organic carbon and beneficial species, thus affecting the growth and yield of plants and contributing to the emission of greenhouse gases (Bisht and Chauhan 2020; Park et al. 2023; Tayoh 2020).

An alternative to reduce the use of chemical fertilizers is the total or partial use of biofertilizers, defined as products that contain live microorganisms or natural compounds derived from organisms such as bacteria and fungi that improve the chemical and biological properties of the soil, stimulate plants growth and restore soil fertility (Pirttilä et al. 2021). Some studies show that biofertilizers do not affect health or pollute the environment. Among biofertilizers, formulations based on photosynthetic organisms, including prokaryotic cyanobacteria and microalgae, have received a lot of interest mainly due to their excellent ability to increase bioavailability, soil fertility, and crop yield (Alvarez et al. 2021; Chittora et al. 2020).

The use of microalgae in the production of biofertilizers has emerged as a contribution of biotechnology and as a promising option due to its multifunctional capacity, photosynthetic efficiency, and ability to supply N in forms available to plants (García-Orellana et al. 2016). Microalgae are considered an organic fertilizer because they can minimize nutrient losses by progressively releasing the macronutrients N, P, and K (Coppens et al. 2016; Mulbry et al. 2007). In this sense, biofertilizers based on microalgae are considered a sustainable, profitable, and environmentally friendly alternative compared to chemical fertilizers, since they not only improve agricultural production but also favor the microbiological niche of the soil and reduce environmental pollution (García-Orellana et al. 2016). Several studies show that some microalgae extracts enhance the growth of crops through plant hormones and proteins (auxins, gibberellins, and cytokines), which work as growth regulators and a high level of macro-micronutrients essential for plants (Tarakhovskaya et al. 2007).

On the other hand, microalgae could influence the soil microbial populations (diversity, community composition, and activity) and have the intrinsic ability to produce a wide variety of bioactive metabolites that have a positive impact on plant growth, as well as help control pests and pathogens (Alvarez et al. 2021). Among these substances, phenolic compounds, carotenoids, terpenoids, polysaccharides, free fatty acids, and phytohormones are of notable interest since they are recognized as plant growth promoters (Gonçalves 2021). Given these properties, microalgae can be applied to improve crop protection and yield production, as they can be a source of nutrients, increase biomass production, or even improve plant tolerance to stress factors, such as drought, salinity, or even pollution (Dmytryk and Chojnacka 2018; Ronga et al. 2019). Regarding this, studies with *Dunaliella* spp. and *Phaeodactylum* spp. have demonstrated their ability to reduce salt stress during the seed germination process of bell peppers (*Capsicum annuum*) (Guzmán-Murillo et al. 2013). Additionally, after their death, they benefit from the development of other soil microorganisms and secrete exopolysaccharides and bioactive substances that are involved in the recovery of soil nutrients and the mobilization of insoluble forms of inorganic phosphates (Chatterjee et al. 2017).

The use of microalgae as biofertilizers has mainly focused on chlorophytes due to their high growth rate and ability to adapt to soil conditions. The microalgae *Chlorella* genus is mainly used due to their specific properties, such as N fixation and the production of metabolites, beneficial for plant development (Ortiz-Moreno et al. 2019). However, different microalgae species have been found to have biostimulant effects on plants, such as *Arthrospira platensis*, *Scenedesmus obliquus*, *Scenedesmus* spp., and *Synechocystis* spp. (Carapinha 2019; Junior et al. 2020). Also, microalgae are a valuable organic carbon source when applied to soil (Alvarez et al. 2021). This is a key aspect in agriculture, considering that the depletion of organic carbon in soil is a significant form of degradation in cropland, leading to decreased land quality and fertility (Stavi and Lal 2015). Moreover, microalgae can assimilate organic carbon into their biomass through photosynthesis, and several strains can release exopolysaccharides, which help as a carbon source and carbon sequestrant, improving soil aggregation and stabilization (Costa et al. 2018). Furthermore, microalgae generally exhibit interesting properties that improve agricultural yield through three distinct modes: biofertilizers, biostimulants, and biopesticides (Gonçalves 2021; Parmar et al. 2023).

17.2.2 Production of Hormones for Plant Growth

Some studies report that microalgae contain plant growth-promoting substances such as auxins, cytokinins, betaines, amino acids, vitamins, polyamines (spermine and spermidine), and polysaccharides (β-glucan) from their primary metabolism (Ronga et al. 2019). Polysaccharides have been described within these metabolites as stimulating tomato crops and improving physical and nutritional parameters (Ronga et al. 2019). Additionally, another group of compounds, such as phytohormones, derived from the secondary metabolism of microalgae, have a biostimulant effect in crops. The main phytohormones produce 86% of auxins (Morgado-González et al. 2015), gibberellins, cytokines, and abscisic acid (Angulo et al. 2014). These phytohormones control cell division, stem elongation, induction of the complex of defensive responses during pathogenesis, initiate seed germination, development of the leaf blade, delay senescence, as well as the induction of elongation growth, differentiation of phloem elements, apical dominance, tropisms, and initiate root formation (Tarakhovskaya et al. 2007). Microalgae also promote the production of chemical substances by the plant that inhibit the development of pathogens, increase competitiveness, induce plant resistance, and improve the response to external factors that can produce stress (Egamberdiyeva and Höflich 2004). Furthermore, microalgae extracts can replace micronutrient foliar fertilizers that function as an N supplement and, therefore, increase nutrient uptake, as well as increase the amount of N in shoot and root tissues (Adamczewska-Sowińska and Uklańska 2009).

17.3 Description of Current and Novel Microalgae Used As Biofertilizers

There are experimental evaluations of microalgae as biofertilizers in different crops of global agricultural importance, such as rice, corn, onion, beans, and chili (Dineshkumar et al. 2019, 2020; Li et al. 2021; Tian et al. 2022). These microalgae are usually defined as prokaryotic microorganisms (cyanobacteria or blue-green algae) and eukaryotes that have a simple structure, with a great capacity for photosynthesis and an adaptation to different environmental conditions to generate potential products for use in different activities, including agriculture (Chiaiese et al. 2018; Kholssi et al. 2022), considering that microalgae are rich in nutrients and can absorb high concentrations of N and P for the synthesis of proteins, and nucleic acids up to 40–60% of their dry weight (Muñoz and Guieysse 2006). The main microalgae used as biofertilizers include microalgae of the Phyla Cyanobacteria and Chlorophyta, with 11 and 6 genera represented, respectively (Table 17.1).

Furthermore, mainly green algae such as *Chlorella* spp. stand out for their experimental research as sources of biofertilizers based on microalgae (Ammar et al. 2022; Dineshkumar et al. 2019, 2020; Li et al. 2021; Tian et al. 2022), these species belong to Phylum *Chlorophyta*, which is characterized by members that inhabit both aquatic (freshwater and marine) and also terrestrial environments (Heimann and Huerlimann 2015). In particular, the *Chlorella* genus is located in soil under symbiotic conditions with lichens or protozoans (Liu and Chen 2016). Besides, in biofertilization studies, the most commonly used species is *Chlorella vulgaris* (Ammar et al. 2022), which is unicellular and has a size of 2–10 μm, is fast growing, has a short reproduction time, and has flexibility in cultivation conditions (Dalіry et al. 2017; Vander Wiel et al. 2017; Wirth et al. 2020). Moreover, under nutrient-limiting and stress conditions, this species accumulates high amounts of lipids and

Table 17.1 Taxonomic classification of the main microalgae used as a source of biofertilizers

Genus	Common name	Phylum	Class	Order	Family
Chlorella	Green algae	Chlorophyta	Trebouxiophyceae	Chlorellales	Chlorellaceae
Scenedesmus	Green algae	Chlorophyta	Chlorophyceae	Sphaeropleales	Scenedesmaceae
Acutodesmus	Green algae	Chlorophyta	Chlorophyceae	Sphaeropleales	Scenedesmaceae
Tetradesmus	Green algae	Chlorophyta	Chlorophyceae	Sphaeropleales	Scenedesmaceae
Dunaliella	Green algae	Chlorophyta	Chlorophyceae	Chlamydomonadales	Dunaliellaceae
Chlorococcum	Green algae	Chlorophyta	Chlorophyceae	Chlamydomonadales	Chlorococcaceae
Spirulina	Cyanobacteria	Cyanobacteria	Cyanophyceae	Spirulinales	Spirulinaceae
Nostoc	Cyanobacteria	Cyanobacteria	Cyanophyceae	Nostocales	Nostocaceae
Anabaena	Cyanobacteria	Cyanobacteria	Cyanophyceae	Nostocales	Nostocaceae
Westiellopsis	Cyanobacteria	Cyanobacteria	Cyanophyceae	Nostocales	Hapalosiphonaceae
Aulosira	Cyanobacteria	Cyanobacteria	Cyanophyceae	Nostocales	Fortieaceae
Scytonema	Cyanobacteria	Cyanobacteria	Cyanophyceae	Nostocales	Scytonemataceae
Tolypothrix	Cyanobacteria	Cyanobacteria	Cyanophyceae	Nostocales	Tolypothrichaceae
Calothrix	Cyanobacteria	Cyanobacteria	Calotrichaeceae	Nostocales	Calotrichaceae
Microcystis	Cyanobacteria	Cyanobacteria	Cyanophyceae	Chroococcales	Microcystaceae
Synechocystis	Cyanobacteria	Cyanobacteria	Cyanophyceae	Synechococcales	Merismopediaceae
Arthrospira	Cyanobacteria	Cyanobacteria	Cyanophyceae	Oscillatoriales	Microcoleaceae

Adapted from NCBI (2023)

stores materials (Wirth et al. 2020). Besides, the *Chlorella* genus requires nutrients such as C, N, P, S, and metals. Carbon is the predominant element of *Chlorella*, which is considered to be 50% of the fraction of elements in the biomass of microalgae (Dolganyuk et al. 2020). On the other hand, carbon dioxide is the primary carbon source for photoautotrophic growth. Due to this, to maintain its rapid growth with an exponential cell density, adding CO_2 is recommended since atmospheric air is insufficient. In addition to the ability to fix carbon dioxide, it removes nutrients such as N and P; therefore, it is considered a candidate for gas mitigation and bioremediation (Liu and Chen 2016). Also, *Chlorella* extracts are usually rich in chlorophyll, amino acids, peptides, potassium, calcium, magnesium, and various vitamins such as E, B1, B2, B6, and C, as well as phytohormones for plant growth. These extracts promote plant growth by stimulating cell division, making it a growth promoter. Recently, it has been considered that the photosynthetic capacity of *Chlorella* is tens of times higher compared to that of plants because it contains a "Chlorella Growth Factor (CGF)" rich in nucleoprotein, nucleic acids, vitamins, amino acids, polysaccharides, enzymes, glycoproteins, and various phytohormones, which could be the reason for their promoting capacity for plant growth (Tian et al. 2022). Currently, the microalgae *Chlorella* is considered to contain significant amounts of N and P (Mutum et al. 2022); in a microalgal consortium with *Scenedesmus* and *Pediastrum*, it presented a dry matter content of N with 4.45%, P with 1.21%, and K with 0.77% (Mückschel et al. 2023).

The *Scenedesmus* genus is another relevant microalga with biofertilizer potential for agriculture. *Scenedesmus* is a rarely single-celled microalgae, mainly for its exponential growth and increased exposure to nutrients and light. However, its culture is frequently colonial, presenting 2–32 cells. This microalga lives in fresh and brackish waters rich in nutrients; hence, it is considered an indicator of pollution (SCB 2023). Moreover, *Scenedesmus* is widely produced for aquaculture and contains particularly high levels of lipids and proteins (Colla et al. 2020). In another study, *Scenedesmus* grown in consortium with *Chlorella variabilis* was applied to corn (*Zea mays*) and soybeans (*Glycine max*), presented positive effects on plant growth, total phenolic content, and flavonoids, as well as in the antioxidant activity and minerals of plants (Gatamaneni-Loganathan et al. 2020). Also, the foliar application of the *Scenedesmus* hydrolyzate was evaluated in an ornamental plant, the garden petunia (*Petunia x hybrida*), finding results in improving plant development, observing promotion of root growth, early flowering and high foliar concentrations of four mineral elements such as P, K, Ca, and Mg (Plaza et al. 2018). Furthermore, *Scenedesmus* also presented a higher concentration of phytohormones such as cytokinins (isopentenyl adenine), gibberellins (GA1), auxins (indoleacetic acid), salicylic and abscisic acids compared to the treatment of the cyanobacteria *Arthrospira* (Plaza et al. 2018).

According to some research, cyanobacteria is another phylum with potential as a biofertilizer. This type of microalgae is photosynthetic prokaryotes in diverse habitats such as aquatic (freshwater and marine), terrestrial, and cryosphere. Also, they are found in symbiosis with different eukaryotic hosts. The group has a complex taxonomy due to its morphological plasticity and ecological flexibility (Gaysina et al. 2019). Moreover, cyanobacteria are a diverse group, Gram-negative, ubiquitous, with a short generation time, and some species are able to fix N. Therefore, these microorganisms are considered an important application in the agricultural sector as sources of biofertilizers that contribute to soil fertility and environmental quality (Kumar et al. 2019; Rai et al. 2019; Singh et al. 2016). In this field, it is known that some cyanobacteria can colonize the roots and shoots of plants; the majority are considered to be atmospheric nitrogen fixers, among which are the genera *Anabaena*, *Nostoc*, and *Oscillatoria* (Ammar et al. 2022). Besides, the *Anabaena* genus has cells shaped like pearls or barrels, forming spores found as plankton in shallow waters and humid soils (Britannica 2016). It presents filaments with specialized cells called heterocysts and akinetes, which are used to fix nitrogen and regenerate cells for future colonies (CFB 2023). This cyanobacterium presents a morphology similar to the *Nostoc* genus; it is distinguished by its habitat, as well as by morphological characters such as indistinct mucilage, less constricted filaments, and akinetes in different places (Britannica 2016). However, the taxonomy of both genera is controversial, so complementary molecular taxonomic approaches are required to correctly identify cyanobacterial species (Makra et al. 2019).

On the other hand, Nostoc spp. are aquatic, endobiotic, symbiotic, and/or terrestrial. They are located in tropical, temperate, and polar environments and are commonly found in gravel and greenhouse hallways, forming biofilms (Laughinghouse et al. 2019). Moreover, cyanobacteria are generally considered good-quality biofertilizers and emerging microorganisms for sustainable agricultural development, as they improve nutrient solubilization and mobility. In addition, they stimulate plant growth due to their ability to produce bioactive compounds, such as phytohormones and other plant growth-regulating macromolecules (Ammar et al. 2022; Elagamey et al. 2023). Recently, it has been estimated that cyanobacteria can contribute 20–30 kg of N per ha as a source of organic matter to the soil (Elagamey et al. 2023). The use of biofertilizers based on cyanobacteria reduces up to 25–40% of the chemical fertilizers used in the soil; due to this, they are consolidated as an alternative to chemical fertilizers (Ammar et al. 2022; Bhuyan et al. 2023).

Furthermore, it is suggested that cyanobacteria improve P bioavailability in plants through solubilization-mobilization of insoluble organic phosphates present in the soil by phosphatase enzymes (Singh et al. 2016). It is also considered that cyanobacteria improve the physical properties of the soil since various components (polysaccharides, peptides, and lipids) are released during their growth in the soil, which helps maintain the soil structure in the form of microaggregates, particularly in pedological conditions of arid and semiarid areas (Jhala et al. 2017). Therefore, due to their N fixation capacity, cyanobacteria are widely used biofertilizers in rice cultivation; this parameter depends on physicochemical and biotic factors (Rai et al. 2019). Cyanobacteria belonging to the Nostocales group represent the order with the most representatives in the search for biofertilizers based on microalgae (Table 17.1); such microalgae are *Anabaena variabilis*, *Nostoc muscorum*, *Aulosira fertilissima*, *Tolypothrix tenuis*, *Calothrix* sp., and *Scytonema* sp. These microalgae are considered efficient biofertilizers for N fixation; particularly, the *Anabaena* and *Nostoc* genera can fix 20–25 kg of N per ha (Elagamey et al. 2023; Jhala et al. 2017).

Other positive effects of cyanobacteria on plants are the activation of the defensive system against soil pathogens, as well as a biocidal effect against various phytopathogens such as bacteria, fungi, and nematodes (Elagamey et al. 2023). Hence, applying biofertilizers based on cyanobacteria attracts the attention of global research as it is environmentally friendly (Rai et al. 2019).

Another cyanobacterium relevant to the agricultural sector is the *Spirulina* genus; it is currently the most produced microalgae and is used as a natural biofertilizer. In addition, it participates in the release of active compounds such as polysaccharides, amino acids, and phytohormones that promote plant growth (Arahou et al. 2023). Moreover, a study in rice (*Oryza* sp.) reported that the addition of *Spirulina platensis* and *Chlorella vulgaris* increased the availability of three macronutrients in the soil (N, P, and K) compared to the control group and also showed an improvement in growth of plants (Dineshkumar et al. 2018). Similar results with these microalgae are reported in the cultivation of corn (*Zea mays*), which positively affects its yield (Dineshkumar et al. 2019).

Besides, in another investigation, the microalgae *Tetradesmus obliquus*, which belongs to *Phylum Chlorophyta*, was isolated from the freshwater environment and showed tolerance to high CO_2 concentrations. Also, it was reported to have multiple potential applications since it is a bioactive compound with anticancer and antioxidant activity and even as a source of biodiesel (Sinha et al. 2021). On the other hand, the multifaceted panorama of microalgae is due to the extraordinary species biodiversity, the improvement of cultivation conditions, and the application of novel molecular techniques. Particularly, genomic studies of microalgae of the genera *Chlorella*, *Chlamydomonas*, *Dunaliella*, and *Phaeodactylum* are proposed as candidates for the synthesis and accumulation of non-indigenous products (Abu-Ghosh et al. 2021).

17.4 Strategies Developed to Obtain Biofertilizers from Microalgae

Regardless of the final product that is desired from microalgae, the main steps reported in the literature to obtain biomass satisfactorily are (1) microalgae cultivation, (2) microalgae biomass harvesting, and (3) microalgae biomass processing (Ammar et al. 2022; Araujo-Abad and Collahuazo-Reinoso 2019; Barbosa et al. 2023; Cao et al. 2023; Deepa et al. 2023; González-Pérez et al. 2022; Mutum et al. 2022; Parmar et al. 2023). The first is the most crucial step because between 20 and 30% of the cost of the process itself is invested in it (Osorio-Reyes et al. 2023). Each stage is described below, emphasizing the various procedures to improve production yields.

17.4.1 Microalgae Cultivation

Before starting the biomass generation from microalgae cultures, the careful selection of the strain(s) must be carried out; these can be either red (Rhodophyta), brown (Phaeophyta), or green (Chlorophyta) (Iqbal et al. 2021). However, the latter are the ones that have been widely reported in the generation of products with agriculture applications. This step is undoubtedly fundamental because different microalgae can more efficiently and particularly generate some metabolites such as lipids, pigments, carbohydrates, and proteins (Araujo-Abad and Collahuazo-Reinoso 2019; Braun and Colla 2023; Mahapatra et al. 2018; Mutum et al. 2022). The isolation of microalgae strains is carried out mainly from the environment, especially those that are considered native, which are tolerant to the presence of some contaminants and hostile conditions present in the environment, as well as due to their diverse adaptive capacity to nourish themselves (autotrophic nutrition, mixotrophic, and heterotrophic). Among the most used microalgae species are *Arthrospira*, *Chlorella*, *Dunaliella*, *Nostoc*, and *Aphanizomenon* (Braun and Colla 2023; Cao et al. 2023; Iqbal et al. 2021; Parmar et al. 2023).

Once the species to be cultivated has been selected, the next step is choosing the type of system used to generate biomass. In this sense, two types of systems are reported for this purpose: (1) the open system that is subdivided into natural waters (ponds, lakes, and lagoons) and artificial ponds (circular and canal-shaped) and (2) the closed system where we find fermenter-type tanks and photobioreactors (PBR) in various modalities such as tubular, laminar and propeller type. Both systems can be suspended or immobilized (biofilms), and the use of combined systems (hybrids) has also been reported (Cao et al. 2023; Osorio-Reyes et al. 2023).

Open culture systems are usually the most widely used because their establishment and maintenance costs are low; however, they have the disadvantage that they are directly exposed to the atmosphere, and they are very susceptible to contamination by various microorganisms (fungi, bacteria, or protozoa). In addition, the temperature of the medium and the quality and quantity of light (photoperiod) to which the cells are exposed are challenging to control since they can be altered by the presence of weather phenomena and attack by predators. Therefore, the efficiency of production is usually low. In contrast, in a closed system (PBR) the conditions to which the crops are exposed are controlled more efficiently, achieving greater performance in the production of high-value goods such as carbohydrates, proteins, pigments, and lipids, although they present the disadvantage of high costs from acquisition to maintenance and cleaning, however, if the process warrants it, they can be considered the best option since the possibility of contamination is reduced to a minimum (Araujo-Abad and Collahuazo-Reinoso 2019; Braun and Colla 2023; González-Pérez et al. 2022).

Among the parameters that must be taken care of during the cultures, regardless of the type of system, are the supply of light (natural or artificial), temperature, pH, the presence of micro (vitamins and trace elements) and macronutrients (N, P, C), salinity, photoperiod, and inoculum volume (Braun and Colla 2023; González-Pérez et al. 2022; Parmar et al. 2023; Ru et al. 2020).

Although some species of microalgae, such as those of the *Dunaliella* genus, support temperatures between 10 and 45 °C, it has been reported that the optimal growth temperatures are between 18 and 35 °C. However, this can vary depending on the final product to be obtained. For example, some reports mentioned that at temperatures close to 40 °C, the synthesis of carotenoids is promoted. However, the growth rate under stressful conditions generally is very low. Moreover, in the case of the pH of the culture medium, optimal conditions are reported in the range of 7 to 9, with an average of 7.5 to 7.6; also, extremophile species grow at extremely low or high pH values.

On the other hand, light/dark photoperiods (natural or artificial) of approximately 12:12 hours are recommended for optimal crop growth, with an average photoperiod of 18 hours. Furthermore, among the nutritional requirements of microalgae are some macroelements such as N, P, and K and some microelements such as B, Cu, Mn, selenium (Se), molybdate (Mo), Cu, and vanadium (V) (Braun and Colla 2023; Cao et al. 2023; González-Pérez et al. 2022; Hussain et al. 2021; Iqbal et al. 2021; Parmar et al. 2023; Win et al. 2018).

Recently, numerous studies have been reported on the use of different types of wastewater, both industrial (wine, meat, dairy, aquaculture, textile, brewing, petrochemical, and livestock, among others) and those of urban origin (domestic and municipal), mainly in open systems, as a sustainable strategy for the microalgae production due to the high presence of usable nutrients and coupled with the bioremediation capacity by reducing or assimilating some toxic elements present such as heavy metals, although a more in-depth evaluation of their possible use because if the toxic elements are not entirely removed, they could be harmful to both the environment and human health (Cao et al. 2023; González-Pérez et al. 2022; Hussain et al. 2021; Parmar et al. 2023; Win et al. 2018).

17.4.2 Microalgal Biomass Harvest

The next step in the biomass production process corresponds to harvesting, which begins with the collection of the microalgae generated; for it, the microalgal cells must be separated from the medium where they developed, seeking to recycle it for reuse as much as possible. In this sense, the harvesting process has been widely documented by various authors through techniques such as centrifugation, filtration, flotation, and flocculation (induced by chemicals and/or bacterial/fungal species). The first is based on using centrifugal and gravitational force to extract biomass from the culture media. However, it is a fast technology that is easy to execute, and with recovery percentages for various species superior to 90%, the investment cost and energy expenditure are high. Also, the cellular interior can be damaged during processing, which consequently generates the wear of delicate cellular components, thus reducing their potential applications. The second is based on the extraction of algal biomass through the use of porous membrane filters of various particle sizes, differentiating into ordinary (>10 µm), microfiltration (0.1–10 µm), ultrafiltration (0.001–0.1 µm) and nanofiltration (0.1 nm–0.001 µm), the application of this technique allows the retention of the biomass in the membrane while allowing the passage of the culture media, which allows a significant cell recovery (>90%). However, some difficulties are reported during its use, such as the tendency for the formation of scale and blockages in the filter, requiring its replacement. The third is a technology whose principle is based on using an air microbubble disperser to float the cells without chemicals, making it highly profitable (minimal operating costs), easy to execute, and with high biomass yields (>95%). Finally, in the fourth, biomass accumulation is carried out through the free-floating of the microalgae cells, forming large particles called flocs induced by a chemical or biological agent that neutralizes the surface charge. This technique, like those mentioned above, has a high recovery efficiency (>96%). Its main disadvantage is the use of chemical flocculating materials, which often present high toxicity, requiring additional treatment processes to eliminate them. Depending on the process, the use in combination of two or more of the described techniques is reported (Ammar et al. 2022; Barbosa et al. 2023; Deepa et al. 2023; Mutum et al. 2022; Osorio-Reyes et al. 2023; Parmar et al. 2023). The appropriate harvesting method is chosen based on the microalgae species used, the density and size of the cells, the conditions of the final product, and the reuse of the growth medium (Deepa et al. 2023).

17.4.3 Microalgal Biomass Processing

The final step in production represents the processing of the biomass separated and obtained during harvest, which normally consists of dehydration or drying, extraction, purification, or formulation of the desired product (Deepa et al. 2023). The processing will depend, therefore, on the intended application, which in this case is agricultural, specifically for the generation of biofertilizers.

In this sense, the dehydration or drying process must be carried out simultaneously with the harvest to prevent the biomass generated from being lost, with three types of drying being mostly reported in the literature: (1) drying in the sun is the simplest and least expensive because, as its name indicates, it is based on the use of solar energy, but it has some drawbacks because the process depends on climatic conditions, involves very long drying periods and requires a considerable size of drying area. In addition, overheating may occur since not regulating it results in changes in some of the properties of the microalgae such as texture, color, and flavor; (2) spray drying generates a fine powder from the spraying of the suspension of small droplets in a large container that is constantly exposed to hot air, among the advantages is the capacity for uninterrupted operation, drying efficiency in a short time and the fact that the product obtained has typically high quality; and finally, (3) freeze-drying is used at the laboratory level to dry microalgae, it is a high-cost process, and as with spray drying, some components can be affected. This method consists of dehydration through a sublimation mechanism from prior freezing to solidify the raw material. Consequently, due to the decrease in humidity, both the solid structure of the microalgae and the quality are preserved (Ammar et al. 2022; Barbosa et al. 2023; Parmar et al. 2023).

The choice of dehydration method will be influenced by the intended application of the final product; it has been reported that spray drying turns out to be the most economically viable option for industrial-type production, and although it has been observed that temperatures above 70 °C can cause the loss or decrease of thermolabile compounds such as some proteins, antioxidants, and secondary metabolites. In the case of the production of biological fertilizers, which contain macro and micronutrients, the drying process does not influence the stability of the raw material to be used for this purpose (González-Pérez et al. 2022; Mutum et al. 2022; Parmar et al. 2023).

Furthermore, depending on the presentation of the final product formulation (powder, granules, or liquid), it is necessary, before taking the process to an industrial scale, to carry out laboratory tests that allow optimizing the process considering, among other aspects, the choice of strain and harvesting microalgae, the design of bioreactors and the choice of the system to use for biomass generation, as well as the use of support materials that guarantee product stability and long shelf life (Araujo-Abad and Collahuazo-Reinoso 2019; Cao et al. 2023; Mahapatra et al. 2018; Mutum et al. 2022; Osorio-Reyes et al. 2023; Ru et al. 2020; Win et al. 2018).

17.5 Future Perspectives on the Use of Microalgae As Biofertilizers in Advanced Agriculture

Although the properties and advantages of using microalgae are well known, the intensive use of these microorganisms as plant growth promoters and soil improvers is still low, unlike other organisms such as macroalgae. However, in recent years, they have gained significant attention worldwide as a biological source for the development of biofertilizers. Parmar et al. (2023) report that the demand for biofertilizers based on microalgae and plant growth stimulants was valued at 9,479 thousand US dollars in 2021 with an annual growth of 7.7% and is expected to rise to 8.7% by 2031. However, there are some challenges to overcome for an industrialization approach and the widespread use of microalgae in agriculture.

One of these challenges is the transition from the laboratory to the field since many strains and/or strain consortia that exhibit excellent in vitro yield perform poorly under field conditions (Mitter et al. 2021). For example, the result of microalgae-bacteria interactions often varies in different studies under different climatic or soil conditions, which constitutes the main problem in implementing this technology (Osorio-Reyes et al. 2023; Solomon et al. 2023). Therefore, it is necessary to develop soil- or crop-specific biofertilizers with strains from the host rhizosphere and native soil and biofertilizer consortia with broad ecological adaptations and wide pH tolerances. In addition, the formulations and vehicles must be stable to environmental fluctuations, such as the use of encapsulation technology and gentler dehydration processes that retain the bioactivity of the formulations (Parmar et al. 2023). The efficiency of biofertilizers is subject to soil properties (moisture, pH, temperature, and organic matter), and soil fertility may not be effectively improved without these favorable factors (Asses et al. 2019).

The success of biofertilizers based on microalgae may depend on the low cost of biomass production and the lower energy footprint of biomass production technologies. In recent years, the generation of microalgae biomass with wastewater treatment has been proposed as an alternative to reduce the cost of microalgae production and contribute to the implementation of a circular economy (Álvarez-González et al. 2023). Besides, different types of wastewater as a source of nutrients to cultivate microalgae must be tested, as they have broad application prospects, thus promoting environmental sustainability.

Generally, wastewater contains abundant organic and inorganic nutrients, such as C, N, and P, which microalgae can effectively utilize. For example, P is a non-renewable resource rich in wastewater; hence, microalgae can use this nutrient and accumulate it to be used later as biofertilizer. However, given the characteristics of the wastewater, the composition of microalgae biofertilizers can vary. Therefore, this creates a challenge to maintain consistent contents and proportions of nutrients in the biofertilizers (Gonçalves et al. 2023). Additionally, depending on the type of wastewater, specific contaminants such as pathogens, heavy metals, and contaminants of emerging concern, such as pharmaceutical and personal care products, are present; hence, the need to carry out studies that evaluate the accumulation of heavy metals and the presence of any of these specific contaminants in the microalgae biomass (Álvarez-González et al. 2023).

Since commercial agriculture requires a large amount of nutrients, a large amount of microalgae biomass is necessary to cover the demand, so using *in vivo* culture as an inoculant is a possible solution to this problem (Guo et al. 2020). Only small amounts of cultured microalgae are necessary as an inoculant for the growth of microalgae and cyanobacteria that will continuously provide the necessary nutrients and compounds for the cultures. The use of cyanobacteria as growth units is reported, which can save between 25 and 75% of N fertilizer, in addition to providing essential elements such as macronutrients, micronutrients, and metabolites beneficial for plant growth (Guo et al. 2020).

Microalgae-based biofertilizers are gaining popularity; however, the high nutrient and water requirements and subsequent processes to obtain biomass make this microalgae-based technology commercially unsustainable (Parmar et al. 2023). There is increasing awareness about the use of microalgae biofertilizers and their benefits. The market potential of microalgae biofertilizers is enormous, but the challenges associated with commercialization must be addressed through extensive field studies and the development of the cost-effectiveness of biofertilizer production technology.

17.6 Conclusions

In recent years, the use of strains and consortia of numerous microalgae as sources of nutrients and agricultural stimulants has been investigated, which exhibit properties that improve agricultural performance through products such as biofertilizers, biostimulants, and biopesticides. Cyanobacteria and Chlorophyta have received much attention due to their ability to increase bioavailability, soil fertility, and crop yields. However, it is necessary to conduct field trials to validate the *in vitro* results on the biofertilizer and/or stimulant activity of extracts derived from microalgae. In addition, the strategies used to obtain biomass from microalgae should be analyzed since the success of biofertilizers based on microalgae may depend on the low cost of biomass production and the lower energy footprint in biomass production technologies. Hence, profitable collection and dehydration processes are necessary to produce valuable microalgal biomass as biofertilizers.

References

Abu-Ghosh S, Dubinsky Z, Verdelho V, Iluz D (2021) Unconventional high-value products from microalgae: a review. Bioresour Technol 329:124895. https://doi.org/10.1016/j.biortech.2021.124895

Adamczewska-Sowińska K, Uklańska CM (2009) Effect of nitrogen fertilization on yield and quality of endive. J Fruit Ornam Plant Res 70(1):193–201. https://doi.org/10.2478/v10032-009-0019-6

Alori ET, Glick BR, Babalola OO (2017) Microbial phosphorus solubilization and its potential for use in sustainable agriculture. Front Microbiol 8:971. https://doi.org/10.3389/fmicb.2017.00971

Alvarez A, Weyers S, Goemann H, Peyton B, Gardner R (2021) Microalgae, soil and plants: a critical review of microalgae as renewable resources for agriculture. Algal Res 54:102200. https://doi.org/10.1016/j.algal.2021.102200

Álvarez-González A, Uggetti E, Serrano L, Gorchs G, Casas ME, Matamoros V, Gonzalez-Flo E, Díez-Montero R (2023) The potential of wastewater grown microalgae for agricultural purposes: contaminants of emerging concern, heavy metals and pathogens assessment. Environ Pollut 324:121399. https://doi.org/10.1016/j.envpol.2023.121399

Ammar EE, Aioub AA, Elesawy AE, Karkour AM, Mouhamed MS, Amer AA, El-Shershaby NA (2022) Algae as Bio-fertilizers: between current situation and future prospective. Saudi J Biol Sci 29(5):3083–3096. https://doi.org/10.1016/j.sjbs.2022.03.020

Angulo VC, Sanfuentes EA, Rodríguez F, Sossa KE (2014) Caracterización de rizobacterias promotoras de crecimiento en plántulas de Eucalyptus nitens. Rev Argent Microbiol 46(4):338–347. https://doi.org/10.1016/S0325-7541(14)70093-8

Arahou F, Lijassi I, Wahby A, Rhazi L, Arahou M, Wahby I (2023) Spirulina-based biostimulants for sustainable agriculture: yield improvement and market trends. Bioenergy Res 16(3):1401–1416. https://doi.org/10.1007/s12155-022-10537-8

Araujo-Abad S, Collahuazo-Reinoso Y (2019) Producción de biofertilizantes a partir de microalgas. Cedamaz 9(2):81–87. https://revistas.unl.edu.ec/index.php/cedamaz/article/view/648

Asses N, Farhat W, Hamdi M, Bouallagui H (2019) Large scale composting of poultry slaughterhouse processing waste: microbial removal and agricultural biofertilizer application. Process Saf Environ Prot 124:128–136. https://doi.org/10.1016/j.psep.2019.02.004

Barbosa M, Inácio LG, Afonso C, Maranhão P (2023) The microalga Dunaliella and its applications: a review. Appl Phycol 4(1):99–120. https://doi.org/10.1080/26388081.2023.2222318

Bhuyan PP, Nayak R, Jena M, Pradhan B (2023) Convoluted role of cyanobacteria as biofertilizer: an insight of sustainable agriculture. Vegetos 36(2):309–321. https://doi.org/10.1007/s42535-022-00415-1

Bisht N, Chauhan PS (2020) Excessive and disproportionate use of chemicals cause soil contamination and nutritional stress. In: Soil contamination-threats and sustainable solutions Vol. 2020, pp 1–10. https://doi.org/10.5772/intechopen.94593

Braun JC, Colla LM (2023) Use of microalgae for the development of biofertilizers and biostimulants. Bioenergy Res 16(1):289–310. https://doi.org/10.1007/s12155-022-10456-8

Britannica (2016) *Anabaena: cyanobacteria*. Encyclopedia Britannica. Retrieved 4 Nov 2023 from https://www.britannica.com/science/Anabaena

Brooks J, Deconinck K, Giner C (2019) Three key challenges facing agriculture and how to start solving them. https://www.oecd.org/agriculture/key-challenges-agriculture-how-solve/

Cao TN-D, Mukhtar H, Le L-T, Tran DP-H, Ngo MTT, Nguyen T-B, Bui X-T (2023) Roles of microalgae-based biofertilizer in sustainability of green agriculture and food-water-energy security nexus. Sci Total Environ 870:161927. https://doi.org/10.1016/j.scitotenv.2023.161927

Carapinha SDSSC (2019) Microalgae cultivated in Swine Wastewater: stimulation of seed growth and biopesticide potential Nova University of Lisbon]. https://run.unl.pt/bitstream/10362/90762/1/Carapinha_2019.pdf

CFB (2023) *Anabaena/Anabaenopsis*. Center for freshwater biology. Retrieved 4 Nov 2023 from http://cfb.unh.edu/CyanoKey/Anabaena.html

Chanda M-J, Merghoub N, El Arroussi H (2019) Microalgae polysaccharides: the new sustainable bioactive products for the development of plant bio-stimulants? World J Microbiol Biotechnol 35(11):177. https://doi.org/10.1007/s11274-019-2745-3

Chandini KR, Kumar R, Prakash O (2019) The impact of chemical fertilizers on our environment and ecosystem. In: A. Publications (ed) Research trends in environmental sciences, vol 3, pp 71–86. https://www.researchgate.net/publication/331132826_The_Impact_of_Chemical_Fertilizers_on_our_Environment_and_Ecosystem

Chatterjee A, Singh S, Agrawal C, Yadav S, Rai R, Rai L (2017) Role of algae as a biofertilizer. In: Algal green chemistry. Elsevier, pp 189–200. https://doi.org/10.1016/B978-0-444-63784-0.00010-2

Chiaiese P, Corrado G, Colla G, Kyriacou MC, Rouphael Y (2018) Renewable sources of plant biostimulation: microalgae as a sustainable means to improve crop performance. Front Plant Sci 9:430391. https://doi.org/10.3389/fpls.2018.01782

Chittora D, Meena M, Barupal T, Swapnil P, Sharma K (2020) Cyanobacteria as a source of biofertilizers for sustainable agriculture. Biochem Biophys Rep 22:100737. https://doi.org/10.1016/j.bbrep.2020.100737

Colla E, Menegotto ALL, Kalschne DL, da Silva-Buzanello RA, Canan C, Drunkler DA (2020) Microalgae: a new and promising source of food. In: Handbook of algal science, technology and medicine. Elsevier, pp 507–518. https://doi.org/10.1016/B978-0-12-818305-2.00032-2

Coppens J, Grunert O, Van Den Hende S, Vanhoutte I, Boon N, Haesaert G, De Gelder L (2016) The use of microalgae as a high-value organic slow-release fertilizer results in tomatoes with increased carotenoid and sugar levels. J Appl Phycol 28:2367–2377. https://doi.org/10.1007/s10811-015-0775-2

Costa OY, Raaijmakers JM, Kuramae EE (2018) Microbial extracellular polymeric substances: ecological function and impact on soil aggregation. Front Microbiol 9:337094. https://doi.org/10.3389/fmicb.2018.01636

Daliry S, Hallajisani A, Mohammadi Roshandeh J, Nouri H, Golzary A (2017) Investigation of optimal condition for Chlorella vulgaris microalgae growth. Glob J Environ Sci Manag 3(2):217–21z. https://doi.org/10.22034/gjesm.2017.03.02.010

Deepa P, Sowndhararajan K, Kim S (2023) A review of the harvesting techniques of microalgae. Water 15(17):3074. https://doi.org/10.3390/w15173074

Dineshkumar R, Kumaravel R, Gopalsamy J, Sikder MNA, Sampathkumar P (2018) Microalgae as bio-fertilizers for rice growth and seed yield productivity. Waste Biomass Valoriz 9:793–800. https://doi.org/10.1007/s12649-017-9873-5

Dineshkumar R, Subramanian J, Gopalsamy J, Jayasingam P, Arumugam A, Kannadasan S, Sampathkumar P (2019) The impact of using microalgae as biofertilizer in maize (Zea mays L.). Waste Biomass Valoriz 10:1101–1110. https://doi.org/10.1007/s12649-017-0123-7

Dineshkumar R, Subramanian J, Arumugam A, Ahamed Rasheeq A, Sampathkumar P (2020) Exploring the microalgae biofertilizer effect on onion cultivation by field experiment. Waste Biomass Valoriz 11:77–87. https://doi.org/10.1007/s12649-018-0466-8

Dmytryk A, Chojnacka K (2018) Algae as fertilizers, biostimulants, and regulators of plant growth. In: Chojnacka PPWK, Schroeder G, Michalak I (eds) Algae biomass: characteristics and applications, vol 8. Springer, Cham, pp 115–122. https://doi.org/10.1007/978-3-319-74703-3_10

Dolganyuk V, Belova D, Babich O, Prosekov A, Ivanova S, Katserov D, Patyukov N, Sukhikh S (2020) Microalgae: a promising source of valuable bioproducts. Biomol Ther 10(8):1153. https://doi.org/10.3390/biom10081153

Egamberdiyeva D, Höflich G (2004) Effect of plant growth-promoting bacteria on growth and nutrient uptake of cotton and pea in a semi-arid region of Uzbekistan. J Arid Environ 56(2):293–301. https://doi.org/10.1016/S0140-1963(03)00050-8

Elagamey E, Abdellatef MA, Flefel HE (2023) Perspective chapter: Cyanobacteria–a futuristic effective tool in sustainable agriculture. In: Cyanobacteria-recent advances and new perspectives. IntechOpen. https://doi.org/10.5772/intechopen.109829

Elakbawy WM, Shanab SM, Shalaby EA (2022) Enhancement of plant growth regulators production from microalgae cultivated in treated sewage wastewater (TSW). BMC Plant Biol 22(1):377. https://doi.org/10.1186/s12870-022-03764-w

FAO (2021) Agricultural production statistics. 2000–2021. Organización de las Naciones Unidas para la Alimentación y la Agricultura. Retrieved 30 Oct 2023 from https://openknowledge.fao.org/items/f769fb2b-a461-4cda-a6f9-4649f89616df

FAO (2022, 16 Oct 2023) The State of World Fisheries and Aquaculture 2022. Organización de las Naciones Unidas para la Alimentación y la Agricultura. Retrieved 16 Oct 2023 from https://openknowledge.fao.org/items/11a4abd8-4e09-4bef-9c12-900fb4605a02

Franco-Morgado M, Amador-Espejo GG, Pérez-Cortés M, Gutiérrez-Uribe JA (2023) Microalgae and cyanobacteria polysaccharides: important link for nutrient recycling and revalorization of agro-industrial wastewater. Appl Food Res 3:100296. https://doi.org/10.1016/j.afres.2023.100296

García-Orellana Y, Soto G, Tafur V, Simbaña A, Tello E, Brito JJ (2016) Efecto de un fertilizante orgánico microalgal en la germinación y crecimiento de plántulas de albahaca (Ocimum basilicum L.). Revista Unellez de Ciencia y Tecnología 34:33–39

Gatamaneni-Loganathan B, Orsat V, Lefsrud M (2020) Utilizing the microalgal biomass of Chlorella variabilis and Scenedesmus obliquus produced from the treatment of synthetic dairy wastewater as a biofertilizer. J Plant Nutr 44(10):1486–1497. https://doi.org/10.1080/01904167.2020.1862191

Gaysina LA, Saraf A, Singh P (2019) Cyanobacteria in diverse habitats. In: Cyanobacteria. Elsevier, pp 1–28. https://doi.org/10.1016/B978-0-12-814667-5.00001-5

Gonçalves AL (2021) The use of microalgae and cyanobacteria in the improvement of agricultural practices: a review on their biofertilising, biostimulating and biopesticide roles. Appl Sci 11(2):871. https://doi.org/10.3390/app11020871

Gonçalves J, Freitas J, Fernandes I, Silva P (2023) Microalgae as biofertilizers: a sustainable way to improve soil fertility and plant growth. Sustain For 15(16):12413. https://doi.org/10.3390/su151612413

González-Pérez BK, Rivas-Castillo AM, Valdez-Calderón A, Gayosso-Morales MA (2022) Microalgae as biostimulants: a new approach in agriculture. World J Microbiol Biotechnol 38(1):4. https://doi.org/10.1007/s11274-021-03192-2

Gosal S, Gill G, Sharma S, Walia S (2018) Soil nutrient status and yield of rice as affected by long-term integrated use of organic and inorganic fertilizers. J Plant Nutr 41(4):539–544. https://doi.org/10.1080/01904167.2017.1392570

Guo S, Wang P, Wang X, Zou M, Liu C, Hao J (2020) Microalgae as biofertilizer in modern agriculture. In: Microalgae biotechnology for food, health and high value products, pp 397–411. https://doi.org/10.1007/978-981-15-0169-2_12

Guzmán-Murillo MA, Ascencio F, Larrinaga-Mayoral JA (2013) Germination and ROS detoxification in bell pepper (Capsicum annuum L.) under NaCl stress and treatment with microalgae extracts. Protoplasma 250:33–42. https://doi.org/10.1007/s00709-011-0369-z

Heimann K, Huerlimann R (2015) Microalgal classification: major classes and genera of commercial microalgal species. In: Handbook of marine microalgae. Elsevier, pp 25–41. https://doi.org/10.1016/B978-0-12-800776-1.00003-0

Hussain F, Shah SZ, Ahmad H, Abubshait SA, Abubshait HA, Laref A, Manikandan A, Kusuma HS, Iqbal M (2021) Microalgae an ecofriendly and sustainable wastewater treatment option: biomass application in biofuel and bio-fertilizer production. A review. Renew Sust Energ Rev 137:110603. https://doi.org/10.1016/j.rser.2020.110603

Iqbal MM, Muhammad G, Aslam MS, Hussain MA, Shafiq Z, Razzaq H (2021) Algal biofertilizer. In: Biofertilizers: study and impact, pp 607–635. https://doi.org/10.1002/9781119724995.ch22

Jhala Y, Panpatte D, Vyas R (2017) Cyanobacteria: source of organic fertilizers for plant growth. In: Microorganisms for Green Revolution: volume 1: microbes for sustainable crop production, pp 253–264. https://doi.org/10.1007/978-981-10-6241-4_13

Junior WGM, Gorgich M, Corrêa PS, Martins AA, Mata TM, Caetano NS (2020) Microalgae for biotechnological applications: cultivation, harvesting and biomass processing. Aquaculture 528:735562. https://doi.org/10.1016/j.aquaculture.2020.735562

Khan BA, Nadeem MA, Nawaz H, Amin MM, Abbasi GH, Nadeem M, Ali M, Ameen M, Javaid MM, Maqbool R (2023) Pesticides: impacts on agriculture productivity, environment, and management strategies. In: Emerging contaminants and plants: interactions, adaptations and remediation technologies. Springer, pp 109–134. https://doi.org/10.1007/978-3-031-22269-6_5

Kholssi R, Lougraimzi H, Grina F, Lorentz JF, Silva I, Castaño-Sánchez O, Marks EA (2022) Green agriculture: a review of the application of micro-and macroalgae and their impact on crop production on soil quality. J Soil Sci Plant Nutr 22(4):4627–4641. https://doi.org/10.1007/s42729-022-00944-3

Krasilnicov P, Taboada MA, Amanullah. (2022) Fertilizer use, soil health and agricultural sustainability. Agriculture 12(4):462. https://doi.org/10.3390/agriculture12040462

Kumar J, Singh D, Tyagi MB, Kumar A (2019) Cyanobacteria: applications in biotechnology. In: Cyanobacteria. Elsevier, pp 327–346. https://doi.org/10.1016/b978-0-12-814667-5.00016-7

Laughinghouse HD, Berthold DE, Marble SC, Saha D (2019) Biología y Manejo de Nostoc (Cyanobacteria) en Viveros e Invernaderos: SS-AGR-431-Span/AG432, 4/2019. EDIS 2019(2). https://doi.org/10.32473/edis-ag432-2019

Li J, Lens PN, Ferrer I, Du Laing G (2021) Evaluation of selenium-enriched microalgae produced on domestic wastewater as biostimulant and biofertilizer for growth of selenium-enriched crops. J Appl Phycol 33:3027–3039. https://doi.org/10.1007/s10811-021-02523-y

Liu J, Chen F (2016) Biology and industrial applications of Chlorella: advances and prospects. In: Microalgae biotechnology, pp 1–35. https://doi.org/10.1007/10_2014_286

Liu Q, Hongwei X, Haijie Y (2021) Impact of fertilizer on crop yield and C:N:P stoichiometry in arid and semi-arid soil. Int J Environ Res Public Health 18(8):4341. https://doi.org/10.3390/ijerph18084341

Mahapatra DM, Chanakya H, Joshi N, Ramachandra T, Murthy G (2018) Algae-based biofertilizers: a biorefinery approach. In: Microorganisms for Green Revolution: volume 2: microbes for sustainable agro-ecosystem, pp 177–196. https://doi.org/10.1007/978-981-10-7146-1_10

Makra N, Gell G, Juhász A, Soós V, Kiss T, Molnár Z, Ördög V, Vörös L, Balázs E (2019) Molecular taxonomic evaluation of Anabaena and Nostoc strains from the Mosonmagyaróvár Algal Culture Collection. S Afr J Bot 124:80–86. https://doi.org/10.1016/j.sajb.2019.03.008

Manjunath M, Kanchan A, Ranjan K, Venkatachalam S, Prasanna R, Ramakrishnan B, Hossain F, Nain L, Shivay YS, Rai AB (2016) Beneficial cyanobacteria and eubacteria synergistically enhance bio availability of soil nutrients and yield of okra. Heliyon 2(2):e00066. https://doi.org/10.1016/j.heliyon.2016.e00066

Mitter EK, Tosi M, Obregón D, Dunfield KE, Germida JJ (2021) Rethinking crop nutrition in times of modern microbiology: innovative biofertilizer technologies. Front Sustain Food Syst 5:606815. https://doi.org/10.3389/fsufs.2021.606815

Moreira JB, Vaz BDS, Cardias BB, Cruz CG, Almeida ACAD, Costa JAV, Morais MGD (2022) Microalgae polysaccharides: an alternative source for food production and sustainable agriculture. Polysaccharides 3(2):441–457. https://doi.org/10.3390/polysaccharides3020027

Morgado-González A, Espinosa-Victoria D, Gómez-Merino FC (2015) Efficiency of plant growth promoting rhizobacteria (PGPR) in sugarcane. Terra Latinoamericana 33(4):321–330

Mückschel F, Ollo E, Glaeser SP, Düring R, Yan F, Velten H, Theilen U, Frei M (2023) Nitrogen use efficiency of microalgae application in wheat compared to mineral fertilizer. J Plant Nutr Soil Sci 186(5):522–531. https://doi.org/10.1002/jpln.202300125

Mukherjee A, Patel J (2020) Seaweed extract: biostimulator of plant defense and plant productivity. Int J Environ Sci Technol 17(1):553–558. https://doi.org/10.1007/s13762-019-02442-z

Mulbry W, Kondrad S, Pizarro C (2007) Biofertilizers from algal treatment of dairy and swine manure effluents: characterization of algal biomass as a slow release fertilizer. J Veg Sci 12(4):107–125. https://doi.org/10.1300/J484V12N04_08

Muñoz R, Guieysse B (2006) Algal–bacterial processes for the treatment of hazardous contaminants: a review. Water Res 40(15):2799–2815. https://doi.org/10.1016/j.watres.2006.06.011

Murata M, Ito Morioka L, Da Silva Marques J, Bosso A, Suguimoto H (2021) What do patents tell us about microalgae in agriculture? AMB Express 11:154. https://doi.org/10.1186/s13568-021-01315-4

Mutum L, Janda T, Ördög V, Molnár Z (2022) Biologia Futura: potential of different forms of microalgae for soil improvement. Biologia Futura 73:1–8. https://doi.org/10.1007/s42977-021-00103-2

NCBI (2023) Taxonomy. National Center for Biotechnology Information. Retrieved 1 Oct 2023 from https://www.ncbi.nlm.nih.gov/taxonomy

ONU (2023) Global challenges: population. Naciones Unidas. Retrieved 29 Oct 2023 from https://www.un.org/es/global-issues/population#:~:text=Una%20poblaci%C3%B3n%20en%20crecimiento,y%202000%20millones%20desde%201998

Ortiz-Moreno ML, Sandoval-Parra KX, Solarte-Murillo LV (2019) Chlorella,≤ un potencial biofertilizante? Orinoquia 23(2):71–78. https://doi.org/10.22579/20112629.582

Osorio-Reyes JG, Valenzuela Amaro HM, Pizaña Aranda JJP, Ramírez Gamboa D, Meléndez Sánchez ER, López Arellanes ME, Castañeda Antonio MD, Coronado Apodaca KG, Gomes Araújo R, Sosa Hernández JE, Melchor-Martínez EM, Iqbal HMN, Parra-Saldivar R, Martínez-Ruiz M (2023) Microalgae-based biotechnology as alternative biofertilizersfor soil enhancement and carbon footprint reduction: advantages and implications. Mar Drugs 21(2). https://doi.org/10.3390/md21020093

Park J-R, Jang Y-H, Kim E-G, Lee G-S, Kim K-M (2023) Nitrogen fertilization causes changes in agricultural characteristics and gas emissions in rice field. Sustain For 15(4):3336. https://doi.org/10.3390/su15043336

Parmar P, Kumar R, Neha Y, Srivatsan V (2023) Microalgae as next generation plant growth additives: functions, applications, challenges and circular bioeconomy based solutions. Front Plant Sci 14:1073546. https://doi.org/10.3389/fpls.2023.1073546

Parwani L, Bhatt M, Singh J (2021) Potential biotechnological applications of cyanobacterial exopolysaccharides. Braz Arch Biol Technol 64:e21200401. https://doi.org/10.1590/1678-4324-2021200401

Penuelas J, Coello F, Sardans J (2023) A better use of fertilizers is needed for global food security and environmental sustainability. Agric Food Secur 12(1):1–9. https://doi.org/10.1186/s40066-023-00409-5

Pirttilä AM, Mohammad Parast Tabas H, Baruah N, Koskimäki JJ (2021) Biofertilizers and biocontrol agents for agriculture: how to identify and develop new potent microbial strains and traits. Microorganisms 9(4):817. https://doi.org/10.3390/microorganisms9040817

Plaza BM, Gómez-Serrano C, Acién-Fernández FG, Jimenez-Becker S (2018) Effect of microalgae hydrolysate foliar application (Arthrospira platensis and Scenedesmus sp.) on Petunia x hybrida growth. J Appl Phycol 30:2359–2365. https://doi.org/10.1007/s10811-018-1427-0

Prisa D, Spagnuolo D (2023) Plant Production with Microalgal Biostimulants. Horticulturae 9(7):829. https://doi.org/10.3390/horticulturae9070829

Rai A, Singh A, Syiem M (2019) Plant growth-promoting abilities in cyanobacteria. In: Cyanobacteria. Elsevier, pp 459–476. https://doi.org/10.1016/B978-0-12-814667-5.00023-4

Rani L, Thapa K, Kanojia N, Sharma N, Singh S, Grewal AS, Srivastav AL, Kaushal J (2021) An extensive review on the consequences of chemical pesticides on human health and environment. J Clean Prod 283:124657. https://doi.org/10.1016/j.jclepro.2020.124657

Ronga D, Biazzi E, Parati K, Carminati D, Carminati E, Tava A (2019) Microalgal biostimulants and biofertilisers in crop productions. Agronomy 9(4):192. https://doi.org/10.3390/agronomy9040192

Ru ITK, Sung YY, Jusoh M, Wahid MEA, Nagappan T (2020) Chlorella vulgaris: a perspective on its potential for combining high biomass with high value bioproducts. Appl Phycol 1(1):2–11. https://doi.org/10.1080/26388081.2020.1715256

Santini G, Biondi N, Rodolfi L, Tredici MR (2021) Plant biostimulants from cyanobacteria: an emerging strategy to improve yields and sustainability in agriculture. Plan Theory 10(4):643. https://doi.org/10.3390/plants10040643

SCB (2023) *Scenedesmus sp.* Sarawak Biodiversity Centre. Retrieved 30 Oct 2023 from https://www.sbc.org.my/component/content/article/866-scenedesmus?catid=187&Itemid=201

Singh JS, Kumar A, Rai AN, Singh DP (2016) Cyanobacteria: a precious bio-resource in agriculture, ecosystem, and environmental sustainability. Front Microbiol 7:186282. https://doi.org/10.3389/fmicb.2016.00529

Sinha A, Goswami G, Kumar R, Das D (2021) A microalgal biorefinery approach for bioactive molecules, biofuel, and biofertilizer using a novel carbon dioxide-tolerant strain Tetradesmus obliquus CT02. Biomass Convers Biorefinery 13:1–14. https://doi.org/10.1007/s13399-021-02098-1

Solomon W, Mutum L, Janda T, Molnár Z (2023) Potential benefit of microalgae and their interaction with bacteria to sustainable crop production. Plant Growth Regul 101(1):53–65. https://doi.org/10.1007/s10725-023-01019-8

Song X, Bo Y, Feng Y, Tan Y, Zhou C, Yan X, Ruan R, Xu Q, Cheng P (2022) Potential applications for multifunctional microalgae in soil improvement. Front Environ Sci 10:1035332. https://doi.org/10.3389/fenvs.2022.1035332

Stavi I, Lal R (2015) Achieving zero net land degradation: challenges and opportunities. J Arid Environ 112:44–51. https://doi.org/10.1016/j.jaridenv.2014.01.016

Tarakhovskaya E, Maslov YI, Shishova M (2007) Phytohormones in algae. Russ J Plant Physiol 54:163–170. https://doi.org/10.1134/S1021443707020021

Tayoh LN (2020) Destruction of soil health and risk of food contamination by application of chemical fertilizer. In: Bauddh K, Kumar S, Singh RP, Korstad J (eds) Ecological and practical applications for sustainable agriculture. Springer, pp 53–64. https://doi.org/10.1007/978-981-15-3372-3_3

Tian S-L, Khan A, Zheng W-N, Song L, Liu J-H, Wang X-Q, Li L (2022) Effects of Chlorella extracts on growth of Capsicum annuum L. seedlings. Sci Rep 12(1):15455. https://doi.org/10.1038/s41598-022-19846-6

Uysal O, Uysal FO, Ekinci K (2015) Evaluation of microalgae as microbial fertilizer. Eur J Sustain Dev 4(2):77. https://doi.org/10.14207/ejsd.2015.v4n2p77

Vander Wiel JB, Mikulicz JD, Boysen MR, Hashemi N, Kalgren P, Nauman LM, Baetzold SJ, Powell GG, He Q, Hashemi NN (2017) Characterization of Chlorella vulgaris and Chlorella protothecoides using multi-pixel photon counters in a 3D focusing optofluidic system. RSC Adv 7(8):4402–4408. https://doi.org/10.1039/c6ra25837a

Wang C, Qi M, Guo J, Zhou C, Yan X, Ruan R, Cheng P (2021) The active phytohormone in microalgae: the characteristics, efficient detection, and their adversity resistance applications. Molecules 27(1):46. https://doi.org/10.3390/molecules27010046

Win TT, Barone GD, Secundo F, Fu P (2018) Algal biofertilizers and plant growth stimulants for sustainable agriculture. Ind Biotechnol 14(4):203–211. https://doi.org/10.1089/ind.2018.0010

Wirth R, Pap B, Böjti T, Shetty P, Lakatos G, Bagi Z, Kovács KL, Maróti G (2020) Chlorella vulgaris and its phycosphere in wastewater: microalgae-bacteria interactions during nutrient removal. Front Bioeng Biotechnol 8:557572. https://doi.org/10.3389/fbioe.2020.557572

Microalgae Biomass As a Sustainable Source of Bioplastics

18

Omar Nateras-Ramírez, Karen Itzel Carrillo-Cañedo, Ernesto Aguilar-Palazuelos, and Perla Rosa Fitch-Vargas

Abstract

Due to the environmental pollution issues associated with non-biodegradable plastics, there is a growing interest in the development of bioplastics. Microalgae biomass has emerged as an excellent option for producing biodegradable materials due to its rapid growth and low nutritional requirements, which means it does not need arable land. Microalgae are a source of various bioactive compounds, such as proteins, carbohydrates, polylactic acid (PLA), and polyhydroxyalkanoates (PHAs), which can be used to form bioplastics. Bioplastics can be formed by blending biomass with conventional polymers or biopolymers, extracting biopolymers of interest, or producing PLA or PHAs through biological processes. Besides, PHAs are biodegradable polyesters, particularly promising due to their similarity to petrochemical-based plastics produced by different microalgae species under environmental and nutrient stress. Likewise, microalgae-based bioplastics can undergo conventional plastics processing techniques such as extrusion, injection molding, and compression molding. However, industrial-scale development of microalgae-based bioplastics remains relatively limited. Therefore, exploring innovative strategies such as biorefinery approaches and genetic engineering is essential to address economic viability challenges hindering the widespread adoption of microalgae-based bioplastics in the market.

Keywords

Microalgae biomass · Bioplastics · Biopolymers · Polyhydroxyalkanoates · Biorefinery

18.1 Introduction

Since conventional petroleum-derived plastics were invented in the twentieth century, they have played an indispensable role in our society due to their low cost, good formability, strength, water resistance, and lightweight. However, despite their numerous applications, mainly in packaging, they contribute to most plastic being discarded as garbage worldwide, reaching degradation times of up to 500 years. The most employed types include polyethylene (PE), polystyrene (PS), polypropylene (PP), polyvinyl chloride (PVC), polyethylene terephthalate (PET) (Dang et al. 2022; Arora et al. 2023; Costa et al. 2019; López Rocha et al. 2020).

In 2019, global plastic production reached 368 million metric tons and is predicted to double within 20 years (Walker and Fequet 2023; Cheah et al. 2023). According to Arora et al. (2023), approximately 34 million metric tons of plastics leak into the aquatic environment annually due to poor waste management, which is expected to increase threefold by 2040. Moreover, it is well-known that prolonged exposure to plastic residues in landfills stimulates their deterioration due to environmental factors. Consequently, plastics break down into smaller particles known as microplastics (<5 mm) and nanoplastics (<100 nm), which persist in the environment for extended periods. These particles are recognized as emerging pollutants threatening the environment and human health (Cheah et al. 2023; Arora et al. 2023). For these reasons, many researchers are looking for sustainable and environmentally friendly resources as alternatives to plastic production.

Moreover, bioplastics differ from petroleum-based plastics as they are either bio-based, biodegradable, or both,

O. Nateras-Ramírez
Instituto Tecnológico de Mazatlán, Instituto Tecnológico de Mexico, Mazatlán, Sinaloa, Mexico

K. I. Carrillo-Cañedo · E. Aguilar-Palazuelos
Posgrado en Ciencia y Tecnología de Alimentos, Facultad de Ciencias Químico-Biológicas, Universidad Autónoma de Sinaloa, Culiacán, Sinaloa, Mexico

P. R. Fitch-Vargas (✉)
Facultad de Ciencias del Mar, Universidad Autónoma de Sinaloa, Mazatlán, Sinaloa, Mexico
e-mail: perlafitch@uas.edu.mx

D. U. Santos Ballardo, S. Rossi (eds.), *Microalgae as Promising Source of Commercial Bioproducts*, Developments in Applied Phycology 14, https://doi.org/10.1007/978-3-031-86433-9_18

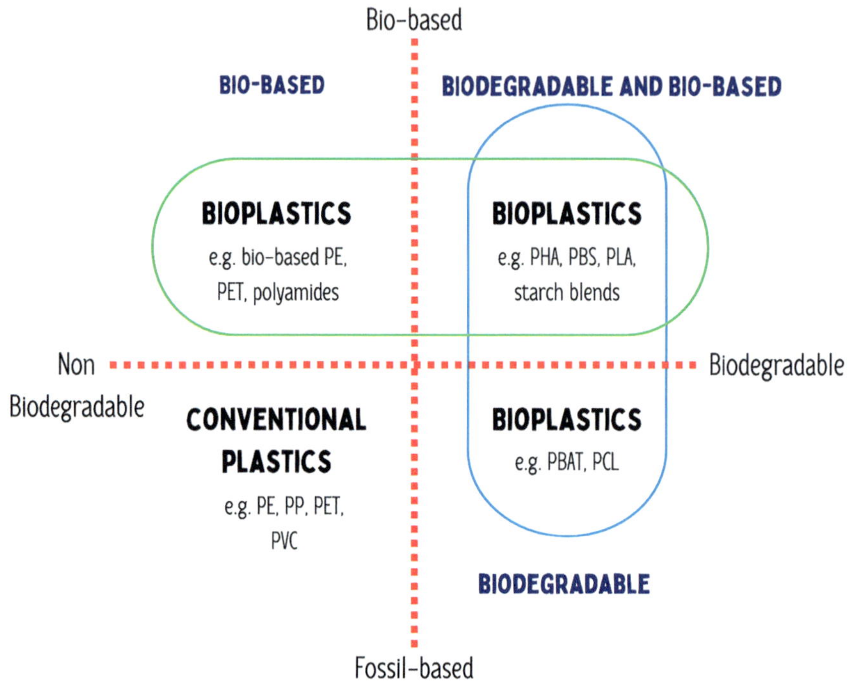

Fig. 18.1 Plastics classification according to bio-based content and biodegradability (European Bioplastics 2022)

produced from renewable biomass sources, agricultural byproducts, algae, and microorganisms (Park and Lee 2022; Madadi et al. 2021). Bioplastics are classified into three groups: bio-based non-biodegradable plastics, bio-based and biodegradable plastics, and plastics based on fossil resources that are also biodegradable (Fig. 18.1) (Onen Cinar et al. 2020; European Bioplastics 2022). These materials enhance resource efficiency, lower carbon footprint and greenhouse gas emissions, and also help to conserve fossil resources (Park and Lee 2022).

According to European Bioplastics (2023), the global bioplastics market is expected to grow from 2.2 million to 6.3 million tons by 2027. Despite the increasing demand of the bioplastics market, they currently account for only 1% of the total annual plastic production (Arora et al. 2023). The main application areas include the packaging, textile, automotive, and construction industries (Park and Lee 2022). Furthermore, bioplastics are commonly made of carbohydrate-rich plants and crops (first-generation feedstock) and waste from lignocellulosic biomass and food industries (second-generation feedstock) (Cheah et al. 2023). Nonetheless, plant-based resources for bioplastic production can lead to environmental challenges and competition with food supplies, which does not make them sustainable in the long term (Ugya et al. 2023; Onen Cinar et al. 2020; Dang et al. 2022). Meanwhile, agricultural wastes are insufficient and limited for bioplastic production (Madadi et al. 2021). Therefore, an increasing interest has been in developing bioplastics based on fast-growing microorganisms such as microalgae (third-generation feedstock).

Microalgae, a third-generation feedstock, have outperformed food crops and lignocellulosic biomass in bioplastic development due to their rapid growth rate, fewer nutrient requirements, and the absence of a need for arable land. They can grow in freshwater, saline water, and wastewater and accumulate high biomass concentrations (Madadi et al. 2021; Chong et al. 2022). Moreover, microalgae bioplastic production contributes to sustainability and the circular economy. Likewise, microalgae can assimilate atmospheric CO_2; therefore, the conversion of microalgae into bioplastics represents a direct means of sequestering atmospheric CO_2 in polymer form (Park and Lee 2022; Onen Cinar et al. 2020).

Microalgae have the potential to produce bioactive compounds such as lipids, proteins, carbohydrates, and polyhydroxyalkanoates (PHAs), which can be utilized to develop bioplastics (Chong et al. 2022). There are three routes for obtaining bioplastics from microalgae: using whole biomass, extracting and utilizing biopolymers, and producing volatile organic acids and polymers (Calijuri et al. 2022). Additionally, it has been demonstrated that microalgae-based bioplastics can undergo conventional plastics processing techniques such as extrusion, injection molding, and compression molding, enabling their industrial-scale production (González-Balderas et al. 2020).

By the above, this chapter aims to promote microalgae as a sustainable, cost-effective, and renewable source for producing microalgae-based bioplastics as an eco-friendly alternative to petroleum-based plastics. It will also analyze microalgae's key characteristics and advantages and the most used species for bioplastic production. Additionally, it will explore the primary biopolymers found in microalgae that hold potential for bioplastic development, the various methods employed in their production, and their characterization techniques.

18.2 Microalgae As a Promising Source for Bioplastic Production

Microalgae are a large group of unicellular photosynthetic microscopic (ranging between 1 and 50 mm in diameters) heterotrophic or autotrophic organisms of which there are approximately 800,000 prokaryotic or eukaryotic species that are also known as microphytes and represent one of the most important of primary producers on the planet. They are present not only in aquatic but also in terrestrial ecosystems like wet soil or moist rocks (Mata et al. 2010; Rizwan et al. 2018; Yousuf 2019), and their reproduction may be sexually or asexually or by a combination of both processes through alternation of generations (Yousuf 2019). They play a crucial role in the food chain, guaranteeing the flow of matter and energy required to maintain heterotrophic organisms in aquatic environments (Andreotti et al. 2020). Therefore, they represent a vast variety of species that can live in a wide range of environments. Microalgae need three major components for growth, including sunlight, water, and carbon sources (Tan et al. 2011; Rizwan et al. 2018). Most species contain chlorophyll; they can transform nutrients from aquatic habitats to produce about 50% of the atmospheric oxygen and generate value-added products (Hadipoor et al. 2021).

They are categorized into four main classes: diatoms, green algae, blue-green algae, and golden algae. Due to simple structures, many species exhibit rapid growth and high productivity (Ruane et al. 2010; Tan et al. 2011). Many algal species have been designated as containing enough material and utilities for industrial application (Table 18.1). Due to their importance, microalgae have been studied worldwide for different purposes, which has generated local collections that have promoted the research of these organisms (Mata et al. 2010).

Table 18.1 Microalgae and its commercial applications

Entry	Microalgae genre	Applications	References
1	*Spirulina*	Supplements, cosmetics	Costa et al. (2019)
		bioplastics	Dianursanti et al. (2019)
2	*Chlorella*	Supplements, cosmetics	Costa et al. (2019)
		bioplastics	Sabathini et al. (2018) and Zeller et al. (2013)
		Biofuels	Mata et al. (2010)
3	*Dunaliella*	Pigments and antioxidant	Costa et al. (2019)
		Bioplastics	Mondal et al. (2021b)
4	*Nannochloropsis*	Bioplastics	Shi et al. (2012) and Wang et al. (2016)
5	*Haematococcus*	Pigments and antioxidants	Costa et al. (2019)

Microalgae contain lipids and fatty acids as membrane components, storage products, metabolites, and energy sources such as proteins, carbohydrates, lipids, and pigments (Tan et al. 2011; Yousuf 2019). Therefore, they represent a biotechnological source of natural compounds with significant commercial potential for bioplastic production (Abdo et al. 2023; González-Balderas et al. 2021; Mondal et al. 2021b).

Microalgae cultivation offers several advantages compared to plants and seeds since they do not need soil to develop and allow non-potable water as a culture medium (Mata et al. 2010; López Rocha et al. 2020). Microalgae has much higher biomass production efficiency per hectare than other crops concerning terrestrial plants (50–70 t ha^{-1} $year^{-1}$ per year vs 10–20 t ha^{-1} $year^{-1}$) (Di Caprio et al. 2019) since they effortlessly double their biomass in 24 h and can be harvested throughout most of the year, giving a regular supply of biomass and less freshwater using (Ruane et al. 2010; Yousuf 2019; Hadipoor et al. 2021). In addition, microalgae cultivation provides very convenient flexibility because they can tolerate a wide range of pH, salinity, and temperature (Yousuf 2019). The high productivity rate of microalgae ensures quick generation of desired products and, more importantly, causes higher capacity and rate of CO_2 fixation (Hadipoor et al. 2021). Microalgae can be grown using sea or brackish water, and microalgal biomass production systems can easily be adapted to various levels of operational or technological skills. According to Mata et al. (2010), microalgae can adopt various types of metabolisms, including autotrophic, heterotrophic, mixotrophic, and photoheterotrophic, and they can shift their metabolism in response to environmental conditions.

18.2.1 Microalgae Culture Systems

Microalgae have been cultivated worldwide for different purposes. Their production occurs in ponds or bioreactors on non-arable land or in a marine environment; it need not compete with food production for land or water. Different microalgae species can be adapted to live in various environmental conditions. Thus, it is possible to find species best suited to local environments or specific growth characteristics, which is not possible to do with other primary producers (Mata et al. 2010).

Several cultivation techniques are currently employed for the large-scale production of microalgae biomass (MB), two of them being the most used: open culture systems (OCS) and photobioreactors (PBRs) (Yousuf 2019; Onen Cinar et al. 2020).

Generally, the efficiency of the culture system is commonly evaluated by microalgal productivity components such as areal productivity (AP) in ton ha^{-1} $year^{-1}$ (units may vary according to the scale, g m^{-2} d^{-1}). The choice of cultivation system depends upon several factors, including the nature of microalgae species, nutrient availability, climate, mode of cultivation (autotrophic/heterotrophic), and the final use of biomass (Rizwan et al. 2018).

OCS represents the simplest and most widespread way to produce large quantities of MB. They consist of pools of different sizes and shapes in which microalgae grow outdoors, stirred by shovels or some mechanical stirring device to promote the homogeneous integration of the biomass. According to Onen Cinar et al. (2020), open systems are considered easy to clean due to their simple design and are generally more durable. Raceway ponds represent the less energy-consuming and most used system for large-scale microalgae production and probably the less expensive option for cultivation, considering the capital and operating costs (Chiaramonti et al. 2013). These systems can reach an AP of 51.1-ton ha^{-1} $year^{-1}$ for *Nannochloropsis* sp. and 32.5-ton ha^{-1} $year^{-1}$ for *Tetraselmis suecica.*

PBRs are closed systems that allow for better control of the culture conditions, helping limit the contamination from other microorganisms. Compared with traditional open ponds, PBRs can be operated at very high concentrations (thanks to the higher surface-to-volume ratio). Typical biomass production rates achievable with PBRs are considered higher than OCS, but a realistic figure can be estimated in 60–70 tons ha^{-1} $year^{-1}$ (Chiaramonti et al. 2013). The significant advantage of closed reactors is their flexibility compared to open reactors. The optimized growth conditions in closed systems enable the cultivation of those microalgae that could not grow in open ponds (Rizwan et al. 2018).

Another microalgae cultivation option that addresses problems such as the time- and energy-consuming harvesting methods is a cultivation system based on microalgal immobilization (Gonçalves et al. 2017). These bioreactors are mainly used to grow microalgae to treat wastewater by cell entrapment within a polymeric matrix, usually alginate and carrageenan, and cell adhesion and biofilm formation on a solid surface (Christenson and Sims 2011). Also, microalgal immobilization wants prolonged stability, enhanced biomass hold-ups, reaction selectivity, increased product yield, and simplification of downstream processing (Nemati and Webb 2011). Nevertheless, this method is not indicated to produce large quantities of biomass.

Some severe limitations must be overcome before the process becomes a realistic alternative, such as the production cost of developing efficient methods for recovering algal biomass (Chiaramonti et al. 2013). Genetic modification and metabolic engineering are likely to have the most significant impact on improving the economics of production of some specific applications, including increasing the biomass yield and growth rate and improving the temperature tolerance of the microalgae (Ruane et al. 2010).

On the other hand, research projects on microalgae have been reorientated toward producing high-value products following the biorefinery concept so that the production of biofuels and bioproducts from microalgae could attain economic feasibility. Three aspects of microalgae production that will strongly influence the future sustainability of algal production are the energy and carbon balance, environmental impacts, and production cost (Yousuf 2019). Some complex compounds in wastewater may benefit or harm the growing process of cultures; in this point, the bacteria that live in association with microalgae may help the enzymatic conversion of compounds into other usable molecules for microalgae (Di Caprio et al. 2019).

Furthermore, microalgal strain improvement involves using genomics, proteomics, and metabolomics to screen for and develop new strains capable of achieving high growth rates (Ruane et al. 2010). This approach suggests using microalgal proteins for human consumption and the production of biopolymers, contributing to the development of bioplastics (Di Caprio et al. 2019).

18.2.2 Key Microalgae Bioactive Compounds with Bioplastic Applications

Microalgae presents a versatile resource for various bioactive compounds, offering solutions to various challenges. A prominent strategy in microalgae cultivation focuses on harnessing their capabilities to produce essential compounds that facilitate the manufacturing of bioplastics (Table 18.2). Only a few microalgae species have undergone extensive research for their bioplastic production potential, with *Chlorella* and *Spirulina* emerging as the most thoroughly investigated genera (Bussa et al. 2019).

Biopolymers produced from microalgae offer the advantage of being non-toxic and can be employed in synthesizing plastics that resemble petroleum-based plastics thanks to their similar mechanical properties. Some examples of these biopolymers are polyhydroxyalkanoates (PHAs) (Costa et al. 2019). PHAs are composed of hydroxyalkanoic acid monomers with a straight or branched chain containing aliphatic or aromatic side groups monomers containing carbon atoms ranging from 4 to 14 (Kaewbai-ngam et al. 2016; Bagatella et al. 2022). PHAs are synthesized from neutral lipids accumulated in the microalgae cells that require control cultivation conditions using vinasse, molasses from

Table 18.2 Overview of the major polymer production in the microalgae

Microalgae genre	Polymer	Conditions	Concentration	References
Spirulina	Biomass (Proteins)	–	–	Zeller et al. (2013)
	P(3HB)	–	–	Costa et al. (2018)
	P(3HB)	Nitrogen deficiency	30.7% (w/w dry biomass)	Costa et al. (2015)
	PLA	30 °C for 30 min	4% (w/w dry biomass)	Niccolai et al. (2019)
Chlorella	P(3HB)	Nutrient limiting conditions	29.5% (w/w dry biomass)	Kumari et al. (2022)
	P(3HB)	pH 7, temperature 30 °C, under sunlight	80% (w/w dry biomass)	Selvaraj et al. (2021)
	PLA	pH 5.5 and temperature 30 °C	98% (g/g)	Nandini et al. (2020)
Synechococcus	PHAs	–	16% (w/w dry biomass)	Cruz et al. (2004)
	PHAs	–	55% (w/w dry biomass)	Kaewbai-ngam et al. (2016)
Phormidium	P(3HB)	Nitrogen deprivation	14.8% (w/w dry biomass)	Kaewbai-ngam et al. (2016)
Calothrix	P(3HB)	Nitrogen deprivation	25.2% (w/w dry biomass)	Kaewbai-ngam et al. (2016)
Scenedesmus	PHAs	Phosphorus deficiency, low glucose, low salinity	29.92% (w/w dry biomass)	García et al. (2021)
	PLA	210 °C, 2 h over a Fe-Sn-Beta catalyst	33.9% (w/w dry biomass)	Xia et al. (2021)
Oscillatoria	P(3HB)	Nitrogen deprivation	15.7% (w/w dry biomass)	Kaewbai-ngam et al. (2016)
Chlamydomonas	Starch	25 °C, under an incident photosynthetic photon flux density of 125 μmol m^{-1} s^{-1}	49% (w/w dry biomass)	Mathiot et al. (2019)
Nannochloropsis	Cellulose	–	34% (w/w dry biomass)	Lee et al. (2018)

sugar production, or even waste frying oil (Dalsasso et al. 2019). Microalgae can store various intracellular compounds for diverse biological functions, with poly 3-hydroxybutyrate P(3HB) being a prominent example found in prokaryotic cyanobacteria. The accumulation of PHAs in cyanobacteria is known to enhance when supplemented with organic carbon sources like glucose, acetate, citrate, xylose, and arabinose, particularly in photoautotrophic conditions (Cassuriaga et al. 2018; Roja et al. 2019).

Besides, PHAs are the most promising biopolymers due to the distinctive array of functional groups present in the side chain, encompassing carboxyl, hydroxyl, methyl esters, epoxy, phenoxy, cyano, nitro, and thiophenoxy groups, along with its noteworthy bioplastic characteristics including melting point, glass transition temperature, and crystallinity. Among the primary sources of PHAs are saturated and unsaturated hydroxyalkanoic acids (HAA). PHAs biopolymers can be categorized into three main types based on the number of carbonyl atoms: short-chain hydroxyalkanoic acids (C3–C5), medium-chain hydroxyalkanoic acids (C6–C14), and long-chain hydroxyalkanoic acids (>C15) (Kumari et al. 2022).

Furthermore, P(3HB) is a biopolymer produced by microorganisms as a strictly isotactic polyester of the monomer 3-(R)-hydroxybutyric acid. Up to now, the fermentation of glucose has been the primary way to produce P(3HB) (Polotti 2020). Also, P(3HB) is a homopolymer of PHAs, biosynthesized by the action of P(3HB) synthase to polymerize acetoacetyl-CoA to 3-hydroxybutyryl-CoA and then to hydroxybutyrate. Unlike PHAs, P(3HB) is mainly accumulated with polyesters that increase their stiffness and brittleness material with an elastic modulus and elongation at break of 1 GPa and below 15%, respectively; with all this, P(3HB) has similar or even better material characteristics to petrochemical-based plastics that possess melting temperature within 110 to 170 °C. P(3HB) has high crystallinity of up to 70% and excellent gas barrier characteristics, which can be degraded or decomposed quickly, making them suitable for various industrial applications (Devadas et al. 2021).

The PHAs extraction process involves cellular disruption treatments and operations for recovering and purifying biopolymers. Chemical, physical, and biological methods, or even a combination, can guarantee a product with high purity and preserved physical and thermal characteristics (Costa et al. 2019). The physical extraction begins by separating the solid material (cells and salts) from the algae culture. Physical rupture processes may be carried out by centrifugation or sonication. Chemical extraction involves solvents such as chloroform (Kaewbai-ngam et al. 2016; García et al. 2021). The purification is a cleaning process with solvents and filters that allow the recovery of most of the polymer (García et al. 2021).

Furthermore, PHAs and P(3HB) are biodegradable polyesters produced by different microalgae species under environmental stress or nutrient limitation (Kaewbai-ngam et al. 2016; Chong et al. 2022). Microorganism's production of these polymers depends on medium conditions, mainly on the composition of the feeding solution and pH (Reis et al. 2003). Moreover, García et al. (2021) found that microalgae *Scenedesmus* sp. presented higher accumulations of PHAs (0.82 vs 29.92% w/w CDW) in stress conditions (P-deficiency, normal N and Fe, low glucose, low salinity) than under conventional conditions. Besides, Costa et al. (2015) noticed that a nitrogen deficiency may improve PHA accumulation (30.7%) in *Spirulina* biomass. Also, Kaewbai-ngam et al. (2016) reported the same effect of nutrient deficiency in different microalgae strains, such as *Synechococcus, Phormidium,* and *Scenedesmus,* where nitrogen and phosphate deprivation improve their P(3HB) accumulation. However, other studies do not necessarily involve specific stress to cause an unusual P(3HB) accumulation, as demonstrated by Selvaraj et al. (2021). They demonstrated that *Chlorella* sp. serves as an excellent reservoir of P(3HB). In pursuit of optimal P(3HB) accumulation, *Chlorella* sp. was cultured under stressful physicochemical conditions: a pH of 7, a temperature of 30 °C, exposure to sunlight, and the presence of 0.2% sodium bicarbonate as the carbon source. Under these parameters, *Chlorella* sp. exhibited an 80% accumulation of P(3HB).

Another valuable bioactive compound derived from microalgae is polylactic acid (PLA), a biodegradable polymer derived from aliphatic polyester that stands on par with fossil-based polymers like PP, PET, or PS. This versatile biopolymer finds applications across various industries, including textiles, agriculture, and medicine, owing to its exceptional biocompatibility (Castro-Aguirre et al. 2016). The MB can serve as a substrate to produce PLA through microbial fermentation processes. Besides, Nandini et al. (2020) used *Chlorella vulgaris* to produce PLA through the fermentation process of *Lactobacillus plantarum* 23, obtaining a maximum lactic acid productivity of 12.59 g L^{-1} h^{-1}. The optimal conditions for this fermentation were pH 5.5 and 30 °C. Under continuous culture conditions with a hydraulic retention time of just 4 h, they attained remarkable yields of 98% g/g of PLA.

Niccolai et al. (2019) used lyophilized biomass of *Artrospira platensis* F&M- C256 (*Spirulina*) as the sole substrate for lactic acid fermentation by the probiotic bacterium *Lactobacillus plantarum* ATCC 8014. The maximum concentration of PLA was reached after 48 h (3.67 ± 0.18 g L^{-1}) with a conversion yield of about 40 mg of lactic acid per gram of dry *A. platensis* biomass. More recently, Xia et al. (2021) converted *Scenedesmus* residue into lactic acid with a high yield (33.9%) under mild reaction conditions (210 °C, 2 h) over a Fe-Sn-Beta catalyst.

On the other hand, another group of microalgae components suitable for bioplastic manufacturing includes polysaccharides. In eukaryotic green algae, cellulose, and hemicellulose are the primary polysaccharides, primarily serving as structural support for the cell wall. Moreover, starch accumulates in response to nutrient scarcity and day/night cycles, reaching up to 50%. A significant issue in starch production from microalgae is that it is hard to extract (Lutzu et al. 2021). Also, starch storage in microalgae can be increased through cell starvation induced by nitrogen or sulfur limitation, as these limitations impact cell metabolism (Mathiot et al. 2019). Different works have reported the use of these polysaccharides from microalgae for the manufacture of bioplastics. For example, Mathiot et al. (2019) developed starch-based bioplastics from microalgae *Chlamydomonas reinhardtii* 11- 32A due its high starch accumulation up to 49%w/w and maximum biomass productivity of 5.07 g L^{-1} obtained in flasks with sulfur-deplete media at 25 °C, under an incident photosynthetic photon flux density (PPFD) of 125 μmol m^{-1} s^{-1}. These results offer opportunities for the development of starch-based bioplastics derived from microalgae.

Furthermore, cellulose is the most abundant biopolymer obtained from nature. This is an essential component of the cell walls of several algal taxa, such as the green algae Chlorophyta (e.g., *Trebouxiophyceae, Ulvophyceae, Chlorophyceae*) and Charophyta (e.g., *Charophyceae*), or the red algae Rhodophyta. For the phylum Ochrophyta, the *Phaeophyceae* class (brown algae) generally reports to contain cellulose and hemicelluloses. In contrast, a more limited number of species from *Xanthophyceae* (yellow-green algae) and *Chrysophyceae* class (golden algae) were found to use cellulose as a structural polysaccharide (Zanchetta et al. 2021). Also, cellulose is more resistant to water and microwave heating than other materials (Rajpoot et al. 2022). Moreover, nanocellulose fibers, derived from MB, possess great strength attributes, boasting a remarkable elastic modulus of up to 220 GPa and tensile strength of almost 3 GPa. These lightweight fibers exhibit a low density of around 1.6 gcm^{-3}. Utilizing nanocellulose fibers in bioplastics enhances their mechanical characteristics without compromising their biodegradability (Asadi et al. 2016). Regarding this, Lee et al. (2018) showcase a novel approach to extracting cellulose nanofibrils (CNFs) from *Nannochloropsis oceanica*. Remarkably, CNFs exhibited an average strength of approximately 3–4 GPa. This strength places them in the same league as joint reinforcing fillers and surpasses the mechanical potency of wood-based CNFs.

18.3 Microalgae-Based Bioplastics

As mentioned earlier, microalgae have the potential to produce bioactive compounds, including lipids, proteins, carbohydrates, PLA, and polyhydroxyalkanoates (PHAs), which can be employed in bioplastic development. There are three primary approaches to producing bioplastics from microalgae (Fig. 18.2). In the first approach, microalgae biomass (MB) can be directly used after harvesting and drying. This is then blended with either synthetic or biologically derived polymers. The second route involves subjecting the MB to an extraction process to recover valuable biopolymers such as cellulose, starch, and proteins. In the third route, MB serves as a substrate for PHAs and PLA biological production through processes like digestion or fermentation (Chong et al. 2022; Calijuri et al. 2022).

18.3.1 Bioplastics Based on Microalgae Biomass

Blending is the most commonly employed method for modifying and enhancing the physical properties of bioplastics, involving mixing multiple polymers to achieve a uniform phase (Cheah et al. 2023). MB can be mixed or blended with non-biodegradable or biodegradable polymers to produce bioplastics with improved mechanical properties (Chong et al. 2022). MB has been recognized as an attractive raw material for bioplastic production due to its high protein content (20% to 70%), rendering it remarkably similar to petroleum-derived plastics (González-Balderas et al. 2020). Moreover, their small size allows them to be transformed into plastic materials without prior treatment, enhancing the cost-effectiveness of scalable production and reducing waste generation (López Rocha et al. 2020).

One of the most significant challenges hampering the widespread production of MB-based bioplastics is the cost associated with their processing. Diverse strategies have been proposed to mitigate these costs, such as using wastewater for microalgae cultivation (Cheah et al. 2023). The bioplastics synthesis from MB and cyanobacteria can be coupled with wastewater treatment, allowing the growth of these organisms without requiring a synthetic culture media while simultaneously treating wastewater (González-Balderas et al. 2020, 2021; López Rocha et al. 2020). This approach can potentially enhance bioplastic's economic viability, positioning them as competitive alternatives to petroleum-based plastics (López-Pacheco et al. 2022).

18.3.1.1 Blending Microalgae Biomass with Non-biodegradable Polymers

MB can be blended with synthetic polymers, extending the duration of microalgae-based plastics with enhanced mechanical properties. Besides, microalgae cells are generally 50 μm in size, enabling them to be well mixed with conventional plastics. Some petroleum-based plastics mixed with MB are PP, PE, and PVA. The compatibility of these

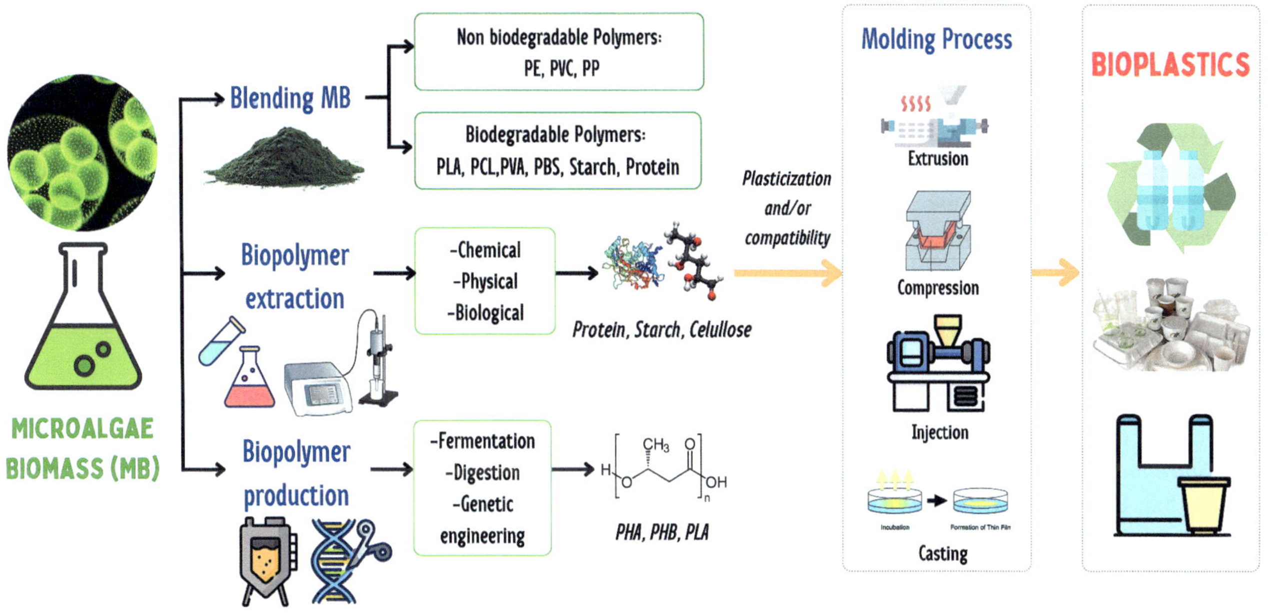

Fig. 18.2 Microalgae-based bioplastics production

Table 18.3 Bioplastics based on microalgae biomass blended with non-biodegradable polymers

Entry	Microalgae genre	Ratio of materials	Plasticizer	Characterization	Products	References
1	*Chlorella*	4:1 biomass to glycerol (0–30%) and UHMW-PE	Glycerol	58% protein	Compressed bioplastic	Zeller et al. (2013)
2	*Chlorella*	20% *Chlorella* and 80% PVC	–	–	PVC-*Chlorella* composite	Zhang et al. (2000)
3	*Spirulina*	1:9 ethylene glycol to *Spirulina* with 20, 35, 50, 65, and 80% of PE-graft-maleic anhydride	Ethylene glycol	33% protein, 20% carbohydrate, 4% fat	Plasticized *Spirulina*	Wang (2014)
4	*Spirulina*	1:9 ethylene glycol to *Spirulina* with 15, 20, and 80% PE	Ethylene glycol	33% protein, 20% carbohydrate, 4% fat	Compressed bioplastic	Wang (2014)
5	*Spirulina*	4:1 biomass to Glycerol (0–30%) and UHMW-PE	Glycerol	57% protein	Compressed bioplastic	Zeller et al. (2013)
6	*Spirulina*	3% *Spirulina*, 7% corn starch and 90% PP	Glycerol	61% protein	Injected bioplastic	Shi et al. (2012)
7	*Nannochloropsis*	14% *Nannochloropsis*, 6% corn starch, and 80% PP	Glycerol	37% protein	Injected bioplastic	Shi et al. (2012)
8	*Nannochloropsis*	50% *Nannochloropsis* and UHMW-PE	–	–	Compressed bioplastic	Wang et al. (2016)
9	*Nannochloropsis*	50% *Nannochloropsis* and 50% PP	–	–	Compressed bioplastic	Wang et al. (2016)

polymers and MB determines the structure and properties of the resulting blends (Cheah et al. 2023; Park and Lee 2022). The polymers must be thermally compatible to prevent the formation of two or more phases and ensure a single glass transition temperature between the respective polymer's glass transition points. Nevertheless, maleic anhydride, starch, and glycerol are often required for blending (Cheah et al. 2023). Table 18.3 shows the array of bioplastics resulting from blending MB with non-biodegradable polymers.

Furthermore, Zeller et al. (2013) produced microalgae-based bioplastics based on MB of *Chlorella* and *Spirulina* with ultra-high molecular weight polyethylene (UHMW-PE). All bioplastics were processed by compression molding for 20 min at 150 °C. *Spirulina* and *Chlorella* showed high protein contents (57 and 58%, respectively) and were plasticized with glycerol (0–30 wt.%) (Table 18.3, Entry 1 and 5). The authors state that *Chlorella* and *Spirulina* have relatively small cells, enabling better dispersion in the matrix when blended with a synthetic polymer. Furthermore, it was found that *Spirulina* presented a higher tensile strength (1.9 MPa) compared to *Chlorella* (1.6 MPa) in the thermoplastic blends, indicating that *Chlorella* had lower compatibility with UHMW-PE; nonetheless, this could be improved by adding a compatibilizer (Chong et al. 2022). On the other hand, Shi et al. (2012) produced bioplastics utilizing *Nannochloropsis* and *Spirulina* with corn starch and polypropylene (PP) (Table 18.3, Entry 6–7). The raw materials were fed into a twin-screw extruder to produce blends, followed by injection molding to make ASTM D638 standard test samples. The tensile strengths of *Nannochloropsis*/starch/PP and *Spirulina*/starch/PP were 17.1 and 19.3 MPa, respectively, comparable to those of neat PE films or polyurethane (Rahman and Miller 2017).

In another study, Wang (2014) reported differences between a blend of *Spirulina* with a UHMW-PE and *Spirulina* compatibilized with maleic acid-modified P (Table 18.3, Entry 3–4). After plasticizing *Spirulina* with ethylene glycol, homogenous phase distribution and high surface adhesion were achieved.

Moreover, bioplastics based on 95% *Spirulina* plasticized with 5% ethylene glycol showed better load and extension characteristics than those without plasticizers. Furthermore, adding a compatibilizer did not lead to any improvement in the bioplastic's mechanical properties. Besides, Zhang et al. (2000) produced bioplastics by blending PVC and *Chlorella* biomass in an 80:20 ratio. PVC and *Chlorella* composite recorded a tensile strength of 30 MPa, whereas PVC without *Chlorella* recorded 50 MPa (Table 18.3, Entry 2). The authors reported that an increased water content in *Chlorella* affected the bioplastic's mechanical properties. However, depending on the intended application of the PVC-*Chlorella* composite, the lower tensile strength could have practical advantages (Rahman and Miller 2017).

18.3.1.2 Blending Microalgae Biomass with Biodegradable Polymers

Blending of MB is essential due to the impossibility of achieving melt formation in the traditional thermoplastic sense (Zeller et al. 2013). Furthermore, mechanical properties such as homogenized interfacial/chemical bonding, high adhesion forces, and tight contact between MB and biodegradable polymers are necessary for bioplastic production, which is solved by incorporating a suitable plasticizer and compatibilizer. Studies have shown that blending MB with biodegradable polymers yields positive results or superior mechanical properties than some non-biodegradable plastics

(Chong et al. 2022). Table 18.4 showcases examples of bioplastics developed by blending MB and biodegradable polymers.

Furthermore, Dianursanti and Khalis (2018) evaluated the compatibilizer ratio (0–6%) effect on the quality of bioplastics derived from PVA and *Chlorella*, finding a notable enhancement in the mechanical properties using 6% maleic anhydride as a compatibilizer (Table 18.4, Entry 3). Similarly, Dianursanti et al. (2019) developed bioplastics based on *Spirulina* and PVA with 6% maleic anhydrate, resulting in comparable tensile strength to conventional plastic bags (Table 18.4, Entry 4).

The compatibilizer incorporation also improved the bioplastic's elongation capability and smoother surface characteristics. For example, in a study by Zhang et al. (2020), a bioplastic film was produced using *Spirulina*, derived from hazardous high-salt-containing microalgal residues and PVA via solvent casting (Table 18.4, Entry 5). The biocomposite film containing 35% PVA exhibited the most favorable mechanical and physicochemical properties, boasting a tensile strength of 22 MPa and a contact angle of 94.48°, indicating excellent water resistance. This study demonstrated that the inorganic salts in the microalgal residue served as fillers, effectively enhancing the film's tensile strength and flexibility.

Moreover, Bulota and Budtova (2015) developed biocomposites based on PLA reinforced with algae (from 0 to 40 wt.%), where larger algae particles positively impacted their mechanical properties (Table 18.4, Entry 13). The PLA biocomposite containing 40 wt.% green algae exhibited a higher Young's modulus than native PLA. This enhancement was attributed to the high cellulose content in the green

Table 18.4 Bioplastics based on microalgae biomass blended with biodegradable polymers

Entry	Microalgae genre	Ratio of materials	Plasticizer	Characterization	Products	References
1	*Chlorella*	5–20 wt.% green microalgae/PLA	–	–	PLA/green microalgae composites	Kodal et al. (2019)
2	*Chlorella*	2.8 g PVA, 2.11 g glycerol, 1 g citric acid and 2.8 g *Chlorella* solution (1:10)	Glycerol and citric acid	–	PVA and *Chlorella* blend biofilm	Sabathini et al. (2018)
3	*Chlorella*	15% glycerol from 5 g from *Chlorella* and 2.5 g PVA with 6% maleic anhydrate	Glycerol and maleic anhydrate	58.5% protein (dry weight basis)	*Chlorella*/PVA composites	Dianursanti and Khalis (2018)
4	*Spirulina*	*Spirulina*, PVA with 6% of maleic anhydride	Glycerol and maleic anhydrate	60% protein	Bioplastic biofilm	Dianursanti et al. (2019)
5	*Spirulina*	3 g *Spirulina* and PVA (0.3, 0.5, 1 and 1.6 g)	Water	–	Bioplastic biofilm	Zhang et al. (2020)
6	*Spirulina*	Polyethylene oxide (1.8–4%), *Spirulina* (0.2–12%)	–	–	Nanofibers for tissue engineering applications	de Morais et al. (2010)
7	*Spirulina*	*Spirulina* (15–50%) and PBS with and without maleic anhydride (6% based on total weight)	Glycerol and maleic anhydride	60% protein (dry weight basis)	PBS/*Spirulina* composites	Zhu et al. (2017)
8	*Spirulina*	Wheat gluten and 10, 20 and 30 php microalgal biomass	Glycerol, octanoic acid and 1,4-butanediol	–	Protein-based composites	Ciapponi et al. (2019)
9	*Dunaliella*	0–10 mg WABR (Washed marine green algae biomass residue) and 490–500 mg PLA	–	19.7 ± 1.6% cellulose	Biomass/PLA biocomposite films	Mondal et al. (2021b)
10	*Dunaliella*	3–4.5%PCL/0.5–2% biomass	–	19.7 ± 1.6% cellulose	Biomass/PCL biocomposite films	Mondal et al. (2021a)
11	*Nannochloropsis*	20% microalgae in corn starch-glycerol solution (1:0:3)	Glycerol	–	Bioplastic biofilm	Fabra et al. (2018)
12	Microalgae consortium: *Scenedesmus* (50%) and *Desmodesmus* (30%)	Microalgae consortium and *Arthrospira* at different ratios of glycerol (50, 55 and 60 wt.%)	Glycerol	48.3% protein	Protein-based bioplastics	López Rocha et al. (2020)
13	Green algae	2–40 wt.% algae/PLA	–	–	Thermoplastic composites	Bulota and Budtova (2015)

algae. However, the authors acknowledge the need for further improvement in the adhesion between the thermoplastic matrix and reinforcement.

Also, Zhu et al. (2017) produced composites using a blend of *Spirulina* and PBS, both with and without a compatibilizer (PBS-g-maleic anhydride) (Table 18.4, Entry 7). The findings revealed that adding PBS-g-maleic anhydride improved tensile strength and Young's modulus, facilitating better *Spirulina* dispersion within the PBS thermoplastic matrix.

Furthermore, Ciapponi et al. (2019) created bioplastics by blending wheat gluten with microalgae as a filler (Table 18.4, Entry 8). The microalgal filler enhanced the bioplastic's mechanical properties when the concentration reached 30 php (parts per hundred) and when 1,4-butanediol was used as a plasticizer. Under these conditions, the tensile stress increased from 3.3 MPa to 4.9 MPa, while Young's modulus increased from 36.5 MPa to 273.1 MPa compared to the unfilled bioplastic.

Fabra et al. (2018) developed thermoplastic corn starch films incorporating diverse microalgae species (*Spirulina, Scenedesmus,* and *Nannochloropsis*) (Table 18.4, Entry 11). The authors examined how adding microalgae affected the film's physicochemical properties. Among the three species studied, *Nannochloropsis* demonstrated the most promising results. It reduced the material's rigidity and enhanced its water vapor and oxygen permeability (OP) compared to films made from native starch.

18.3.2 Bioplastics Based on Biopolymers Extracted from Microalgae Biomass

Microalgae can be grown under various conditions, including heterotrophic, photoautotrophic, or mixotrophic, and they can produce significant amounts of diverse bioactive compounds, depending on the species, such as cellulose, protein, starch, lipids, and lactic acid. These biopolymers hold great potential for bioplastic production; however, an effective extraction process is essential to enhance their bioavailability (Chong et al. 2022). Typically, the extraction process of the desired biopolymer involves treating the wet biomass and includes stages such as destabilization, cell rupture, recovery, and purification (Calijuri et al. 2022). Nevertheless, it has been reported that these extraction processes are costly and require a significant amount of energy and solvents.

On the other hand, there is growing interest in adopting a comprehensive biorefinery process based on the principles of a circular economy to reduce the production costs of microalgae-based biopolymers and bioplastics (López-Pacheco et al. 2022). This approach is proposed mainly because microalgae have the potential to generate multiple high-value products simultaneously, all while remaining sustainable and saving on energy, time, and costs (Arora et al. 2023). The main challenge is identifying the most productive microalgae strains to reduce production costs and enhance competitiveness with bacterial and conventional plastics (Costa et al. 2019).

Some bioplastics based on biopolymers extracted from MB are presented in Table 18.5. Furthermore, Mathiot et al. (2019) analyzed 10 microalgae strains for intracellular starch production. Among these strains, *Chlamydomonas reinhardtii* 11-32A demonstrated the highest starch production, reaching 49 wt.%, in a sulfur-deprived medium (Table 18.5, Entry 1). Moreover, the starch extracted from microalgae was plasticized with glycerol using a twin-screw extruder, introducing a novel approach to bioplastic production. Also, González-Balderas et al. (2021) employed residual MB cultivated in different wastewater sources to develop composites (Table 18.5, Entry 2–3). *Desmodesmus* sp. and *Tetradesmus obliquus* biomass underwent an ultrasound pretreatment to facilitate protein extraction. Once the highest protein content was extracted from these two residues, each was mixed with polycaprolactone (PCL). As a result, it was obtained that the biomass ultrasound pretreatment promoted a favorable synergy between the proteins and PCL, resulting in biocomposites with improved viscoelastic and mechanical properties as mold temperature increased.

According to the literature, the biomass of some microalgae strains can present up to 30% cellulose after oil extraction. It is known that the microalgae cell wall is reinforced by a matrix of microfibrils composed of cellulose (Chong et al. 2022). This has led to the recent utilization of cellulose extracted from microalgae as a reinforcing agent to enhance the mechanical properties of bioplastics (Ugya et al. 2023). However, it is essential to mention that cellulose derived from microalgae may exhibit thermal stability, compatibility, and moisture absorption limitations. These challenges can be addressed by incorporating cellulose into the thermoplastic matrix, such as nanofibers, cellulose nanocrystals (CNCs), and microcrystals (Madadi et al. 2021). Besides, Kassab et al. (2019) developed biocomposite films based on biopolymeric matrices reinforced with CNCs extracted from various lignocellulosic materials, including red algae waste (Table 18.5, Entry 4). Incorporating CNCs into a thermoplastic matrix resulted in films with significantly improved tensile strength and elastic modulus due to their homogeneous dispersion and favorable interfacial interactions.

On the other hand, Mondal et al. (2021b) isolated CNCs from *Dunaliella tertiolecta* green algae waste. These CNCs were then used to produce nanocomposites by blending them with PLA (0–2 wt.%) via casting technique (Table 18.5, Entry 5). The CNCs were effective nucleating and crystallizing agents, leading to a 30% increase in crystallinity in the PLA nanocomposites. Additionally, Mondal et al. (2021a) investigated the effectiveness of CNCs extracted from *Dunaliella tertiolecta* in synthesized poly(ε-caprolactone)

Table 18.5 Bioplastics based on biopolymers extracted from microalgae biomass

Entry	Microalgae genre	Biopolymer and extraction method	Characterization	Ratio of materials	Plasticizer	Products	References
1	*Chlamydomonas*	Starch by cultivation in a reactor with a controlled environment	49% w/w Starch	Starch-rich microalgal biomass with 30% w/w glycerol	Glycerol	Microalgae starch-based bioplastics	Mathiot et al. (2019)
2	*Desmodesmus*	Protein by ultrasound pretreatment	26 ± 0.5% protein, 11.2 ± 0.4% lipids, 14.3 ± 0.9% carbohydrates	48% residual algae/32% glycerol/20% polycaprolactone (PCL)	Glycerol	PCL/algae biocomposites	González-Balderas et al. (2021)
3	*Tetradesmus*	Protein by ultrasound pretreatment	28.5 ± 0.5% protein, 7.4 ± 0.9% lipids, 11.3 ± 0.3% carbohydrates	48% residual algae/32% glycerol/20% polycaprolactone x(PCL)	Glycerol	PCL/algae biocomposites	González-Balderas et al. (2021)
4	Unpurified algae waste	CNCs by acid hydrolysis	–	PVA/CNCs (1, 2, 5 y 8%)	Water	Nanocomposite films	Kassab et al. (2019)
5	*Dunaliella*	CNCs by acid hydrolysis	Yield 20–30%	490–500 mg PLA/0–10 mg CNCs	Chloroform	CNCs/PLA biocomposite films	Mondal et al. (2021b)
6	*Dunaliella*	CNCs by acid hydrolysis	Yield 20–30%	3–5% PCL/0.5–2% CNCs	Water, acetone, and chloroform	CNCs/PCL biocomposite films	Mondal et al. (2021a)

(Table 18.5, Entry 6). Biocomposites reinforced with CNCs improved thermal stability and mechanical properties compared to biocomposites produced with the whole biomass. Despite the promising properties of microalgal CNCs, their primary application has been producing reinforcing fillers rather than developing bioplastics (Madadi et al. 2021).

On the other hand, polylactic acid (PLA) is a thermoplastic polyester produced by chemical synthesis or bacteria fermentation. PLA exhibits mechanical properties like PET's, making it a suitable material for food packaging. Commonly, PLA is derived from sugarcane bagasse and corn plants, which are not sustainable due to the need for agricultural land for crops. In contrast, microalgae offer a promising alternative, as they do not need arable land, have a concise harvesting cycle, and have high starch content ideal for the fermentative production of lactic acid (Chong et al. 2022; López-Pacheco et al. 2022). Nonetheless, despite the existing research on producing PLA through the fermentation of MB, there is an absence of studies focusing on the production of bioplastics using microalgae-derived PLA. Hence, studying these photosynthetic organisms is essential, as they represent a viable option for PLA production, a biopolymer with significant potential for bioplastic manufacturing, primarily due to its biodegradability.

18.3.3 PHA-Based Bioplastics

Polyhydroxyalkanoates (PHAs) are polyesters produced by bacterial and algal cells, usually when entering the stationary growth phase. PHAs form intracellular inclusions that can be synthesized to store carbon and energy, reaching 80% of cell weight. PHAs have gained attention as suitable substitutes for petrochemical-based plastics such as PP, offering similar physical properties and processability through extrusion and injection molding methods, with complete biodegradability (Madadi et al. 2021; Costa et al. 2019). PHAs are characterized by high hydrophobicity, biodegradability, and biocompatibility. They are considered inert, non-toxic, and indefinitely stable in the air. PHAs can exhibit elastomeric and thermoplastic properties, high purity within cells, high resistance to UV light degradation, and low solvent susceptibility. These properties make them versatile materials for diverse applications, including fiber and foam materials, packaging films for shopping bags, containers, utensils, drug delivery carriers, and medical implants (Madadi et al. 2021; Chong et al. 2022).

Typically, heterotrophic bacteria produce PHAs in large fermenters; however, their production cost is relatively high compared to petroleum-based plastics. An alternative approach to PHA production involves cellular biomass or prokaryotic microalgae, specifically cyanobacteria. As a part of phytoplankton, they produce biomass using only light and atmospheric CO_2 as energy sources, reducing production costs. Moreover, it is reported that PHAs can be synthesized under environmental stress or nutrient limitations. Hence, it is fundamental to understand the properties of the obtained biopolymers and verify their potential industrial applications (Costa et al. 2019; Chong et al. 2022).

Tensile strength, extension to break, melting temperature (T_m), glass transition temperature (T_g), crystallinity, and molecular mass are properties used to understand the PHAs' physical behavior. Usually, PHAs exhibit a T_m from 50 to 180 °C and crystallinity from 30 to 70%, depending on the polymer composition (Rehm 2010). With 60–80% crystallinity, PHAs behave as a rigid material, while PHAs with 30–40% crystallinity are flexible. Therefore, a lower degree of PHAs crystallinity is essential to enhance their industrial processing (Madadi et al. 2021). Additionally, biodegradable polymers such as starch, cellulose, lignin, PLA, and PCL are commonly blended with PHAs to improve processing and physical properties (Li et al. 2016). Several authors have explored blends between PHAs or P(3HB) with cellulose derivatives (e.g., ethyl cellulose, cellulose propionate, and cellulose acetate butyrate) to reduce crystallinity. Also, blending PHAs with plasticizers is critical in enhancing the bioplastics' performance. Plasticization is achieved through their integration with raw polymers, which weakens the hydrogen bonds and dispersion forces (Mekonnen et al. 2013).

The most common short-chain-length PHAs produced by cyanobacteria and the most extensively researched and commercially available is poly 3-hydroxybutyrate P(3HB). This biopolymer has demonstrated better biodegradability, recyclability, plasticizing capacity, biocompatibility, and utility than petroleum-based plastics (Madadi et al. 2021). Moreover, P(3HB) exhibits thermoplastic properties like PP. However, processing P(3HB) for bioplastic is complex due to its melting temperature nearly to its degradation temperature (180 and 200 °C, respectively) and lower mechanical properties than PP. Consequently, P(3HB) copolymerization with other monomers can enhance flexibility and decrease melting point and crystallinity (Dang et al. 2022).

Abdo and Ali (2019) explored using MB as a resource for P(3HB) synthesis (Table 18.6, Entry 1–2). They obtained the MB from a high-rate algal pond and found that the strain *Microcystis aeruginosa* presented the highest P(3HB) production, with a yield of 0.49 ± 0.5 mg/mL. The algal bioplastic demonstrated a good plasticizing capacity, reaching 53%. Therefore, their study suggests that using an algal pond can offer an excellent and cost-effective option for P(3HB) production and bioplastic manufacturing. Furthermore, Abdo et al. (2023) prepared a biodegradable packaging film using a polyurethane (PU) nanocomposite immersed with P(3HB) extracted from *Microcystis* sp. in the presence of CuO nanoparticles (CuO-NPs) (1, 2, and 4%) (Table 18.6, Entry 3) and employing 40 wt.% P(3HB) improvements in the film's tensile properties were recorded. Furthermore, the contact angle increased from 77° to 87° with the rising CuO-NPs' concentration, indicating enhanced surface hydrophobicity.

Also, Bhati and Mallick (2012) synthesized a P(3HB-co-3 HV) co-polymer by *Nostoc muscorum Agardh* under propionate-supplemented conditions (Table 18.6, Entry 4). They achieved a co-polymer composition of (3HB-co-3 HV) from 58% to 60%. The films' material properties were comparable to commercial polymers, indicating their potential applications in different fields. Moreover, Monshupanee et al. (2016) harvested *Chlorogloea fritschii*, from which the biomass instantaneously swiftly settled to the bottom of a liquid culture to produce P(3HB) through a two-stage cultivation strategy (Table 18.6, Entry 5). By optimizing this process, they achieved a mass conversion efficiency of acetate substrate to P(3HB) at 51 ± 7% (w/w). These findings indicate that this system shows a viable alternative for producing biopolymers and bioplastics.

To fully consider microalgae PHAs as a potential alternative for the development of bioplastics, several challenges

Table 18.6 PHA-based bioplastics

Entry	Microalgae specie	Biopolymer	Ratio of materials	Production method	Characterization	References
1	*Microcystis aeruginosa*	P(3HB)	2.25 g sorbitol, 2.25 g gelatin, 2.25 g *Microcystisflos*, and 75 mL of 2% glycerol solution	Casting method	P(3HB) extraction: 0.49 ± 0.5 mg mL^{-1}	Abdo and Ali (2019)
2	*Haematococcus pluvialis*	P(3HB)	2.25 g sorbitol, 2.25 g gelatin, 2.25 g *Haematococcus pluvialis*, and 75 mL of 2% glycerol solution	Casting method	P(3HB) extraction: 0.39 ± 0.42 mg mL^{-1}	Abdo and Ali (2019)
3	*Microcystis* sp.	P(3HB)	Blend 40:60 of PU solution and P(3HB) solution with CuO (1, 2 and 4%)	Casting method	–	Abdo et al. (2023)
4	*Nostoc muscorum A*	P(3HB)	–	Casting method	P(3HB) accumulation: 60%	Bhati and Mallick (2012)
5	*Chlorogloea fritschii*	P(3HB)	–	–	Mass conversion efficiency: 51 ± 7% (w/w)	Monshupanee et al. (2016)

must be overcome, such as poor mechanical strength, versatility, and limited production compared to conventional plastics (Chong et al. 2022). Despite these drawbacks and the fact that commercial production of bioplastics from microalgae PHAs is not yet widespread, numerous researchers and companies are actively advancing this goal (Costa et al. 2019). On the other hand, the production costs for purifying PHAs from MB are very high. This issue could be addressed by maximizing output through genetically engineered microalgae (Cheah et al. 2023). Moreover, according to diverse authors, a hybrid biorefinery may offer a more economically attractive solution (Chew et al. 2017; López-Pacheco et al. 2022; Cheah et al. 2023). As a result, microalgae can fit into most integrated biorefinery projects, as their primary components can be targeted and optimized to yield a greater volume of PHAs (Costa et al. 2019).

18.4 Techniques for Producing Bioplastics Based on Microalgae

Regarding the processing of bioplastics based on microalgae, most of the work found in the present review focuses on obtaining formulations to obtain evaluation specimens to analyze their mechanical and stability properties and not on commonly used products such as those generally obtained using conventional plastics. The processes in the manufacture of bioplastics necessarily require that the composition to be used go through a fluid state achieved either by subjecting the material to heating and high pressures through the addition of ingredients (plasticizers, stabilizers, compatibilizers) and additives; the material can be made to assume the desired shape (Brown and Ives 1981). Compression Molding, Injection Molding, Extrusion, and Casting Techniques are the most important processes.

18.4.1 Injection Molding

Injection molding is a process suitable for high-consumption parts. The raw material can be transformed into a finished product in a single stage. Obtaining pieces of different weights and complicated geometries is feasible throughout the injection process. The most important characteristics of the injection process are the following: (a) the material is obtained in a single step; (b) little or no final work is needed on the piece obtained; (c) the process is fully automatable; (d) the manufacturing conditions are easily reproducible; and (e) the finished parts are of high quality (Robertson 2008).

The injection machine is the leading equipment for this process, and the peripheral equipment (Fig. 18.3) is divided into plasticizing, process control, and closing units (Ketabchi et al. 2015). The injection unit is responsible for plasticizing, homogenizing, transporting, dosing, and injecting the material into the mold; this machine is very similar to an extruder, having a screw inside a hot cylinder, which incorporates material from a hopper. The control injection panel contains the elements responsible for adjusting the most important operating parameters. The clamping unit can be compared to a press placed in a vertical position and is responsible for applying the closing force necessary to keep the mold closed, thus counteracting the high pressures reached inside the mold at this stage of the injection process. Moreover, the camping unit carries out the mold's opening and closing movements and expulsions of the injected part (Gordillo 2000).

Injection molding has primarily been applied to produce microalgae-based bioplastics when blended with other conventional polymers. Regarding this, Torres et al. (2015) employed a mixture of poly(butylene adipate-co-terephthalate) and microalgae biomass (MB) to produce bio-

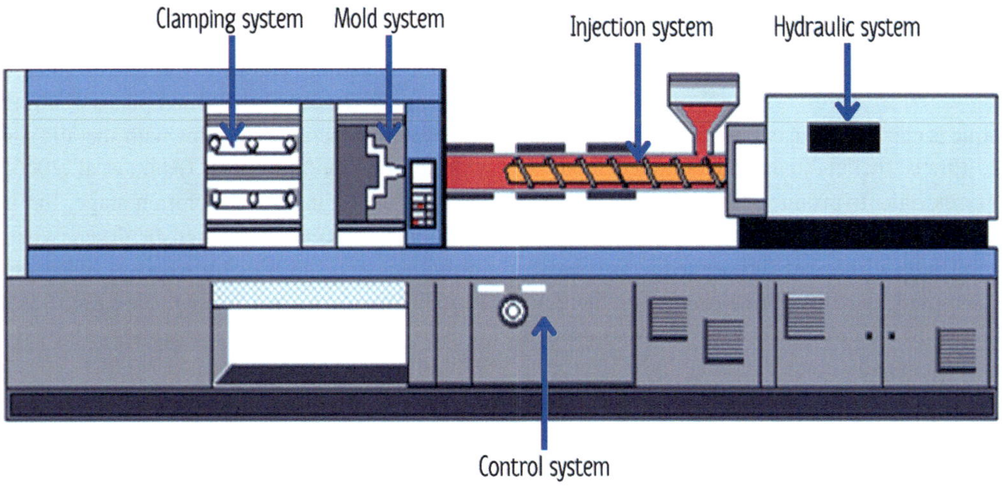

Fig. 18.3 Injection machine and components

composites using injection molding technology and urea and glycerol as plasticizers. They found that a mixture of 20% MB generated bioplastics with improved mechanical properties (Young's modulus and elongation). This could be an alternative to applying them as agricultural films with rapid reintegration into the soil and reduced total production cost. Moreover, Garrido et al. (2016) evaluated the potential of the injection molding process to manufacture bioplastics through mixtures of soy protein and red algae *Gelidium sesquipedale* (Rodophyta), used as filling material. The authors found that using red algae improved the mechanical and optical properties of the bioplastics, reducing the production cost by using waste material and thereby reducing the environmental impact by eliminating the presence of the algae from the coasts.

On the other hand, some bioplastics have also been obtained from different strains that are an environmental problem. For example, in Texcoco Lake in Mexico, a bioplastic was produced from two microalgae, *Desmodesmus* sp. and T. *Obliquus*, mixed with glycerol (ratio 60 protein/40 plasticizer). In the injection molding process, the effect of temperature was evaluated, finding that the conventional polymer processing process was sufficient to modify the protein structure and carbohydrates from the MB, generating sufficient interactions or cross-links to produce bioplastics with suitable physical and mechanical properties (González-Balderas et al. 2020).

18.4.2 Compression Molding

Compression molding is one of the oldest methods used for manufacturing biocomposites. This process is a high-pressure, high-volume molding technique for producing high-strength products. Many production lines have chosen compression molding to produce parts due to its low cost, short cycle times, high production volume, dimensional accuracy, improved impact resistance, uniform shrinkage due to uniform flow, and uniform density (Ketabchi et al. 2015).

This technique is carried out in two stages: preheating and pressurization. Firstly, pellets must be obtained from an internal mixer or extruder to produce biocomposites by compression molding. The pellets are placed in a cavity of the paired mold, and after closing it (bringing the two halves together), pressure and heat are applied to squeeze the granules, filling the rest of the cavity. Then, the mold is opened, and the part is subsequently ejected. Generally, this process operates at a pressure range of 14–20 MPa and 150–190 °C for temperature. There are two traditional types of compression molding: "sheet molding compound" and "bulk molding compound" (Ketabchi et al. 2015). Besides, fiber density, strength, and orientation are material properties that can be affected by the process. The application of extreme pressure can cause fracture. Furthermore, the viscosity of the thermoplastic matrix during pressing and heating must be carefully controlled to ensure that the matrix is impregnated between the fibers. Suitable quality composites can be produced by controlling viscosity, pressure, holding time, and temperature (Pickering et al. 2016).

There are several investigations where compression molding has been used to generate bioplastics based on MB. Otsuki et al. (2004) developed a biocomposite made from *Chlorella* and chemically modified PE to improve the interaction with the algae, finding satisfactory mechanical properties through solid phase reactions. Similar works using protein-rich MB, such as *Chlorella* vulgaris and *Spirulina platensis*, studied their potential for developing bioplastics and thermoplastic mixtures (Zeller et al. 2013). This work found that depending on the microalgae source, the behavior as a bioplastic varies, observing that *Chlorella* presented a better thermoplastic behavior than *Spirulina* when used individually. Nonetheless, they presented technical limitations typical of biopolymers during their processing, so mixing with conventional plastics such as PE was necessary. In this case, the one that presented the best interaction was *Spirulina*. However, if compatibilizers were added to the mixture, *Chlorella* had the best results. Hence, more studies are necessary to determine the process conditions and formulations to obtain a bioplastic with the best possible properties.

18.4.3 Casting Method

The casting method (solvent casting method) is a molding method mainly used to form biopolymer-based films. This process can be divided into three stages: solubilization, molding, and evaporation. In solubilization, the biopolymer is placed in contact with a solvent, which allows a solution to form with a homogeneous distribution of the materials. The most used solvents are water and organic solvents such as ethanol, methanol, and petroleum ether. In casting, the solution is poured onto a preformed plate. Working with a mold ensures the generation of films with specific shapes, ideal for working on laboratory tests (Rhim et al. 2006; Jensen et al. 2015). Finally, in the evaporation stage, the mold with the solution is placed in an oven or dryer, where the solvent evaporates and the film forms. It is subsequently removed from the mold for use (Suhag et al. 2020; Koide et al. 2013).

Various authors have generated films by the casting method using MB. Sabathini et al. (2018) produced films of *Chlorella* and PVA by casting method and ultrasonically homogenization, using water as a solvent, placing them on glass plates, and drying them for 24 h. They studied the physical and mechanical properties of the bioplastic, finding that by this method, a more compact film is created with a smooth surface and without pores, suggesting a potential use as an eco-sustainable plastic material for food packaging.

Moreover, Zhang et al. (2020) added that a sustainable solution against plastic pollution is using bioplastic films generated from *Spirulina* sp. Likewise, some authors have shown that using microalgae as raw material to produce bioplastics can generate reusable and biodegradable bags. Also, Kadell and Callychurn (2023) investigated the use of algae, mainly *Gracilaria Salicornia* and *Ulva Lactuca*, as potential raw material to produce reusable bioplastic bags, using the casting method to be able to carry out tests at the laboratory level. They found that the mechanical properties of the material generated from algae are like conventional plastic bags, highlighting that some of their treatments had much higher tensile strength, which confirms their potential use.

18.4.4 Extrusion

The extrusion process is any transformation operation in which a material is melted and forced to pass through a die to produce an object of constant cross-section and indefinite length (Osswald 1998). The materials used in this process can be of various types: plastics, metals, ceramics, or foods, and therefore, a wide variety of products can be obtained, such as frames, rods, pipes, and pasta, among others. From the point of view of plastics and bioplastics, the most used extruder is the single screw extruder (Wilkinson and Ryan 1998). The main components that make up a single screw extruder are the screw, cylinder or barrel, feed throat or inlet, feed hopper, die, and transmission (White 1991). The raw material is placed in the feeding hopper and joined to the barrel by the feeding throat. The material enters and is transported by a screw, where it will be subjected to high pressures and temperatures that will allow it to be melted. The screw forces the material to be compressed at the exit of the extruder, where the exit die is located, and expels it (Crawford and Martin 2020; Hensen and Berghaus 1988).

In extrusion, the raw material is generally fed in solid form and comes out molten. In the industry, in some cases, the raw material enters the extruder already molten from a reactor, functioning as a pump, generating the appropriate pressure so that the polymer can pass through a die with an already defined shape. On some occasions, extrusion is a previous stage of mixing and melting, where the material is taken to be formed by other processes such as blowing or thermoforming (Rosato 2012).

Various authors have used extrusion as a method of forming biomaterials from microalgae as a proposal to provide a solution to the pollution produced by plastics. Torres et al. (2015) used MB and poly(butylene adipate-co-terephthalate) to produce bioplastics. Urea and glycerol were used as plasticizers. It was found that the best extrusion results had 20% microalgae in their mixture. Moreover, Mathiot et al. (2019) studied the plasticization of 10 microalgae strains to form bioplastics using the extrusion method. Microstructural studies confirmed the ability of these microalgae to be homogeneously plasticized and, therefore, the possibility of being used for the development of bioplastics.

18.5 Characterization Methods of Bioplastics Based on Microalgae Biomass

Once the bioplastic is developed, it is necessary to conduct different tests to determine its performance and investigate its structural and molecular properties. The most favorable properties, that is, the optimal properties, largely depend on the application that the product is to be given, as well as the raw material from which said bioplastic is made (Pelleg 2013). However, different methods can be used to carry out different characterizations; the most used are determining mechanical, microstructural, thermal, and physicochemical properties. Through these tests, it is possible to know the bioplastic's behavior and obtain valuable information on the quality and performance of bioplastics to facilitate their development and application (Onen Cinar et al. 2020).

18.5.1 Mechanical Properties

Mechanical properties refer to how materials respond to external forces or loads applied to them and are essential to determine biocomposites' resistance, elasticity, durability, and mechanical behavior in various applications. These properties vary significantly depending on the material's composition, structure, and processing. Therefore, understanding and fine-tuning these properties is essential to tailor biocomposites for specific applications, such as medical devices, packaging, cosmetic products, and more (Pelleg 2013). Some authors state that protein-rich materials, such as microalgae, have poor mechanical properties because the proteins participate in intermolecular interactions that cause them to form complex structures by joining to form microfibers, which reduces the interaction between the polymer matrix and the additive (Torres et al. 2015).

Tensile stress measures the force per unit area applied to a material when stretched out of traction. It is essential to understand how materials respond to traction and their resistance to deformation under tensile loading (Gor et al. 2021). Regarding this, Wang (2014) studied the tensile properties test for *Spirulina* bioplastics, concluding that the best blend was the one containing 95% *Spirulina* and 5% ethylene glycol since that bioplastic had good load and extension characteristics if used compared with the one that had 90% *Spirulina* and 10% ethylene glycol. In addition, Gozan and Noviasari (2018) evaluated the effect of glycerol in bioplastics based on *Spirulina platensis* and PVA. They reported that plasticizer concentration was inversely proportional to the tensile stress since each time it added 5% more glycerol, the tensile strength decreased by 1.2 kgf/cm^2. This decrease was attributed to replacing the intermolecular bond between PVA with an intermolecular bond between PVA and glycerol. Likewise, they compared a conventional plastic bag with a microalgae-based bioplastic, recording tensile stress of 26.4 kgf/cm^2 and 30.8 kgf/cm^2, respectively, concluding that the microalgae-based bioplastic film could be used as a conventional plastic bag.

Elongation is an essential property of materials and is related to their ability to deform under load without breaking. It is a critical property in materials science and engineering, affecting the choice of materials for specific applications (Zeller et al. 2013). Abdo and Ali (2019) developed bioplastics from microalgae-derived P(3HB), which exhibited superior plasticizing capability compared to conventional petroleum-based plastics. The small particle size of the MB contributed to improved distribution in the medium when combined with plasticizers and gelatin, resulting in favorable mechanical properties, including enhanced elongation. On the other hand, Zeller et al. (2013) produced bioplastics based on *Spirulina* and *Chlorella* by compression molding, reporting that *Spirulina* bioplastic presented a better extension than *Chlorella* at the same ratio of ingredients (80:20) due to better particle dispersion.

Young's modulus is also known as the modulus of elasticity, a physical property that describes the material's stiffness. It measures the relationship between the stress applied to a material and the resulting elastic deformation. Materials with a high Young's modulus are rigid and do not deform easily, while materials with low modulus are more flexible and deformable (Li et al. 1999). Wang (2014) measured the mechanical properties of a bioplastic based on *Spirulina* microalgae plasticized with ethylene glycol, reported that with increasing plasticizer, the bioplastic became less rigid, and therefore the modulus values decreased.

18.5.2 Microstructural Properties

Microstructural properties refer to the characteristics and organization of components at the microscopic level within a material or substance. These properties are essential for understanding and predicting material's behavior and macroscopic properties (Onen Cinar et al. 2020). Scanning electron microscopy (SEM) is an advanced microscopy technique that uses an electron beam instead of visible light to obtain high-resolution images of samples at the microscopic level. SEM uses electromagnetic lenses to focus the electron beam onto the sample surface. This technique is widely used in scientific research, industry, and various fields to study microscopic structures on various samples, such as materials, cells, tissues, and more. It provides a detailed view of the morphology and structure of the samples, which is essential to understanding the behavior and properties at the microscopic level (Smith and Oatley 1955).

With a *Chlorella* and PE bioplastic, Otsuki et al. (2004) reported that SEM images show that the *Chlorella* biomass was firmly embedded in the modified PE matrix without air spaces. This technique revealed stronger interactions between *Chlorella* and the matrix than within the microalgal cells. These SEM findings provided insights into the tensile properties of the composite. Furthermore, Dianursanti and Khalis (2018) worked with *Chlorella vulgaris* and a compatibilizer to generate a bioplastic. SEM analysis revealed differences between the film with and without maleic anhydride. The film without maleic anhydride exhibited a very rough surface with many pores. The SEM results led them to conclude that adding maleic anhydride as a compatibilizer to the plastic mixture improved the solubilization of *Chlorella* granules and PVA. This enhancement resulted in better cohesiveness of the film's microstructure, increased homogeneity, compactness, and a denser film surface.

On the other hand, Fourier-transform infrared spectroscopy (FTIR) is a technique valuable for identifying organic and inorganic compounds, characterizing polymers, analyzing biological samples, and controlling quality in pharmaceutical and food industries, chemicals, and materials (Dutta 2017). Otsuki et al. (2004) confirmed the successful formation of a biocomposite of *Chlorella* and PE modified with maleic anhydride since the FTIR spectra showed characteristic bands of grafted succinic anhydride groups and their hydrolysis derivatives. In addition, the biocomposite's FTIR spectra showed a lower intensity absorption at 3040 cm^{-1} compared to *Chlorella;* a band disappeared at 1792 cm^{-1}, and a new band attributed to ester groups appeared at 1735 cm^{-1}. These findings suggest that the maleic acid groups grafted to the PE bind to the OH groups of *Chlorella* by esterification. In contrast, Wang (2014) developed a bioplastic based on *Spirulina* and PE modified with maleic anhydride, commenting that FTIR spectra effectively demonstrate structural changes in proteins. They emphasize the presence of amide I and II peaks, with amide I being particularly intense and helpful in analyzing the protein's secondary structure. Higher intensity peaks were observed for the mentioned amides in the bioplastic's spectra. This increase in intensity can be attributed to their formation in the compatibilizer presence.

18.5.3 Thermal Properties

Thermal analysis of biocomposites is crucial to understanding and optimizing their thermal properties, influencing their applicability in various industries. This testing allows the determination of optimum processing temperatures, such as those for injection or extrusion molding, ensuring efficient and high-quality biocomposite manufacturing. Furthermore, it provides insights into the biocomposite's shelf life and how it degrades over time when heat exposure (Ziman 1967). Thermogravimetric analysis (TGA) and differential scanning calorimetry (DSC) are commonly used to determine thermal stability and analyze the composition of biocomposites.

TGA is a characterization technique employed to study a sample's weight variation as a function of temperature or time while it undergoes controlled heating. TGA measures how a sample's mass changes as it is heated (Coats and Redfern 1963). Sabathini et al. (2018) assessed the thermal stability of *Chlorella* and PVA-based bioplastic according to the ASTM D1131-14. They found that adding *Chlorella* biomass enhanced the compactness of the cross-links within the material, as evidenced by the increased residue levels after heating. Similarly, they found that ultrasonic pretreatment of the *Chlorella*-PVA bioplastic film led to even higher residue levels, demonstrating that ultrasonication promoted the formation of denser cross-links between microalgae and PVA, resulting in an enhanced bioplastic film.

On the other hand, Zeller et al. (2013) achieved bioplastics based on *Spirulina* and *Chlorella* thermally stable since formulations with lower glycerol exhibited degradation temperatures around 350–475 °C, indicating stability. Conversely, formulations containing higher glycerol levels displayed lower degradation temperatures. This decrease is attributed to glycerol interfering with the interactions between carbohydrate and protein molecules within the thermoplastic matrix, rendering them more susceptible to degradation.

DSC is a thermal analysis technique used to study changes in a sample's thermal properties as it undergoes heating or cooling. Unlike TGA, which measures a sample's weight change during heating, DSC focuses on measuring changes in heat flux associated with thermal transitions, such as melting, crystallization, glass transition, or chemical reactions. Zeller et al. (2013) studied the effect of *Spirulina* and *Chlorella* content on glycerol-plasticized bioplastics. The DSC test found that high glycerol content and significant plasticization replaced protein-protein interactions primarily with protein-plasticizer interactions. This shift prevented the formation of protein macromolecule structures, resulting in the loss of plastic-like characteristics. As a result, they conclude that less than 25% plasticization is needed for *Spirulina* and *Chlorella* to preserve their plastic-like characteristics. Zhang et al. (2020) produced bioplastic films from *Spirulina* biomass, treating them under alkaline and acidic conditions. Through DSC, they verified the effect of inorganic salts, which performed as fillers under alkaline conditions, mainly in an amorphous form, while under acid conditions, they acted as crosslinkers between microalgae polymers and PVA.

18.5.4 Physicochemical Properties

The bioplastics' physicochemical properties are characteristics that merge the physical and chemical aspects of these materials. Understanding and evaluating these properties is essential for comprehending the bioplastic's behavior, which varies according to their composition and specific processing. Moreover, this knowledge is essential for designing and employing bioplastics in various packaging, agriculture, medicine, and other industries. According to the bibliography consulted, water vapor permeability (WVP), oxygen permeability (OP), and transparency are techniques often emphasized in this category.

WVP is the mass loss over time normalized by the sample's cross-sectional area. In a study by Ciapponi et al. (2019), different tests were carried out to observe the sensitivity to water of bioplastics based on wheat gluten with microalgae as a filler. The WVP test results indicated that adding 20 php of biomass did not lead to a significant difference compared to the gluten film without microalgae. In contrast, Fabra et al. (2018) found that corn starch films containing microalgae (*Nannochloropsis, Spirulina,* and *Scenedesmus*) had significantly lower permeability, with an average reduction of 54% compared to pure corn starch film. This reduction in *Spirulina* and *Scenedesmus* films can be attributed to cell wall rupture during processing, allowing lipid extraction and making the material more hydrophobic. In the case of the *Nannochloropsis* films, the well-dispersed crystalline microalgae may have acted as a filling and blocking agent, reducing WVP through the film.

On the other hand, OP testing is vital for evaluating bioplastic performance, mainly when the material is intended for food packaging applications. It measures the oxygen transmission rate (Onen Cinar et al. 2020). Fabra et al. (2018) developed corn starch films incorporating various microalgae species (*Nannochloropsis, Spirulina,* and *Scenedesmus*). The *Nannochloropsis* film, characterized by well-dispersed microalgae, reduced OP values by up to 72%, probably due to its role as a filler within the polymer network. Conversely, *Spirulina* and *Scenedesmus* films, which probably contained lipids that were released into the cell wall, their hydrophobic nature may have increased oxygen permeability due to their chemical affinity, favoring gas solubility in the biopolymer matrix. Hence, the physicochemical composition of microalgae-based bioplastics is crucial to define their applications, as some types can enhance barrier properties while others may diminish them.

18.6 Conclusions

Microalgae, considered a third-generation feedstock, have outperformed lignocellulosic biomass in bioplastic development due to their rapid growth rate, absence of a need for arable land, and reduced nutrient requirements. They can grow in freshwater, saline water, and wastewater, accumulating high biomass concentrations. Therefore, microalgae bioplastic production could contribute significantly to sustainability and the circular economy.

Chlorella and *Spirulina* are among the most used microalgae species for producing bioplastics and thermoplastic mixtures. It has been reported that microalgae can potentially produce bioactive compounds of interest for bioplastic manufacturing, including proteins, cellulose, starch, PLA, and PHAs. Therefore, there are three approaches to developing bioplastics from microalgae biomass (MB): blending MB with conventional polymers or biopolymers, extracting a desired biopolymer from MB, or using biomass to produce, mainly PHAs through biological processes or genetic engineering. PHAs are the most promising microalgae-derived biopolymers due to their similarity to petroleum-derived plastics; it has been reported that biomass derived from microalgae biorefinery can generate PHAs under specific nutrient limitation conditions.

Furthermore, algal-based bioplastics are projected to find applications across various industries owing to their eco-friendliness, biodegradability, and excellent mechanical properties, such as high durability and rigidity. Microalgae-based bioplastics can undergo conventional plastic processing techniques such as injection molding, compression molding, and extrusion. Nonetheless, much of the existing research focuses on formulating materials for evaluation purposes, analyzing their mechanical and stability properties, rather than producing commonly used products like those derived from conventional plastics. Different testing methods, such as thermal analytical methods, microstructural techniques, mechanical testing, and physicochemical characterization, have assessed these materials. These methods are crucial in analyzing bioplastics' performance and structural and molecular properties, thereby defining their potential applications. Consequently, there is a pressing need for continued research and development efforts in this field to enhance further the feasibility of utilizing algal-based bioplastics for environmental sustainability.

References

Abdo SM, Ali GH (2019) Analysis of polyhydroxybutrate and bioplastic production from microalgae. Bull Natl Res Centre Biomass Convers 43(1):1–4. https://doi.org/10.1186/s42269-019-0135-5

Abdo SM, Youssef AM, El-Liethy MA, Ali GH (2023) Preparation of simple biodegradable, nontoxic, and antimicrobial PHB/PU/CuO bionanocomposites for safely use as bioplastic material packaging. Biomass Convers Biorefinery 1-11:28673. https://doi.org/10.1007/s13399-022-03591-x

Andreotti V, Solimeno A, Rossi S, Ficara E, Marazzi F, Mezzanotte V, García J (2020) Bioremediation of aquaculture wastewater with the microalgae *Tetraselmis suecica*: semi-continuous experiments, simulation and photo-respirometric tests. Sci Total Environ 738:139859. https://doi.org/10.1016/j.scitotenv.2020.139859

Arora Y, Sharma S, Sharma V (2023) Microalgae in bioplastic production: a comprehensive review. Arab J Sci Eng:1–17. https://doi.org/10.1007/s13369-023-07871-0

Asadi A, Miller M, Sultana S, Moon RJ, Kalaitzidou K (2016) Introducing cellulose nanocrystals in sheet molding compounds (SMC). Compos A: Appl Sci Manuf 88:206–215. https://doi.org/10.1016/j.compositesa.2016.05.033

Bagatella S, Ciapponi R, Ficara E, Frison N, Turri S (2022) Production and characterization of polyhydroxyalkanoates from wastewater via mixed microbial cultures and microalgae. Sustain For 14:1–19. https://doi.org/10.3390/su14063704

Bhati R, Mallick N (2012) Production and characterization of poly (3-hydroxybutyrate-co-3-hydroxyvalerate) co-polymer by a N2-fixing cyanobacterium, *Nostoc muscorum* Agardh. J Chem Technol Biotechnol 87(4):505–512. https://doi.org/10.3390/su14063704

Brown R, Ives GC (1981) Handbook of plastics test methods

Bulota M, Budtova T (2015) PLA/algae composites: morphology and mechanical properties. Compos Part A Appl Sci Manuf 73:109–115. https://doi.org/10.1016/j.compositesa.2015.03.001

Bussa M, Eisen A, Zollfrank C, Röder H (2019) Life cycle assessment of microalgae products: state of the art and their potential for the production of polylactic acid. J Clean Prod 213:1299–1312. https://doi.org/10.1016/j.jclepro.2018.12.048

Calijuri ML, Silva TA, Magalhães IB, de Paula Pereira ASA, Marangon BB, de Assis LR, Lorentz JF (2022) Bioproducts from microalgae biomass: technology, sustainability, challenges and opportunities. Chemosphere 305:135508. https://doi.org/10.1016/j.chemosphere.2022.135508

Cassuriaga APA, Freitas BCB, Morais MG, Costa JAV (2018) Innovative polyhydroxybutyrate production by *Chlorella fusca* grown with pentoses. Bioresour Technol 265:456–463. https://doi.org/10.1016/j.biortech.2018.06.026

Castro-Aguirre E, Iñiguez-Franco F, Samsudin H, Fang X, Auras R (2016) Poly(lactic acid) – mass production, processing, industrial applications, and end of life. Adv Drug Deliv Rev 107:333–366. https://doi.org/10.1016/j.addr.2016.03.010

Cheah WY, Er AC, Aiyub K, Yasin NHM, Ngan SL, Chew KW, Khoo KS, Ling TC, Juan JC, Ma Z (2023) Current status and perspectives of algae-based bioplastics: a reviewed potential for sustainability. Algal Res 71:103078. https://doi.org/10.1016/j.algal.2023.103078

Chew KW, Yap JY, Show PL, Suan NH, Juan JC, Ling TC, Lee D-J, Chang J-S (2017) Microalgae biorefinery: high value products perspectives. Bioresour Technol 229:53–62. https://doi.org/10.1016/j.biortech.2017.01.006

Chiaramonti D, Prussi M, Casini D, Tredici MR, Rodolfi L, Bassi N, Zittelli GC, Bondioli P (2013) Review of energy balance in raceway ponds for microalgae cultivation: re-thinking a traditional system is possible. Appl Energy 102:101–111. https://doi.org/10.1016/j.apenergy.2012.07.040

Chong JWR, Tan X, Khoo KS, Ng HS, Jonglertjunya W, Yew GY, Show PL (2022) Microalgae-based bioplastics: future solution towards mitigation of plastic wastes. Environ Res 206:112620. https://doi.org/10.1016/j.envres.2021.112620

Christenson L, Sims R (2011) Production and harvesting of microalgae for wastewater treatment, biofuels, and bioproducts. Biotechnol Adv 29:686–702. https://doi.org/10.1016/j.biotechadv.2011.05.015

Ciapponi R, Turri S, Levi M (2019) Mechanical reinforcement by microalgal biofiller in novel thermoplastic biocompounds from plasticized gluten. Materials 12(9):1476. https://doi.org/10.3390/ma12091476

Coats A, Redfern J (1963) Thermogravimetric analysis. A review. Analyst 88(1053):906–924

Costa V, Silva C, Ana L, Costa JA, Morais M (2015) Polyhydroxybutyrate production by *Spirulina* sp. LEB 18 grown under different nutrient concentrations. Afr J Microbiol Res 9:1586–1594. https://doi.org/10.5897/AJMR2015.7530

Costa SS, Miranda AL, Andrade BB, de Jesus AD, Souza CO, de Morais MG, Costa JAV, Druzian JI (2018) Influence of nitrogen on growth, biomass composition, production, and properties of polyhydroxyalkanoates (PHAs) by microalgae. Int J Biol Macromol 116:552–562. https://doi.org/10.1016/j.ijbiomac.2018.05.064

Costa SS, Miranda AL, de Morais MG, Costa JAV, Druzian JI (2019) Microalgae as source of polyhydroxyalkanoates (PHAs) – a review. Int J Biol Macromol 131:536–547. https://doi.org/10.1016/j.ijbiomac.2019.03.099

Crawford RJ, Martin PJ (2020) Plastics engineering. Butterworth-Heinemann

Cruz CCV, da Costa ACA, Henriques CA, Luna AS (2004) Kinetic modeling and equilibrium studies during cadmium biosorption by dead *Sargassum* sp. biomass. Bioresour Technol 91:249–257. https://doi.org/10.1016/S0960-8524(03)00194-9

Dalsasso RR, Pavan FA, Bordignon SE, Aragão GMF, Poletto P (2019) Polyhydroxybutyrate (PHB) production by *Cupriavidus necator* from sugarcane vinasse and molasses as mixed substrate. Process Biochem 85:12–18. https://doi.org/10.1016/j.procbio.2019.07.007

Dang B-T, Bui X-T, Tran DP, Ngo HH, Nghiem LD, Nguyen P-T, Nguyen HH, Lin C, Lin KYA, Varjani S (2022) Current application of algae derivatives for bioplastic production: a review. Bioresour Technol 347:126698. https://doi.org/10.1016/j.biortech.2022.126698

de Morais MG, Stillings C, Dersch R, Rudisile M, Pranke P, Costa JAV, Wendorff J (2010) Preparation of nanofibers containing the microalga *Spirulina* (*Arthrospira*). Bioresour Technol 101(8):2872–2876. https://doi.org/10.1016/j.biortech.2009.11.059

Devadas VV, Khoo KS, Chia WY, Chew KW, Munawaroh HSH, Lam MK, Lim JW, Ho YC, Lee KT, Show PL (2021) Algae biopolymer towards sustainable circular economy. Bioresour Technol 325:124702. https://doi.org/10.1016/j.biortech.2021.124702

Di Caprio F, Altimari P, Pagnanelli F (2019) New strategies enhancing feasibility of microalgal cultivations. Stud Surf Sci Catal 179:287–316. https://doi.org/10.1016/B978-0-444-64337-7.00016-1

Dianursanti CN, Khalis SA (2018) The effect of compatibilizer addition on *Chlorella vulgaris* microalgae utilization as a mixture for bioplastic. In: E3S web of conferences. EDP Sciences, p 03047. https://doi.org/10.1051/e3sconf/20186703047

Dianursanti N, Citra WL, Misri G (2019) Effect of compatibilizer addition in *Spirulina platensis* based bioplastic production. In: AIP conference proceedings, vol 1. AIP Publishing. https://doi.org/10.1063/1.5096716

Dutta A (2017) Fourier transform infrared spectroscopy. Spectrosc Methods Nanomater Charact:73–93. https://doi.org/10.1016/B978-0-323-46140-5.00004-2

European Bioplastics (2022) Fact sheet, European bioplastics: what are the bioplastics? European Bioplastics

European Bioplastics (2023) Bioplastics market data. Accessed 30 Aug 2023

Fabra MJ, Martínez-Sanz M, Gómez-Mascaraque LG, Gavara R, López-Rubio A (2018) Structural and physicochemical characterization of thermoplastic corn starch films containing microalgae. Carbohydr Polym 186:184–191. https://doi.org/10.1016/j.carbpol.2018.01.039

García G, Sosa-Hernández JE, Rodas-Zuluaga LI, Castillo-Zacarías C, Iqbal H, Parra-Saldívar R (2021) Accumulation of PHA in the microalgae *Scenedesmus* sp. Under nutrient-deficient conditions. Polymers 13:1–23. https://doi.org/10.3390/polym13010131

Garrido T, Penalba M, de la Caba K, Guerrero P (2016) Injection-manufactured biocomposites from extruded soy protein with algae waste as a filler. Compos Part B 86:197–202. https://doi.org/10.1016/j.compositesb.2015.09.058

Gonçalves AL, Pires JCM, Simões M (2017) A review on the use of microalgal consortia for wastewater treatment. Algal Res 24:403–415. https://doi.org/10.1016/j.algal.2016.11.008

González-Balderas R, Felix M, Bengoechea C, Guerrero A, Ledesma MO (2020) Influence of mold temperature on the properties of wastewater-grown microalgae-based plastics processed by injection molding. Algal Res 51:102055. https://doi.org/10.1016/j.algal.2020.102055

González-Balderas R, Felix M, Bengoechea C, Ledesma MO, Guerrero A, Velasquez-Orta S (2021) Development of composites based on residual microalgae biomass cultivated in wastewater. Eur Polym J 160:110766. https://doi.org/10.1016/j.eurpolymj.2021.110766

Gor M, Soni H, Wankhede V, Sahlot P, Grzelak K, Szachgluchowicz I, Kluczyński J (2021) A critical review on effect of process parameters on mechanical and microstructural properties of powder-bed fusion additive manufacturing of ss316l. Materials 14(21):6527. https://doi.org/10.3390/ma14216527

Gordillo A (2000) Influencia de la inyección sobre el comportamiento mecánico de compuestos de polipropileno. Tesis doctoral. UPC

Gozan M, Noviasari C (2018) The effect of glycerol addition as plasticizer in *Spirulina platensis* based bioplastic. In: E3S web of conferences. EDP Sciences, p 03048. https://doi.org/10.1051/e3sconf/20186703048

Hadipoor M, Keivanimehr F, Baghban A, Ganjali MR, Habibzadeh S (2021) Carbon dioxide as a main source of air pollution: prospective and current trends to control. In: Sorbents materials for controlling environmental pollution. Elsevier. https://doi.org/10.1016/B978-0-12-820042-1.00004-3

Hensen F, Berghaus U (1988) Plastics extrusion technology. Hanser Publishers

Jensen A, Lim L-T, Barbut S, Marcone M (2015) Development and characterization of soy protein films incorporated with cellulose fibers using a hot surface casting technique. LWT-Food Sci Technol 60(1):162–170. https://doi.org/10.1016/j.lwt.2014.09.027

Kadell M, Callychurn D (2023) An investigation on the use algae-based material for the production of reusable bioplastic bags: a Mauritian case study. Clean Mater 9:100201. https://doi.org/10.1016/j.clema.2023.100201

Kaewbai-ngam A, Incharoensakdi A, Monshupanee T (2016) Increased accumulation of polyhydroxybutyrate in divergent cyanobacteria under nutrient-deprived photoautotrophy: An efficient conversion of solar energy and carbon dioxide to polyhydroxybutyrate by *Calothrix scytonemicola* TISTR 8095. Bioresour Technol 212:342–347. https://doi.org/10.1016/j.biortech.2016.04.035

Kassab Z, Hannache H, El Achaby M (2019) Isolation of cellulose nanocrystals from various lignocellulosic materials: physico-chemical characterization and application in polymer composites development. Mater Today Proc 13:964–973. https://doi.org/10.1016/j.matpr.2019.04.061

Ketabchi MR, Hoque ME, Siddiqui MK (2015) Critical concerns on manufacturing processes of natural fibre reinforced polymer composites. In: Manufacturing of natural fibre reinforced polymer composites. Springer, pp 125–138. https://doi.org/10.1007/978-3-319-07944-8_6

Kodal M, Kutluk T, Yarici T, Kapucu N, Ozkoc G (2019) Non-isothermal crystallization kinetics of green microalgae filled poly (lactic acid). Acta Phys Pol A 136(6):974–982. https://doi.org/10.12693/APhysPolA.136.974

Koide Y, Ikake H, Muroga Y, Shimizu S (2013) Effect of the cast-solvent on the morphology of cast films formed with a mixture of stereoisomeric poly (lactic acids). Polym J 45(6):645–650. https://doi.org/10.1038/pj.2012.192

Kumari P, Ravi Kiran B, Venkata Mohan S (2022) Polyhydroxybutyrate production by *Chlorella sorokiniana* SVMIICT8 under nutrient-deprived mixotrophy. Bioresour Technol 354:127135. https://doi.org/10.1016/j.biortech.2022.127135

Lee HR, Kim KH, Mun SC, Chang YK, Choi SQ (2018) A new method to produce cellulose nanofibrils from microalgae and the measurement of their mechanical strength. Carbohydr Polym 180:276–285. https://doi.org/10.1016/j.carbpol.2017.09.104

Li G, Zhao Y, Pang S-S, Li Y (1999) Effective Young's modulus estimation of concrete. Cem Concr Res 29(9):1455–1462. https://doi.org/10.1016/S0008-8846(99)00119-2

Li Z, Yang J, Loh XJ (2016) Polyhydroxyalkanoates: opening doors for a sustainable future. NPG Asia Mater 8(4):e265–e265. https://doi.org/10.1038/am.2016.48

López Rocha CJ, Álvarez-Castillo E, Estrada Yáñez MR, Bengoechea C, Guerrero A, Orta Ledesma MT (2020) Development of bioplastics from a microalgae consortium from wastewater. J Environ Manag 263:110353. https://doi.org/10.1016/j.jenvman.2020.110353

López-Pacheco IY, Rodas-Zuluaga LI, Cuellar-Bermudez SP, Hidalgo-Vázquez E, Molina-Vazquez A, Araújo RG, Martínez-Ruiz M, Varjani S, Barceló D, Iqbal HM (2022) Revalorization of microalgae biomass for synergistic interaction and sustainable applications: bioplastic generation. Mar Drugs 20(10):601. https://doi.org/10.3390/md20100601

Lutzu GA, Ciurli A, Chiellini C, Di Caprio F, Concas A, Dunford NT (2021) Latest developments in wastewater treatment and biopolymer production by microalgae. J Environ Chem Eng 9:104926. https://doi.org/10.1016/j.jece.2020.104926

Madadi R, Maljaee H, Serafim LS, Ventura SP (2021) Microalgae as contributors to produce biopolymers. Mar Drugs 19(8):466. https://doi.org/10.3390/md19080466

Mata TM, Martins AA, Caetano NS (2010) Microalgae for biodiesel production and other applications: a review. Renew Sust Energ Rev 14:217–232. https://doi.org/10.1016/j.rser.2009.07.020

Mathiot C, Ponge P, Gallard B, Sassi J-F, Delrue F, Le Moigne N (2019) Microalgae starch-based bioplastics: screening of ten strains and plasticization of unfractionated microalgae by extrusion. Carbohydr Polym 208:142–151. https://doi.org/10.1016/j.carbpol.2018.12.057

Mekonnen T, Mussone P, Khalil H, Bressler D (2013) Progress in bio-based plastics and plasticizing modifications. J Mater Chem A 1(43):13379–13398. https://doi.org/10.1039/C3TA12555F

Mondal K, Bhagabati P, Goud VV, Sakurai S, Katiyar V (2021a) Utilization of microalgae residue and isolated cellulose nanocrystals: a study on crystallization kinetics of poly (ε-caprolactone) bio-composites. Int J Biol Macromol 191:521–530. https://doi.org/10.1016/j.ijbiomac.2021.09.114

Mondal K, Sakurai S, Okahisa Y, Goud VV, Katiyar V (2021b) Effect of cellulose nanocrystals derived from *Dunaliella tertiolecta* marine green algae residue on crystallization behaviour of poly (lactic acid). Carbohydr Polym 261:117881. https://doi.org/10.1016/j.carbpol.2021.117881

Monshupanee T, Nimdach P, Incharoensakdi A (2016) Two-stage (photoautotrophy and heterotrophy) cultivation enables efficient production of bioplastic poly-3-hydroxybutyrate in auto-sedimenting cyanobacterium. Sci Rep 6(1):37121. https://doi.org/10.1038/srep37121

Nandini A, Nagarajan D, Chang J-s (2020) Production of lactic acid from microalgal biomass *Chlorella vulgar* ESP-31 as a feedstock using PVA immobilized bacteria L. Plantarum 23(2020):165–169. https://doi.org/10.11594/nstp.2020.0525

Nemati M, Webb C (2011) Immobilized cell bioreactors. In: Moo-Young M (ed) Comprehensive biotechnology, 3rd edn. Pergamon, Oxford, pp 489–504. https://doi.org/10.1016/B978-0-444-64046-8.00084-7

Niccolai A, Shannon E, Abu-Ghannam N, Biondi N, Rodolfi L, Tredici MR (2019) Lactic acid fermentation of *Arthrospira platensis* (*Spirulina*) biomass for probiotic-based products. J Appl Phycol 31:1077–1083. https://doi.org/10.1007/s10811-018-1602-3

Onen Cinar S, Chong ZK, Kucuker MA, Wieczorek N, Cengiz U, Kuchta K (2020) Bioplastic production from microalgae: a review. Int J Environ Res Public Health 17(11):3842. https://doi.org/10.3390/ijerph17113842

Osswald TA (1998) Polymer processing fundamentals

Otsuki T, Zhang F, Kabeya H, Hirotsu T (2004) Synthesis and tensile properties of a novel composite of *Chlorella* and polyethylene. J Appl Polym Sci 92(2):812–816. https://doi.org/10.1002/app.13650

Park Y-K, Lee J (2022) Achievements in the production of bioplastics from microalgae. Phytochem Rev:1–19. https://doi.org/10.1007/s11101-021-09788-8

Pelleg J (2013) Mechanical properties of materials. In: Solid mechanics and its applications, vol 190. Springer. https://doi.org/10.1007/978-94-007-4342-7

Pickering KL, Efendy MA, Le TM (2016) A review of recent developments in natural fibre composites and their mechanical performance. Compos A: Appl Sci Manuf 83:98–112. https://doi.org/10.1016/j.compositesa.2015.08.038

Polotti G (2020) Perspectives from industry. In: Advances in polymer reaction engineering, vol 56. Academic. https://doi.org/10.1016/bs.ache.2020.07.003

Rahman A, Miller C (2017) Microalgae as a source of bioplastics. In: Algal green chemistry. Elsevier, pp 121–138. https://doi.org/10.1016/B978-0-444-63784-0.00006-0

Rajpoot AS, Choudhary T, Chelladurai H, Nath Verma T, Shende V (2022) A comprehensive review on bioplastic production from microalgae. Mater Today Proc 56:171–178. https://doi.org/10.1016/j.matpr.2022.01.060

Rehm BH (2010) Bacterial polymers: biosynthesis, modifications and applications. Nat Rev Microbiol 8(8):578–592. https://doi.org/10.1038/nrmicro2354

Reis MAM, Serafim LS, Lemos PC, Ramos AM, Aguiar FR, Van Loosdrecht MCM (2003) Production of polyhydroxyalkanoates by mixed microbial cultures. Bioprocess Biosyst Eng 25:377–385. https://doi.org/10.1007/s00449-003-0322-4

Rhim JW, Mohanty AK, Singh SP, Ng PK (2006) Effect of the processing methods on the performance of polylactide films: thermocompression versus solvent casting. J Appl Polym Sci 101(6):3736–3742. https://doi.org/10.1002/app.23403

Rizwan M, Mujtaba G, Memon SA, Lee K, Rashid N (2018) Exploring the potential of microalgae for new biotechnology applications and beyond: a review. Renew Sust Energ Rev 92:394–404. https://doi.org/10.1016/j.rser.2018.04.034

Robertson GL (2008) State-of-the-art biobased food packaging materials. In: Environmentally compatible food packaging, vol 1. Woodhead Publishing, pp 3–18

Roja K, Ruben Sudhakar D, Anto S, Mathimani T (2019) Extraction and characterization of polyhydroxyalkanoates from marine green alga and cyanobacteria. Biocatal Agric Biotechnol 22:101358. https://doi.org/10.1016/j.bcab.2019.101358

Rosato DV (2012) Plastics processing data handbook. Springer

Ruane J, Sonnino A, Agostini A (2010) Bioenergy and the potential contribution of agricultural biotechnologies in developing countries. Biomass Bioenergy 34:1427–1439. https://doi.org/10.1016/j.biombioe.2010.04.011

Sabathini H, Windiani L, Gozan M (2018) Mechanical physicial properties of *chlorella*-PVA based bioplastic with ultrasonic homogenizer. In: E3S web of conferences. EDP Sciences, p 03046. https://doi.org/10.1051/e3sconf/20186703046

Selvaraj K, Vishvanathan N, Dhandapani R (2021) Screening, optimization and characterization of poly hydroxy butyrate from fresh water microalgal isolates. Int J Biobased Plast 3:139–162. https://doi.org/10.1080/24759651.2021.1926621

Shi B, Wideman G, Wang JH (2012) A new approach of BioCO2 fixation by thermoplastic processing of microalgae. J Polym Environ 20(1):124–131. https://doi.org/10.1007/s10924-011-0329-x

Smith K, Oatley C (1955) The scanning electron microscope and its fields of application. Br J Appl Phys 6(11):391. https://doi.org/10.1088/0508-3443/6/11/304

Suhag R, Kumar N, Petkoska AT, Upadhyay A (2020) Film formation and deposition methods of edible coating on food products: a review. Food Res Int 136:109582. https://doi.org/10.1016/j.foodres.2020.109582

Tan T, Yu J, Shang F (2011) Biorefinery engineering. In: Moo-Young M (ed) Comprehensive biotechnology, 2nd edn. Academic, Burlington, pp 815–828. https://doi.org/10.1016/B978-0-08-088504-9.00138-0

Torres S, Navia R, Campbell Murdy R, Cooke P, Misra M, Mohanty AK (2015) Green composites from residual microalgae biomass and poly (butylene adipate-co-terephthalate): processing and plasticization. ACS Sustain Chem Eng 3(4):614–624. https://doi.org/10.1021/sc500753h

Ugya YA, Chen H, Sheng Y, Ajibade FO, Wang Q (2023) A review of microalgae biofilm as an eco-friendly approach to bioplastics, promoting environmental sustainability. Environ Res:116833. https://doi.org/10.1016/j.envres.2023.116833

Walker TR, Fequet L (2023) Current trends of unsustainable plastic production and micro (nano) plastic pollution. TrAC Trends Anal Chem:116984. https://doi.org/10.1016/j.trac.2023.116984

Wang K (2014) Bio-plastic potential of spirulina microalgae. University of Georgia

Wang K, Mandal A, Ayton E, Hunt R, Zeller M, Sharma S (2016) Modification of protein rich algal-biomass to form bioplastics and odor removal. In: Press A (ed) Protein byproducts. Elsevier, pp 107–117. https://doi.org/10.1016/B978-0-12-802391-4.00006-9

White JL (1991) Twin screw extrusion: technology and principles. Hanser Publishers

Wilkinson AN, Ryan AJ (1998) Polymer processing and structure development. Springer

Xia M, Shen Z, Gu M, Chen W, Dong W, Zhang Y (2021) Efficient catalytic conversion of microalgae residue solid waste into lactic acid over a Fe-Sn-Beta catalyst. Sci Total Environ 771:144891. https://doi.org/10.1016/j.scitotenv.2020.144891

Yousuf A (2019) Fundamentals of microalgae cultivation. In: Press A (ed) Microalgae cultivation for biofuels production. Elsevier, pp 1–9. https://doi.org/10.1016/B978-0-12-817536-1.00001-1

Zanchetta E, Damergi E, Patel B, Borgmeyer T, Pick H, Pulgarin A, Ludwig C (2021) Algal cellulose, production and potential use in plastics: challenges and opportunities. Algal Res 56:102288. https://doi.org/10.1016/j.algal.2021.102288

Zeller MA, Hunt R, Jones A, Sharma S (2013) Bioplastics and their thermoplastic blends from *Spirulina* and *Chlorella* microalgae. J Appl Polym Sci 130(5):3263–3275. https://doi.org/10.1002/app.39559

Zhang F, Kabeya H, Kitagawa R, Hirotsu T, Yamashita M, Otsuki T (2000) An exploratory research of PVC-*Chlorella* composite material (PCCM) as effective utilization of *Chlorella* biologically fixing CO2. J Mater Sci 35:2603–2609. https://doi.org/10.1023/A:1004779415778

Zhang C, Wang C, Cao G, Wang D, Ho S-H (2020) A sustainable solution to plastics pollution: an eco-friendly bioplastic film production from high-salt contained *Spirulina* sp. residues. J Hazard Mater 388:121773. https://doi.org/10.1016/j.jhazmat.2019.121773

Zhu N, Ye M, Shi D, Chen M (2017) Reactive compatibilization of biodegradable poly (butylene succinate)/*Spirulina* microalgae composites. Macromol Res 25:165–171. https://doi.org/10.1007/s13233-017-5025-9

Ziman J (1967) The thermal properties of materials. Sci Am 217(3):180–193

Enzymes Bio-factories Developed Through Microalgae

19

Gissel Daniela Rios-Herrera, Idalia Osuna-Ruiz, Jesús Aarón Salazar-Leyva, Emmanuel Martínez-Montaño, Israel Benítez-García, and Jorge Manuel Sandoval-Gallardo

Abstract

Microalgae comprise a diverse group of eukaryotic photosynthetic microorganisms that have stood out for their abilities to generate a large number of organic compounds, leading them to be classified as bio-factories. Therefore, they have aroused the interest of industries such as pharmaceuticals, food processing, cosmetics, energy, agro-industrial, aquaculture, etc., and they are used in applications of high commercial value. Its use stands out in the production of energy, wastewater purification, food supplements, vitamins, fatty acids, pigments, and others. However, it is necessary to identify the correct microalgae, as well as the appropriate growth conditions, to be able to bring them to production on an industrial scale. Currently, lines of research have been developed around microalgae focused on studying the enzymes involved in the various biological processes they carry out to produce compounds of industrial interest. In this sense, a large number of enzymes have been identified, and different activities that can be applied in various agricultural, food, and energy processes have been shown.

Keywords

Microalgal biotechnology · Enzymes · Genetic engineering · Microalgae cultures

19.1 Introduction

Enzymes are biological catalysts that accelerate biochemical reactions in living organisms and can be extracted from cells and subsequently used to catalyze a wide range of commercially important processes (Robinson 2015). Currently, enzyme-based technology is important and, in some cases, essential—for many industrial applications because enzymes possess high activity and selectivity, even at very low concentrations and wide ranges of temperatures and pH values, resulting in fewer side effects and unwanted by-products (Ferraro et al. 2010).

Different important enzymatic functions could be applied in the food industry, production of biofuels, formulation of detergents and various cleaning products, and even in environmental applications such as bioremediation (Osorio-Reyes et al. 2023; Alvarez et al. 2021; Righini and Roberti 2019). In recent years, the use of biocatalysts has increased in different industrial processes; however, the current request for enzymes is high, and production is still not sufficient to satisfy industrial demands (Mogharabi and Faramarzi 2016), which has led to the search for new sources of enzymes.

Additionally, various investigations have shown that microalgae are more stable than plants and animals for isolating enzymes since microalgae can grow with low nutritional requirements and produce high amounts of biomass; however, microalgae are not currently used to produce

G. D. Rios-Herrera (✉)
Estancias Posdoctorales por México SECIHTI, Secretaría de Ciencia, Humanidades, Tecnología e Innovación, Ciudad de México, México/Universidad Politécnica de Sinaloa (UPSIN), Mazatlán, Mexico

Facultad de Ingeniería y Tecnología de Mazatlán, Universidad Autónoma de Sinaloa, Mazatlán, Sinaloa, Mexico
e-mail: gissel.rios@upsin.edu.mx

I. Osuna-Ruiz · J. A. Salazar-Leyva · I. Benítez-García
Maestría en Ciencias Aplicadas, Unidad Académica de Ingeniería en Biotecnología, Universidad Politécnica de Sinaloa (UPSIN), Mazatlán, Sinaloa, Mexico

E. Martínez-Montaño
Maestría en Ciencias Aplicadas, Unidad Académica de Ingeniería en Biotecnología, Universidad Politécnica de Sinaloa (UPSIN), Mazatlán, Sinaloa, Mexico

Investigadores por México SECIHTI, Secretaría de Ciencia, Humanidades, Tecnología e Innovación, Ciudad de México, Mexico

J. M. Sandoval-Gallardo
Facultad de Ingeniería y Tecnología de Mazatlán, Universidad Autónoma de Sinaloa, Mazatlán, Sinaloa, Mexico

D. U. Santos Ballardo, S. Rossi (eds.), *Microalgae as Promising Source of Commercial Bioproducts*, Developments in Applied Phycology 14, https://doi.org/10.1007/978-3-031-86433-9_19

enzymes at an industrial scale, although some studies have reported the ability of microalgal cells to synthesize different types of enzymes. Therefore, microalgae represent a promising source of enzyme production (Spier et al. 2020).

In this sense, this chapter focuses on microalgal enzymes and their commercial and/or experimental use, along with their role in different industrial applications; also, the importance of generating and developing microalgae as biofactories is discussed. Different classes of enzymes, including hydrolases, oxidoreductases, and lyases, are mentioned in published studies and are described below.

19.2 Microalgae Enzyme Applications and Their Importance Like Bio-factories

Therefore, large-scale production of enzymes isolated from microalgae has not been reported; however, microalgae have been found to possess a high capacity to synthesize enzymes for industrial applications, such as xylanases, galactosidases, phytases, and glucanases (Table 19.1). The demand for these biomolecules has increased over recent years due to the wide range of potential applications in the textile, pharmaceutical, food, and biofuel industries.

19.2.1 Cell Wall-Degrading Enzymes

Strategies that damage the structural integrity of the cell wall (a resistant and rigid layer located on the outside of the plasma membrane of plant, fungal, algal, bacterial, and archaeal cells) must be implemented in various biotechnological processes. Using various solvents in combination with temperature, pressure, and friction usually results in environmentally unfriendly wastes and, in some cases, causes the loss and/or modification of the intracellular components required in subsequent stages. Thus, innovations in cell disruption increasingly include hydrolytic enzymes capable of breaking down the cell wall polymers via the cleavage of specific bonds; these processes are less aggressive, enabling increased process efficiency with a lower environmental impact.

Microalgae are natural sources of cell wall-degrading enzymes since they are aquatic organisms with genes that encode polysaccharide-metabolizing enzymes. Cellulose is one of the most common cell wall components; therefore, one of the most relevant classes of enzymes for cell wall degradation is cellulases, namely, exo-β-glucanase (EC 3.2.1.4), endo-β-glucanase (EC 3.21.91), and β-glucosidase/cellobiose (EC 3.2.1.21). In silico studies have shown microalgae's potential for producing cellulases (Guerriero et al. 2018). However, research focused on using microalgae as sources of cell wall-degrading enzymes is scarce.

Some examples include a study related to cellulose degradation and assimilation by *Chlamydomonas reinhardtii* grown under light conditions with limiting carbon dioxide (CO_2); it was capable of secreting endo-β-1,4-glucanases and thus can digest cellulose that is exogenously available in the medium and then absorb and assimilate the cellobiose produced (Blifernez-Klassen et al. 2012). Furthermore, some strains of microalgae can be genetically modified to stably express hydrolytic enzymes, which can be used to hydrolyze polysaccharides; one example is *Dunaliella tertiolecta*, which has been modified to produce recombinant enzymes with xylanase, α-galactosidase, phytase, phosphate anhydrolase, and β-mannanase activities (Georgianna et al. 2013). Likewise, novel enzymes with unique structural characteristics that make them potentially attractive for biotechnological applications have been identified; for example, Dong et al. (2021) identified a β-glucosidase in *Nannochloropsis oceanica* that exhibits high activity on laminaribiose.

Moreover, Demuez et al. (2015) proposed the autolysis of microalgae as an alternative method of enzymatic cell disruption. Autolysis occurs naturally during certain events and at particular sites on the cell wall and involves endogenous enzymes that act in specific stages of the life cycle; for example, during fusion or fission, autolysins (endopeptidases) act; during asexual reproduction via spores, sporangins (glycoproteases) act; during sexual reproduction, gametolysins (metalloproteases) degrade the walls of gametes during mating; and during apoptosis, cell wall lysis is associated with caspase enzymes.

Besides, it has been reported the activity increments of some of these lysis enzymes; for example, caspase activity was induced in *Chlamydomonas reinhardtii* via abiotic stress with hydrogen peroxide (Vavilala et al. 2015), and such induction was carried out using dark conditions and/or oxidative stress in *Dunaliella tertiolecta* cultures (Segovia et al. 2003), using environmental stress in *Dunaliella viridis* (Jiménez et al. 2009), and employing heat stress and high temperatures in *Chlorella saccharophila* (Zuppini et al. 2007).

On the other hand, for *Ankistrodesmus densus*, the activation of meta-caspases related to vacuolar cell death has been observed via induction with heat stress and a lack of light and nitrogen (Barreto Filho et al. 2022). Recently, a protocol that uses *Chlamydomonas reinhardtii* to produce autolysin has been published (Findinier 2023).

The search for enzymes capable of degrading cell walls is currently a subject of global interest due to multiple potential environmentally friendly biotechnological applications; in this sense, microalgae represent a promising source for the production of these biocatalysts, mainly for the advantages that they offer in terms of optimization, manipulation, and handling.

Table 19.1 Microalgae enzyme applications

Enzyme	Microalgae	Molecular/catalytic features of the enzyme	Extraction/purification method of enzyme	Author
Cell wall disruption				
β-Glucosidase, recombinant (BGLN1)	*Nannochloropsis oceanica*	Higher activity against laminaribiose than other oligosaccharides; partially inhibited by metal ions; stability by residues N-terminus; has a more acidic substrate binding pocket than other β-glucosidases	Recovered from *Escherichia coli* BL21 (DE3), by cells lysis with ultrasonication and centrifugation/purified by Sepharose IIP column (50 mM Tris-HCl, pH 8.0, 500 mM NaCl, 250 mM imidazole), then Q-Sepharose column (50 mM Tris-HCl, pH 8.0, 1 M NaCl), and Superdex 75 gel filtration column (10 mM Tris-HCl, pH 8.0, 100 mM NaCl)	Dong et al. (2022)
β-1,4-endoglucanase	*Chlamydomonas reinhardtii*	It contains a catalytic domain having a (glycosyl hydrolase family) GHF5-specific signature, a short stretch of proline and/or hydroxy amino acids termed a linker sequence, and a cellulose-binding module	Microalgae cells were disrupted by sonication on ice; Cell lysates were centrifuged at 16,000 g for 5 min to separate the soluble and insoluble protein fraction; after concentrated and washed, they were eluted in non-reducing SDS–polyacrylamide gel electrophoresis	Blifernez-Klassen et al. (2012)
Starch hydrolysis				
α-Amylase	*Chlamydomonas reinhardtii*	The optimum pH and temperature of thermophilic amylase (arAmyBH) were 5.5 and 60 °C, respectively in vivo at an elevated temperature	The harvested cells were centrifugated, resuspended in lysis buffer, and broken by sonication. After that, the crude extract was centrifuged, and the supernatant was mixed with pre-equilibrated Ni-NTA slurry. The lysate-Ni-NTA mixture was loaded into a column and the His6-arAmyBH was eluted with a buffer containing 200 Mm imidazole. The active fractions were dialyzed and then concentrated using an Amicon Ultra-15 30 K filtration device	Wang et al. (2015)
Lactose hydrolysis				
β-Galactosidase	*Nannochloropsis limnetica*	The highest enzymatic activity in cultures was 40.84 ± 0.23 U/L on day 8 of cultivation	Extracellular β-galactosidase was centrifuged at 4000 rpm for 15 min; the resulting supernatant was filtered through 0.22 μm nylon membrane to remove any biomass	Li et al. (2024)
Phytate degradation				
Phytase	*Chlamydomonas reinhardtii*	The best-engineered strain achieved up to 34 U/g dry weight at 37 °C and pH 2.5	Cells were collected by centrifugation and frozen in liquid nitrogen. Frozen cells were homogenized in protein extraction buffer supplemented with β-mercaptoethanol, aprotinin, leupeptin, and PMSF, and supernatant was collected	Peraza-Echeverria et al. (2021)
	Phaeodactylum tricornutum	The best-engineered strain achieved up to 40,000 phytase activity units (FTU)/g soluble protein, at 37 °C and pH 2.5	Microalgae were harvested by centrifugation. Then, pellets were resuspended, and after that, lysed by sonication. Lysates were clarified by centrifugation (12,000 rcf, 10 min; 16,000 g, 5 min), and recovered supernatants stored on ice	Pudney et al. (2019)
	Schizochytrium sp.	A crude microalgae lysate shows a phytase activity of 1.77 U/mL. Further chromatographic purification, the recombinant phytase exhibited a specific activity of 15.59 U/mg, at pH of 5.5, and temperature of 35 °C	Harvested recombinant *Schizochytrium* sp. were resuspended with Tris-buffer plus 1 mM PMSF, then lysed, centrifuged, and filtered (0.45 μm). Purification of the recombinant phytase with a His tag was carried out on 20 ml of Ni-NTA affinity chromatography matrix connected to the FPLC	Singarapu et al. (2023)

(continued)

Table 19.1 (continued)

Enzyme	Microalgae	Molecular/catalytic features of the enzyme	Extraction/purification method of enzyme	Author
Fibrin degradation				
	Chlorella vulgaris	Crude extract exhibited an elevated specific activity (662.6 U/mg) to degrade fibrin directly	Lyophilized microalgae were homogenized in phosphate buffer 0.1 M, pH 7.0. The homogenate was centrifuged at 15,000 rpm for 10 min at 4 °C with no further purification steps	Páblo et al. (2017)
Fibrinolytic enzyme	*Tetradesmus obliquus*	A protein band around 97 kDa was detected in the purified enzymatic extract and exhibited a specific activity of 1221 U/mg A high degradation of fibrin was observed by the plate method, obtaining a clear zone of 82 mm^2	Microalgae cells were concentrated and lysed by two different methods (sonication or homogenization) and were precipitated via acetone or ammonium sulfate additions and purified using ion exchange chromatography	Moura et al. (2023)
Increase the concentration of TAGs				
Diacylglycerol acyltransferase (DGAT2)	*Neochloris oleoabundans*	Is responsible for catalyzing the final reaction in the glycerol phosphate pathway of triglycerol synthesis	The yeast cells suspended in methanol:chloroform (2:1 v/v) were lysed using 0.5 mm glass beads with vortexing. The cell lysate was obtained by centrifugation. Total lipid was extracted from cell lysate using 50 mM citric acid:chloroform (2:3 v/v), and chloroform was evaporated using N_2 gas. The total lipid was kept at −20 °C until use. To separate lipid classes by thin-layer chromatography (TLC), the total lipid was dissolved in chloroform and applied to Silica gel 60 F254	Vingiani et al. (2019) Chungjatupornchai and Watcharawipas (2015)
Increased lipid synthesis				
Glycerol-3-phosphate dehydrogenase	*Chlamydomonas reinhardtii*	Are commonly involved in glycerolipid metabolism and in the preservation of cellular redox status by consuming NAD(P)H and regenerating $NAD(P)^+$	Has been purified to homogeneity using blue Sepharose and hydroxylapatite chromatography, and monomeric molecular mass of GPDH has been estimated using SDS—PAGE. In contrast, the dimeric mass was estimated using non-denaturing PAGE.	Wang et al. (2015)
Glucose-6-phosphate dehydrogenase	*Phaeodactylum tricornutum*	G6PD increased by 4.4-fold, and G6PD enzyme activity increased by more than 3.1-fold with enhanced NADPH production	Chromatography on the yellow dye Remazol Brilliant Gelb GL to highly purify the cytoplasmic isoenzyme of glucose-6-phosphate dehydrogenase	Xue et al. (2017)
Biocontrol				
β-1,4-Endoglucanase	*Anabaena laxa* RPAN8	Hydrolyze the β-1,4-glycosidic bonds in cellulose, breaking down the complex carbohydrate into simpler sugars like cellobiose and glucose	Cells were harvested by centrifugation and lysed in a buffer containing lysozyme and benzonase. The lysate was then centrifuged to remove debris, and the soluble protein was purified using Ni-NTA kit, Quiagen, USA affinity chromatography. The protein was eluted with a high concentration of phosphate buffer (5 mM) supplemented with NaCl (300 mM) and imidazole (20 mM). The final protein was eluted in a very high concentration of imidazole (500 mM), keeping a similar concentration of phosphate buffer and NaCl. Protein concentration was determined using the Bradford method, and molecular weight was assessed by SDS-PAGE, visualized with Coomassie Brilliant Blue staining	Gupta et al. (2011)

(continued)

Table 19.1 (continued)

Enzyme	Microalgae	Molecular/catalytic features of the enzyme	Extraction/purification method of enzyme	Author
β-1,3-glucanase	*Anabaena laxa*	Degradation to produce (1 → 3)-linked β-glucan oligosaccharides with pharmacological activity properties	Cells were harvested by centrifugation and were filtered through a 0.22-μm filter the crude enzyme was precipitated from the supernatant via ammonium sulfate saturation (0–40%, 0–50%, 0–60%, 0–70%, 0–80% saturation) and dialysate at Superdex75PG column (1.6 × 64.6 cm, Beijing Huideyi Technology Co., Ltd., Beijing, China), and purified using ion exchange chromatography	Prasanna et al. (2013)

19.2.2 Food Industry

Several studies have considered different species of microalgae as authentic enzyme-producing bio-factories. Enzymes can be obtained from intracellular contents or extracellular liquids and, once isolated, show biochemical and catalytic properties that make them of industrial interest; however, their application on a commercial scale has not yet been implemented. Despite the diversity of microalgal enzymes studied to date, their potential for application in the animal and human food industries has been reported only for specific subclasses of enzymes, notably glycosidases and phytases.

Moreover, glycosidases (EC 3.2.1) catalyze the breaking of the glycosidic bonds present in polysaccharides and oligosaccharides, consequently producing smaller carbohydrates. The enzymes in this group are diverse; however, only a few glycosidases produced by microalgae with potential use in the food industry have been reported. Examples of these are the enzymes α-amylase (EC 3.2.1.1), β-amylase (EC 3.2.1.2), β-glucosidase (EC 3.2.1.21), and β-galactosidase (EC 3.2.1.23).

Some enzymes of the α-amylase class have been isolated and studied in microalgae, such as *Chlorella vulgaris* (Ben Hlima et al. 2021). Moreover, an interesting microalgal amylase was reported by Wang et al. (2015), who introduced recombinant genes from the hyperthermophilic bacterium *Thermotoga neapolitana* into the chloroplasts of the microalgae *Chlamydomonas reinhardtii*, then expressed, accumulated, and subsequently purified an α-amylase (arAmyBH) via Ni-NTA purification. This enzyme completely transformed starch into reducing sugars even at temperatures above 60 °C at an optimum pH of 5.5. This biochemical property of arAmyBH has garnered interest for application in industrial high-temperature food processes, such as those used in the brewing industry.

On the other hand, an extracellular β-galactosidase enzyme was recovered from the aqueous fraction of the freshwater microalgae *Nannochloropsis limnetica* grown in whey-powder media (Li et al. 2024). According to this study, the high activity of this β-galactosidase makes it potentially valuable for the bioremediation of effluents from the dairy industry; however, this enzyme could also be used as a dietary supplement for those who suffer from lactose intolerance.

Furthermore, phytates (phytic acid bound to a mineral) are the primary form in which some plants (e.g., leguminous, oilseeds, cereals, and nuts) store phosphorus (approximately 70% of the total phosphates). Phytates are considered antinutritional factors in human and animal nutrition because they bind strongly to cations such as Ca, Fe, Mn, K, Zn, and Mg, thereby reducing their bioavailability. One strategy to reduce phytate contents is using phytases (EC 3.1.3.8 and EC 3.1.3.26), which hydrolyze the organic phosphate portion, releasing the cations (Handa et al. 2020). Thus, it is a conventional practice to use phytases of microbial origin as food additives to increase phosphorus viability and reduce phytate contents in the fecal matter and, in turn, the discharge of inorganic phosphate into the environment, consequently decreasing environmental eutrophication.

Several studies have reported the production of phytases by microalgae via recombinant technology. Of particular interest is creating microalgae that lack cell walls by inserting microbial phytase genes into the plastid genome via transformation. Once the transgenic microalgae express the microbial phytase and the enzyme accumulates in the organelles, the microalgae can be used as a food supplement for different species of monogastric animals without the need for protein purification (Erpel et al. 2016).

A microalga that is commonly transformed using recombinant technology is the green microalgae *Chlamydomonas reinhardtii* (Masi et al. 2023). Good results have been obtained after the insertion of ruminal bacterial phytase genes (PhyAsr) into the chloroplasts of this microalga, with high phytase activity observed (34 U/g dry weight) (Peraza-Echeverria et al. 2021). Moreover, an *Escherichia coli* phytase gene (AppA) was also inserted into the chloroplast genome of *Chlamydomonas reinhardtii* (Yoon et al. 2011). Once the gene was expressed, the spray-dried whole-cell microalgal lysate (termed Chlasate) registered an activity of

10 phytase units per gram at a pH of 4.5 at 37 °C. In addition, including the transgenic Chlasate in a diet for young broilers significantly reduced phytate excretion in the bird's feces (Yoon et al. 2011).

An advantage of the production of recombinant phytases through the use of microalgae is the co-production of long-chain omega-3 polyunsaturated fatty acids (LC-PUFAs), such as eicosapentaenoic (EPA) and docosahexaenoic (DHA) acids. For example, for the marine diatom *Phaeodactylum tricornutum*, in which the accumulation of high levels of EPA and DHA was promoted together with recombinant phytases from the fungus *Aspergillus niger* (PhyA) or the bacterium *E. coli* (AppA) (Pudney et al. 2019). In this study, the engineered strain that showed the best results produced more than 40,000 units of phytase activity per gram of soluble protein, such that, according to the authors, one gram of soluble protein is sufficient to treat 20 kg of Atlantic salmon feed. Also, the levels of omega-3 LC-PUFAs were increased.

Furthermore, the cloning and expression of phytase genes from *Bacillus* sp. (DS11) in *Schizochytrium* sp. and the subsequent purification of the phytase using Ni-NTA affinity chromatography yielded a specific activity of 15.59 U/mg (Singarapu et al. 2023). The enzymatic activity reported for this phytase gives it the potential to be used in feed for poultry and pigs, including the wet microalgae at 1.80% (w/w) of the diet would completely cover the recommended phytase and even DHA requirements.

19.2.3 Pharmaceutical Industry: Microalgae As a Source of Antithrombotic Agents

According to the World Health Organization, cardiovascular diseases (CVDs) are the leading cause of death worldwide. In this sense, in 2021 alone, CVD-related deaths represented approximately one-third of all deaths worldwide (Roth et al. 2020). Thrombosis can be understood as the accumulation of fibrin clots in blood vessels and is recognized as one of the main causes associated with cardiovascular disorders, such as acute myocardial infarction, ischemic heart disease, high blood pressure, and stroke (Moura et al. 2023; Wendelboe and Raskob 2016). The formation of fibrin clots is a natural response where the body seeks to seal and repair the breakage of blood vessels. However, if there is an imbalance in fibrin formation and degradation, fibrin can accumulate irregularly in blood vessels and trigger thrombosis (Hazare et al. 2024; Sharma et al. 2021).

The pharmaceutical industry has developed strategies to combat thrombosis, such as antiplatelet drugs (aspirin and dipyridamole), anticoagulants (heparin and warfarin), and surgical operations; however, such strategies carry side effects, notably hemorrhage (Singh et al. 2023). Fibrinolytic enzymes are emerging as promising medications for thrombosis, as they can hydrolyze insoluble fibrin fibers, lyse thrombi, and restore blood flow to ischemic areas (Huang et al. 2013).

Interestingly, various investigations suggest that fibrinolytic enzymes can be produced in masse because they are secreted by a wide range of microorganisms, including bacteria, filamentous fungi, and microalgae (Sharma et al. 2021). Although microalgae have been widely studied for their ability to produce diverse bioactive metabolites, studies related to the biosynthesis of fibrinolytic enzymes from microalgae are limited. In this context, bio-factories based on microalgae can help mitigate CVDs and thus contribute biotechnological solutions to health problems of global impact.

Besides, fibrinolytic enzymes are hydrolases and are, therefore, members of the protease family (EC 3.4). Considering their catalytic mechanism at the active site level, fibrinolytic enzymes are classified as metalloproteases, serine proteases, or serine metalloproteases (Barzkar et al. 2022; Hazare et al. 2024). The production of fibrinolytic enzymes using microalgae is of great interest because these organisms possess high photosynthetic efficiency, high growth rates, resistance to various contaminants, and the ability to be cultivated under different growth conditions (i.e., autotrophic, heterotrophic, and mixotrophic growth) (Silva et al. 2023). Moreover, microalgae grown especially under mixotrophic conditions can produce excellent enzyme yields using different carbon sources, which can come from organic wastes and by-products of the agri-food industry. As an example, Páblo et al. (2017) used a mixture of glycerol and corn steep liquor (CSL), which is a by-product resulting from the corn wet milling industry, as nutrients to produce fibrinolytic enzymes from the microalgae *Chlorella vulgaris*. These authors reported maximal production of fibrinolytic protease (21.7 U/mL) when 1.1% and 0.9% glycerol and CSL were used as nutrients. Similarly, in the same study, the protease produced exhibited a specific activity of 662.6 U/mg, indicating its high ability to degrade fibrin directly. This promising result led to another scientific study in which it was possible to produce and purify a fibrinolytic enzyme from *C. vulgaris* cultivated under mixotrophic conditions with CSL added as a nutrient. The pure enzyme had a molecular weight of 45 kDa and exhibited in vitro thrombolytic activity (25.6% in 90 min of reaction) and the ability to cause red blood cell lysis. Furthermore, fibrinolytic activity was enhanced in the presence of Fe^{2+} and inhibited by phenylmethanesulfonyl fluoride (PMSF) and ethylenediamine tetraacetic acid (EDTA), the pure enzyme may be a serine metalloprotease (Silva et al. 2018).

In a recent study, the production of fibrinolytic enzymes was evaluated by cultivating the green microalgae *Tetradesmus obliquus* under different cultivation conditions;

different methods of cell lysis were also tested to extract the enzymes, and different protein precipitation procedures (acetone or ammonium sulfate) were used to establish their effects on the production of purified fibrinolytic enzymes (Moura et al. 2023). These authors noted that cell growth conditions have an important influence on the production of microalgal enzymes. When cultivated under mixotrophic conditions, *Tetradesmus obliquus* produced the highest amount of protease (when 0.25% CSL was used), followed by autotrophic and mixotrophic under 0.50% CSL conditions.

On the other hand, cellular lysis, as a homogenization method, was the most efficient method for extracting proteases and yielded 430 U/mg of fibrinolytic activity. The most efficient protein precipitation method involved acetone, resulting in a high enzyme yield (80%) and the highest fibrinolytic activity (625 U/mg). The enzyme purified from *T. obliquus* cells exhibited a specific fibrinolytic activity of 1221 U/mg, which was higher than that presented by extracts of *Bacillus* bacteria, the conventional sources of fibrinolytic enzymes (Barzkar et al. 2022).

Considering that microalgae are relatively inexpensive to culture, have the ability to grow with substrates from agro-industrial waste, and are feasible to scale up, the production of fibrinolytic enzymes through enzyme bio-factories based on microalgae is promising, opening a window of application in the pharmaceutical industry, especially in the development of alternative antithrombotic agents.

19.2.4 Biofuel Production

Population and technological growth worldwide have caused an increase in the demand for energy resources. However, coal and oil derivatives have provided energy to various sectors of the population, and they have also contributed to the alarming increase in the concentration of greenhouse gases in the environment; the use of fossil fuels is reported to release approximately 30 million tons per year of carbon dioxide (CO_2) alone (Hossain et al. 2020).

Regardless, research, legislation, standards, and the implementation of new technologies that contribute to reducing greenhouse gases (CO_2, CH_4, N_2O, and fluorinated gases) have been applied gradually and, in some cases, inefficiently. Besides, these attempts are mostly in countries that depend excessively on the industrial sector and those without sufficient resources; due to these incomplete transitions, these polluting gases have not decreased as society needs (Lamb et al. 2022).

In addition, it is well known that fossil fuels are not renewable resources, and together with the growth of countries around the world, the demand for these non-renewable fuels strongly grows each year; this increasing demand causes increments in the cost of fuels, which has caused a catastrophic effect on different economies, leading to generalized growths in inflation for all productive sectors.

To combat these problems, various technologies have been developed in recent decades to allow more renewable and sustainable energy generation, such as solar, geothermal, hydroelectric, wind energy, nuclear energy, and bioenergy (Anto et al. 2020).

In 2018, the International Energy Agency (IEA) reported a substantial increase in bioenergy generation worldwide, with the production of 154 billion liters, with an annual growth expectation of 6–7%, reaching 25% in 2024 (de Carvalho Silvello et al. 2022). In recent years, bioenergy has accounted for approximately 10% of the primary energy supply worldwide, and it is expected that it may cover more than 60% in the coming decades; therefore, diverse bioenergetic resources, including lignocellulosic biomass, woody plants, grass or herbaceous plants, starch crops, sugarcane crops, oilseeds (sunflower, soybean, corn, etc.), and microalgal cultures (Duarah et al. 2022), will be used.

In this sense, bioenergy is obtained from various types of biomass, which are used to produce biofuels, such as biodiesel, bioethanol, and biohydrogen (Thirumalaivasan et al. 2024). Generally, biofuels are classified as first-, second-, and third-generation; however, a fourth-generation class has been suggested recently. First-generation methods are based on food crops such as sugarcane, sugar beets, canola, and animal and vegetable oils. Second-generation methods do not use food crops but focus mainly on lignocellulosic biomass, woody plants, etc. Third-generation methods use microalgae as the raw material and are considered to have the ability to essentially complement and/or replace petroleum-based fuels since they produce high concentrations of oils, do not require the use of farmland, reproduce easily, and use CO_2, nitrogen, and phosphorus as nutrients; they can also grow in water with high salinity and even in residual waters. Finally, fourth-generation methods use molecular biology tools to introduce genetic modifications, mainly in microalgae and other microorganisms, to increase the concentration of lipids without reducing biomass yield (Gilmour 2019; Shahidi et al. 2019).

Microalgae are photosynthetic microorganisms that vary morphologically, can be found in almost all habitats, and notably have the ability to produce reserve energy compounds, such as lipids and polysaccharides, at concentrations that vary by species, a factor that should be considered for the production of biofuels (de Carvalho Silvello et al. 2022).

One of the main characteristics of microalgae is that they accumulate large amounts of lipids, with unsaturated fatty acids (PUFAs) and triglycerides (TAGs) being of most significant commercial interest; however, TAGs are of greater interest for biofuel production and can be produced by transesterification reactions catalyzed by acids or bases (Sun et al.

2019). The biosynthesis of TAGs in microalgae follows the Kennedy route, which begins with the precursor sn-1,2-diacylglycerol (DAG) and ends with the enzyme acyl-CoA: diacylglycerol acyltransferase (DGAT); also, three independent groups, DGAT1, DGAT2, and DGAT3, have been identified (Jothibasu et al. 2021).

Moreover, studies of the enzymes produced by microalgae relevant to biofuel production have focused on the overexpression of endogenous genes encoding different enzymes involved or related to the synthesis of lipids or carbohydrates. By inducing DGAT2 overexpression in some microalgae (*Phaeodactylum tricornutum*, *Thalassiosira pseudonana*, *Neochloris oleoabundans*, and *Nannochloropsis oceanica*), the concentration of TAGs was significantly increased, and even the overexpression of other isoforms of DGAT2 (NoDGAT2A, 2C and 2D) in *N. oceanica* also contributed to increased lipid production, which is ultimately reflected in increased yields during the biodiesel production from these microalgae (Vingiani et al. 2019).

In another study where an increase in the concentration of the enzyme NAD(H) kinase was induced in *Chlorella pyrenoidosa*, the microalga was able to synthesize 110% more in total lipids; besides, in *Chlamydomonas reinhardtii*, production of the enzymes lysophosphatidic acyltransferase (c-LPAAT) and glycerol-3-phosphate dehydrogenase (GPD1) was increased, which led to a 45–68% increase in the total lipid concentration. Also, Xue et al. (2017) reported that the overexpression of glucose-6-phosphate dehydrogenase in *Phaeodactylum tricornutum* increased lipid synthesis.

On the other hand, some microalgae also produce high concentrations of carbohydrates, which can be used to produce bioethanol. Some species have been reported as sources of bioethanol, reaching contents between 50% and 70% of their total dry weight; the most used strains are *Dunaliella salina* (57–69.7%), *Spirulina platensis* (60%), *Chlorella* sp. (56%), and *Scenedesmus obliquus* (55.4%) (Efremenko et al. 2012; Sajjadi et al. 2018; Hossain and Mahlia 2019).

However, studies of microalgal enzymes involved in the synthesis or transformation of carbohydrates are limited compared to those in other organisms, such as plants, bacteria, or yeasts, possibly due to the complexity of the metabolic pathways and the diverse nature of microalgae making them more difficult to study (Alam et al. 2020). Despite these limitations, some microalgal enzymes have been reported to be involved in the generation and/or transformation of carbohydrates; for example, the overexpression of the enzyme glucosyltransferase is related to the accumulation and production of some sugars or some cellulases, which may improve the yield and production of bioethanol.

Therefore, the potential use of microalgae as biofuel biofactories can be highlighted, mainly because different authors have demonstrated their potential in the area of biofuel production; however, many limitations can be addressed through the study and development of new technologies to create a highly competitive industry (Park et al. 2019).

19.2.5 Biocontrol Agents

Every year, a variety of pathogens, such as viruses, bacteria, fungi, nematodes, and insects, cause enormous crop losses in both the preharvest and postharvest stages (Jamiołkowska 2020). Attacks by plant pathogens are very persistent, resulting in direct and indirect losses of 40 billion dollars in global value (Kursa et al. 2022). Although the application of synthetic pesticides has increased agricultural productivity in recent decades, excessive use has also caused detrimental impacts on the environment worldwide (Lengai et al. 2020). Therefore, environmentally friendly alternatives using plant extracts, essential oils, and beneficial microorganisms have been sought since they tend to be less toxic to the environment and human health than synthetic pesticides are. These natural products break down faster, reducing soil and water contamination, and their use reduces resistance to pathogens, a common problem with synthetic pesticides (Pandit et al. 2022).

Furthermore, biological control agents such as fungi, bacteria, viruses, and insects are used worldwide to reduce the populations of pests that affect crops (Viera-Arroyo et al. 2020). However, new agents for biological control continue to be explored, and in this sense, microalgae exhibit a series of distinctive characteristics that can help address this difficult agricultural problem.

Microalgae include different types of photosynthetic microorganisms, including cyanobacteria (prokaryotic) and green algae (eukaryotic) (Osorio-Reyes et al. 2023). Microalgae improve soil fertility and contribute to the growth and protection of plants against phytopathogens and pests, offering an alternative to reduce dependence on pesticides and chemical fertilizers in agricultural environments (Alvarez et al. 2021; Osorio-Reyes et al. 2023).

Microalgae contain bioactive compounds for biological control, particularly hydrolytic enzymes with antifungal activity (Prasanna et al. 2008). Therefore, there is a general view of the great potential of microalgae to synthesize enzymes as adjuvants for biological control (Holajjer et al. 2013).

Moreover, enzymes derived from photosynthetic microorganisms include cellulases, galactosidases, proteases, chitinases, and amylases, which can directly contribute to fungicidal activity by breaking down the cell walls of phytopathogenic fungi (Brasil et al. 2017; Alvarez et al. 2021). For example, Prasanna et al. (2008) evaluated the effects of cellular extracts of *Anabaena* hydrolytic enzymes (chitosanase and xylanase) against fungi and demonstrated that cyanobacterial extracts are potential candidates for the development of biological control agents against phytopathogenic

Table 19.2 Antifungal activity of enzymes extracted from microalgae for biological control

Enzyme	Microalgae	Biological activity	Reference
Hydrolytic enzymes, chitosanase, and xylanase	*Cyanobacteria*	Fungicidal	Prasanna et al. (2008)
	Anabaena spp.		
Peroxidase and endoglucanase	*Anabaena laxa* and *Calothrix elenkinii*		Kumar et al. (2013)
Chitosanase, β-1,3-glucanase)	*Anabaena laxa* or *A. variabilis*		Prasanna et al. (2013)

fungi such as *Aspergillus candida*, *Pythium aphanidermatum*, *Alternaria solani*, and *Fusarium solani*. Similarly, Gupta et al. (2011) and Natarajan et al. (2013) characterized endoglucanases from *Anabaena laxa* and *Calothrix elenkinii*, respectively, and reported that the enzymes showed fungicidal activity, inducing structural deformities in the cell wall of the mycelium of *Pythium aphanidermatum*.

The studies in Table 19.2 show that enzymes such as endoglucanases, chitinases, and chitosanases can be used as biocontrol agents to act as fungicides. In addition, the genetic diversity of algae has been explored to efficiently use photosynthetic microorganisms as recombinant enzyme bio-factories that are useful for the agroindustry (Brasil et al. 2017).

Finally, encouraging natural pesticides can promote sustainable agricultural practices, aligning with global efforts to reduce dependence on synthetic chemicals and move toward greener and safer food production. In addition, these findings help better elucidate how plants respond to and associate with cyanobacteria, which may allow the creation and optimization of these associations for specific benefits, potentially improving aspects such as plant health and disease resistance.

19.3 Advantages and Disadvantages of Obtaining Enzymes from Microalgae

Microalgal bio-factories represent a promising development area; however, there are various limitations, such as high biomass production costs and difficulties obtaining and purifying the enzymes of interest. It is important to note that it is not currently possible to compete with bacterial or animal systems to produce industrial enzymes due to the limited knowledge available, which leaves a wide field of study to be investigated. In this sense, advances related to optimizing large-scale phototrophic growth systems and developing powerful genetic tools to increase high levels of enzyme expression have occurred. However, significant obstacles still exist that must be overcome (Fig. 19.1).

19.4 Conclusions and Perspectives

The variety of enzymes that can be synthesized from microalgae shows how versatile they are for use as bio-factories. This source is still poorly explored compared with other organisms; however, enzymes that can be applied in the chemical, pharmaceutical, and food industries have been discovered.

Developing new strategies and methods to induce and optimize the production of microalgal enzymes, including bio-fabrication, is an important approach for expanding and industrializing potential enzymes. Some significant challenges include increasing the enzyme concentration and reducing the cost of subsequent operations. However, advances in engineering, genetics, and synthetic biology, coupled with using microalgal biomass to produce various bioproducts, offer promising possibilities to overcome these limitations and achieve the profitable production of industrial enzymes from microalgae.

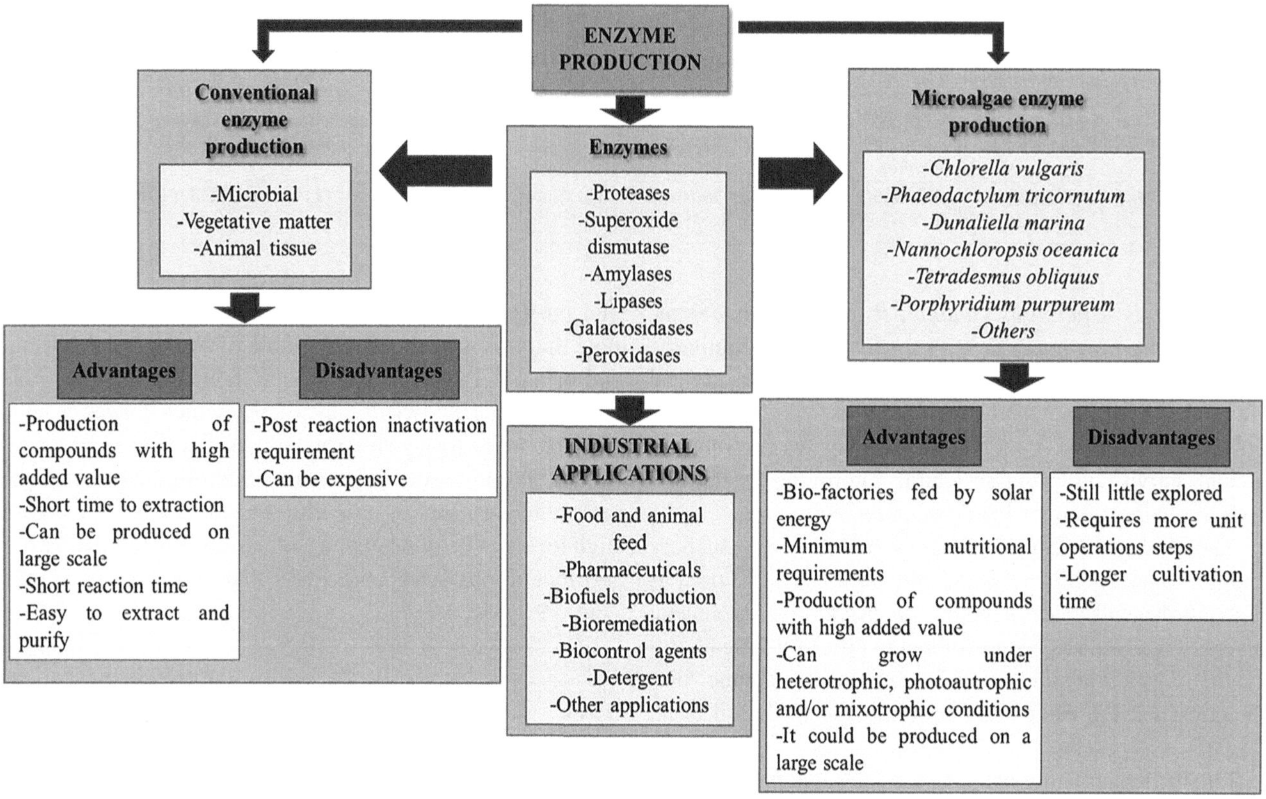

Fig. 19.1 Advantages and disadvantages of conventional enzyme and microalgal enzyme production

References

Alam MA, Xu J-L, Wang Z (2020) Microalgae biotechnology for food, health and high value products. Springer

Alvarez AL, Weyers SL, Goemann HM, Peyton BM, Gardner RD (2021) Microalgae, soil and plants: a critical review of microalgae as renewable resources for agriculture. Algal Res 54:102200

Anto S, Mukherjee SS, Muthappa R, Mathimani T, Deviram G, Kumar SS, Verma TN, Pugazhendhi A (2020) Algae as green energy reserve: technological outlook on biofuel production. Chemosphere 242:125079

Barreto Filho MM, Durand PM, Andolfato NE, Jordaan A, Sarmento H, Bagatini IL (2022) Programmed cell death in the coccoid green microalga Ankistrodesmus densus Korshikov (Sphaeropleales, Selenastraceae). Eur J Phycol 57(2):193–206. https://doi.org/10.1080/09670262.2021.1938240

Barzkar N, Jahromi ST, Vianello F (2022) Marine microbial fibrinolytic enzymes: an overview of source, production, biochemical properties and thrombolytic activity. Mar Drugs 20(1):46

Ben Hlima H, Karray A, Dammak M, Elleuch F, Michaud P, Fendri I, Abdelkafi S (2021) Production and structure prediction of amylases from Chlorella vulgaris. Environ Sci Pollut Res 28(37):51046–51059

Blifernez-Klassen O, Klassen V, Doebbe A, Kersting K, Grimm P, Wobbe L, Kruse O (2012) Cellulose degradation and assimilation by the unicellular phototrophic eukaryote Chlamydomonas reinhardtii. Nat Commun 3(1):1214

Brasil BSAF, de Siqueira FG, Salum TFC, Zanette CM, Spier MR (2017) Microalgae and cyanobacteria as enzyme biofactories. Algal Res 25:76–89

Chungjatupornchai W, Watcharawipas A (2015) Diacylglycerol acyltransferase type 2 cDNA from the oleaginous microalga Neochloris oleoabundans: cloning and functional characterization. J Appl Phycol 27:1499–1507

de Carvalho Silvello MA, Gonçalves IS, Azambuja SPH, Costa SS, Silva PGP, Santos LO, Goldbeck R (2022) Microalgae-based carbohydrates: a green innovative source of bioenergy. Bioresour Technol 344:126304

Demuez M, Mahdy A, Tomás-Pejó E, González-Fernández C, Ballesteros M (2015) Enzymatic cell disruption of microalgae biomass in biorefinery processes. Biotechnol Bioeng 112(10):1955–1966

Dong S, Liu Y-J, Zhou H, Xiao Y, Xu J, Cui Q, Wang X, Feng Y (2021) Structural insight into a GH1 β-glucosidase from the oleaginous microalga, Nannochloropsis oceanica. Int J Biol Macromol 170:196–206

Dong F-k, Quan X-g, Wang Q-b, Liu Z-m, Cui T, Wang W-j, Tang D-m, Zhang R-m, Zhang C, Wang H-y, Ren Q (2022) Purification, structural characterization, and anticoagulant activity evaluation of chondroitin sulfate from codfish (Gadus macrocephalus) bones. Int J Biol Macromol 210:759–767. https://doi.org/10.1016/j.ijbiomac.2022.05.001

Duarah P, Haldar D, Patel AK, Dong C-D, Singhania RR, Purkait MK (2022) A review on global perspectives of sustainable development in bioenergy generation. Bioresour Technol 348:126791

Efremenko E, Nikolskaya A, Lyagin I, Senko O, Makhlis T, Stepanov N, Maslova O, Mamedova F, Varfolomeev S (2012) Production of biofuels from pretreated microalgae biomass by anaerobic fermentation with immobilized Clostridium acetobutylicum cells. Bioresour Technol 114:342–348

Erpel F, Restovic F, Arce-Johnson P (2016) Development of phytase-expressing Chlamydomonas reinhardtii for monogastric animal nutrition. BMC Biotechnol 16:1–7

Ferraro V, Cruz IB, Jorge RF, Malcata FX, Pintado ME, Castro PM (2010) Valorisation of natural extracts from marine source focused on marine by-products: a review. Food Res Int 43(9):2221–2233

Findinier J (2023) Autolysin production from Chlamydomonas reinhardtii. Bio Protoc 13(13):e4705. https://doi.org/10.21769/BioProtoc.4705

Georgianna DR, Hannon MJ, Marcuschi M, Wu S, Botsch K, Lewis AJ, Hyun J, Mendez M, Mayfield SP (2013) Production of recombinant enzymes in the marine alga Dunaliella tertiolecta. Algal Res 2(1):2–9. https://doi.org/10.1016/j.algal.2012.10.004

Gilmour DJ (2019) Microalgae for biofuel production. Adv Appl Microbiol 109:1–30

Guerriero G, Sergeant K, Legay S, Hausman J-F, Cauchie H-M, Ahmad I, Siddiqui KS (2018) Novel insights from comparative in silico analysis of green microalgal cellulases. Int J Mol Sci 19(6):1782

Gupta V, Natarajan C, Kumar K, Prasanna R (2011) Identification and characterization of endoglucanases for fungicidal activity in Anabaena laxa (Cyanobacteria). J Appl Phycol 23:73–81

Handa V, Sharma D, Kaur A, Arya SK (2020) Biotechnological applications of microbial phytase and phytic acid in food and feed industries. Biocatal Agric Biotechnol 25:101600

Hazare C, Bhagwat P, Singh S, Pillai S (2024) Diverse origins of fibrinolytic enzymes: a comprehensive review. Heliyon 10(5):e26668

Holajjer P, Kamra A, Gaur H, Manjunath M (2013) Potential of cyanobacteria for biorational management of plant parasitic nematodes: a review. Crop Prot 53:147–151

Hossain N, Mahlia TMI (2019) Progress in physicochemical parameters of microalgae cultivation for biofuel production. Crit Rev Biotechnol 39(6):835–859

Hossain SZ, Razzak SA, Al-Shater AF, Moniruzzaman M, Hossain MM (2020) Recent advances in enzymatic conversion of microalgal lipids into biodiesel. Energy Fuel 34(6):6735–6750

Huang S, Pan S, Chen G, Huang S, Zhang Z, Li Y, Liang Z (2013) Biochemical characteristics of a fibrinolytic enzyme purified from a marine bacterium, Bacillus subtilis HQS-3. Int J Biol Macromol 62:124–130

Jamiołkowska A (2020) Natural compounds as elicitors of plant resistance against diseases and new biocontrol strategies. Agronomy 10(2):173

Jiménez C, Capasso JM, Edelstein CL, Rivard CJ, Lucia S, Breusegem S, Berl T, Segovia M (2009) Different ways to die: cell death modes of the unicellular chlorophyte Dunaliella viridis exposed to various environmental stresses are mediated by the caspase-like activity DEVDase. J Exp Bot 60(3):815–828. https://doi.org/10.1093/jxb/ern330

Jothibasu K, Dhar D, Rakesh S (2021) Recent developments in microalgal genome editing for enhancing lipid accumulation and biofuel recovery. Biomass Bioenergy 150:106093

Kumar M, Prasanna R, Bidyarani N, Babu S, Mishra BK, Kumar A, Adak A, Jauhari S, Yadav K, Singh R (2013) Evaluating the plant growth promoting ability of thermotolerant bacteria and cyanobacteria and their interactions with seed spice crops. Sci Hortic 164:94–101

Kursa W, Jamiołkowska A, Wyrostek J, Kowalski R (2022) Antifungal effect of plant extracts on the growth of the cereal pathogen Fusarium spp. – an in vitro study. Agronomy 12(12):3204

Lamb WF, Wiedmann T, Pongratz J, Andrew R, Crippa M, Olivier JG, Wiedenhofer D, Mattioli G, Al Khourdajie A, House J (2022) Corrigendum: a review of trends and drivers of greenhouse gas emissions by sector from 1990 to 2018 (2021 Environ. Res. Lett. 16 073005). Environ Res Lett 17(4):049502

Lengai GM, Muthomi JW, Mbega ER (2020) Phytochemical activity and role of botanical pesticides in pest management for sustainable agricultural crop production. Sci Afr 7:e00239

Li Y, Miros S, Eckhardt H-G, Blanco A, Mulcahy S, Tiwari BK, Halim R (2024) Freshwater microalgae Nannochloropsis limnetica for the production of β-galactosidase from whey powder. Sci Rep 14(1):14346

Masi A, Leonelli F, Scognamiglio V, Gasperuzzo G, Antonacci A, Terzidis MA (2023) Chlamydomonas reinhardtii: a factory of nutraceutical and food supplements for human health. Molecules 28(3):1185

Mogharabi M, Faramarzi MA (2016) Are algae the future source of enzymes? Trend Peptide Protein Sci 1(1):1–6

Moura YAS, De Souza ATV, Silva PEDCE, Da Silva MM, Porto ALF, Bezerra RP (2023) Fibrinolytic enzyme from green microalgae: a new potential drug for thrombolytic therapy? Phycology 3(4):436–446

Natarajan C, Gupta V, Kumar K, Prasanna R (2013) Molecular characterization of a fungicidal endoglucanase from the cyanobacterium Calothrix elenkinii. Biochem Genet 51:766–779

Osorio-Reyes JG, Valenzuela-Amaro HM, Pizaña-Aranda JJP, Ramírez-Gamboa D, Meléndez-Sánchez ER, López-Arellanes ME, Castañeda-Antonio MD, Coronado-Apodaca KG, Gomes Araújo R, Sosa-Hernández JE (2023) Microalgae-based biotechnology as alternative biofertilizers for soil enhancement and carbon footprint reduction: Advantages and implications. Mar Drugs 21(2):93

Páblo E, de Souza F, de Barros R, Marques D, Porto A, Bezerra R (2017) Enhanced production of fibrinolytic protease from microalgae Chlorella vulgaris using glycerol and corn steep liquor as nutrient. Ann Microbiol Res 1(1):9–19

Pandit MA, Kumar J, Gulati S, Bhandari N, Mehta P, Katyal R, Rawat CD, Mishra V, Kaur J (2022) Major biological control strategies for plant pathogens. Pathogens 11(2):273

Park S, Nguyen THT, Jin E (2019) Improving lipid production by strain development in microalgae: strategies, challenges and perspectives. Bioresour Technol 292:121953

Peraza-Echeverria S, Bernardo-Candelero S, Baas-Espinola FM, Puch-Hau C, Rivera-Solís RA, Echevarría-Machado I, Borges-Argáez IC, Herrera-Valencia VA (2021) Production of a ruminal bacterial phytase in the green microalga Chlamydomonas reinhardtii with potential applications in monogastric animal feed. Bioresour Technol Rep 14:100660

Prasanna R, Nain L, Tripathi R, Gupta V, Chaudhary V, Middha S, Joshi M, Ancha R, Kaushik BD (2008) Evaluation of fungicidal activity of extracellular filtrates of cyanobacteria–possible role of hydrolytic enzymes. J Basic Microbiol 48(3):186–194

Prasanna R, Chaudhary V, Gupta V, Babu S, Kumar A, Singh R, Shivay YS, Nain L (2013) Cyanobacteria mediated plant growth promotion and bioprotection against Fusarium wilt in tomato. Eur J Plant Pathol 136:337–353

Pudney A, Gandini C, Economou CK, Smith R, Goddard P, Napier JA, Spicer A, Sayanova O (2019) Multifunctionalizing the marine diatom Phaeodactylum tricornutum for sustainable co-production of omega-3 long chain polyunsaturated fatty acids and recombinant phytase. Sci Rep 9(1):11444

Righini H, Roberti R (2019) Algae and cyanobacteria as biocontrol agents of fungal plant pathogens. Plant Microbe Interface: 219–238

Robinson PK (2015) Enzymes: principles and biotechnological applications. Essays Biochem 59:1

Roth GA, Mensah GA, Johnson CO, Addolorato G, Ammirati E, Baddour LM, Barengo NC, Beaton AZ, Benjamin EJ, Benziger CP (2020) Global burden of cardiovascular diseases and risk factors, 1990–2019: update from the GBD 2019 study. J Am Coll Cardiol 76(25):2982–3021

Sajjadi B, Chen W-Y, Raman AAA, Ibrahim S (2018) Microalgae lipid and biomass for biofuel production: a comprehensive review on lipid enhancement strategies and their effects on fatty acid composition. Renew Sust Energ Rev 97:200–232

Segovia MA, Haramaty L, Berges JA, Falkowski PG (2003) Cell death in the unicellular chlorophyte Dunaliella tertiolecta. A hypothesis on the evolution of apoptosis in higher plants and metazoans. Plant Physiol 132(1):99–105. https://doi.org/10.1104/pp.102.017129

Shahidi F, Varatharajan V, Peng H, Senadheera R (2019) Utilization of marine by-products for the recovery of value-added products. J Food Bioact 6:10. https://doi.org/10.31665/JFB.2019.6184

Sharma C, Osmolovskiy A, Singh R (2021) Microbial fibrinolytic enzymes as anti-thrombotics: production, characterisation and prodigious biopharmaceutical applications. Pharmaceutics 13(11):1880

Silva PEDCE, de Barros RC, Albuquerque WWC, Brandão RMP, Bezerra RP, Porto ALF (2018) In vitro thrombolytic activity of a purified fibrinolytic enzyme from Chlorella vulgaris. J Chromatogr B 1092:524–529

Silva G, Cerqueira K, Rodrigues J, Silva K, Coelho D, Souza R (2023) Cultivation of microalgae Chlorella vulgaris in open reactor for bioethanol production. Phycology 3(2):325–336

Singarapu RK, Reddy BRB, Rajesh M (2023) Heterologous expression of phytase in Schizochytrium sp. as a fortified feed additive for the Livestock industry. J Appl Nat Sci 15(3):1245–1253

Singh R, Gautam P, Sharma C, Osmolovskiy A (2023) Fibrin and fibrinolytic enzyme cascade in thrombosis: unravelling the role. Life 13(11):2196

Spier MR, Peron-Schlosser B, Paludo LC, Gallo-García LA, Zanette CM (2020) Microalgae as enzymes biofactories. In: Handbook of microalgae-based processes and products. Elsevier, pp 687–706

Sun X-M, Ren L-J, Zhao Q-Y, Ji X-J, Huang H (2019) Enhancement of lipid accumulation in microalgae by metabolic engineering. Biochimica et Biophysica Acta (BBA) Mol Cell Biol Lipids 1864(4):552–566

Thirumalaivasan N, Nangan S, Kanagaraj K, Rajendran S (2024) Assessment of Sustainability and environmental impacts of renewable energies: focusing on biogas and biohydrogen (biofuels) production, vol 189. Process Saf Environ Prot, p 467

Vavilala SL, Gawde KK, Sinha M, D'Souza JS (2015) Programmed cell death is induced by hydrogen peroxide but not by excessive ionic stress of sodium chloride in the unicellular green alga Chlamydomonas reinhardtii. Eur J Phycol 50(4):422–438. https://doi.org/10.1080/09670262.2015.1070437

Viera-Arroyo WF, Tello-Torres CM, Martínez-Salinas AA, Navia-Santillán DF, Medina-Rivera LA, Delgado-Párraga AG, Perdomo-Quispe CE, Pincay-Verdezoto AK, Báez-Cevallos FJ, Vásquez-Castillo WA (2020) Biological control: a tool for sustainable agriculture, a point of view of its benefits in Ecuador. J Selva Andina Biosphere 8(2):128–149

Vingiani GM, De Luca P, Ianora A, Dobson AD, Lauritano C (2019) Microalgal enzymes with biotechnological applications. Mar Drugs 17(8):459

Wang X, Ruan Z, Boileau D, Sears BB, Liu Y, Liao W (2015) Transgenic expression of a bacterial thermophilic amylase in the Chlamydomonas reinhardtii chloroplast to facilitate algal biofuel production. Bioenergy Res 8:527–536

Wendelboe AM, Raskob GE (2016) Global burden of thrombosis: epidemiologic aspects. Circ Res 118(9):1340–1347

Xue J, Balamurugan S, Li D-W, Liu Y-H, Zeng H, Wang L, Yang W-D, Liu J-S, Li H-Y (2017) Glucose-6-phosphate dehydrogenase as a target for highly efficient fatty acid biosynthesis in microalgae by enhancing NADPH supply. Metab Eng 41:212–221

Yoon S-M, Kim SY, Li KF, Yoon BH, Choe S, Kuo MM-C (2011) Transgenic microalgae expressing Escherichia coli AppA phytase as feed additive to reduce phytate excretion in the manure of young broiler chicks. Appl Microbiol Biotechnol 91:553–563

Zuppini A, Andreoli C, Baldan B (2007) Heat stress: an inducer of programmed cell death in Chlorella saccharophila. Plant Cell Physiol 48(7):1000–1009

Microalgal Applications in Nanotechnology: Potential for Bioproducts Obtention and Nano-Compounds Synthesis

20

Jennifer Ortiz-Letechipia,
Jacqueline Stephanie Ortiz-Letechipia,
Salvador Gómez-Jiménez, Julián González-Trinidad,
and Ada Rebeca Contreras-Rodríguez

Abstract

Microalgae are being studied due to their potential for a wide range of applications, such as the ability to capture and sequester carbon dioxide and convert it into organic compounds used to produce various products. In recent years, the combination of microalgae and nanotechnology has led to the development of innovative approaches like creating high-value products and has been studied in several different ways across various industries, such as the production of biodiesel and bioethanol, to name a few. This work aims to provide a comprehensive review of the main aspects of microalgal applications in nanotechnology, mainly as a potential for bioproduct obtention and nano-compounds synthesis.

Keywords

Nanotechnology · Green synthesis · Bioproducts · Nano-compounds

J. Ortiz-Letechipia
Unidad Académica de Ciencias Químicas, Universidad Autónoma de Zacatecas, Zacatecas, Mexico

J. S. Ortiz-Letechipia
Doctorado en Ingeniería para la Innovación Tecnológica, Unidad Académica de Ingeniería Eléctrica, Universidad Autónoma de Zacatecas, Zacatecas, Mexico

S. Gómez-Jiménez
Maestría en Ciencia e Ingeniería de Materiales de la Unidad Académica de Ingeniería I, Universidad Autónoma de Zacatecas, Zacatecas, Mexico

J. González-Trinidad · A. R. Contreras-Rodríguez (✉)
Unidad Académica de Ciencia y Tecnología de la Luz y la Materia, Universidad Autónoma de Zacatecas, Zacatecas, Mexico
e-mail: rebeca.contreras@uaz.edu.mx

20.1 Introduction

Microalgae are unicellular microorganisms that require light, water, carbon dioxide (CO_2), and inorganic nutrients for growth and reproduction. These organisms contain lipids and fatty acids as membrane components, storage products, and metabolites (Fonseca et al. 2020). They are considered one of the most diverse groups of organisms on Earth, with over 100,000 known species. Usually, they are single-celled, colony-forming, or filamentous photosynthetic microorganisms belonging to several algal divisions, such as Chlorophyta, Charophyta, and Bacillariophyta (Dahoumane et al. 2016).

Besides, microalgae have been studied because of their potential for a wide range of applications and their ability to capture and sequester carbon dioxide (CO_2) from the atmosphere (Barsanti and Gualtieri 2007). Through photosynthesis, microalgae convert CO_2 into organic compounds, which can produce different high-value products (Wijffels and Barbosa 2010).

On the other hand, nanotechnology refers to manipulating different compounds at the nanoscale, typically between 1 and 100 nm. It involves working at the atomic, molecular, and supramolecular levels to create and use materials, devices, and systems with fundamentally new properties and functions because of their small structure. It has been applied in different fields, including electronics, medicine, and environmental science (Mahdia et al. 2021).

In 1959, Richard Feynman, in his talk "*There is Plenty of Room at the Bottom*," proposed the idea of manipulating materials on an atomic and molecular scale, laying the foundation for nanotechnology's development (Feynman 1959). Moreover, Norio Taniguchi, in his 1974 research, coined the term "nanotechnology" to describe the manipulation of materials at the nanoscale (Taniguchi 1974). Later, Richard Smalley and colleagues, in 1985 at Rice University, discov-

D. U. Santos Ballardo, S. Rossi (eds.), *Microalgae as Promising Source of Commercial Bioproducts*, Developments in Applied Phycology 14, https://doi.org/10.1007/978-3-031-86433-9_20

ered a new form of carbon known as buckminsterfullerene or C60, a nanoparticle (NP) with a unique structure and properties (Kroto et al. 1985). Besides, in the 90 s, Sumio Iijima discovered carbon nanotubes and cylindrical carbon nanoparticles with various applications in electronics, materials science, and other fields (Iijima 1991).

The combination of microalgae and nanotechnology has led to the development of innovative approaches like creating high-value products and has been studied in several different ways across various industries. Some of the most relevant applications are: biodiesel production, using nanotechnology for lipid accumulation and extraction in microalgae-based biodiesel production (Li et al. 2020; Santos-Ballardo et al. 2015, 2016a, b); bioethanol production, where microalgae could be used as a source of biomass for its production (Condor et al. 2022); nanosensing and nanofabrication, since the microalgae biosensor chip that has been developed for pesticide sensing by incorporating microarray oxygen sensors (Kashem et al. 2019); cultivation and harvesting, applied to enhance lipid accumulation, biomass yields, and carbon fixation in microalgae cultivation (Kurniawan et al. 2022); biomedical applications, used for drug delivery and imaging (Wu et al. 2021); environmental remediation through the removal of heavy metals and other pollutants from wastewater and contaminated soils (Cameron et al. 2018); agriculture with microalgae-based products, such as biofertilizers, that can be used to improve crop yields and soil health (Çakirsoy et al. 2022); aquaculture, using microalgae as a source of feed for fish and other aquatic organisms (Guedes and Malcata 2012); microalgae-based products can be used in cosmetics and skincare products for their antioxidant and anti-inflammatory properties (Martínez-Ruiz et al. 2022); microalgae-based products, such as spirulina and chlorella, which are rich in nutrients and can be used as a source of protein, vitamins, and minerals in food and supplements (Dalla Costa et al. 2022); and microalgae-based polymers, synthesized using nanotechnology, which have potential as sustainable and biodegradable alternatives to traditional plastics (Martínez-Ruiz et al. 2022).

On the other hand, nanotechnology has become a valuable tool in the current era because it has explored new fields of study that did not exist before. If combined with other branches, such as biology and the use of microalgae, it can become a very versatile tool with the potential to be used in countless applications in everyday life to improve the quality of life of human beings.

This chapter aims to comprehensively review the main aspects of microalgal applications in nanotechnology, mainly as a potential for bioproduct obtention and nano-compounds synthesis.

20.2 Microalgal Applications in Nanotechnology

In recent years, nanotechnology has combined with microalgae efficiently. Their success in different research has positioned them in a wide range of applications. Some of these are as follows:

Nanoparticle synthesis, biosensors, drug-delivery systems, environmental remediation, photocatalysis, nanofabrication templates, nanocomposite materials, nanofiltration membranes, energy harvesting, antimicrobial nanomaterials, nanosensors for environmental monitoring, photonic devices, nanomedicine, nanowire fabrication, nanoporous materials, nanofluids, nanocarriers for gene delivery, environmental monitoring using nanobiosensors, nanocatalysts, nanophotonics, nanomaterials for water treatment, nanoscale imaging agents, nanoscale sensors for food safety, self-cleaning surfaces, carbon capture and utilization, nanoparticle-based imaging contrast agents, nanoscale electronic devices, nanobiofertilizers, nanoscale wound dressings, nanotechnology-based cosmetics, etc.

Besides, microalgae have also gained significant attention in recent years as potential nanocarriers and drug-delivery systems, mainly because they can produce different bioactive molecules, including carotenoids, polysaccharides, vitamins, and lipids. Furthermore, they have potential uses as cosmetic, antioxidant, anti-inflammatory, antitumor, anticancer, antimicrobial, antiviral, and anti-allergy agents (Edis et al. 2021). Moreover, microalgae pigments also have health benefits, such as preventing cancer, cardiac diseases, neurological disorders, and eye diseases (Ramana et al. 2017). They can also produce pharmaceuticals and therapeutic proteins for subunit vaccines, monoclonal antibodies, antimicrobial agents, etc. (Yan et al. 2016). On the other hand, the small size of microalgae allows efficient cellular uptake and distribution within the body, and its large surface area-to-volume ratio enhances its capacity to carry and deliver drugs. The surface of microalgae can be easily modified to enhance their drug loading and targeting capabilities (Edis et al. 2021). These single-celled organisms possess several properties that make them suitable for these applications and are generally considered safe and biocompatible with the human body. They have been extensively studied for their potential health benefits as nutritional supplements and have shown low toxicity profiles (Khavari et al. 2021).

Also, due to their properties, microalgae are considered biodegradable and environmentally friendly. Furthermore, after drug delivery, microalgae can be easily metabolized or eliminated from the body, reducing the potential for long-term accumulation. Additionally, microalgae possess natural capabilities for bioremediation, which can be harnessed to eliminate toxins or pollutants from the environment (Khavari et al. 2021). On the other hand, these microorganisms can be cultivated relatively rapidly and cost-effectively using photobioreactors or open ponds. Different authors have reported high growth rates and the possibility of culturing in non-arable land using sunlight and carbon dioxide, reducing the reliance on scarce resources and minimizing production costs (Khavari et al. 2021).

Some examples of recent studies in this area are: Ulvan, a bioactive marine sulfated polysaccharide isolated from green algae, has been incorporated into liposomes to be evaluated for their antibacterial activity against *Staphylococcus aureus* (Tziveleka et al. 2022); also, sulfated polysaccharides isolated from marine algae and bacteria have been explored for their potential as nanocarriers and stimuli-responsive drug-delivery systems for various applications, including cancer therapy (Avramović et al. 2020); besides, topical ophthalmic nanotherapies in the treatment of intraocular diseases with lipid-based nanosystems have shown increased efficiency and feasibility in topical formulations (Navarro-Partida et al. 2021).

In summary, microalgae and sulfated polysaccharides from marine algae and bacteria have shown potential as nanocarriers and drug-delivery systems for various applications, including antibacterial and cancer therapy. Liposomes and polymeric nanocarriers have been designed to improve drug-delivery efficiency, specificity, and control. However, further research is needed to optimize their design, evaluate their safety, and assess their performance in vivo. Nonetheless, these microscopic organisms offer exciting possibilities for developing innovative and sustainable drug delivery platforms.

20.3 Bioproducts Obtained from Microalgae in Nanotechnology

Microalgae are part of the sustainable food sources, providing global food security while counteracting the challenges related to global warming (Aratboni et al. 2023).

Besides, microalgae can substitute non-renewable energy sources and non-biological-based products. They produce valuable biological resources such as lipids, pigments, carbohydrates, and proteins capable of entering multiple industries (Dolganyuk et al. 2020). These can be used to generate products such as protein-rich animal feed in aquaculture, high-value products, namely, polyunsaturated fatty acids, bioactive and functional pigments and natural colorants, health foods, cosmetics, and pharmaceuticals (Paniagua-Michel et al. 2015).

In the medical field, microalgae have multiple applications; the influence of secondary metabolites of microalgae in the synthesis of nanoparticles are applied as therapeutics, and the use of isolated compounds on the surface of nanoparticles to combat diseases has also been addressed (Sidorowicz et al. 2023). Among the medical uses is the application of spirulina (also called Arthrospira), a type of microfilamentous cyanobacteria that is considered a sustainable and environmentally friendly microalgae (Dai et al. 2022), designated by the U.S. Food and Drug Administration (FDA) as "generally recognized as safe" (GRAS) for human consumption, without serious health risks according to the Dietary Supplement Information Expert Committee (DSI-EC) (Marles et al. 2011).

On the other hand, the production of biomass consisting of numerous nutrients can also be derived from the cultivation of microalgae, which form an efficient production of renewable elements such as biopolymers (Rueda et al. 2020), biofertilizers (Khan et al. 2019), and natural pigments (Chakravorty et al. 2023).

The main bioproducts obtained by optimizing the hydrolyzed algae biomass through microbial fermentation are bioalcohol, biogas, biohydrogen as energy generators, and organic acids and biopolymers as biochemicals (Dutta et al. 2023; Santos-Ballardo et al., 2015).

Some of the advantages of microalgae-based bioproducts are economical (because they omit the use of a bioreactor and additional nutrients, unlike bacteria, they have a high biomass production and are readily available throughout the year (Thiruchelvi et al. 2021), in turn, algae do not compete with food sources, they can survive in a hostile environment, they can restore water quality by treating wastewater, and they help remove carbon dioxide from the environment by using it as a source of nutrients for biomass production (Chia et al. 2020; Khoo et al. 2020).

20.4 Microalgal Nanoparticles: Synthesis and Properties

Ecological synthesis of nanoparticles is a method that aims to produce nanoparticles with unique biological, physical, and chemical properties; this method has the advantage of being environmentally friendly (Ali Anvar et al. 2023). Some organisms, such as algae and microalgae, contribute to the green synthesis of pure nanoparticles (Velusamy et al. 2016). These biological systems act as regenerative agents as inhibitors in the production of intracellular or extracellular nanoparticles; in terms of functional groups (hydroxyl, carbonyl, amine, and thiol), biological organism's proteins and peptides reduce the cutting-edge salt to nanoparticles. This synthesis is simple, repeatable, and constant (Ali et al. 2023).

The application of cyanobacteria within green synthesis is increasing due to their rich sources of secondary metabolites, pigments, proteins, and peptides that act as the basis of bio nanofactories (Sahoo et al. 2020).

The natural substances of microalgae act as a protective or reducing agent and stabilizer in metal salts of metal, metal oxide, or bimetallic NP conversion. NPs usable for algae can be intracellular or extracellular, depending on the area of the NPs (AlNadhari et al. 2021).

The synthesis of nanoparticles obtained from microalgae is exemplified by the use of aqueous extracts of two cyanobacteria (*Synechococcus elongatus* and *Microcystis aeruginosa*) and different microalgae (*Synechococcus elongatus, Microcystis aeruginosa, Coelastrum astroideum, Desmodesmus armatus, Cosmarium punctulatum,* and *Klebsormidium flaccidum*) for the biosynthesis of silver nanoparticles (AgNP) (Moraes et al. 2021). The synthesized silver nanoparticles act as an effective antimicrobial agent and proved to be an alternative for developing new antimicrobial agents to combat the problem of resistance (Sudha et al. 2013).

Microalgae can also remove heavy metals from wastewater by producing microalgae-based carbon-encapsulated iron nanoparticles (ME-nFe) (Mantovani et al. 2023).

Another product obtained from microalgae (*Stoechospermum marginatum*) is the biological synthesis of gold nanoparticles; the formation of the particles can be observed in just 10 min. They have several properties, one of them being photoluminescence. They are spherical, and some are hexagonal and triangular, with a size ranging from 18.7 to 93.7 nm. A positive effect of biologically synthesized gold nanoparticles is their efficacy against bacterial pathogens, an alternative to counteract infectious diseases (Rajathi et al. 2012).

The outstanding advantages of large-scale synthesis of algae-derived nanoparticles include their ease of growth and ability to survive extreme pH and temperature conditions, making them promising low-cost green machines (Jacob et al. 2021).

20.5 Exploring Microalgae for Biofuel Production at the Nanoscale

According to Krugman et al. (2013), a resource is anything used to do something else. However, there are not enough resources to use in all the ways that society wants; in this sense, the resources are considered scarce. Within these scarce resources, we can include natural ones, such as minerals, wood, or oil, which arise from the physical environment. However, the number of human resources is also considered limited, and in a growing world economy with an accelerated demographic explosion, resources such as clean air and water have become vital but limited.

In the last 20 years, due to the nature of the energy resources, an important discussion of the need for renewable and sustainable alternatives for energy sources has been in the spotlight (Pattarkine and Pattarkine 2012). There is a discussion of clean energy and technological power that will provide strength to maintain production processes with the least possible environmental impact. It will be necessary to add the virtues of the fourth scientific-technological revolution that will make the abundance infinite. It also insists that the damage caused by the profit motive of economic growth models (the economic footprint) to the base that sustains life can be reversed. Currently, compensatory alternatives to the biosphere are being sought. Solutions to the pollution of the oceans by plastics or nuclear waste are being explored. Today, there is concern about the atmospheric integrity impacted by polluting gasses, the scarcity of fossil fuels as the primary source of energy, or the slowness to make improvements in the integration of clean water into its natural cycle as well as the indiscriminate use of pesticides, genetically modified seeds, industrial food production, deforestation, etc.

Undoubtedly, challenges require linking two central elements: economy and nature. The logic of economic benefit cannot be stopped, nor should it be obtained at the cost of increasing externalities. The challenge is to find a balance between the two, trying to achieve empathy with global plans and strategies such as the millennium development goals, others focused on sustainable production and green development, etc., based on a proposal where the factors of production, biology, physics, and chemistry contribute to greater energy efficiency and to develop and value renewable energies.

For this purpose, researchers are seeking alternatives to renewable sources such as solar, hydro, ocean, geothermal, wind, and bioenergy (Hossain et al. 2020). Bioenergy technologies can provide an attractive solution to these issues (Pattarkine and Pattarkine 2012). One of the most promising is substituting fossil fuel-based diesel oil with biofuels (biogas, bioethanol, but mainly biodiesel) (Hossain et al. 2020). A biofuel is any solid, liquid, or gaseous compound derived from biological materials or biomass. The biological material or organic matter converted to biofuels includes food crops, dedicated bioenergy crops, agricultural residues, wood/forestry waste, animal manure, microalgae, etc. The most primary common types of biofuels that are produced commercially are ethanol/gasoline blends and biodiesel/diesel blends (Lawan Muhammad 2018). Biodiesel can be used pure or in mixtures with diesel in diesel cycle engines, being, in some ways, comparatively less polluting than diesel (Karthikeyan et al. 2020).

Biofuels can be categorized into current/conventional or advanced biofuels. Current or conventional bioenergy is often called first-generation, whereas advanced bioenergy refers to second and third-generation biofuels (Lawan Muhammad 2018).

They typically have characteristics similar to petrodiesel's, such as viscosity, energy content, cetane number, and phase changes. Due to this, it can be easily blended with petrodiesel and can be used in conventional diesel engines without modification. This process's main advantage includes reducing greenhouse gas emissions, and the increase of lubricity and cetane ignition (Hossain et al. 2020).

The fuels obtained from biomass are essential, and there is a great expectation for the consolidation of microalgae as a viable source of biomass to produce them.

The biomass of microalgae produces three major biochemical components by de novo synthesis consisting of carbohydrates, proteins, and lipids (natural oils). These microorganisms are known to synthesize and rapidly accumulate higher amounts of lipids than terrestrial plants due to their high growth rates, concomitantly by alteration of the lipid biosynthetic pathways for storage as neutral lipids. The lipid yields and growth rates vary significantly among different species. Microalgal oil and spent biomass offered as good potential sources as biodiesel feedstocks. Because lipids contain twice the energy stored per carbon atoms than carbohydrates, which translates directly into a twofold increase in fuel energy content, lipid-rich microalgae biomass feedstock has been the focus of attention for the scientific community for an extended period for the production of alternate sustainable energy to resolve the world energy fossil fuel depletion and its harmful effect on the environment (Rawat et al. 2013).

The biodiesel production process from microalgae generally consists of five phases: cultivation, harvesting, drying, lipid extraction, and transesterification (Collotta et al. 2016). The first begins by growing microalgae in carefully controlled environments, utilizing photobioreactors or open ponds. These environments are optimized for light exposure, nutrient supply, temperature, and pH levels to facilitate robust growth and maximize lipid accumulation within the microalgae cells. The second phase, harvesting, occurs once the microalgae culture reaches maturity, typically after a few days to weeks, depending on the species and growth conditions. Harvesting methods can include centrifugation, filtration, flocculation, or gravity sedimentation, aiming to separate the microalgae biomass from the growth medium (Collotta et al. 2016). Subsequently, the harvested biomass is processed through drying, the third phase, to reduce water content, which is vital for efficient lipid extraction. The fourth phase involves lipid extraction, where the lipids (oils) are separated from the microalgae biomass. This can be achieved through various methods, such as solvent extraction, mechanical pressing, or supercritical fluid extraction. These lipids, which contain triglycerides, are then converted into biodiesel in the fifth phase through a chemical process called transesterification. This involves reacting the lipids with an alcohol in the presence of a catalyst to produce biodiesel (fatty acid methyl or ethyl esters) and glycerol. Overall, this process showcases the potential of microalgae in generating biodiesel, a promising and sustainable alternative to conventional fossil fuels (Collotta et al. 2016).

To accelerate the biofuel yield and improve the efficiency of biofuel utilization in petrol and diesel, nanotechnology has been initiated via nano additives such as nano-magnets, nano-crystals, nano-fibers, nano-droplets, and others (Hossain et al. 2019). The effects of nanoparticle addition on microalgal growth, lipid production, CO_2 absorption and sequestration, light conversion, toxicity, and other factors have been studied (Vargas-Estrada et al. 2020).

Thus, microalgae are a sustainable resource for third-generation biodiesel, but large-scale production is limited due to lipid extraction, biomass productivity, and harvesting challenges. However, recent cultivation, harvesting technologies, and environmental pollution control advances can offer a promising roadmap for cost-effective biodiesel production (Yin et al. 2020).

The results suggested that the new process, which combined bioengineering and transesterification, was a feasible and effective method for producing high-quality biodiesel from microalgal oil (Miao and Wu 2006).

20.6 Biocompatible Nanomaterials from Microalgae

Biocompatible nanomaterials can be used in synergy with microalgae to enhance their functionalities and applications. These nanomaterials can serve as carriers, sensors, modifiers, and catalysts, among other functions. When combined, they can improve efficiency, selectivity, and overall performance. They can represent a fascinating field of research at the intersection of nanotechnology and biotechnology. By taking advantage of their unique properties, the potential of nanomaterials that can be obtained from microalgae has been explored for various applications, such as medicine, environmental remediation, and energy storage. Microalgae can be manipulated to produce nanoparticles or nanoscale materials through various techniques, such as biomineralization, chemical synthesis, and genetic engineering. These nanoparticles can comprise metals, metal oxides, or organic compounds (Sidorowicz et al. 2023).

Examples of materials are metallic nanoparticles such as gold, silver, and platinum. These nanoparticles can enhance the photosynthetic activity of microalgae by absorbing and utilizing specific wavelengths of light more efficiently. They can also act as catalysts in specific reactions, such as CO_2 fixation (Li et al. 2021).

One of the main advantages of these microalgae-derived nanomaterial composites is their inherent biocompatibility. Since living organisms produce these materials, they exhibit lower toxicity and better compatibility with biological systems than synthetic nanomaterials. Due to these properties, they have shown promise in various applications. However, the ones that stand out the most are medical, as they can be used in drug-delivery systems, imaging agents, and even regenerative medicine thanks to their biocompatibility (Sidorowicz et al. 2023).

Besides, microalgae can be combined with nanosensors to detect changes in environmental parameters such as pH, temperature, and pollutant concentrations. They can also provide real-time data for environmental monitoring and assessment (Patwari et al. 2023).

According to several studies, carbon-based nanomaterials in microalgae for the sustainable production of bioactive compounds have shown better biocompatibility than others. However, the exact mechanism by which carbon-based nanomaterials improve the biocompatibility of microalgae is not yet clear. The unique properties of carbon-based nanomaterials, such as their large surface-to-volume ratio and optical properties, may contribute to improved biocompatibility of microalgae. Further research is needed to fully understand this phenomenon's mechanism (Agarwal et al. 2022).

Considering these nanomaterial's compatibility, toxicity, and long-term effects on microalgae and the surrounding environment is important. Ethical and regulatory considerations also play a role in developing these systems.

20.7 Microalgae in Nanotechnology for Environmental Remediation

The improper use of natural resources has led us to create pollution problems that threaten the planet. Various toxic chemical substances and elements are harmful not only to humans but also to all species that inhabit ecosystems. That is why it is necessary to implement measures to improve and clean the nature of these compounds.

Throughout history, various alternatives have been created that can help remediate the natural resources that have been affected. One of these tools is microalgae in nanotechnology.

Microalgae have been gaining attention as a promising environmental tool for its remediation. The properties, characteristics, and capabilities of these materials are the best advantages for removing toxic materials. Some of these are their high surface area, the ability to sequester metals and toxins, their photosynthetic activity, and the small size of their particles. They offer a cost-effective approach in different fields (Dwivedi and Ahmad 2021).

There are multiple applications in nanomaterials for environmental remediation. Microalgae can absorb and accumulate heavy metals from water, soil, and sediments. Nanoparticles can help improve the algae's metal binding capacity and increase the overall efficiency of the remediation process (Danouche et al. 2021). Some studies have reviewed the potential of removing toxic metals like arsenic (As), cadmium (Cd), chromium (Cr), lead (Pb), and mercury (Hg) (Leong and Chang 2020).

These also can break down and metabolize certain toxic pollutants, including organic compounds and hydrocarbons, and waste from agriculture. Introducing nanoparticles with catalytic properties can enhance their ability to effectively degrade these pollutants (Hidangmayum et al. 2022).

Microalgae-based systems integrated with nanomaterials can treat wastewater (Sonali et al. 2023). They help in nutrient removal and organic matter degradation, while nanomaterials enhance the removal of recalcitrant pollutants and facilitate the separation of algal biomass from treated water (Goh et al. 2023; Abdelfattah et al. 2022). Microalgae have been used to remove heavy metals such as copper and zinc from secondary effluent from wastewater treatment plants. They can be used as adsorbents for wastewater treatment, soil remediation, and heavy metal removal (Goswami et al., 2022).

On the other hand, microalgae can serve as biosensors to detect and monitor environmental pollutants, including heavy metals and organic compounds. When combined with nanoparticles, the sensitivity and selectivity can be improved, allowing for real-time and accurate pollutant detection (Ferro et al. 2012).

Furthermore, microalgae's carbon uptake efficiency can be enhanced by incorporating nanomaterials, such as carbon nanotubes or graphene. The captured carbon can then be utilized for various purposes, such as biofuel production or bio-based materials (Argarwal et al. 2022).

Moreover, nanomaterials applied in controlled doses can stimulate the growth and productivity of microalgae. By providing essential nutrients or improving light absorption, nanoparticles can enhance algal biomass yield, making remediation more efficient (Dev Sarkar et al. 2021).

It is of utmost importance to consider that although microalgae-based nanotechnology holds great promise for environmental remediation, there are also challenges and drawbacks that still need to be addressed. Some of these include the potential toxicity issues of nanoparticles and their environmental impact. For this reason, further research is needed to optimize these technologies and ensure their safe and practical application in real situations.

20.8 Conclusions and Future Perspectives of Microalgal Applications in Nanotechnology

Microalgae are single-celled organisms that can do amazing things, and scientists have been studying how to use them in conjunction with nanotechnology. This field involves working with tiny materials and structures. Some of the challenges still facing this technique are mentioned below.

Although microalgae can be grown in small quantities, large-scale production for nanotechnology applications can be challenging, so scientists are tasked with figuring out how to grow enough microalgae efficiently. Once the microalgae are grown, they must be harvested and processed to extract their useful components. This can be complicated and require special techniques.

Keeping the microalgae pure and free of unwanted contaminants is important for nanotechnology applications, as these could affect the properties of the materials produced.

Nanotechnology requires precise control of the materials used. Achieving consistent results with microalgae-based materials could be challenging due to their natural variability; any biological product has this, which complicates their study.

On the other hand, microalgae can be sensitive to changes in their environment, such as temperature and light. To use them in nanotechnology, researchers have to ensure that the properties of the microalgae remain constant. As for the prospects of this technique, the following are mentioned: microalgae could create new nanomaterials for various applications, such as electronics, medicine, etc.

They are excellent for removing pollutants from water and air. They could be used to develop nanomaterials that clean up pollutants present in the environment that other existing techniques cannot remove. They can also be used when integrated into nanosensors that detect specific substances. This could have applications in environmental monitoring, food safety, disease detection, etc.

Microalgae can also produce biofuels and together with nanotechnology, they could help improve the efficiency of converting this into biofuels that provide energy. They can also be used as compounds that have potential medical uses, and mixed with nanotechnology, they could help to develop targeted drug-delivery systems with specific release. These are just some of the applications that microalgae could have and where they can be explored to improve the compounds we already know. The future looks promising, and this area is expected to develop to its full potential.

References

Abdelfattah A, Ali SS, Ramadan H, El-Aswar EI, Eltawab R, Ho SH, Elsamahy T, Li S, El-Sheekh MM, Schagerl M, Kornaros M, Sun J (2022) Microalgae-based wastewater treatment: mechanisms, challenges, recent advances, and future prospects. Environ Sci Ecotechnol 13:100205. https://doi.org/10.1016/j.ese.2022.100205. PMID: 36247722; PMCID: PMC9557874

Agarwal A, Jeevanandham S, Sangam S, Chakraborty A, Mukherjee M (2022) Exploring the role of carbon-based nanomaterials in microalgae for the sustainable production of bioactive compounds and beyond. ACS Omega 7(26):22061–22072. https://doi.org/10.1021/acsomega.2c01009. PMID: 35811909; PMCID: PMC9260754

Agarwal, K., Shah, M. P., & Verma, P. (2022). Bioremediation of heavy metals from wastewater: A current perspective on microalgae-based future. Letters in Applied Microbiology, 75(4), 701–717. https://doi.org/10.1111/lam.13564

Ali SA, Nowruzi B, Afshari G (2023) A review of the application of nanoparticles biosynthesized by microalgae and cyanobacteria in medical and veterinary sciences. Iran J Vet Med 17(1). https://doi.org/10.32598/IJVM.17.1.1005309

AlNadhari S, Al-Enazi NM, Alshehrei F, Ameen F (2021) A review on biogenic synthesis of metal nanoparticles using marine algae and its applications. Environ Res 194:110672. https://doi.org/10.1016/j.envres.2020.110672

Aratboni HA, Rafiei N, Allaf MM, Abedini S, Rasheed RN, Seif A et al (2023) Nanotechnology: an outstanding tool for increasing and better exploitation of microalgae valuable compounds. Algal Res 71:103019. https://doi.org/10.1016/j.algal.2023.103019

Avramović N, Mandić B, Savić-Radojević A, Simić T (2020) Polymeric nanocarriers of drug delivery systems in cancer therapy. Pharmaceutics 12(4):298. https://doi.org/10.3390/pharmaceutics12040298. PMID: 32218326; PMCID: PMC7238125

Barsanti L, Gualtieri P (2007) Algae: anatomy, biochemistry, and biotechnology. CRC Press. J Phycol 43(2). https://doi.org/10.1111/j.1529-8817.2007.00335.x

Çakirsoy I, Miyamoto T, Ohtake N (2022) Physiology of microalgae and their application to sustainable agriculture: a mini-review. Front Plant Sci 13:1005991. https://doi.org/10.3389/fpls.2022.1005991. PMID: 36466259; PMCID: PMC9712798.ID: 30065883; PMCID: PMC6064205

Cameron H, Mata MT, Riquelme C (2018) The effect of heavy metals on the viability of Tetraselmis marina AC16-MESO and an evaluation of the potential use of this microalga in bioremediation. PeerJ 6:e5295. https://doi.org/10.7717/peerj.5295

Chakravorty M, Nanda M, Bisht B, Sharma R, Kumar S, Mishra A et al (2023) Heavy metal tolerance in microalgae: detoxification mechanisms and applications. Aquat Toxicol 260:106555. https://doi.org/10.1016/j.aquatox.2023.106555

Chia WY, Tang DYY, Khoo KS, Lup ANK, Chew KW (2020) Nature's fight against plastic pollution: algae for plastic biodegradation and bioplastics production. Environ Sci Ecotechnol 4:100065. https://doi.org/10.1016/j.ese.2020.100065

Condor BE, de Luna MDG, Chang YH, Chen JH, Leong YK, Chen PT, Chen CY, Lee DJ, Chang JS (2022) Bioethanol production from microalgae biomass at high-solids loadings. Bioresour Technol 363:128002. https://doi.org/10.1016/j.biortech.2022.128002. Epub 2022 Sep 22

Collotta Leonardo, Busi Pascale, Champagne Warren, Mabee Giuseppe, Tomasoni Marco, Alberti (2016) Evaluating microalgae-to-energy -systems: different approaches to life cycle assessment(LCA) studiesAbstract Biofuels Bioproducts and Biorefining 10(6) 883-895 10.1002/bbb.2016.10.issue-6 10.1002/bbb.1713

Dahoumane SA, Mechouet M, Álvarez FJ, Agathos SN, Jeffryes CS (2016) Microalgae: an outstanding tool in nanotechnology. Bionatura 1(4):196–120. https://doi.org/10.21931/RB/2016.01.04.7

Dai N, Wang Q, Xu B, Chen H (2022) Remarkable natural biological resource of algae for medical applications. Front Mar Sci 9:912924. https://doi.org/10.3389/fmars.2022.912924

Dalla Costa V, Filippini R, Zusso M, Caniato R, Piovan A (2022) Monitoring of Spirulina flakes and powders from Italian companies. Molecules 27(10):3155. https://doi.org/10.3390/molecules27103155. PMID: 35630631; PMCID: PMC9143159

Danouche M, El Ghachtouli N, El Arroussi H (2021) Phycoremediation mechanisms of heavy metals using living green microalgae: physicochemical and molecular approaches for enhancing selectivity and removal capacity. Heliyon 7(7):e07609. https://doi.org/10.1016/j.heliyon.2021.e07609. PMID: 34355100; PMCID: PMC8322293

Dev Sarkar R, Singh HB, Chandra KM (2021) Enhanced lipid accumulation in microalgae through nanoparticle-mediated approach, for biodiesel production: a mini-review. Heliyon 7(9):e08057. https://doi.org/10.1016/j.heliyon.2021.e08057. PMID: 34622062; PMCID: PMC8481968

Dolganyuk V, Belova D, Babich O, Prosekov A, Ivanova S, Katserov D, Patyukov N, Sukhikh S (2020) Microalgae: a promising source of valuable bioproducts. Biomol Ther 10(8):1153. https://doi.org/10.3390/biom10081153

Dutta N, Kundu P, Lee JTE, Bhattacharya S (2023) Implementation and optimization of algal biomass in value-added products recovery: a step towards algae-based green economy. Hydrobiology 2(2):326–346. https://doi.org/10.3390/hydrobiology2020021

Dwivedi S, Ahmad IZ (2021) Microalgal nanotechnology for the remediation of environmental pollutants. In: Thangadurai D, Sangeetha J, Prasad R (eds) Bioprospecting algae for nanosized materials. Nanotechnology in the life sciences. Springer, Cham. https://doi.org/10.1007/978-3-030-81557-8_17

Edis Z, Wang J, Waqas MK, Ijaz M, Ijaz M (2021) Nanocarriers-mediated drug delivery systems for anticancer agents: an overview and perspectives. Int J Nanomed 16:1313–1330. https://doi.org/10.2147/IJN.S289443. Erratum in: Int J Nanomedicine Jul 27;16:5099. PMID: 33628022; PMCID: PMC7898224

Ferro Y, Perullini M, Jobbagy M, Bilmes SA, Durrieu C (2012) Development of a biosensor for environmental monitoring based on microalgae immobilized in silica hydrogels. Sensors 12:16879–16891. https://doi.org/10.3390/s121216879

Richard Feynman (1959) "There's Plenty of Room at the Bottom," a talk given by Feynman at the American Physical Society meeting at Caltech. Available at: https://www.zyvex.com/nanotech/feynman.html

Fonseca E, Santini J, Filho WP, Bortoluzzi LD, Vieira WD, Souza DD, Fernandes LD, Sagrillo MR (2020) Use of microalgae and potential nanotechnological application: a review. Disciplinarum Scientia – Ciências Naturais e Tecnológicas. https://doi.org/10.37779/nt.v21i3.3553

Goh PS, Lau WJ, Ismail AF, Samawati Z, Liang YY, Kanakaraju D (2023) Microalgae-enabled wastewater treatment: a sustainable strategy for bioremediation of pesticides. Water 15:70. https://doi.org/10.3390/w15010070

Goswami RK, Agrawal K, Shah MP, Verma P (2022) Bioremediation of heavy metals from wastewater: a current perspective on microalgae-based future. Lett Appl Microbiol 75(4):701–717. https://doi.org/10.1111/lam.13564

Guedes AC, Malcata FX (2012) Nutritional value and uses of microalgae in aquaculture. InTech. https://doi.org/10.5772/30576

Hidangmayum A, Debnath A, Guru A (2022) Mechanistic and recent updates in nano-bioremediation for developing green technology to alleviate agricultural contaminants. Int J Environ Sci Technol. https://doi.org/10.1007/s13762-022-04560-7

Hossain, N., Zaini, J., Mahlia, T. M. I., & Azad, A. K. (2019). Elemental, morphological and thermal analysis of mixed microalgae species from drain water. Renewable Energy, 131, 617-624. https://doi.org/10.1016/j.renene.2018.07.082

Hossain N., Sabzoi, Nizamuddin Gregory, Griffin Periasamy, Selvakannan Nabisab Mujawar, Mubarak Teuku Meurah Indra, Mahlia (2020) Synthesis and characterization of rice husk biochar via hydrothermal carbonization for wastewater treatment and biofuel production Abstract Scientific Reports 10(1) 10.1038/s41598-020-75936-3

Iijima S (1991) Helical microtubules of graphitic carbon. Nature 354(6348):56–58. https://www.nature.com/articles/354056a0

Jacob JM, Ravindran R, Narayanan M, Samuel SM, Pugazhendhi A, Kumar G (2021) Microalgae: a prospective low cost green alternative for nanoparticle synthesis. Curr Opin Environ Sci Health 20:100163. https://doi.org/10.1016/j.coesh.2019.12.005

Karthikeyan, S., Periyasamy, M., & Prathima, A. (2020). Biodiesel from microalgae: Environmental aspects. Materials Today: Proceedings, 33(7), 3664-3667. https://doi.org/10.1016/j.matpr.2020.04.746

Kashem MA, Kimoto K, Iribe Y, Suzuki M (2019) Development of microalgae biosensor chip by incorporating microarray oxygen sensor for pesticides sensing. Biosensors (Basel) 9(4):133. https://doi.org/10.3390/bios9040133. PMID: 31726653; PMCID: PMC6956216

Khan SA, Sharma GK, Malla FA, Kumar A, Gupta N (2019) Microalgae based biofertilizers: a biorefinery approach to phycoremediate wastewater and harvest biodiesel and manure. J Clean Prod 211:1412–1419. https://doi.org/10.1016/j.jclepro.2018.11.281

Khavari F, Saidijam M, Taheri M (2021) Microalgae: therapeutic potentials and applications. Mol Biol Rep 48:4757–4765. https://doi.org/10.1007/s11033-021-06422-w

Khoo KS, Chia WY, Tang DYY, Show PL, Chew KW, Chen WH (2020) Nanomaterials utilization in biomass for biofuel and bioenergy production. Energies 13(4):892. https://doi.org/10.3390/en13040892

Kroto HW, Heath JR, O'Brien SC, Curl RF, Smalley RE (1985) C60: Buckminsterfullerene. Nature 318(6042):162–163. https://www.nature.com/articles/318162a0

Krugman PR, Wells R, Graddy K (2013) Microeconomics, 3rd edn. Worth Publishers, Inc. ISBN: 9781429278508. Worth Publishers, Inc, New York

Kurniawan SB, Ahmad A, Imron MF, Sheikh Abdullah SR, Othman AR, Hasan HA (2022) Potential of microalgae cultivation using nutrient-rich wastewater and harvesting performance by biocoagulants/bioflocculants: mechanism, multi-conversion of biomass into valuable products, and future challenges. J Clean Prod 365:132806. https://doi.org/10.1016/j.jclepro.2022.132806

Lawan Muhammad, Ibrahim Muhammad Shamsuddin, Abdurrahman Danjuma, Rabi'u Sulaiman Musawa, Umar Hafsat Dembo. (2018). Biofuels as the Starring Substitute to Fossil Fuels. Petroleum Science and Engineering, 2(1), 44-49. https://doi.org/10.11648/j.pse.20180201.17

Leong YK, Chang JS (2020) Bioremediation of heavy metals using microalgae: recent advances and mechanisms. Bioresour Technol 303:122886. https://doi.org/10.1016/j.biortech.2020.122886

Li D, Du W, Fu W, Cao X (2020) A quick look back at the microalgal biofuel patents: rise and fall. Front Bioeng Biotechnol 8:1035. https://doi.org/10.3389/fbioe.2020.01035. PMID: 32984291; PMCID: PMC7479055

Li Y, Liu X, Shinde S, Wang J, Zhang P (2021) Impacts of micro- and nanoplastics on photosynthesis activities of photoautotrophs: a mini-review. Front Microbiol 12:773226. https://doi.org/10.3389/fmicb.2021.773226. PMID: 34899657; PMCID: PMC8660080

Mahdia DB, Majeed DZ, Hussein ES, Abbas EZ, Hasan É, Hamood H (2021) Application of nanotechnology in geotechnical engineering: a review. Saudi J Eng Technol 6:143–151. https://doi.org/10.36348/sjet.2021.v06i06.005

Mantovani M, Collina E, Lasagni M, Marazzi F, Mezzanotte V (2023) Production of microalgal-based carbon encapsulated iron nanoparticles (ME-nFe) to remove heavy metals in wastewater. Environ Sci Pollut Res 30(3):6730–6745. https://doi.org/10.1007/s11356-022-22506-x

Marles RJ, Barrett ML, Barnes J, Chavez ML, Gardiner P, Ko R (2011) United States pharmacopeia safety evaluation of Spirulina. Crit Rev Food Sci Nutr 51:593–604. https://doi.org/10.1080/10408391003721719

Martínez-Ruiz M, Martínez-González CA, Kim DH, Santiesteban-Romero B, Reyes-Pardo H, Villaseñor-Zepeda KR, Meléndez-Sánchez ER, Ramírez-Gamboa D, Díaz-Zamorano AL, Sosa-Hernández JE, Coronado Apodaca KG, Gámez-Méndez AM, Iqbal HMN, Parra-Saldivar R (2022) Microalgae bioactive compounds to topical applications products-a review. Molecules 27(11):3512. https://doi.org/10.3390/molecules27113512. PMID: 35684447; PMCID: PMC9182589

Miao X, Wu Q (2006) Biodiesel production from heterotrophic microalgal oil. Bioresour Technol 97(6):841–846. https://doi.org/10.1016/J.BIORTECH.2005.04.008

Moraes LC, Figueiredo RC, Ribeiro-Andrade R, Pontes-Silva AV, Arantes ML, Giani A, Figueredo CC (2021) High diversity of microalgae as a tool for the synthesis of different silver nanoparticles: a species-specific green synthesis. Colloids Interface Sci Commun 42:100420. https://doi.org/10.1016/j.colcom.2021.100420

Navarro-Partida J, Castro-Castaneda CR, Santa Cruz-Pavlovich FJ, Aceves-Franco LA, Guy TO, Santos A (2021) Lipid-based nanocarriers as topical drug delivery systems for intraocular diseases. Pharmaceutics 13(5):678. https://doi.org/10.3390/pharmaceutics13050678. PMID: 34065059; PMCID: PMC8151015

Pattarkine Vikram M., Pattarkine Richard, Gordon Joseph, Seckbach (2012) The Science of Algal Fuels Nanotechnology for Algal Biofuels Springer Netherlands Dordrecht 147-163. 10.1007/978-94-007-5110-1

Paniagua-Michel JDJ, Morales Guerrero E, Olmos Soto J (2015) Microalgal biotechnology: biofuels and bioproducts. In: Springer handbook of marine biotechnology. Springer, pp 1355–1370

Patwari P, Pruckner F, Fabris M (2023) Biosensors in microalgae: a roadmap for new opportunities in synthetic biology and biotechnology. Biotechnol Adv 68:108221. https://doi.org/10.1016/j.biotechadv.2023.108221

Rajathi FAA, Parthiban C, Kumar VG, Anantharaman P (2012) Biosynthesis of antibacterial gold nanoparticles using brown alga, Stoechospermum marginatum (kützing). Spectrochim Acta A Mol Biomol Spectrosc 99:166–173. https://doi.org/10.1016/j.saa.2012.08.081

Ramana KV, Xavier JR, Sharma RK (2017) Recent trends in pharmaceutical biotechnology. Pharm Biotechnol Curr Res 1(1):5

Rawat, I., Ranjith Kumar, R., Mutanda, T., & Bux, F. (2013). Biodiesel from microalgae: A critical evaluation from laboratory to large scale production. Applied Energy, 103, 444-467. https://doi.org/10.1016/j.apenergy.2012.10.004

Rueda E, García Galán MJ, Ortiz A, Uggetti E, Carretero J, García J, Díez-Montero R (2020) Bioremediation of agricultural runoff and biopolymers production from cyanobacteria cultured in demonstrative full-scale photobioreactors. Process Saf Environ Prot 139:241–250. https://doi.org/10.1016/j.psep.2020.03.035

Sahoo CR, Maharana S, Mandhata CP, Bishoyi AK, Paidesetty SK, Padhy RN (2020) Biogenic silver nanoparticle synthesis with cyanobacterium Chroococcus minutus isolated from Baliharachandi Sea-mouth, Odisha, and in vitro antibacterial activity. Saudi J Biol Sci 27(6):1580–1586. https://doi.org/10.1016/j.sjbs.2020.03.020

Santos-Ballardo DU, Font-Segura X, Ferrer AS, Barrena R, Rossi S, Valdez-Ortiz A (2015) Valorisation of biodiesel production wastes: anaerobic digestion of residual Tetraselmis suecica biomass and co-digestion with glycerol. Waste Manag Res 33(3):250–257. https://doi.org/10.1177/0734242X15572182

Santos-Ballardo DU, Rendón-Unceta MD, Rossi S (2016a) Effects of outdoor cultures on the growth and lipid production of Phaeodactylum tricornutum using closed photobioreactors. World J Microbiol Biotechnol 32:128. https://doi.org/10.1007/s11274-016-2089-1

Santos-Ballardo DU, Rossi S, Reyes-Moreno C (2016b) Microalgae potential as a biogas source: current status, restraints and future trends. Rev Environ Sci Biotechnol 15:243–264. https://doi.org/10.1007/s11157-016-9392-z

Sidorowicz A, Fais G, Casula M, Borselli M, Giannaccare G, Locci AM, Lai N, Orrù R, Cao G, Concas A (2023) Nanoparticles from microalgae and their biomedical applications. Mar Drugs 21(6):352. https://doi.org/10.3390/md21060352. PMID: 37367677; PMCID: PMC10301525

Sonali J, Isabella M, Krishnaswamy G, Veena (2023) Use of microalgae for the removal of emerging contaminants from wastewater. In: Shah MP (ed) Microbial degradation and detoxification of pollutants. De Gruyter, Berlin, Boston, pp 189–204. https://doi.org/10.1515/9783110743623-009

Sudha SS, Rajamanickam K, Rengaramanujam J (2013) Microalgae mediated synthesis of silver nanoparticles and their antibacterial activity against pathogenic bacteria

Taniguchi N (1974) On the basic concept of 'nanotechnology'. Proceedings of the International Conference on Production Engineering, Tokyo, 18–23 August 1974, vol 2, pp 18–23. https://iopscience.iop.org/article/10.1088/1468-6996/6/3/310

Thiruchelvi R, Das A, Sikdar E (2021) Bioplastics as better alternative to petro plastic. Mater Today Proc 37:1634–1639. https://doi.org/10.1016/j.matpr.2020.07.176

Tziveleka LA, Pippa N, Ioannou E, Demetzos C, Roussis V (2022) Development of ulvan-containing liposomes as antibacterial drug delivery platforms. J Funct Biomater 13(4):186. https://doi.org/10.3390/jfb13040186. PMID: 36278655; PMCID: PMC9589965

Vargas-Estrada, L., Torres-Arellano, S., Longoria, A., Arias, D. M., Okoye, P. U., & Sebastian, P. J. (2020). Role of nanoparticles on microalgal cultivation: A review. Fuel, 280, 118598. https://doi.org/10.1016/j.fuel.2020.118598

Velusamy P, Kumar GV, Jeyanthi V, Das J, Pachaiappan R (2016) Bio-inspired green nanoparticles: synthesis, mechanism, and antibacterial application. Toxicol Res 32(2):95–102. https://doi.org/10.5487/TR.2016.32.2.095

Wijffels RH, Barbosa MJ (2010) An outlook on microalgal biofuels. Science 329(5993):796–799. https://doi.org/10.1126/science.1189003

Wu J, Gu X, Yang D, Xu S, Wang S, Chen X, Wang Z (2021) Bioactive substances and potentiality of marine microalgae. Food Sci Nutr 9(9):5279–5292. https://doi.org/10.1002/fsn3.2471. PMID: 34532034; PMCID: PMC8441504

Yan N et al (2016) The potential for microalgae as bioreactors to produce pharmaceuticals. Int J Mol Sci 17(6):962. https://doi.org/10.3390/ijms17060962

Yin Z, Zhu L, Li S, Hu T, Chu R, Mo F, Hu D, Liu C, Li B (2020) A comprehensive review on cultivation and harvesting of microalgae for biodiesel production: environmental pollution control and future directions. Bioresour Technol 301:122804. https://doi.org/10.1016/j.biortech.2020.122804

Part VII

Techno-economical Approaches Applied to Microalgal Technologies and Products

Microalgal Biorefineries: Advances and Techno-Economic Challenges

21

Andrea de Jesús Gárate-Osuna, David Ulises Santos-Ballardo, Miguel Betancourt-Lozano, and Karla Denisse Luna-Avelar

Abstract

Microalgal biorefineries represent a promising frontier in sustainable biofuel production and the generation of value-added bioproducts. This chapter explores the biorefinery concept, emphasizing microalgae's unique potential as a versatile feedstock due to its high productivity, adaptability, and ability to accumulate different valuable compounds. It highlights microalgal applications in biofuel production, nutraceuticals, biomedicine, and agrochemicals while addressing techno-economic and environmental challenges. Key barriers include high operational costs, low biomass production efficiency, and the energy-intensive nature of cultivation and processing. The chapter discusses advancements in cultivation techniques, harvesting technologies, and bioproduct extraction methods that aim to optimize resource utilization and economic viability. Furthermore, integrating tools like life cycle assessment (LCA) and techno-economic analysis (TEA) is essential for evaluating and improving sustainability metrics. Finally, innovations in zero-waste biorefining and coupling CO_2 capture with biofuel production underscore the potential of microalgal biorefineries to contribute significantly to global renewable energy and sustainable development goals.

A. de. J. Gárate-Osuna
Programa Regional de Posgrado en Biotecnología, Facultad de Ciencias Químico-Biológicas, Universidad Autónoma de Sinaloa, Culiacan, Sinaloa, Mexico

D. U. Santos-Ballardo · M. Betancourt-Lozano
Laboratorio de Ecotoxicología y Bioprocesos, Centro de Investigación en Alimentación y Desarrollo Unidad Mazatlán, Mazatlán, Sinaloa, Mexico

K. D. Luna-Avelar (✉)
Estancias posdoctorales por México, SECIHTI, Mexico City, Mexico

Laboratorio de Ecotoxicología y Bioprocesos, Centro de Investigación en Alimentación y Desarrollo Unidad Mazatlán, Mazatlan, Sinaloa, Mexico
e-mail: karla.luna@ciad.mx

Keywords

Microalgal biomass · Valuable bioproducts · Biorefinery · Limitations and challenges

21.1 Introduction to Microalgae Biorefinery

21.1.1 Biorefinery Concept

The biorefinery concept can be described as similar to a conventional biorefinery (oil refineries), because the processes are almost the same, only with two important differences: (1) the raw materials that biorefinery uses are from some sustainable source, or raw materials that are used for another principal bioproduct obtaining, and a biorefinery can valorize the residue to obtain other additional products, and (2) the processes involved in biorefineries can be sustainable, because it can use their bioproducts, for example, bioenergy in thermal processes (González-Delgado and Kafarov 2011; Posada and Osseweijer 2016).

These biorefineries are a significant promise to help mitigate greenhouse gases (GHG), the principal protagonists of climate change, typically from fossil power plants (Chew et al. 2017). There are many types or configurations of biorefineries. Still, most designs have one main product to produce, such as biofuels (biogas, biodiesel, bioethanol, and biohydrogen). Then, on the second line, bioproducts can be other value-added molecules (i.e., antioxidants, proteins, amino acids, etc.). Besides, the products made in the biorefinery can be used in themselves; for example, the biogas obtained in the processes can be applied in some internal processes of the biorefinery that need heat to obtain another product (Gárate-Osuna et al. 2022; González-Delgado and Kafarov 2011). Figure 21.1 describes the main idea of a biorefinery related to production and market value.

Some raw materials are already used in biorefineries. For example, there are agricultural materials (crop wastes, pruning residues), livestock (manure from cows, pigs, poultry,

D. U. Santos Ballardo, S. Rossi (eds.), *Microalgae as Promising Source of Commercial Bioproducts*, Developments in Applied Phycology 14, https://doi.org/10.1007/978-3-031-86433-9_21

etc.), and industry (industrial wastes such as fruits and vegetables (Gárate-Osuna et al. 2022; Junghare et al. 2023). These materials are currently used since the cultivation conditions are known, and estimates of bioproduct generation are already known. However, a material that has gained interest in recent years for application in biorefineries is microalga biomass (Chhandama et al. 2023; Figueroa-Torres and Theodoropoulos 2023; Okeke et al. 2022).

Table 21.1 shows some examples of biorefinery proposals and research; the most recent investigations are for using microalgae (green, brown, and red). This is mainly because microalgae have many compounds that can be transformed into high and low value-added products.

Besides, microalgae can have high yields in small areas, compounds like lipids, proteins, and pigments can be obtained from it; and then, it can be transformed into bioproducts such as bioenergy (biofuels), antioxidants, fatty acids (FA), vitamins, nutricosmetics, and fish food, among others. (Branco-Vieira et al. 2017; Francavilla et al. 2015; Guldhe et al. 2017; Mussgnug et al. 2010).

21.1.2 Microalgae As a Source of Biorefineries

Microalgae are photosynthetic procaryote microorganisms present in different types of water (freshwater, seawater, and wastewater). It contains chlorophyll "*a*" and other photosynthetic pigments, and they are capable of oxygenic photosynthesis (Masojídek et al. 2013). Different studies mention that microalgae have a high versatility over the common crops to obtain biofuels and bioproducts, including pigments, lipids, amino acids, proteins, etc. (Choi et al. 2021; Takahashi 2021; Wicker et al. 2021). Microalgae have this versatility for some important reasons: (1) they can accumulate different value-added compounds, which can be transformed through industrial processes into bioproducts, (2) they can fix CO_2, (3) microalgae have high growth rates, which leads to (4) accumulate large amounts of biomass in relatively small spaces, (5) they can be cultivated indoor conditions (e.g., photobioreactors), or in outdoor conditions (e.g., raceways), and (6) microalgae can adapt to changes in their growth conditions, such as nutritional or lighting needs, making it a characteristic that can be used to take advantage of stressful conditions for improving the accumulation of compounds like lipids or pigments (Santos-Ballardo et al. 2016; Zhang et al. 2022).

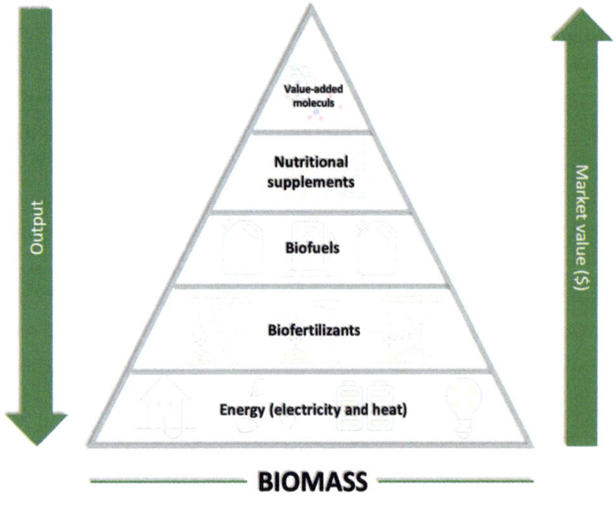

Fig. 21.1 Biorefineries pyramid. Relation of volume of production and market value

In recent years, microalgae and other cyanobacteria such as diatoms and euglenoids could be potential microorganisms to obtain products like biofuels, under the argument that the fossil demands these days are higher than in the past years, and the tendency is still increasing (Wicker et al. 2021). The International Energy Agency (IEA) (2020), highlights that biofuels should not only have a targeted application in common-use transport like cars and buses but should also be applied in "*heavy transportation*" such as airplanes and cargo ships. This is mainly because fossil fuel production is not enough for society's demands, and the distribution is not equal across the different countries of the world.

21.1.2.1 Value-Added Bioproducts from Microalgae Biomass

Microalgae have had an important role as potential biofuel producers since microalgae-based biofuels present numerous benefits in terms of economic growth. In this sense, developing a biofuel industry based on microalgal biomass can provide many socio-economic advantages that contribute to a sustainable outcome for the population. However, to get an economic balance and make microalgae biorefineries a competitive idea, obtaining compounds of higher value in the market (Gárate-Osuna et al. 2022; Mahmood et al. 2023). These value-added bioproducts include functional foods and food supplements, antioxidant compounds, nutricosmetics, and biomedicines, among others.

Furthermore, fatty acids (FA) are necessary in the human diet. However, the amounts that the average population obtains from foods, such as fish, are relatively low, most in the occidental population. Low consumption of fatty acids can trigger chronic degenerative diseases (e.g., antihypertension, neurological deficiencies, inflammatory problems, etc.) (de Carvalho and Caramujo 2018). The consumption of fatty acids such as DHA and EPA (docosahexaenoic and eicosatetraenoic, respectively), present in microalgae, are essential in the diet of children (Van Den Hoek et al. 1995; Vanthoor-Koopmans et al. 2013). Also, pigments are used as functional supplements and antioxidant compounds, and these have been the subject of interest to the food industries; the research is to find colorants that meet current demands and do not have negative repercussions on health. Some pigments used

Table 21.1 Examples of biorefinery studies from different biomass resources

Biomass	Processes	Bioproducts	Goal description	Reference
Lignocellulosic biomass of Napier grass (*Pennisetum purpureum*)	Anaerobic digestion and thermochemical conversion	Biomethane and solid biochar	An innovative decentralized biorefinery of lignocellulosic feedstocks through the integration of anaerobic digestion with thermochemical conversion	Sawatdeenarunat et al. (2018)
Fruit waste from wine and cider industries	Extraction processes, pyrolysis, and gasification.	Bioactive compounds and value-added chemical products (*biochar*).	The biorefinery concept to re-valorize fruit waste contributes to a circular economy, with the application of the 3R approach of "reduce, recycle, and reuse".	Sette et al. (2020)
Wood lignin	Reductive catalytic fractionation of lignocellulose, hydroprocessing, and dealkylation	Phenol, propylene, phenolic oligomers, and bioethanol.	Proposal of an integrated biorefinery for the production of phenol, propylene, useful phenolic oligomers, and bioethanol from an amenable carbohydrate pulp of wood lignin.	Liao et al. (2020)
Apple orchard waste (AOW)	Thermochemical and biological processes.	Biofuels, bioenergy, chemicals, fertilizers.	A theoretical proposal of the potential use of AOW, it is for the bioeconomy of the apple farms, to make it sustainable and add valor to the industrial waste.	Qin et al. (2021)
Spent coffee grounds (SCG)	Microwave-assisted deep eutectic solvent extraction, and enzymatic hydrolysis and fermentation of extracted SCG.	Antioxidant compounds and biobutanol	As the second most consumed beverage in the world, the coffee industry has a lot of unvalued waste, which can be used under the biorefinery concept.	López-Linares et al. (2021)
Microalgal biomass	Acid hydrolysis (AH) and solvent extraction (SX).	Biofuels and value-added products.	Biorefineries are a potential and competitive industry. The study shows a techno-economic vision of the use of microalgae to produce biofuels such as bioethanol, biobutanol, and subproducts such as glycerol.	Figueroa-Torres and Theodoropoulos (2023)

in food, such as fucoxanthin, diatoxanthin, β-carotene, and astaxanthin, are found naturally in vegetables such as carrots; however, some species of microalgae (e.g., *Haematococcus pluvialis*) are capable of accumulating large quantities and could be used in commercial products (Chew et al. 2017; Zhou et al. 2015).

On the other hand, nutricosmetic products have gained considerable prominence in recent years since they are categorized as food supplements. Notably, nutricosmetics offer distinct advantages as they are in charge of beauty care, targeting aspects like hair, nails, and skin. Due to this, different products labeled "anti-aging" have emerged, primarily combating aging induced by oxidative cellular stress. This biological process requires the participation of enzymes such as elastase, hyaluronidase, and collagenase for mitigation. Several microalgae species have emerged as potential candidates for commercialization under this paradigm (Dini and Laneri 2019; Masaki 2010; Norzagaray-Valenzuela et al. 2016).

Moreover, microalgae have been extensively investigated for their potential applications in biomedicine. Currently, they are being utilized as dietary supplements to aid in preventing and managing health issues. Noteworthy examples include their demonstrated anti-cancer, anti-inflammatory, and anti-obesity properties (Hassan et al. 2021; Li et al. 2020). Certain bioactive compounds like chlorophyll and carotenoids have shown promise in disrupting the formation of cancer cells (Ambriz-Perez et al. 2017; Chew et al. 2017; Ferruzzi and Blakeslee 2007). Additionally, various pigments and polyunsaturated fatty acids (PUFAs) derived from microalgae have been found to mitigate inflammatory processes (Chew et al. 2017; Robertson et al. 2015). In recent times, there has been a focus on investigating antihypertensive activity, given the escalating prevalence of hypertension worldwide. One approach involves exploring peptides that can inhibit angiotensin-converting enzyme (ACE), which could aid in managing this condition effectively (Medina-Godoy et al. 2012). Norzagaray-Valenzuela et al. (2016) studied the *Dunaliella tertiolecta* species as a potential candidate for developing supplements with antihypertensive activity.

In addition to the use of microalgal biomass as a valuable biological resource for industries such as the food sector, the pharmaceutical industry, and biofuel production, microalgae are presented as a promising and sustainable alternative to agricultural products (Guo et al. 2020; Cakirsoy et al. 2022). Because of the ability of these microorganisms to cycle nutrients, they can assimilate and convert essential nutrients into forms readily available for plants. Additionally, their physiology influences the effectiveness and characteristics of microalga-based fertilizers; also, due to their capacity for bioactive substance accumulation, they can produce phytohormones, which promote the growth of plants, and they are involved in their physiological processes. In this sense, agrochemicals from microalgal biomass offer various benefits to improve soil fertility, optimize nutrient management, and reduce dependence on synthetic fertilizers (Cakirsoy et al. 2022; Gonçalves et al. 2023).

As a result, microalgae are considered microorganisms that can be used under the biorefinery scheme. Besides, being a green and sustainable alternative, it could also be technically and economically viable by optimizing processes and resources (Fal et al. 2023).

21.1.3 Microalgal Biorefineries

A microalgae biorefinery entails cultivating one or multiple microalgae species to obtain biomass. Subsequently, employing diverse processes, a considerable quantity of different compounds (e.g., pigments, lipids, proteins, etc.) can be extracted from this biomass to be further transformed into valuable products for the market (Gárate-Osuna et al. 2022), an example of microalgae biorefinery is shown in Fig. 21.2.

The main advantages of using microalgae under a biorefinery scheme are their ability to fix CO_2 efficiently in the photosynthetic process and their remarkable ability to adapt to any environmental conditions, including ultraviolet radiation, osmotic pressure, salt water, and even wastewater (Junghare et al. 2023).

However, the biomass productivity levels will depend on evaluating the most promising way to fix and use CO_2. This assessment involves investments in necessary cultivation equipment, adaptations to achieve stress-inducing conditions, or supplying the necessary nutrients for their development and favorable value-added compound accumulation (Chew et al. 2017; Ferreira and Gouveia 2020; Wichaphian et al. 2024).

The use of microalgal biomass in biofuel biorefinery is the most studied topic; it is supported by the existence of techniques such as anaerobic digestion, transesterification, and photobiological conversion for turning microalgal biomass into renewable biofuels, like methane, biodiesel, and hydrogen, respectively. Although biofuel production from microalgae is not a new concept, it is currently being investigated more earnestly due to the demand for fossil fuels and the increasing prices (Javed et al. 2019). For example, biodiesel can be produced owing to its abundant oil content and the microalgae's ability to accumulate lipids depending on some cultivation conditions, such as light, temperature, and salinity, which also influence microalgal growth (Mahmood et al. 2023). However, the production of biofuel through microalgae is obstructed by many commercialization challenges, mainly operational and capital costs, which represent about 50% of the investment when a photobioreactor is installed, including nutrients and chemicals costs as salaries. Using open ponds as an alternative for photobioreactors is anticipated to decrease capital investment costs substantially. Additionally, the employment of oil-rich microalgae species, coupled with cultivation in wastewater and the incorporation of industrial recycled fractions, is expected to enhance the economic viability of biodiesel production (Özçimen et al. 2018).

Despite the challenges of biofuel production, research is still developing strategies for improving the quantity and quality of bioproducts to advance more techno-economically viable production options. Furthermore, Wichaphian et al. (2024) integrate *Chlorella* biomass for zero-waste biorefining; they optimize the lipid extraction process using an ultrasonic-assisted deep eutectic solvent. This results in lipid recovery with biodiesel-quality lipids of 139.52 mg/g of microalgal biomass; moreover, it boasts a high calorific value and favorable ignition characteristics. Meanwhile, de-oiled residual microalgal biomass was evaluated as a soil promoter on lettuce growth, enhancing photosynthetic pigments, and the community structure showed alterations in microbial communities, indicating both positive and negative effects on lettuce growth. Therefore, *Chlorella* microalgal biomass was used simultaneously to generate biodiesel and biofertilizer into a zero-waste biorefining concept.

On the other hand, after varying CO_2 feeding concentration from 0.04% to 40%, it was found that cyanobacterium *Leptolyngbya* sp. KC45 was proposed as a highly efficient photocatalyst, and it was able to produce 0.182 g L^{-1} day^{-1} of microalgal biomass at a CO_2 concentration of 10% under a photoautotrophic culture. With the biorefinery co-products processing, according to these data, *Leptolyngbya* sp. KC45 can convert between 1875 and 1883 kg of waste CO_2 into one ton of cyanobacterial biomass, meaning a potential production of 93–402 kg of antioxidative and anti-cancer fluorescent pigments as well as between 26 and 115 kg of lipid, which is traduced until 112 kg of biodiesel (Pekkoh et al. 2023).

According to the above-mentioned, a biorefinery that involves biofuel and value-added product recovery has higher feasibility in terms of profit. Moreover, the CO_2 capture and biorefinery could significantly contribute to sustainable development goals (Rajesh Banu et al. 2020; Pekkoh et al. 2023).

Table 21.2 shows information about diverse investigations, wherein they proffer varied biorefinery proposals using distinct microalgae species. The authors of those researchers provided regarding the broad spectrum of bioproducts amenable to extraction from different microalgae species strains. Furthermore, the processes under which the microalgae biomass was subjected, and the rationale for choosing microal-

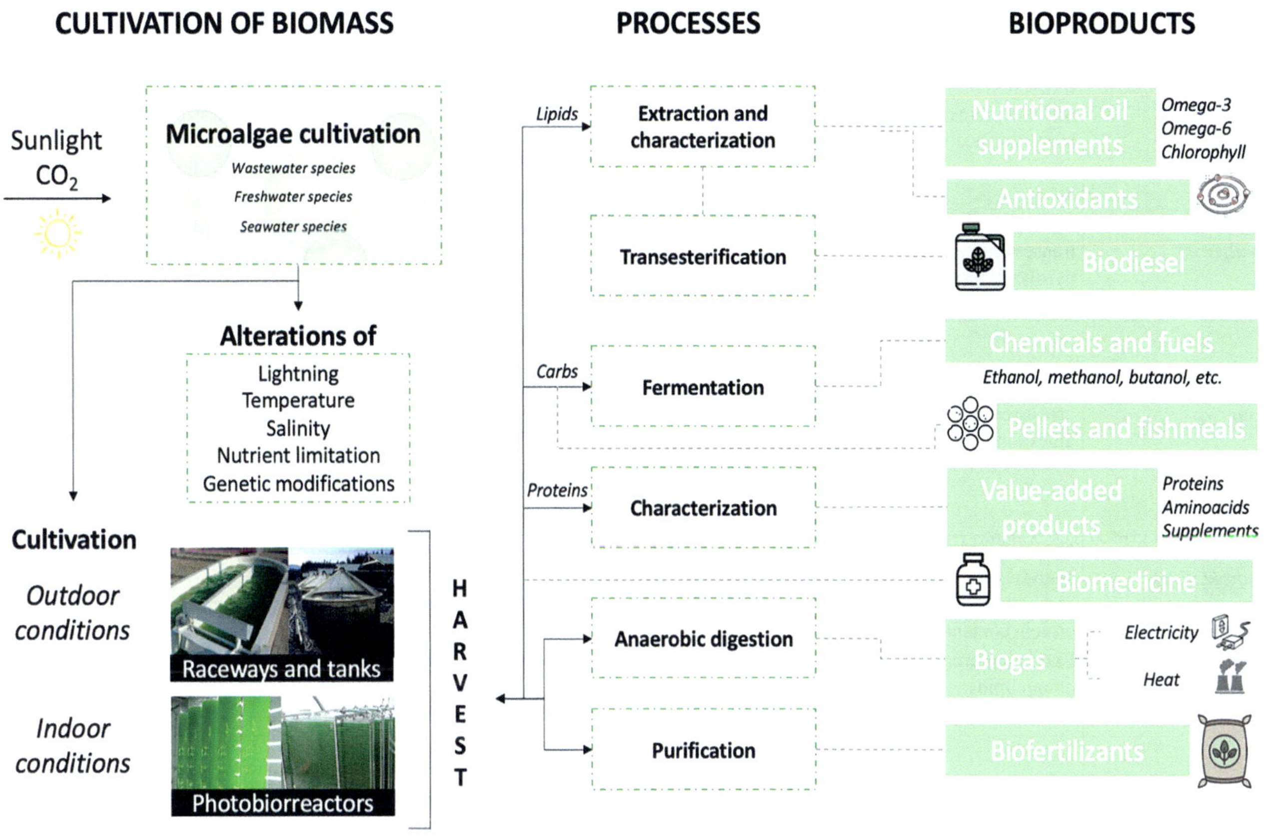

Fig. 21.2 Proposed scheme of microalgal biorefinery

gae are shown. Due to these different studies and the current investigations, it has been demonstrated that under the right conditions and ideal cultivation, microalgal biomass is capable of partially supplying the current demands of products such as biofuels (biogas, biodiesel, bioethanol, biohydrogen, etc.), biochemicals, biofertilizers and helps improve the food supplement and biomedicine industries through green and sustainable processes.

21.1.4 Limitations and Challenges of Microalgal Biorefinery

One of the principal addresses concerning the application and utilization of microalgae in the biorefinery concept revolves around the range of challenges that they confront. Microalgae entail distinct considerations compared to well-established crops, over which we exert more control and possess deeper comprehension. Furthermore, ensuring the growth of microalgae without interruption demands vigilance against external contaminants such as bacteria or potentially threatening microorganisms (Chew et al. 2017; Ferreira and Gouveia 2020).

Additionally, selecting microalgae species is important to engender a favorable technical-economic equilibrium. It is important to highlight that not all biomass obtained from microalgae cultivation is equally suitable for deriving diverse bioproducts. In essence, the blueprint of a microalgae-based biorefinery necessitates meticulous and strategic design considerations to achieve competitiveness options prevailing technologies in current practice (Yen et al. 2013). In this sense, it is important to continue researching this topic, because of the challenges that microalgae biorefinery presents.

Numerous economic, ecological, sociological, and continuing problems exist during the microalgal biorefinery establishment. However, the current bottlenecks of an algal biorefinery are technological issues such as cultivation, harvesting, and compound extraction. However, the acceptance in the market of the traditional economy based on fossil fuel, the availability and composition of the feedstock, meeting the market demand, the efficiency in the product recovery, techno-economic viability, and sustainability are the key difficulties facing biorefineries today (Asghar et al. 2022). In this sense, microalgae could be employed in extracting high-value products and taking advantage of their multiple components value, which maximizes the production output from a single biological material and turns it into an attractive option for the application in biorefinery scheme (Chandra et al. 2019).

Table 21.2 Examples of microalgae biorefineries: different technical and theoretical studies

Microalga	Processes	Bioproducts	Description	Reference
Nannochloropsis sp.	Extraction processes for oils and pigments and transesterification	Biodiesel, biohydrogen, and pigments (carotenoids)	This microalga has a high lipid and pigment content; in addition to possessing characteristics such as fast doubling time and high biomass productivity, it is a candidate for use under the biorefinery concept.	Nobre et al. (2013)
Chlorella vulgaris	Lipid extraction, transesterification, fast pyrolysis, fermentation and anaerobic digestion	Biodiesel, glycerol, bio-oil, bioethanol and biogas	Theoretical proposal of a superstructure for the production of biodiesel from microalgal biomass, the processing of microalgae residue, and recycling of the water and solvents, which includes mass and energy balance constraints and objective function.	Rizwan et al. (2015)
Tetraselmis suecica	Lipid extraction processes, anaerobic digestion, and co-digestion.	Fatty Acids for biodiesel and residual biomass for biomethane.	The investigation shows residual biomass can be used after the lipid extraction as a substrate for anaerobic digestion. Furthermore, glycerol it is a second product obtained from transesterification, and it can be used as co-substrate to improve anaerobic digestion.	Santos-Ballardo et al. (2016)
Chlorella sorokiniana	Nutrient removal and analysis of lipid, carbohydrates, and protein content.	Bioremediation of wastewater and biofuels	By cultivating it in wastewater, heavy metals are removed, and the biomass has ideal characteristics (i.e., high lipid content) for biofuel production.	Guldhe et al. (2017)
Phaeodactylum tricornutum	Extraction processes to obtain lipids.	Biodiesel	The research shows that this diatomaceous is an alternative capable of obtaining large quantities of biomass and lipids with suitable characteristics for transformation to biodiesel, complying with the regulatory standards of the current regulations.	Branco-Vieira et al. (2017)
Scenedesmus obliquus	Dark fermentation, pyrolysis, and subcritical water extraction	BioH_2, bio-oil, biochar, biogas, bioactive compounds and biofertilizers	In the investigation, this microalga was used to treat brewery industry wastewater. This bioremediation helped to obtain biomass rich in different compounds to obtain a wide variety of high-value-added bioproducts.	Ferreira et al. (2019)
Haematococcus pluvialis	Extraction processes, fermentation, and anaerobic digestion.	*Superantioxidants* (astaxanthin) and biofuels (ethanol and biogas)	A "cascade" biorefinery scheme is proposed, where antioxidants are obtained first, and biofuels are obtained from the residual biomass. Under this argument, this microalga is a candidate for obtaining value-added bioproducts.	Hosseini et al. (2020)
Dunaliella tertiolecta	Extraction processes and anaerobic digestion.	Fatty acids for biodiesel and residual biomass for biogas and bioproducts	Under the biorefinery concept, it is argued that lipids suitable for conversion to biodiesel can be obtained from the first line, to later obtain residual biomass that has potential for other bioproducts such as biogas and antioxidants.	Gárate-Osuna et al. (2022)
Nitzschia palea	Lipid extraction processes and characterization of lipids.	Biodiesel	Cultivation under outdoor conditions is promising for this microalga, rendering it a more favorable purpose from a technical-economic standpoint. The microalga exhibits compounds with favorable properties for transformation into biodiesel.	Touliabah et al. (2020)
Chromochloris zofingiensis	Extraction processes, enzymatic hydrolysis, and anaerobic digestion.	Astaxanthin, ethanol, and methane.	Stressful conditions were used in this study to obtain a rich carbohydrate and astaxanthin biomass. After that, the residual biomass was used to obtain ethanol, and finally, a mix of biomasses was used to obtain biogas. This "cascade" biorefinery framework has shown promising results.	Mirzaei et al. (2022)
Chlorella sp. microalga	Alkaline saponification treatments.	Proteins, chlorophyllins, and carotenoids (lutein).	Microalgae biorefineries could be a competitive industry, as bioproducts can be obtained under a "cascade" concept. This study is a biorefinery simulation using a microalga from several value-added compounds.	Karan et al. (2023)

(continued)

Table 21.2 (continued)

Microalga	Processes	Bioproducts	Description	Reference
Leptolyngbya sp. KC45	Photoautotrophic culture, fluorescent pigment extraction, and acid-transesterification	CO_2 capture, extraction of antioxidative and anti-cancer pigments, biodiesel	Using cyanobacteria for CO_2 capturing and biorefinery contributes to sustainable development goals and leads to a sustainable society.	Pekkoh et al. (2023)
Chlorella pyrenoidosa	Ultrasonic-assisted deep eutectic solvent pretreatment, transesterification	Biodiesel, de-oiled microalgal biomass (DMB) as pigments source or biofertilizer	Integration of *Chlorella* biomass for zero-waste biorefining, using a Response Surface Methodology for the lipid extraction and DMB as biofertilizer for the cultivation of baby lettuce.	Wichaphian et al. (2024)

21.1.4.1 Microalgal Cultivation and Biomass Harvesting

The major problem associated with the production of microalgal bioproducts is the low biomass production and the small size of cells; therefore, when they are cultured in a liquid medium, the harvesting process of microalgae is very costly (Udayan et al. 2021). One of the most significant limitations in microalgal mass cultivation is creating a cost-effective production system and one way to solve the difficulties of using microalgae for biorefinery could be increasing their growth rate to compensate for their low cell density and difficulties in harvesting (Hu, 2019).

Mostly, microalgae grow on a salty culture media with critical influencing micro-elements such as nitrogen, phosphorus, sulfur, and magnesium; the main trace minerals in culture media are iron, manganese, cobalt, copper, and molybdenum. Although the nutrients in the cultivation medium are related to the microalgal composition, the light intensity and pH value also influence microalgae's metabolism and cell composition. Furthermore, available minerals, vitamins, and hormones affect algal cells' size, growth rate, and biochemical composition (Aslam et al. 2021; de Carvalho et al. 2019).

For example, nitrogen starvation is the most widely used strategy to improve lipid accumulation. It could lead to reduced cell division, but it promotes the synthesis of more neutral lipids instead of membrane lipids required for cell wall formation. In this way, lipid accumulation is increased, which is frequently required for the commercialization of biofuels (Singh et al. 2016). Also, iron and magnesium ions increase lipid accumulation, reducing the time for the onset of the stationary phase of microalgal growth and thereby increasing lipid accumulation once they promote the activity of the acetyl-CoA enzyme that regulates the first obligating step of microalgal lipid biosynthesis (Begum et al. 2016). On the other hand, other micronutrients can produce changes in pigment metabolic pathways, generally low or even trace amounts are needed. The prominent produced microalgal pigments, in addition to chlorophylls, include carotenes, xanthophylls, and phycobiliproteins; they are produced by the conversion of sunlight into chemical energy through photosynthesis. In this sense, magnesium ions are essential for chlorophyll biosynthesis, and they are also related to the pigment metabolic pathway as the cofactor for critical enzymes, which is involved in the tricarboxylic acid cycle and other microalgal metabolic pathways; also, iron supplementation can enhance the chlorophyll production, and copper and zinc could serve as inducers in astaxanthin production (Lu et al. 2021; Sun et al. 2023). The valuable pigments that microalgae produce and accumulate are attractive for industrial production; moreover, the photosynthetic nature of these microorganisms could also reduce the cost of production because they do not need carbon or nitrogen supplementation and can fixate nitrogen compared to heterotrophic organisms (Pagels et al. 2020). In addition to the above, several studies are still being done to find a diverse range of microalgal cultivation techniques that can offer differing control over growth and product yield. For example, some methods can be used to achieve high biomass production, including substance growth-helping, such as nutrient supplementation or mixotrophic cultivation using cheaper low-carbon sources and industrial and municipal wastewater as a cultivation medium (Udayan et al. 2021).

In this sense, have been shown that the study of cultivation methods for high biomass production is one strategy for decreasing the cost of microalgae utilization due to the cell density and biomass being important factors for microalgae large-scale production and application (Zhan et al. 2017).

For some microalgae that can grow under mixotrophic conditions, biomass accumulation is improved by cultivation in which the carbon sources are assimilated with CO_2, and photosynthetic and respiratory metabolisms are active (Meng et al. 2020). This type of cultivation presents advantages under the other shown in Table 21.3 and is presented as a good strategy to promote a high growth rate and biochemical accumulation simultaneously. However, it is still more challenging to ensure the high productivity of microalgae in large-scale production. It has several practical limitations regarding the culture of microalgae on a larger scale. One of the oldest microalgae cultivation techniques is microalgal culture, produced from naturally occurring sources. In other words, the cultivation method relies entirely on naturally

Table 21.3 Different types of microalgae cultivation

Cultivation type	Description	Energy	Carbon source	Advantages	Disadvantages
Photoautotrophic	The process of photosynthesis makes chemical energy.	Obligatory light.	CO_2.	Appropriate for large-scale algal biomass production. Low cost.	Low growth rates and biomass production. Necessity for specialized bioreactors, specifically closed photobioreactors. Significant dependence on environmental conditions.
Heterotrophic	Carbon materials are used as a source of energy to promote algal growth.	Not obligatory light.	Organic carbon sources.	Appropriate for large-scale microalgal biomass production. High biomass productivity and lipid accumulation. Bioreactor designs with minimal constraints simplified scaling-up processes. Potential to modify biomass composition by altering the culture medium and organic substrates.	A limited range of microalgal species grow heterotrophically. There are elevated costs of organic substrates and energy requirements. The necessity for sterile media. Growth inhibition by excessive concentrations of organic substrates. Inhibition of synthesis metabolites induced by light.
Photoheterotrophic	Microalgae require light, needed to utilize the organic compounds as a source of carbon.	Obligatory light.	Organic carbon.	High biomass productivity.	High costs associated with equipment and substrates. Low carbohydrate and lipid content. Concurrent requirement for both light and organic carbon. Risk of contamination.
Mixotrophic	Microalgae exploit equally the inorganic and organic sources of carbon.	Not obligatory light.	Inorganic (CO_2) and organic carbon sources.	Increased growth rate, biomass, carbohydrate, and lipid accumulation. Prolonged exponential growth phase. Decreased biomass loss due to respiration during dark periods. Mitigation of photoinhibitory effects. Flexibility to switch between photoautotrophic and heterotrophic regimens as needed.	Limited microalgal species are capable of mixotrophic growth. Higher energy costs. Requirement for sterile media susceptible to contamination.

occurring water bodies to supply the conditions and nutrients necessary for microalgae growth. The cultivation method completely depends on naturally occurring water bodies to provide the right conditions and nutrients for the growth of microalgae (Tan et al. 2020).

Concerning biomass harvesting, all techniques entail the elimination of liquid media from biomass for subsequent biocomponent extraction. It is considered one of the main parts of microalgae processing, responsible for almost 30% of the total production cost, while for outdoor microalgal biomass production, it is calculated that 90% of device costs originate from harvesting (Mathimani and Mallick 2018; Show et al. 2020). Currently, a range of established harvesting techniques are available, with gravity sedimentation being the most conventional and cost-effective method. However, this technique is unsuitable for small-sized microalgal cells and is characterized by a slow sedimentation rate, necessitating a substantial operational footprint (Cheirsilp et al. 2020; Mathimani and Mallick 2018).

Other options include physical, chemical, and biological methods for harvesting microalgal biomass, such as centrifugation, filtration, flotation, chemical flocculation, the use of organic compounds, the application of certain microorganisms (including fungi and bacteria), self-flocculation, electroflocculation, and the utilization of biofilm reactors. In biofilm reactors, effluents are recirculated over a solid support material, allowing cells to attach to the surface over time to concentrate the biomass (Cheirsilp et al. 2020; Deconinck et al. 2018; Ferreira et al. 2019). One of the emerging methods for biomass recovery involves using magnetic particles to flocculate, which can be tagged or adsorbed by microalgal cells. Subsequently, an external magnetic force is employed to facilitate separation. Despite the rapidity of this separation process, challenges remain, including post-harvest separation, the recycling of nanoparticles, practical applicability, and the production of uncontaminated microalgal biomass, which are significant bottlenecks (Mathimani and Mallick 2018).

Technologies based on flocculation with organic, magnetic compounds and electrofocculation, are more successful. However, no single harvesting process has demonstrated superior feasibility for large-scale commercial applications. Consequently, future research and development should emphasize the sequential integration of cost-effective concentration and separation techniques (Deconinck et al. 2018; Fuad et al. 2018).

Singh and Patidar (2018) highlight the flocculation, centrifugation, and filtration on the top of harvesting techniques for applications such as biofuel, high-value products extraction, and human and animal food, according to these six important criteria: (1) biomass quantity, (2) biomass quality, (3) cost, (4) processing time, (5) species-specific, and (6) toxicity. These techniques can be employed individually or in combination to enhance harvesting efficiency. However, selecting a specific combination of harvesting techniques will depend on the microalgal species, the desired final biocompound concentration, and quality.

21.1.4.2 Biocomponent Extraction from Microalgal Biomass

The harvesting and extracting of valuable components are expensive in terms of capital and maintenance costs. Separating various components without significant loss of other constituents is critical in a microalgae biorefinery. This challenge can be addressed by employing scalable, cost-effective, and energy-efficient separation techniques alongside the development of efficient large-scale cultivation systems and by utilizing all components of microalgae, not only the lipids employed for biodiesel production but also the lipid-extracted residue (microalgal residual biomass). To enhance the economic feasibility of algae biofuel production, it is possible to produce value-added products, including other biofuels, valuable chemicals, and other useful industrial intermediates (Rizwan et al. 2015; Udayan et al. 2021). A critical factor in developing a microalgal biorefinery is the selection of appropriate algal strains that exhibit rapid growth to facilitate high biomass density accumulation. Additionally, it is essential to consider microalgal cells' mechanical and physicochemical properties when selecting a suitable process-oriented algal strain.

The initial challenge is the cell disruption for the recovering value-added compounds since microalgae present a distinctive limitation due to their robust and complex cell walls, which exhibit varying compositions across different species; therefore, it is crucial for maximizing product recovery and keeping prices low (Diankristanti et al. 2024). In this sense, a variety of disruption methods are available, which can be broadly categorized into (i) physical methods, such as bead milling, high-pressure homogenization, microwave treatment, and ultrasonication; (ii) chemical methods, such as alkaline treatment, surfactant disruption, and osmotic shock; and (iii) enzymatic methods (Lee et al. 2020). Subsequently, for biofuel production, dry extraction mainly focuses on pyrolysis and combustion techniques. However, the energy recovery and environmental feasibility of these methods are inferior to those of wet extraction techniques, which emerge as the most promising option for producing microalgal biofuels in a biorefinery (Khoo et al. 2019).

On the other hand, Wu and Chang (2019) presented various scenarios in a microalgae biorefinery scheme for biofuel production; they pointed out the possibility of generating bio-oil and syngas by pyrolysis and/or hydrothermal liquefaction process (thermochemical conversion); biodiesel by transesterification process (chemical conversion); bioalcohols and biogas by fermentation and anaerobic digestion processes (biochemical conversion). Moreover, these sequential processing steps for producing different biofuels could generate other byproducts, such as glycerol, biochar, and aqueous products (digestate), which can be used as fertilizer.

Other researchers have noticed the extraction of pigments, which are utilized in various nutraceutical, pharmaceutical, and cosmeceutical industries (Yadav et al. 2022). For example, astaxanthin, a kind of carotenoid, is recovered through conventional solvent extraction techniques, using either alcohols or concentrated acids and alkalis (Ambati et al. 2014; Panis and Carreon 2016); recently, magnetic-assisted extraction, supercritical carbon dioxide and ionic liquid-based extraction have developed as emerging technologies for the industrial biorefinery of microalgae, they are presented as an alternative greener solvent but they require more intensive studies to optimize the astaxanthin recovering (Cheng et al. 2018; Liu et al. 2018; Zhao et al. 2016). According to this, optimization of the chemistry-based alkaline saponification treatment was performed, and the parameters: 0.4% NaOH, 45 °C, 0.5 h, allowed to extract proteins, chlorophyllins, and lutein from microalgae *Chlorella* sp. Moreover, using ethanol solvent, it is possible to extract astaxanthin from *Chromochloris zofingiensis*, a green microalga. Also, waste biomass from astaxanthin production is a promise for generating bioethanol due to its high carbohydrate content (54.7%) and biomethane due to its increasing digestibility (Karan et al. 2023; Mirzaei et al. 2022). Fundamentally, system integration technology seeks to optimize the utilization of energy and materials within these sequential processes to enhance the economic viability and minimize environmental impacts.

21.1.5 Tools for Analyzing the Feasibility of Microalgal Biorefinery

The microalgae biorefinery field presents significant challenges across both environmental and economic dimensions; to address this concern, integrating suitable technologies within the biorefinery framework is essential for the sustainable production of microalgal biofuels at competitive prices. In this sense, life cycle assessment (LCA) is a certified scientific methodology for evaluating a product's environmental impacts, human health effects, and resource utilization. This approach has proven instrumental in identifying bottlenecks that hinder the successful commercialization of microalgal technologies (Gifuni et al. 2019). Life cycle accounting encompasses emissions from various stages, including algae pond cultivation, harvesting, downstream processes, fuel transportation and distribution, and the end use of biofuels. Techno-economic analysis (TEA) is regarded as a fundamental tool for assessing the feasibility of new processes, with its system boundaries incorporating the costs associated with downstream processes, fuel upgrading, utilities, and transportation (Beal et al. 2015; Ou et al. 2015; Richardson et al. 2014). The integration of comparative LCA and TEA methodologies facilitates a comprehensive evaluation of sustainability and economic performance. Consequently, the coupling of TEA with LCA has emerged as an accepted approach for the sustainable assessment of microalgal biofuel production (Cruce et al. 2021; Ubando et al. 2022).

The use of these tools has demonstrated that microalgae cultivation with separated nutrients: wastewater influents, municipal wastewater, and human urine, is more effective compared to the culture with non-separated nutrient resources to reduce the environmental impacts and increase the commercial potential of biomethane, bio-oil, biogas, biochar and composting fertilizers (Li et al. 2022). Also, the bio-crude from microalgal hydrothermal liquefaction has been the subject of study of these tools; despite operating temperatures below 350 °C can modestly ease energy duties and attempt to improve performance metrics, the economic viability of this process is ultimately limited by fuel yield and biomass cost. However, the GHG emissions indicated a significant reduction in this kind of emissions compared to petroleum-based fuel production. It is important to particularize the application of hydrothermal liquefaction, a technology that has shown success in the commercialization of microalgal biorefinery (Masoumi and Dalai 2021; Ubando et al. 2022). In future work, it is proposed that continuous-scale hydrothermal liquefaction research be conducted to enhance process models with experimental data, thereby reducing uncertainty in critical performance areas such as product yields, nutrient recyclability, and seasonal variations in microalgal culture. Additionally, there exists an opportunity to advance modeling efforts by integrating hydrothermal liquefaction with other conversion technologies to demonstrate better the viability of fuels and various bioproducts (Chen and Quinn 2021; Sills et al. 2020).

21.2 Concluding Remarks

In conclusion, microalgal biomass holds significant potential for biofuel biorefineries; however, several stages in bioproduct generation remain bottlenecks in this process. To ensure the microalgal industry's economic feasibility, energy cost control, and environmental sustainability, it is crucial to optimize the downstream routes. Reducing energy costs associated with harvesting, extraction, and conversion presents a viable strategy to enhance the sector's economic viability. Additionally, efficiently utilizing byproducts and upgrading high-value biochemicals from microalgal cells can substantially increase revenue streams. Potential improvement avenues include producing high-value co-products, enhancing biomass productivity, and developing automated technologies to minimize labor costs. Furthermore, designing cost-effective production systems, integrating algal cultivation with low-cost renewable energy, and leveraging atmospheric or waste-derived carbon, nitrogen, and phosphorus can significantly improve microalgal biofuel production's overall efficiency and sustainability. In this sense, it helps technologies, such as the coupling of TEA with LCA, to understand the evaluation of sustainability and economic performance.

References

Ambati RR, Phang SM, Ravi S, Aswathanarayana RG (2014) Astaxanthin: sources, extraction, stability, biological activities and its commercial applications–a review. Mar Drugs 12(1):128–152

Ambriz-Perez DL, Ramos-Payan R, Angulo-Escalante MA, Leon-Felix J, Delgado-Vargas F, Heredia JB (2017) Antioxidant capacity and protective role of methanolic extracts of jatropha platyphylla against oxidative stress on macrophage cells. Int J Pharma Bio Sci 8(1):245–253

Asghar A, Sairash S, Hussain N, Baqar Z, Sumrin A, Bilal M (2022) Current challenges of biomass refinery and prospects of emerging technologies for sustainable bioproducts and bioeconomy. Biofuels Bioprod Biorefin 16(6):1478–1494

Aslam A, Rasul S, Bahadar A, Hossain N, Saleem M, Hussain S, Rasool L, Manzoor H (2021) Effect of micronutrient and hormone on microalgae growth assessment for biofuel feedstock. Sustain For 13(9):5035

Beal CM, Gerber LN, Sills DL, Huntley ME, Machesky SC, Walsh MJ, Tester JW, Archibald I, Granados J, Greene CH (2015) Algal biofuel production for fuels and feed in a 100-ha facility: a comprehensive techno-economic analysis and life cycle assessment. Algal Res 10:266–279

Begum H, Yusoff FM, Banerjee S, Khatoon H, Shariff M (2016) Availability and utilization of pigments from microalgae. Crit Rev Food Sci Nutr 56(13):2209–2222

Branco-Vieira M, San Martin S, Agurto C, Santos M, Freitas M, Mata T, Martins A, Caetano N (2017) Potential of phaeodactylum tricornutum for biodiesel production under natural conditions in Chile. Energies 11(1):54

Cakirsoy I, Miyamoto T, Ohtake N (2022) Physiology of microalgae and their application to sustainable agriculture: a mini-review. Front Plant Sci 13:1005991

Chandra R, Iqbal HMN, Vishal G, Lee HS, Nagra S (2019) Algal biorefinery: a sustainable approach to valorize algal-based biomass towards multiple product recovery. Bioresour Technol 278:346–359

Cheirsilp B, Srinuanpan S, Mandik YI (2020) Efficient harvesting of microalgal biomass and direct conversion of microalgal lipids into biodiesel. In: Microalgae cultivation for biofuels production. Academic Press, pp 83–96

Chen PH, Quinn JC (2021) Microalgae to biofuels through hydrothermal liquefaction: open-source techno economic analysis and life cycle assessment. Appl Energy 289:116613

Cheng X, Qi Z, Burdyny T, Kong T, Sinton D (2018) Low pressure supercritical CO(2) extraction of astaxanthin from Haematococcus pluvialis demonstrated on a microfluidic chip. Bioresour Technol 250:481–485

Chew KW, Yap JY, Show PL, Suan NH, Juan JC, Ling TC, Lee DJ, Chang JS (2017) Microalgae biorefinery: high value products perspectives. Bioresour Technol 229:53–62

Chhandama MVL, Rai PK, Lalawmpuii. (2023) Coupling bioremediation and biorefinery prospects of microalgae for circular economy. Bioresour Technol Rep 22:101479

Choi HI, Hwang S-W, Kumar Patel A, Sim SJ (2021) Microalgal biorefinery: a sustainable technology toward circular bioeconomy and microalgal biomass valorization. In: Pandey RDTA, Varjani S (eds) Biomass, biofuels, biochemicals. Elsevier, pp 323–350

Cruce JR, Beattie A, Chen P, Quiroz D, Somers M, Compton S, DeRose K, Beckstrom B, Quinn JC (2021) Driving toward sustainable algal fuels: a harmonization of techno-economic and life cycle assessments. Algal Res 54:102169

de Carvalho CCCR, Caramujo MJ (2018) The various roles of fatty acids. Molecules 23(10):2583

de Carvalho JC, Sydney EB, Assú Tessari LF, Soccol CR (2019) Culture media for mass production of microalgae. In: Biofuels from algae. Elsevier, pp 33–50

Deconinck N, Muylaert K, Ivens W, Vandamme D (2018) Innovative harvesting processes for microalgae biomass production: a perspective from patent literature. Algal Res 31:469–477

Diankristanti PA, Ho NHE, Chen J-H, Nagarajan D, Chen C-Y, Hsieh Y-M, Ng IS, Chang J-S (2024) Unlocking the potential of microalgae as sustainable bioresources from up to downstream processing: a critical review. Chem Eng J 488:151124

Dini I, Laneri S (2019) Nutricosmetics: a brief overview. Phytother Res 33(12):3054–3063

Fal S, Smouni A, Arroussi HE (2023) Integrated microalgae-based biorefinery for wastewater treatment, industrial CO_2 sequestration and microalgal biomass valorization: a circular bioeconomy approach. Environ Adv 12:100365

Ferreira A, Gouveia L (2020) Microalgal biorefineries. In: Handbook of microalgae-based processes and products, pp 771–798

Ferreira A, Ribeiro B, Ferreira AF, Tavares MLA, Vladic J, Vidović S, Cvetkovic D, Melkonyan L, Avetisova G, Goginyan V, Gouveia L (2019) Scenedesmus obliquusmicroalga-based biorefinery – from brewery effluent to bioactive compounds, biofuels and biofertilizers – aiming at a circular bioeconomy. Biofuels Bioprod Biorefin 13(5):1169–1186

Ferruzzi MG, Blakeslee J (2007) Digestion, absorption, and cancer preventative activity of dietary chlorophyll derivatives. Nutr Res 27(1):1–12

Figueroa-Torres GM, Theodoropoulos C (2023) Techno-economic analysis of a microalgae-based biorefinery network for biofuels and value-added products. Bioresour Technol Rep 23:101524

Francavilla M, Kamaterou P, Intini S, Monteleone M, Zabaniotou A (2015) Cascading microalgae biorefinery: fast pyrolysis of Dunaliella tertiolecta lipid extracted-residue. Algal Res 11:184–193

Fuad N, Omar R, Kamarudin S, Harun R, Idris A, Wan Azlina WAKG (2018) Mass harvesting of marine microalgae using different techniques. Food Bioprod Process 112:169–184

Gárate-Osuna ADJ, Libna A-PD, Valdez-Ortiz A, Santos-Ballardo DU (2022) Microalgae as a promising alternative for development of biorefineries: main technological and economical challenges. Arquivos de Ciências do Mar 55(Especial):369–412

Gifuni I, Pollio A, Safi C, Marzocchella A, Olivieri G (2019) Current bottlenecks and challenges of the microalgal biorefinery. Trends Biotechnol 37(3):242–252

Gonçalves J, Freitas J, Fernandes I, Silva P (2023) Microalgae as biofertilizers: a sustainable way to improve soil fertility and plant growth. Sustain For 15(16):12413

González-Delgado ÁD, Kafarov V (2011) Microalgae based biorefinery: issues to consider. CTF Cienc Tecnol Futuro 4(4):05–21

Guldhe A, Ansari FA, Singh P, Bux F (2017) Heterotrophic cultivation of microalgae using aquaculture wastewater: a biorefinery concept for biomass production and nutrient remediation. Ecol Eng 99:47–53

Guo S, Wang P, Wang X, Zou M, Liu C, Hao J (2020) Microalgae as biofertilizer in modern. In: Asraful Alam J-LXM (ed) Microalgae biotechnology for food, health and high value products. Springer Singapore, Singapore, pp 397–413

Hassan ME, El-Sayed AEB, Abdel-Wahhab MA (2021) Screening of the bioactive compounds in Amphora coffeaeformis extract and evaluating its protective effects against deltamethrin toxicity in rats. Environ Sci Pollut Res Int 28(12):15185–15195

Hosseini A, Jazini M, Mahdieh M, Karimi K (2020) Efficient supe rantioxidant and biofuel production from microalga Haematococcus pluvialis via a biorefinery approach. Bioresour Technol 306:123100

Hu IC (2019) Production of potential coproducts from microalgae. In: Biofuels from Algae. Elsevier, pp 345–358

IEA (2020) Global energy review. The impacts of the Covid-19 crisis on global energy demand and CO_2 emissions Paris

Javed F, Aslam M, Rashid N, Shamair Z, Khan AL, Yasin M, Fazal T, Hafeez A, Rehman F, Rehman MSU, Khan Z, Iqbal J, Bazmi AA (2019) Microalgae-based biofuels, resource recovery and wastewater treatment: a pathway towards sustainable biorefinery. Fuel 255:115826

Junghare M, Saxena S, Ingle AP, Moharil MP (2023) Biorefineries: current scenario, feedstocks, challenges, and future perspectives. In: Ingle AP (ed) Nanotechnology for biorefinery. Elsevier, pp 1–25

Karan H, Roles J, Hankamer B, Ross IL (2023) Targeting greens and yellows: a solar biorefinery analysis for the microalgae-based co-production of pigments, proteins, and fuel. Algal Res 74:103187

Khoo CG, Dasan YK, Lam MK, Lee KT (2019) Algae biorefinery: review on a broad spectrum of downstream processes and products. Bioresour Technol 292:121964

Lee SY, Khoiroh I, Vo D-VN, Senthil Kumar P, Show PL (2020) Techniques of lipid extraction from microalgae for biofuel production: a review. Environ Chem Lett 19(1):231–251

Li Y, Liu L, Sun P, Zhang Y, Wu T, Sun H, Cheng KW, Chen F (2020) Fucoxanthinol from the Diatom Nitzschia Laevis ameliorates neuroinflammatory responses in lipopolysaccharide-stimulated BV-2 microglia. Mar Drugs 18(2):116

Li P, Luo Y, Yuan X (2022) Life cycle and techno-economic assessment of source-separated wastewater-integrated microalgae biofuel production plant: a nutrient organization approach. Bioresour Technol 344(Pt B):126230

Liao Y, Koelewijn SF, Van den Bossche G, Van Aelst J, Van den Bosch S, Renders T, Navare K, Nicolaï T, Van Aelst K, Maesen M, Matsushima H, Thevelein J, Van Acker K, Lagrain B, Verboekend D, Sels BF (2020) A sustainable wood biorefinery for low–carbon footprint chemicals production. Sci 367 (6484):1385-1390

Liu ZW, Yue Z, Zeng XA, Cheng JH, Aadil RM (2018) Ionic liquid as an effective solvent for cell wall deconstructing through astaxanthin extraction from Haematococcus pluvialis. Int J Food Sci Technol 54(2):583–590

López-Linares JC, García-Cubero MT, Coca M, Lucas S (2021) A biorefinery approach for the valorization of spent coffee grounds to produce antioxidant compounds and biobutanol. Biomass Bioener 147:106026

Lu Q, Li H, Zou Y, Liu H, Yang L (2021) Astaxanthin as a microalgal metabolite for aquaculture: a review on the synthetic mechanisms, production techniques, and practical application. Algal Res 54:102178

Mahmood T, Hussain N, Shahbaz A, Mulla SI, Iqbal HMN, Bilal M (2023) Sustainable production of biofuels from the algae-derived biomass. Bioprocess Biosyst Eng 46(8):1077–1097

Masaki H (2010) Role of antioxidants in the skin: anti-aging effects. J Dermatol Sci 58(2):85–90

Masojídek J, Torzillo G, Koblízek M (2013) Photosynthesis in microalgae. In: Hu ARAQ (ed) Handbook of microalgal culture: applied phycology and biotechnology. Blackwell Publishing Ltd

Masoumi S, Dalai AK (2021) Techno-economic and life cycle analysis of biofuel production via hydrothermal liquefaction of microalgae in a methanol-water system and catalytic hydrotreatment using hydrochar as a catalyst support. Biomass Bioenergy 151:106168

Mathimani T, Mallick N (2018) A comprehensive review on harvesting of microalgae for biodiesel – key challenges and future directions. Renew Sust Energ Rev 91:1103–1120

Medina-Godoy S, Ambriz-Perez DL, Fuentes-Gutierrez CI, German-Baez LJ, Gutierrez-Dorado R, Reyes-Moreno C, Valdez-Ortiz A (2012) Angiotensin-converting enzyme inhibitory and antioxidative activities and functional characterization of protein hydrolysates of hard-to-cook chickpeas. J Sci Food Agric 92(9):1974–1981

Meng TK, Kassim MA, Cheirsilp B (2020) Mixotrophic cultivation: biomass and biochemical biosynthesis for biofuel production. In: Microalgae cultivation for biofuels production. Academic Press, pp 51–67

Mirzaei D, Jazini M, Rahimi M, Mahdieh M, Karimi K (2022) Production of astaxanthin, ethanol and methane from Chromochloris zofingiensis microalga in an integrated biorefinery. Algal Res 68:102905

Mussgnug JH, Klassen V, Schluter A, Kruse O (2010) Microalgae as substrates for fermentative biogas production in a combined biorefinery concept. J Biotechnol 150(1):51–56

Nobre BP, Villalobos F, Barragan BE, Oliveira AC, Batista AP, Marques PA, Mendes RL, Sovova H, Palavra AF, Gouveia L (2013) A biorefinery from Nannochloropsis sp. microalga–extraction of oils and pigments. Production of biohydrogen from the leftover biomass. Bioresour Technol 135:128–136

Norzagaray-Valenzuela CD, Valdez-Ortiz A, Shelton LM, Jiménez-Edeza M, Rivera-López J, Valdez-Flores MA, Germán-Báez LJ (2016) Residual biomasses and protein hydrolysates of three green microalgae species exhibit antioxidant and anti-aging activity. J Appl Phycol 29(1):189–198

Okeke ES, Ejeromedoghene O, Okoye CO, Ezeorba TPC, Nyaruaba R, Ikechukwu CK, Oladipo A, Orege JI (2022) Microalgae biorefinery: an integrated route for the sustainable production of high-value-added products. Energy Convers Manag X 16:100323

Ou L, Thilakaratne R, Brown RC, Wright MM (2015) Techno-economic analysis of transportation fuels from defatted microalgae via hydrothermal liquefaction and hydroprocessing. Biomass Bioenergy 72:45–54

Özçimen D, İnan B, Koçer AT, Vehapi M (2018) Bioeconomic assessment of microalgal production. In: Microalgal biotechnology. IntechOpen

Pagels F, Salvaterra D, Amaro HM, Guedes AC (2020) Pigments from microalgae. In: Handbook of microalgae-based processes and products. Springer, pp 465–492

Panis G, Carreon JR (2016) Commercial astaxanthin production derived by green alga Haematococcus pluvialis: a microalgae process model and a techno-economic assessment all through production line. Algal Res 18:175–190

Pekkoh J, Duangjan K, Phinyo K, Kaewkod T, Ruangrit K, Thurakit T, Pumas C, Pathom-aree W, Cheirsilp B, Gu W, Wang G, Chaichana C, Srinuanpan S (2023) Turning waste CO2 into value-added biorefinery co-products using cyanobacterium Leptolyngbya sp. KC45 as a highly efficient living photocatalyst. Chem Eng J 460:141765

Posada JA, Osseweijer P (2016) Socioeconomic and environmental considerations for sustainable supply and fractionation of Lignocellulosic biomass in a biorefinery context. In: Biomass fractionation technologies for a Lignocellulosic feedstock based biorefinery. Elsevier Inc, pp 611–631

Rajesh Banu J, Preethi S, Kavitha MG, Kumar G (2020) Microalgae based biorefinery promoting circular bioeconomy-techno economic and life-cycle analysis. Bioresour Technol 302:122822

Richardson JW, Johnson MD, Lacey R, Oyler J, Capareda S (2014) Harvesting and extraction technology contributions to algae biofuels economic viability. Algal Res 5:70–78

Rizwan M, Lee JH, Gani R (2015) Optimal design of microalgae-based biorefinery: economics, opportunities and challenges. Appl Energy 150:69–79

Robertson RC, Guiheneuf F, Bahar B, Schmid M, Stengel DB, Fitzgerald GF, Ross RP, Stanton C (2015) The anti-inflammatory effect of algae-derived lipid extracts on Lipopolysaccharide (LPS)-stimulated human THP-1 macrophages. Mar Drugs 13(8):5402–5424

Santos-Ballardo DU, Rossi S, Valdez-Ortiz A (2016) Microalgas como fuentes de energía. In: Española A (ed) Energía verde a partir de microalgas. Omniscriptum GmbH, Saarbrücken, pp 40–64

Sawatdeenarunat C, Nam H, Adhikari S, Sung S, Khanal SK (2018) Decentralized biorefinery for lignocellulosic biomass: Integrating anaerobic digestion with thermochemical conversion. Bioresour Technol 250:140–147

Sette P, Fernandez A, Soria J, Rodriguez R, Salvatori D, Mazza G (2020) Integral valorization of fruit waste from wine and cider industries. J Clean Prod 242:118486

Sills DL, Van Doren LG, Beal C, Raynor E (2020) The effect of functional unit and co-product handling methods on life cycle assessment of an algal biorefinery. Algal Res 46:101770

Singh G, Patidar SK (2018) Microalgae harvesting techniques: a review. J Environ Manag 217:499–508

Singh P, Kumari S, Guldhe A, Misra R, Rawat I, Bux F (2016) Trends and novel strategies for enhancing lipid accumulation and quality in microalgae. Renew Sust Energ Rev 55:1–16

Sun H, Wang Y, He Y, Liu B, Mou H, Chen F, Yang S (2023) Microalgae-derived pigments for the food industry. Mar Drugs 21(2):82

Takahashi T (2021) Usefulness of a microalgal biorefinery in conversion to chemical products and recent technology in automatic evaluation of microalgae. Curr Opin Green Sustain Chem 27:100410

Tan JS, Lee SY, Chew KW, Lam MK, Lim JW, Ho SH, Show PL (2020) A review on microalgae cultivation and harvesting, and their biomass extraction processing using ionic liquids. Bioengineered 11(1):116–129

Touliabah HE, Abdel-Hamid MI, Almutairi AW (2020) Long-term monitoring of the biomass and production of lipids by Nitzschia palca for biodiesel production. Saudi J Biol Sci 27(8):2038–2046

Qin S, Shekher Giri B, Kumar Patel A, Sar T, Liu H, Chen H, Taherzadeh MJ (2021) Resource recovery and biorefinery potential of apple orchard waste in the circular bioeconomy. Bioresour Technol 321:124496

Ubando AT, Anderson SNE, Chen WH, Culaba AB, Kwon EE (2022) Life cycle assessment of microalgal biorefinery: a state-of-the-art review. Bioresour Technol 360:127615

Udayan A, Pandey AK, Sharma P, Sreekumar N, Kumar S (2021) Emerging industrial applications of microalgae: challenges and future perspectives. Syst Microbiol Biomanufacturing 1(4):411–431

Van Den Hoek C, Mann DG, Jahns HM (1995) Algae. In: An introduction to phycology. Cambridge University Press, Cambridge, p xiv+623. ISBN: 0 521 30419 9 (hardback); 0 051 31687 1 (paperback). Price: £70.00 (hard); £24.95 (soft). Eur J Phycol 32 (2):203–205

Vanthoor-Koopmans M, Wijffels RH, Barbosa MJ, Eppink MH (2013) Biorefinery of microalgae for food and fuel. Bioresour Technol 135:142–149

Wichaphian A, Kaewman N, Pathom-Aree W, Phinyo K, Pekkoh J, Chromkaew Y, Cheirsilp B, Srinuanpan S (2024) Zero-waste biorefining co products from ultrasonically assisted deep eutectic solvent-pretreated Chlorella biomass: sustainable production of biodiesel and bio-fertilizer. Bioresour Technol 408:131163

Wicker RJ, Kumar G, Khan E, Bhatnagar A (2021) Emergent green technologies for cost-effective valorization of microalgal biomass to renewable fuel products under a biorefinery scheme. Chem Eng J 415:128932

Wu W, Chang JS (2019) Integrated algal biorefineries from process systems engineering aspects: a review. Bioresour Technol 291:121939

Yadav K, Vasistha S, Nawkarkar P, Kumar S, Rai MP (2022) Algal biorefinery culminating multiple value-added products: recent advances, emerging trends, opportunities, and challenges. 3 Biotech 12(10):244

Yen HW, Hu IC, Chen CY, Ho SH, Lee DJ, Chang JS (2013) Microalgae-based biorefinery–from biofuels to natural products. Bioresour Technol 135:166–174

Zhan J, Rong J, Wang Q (2017) Mixotrophic cultivation, a preferable microalgae cultivation mode for biomass/bioenergy production, and bioremediation, advances and prospect. Int J Hydrog Energy 42(12):8505–8517

Zhang S, Zhang L, Xu G, Li F, Li X (2022) A review on biodiesel production from microalgae: influencing parameters and recent advanced technologies. Front Microbiol 13:970028

Zhao X, Zhang X, Fu L, Zhu H, Zhang B (2016) Effect of extraction and drying methods on antioxidant activity of astaxanthin from Haematococcus pluvialis. Food Bioprod Process 99:197–203

Zhou Q, Zhang P, Zhang G, Peng M (2015) Biomass and pigments production in photosynthetic bacteria wastewater treatment: effects of photoperiod. Bioresour Technol 190:196–200

Life-Cycle Assessment of Microalgae Technology and Products

22

Oscar M. Hernández-Calderón,
María G. Laguna-Martínez, Eusiel Rubio-Castro,
Erika Y. Rios-Iribe, Marcos D. González-Llanes,
Maritza E. Cervantes-Gaxiola,
and Jesús R. Ortiz-del-Castillo

Abstract

This chapter provides a comprehensive overview of the Life Cycle Assessment (LCA) applied to microalgae-based technologies and products, emphasizing their potential as a sustainable alternative to conventional fossil fuels. Microalgae, recognized for their high efficiency in biomass production and the ability to be cultivated in environments unsuitable for traditional agriculture, offer significant environmental sustainability advantages. The chapter explores the environmental, economic, and social impacts associated with biofuel and bioproduct production and the integration of microalgae in wastewater treatment. It highlights the need for process optimization to maximize energy efficiency and reduce greenhouse gas (GHG) emissions. Furthermore, commonly used LCA methodologies are analyzed, identifying key areas for improvement and standardization to enhance the commercial viability and widespread adoption of microalgae-based technologies. The insights gained from these assessments provide crucial information for strategic decision-making in the transition toward a circular, low-carbon economy, where microalgae play a crucial role in climate change mitigation and global sustainability.

Keywords

Microalgae · Life-Cycle Assessment (LCA) · Environmental sustainability · Microalgae-based technologies

22.1 Introduction

Life-Cycle Assessment (LCA) is a widely recognized and utilized methodology for evaluating the environmental impacts of products, processes, and services across their entire life cycle, from raw material extraction to final disposal. This approach enables identifying and quantifying material and energy inputs and outputs at each stage, along with their environmental effects. The primary goal of LCA is to provide a robust foundation for decision-making aimed at reducing environmental impacts and enhancing the sustainability of human activities.

Microalgae-based technologies have attracted significant interest due to their potential to mitigate the environmental impacts of fossil resource production and consumption. As photosynthetic organisms, microalgae capture carbon dioxide (CO_2) from the atmosphere, helping to reduce greenhouse gas (GHG) emissions. Furthermore, these technologies facilitate the production of diverse products, including biofuels, biofertilizers, high-value compounds, and animal feed, many of which can replace their more energy-intensive fossil or chemically derived counterparts.

Moreover, microalgae are increasingly being utilized to treat industrial and domestic wastewater effluents. In some cases, these processes help clean the wastewater and produce valuable compounds, adding an extra layer of sustainability to the technology.

Integrating microalgae processes in biorefineries, where biomass utilization is optimized to produce multiple value-added products, exemplifies how these technologies can support sustainability goals. For instance, the production of

O. M. Hernández-Calderón (✉) · M. G. Laguna-Martínez
E. Rubio-Castro · M. D. González-Llanes ·
M. E. Cervantes-Gaxiola · J. R. Ortiz-del-Castillo
Department of Chemical Engineering, Facultad de Ciencias Químico-Biológicas, Universidad Autónoma de Sinaloa, Culiacán, Sinaloa, Mexico
e-mail: omhc@uas.edu.mx

E. Y. Rios-Iribe
Programa de Posgrado Integral en Biotecnología, Facultad de Ciencias Químico Biológicas, Universidad Autónoma de Sinaloa, Culiacán, Sinaloa, México

D. U. Santos Ballardo, S. Rossi (eds.), *Microalgae as Promising Source of Commercial Bioproducts*, Developments in Applied Phycology 14, https://doi.org/10.1007/978-3-031-86433-9_22

biodiesel from microalgal oil, coupled with the extraction of co-products such as proteins, organic acids, pigments, and other compounds, maximizes resource efficiency while minimizing waste.

LCA plays a crucial role in evaluating these technologies, offering a detailed analysis of their environmental impacts and helping identify best practices to minimize them. Through LCA studies, it is possible to compare various cultivation system designs, biomass extraction methods, and co-product strategies to determine the most sustainable options. These studies have demonstrated that, under certain conditions, microalgae can significantly reduce CO_2 emissions, land use, and water demand compared to conventional technologies.

However, the efficiency and environmental benefits of microalgae technologies are highly dependent on factors such as the cultivation method, production system design, geographic location, and scale of operation. The variability in these factors makes LCA essential for evaluating the specific conditions under which microalgae can achieve true sustainability.

This chapter discusses the opportunities and challenges that microalgae-based technologies and products face in contributing to sustainability, exploring the environmental, economic, and social implications across the entire life cycle, from cultivation to downstream processing. Additionally, it provides a historical overview of the role LCA has played in optimizing key variables for microalgae systems. The chapter also highlights specific limitations and potential areas for improving current LCA practices. Furthermore, it discusses how LCA can be an effective tool for decision support and product innovation when combined with spatialized inventories, methodological updates, and new data science solutions.

22.2 Microalgae Technology for Sustainability

In the context of microalgae-based technologies and products, microalgae represent a promising alternative to the unsustainability of fossil fuels, which constitute a significant part of global primary energy consumption. Historically, the continued use of fossil fuels has been recognized as unsustainable due to resource depletion and greenhouse gas (GHG) emissions, which contribute significantly to the increase in atmospheric CO_2 concentration, exceeding the "dangerously high" threshold of 450 ppm CO_2-e. This reality has driven the development of mitigation techniques and the adoption of policies promoting renewable energy sources, such as microalgae-based technologies, which can sequester atmospheric CO_2, reduce dependence on fossil reserves, and maintain environmental and economic sustainability (Zaimes and Khanna 2015; Shimako et al. 2016; Gholkar et al. 2021).

This recent line of research has focused on fuel products derived from microalgae to address the growing energy demand while reducing both consumption and dependence on fossil fuels. Most studies aim to provide local energy resources and mitigate the effects of climate change while reducing greenhouse gas (GHG) emissions (Rajesh Banu et al. 2020). Fuel products derived from microalgae are presented as a viable solution to the challenges associated with energy production and consumption. As a renewable resource, these products can generate heat, electricity, and transportation fuels, with the widely recognized advantage of requiring minimal modifications to existing technologies. This positions them as a significant opportunity to enhance sustainability and reduce GHG emissions (Guo et al. 2016). Many developed and developing countries have identified microalgae-derived biofuels as a critical tool for reducing dependence on foreign oil, decreasing GHG emissions, and achieving rural development goals. Some microalgae fuel products have been proposed to displace traditional fuels, mitigate vulnerabilities in the energy sector, and contribute to global energy security. Research aimed at improving the production of these products has accelerated for ecological and economic reasons. Sustainable development involves meeting current needs using resources available within a production system. Energy conversion, utilization, and access are fundamental aspects of many significant challenges related to sustainability, environmental quality, security, and poverty (European Commission: Joint Research Centre et al. 2020; Hansen et al. 2022). The World Commission on Environment and Development defined "sustainability" as "development that meets the needs of the present without compromising the ability of future generations to meet their own needs" (Hariem Brundtland 1985).

The sustainability of products or technologies must be evaluated across three dimensions: economic, environmental, and social (Gholkar et al. 2021; Rajesh Banu et al. 2020; Drejeris and Oželienė 2019).

The economic dimension encompasses cost competitiveness, energy security, self-sufficiency, and financial investment. For microalgae-based technologies to be viable, they must demonstrate cost competitiveness compared to conventional alternatives while contributing to energy security and self-sufficiency. Additionally, these technologies should attract sufficient financial investments to support ongoing development and expansion.

While high initial production costs and energy inputs for microalgae cultivation are valid concerns, it is essential to recognize the long-term potential of this field, as well as the technological advances that enable rapid biomass production and higher oil yields per acre, which may justify the initial investment. Operating costs can be further reduced by incorporating waste streams and renewable energy sources that align with sustainable practices. As previously mentioned, the environmental dimension typically assesses factors such as GHG emissions, life-cycle carbon balance, air and water quality, soil quality, water use, and the impacts on biodiversity and land use. In this context, microalgae-based technologies and products have the potential to reduce emissions, improve carbon balance, and minimize negative impacts on biodiversity, thereby contributing to environmental sustainability (Verdelho Vieira et al. 2022; Merlo et al. 2021).

Finally, achieving sustainability in microalgae cultivation requires a global perspective encompassing economic, social, and environmental aspects. The social dimension, however, is more complex to characterize due to its intricacies, yet numerous strategies have been introduced to address this challenge. One approach seeks to explain social sustainability through the impact of habitat management, highlighting that strategic planning is crucial for diversifying resources from agricultural activities. Cultivation systems should be designed to complement local farming practices, utilize marginal lands, and integrate rural waste management systems. Consequently, community participation is essential for effectively addressing social implications (García-Llorente et al. 2016; Ahmadzai et al. 2021). These three interrelated objectives must be balanced to ensure the sustainability of microalgae-derived technologies and products (Renuka et al. 2018).

22.3 LCA in Microalgae Technologies: A Survey

This section provides a comprehensive overview of the application of LCA in microalgae technologies, drawing from a review of over a hundred research papers. The most significant studies have been selected to highlight key advancements and persistent challenges in this field. The section is organized into four subtopics: Biofuels, Bioproducts, Wastewater Treatment, and Process Characteristics. Each subtopic examines how LCA has been utilized to critically assess and enhance the environmental sustainability of microalgae-based systems, tracing the historical evolution of these technologies. Through this detailed review, we aim to underscore the significant developments that have redefined current understanding and shaped the future trajectory of microalgae in sustainable applications.

22.3.1 Biofuels

22.3.1.1 Methane Production from Microalgae

This study employed LCA to examine the environmental impacts of producing methane from *Chlorella vulgaris* microalgae. The process involves the anaerobic digestion of raw microalgae biomass to produce methane, which also facilitates the recycling of key nutrients such as nitrogen, phosphorus, and potassium. The LCA revealed that environmental impacts are significantly influenced by the high electricity consumption required for operations such as mixing and circulation throughout various production steps. Improvements in these areas, such as reducing the energy needed to mix and optimize the anaerobic digestion process under controlled conditions, could significantly reduce the environmental footprint. Despite these energy challenges, the study suggests that methane production from microalgae is a competitive alternative to other biofuels, such as algal biodiesel and first-generation bio-diesels (Collet et al. 2011).

22.3.1.2 Impact of Lipid Content and Nitrogen on Microalgae-Based Fuel

The LCA of microalgae-based aviation fuel indicates that increasing lipid content in microalgae can significantly reduce total energy input. However, this increase in lipid content is accompanied by higher fossil fuel consumption and greenhouse gas (GHG) emissions. This trade-off arises from the relationship between lipid content and specific productivity, where higher lipid levels are associated with lower productivity, necessitating more intensive cultivation processes. Additionally, varying nitrogen nutrient levels influence both lipid content and the environmental impact, with lower nitrogen recovery efficiencies exacerbating GHG emissions. The study emphasizes the need to balance lipid content and specific productivity to optimize the sustainability of microalgae-based aviation fuels (Guo et al. 2016).

22.3.1.3 Biomethane Production from Microalgae Cultivated in Wastewater

Maga (2017) conducted a study on producing biomethane from microalgae cultivated in municipal wastewater as part of the EU-funded All-Gas project. The study analyzed the entire process chain, from microalgae cultivation in open ponds (OP) to biomethane production and its use as vehicle fuel. The LCA demonstrated significant environmental benefits, particularly in climate protection and fossil resource conservation, compared to conventional wastewater treatment and using compressed natural gas (CNG) as fuel. However, the study also identified challenges, including higher impacts in categories such as particulate matter formation and water depletion. Despite these drawbacks, the results support further development and optimization of this technology as a sustainable alternative in biofuel production.

22.3.1.4 Microalgae-Based Biodiesel Production

Raghuvanshi et al. (2018) conducted an LCA to evaluate the environmental impacts of biodiesel production from microalgae cultivated in wastewater compared to freshwater. The study employed a cradle-to-gate approach to analyze energy consumption and environmental impacts throughout biodiesel production, including stages such as cultivation, flocculation, centrifugation (CT), extraction, and transesterification. The findings indicated that using wastewater for microalgae cultivation offers significant energy savings compared to freshwater, particularly during the cultivation phase. However, the extraction process was identified as the most energy-intensive and environmentally challenging stage for both water sources, primarily due to chemical solvents. The study highlights the sustainability benefits of incorporating wastewater into biodiesel production while identifying key areas for improvement in the extraction process.

22.3.1.5 Comparative Environmental Impact of Microalgae-to-Biofuel Chains

The LCA of upgraded microalgae-to-biofuel chains revealed that the microalgae-to-butanol chain is more environmentally friendly than the microalgae-to-diesel chain. The analysis demonstrated that the microalgae-to-butanol process achieved a 54.4% lower ReCiPe endpoint score compared to the microalgae-to-diesel process, primarily due to lower impacts in categories such as Climate Change Human Health, Climate Change Ecosystems, and Fossil Depletion. Furthermore, integrating combined heat and power (CHP) systems and solvent reuse in the microalgae-to-butanol chain further minimized the environmental footprint. However, both processes face limitations due to the significant energy demands associated with cultivation, pretreatment, and biofuel production, which continue to pose challenges for the overall sustainability of microalgae-based biofuels (Wu et al. 2019).

22.3.1.6 Bioethanol Production from Microalgae

The LCA of bioethanol production from microalgae in Brunei Darussalam, a tropical country, demonstrated significant environmental benefits, including a low-carbon footprint and efficient water and land use. The study revealed a favorable net energy balance, with a net energy ratio (NER) of 0.45, indicating that bioethanol produced exceeded the energy consumed in the process. Using photobioreactors (PBR) and open ponds (OP) for microalgae cultivation showcased the advantages of utilizing coastal areas and seawater, minimizing the need for freshwater resources. Additionally, integrating wastewater from industrial and agricultural sources as nutrient inputs further enhanced the sustainability of the process, reducing reliance on synthetic fertilizers and contributing to effective waste management (Hossain et al. 2019).

22.3.1.7 Bio-Crude Production from Marine Microalgae

The study by Al-Jabri et al. (2022) utilizes LCA to evaluate the environmental impacts of bio-crude production from various marine microalgae species. *Tetraselmis* sp. is highlighted for its high biomass yield and favorable energy balance, positioning it as an optimal candidate for sustainable bio-crude production. The analysis reveals that *Tetraselmis* sp. provides better net energy output and contributes to lower greenhouse gas emissions compared to other species. Moreover, the proximity of cultivation sites to CO_2 sources is crucial, as the LCA demonstrates that reducing CO_2 transport distances significantly decreases energy consumption and the overall environmental footprint of the production process.

22.3.1.8 Potential of Microalgae for CO_2 Sequestration and Biofuel Production

Microalgae, particularly *Chlorella*, *Scenedesmus*, and *Euglena*, have significant potential for CO_2 bio-sequestration by converting CO_2 and water into organic compounds through photosynthesis. This carbon fixation process is driven by solar energy, making microalgae a promising tool for carbon mitigation strategies. The resulting biomass can also be utilized to produce various biofuels, including biodiesel, bioethanol, biohydrogen, and other value-added products. Despite these advantages, the microalgae industry remains in its early stages, facing challenges related to production scalability and the need for further technological advancements. The study emphasizes that achieving large-scale application will require improvements in the efficiency of carbon fixation processes and developing a robust biorefinery infrastructure to fully capitalize on the environmental benefits of microalgae-based biofuels (Li et al. 2023).

22.3.2 Bioproducts

22.3.2.1 Microalgae Oil Production

Benemann et al. (2012) conducted an LCA to evaluate the environmental impacts of oil production from microalgae cultivated in large open ponds. The study found that microalgae oil production could be achieved with minimal fossil energy inputs and negligible net greenhouse gas (GHG) emissions, provided those realistic assumptions regarding productivity, scale, and operational efficiency are met. The findings underscore the importance of optimizing energy use and operational practices to minimize the environmental footprint of microalgae-based biofuels. The study also highlights the necessity of detailed engineering designs and systems analyses to accurately assess energy inputs and mass balances, which are crucial for achieving sustainable large-scale microalgae oil production.

22.3.2.2 Astaxanthin Production from Microalgae

An LCA revealed that electricity consumption is the primary driver of environmental burdens during the pilot-scale production of astaxanthin from *Haematococcus pluvialis*. This impact is particularly pronounced during the cultivation phase, where artificial lighting is extensively used. Although sunlight was considered an alternative to reduce energy use, the corresponding decrease in biomass productivity limited the potential environmental benefits. The study concludes that optimizing photobioreactor configurations and investigating energy-efficient alternatives are crucial steps toward reducing the environmental footprint of astaxanthin production at a pilot scale (Pérez-López et al. 2014a).

22.3.2.3 Bioactive Compounds Production from *Tetraselmis suecica*

An LCA of bioactive compounds production from *Tetraselmis suecica* at a pilot scale identified the inoculation and culturing stages as the largest contributors to environmental impacts, mainly due to high electricity consumption. The process includes five main stages: sterilization, inoculation and culturing, harvesting, extraction, and anaerobic digestion of the residual algal paste. The key bioactive compounds produced are polyunsaturated fatty acids (PUFAs), α-tocopherol, chlorophyll, β-carotenoid, and polyphenols. Using alternative nitrogen sources, such as inorganic and organic fertilizers, demonstrated considerable environmental benefits over sodium nitrate, indicating opportunities for reducing the overall environmental footprint. Additionally, integrating anaerobic digestion of the residual algal paste as a nutrient source further enhanced sustainability by minimizing reliance on synthetic fertilizers (Pérez-López et al. 2014b).

22.3.2.4 Cultivation for Food and Feed

A study by Smetana et al. (2017) employed the LCA approach to compare microalgae production's environmental impacts under autotrophic and heterotrophic conditions. The findings suggest that protein powders derived from microalgae typically have a higher environmental impact than traditional protein concentrates. However, the extrusion of high-moisture products from heterotrophically cultivated *Chlorella vulgaris* proved to be more sustainable than pork and beef production. Moreover, using hydrolyzed food waste as a carbon source for producing *Chlorella pyrenoidosa* significantly reduced environmental impacts, positioning it as one of the most sustainable protein sources evaluated in the study.

22.3.2.5 Environmental Impact of Microalgae-Based Biofertilizer

LCA trials indicated that microalgae-based biofertilizer, though promising, exhibited higher environmental impacts compared to conventional triple superphosphate (TSP) fertilizer across several categories, particularly in terms of climate change and terrestrial ecotoxicity. The study highlighted that algal biomass production and processing were the primary contributors to these impacts. However, optimizing energy sources by integrating photovoltaic panels and improving biomass separation and drying methods could significantly reduce these environmental burdens. Despite the higher overall impacts, the potential for nutrient recovery and reduced reliance on non-renewable phosphate resources underscore the need for further research and technological advancements in this field (Castro et al. 2020).

22.3.2.6 Microalgae and Fish for Nutrient Production

Schade et al. (2020) conducted an extensive LCA to evaluate the environmental impacts of producing essential nutrients such as eicosapentaenoic acid (EPA), docosahexaenoic acid (DHA), and protein from microalgae species like *Nannochloropsis* sp. and *Phaeodactylum tricornutum*. The study compared these findings with traditional nutrient sources, including wild capture and aquaculture fish. Several scenarios were analyzed, including photobioreactor materials, tube diameters, and cultivation season length. Notably, the results showed that microalgae production, particularly when considering CO_2 sequestration as an avoided burden, often had lower or comparable environmental impacts than fish fillets. Even when accounting for the CO_2 burden, microalgae cultivation demonstrated similar or reduced impacts compared to aquaculture. These findings highlight the potential of microalgae as a sustainable alternative to traditional fish sources, especially in efforts to alleviate the ecological strain on marine environments.

22.3.2.7 Impact of Microalgae Biomass as a Soil Biostimulant

This study utilized an LCA to evaluate the environmental impacts of using microalgae biomass cultivated in agro-industrial effluents as a soil biostimulant for corn cultivation. The analysis included six scenarios, varying soil types, and functional units. The results identified biomass harvesting and supplementary soil fertilization as the main contributors to the carbon footprint. A significant reduction in this footprint was observed when chemical coagulation for biomass harvesting was replaced with gravitational sedimentation. This change also reduced freshwater ecotoxicity, terrestrial ecotoxicity, and terrestrial acidification. The study concluded that applying microalgae biomass to soil is an environmentally beneficial strategy, offering a sustainable solution for effluent treatment and biostimulant production (Silva et al. 2022).

22.3.2.8 Evaluating Algae Oil as a Sustainable Fish Oil Substitute

The LCA demonstrated that algae oil rich in docosahexaenoic acid (DHA) significantly reduces environmental impacts compared to traditional fish oil, particularly regarding global warming, terrestrial acidification, freshwater eutrophication, and land use. The production process utilized wastewater, further lowering the environmental footprint. The study highlights the potential of algae oil as a sustainable alternative in aquaculture, provided continued technological advancements and strategic investments are made (Bartek et al. 2021).

22.3.2.9 Omega-3 Fatty Acid Production

Togarcheti and Padamati (2021) underscore the importance of LCA in assessing the environmental impacts of producing eicosapentaenoic acid (EPA) and docosahexaenoic acid (DHA) from microalgae and farmed fish. Their study rigorously compares the primary energy demands and environmental burdens of microalgae cultivated via photoautotrophic and heterotrophic methods with fish oil extracted from farmed fish. The results indicate that heterotrophic microalgae cultivation significantly reduces environmental impacts compared to photoautotrophic cultivation and fish farming. This suggests that heterotrophic microalgae-based EPA and DHA production could offer a more sustainable alternative to conventional fish oil, potentially alleviating the environmental pressures on marine ecosystems.

22.3.2.10 Cultivating *Galdieria sulphuraria* on Food Waste

Thielemann et al. (2021) conducted a detailed LCA to assess the environmental impacts of cultivating *Galdieria sulphuraria*, a heterotrophic microalga, using food waste as a nutrient source to produce protein-rich biomass. The study evaluated the sustainability of food waste hydrolysate as a substrate compared to conventional glucose-based alternatives. The findings revealed an 11% higher carbon footprint when using food waste hydrolysate, mainly due to the energy demands of the hydrolysis process. Additionally, the lyophilization contributed to 54% of the total greenhouse gas emissions. The study highlights the importance of optimizing both the hydrolysis and drying stages to improve the overall sustainability of this production method.

22.3.2.11 Performance and Product Viability of Autotrophic Microalgae

Zhang et al. (2022) conducted a detailed LCA to evaluate the environmental impacts of producing food, total fatty acids (TFA), and bio-oil from autotrophically cultivated microalgae. The study found that microalgae-derived products, such as TFA and algal powder, exhibited lower environmental impacts than conventional agricultural products, particularly regarding global warming potential and resource depletion. However, bio-oil production had higher environmental impacts due to the energy-intensive nature of the process. The research emphasized integrating renewable energy sources and optimizing cultivation methods to mitigate these impacts. The study suggests that selecting appropriate algal strains and improving production efficiency are essential for achieving sustainable and economically viable outcomes.

22.3.2.12 Biomass, Glycerol, and Beta-Carotene Production Using Dunaliella

The LCA of cultivating *Dunaliella salina* and *Dunaliella tertiolecta* for biomass, glycerol, and beta-carotene production revealed that the choice of cultivation medium plays a crucial role in environmental outcomes. Scenarios with optimized nitrogen and magnesium concentrations resulted in reduced environmental impacts, particularly in global warming, non-renewable energy use, and respiratory inorganics. Using brine from desalination plants as a salt source also lowered carbon emissions. This highlights the potential for integrating microalgae cultivation with CO_2 capture and brine utilization for a more sustainable production system. However, the study acknowledged that emissions were only calculated based on components in the artificial media, excluding the contributions of electricity, infrastructure, and other critical aspects of microalgae cultivation and bioproduct recovery (de Souza Celente et al. 2023).

22.3.2.13 Microalgae-Derived Biofuels and Biofertilizers

Castro et al. (2023) conducted a detailed LCA and techno-economic analysis to evaluate the production of solid biofuel and biofertilizer from microalgae biomass cultivated in wastewater. The study explored two operating conditions for hydrothermal carbonization (HTC): one optimized for solid biofuel and the other for biofertilizer production. The LCA indicated biofertilizer production had lower environmental impacts than solid biofuel in categories such as freshwater eutrophication, fossil resource scarcity, and global warming potential. From an economic standpoint, biofertilizer production was more favorable, showing a positive net present value (NPV) and a shorter payback period. In contrast, solid biofuel production was economically unfeasible due to its negative NPV and longer payback period. The findings underscore the importance of selecting appropriate valorization pathways to maximize environmental and economic benefits in microalgae biomass utilization.

22.3.3 Wastewater Treatment

22.3.3.1 Microalgae-Based Wastewater Treatment Systems

Colzi Lopes et al. (2018) performed a comprehensive techno-environmental analysis of a medium-scale wastewater treatment system using high-rate algal ponds (HRAPs). The study evaluated mass and energy balances and conducted an LCA across two biogas utilization scenarios: one focused on heat and electricity generation and the other on producing heat, electricity, and biomethane. The findings showed that electricity consumption during the secondary treatment stage was significantly lower than in traditional activated sludge systems. Furthermore, substituting conventional energy with that produced by the microalgae-based system resulted in an improved life-cycle performance. The study also emphasized that the choice between the biogas use scenarios depends on the specific techno-environmental goals of the decision-makers, highlighting the flexibility of microalgae-based systems for wastewater treatment.

22.3.3.2 Integrating Microalgae in Wastewater Treatment

Tua et al. (2021) explored integrating a microalgae-based side-stream process within an existing wastewater treatment plant (WWTP) in northern Italy. The study focused on using microalgae to treat nutrient-rich centers from sludge dewatering and capturing CO_2 from the flue gas of the plant's combined heat and power (CHP) unit. The harvested algal biomass was reintroduced into the system for biogas production via anaerobic digestion. The results demonstrated environmental benefits, especially through electricity savings from the algal process. However, the study pointed out optimization challenges in ammonia stripping and improving the valorization of algal biomass to maximize the environmental advantages of this technology.

22.3.3.3 Biomass Production in Wastewater Systems

This study utilized LCA to evaluate the environmental impacts of producing algal biomass in high-rate ponds using domestic effluent as the culture medium. Seven systems were modeled, incorporating various approaches, including ultraviolet disinfection of effluent, supplementation with industrial CO_2, exhaust gas from gasoline combustion, and coupling with a biofilm reactor. The LCA findings indicated that using wastewater as a source of water and nutrients significantly reduced environmental impacts, with some systems even exhibiting negative impact values. However, eutrophication and toxicity-related impacts were notably higher. The system utilizing CO_2 from exhaust gas performed best across most environmental categories. In contrast, combining industrial CO_2 supply with hybrid systems (HS) to enhance biomass productivity often resulted in increased environmental impacts. Additionally, the LCA identified opportunities for improvement, such as optimizing CO_2 sources and exploring alternative support materials for biofilm reactors to enhance the sustainability of these systems (Magalhães et al. 2021).

22.3.3.4 Environmental Benefits of Using Wastewater in Microalgae Systems

The LCA revealed that microalgae systems utilizing wastewater, particularly from the food industry, significantly reduce environmental impacts compared to conventional systems that rely on standard growth media. These systems facilitate the treatment of urban and industrial wastewater while recovering valuable bioproducts, such as natural pigments, biofertilizers, and biogas. The scenario using food-industry wastewater emerged as the most promising due to the higher quality of the effluent, which supports more efficient recovery of bioproducts and the cultivation of single-species microalgae like *Spirulina*. This approach reduces the need for chemical inputs and capitalizes on wastewater as a nutrient source, promoting a more sustainable and circular bioeconomy within the water sector (Arashiro et al. 2022).

22.3.3.5 Cultivating Microalgae in Adjusted Wastewater for Biofuel Production

Li et al. (2022) examined the cultivation of microalgae using nutrient-adjusted wastewater to enhance biofuel production while addressing environmental impacts. By optimizing the nutrient content of wastewater, the process significantly improved microalgal biomass yield and reduced energy demands during pyrolysis. The LCA conducted in the study demonstrated that this approach not only increases biomass production but also lowers energy consumption and reduces the environmental impacts associated with the process, compared to conventional cultivation methods. The findings indicate that upcycling nutrients from wastewater is an effective strategy for optimizing biofuel and other high-value compounds production from microalgae. This method provides a sustainable pathway for integrating microalgae cultivation into wastewater management systems.

22.3.3.6 Microalgae Production Systems in Urban-Industrial Symbiosis

Pechsiri et al. (2023) conducted a comprehensive LCA to compare the environmental impacts of conventional open-reactor systems and a novel closed photobioreactor for microalgae production. Under different operational scenarios, the study evaluated these systems across various impact categories, including climate change and eutrophication. One key scenario involved urban-industrial symbiosis, where the reactors used wastewater and incinerator flue gases as inputs. The results indicated that the novel closed photobioreactor demonstrated competitive performance, particularly when integrated into urban-industrial symbiosis, significantly reducing environmental impacts, especially when powered by renewable energy. In contrast, conventional open reactors had lower energy demands due to their simpler design and reliance on natural sunlight. The study highlights the potential of the closed photobioreactor to mitigate environmental burdens, particularly in densely populated urban settings where limited space and waste streams can be effectively utilized.

22.3.3.7 Carotenoid Production from Wastewater-Grown Microalgae

Ferreira et al. (2024) conducted a comprehensive LCA and techno-economic analysis to evaluate the environmental impacts and economic viability of producing carotenoids, specifically lutein and beta-carotene, from microalgae cultivated in wastewater. The study compared two cultivation systems—a high-rate algal pond (HRAP) and a hybrid system (HS) along with two extraction methods: ethyl acetate solvent extraction and supercritical CO_2 extraction. The results revealed that although supercritical CO_2 extraction offered better environmental and economic performance, it required more energy than solvent extraction. Among the cultivation systems, the HRAP, combined with supercritical CO_2 extraction, emerged as the most efficient, exhibiting lower environmental impacts and production costs than the HS system. Significant environmental impacts were tied to electricity consumption and the use of ethyl acetate, particularly in the categories of human carcinogenic toxicity and fossil resource depletion. The study suggests that it is essential to optimize both the cultivation and extraction processes to improve the sustainability of carotenoid production from microalgae. This includes implementing solvent recovery techniques and exploring greener extraction methods.

22.3.4 Process Characteristics

22.3.4.1 Microalgal Biomass Production Systems

Jorquera et al. (2010) conducted a comparative LCA to evaluate the energy efficiency of producing microalgal biomass using open raceway ponds versus two types of photobioreactors: horizontal tubular and flat-plate designs. The study calculated each system's net energy ratio (NER), revealing that raceway ponds were the most energy-efficient option, with the most favorable NER. In contrast, horizontal tubular photobioreactors were found to be economically unfeasible due to their high energy consumption, resulting in an NER below one. Flat-plate photobioreactors demonstrated better energy performance than tubular systems but still lagged behind raceway ponds in overall efficiency. The study emphasized that improving the lipid content of microalgae could substantially enhance the NER for all systems, increasing their competitiveness with conventional fossil fuels. This research underscores the critical role of selecting appropriate cultivation systems to optimize the energy efficiency of microalgal biomass production.

22.3.4.2 Sustainability Insights for Microalgae Photobioreactors

The LCA of microalgae biofuel production has identified key factors that influence the sustainability of photobioreactor (PBR) technologies. One critical finding is that utilizing atmospheric CO_2 in compact PBRs, rather than fossil-derived CO_2, significantly reduces greenhouse gas emissions. The analysis also revealed that construction materials such as steel and PVC contribute substantially to the overall environmental impact. Consequently, optimizing material selection and reducing the use of non-renewable resources are essential steps toward improving the sustainability of these systems. These findings underscore the importance of a comprehensive approach considering the entire life cycle of microalgae biomass production to maximize its viability as a renewable energy source (Silva et al. 2015).

22.3.4.3 Assessment of Algae-Derived Biodiesel and Co-Products

Gnansounou and Kenthorai Raman (2016) conducted an LCA to evaluate the environmental and economic impacts of producing biodiesel and co-products, such as proteins and succinic acid, from algae. The study revealed that algae-derived biodiesel and its co-products significantly reduce greenhouse gas emissions and land use compared to traditional fossil fuel systems. Notably, the co-products, particularly succinic acid, are crucial in enhancing the sustainability and economic feasibility of algae biorefineries. The research highlights that optimizing the production processes for these co-products is key to achieving a more sustainable and economically viable algae-based biodiesel industry.

22.3.4.4 Assessing the Environmental Impact of Microalgae Pyrolysis

Wang et al. (2017) conducted an in-depth study on the effects of various pretreatment methods on microalgae pyrolysis, focusing on the process's kinetics, bio-crude yield, and quality, as well as the LCA of the process. The research examined the environmental impacts of producing biofuels through pyrolysis, mainly how these impacts are influenced by pretreatment strategies such as microwave, acid heating, and ultrasonic treatment. Although pretreatments can enhance bio-crude quality, the study found that they typically lead to increased energy consumption and higher greenhouse gas (GHG) emissions. The LCA results showed that most pretreatment methods, except for acid heating, significantly raised energy input and environmental burdens, making these processes less favorable from a sustainability perspective. This study highlights the importance of optimizing pretreatment strategies to improve bio-crude quality without excessively increasing environmental impacts.

22.3.4.5 Impact of System Design on Biofuel Production from Microalgae

Zhang and Kendall (2019) conducted a detailed evaluation of the environmental performance of biofuels derived from microalgae, emphasizing the role of system design and co-product management strategies. The study applied LCA to compare two key production pathways: biodiesel via lipid extraction (LE) and renewable diesel through hydrothermal liquefaction (HTL). Additionally, various co-product allocation methods, including economic and energy-based approaches, were analyzed to assess their influence on carbon intensity and energy consumption. The findings indicate that the carbon intensity estimates are more accurate when co-products contribute minimally to the biofuel's overall performance or are internally recycled.

22.3.4.6 Industrial-Scale Biohythane Production

This study utilized LCA to comprehensively evaluate the environmental impacts of producing biohythane (a blend of biohydrogen and biomethane) on an industrial scale, using a two-stage fermentation process that integrates microalgae and food waste. The results identified biomass pretreatment and microalgae cultivation as the most energy-intensive stages, significantly contributing to the overall environmental footprint. Electricity consumption, CO_2 release during pressurized water processes, and energy recovery operations largely drove greenhouse gas emissions. Additionally, the study emphasized the system's ability to recycle a substantial portion of critical nutrients, including carbon, nitrogen, and phosphorus, which are essential for sustaining microalgae growth. These insights highlight the importance of improving energy efficiency and enhancing nutrient recycling to mitigate the environmental impacts of biohythane production (Sun et al. 2019).

22.3.4.7 Microalgae Cultivation in Tubular Photobioreactors

Schade and Meier (2020) conducted an LCA to evaluate the environmental impacts of cultivating various microalgae species in tubular photobioreactors. The study focused on factors such as tube material, diameter, and the length of the cultivation season in a cold-weather climate. Extending the cultivation season, even with lower productivity during spring and fall, consistently resulted in lower environmental impacts than shorter periods. Additionally, the choice of tube material played a key role, with acrylic glass showing higher environmental impacts than borosilicate glass. Key processes were critical to the system's footprint, including hydrogen peroxide for tube cleaning, nitrogen fertilizer application, and electricity consumption for mixing, centrifugation, and drying. These findings underscore the importance of optimizing photobioreactors' design and operational parameters to minimize their environmental impact, particularly in microalgae production for food applications.

22.3.4.8 Microalgae Cultivation with Different CO_2 Sources

A comprehensive LCA was conducted to evaluate the environmental impacts of cultivating *Phaeodactylum tricornutum* for non-energy purposes using different sources of carbon dioxide. The study compared synthetic CO_2 with CO_2 derived from a biogas upgrading process within an industrial symbiosis network. Both the cultivation and freeze-drying stages contributed significantly to the overall environmental impacts. However, the environmental performance improved markedly across all impact categories when waste-derived CO_2 was used instead of synthetic CO_2. Additionally, integrating renewable energy sources was identified as a key opportunity to enhance the sustainability of processes further. These findings highlight the substantial environmental benefits of leveraging industrial by-products and renewable energy in microalgae cultivation (Porcelli et al. 2020).

22.3.4.9 Considerations for Microalgae-Derived Biofuels

Ali (2022) conducted an LCA to evaluate the environmental impacts and economic viability of producing biofuels from microalgae. The study emphasizes the importance of utilizing wastewater, minimizing freshwater usage, and employing cost-effective infrastructure to lower production costs and reduce competition with arable lands. This approach is increasingly relevant given rising crude oil prices and the global interest in algal-based biofuels, particularly biodiesel, which can be cultivated in smaller land areas unsuitable for agriculture. The LCA results showed that microalgae-derived biofuels significantly reduce greenhouse gas emissions. However, the environmental advantages are closely linked to energy use efficiency and resource management within the production process. The study underscores the need to optimize these processes to improve microalgae-based biofuels' sustainability and economic feasibility, especially for large-scale production.

22.3.4.10 Microalgae Harvesting Strategies for Biogas Production

Wei et al. (2023) conducted a comprehensive LCA to evaluate the energy conversion characteristics and greenhouse gas (GHG) emissions of biogas production from microalgae, comparing four different harvesting strategies: centrifugation (CT), sedimentation followed by centrifugation (SD-CT), flocculation with centrifugation (FC-CT), and flocculation with electro-flotation (FC-EF). The findings revealed that the FC-EF method delivered the best overall performance, achieving the lowest net energy ratio (NER) and the least GHG emissions. Conversely, the CT strategy resulted in the highest energy consumption and GHG emissions, making it the least sustainable option. The study highlights the importance of harvesting efficiency and energy input as key factors influencing the sustainability of microalgae-based biogas production. Flocculation and electro-flotation emerged as the most effective techniques for large-scale applications.

22.3.4.11 Microalgae-Based Biofuel Production

Marangon et al. (2023) provide a comprehensive review of emerging biofuels derived from microalgae, including biohydrogen, biokerosene, and biomethane. The study underscores the versatility of microalgae biomass, which can be converted into various fuels through thermochemical and biochemical processes. Notably, cultivating microalgae in wastewater presents significant environmental and economic advantages, as this approach reduces production costs and enhances sustainability by recycling nutrients and capturing carbon dioxide. The LCA models presented in the review highlight the challenges of scaling up microalgae-based biofuel technologies. Key challenges include improving biomass growth rates and optimizing conversion processes to enhance efficiency. Additionally, the authors emphasize the integration of renewable energy sources as a crucial strategy for boosting the sustainability of these systems. This integration, combined with industrial by-products, offers substantial environmental benefits and strengthens the case for microalgae as a sustainable biofuel source.

22.3.4.12 Industrial-Scale Microalgae Production

An LCA was instrumental in evaluating the environmental impacts of an industrial-scale microalgae cultivation facility. This case study, based on primary data from the production of *Chlorella vulgaris* in vertically stacked horizontal photobioreactors, provided an in-depth analysis of the entire production process. The LCA revealed that the cultivation stage contributed the most to the environmental footprint, primarily due to the use of chemicals (nutrients and cleaning agents) and the high electricity demand for key operations such as agitation and thermoregulation. Compared to existing studies, the facility's environmental performance was considered moderate. This finding highlights the importance of using full-scale operational data to provide a realistic assessment. The study underscores the importance of exploring alternative strategies, including industrial symbiosis and circular bioeconomy, to further reduce the environmental footprint and enhance the sustainability of large-scale microalgae production (Gurreri et al. 2024).

22.4 Life-Cycle Assessment Methodology

As previously discussed, LCA evaluates the environmental and social impacts and resource consumption associated with a technology, product, or service throughout its life cycle, providing valuable insights for decision-making and the development of sustainable applications (Gurreri et al. 2024; Jorgensen 2016). In microalgae-based technologies, LCA covers all stages of the value chain: cultivation, harvesting, processing, transportation, manufacturing, and finally, disposal. By considering the entire life cycle, LCA ensures that impacts are comprehensively assessed and balanced, promoting responsible, long-term solutions (Shimako et al. 2016; Verdelho Vieira et al. 2022; Khan et al. 2018).

With its detailed inventory of inputs and outputs, LCA helps identify environmental and social hotspots, allowing for the optimization of the production processes. This, in turn, enhances the sustainability of these technologies from the cultivation stage to end-use (Smetana et al. 2017; Reijnders 2020; Aitken et al. 2014; Sprecher et al. 2014; Dewulf et al. 2015; Grimaldi et al. 2021; Lardon et al. 2009; Garlapati et al. 2019).

Figure 22.1 shows that the methodological framework for conducting an LCA comprises several vital steps, each with its challenges. The analysis begins by defining the goal and scope, with the principal objectives for microalgae technology and products, often reducing greenhouse gas emissions or optimizing water use. The challenges in this phase involve defining the functional unit, selecting system boundaries, and considering social and economic impacts, including alternative biomass conversion scenarios.

Life-Cycle Inventory: Data on the inputs and outputs associated with microalgae production, typically related to material and energy balances, are compiled. This phase also addresses allocating impacts among co-products, managing insignificant contributions, and handling regional variability.

Life-Cycle Impact Assessment (LCIA): In this phase, the inputs and outputs are quantified, particularly their emissions and social and environmental implications, which are commonly addressed through indicators. The challenges here involve selecting appropriate impact categories, dealing with spatial variation and environmental sensitivity, and establishing a suitable temporal horizon.

Interpretation: The results are interpreted to identify opportunities for improvement and inform decision-making on optimizing microalgae-based technologies, considering weighting, valuation, and uncertainty.

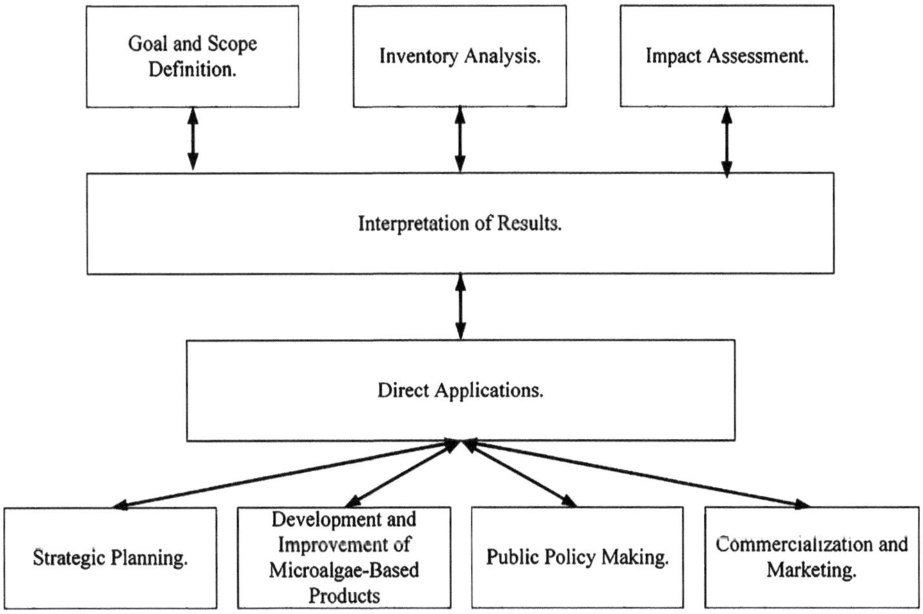

Fig. 22.1 The LCA framework. (Adapted from Dey and Ray (2022))

Data availability and quality are critical throughout the LCA process to ensure accurate and robust assessments. By overcoming these challenges, LCA thoroughly evaluates microalgae-based technologies' environmental and social impacts, guiding more sustainable decisions (Smetana et al. 2017; Garlapati et al. 2019; Soh et al. 2014; Deprá et al. 2020; Morales et al. 2019; Guieysse and Plouviez 2021; Reap et al. 2008).

As described in this chapter, holistic assessment of microalgae-based technologies and products depends on a large-scale evaluation. In the following section, life-cycle analysis, phase by phase, will be discussed.

22.4.1 The Objective of LCA

As previously discussed, the first step in LCA is to define the study's objective, which is essential for guiding the methodological approach and ensuring that the conclusions are relevant. Traditionally, the primary goal of microalgae-derived technologies and products has been to evaluate environmental impacts and quantify ecological benefits compared to conventional alternatives, such as fossil fuels or chemical fertilizers. This is typically done using eco-indicators like CO_2-e (Arashiro et al. 2022), a standard metric for comparing greenhouse gases in terms of their global warming potential by converting them into equivalent amounts. This provides a relatable way to assess the contributions of different activities and materials to global warming.

The results of these studies are essential for policymakers, industries, and consumers to identify and implement the most sustainable and efficient technological options and production methods (Zaimes and Khanna 2015; Ali 2022; de Souza et al. 2019; Singh and Olsen 2011).

Regarding social impact, the objective has been to advance health and safety in occupation areas, protect human rights at source points, enhance local working conditions, minimize detrimental effects on host communities, and provide access to resources (Josa and Garfí 2023). Including social considerations in LCAs, alongside environmental and economic dimensions, enhances the comprehensive sustainability evaluation of microalgae production. This integration is essential for decision-making and the future acceptability of algal technologies. Finally, it will facilitate the transition to a circular economy by delivering benefits that reduce greenhouse gas emissions and enhance resource productivity (Maga 2017; Ali 2022; Singh and Olsen 2011; Fernández et al. 2021; Giwa et al. 2018).

22.4.2 Scope of LCA

The first step in any LCA is to establish the scope, which defines temporal, geographical, scale, technological, and informational boundaries, as well as the limits of the study and the economic processes involved. This is particularly important for processes that require environmental intervention through impact categories, which typically define an indicator used as an assessment tool (Arashiro et al. 2022; Reap et al. 2008; Collotta et al. 2016; Collet et al. 2015; Hosseinzadeh-Bandbafha et al. 2020; Calijuri et al. 2023). In other words, limiting the scope of an LCA is essential to clarify what will be assessed and how the analysis will be designed and conducted to achieve its intended purpose. This encompasses the following crucial components: functional unit, system boundaries, and methodological approaches.

In microalgae-based technologies, a well-defined scope enables a practical assessment of cultivation systems (e.g., photobioreactors vs. open ponds) and a detailed investigation of processing pathways and end-user applications. This integrated approach ensures a comprehensive LCA that considers environmental and potentially economic and social impacts, providing valuable insights for decision-making processes that support the development of microalgae-based solutions (Al-Jabri et al. 2022; Arashiro et al. 2022; Lardon et al. 2009; Collotta et al. 2019). The scale of the systems studied, ranging from pilot to industrial levels, also influences the results, particularly regarding scalability and commercial potential (Zaimes and Khanna 2015; Hossain et al. 2019; Ali 2022; Collotta et al. 2016; Calijuri et al. 2023; Yang et al. 2011; Swathe Sriee et al. 2024).

It is important to note that the LCA process is iterative, meaning that system boundaries and impact categories may need to be adjusted as more baseline data becomes available. For example, when evaluating microalgae-derived fuels, it is crucial to consider these system boundaries and impact categories (e.g., GHG emissions or water footprint), mainly if the goal is environmental sustainability. Additionally, all valorization pathways must account for co-products (such as biofertilizers and bioplastics), as these can significantly influence LCA results. Controllable inputs are essential to ensure the sustainability and economic viability of microalgae processes (Maga 2017; Reijnders 2020; de Souza et al. 2019; Calijuri et al. 2023; Bussa et al. 2020; Sangma and Chalie-u 2023; Grierson et al. 2013). Moreover, the adaptability of these technologies to function effectively depends on their alignment with the current economic landscape. While increasing the scope, boundaries, and depth of a Life-Cycle Assessment (LCA) can yield more reliable results, the volume of data required and the computational power and resources needed to manage this complexity can become unmanageable. Thus, a balance must be struck between these factors. It is important to consider how detailed the analysis should be. A highly detailed approach may offer valuable insights but can also become impractical. Striking this key balance will determine whether the deployment and scaling of microalgae technologies can effectively integrate into established supply chains or be incorporated into existing industrial operations.

22.4.2.1 Functional Unit

The functional unit in any LCA defines the primary function of the product system and how it will be measured. In microalgae LCA studies, the functional unit varies depending on the focus of the study: it may be expressed in terms of energy, such as kilowatt-hours when evaluating algae-derived biofuels, or in terms of nutrient supply, such as kilograms of nitrogen when assessing produced fertilizers from algae (Morales et al. 2019; Collet et al. 2015; Hosseinzadeh-Bandbafha et al. 2020; Grierson et al. 2013; Sills et al. 2020; Cooney et al. 2011).

Ideally, this functional unit should be as clear-cut as possible and designed to facilitate comparisons across technologies, particularly when assessing environmental impacts or evaluating against conventional alternatives like fossil fuels or chemical fertilizers. In biofuel applications, the functional unit relates to the distance traveled per liter of fuel. For bioplastics, the unit typically involves the mass of the product, such as kilograms or tons of bioplastic produced. Biofertilizer functional units often depend on the nutrients provided, such as kilograms of nitrogen or other key elements. In carbon capture applications, the quantity of CO_2 captured serves as a standard functional unit. For food and nutritional supplements, both the mass of the product and its specific nutrient content are important.

Additionally, the mass of biomass is often the chosen unit for evaluating general biomass production. These functional units enable comparisons between algae-based technologies and conventional alternatives, allowing for a structured and measurable assessment of their environmental, economic, and social effects (Sills et al. 2020). Recent studies emphasize the importance of adapting the unit of analysis to specific production conditions while making further adjustments during the analysis to account for the system's complexities. This flexible approach ensures that the results of LCA are rational and beneficial for making sustainability decisions regarding microalgae (Bradley et al. 2023).

22.4.2.2 System Boundaries

Well-defined boundaries enable comparisons and measuring LCA results while keeping the analysis focused. To determine which aspects of the LCA to examine, a "life-cycle model" must be selected. This could be cradle-to-gate, cradle-to-costumer, cradle-to-grave, or cradle-to-cradle. As shown in Fig. 22.2, this choice delineates what materials, transportation, energy use, waste streams, and process emissions (inventory) are included in the product system. Therefore, the data collected in the inventory depends on the system boundaries defined for the analysis. If boundaries are not accurate or consistent, it may lead to discrepancies that affect the validity of the analysis. For instance, a complete picture is obtained when comparing fuels from microalgae with fossil fuels by covering all life-cycle stages. This means considering everything from growing the feedstock to burning the final fuel. Additionally, whether co-products and by-products are included or excluded can significantly alter the results (Zaimes and Khanna 2015; Ali 2022; de Souza et al. 2019; Collotta et al. 2019; Collet et al. 2014a; Saranya and Ramachandra 2020; Ott et al. 2023).

When conducting comparative analyses of production systems, assessing differences in energy consumption and emissions within the defined system boundaries is essential.

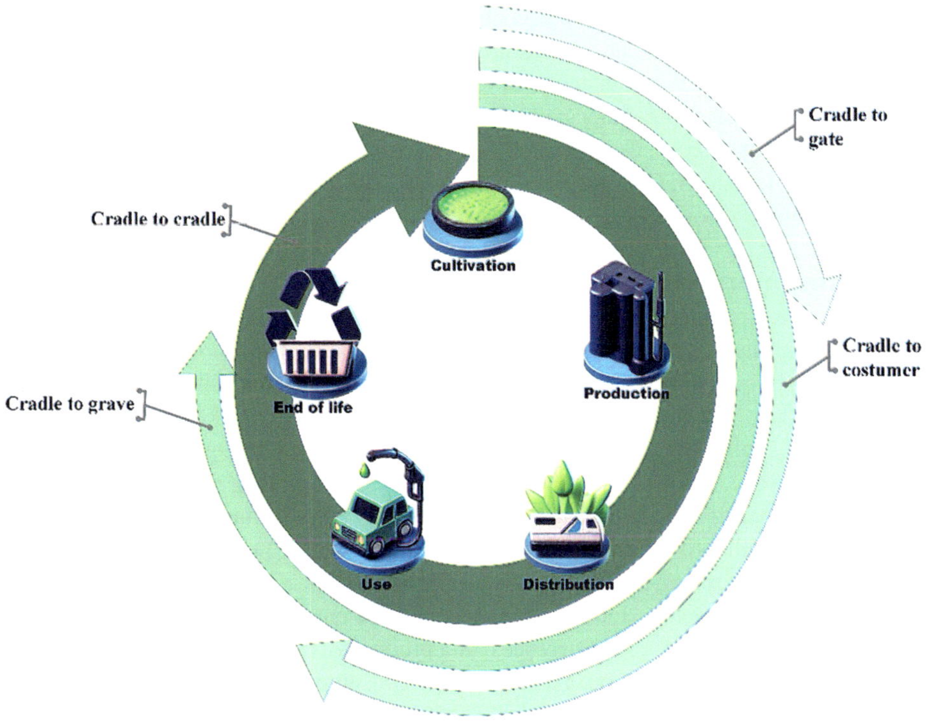

Fig. 22.2 Product life-cycle models. (Adapted from Suppipat et al. (2023))

For example, photobioreactors offer greater control and higher productivity but are associated with significant energy demands. In contrast, open ponds present a more energy-efficient alternative but are more vulnerable to environmental fluctuations (Calijuri et al. 2023; Craggs et al. 2014; Adhikari and Fedler 2020; Amahmoud et al. 2022; Hellweg et al. 2023). A straightforward and flexible approach to defining system boundaries based on an appropriate life-cycle model enhances the accuracy and reliability of LCA studies. This method aids in optimizing and developing eco-friendly microalgae systems, facilitating informed decision-making and driving technological advancements (Zaimes and Khanna 2015; Taelman et al. 2013).

22.4.2.3 Reference System

The reference system is a critical factor in any LCA study, as it compares the environmental impacts of a product, technology, or process with conventional alternatives. This comparison highlights the advantages and disadvantages of each option. For biofuels derived from microalgae, fossil fuels like diesel often serve as the reference system, while for microalgae-based biofertilizers, commonly used chemical fertilizers may be the reference. Alternatively, the reference system can also include other products and technologies, creating a benchmark between microalgae products and the different technologies employed in their processing. For instance, when comparing various microalgae-based fertilizers, the impacts on eutrophication and acidification should be carefully evaluated to understand each option's pros and cons. The less optimal option is the reference system in this context.

The value of an LCA lies in its ability to compare results, as it assesses impacts through eco-indicators or social indicators, which quantify and clarify the level of sustainability of a given process or product. By analyzing these metrics, we can determine how environmentally or socially sustainable one option is relative to another, facilitating more informed decision-making. These indicators provide the context for understanding relative impacts, guiding better decisions regarding the sustainability of different processes and products. LCA enables the comparison of key environmental impacts, such as land use, greenhouse gas emissions, and water consumption. As a result, it helps determine whether a new product or technology offers environmental benefits over its competitors (Renuka et al. 2018; Morales et al. 2019; Collet et al. 2015; Gonçalves et al. 2023).

LCA methods convert raw emissions and resource data into well-defined impact metrics that categorize and quantify environmental effects, such as global warming potential and ozone depletion. Methodological approaches like ReCiPe, IMPACT World+, IMPACT 2002+, and Environmental Prices provide structured frameworks to assess environmental footprints and define indicators, enabling comparative studies across different systems. These impact categories are crucial when comparing fuels derived from microalgae with fossil fuels or other competing alternatives.

The selection of the reference system and LCA methods significantly influences perceptions of microalgae products' sustainability and commercial viability. A comprehensive assessment of the reference system and the resulting LCA outcomes enables informed decision-making regarding microalgae technologies. This approach also facilitates the prioritization of different environmental impacts, focusing on the most relevant ones and transforming raw data into actionable insights for business and policy decisions (Maga 2017; Hossain et al. 2019; Smetana et al. 2017; Reijnders 2020; de Souza et al. 2019; Collet et al. 2015; Grierson et al. 2013; Ketzer et al. 2018; Collotta et al. 2017).

22.4.3 Inventory Analysis

Inventory analysis is the most crucial step in an LCA, as it quantifies material and energy flows within a well-defined system boundary. In a microalgae LCA study, the inventory is compiled by collecting data on water consumption, nutrients, energy usage, and emissions at each life-cycle stage. Accurate and context-specific inventory data are essential for reliable LCA results (Swathe Sriee et al. 2024).

Microalgae processes typically rely on adapted or specialized databases to capture local conditions and system-specific variables, such as differences in productivity between photobioreactors and open ponds, energy consumption, and CO_2 emissions (Hosseinzadeh-Bandbafha et al. 2020; Taelman et al. 2013).

Compared to conventional technologies like fossil fuel or chemical fertilizer systems, inventory data quality is important when assessing microalgae systems. Detailed data are essential for drawing valid comparisons given the wide variability in productivity, cultivation conditions, energy use, and emissions (Reijnders 2020; Collet et al. 2015; Bradley et al. 2023; Schade and Meier 2019). Life-Cycle Inventory (LCI) data can be obtained through primary data collection, which is highly accurate but resource-intensive, or from secondary data available in LCI databases (Kalverkamp et al. 2020). Some of the most notable databases include Ecoinvent, GaBi, and the Product Environmental Footprint (PEF) database, which offer datasets tailored to specific regions, sectors, and LCA standards. The correct choice of the database depends on the geographical context, the LCA methodology applied, and the relevance of the data to the study. LCI databases play a crucial role in filling data gaps, helping to ensure robust and accurate LCA results. Selecting the appropriate database, based on region, methodology, and product category, is a key step toward conducting meaningful assessments and making informed decisions in the microalgae industry.

22.4.4 Impact Assessment

In this phase, the raw data collected during the inventory is transformed into quantified environmental impact indicators within specific impact categories, following the selected methodology. This process links the interactions of a product system or technology with the environment to its potential impacts on human health, ecosystems, and resource depletion. Properly selecting impact categories and assessment methodologies is crucial to ensure that the LCA results are relevant for strategic decision-making and public policy (Silva et al. 2022; Arashiro et al. 2022; Sangma and Chalie-u 2023; Martinez-Porchas and Martinez-Cordova 2012).

The Impact Assessment translates material and energy flows—exchanges between the product system and the environment—into impact categories. Characterization factors standardize the effect of each substance within these categories, allowing for comparisons across different processes. For example, methane (CH_4) has a greenhouse effect 34 times greater than that of carbon dioxide (CO_2), meaning that 1 kg of methane is equivalent to 34 kg of CO_2 in terms of climate change. This process generates impact metrics that quantify a product's environmental footprint. Once the impact data is generated, the next step involves normalizing and weighting those results, which enables comparisons and prioritization of various impact categories based on their relative importance. For instance, biofertilizers derived from microalgae are efficient in nutrient use and reduce the risk of eutrophication, whereas chemical fertilizers may lead to nutrient loss (Hellweg et al. 2023). These impact categories facilitate comprehensive comparisons between options, supporting sustainable decision-making.

Life-Cycle Impact Assessment (LCIA) methods are predefined models identifying impact categories and the characterization models used in environmental impact quantification. The choice of an LCIA method depends on the LCA standard applied to the product or industry under study. These methods influence how elementary flows are categorized, and impact metrics are calculated, ensuring that assessments comply with relevant standards for meaningful outcomes. A thorough understanding of this phase is essential for anyone involved in LCA, as it bridges the gap between raw data and actionable environmental insights, guiding more sustainable practices across industries (Swathe Sriee et al. 2024; Amahmoud et al. 2022; Hellweg et al. 2023; Ghoroghi et al. 2022).

22.4.5 Interpretation of Results in Microalgae LCA

The final critical activity in an LCA is interpretation, where the impact-assessment outcomes are reviewed, and decisions are derived from those conclusions. This process involves evaluating technological and raw material choices and identifying opportunities for process improvement (Ali 2022; Bradley et al. 2023). It is essential to approach the conclusions of this phase with caution, considering the limitations of the available data and the potential for variation. To address these concerns, sensitivity analysis is a key approach that helps identify critical parameters influencing LCA results. This analysis highlights areas where data approximations or improvements are needed to ensure the accuracy of the LCA findings (European Commission: Joint Research Centre et al. 2020; Drejeris and Oželienė 2019; Fritzsche et al. 2018).

Since interpretation also identifies opportunities for innovation and further research, integrating more detailed approaches enhances the potential for sustainability in microalgae-based processes (de Souza et al. 2019; Singh and Olsen 2011; Patil et al. 2020). This stage is iterative because all phases of the LCA are interlinked, meaning that each analysis refines the previous one, resulting in progressively more precise studies. Much like fine-tuning an instrument, each iteration of the LCA hones in on the sensitivity analysis's most significant variables and phases, increasing accuracy over time. Even with variable results, the LCA provides crucial insights into the impacts of processes and products, offering a systematic way to evaluate complex concepts like sustainability. Without it, decision-makers would be left guessing, potentially making choices that could worsen environmental harm. While the results may not always be ideal, they remain valuable for driving better decisions and fostering innovation.

22.4.6 Life-Cycle Assessment Methodologies

This chapter has highlighted the variability in methodologies across LCA studies, mentioning indicators such as CO_2-e, classified as midpoint indicators. These indicators quantify impacts but do not explain the underlying mechanisms; for instance, a higher CO_2-e value indicates a greater impact but does not specify how much temperature will rise, only that global warming will increase. While midpoint indicators are more straightforward to calculate, they may overlook significant effects on health or ecosystems. In contrast, endpoint indicators focus on final impacts, such as disability-adjusted life years (DALYs) for health, providing a more comprehensive view of the impacts but with higher uncertainty. The choice between midpoint and endpoint indicators in LCA depends on the study's objectives, available data, and the intended audience. The ideal approach combines both indicators for decision-makers to fully understand the impacts (Hellweg et al. 2023; Ghoroghi et al. 2022; Collet et al. 2014b; Dong et al. 2018). A mix of these indicators is commonly used to evaluate microalgae products and technologies, including tools such as Eco-indicator 99, carbon dioxide equivalent emissions, energy consumption, water footprint, and eutrophication metrics (Reijnders 2020; Hellweg et al. 2023; Efroymson and Dale 2015; Verones et al. 2020; Bulle et al. 2019; Ramos Huarachi et al. 2020).

Choosing the appropriate LCA methodology for microalgae studies depends on specific objectives, system characteristics, and regional factors. The same indicator (e.g., water footprint) may yield different values depending on the methodologies and databases used. As shown in Table 22.1, more rigorous methods, such as ReCiPe2016 or IMPACT World+, can incorporate detailed geographic and temporal data but may be challenging to implement for initial assessments or when evaluating a wide range of alternatives. The challenge with these methods lies in obtaining and processing large amounts of data, which can be a limiting factor. Maintaining a consistent level of detail across all LCA phases is crucial. Due to these challenges, older approaches, like Eco-indicator-99, have remained relevant and continue to be used for analysis and decision-making support. These methods also serve as drivers for change by providing valuable environmental insights and guiding strategic choices (Smetana et al. 2017; de Souza et al. 2019; Collet et al. 2015; Collotta et al. 2017; Sala et al. 2022; Csiszar et al. 2016).

Due to the complexity of LCA implementation, software tools are essential for evaluating the environmental impacts of products and processes. Widely used tools include SimaPro, GaBi, openLCA, and Umberto, which are popular in growing industries because they provide critical insights into ecological impacts, helping maintain sustainably managed production (Lopes Silva et al. 2019).

Different LCA tools can yield distinct results even when using the same methods, as they rely on varying data sources. Regional factors and energy usage are the primary sources of bias. Additionally, the diversity of the supported impact-assessment methods (e.g., ReCiPe, ILCD, CML) characterizes impacts differently, contributing to variability. In summary, the precision and reliability of LCA results depend on how the methods are constructed, the protocols used to interpret data, and the frequency of database updates. Despite these challenges, LCA tools are crucial for conducting environmental assessments and guiding industry decisions.

22.5 Challenges and Future Perspectives in Microalgae Life-Cycle Assessments

As analyzed in the chapter, several key factors in Life-Cycle Assessment (LCA) for microalgae are crucial for the future. LCA studies consistently highlight high energy consumption, particularly in cultivation, mixing, extraction, and anaerobic digestion, as a significant environmental challenge in producing microalgae-based biofuels and bioproducts. However, integrating renewable energy sources and recycling industrial waste streams, such as wastewater and CO_2, can significantly enhance energy efficiency.

Although microalgae biofuels—such as methane, biodiesel, bioethanol, and bio-crude—present environmental

Table 22.1 Comparison of some LCA methodologies (Bulle et al. 2019; Rybaczewska-Błażejowska and Jezierski 2024; Oele and Dolfing 2019)

LCIA method or indicator	Approach	Impact category coverage	Notes
ReCiPe 2016	Midpoint and endpoint	18 impact categories at midpoint, 3 areas of protection at endpoint	Detailed categories are standardized but can become complex due to dual approach
LCD 2011	Midpoint	16 impact categories at midpoint	Provides a standardized and harmonized approach and is widely accepted in Europe
CML-IA Baseline	Midpoint	11 impact categories at midpoint	Simple and well-established
IMPACT 2002+	Midpoint and endpoint	15 impact categories at midpoint, 4 damage categories at endpoint	Broader coverage of damage categories (human health, ecosystem, resources) and less geographical specificity
IMPACT World+	Midpoint and endpoint	18 impact categories at midpoint, 21 damage categories at endpoint	Provides regional differentiation, a more detailed focus on water and carbon footprints, and is more complex to implement due to high geographical specificity and larger dataset requirements
Eco-indicator 95	Endpoint	Focus on three main damage categories: human health, ecosystem quality, and resource depletion	Easy to interpret and communicate results, useful for decision-making and loss of detail at the midpoint level
Eco-indicator 99	Endpoint	Focus on three damage categories with multiple subcategories	Comprehensive endpoint approach, widely used Flexible for different value choices, loss of detail at the midpoint level, value-based judgments, and no regional differentiation
EDIP 2003	Midpoint	Covers several impact categories such as global warming, acidification, eutrophication, and human toxicity	Suitable for industrial applications, widely used in Europe, and lacks regional differentiation

benefits, their scalability is constrained by the energy-intensive processes. Reducing the environmental impact and improving cost-effectiveness will require advancements in photobioreactor design, carbon fixation pathways, and more efficient harvesting methods. Furthermore, the environmental footprint of microalgae cultivation can be optimized by employing different techniques that enhance both carbon sequestration and biomass production. Ultimately, continued research and technological innovation are vital to fully realizing microalgae's environmental and economic potential within the economy.

As the field of microalgae LCA continues to evolve, it presents both opportunities and challenges. While LCA's flexibility allows adaptation to various research contexts, it can also result in inconsistencies, complicating comparisons across studies. Standardizing these methods is crucial for generating more meaningful insights from microalgae LCA research. Additionally, incorporating regional data and specificity into models will enhance the accuracy and relevance of findings, as resource availability and regulations vary significantly by location, directly influencing variables in microalgae systems. As the field progresses, future studies should explore new conversion methods, alternative nutrient sources, and improving scalability. Emphasizing energy efficiency and nutrient recovery processes will further contribute to a circular economy, paving the way toward more sustainable practices.

As mentioned, accurate environmental assessments increasingly depend on detailed and reliable data. While traditional methods based on outdated eco-indicators remain popular due to their ease of implementation, they often fail to capture the full complexity of modern environmental challenges. Advances in data science are helping to address these limitations. Data-driven LCA models will offer more accurate, context-specific insights, driving sustainability forward and pushing the boundaries of current LCA methodologies. New statistical methods and predictive analytics are transforming environmental modeling, making it more dynamic and precise. LCA's future lies in integrating data science and circular economy principles. LCA will be crucial in designing and optimizing circular strategies as the world shifts toward more sustainable decisions.

References

Adhikari K, Fedler CB (2020) Pond-In-Pond: an alternative system for wastewater treatment for reuse. J Environ Chem Eng 8(2):103523

Ahmadzai H, Tutundjian S, Elouafi I (2021) Policies for sustainable agriculture and livelihood in marginal lands: a review. Sustainability 13. https://doi.org/10.3390/su13168692

Aitken D et al (2014) Life cycle assessment of macroalgae cultivation and processing for biofuel production. J Clean Prod 75:45–56

Ali EM (2022) Chapter 23 – Life cycle assessment for microalgae-derived biofuels. In: El-Sheekh M, Abomohra AE-F (eds) Handbook of algal biofuels. Elsevier, pp 523–545

Al-Jabri H et al (2022) A comparison of bio-crude oil production from five marine microalgae – using life cycle analysis. Energy 251:123954

Amahmoud A, El Attar MM, Meleishy A (2022) The evolution of life cycle assessment approach: a review of past and future prospects. IOP Conf Ser: Earth Environ Sci 992(1):012002

Arashiro LT et al (2022) Life cycle assessment of microalgae systems for wastewater treatment and bioproducts recovery: natural pigments, biofertilizer and biogas. Sci Total Environ 847:157615

Bartek L et al (2021) Life cycle assessment of fish oil substitute produced by microalgae using food waste. Sustain Prod Consump 27:2002–2021

Benemann J, Woertz I, Lundquist T (2012) Life cycle assessment for microalgae oil production. Disruptive Sci Technol 1(2):68–78

Bradley T et al (2023) Life cycle assessment of microalgae-derived biodiesel. Int J Life Cycle Assess 28(5):590–609

Bulle C et al (2019) IMPACT World+: a globally regionalized life cycle impact assessment method. Int J Life Cycle Assess 24(9):1653–1674

Bussa M, Zollfrank C, Röder H (2020) Life-cycle assessment and geospatial analysis of integrating microalgae cultivation into a regional economy. J Clean Prod 243:118630

Calijuri ML et al (2023) Chapter 9 – Life cycle assessment of microalgal biomass for valorization. In: Bandh SA, Malla FA (eds) Valorization of microalgal biomass and wastewater treatment. Elsevier, pp 179 196

Castro JDS et al (2020) Microalgae based biofertilizer: a life cycle approach. Sci Total Environ 724:138138

Castro JS et al (2023) Life cycle assessment and techno-economic analysis for biofuel and biofertilizer recovery as by-products from microalgae. Renew Sust Energ Rev 187:113781

Collet P et al (2011) Life-cycle assessment of microalgae culture coupled to biogas production. Bioresour Technol 102(1):207–214

Collet P et al (2014a) Biodiesel from microalgae – life cycle assessment and recommendations for potential improvements. Renew Energy 71:525–533

Collet P et al (2014b) Chapter 13 – Life-cycle assessment of microalgal-based biofuels. In: Pandey A et al (eds) Biofuels from algae. Elsevier, Amsterdam, pp 287–312

Collet P et al (2015) Recommendations for life cycle assessment of algal fuels. Appl Energy 154:1089–1102

Collotta M et al (2016) Evaluating microalgae-to-energy -systems: different approaches to life cycle assessment (LCA) studies. Biofuels Bioprod Biorefin 10(6):883–895

Collotta M et al (2017) Comparative LCA of flocculation for the harvesting of microalgae for biofuels production. Procedia CIRP 61:756–760

Collotta M et al (2019) Life cycle analysis of the production of biodiesel from microalgae. In: Basosi R et al (eds) Life cycle assessment of energy systems and sustainable energy technologies: the Italian experience. Springer International Publishing, Cham, pp 155–169

Colzi Lopes A et al (2018) Energy balance and life cycle assessment of a microalgae-based wastewater treatment plant: a focus on alternative biogas uses. Bioresour Technol 270:138–146

Cooney MJ, Young G, Pate R (2011) Bio-oil from photosynthetic microalgae: case study. Bioresour Technol 102(1):166–177

Craggs R et al (2014) High rate algal pond systems for low-energy wastewater treatment, nutrient recovery and energy production. N Z J Bot 52(1):60–73

Csiszar SA et al (2016) Conceptual framework to extend life cycle assessment using near-field human exposure modeling and high-throughput tools for chemicals. Environ Sci Technol 50(21):11922–11934

de Souza Celente G et al (2023) Life cycle assessment of microalgal cultivation medium: biomass, glycerol, and beta-carotene production by Dunaliella salina and Dunaliella tertiolecta. Int J Life Cycle Assess:1–14

de Souza MP et al (2019) Microalgae and clean technologies: a review. CLEAN – Soil, Air Water 47(11):1800380

Deprá MC et al (2020) Environmental impacts on commercial microalgae-based products: sustainability metrics and indicators. Algal Res 51:102056

Dewulf J et al (2015) Toward an overall analytical framework for the integrated sustainability assessment of the production and supply of raw materials and primary energy carriers. J Ind Ecol 19(6):963–977

Dey P, Ray S (2022) Chapter 32 – Environmental impacts of biofuels and their blends: a case study on waste vegetable oil-derived biofuel blends. In: Gurunathan B, Sahadevan R (eds) Biofuels and bioenergy. Elsevier, pp 755–770

Dong Y et al (2018) Environmental sustainable decision making– the need and obstacles for integration of LCA into decision analysis. Environ Sci Pol 87:33–44

Drejeris R, Oželienė D (2019) New approach to the technological aspect of corporate sustainable development. Bus Theory Pract 20:363–371

Efroymson RA, Dale VH (2015) Environmental indicators for sustainable production of algal biofuels. Ecol Indic 49:1–13

European Commission: Joint Research Centre et al (2020) Future transitions for the bioeconomy towards sustainable development and a climate-neutral economy – knowledge synthesis – final report. Publications Office

Fernández FGA et al (2021) The role of microalgae in the bioeconomy. New Biotechnol 61:99–107

Ferreira J et al (2024) Carotenoids from wastewater-grown microalgae biomass: life cycle assessment and techno-economical analysis. J Clean Prod 434:140526

Fritzsche K, Niehoff S, G. (2018) Beier Industry 4.0 and climate change—exploring the science-policy gap. Sustainability 10. https://doi.org/10.3390/su10124511

García-Llorente M et al (2016) Social farming in the promotion of social-ecological sustainability in rural and periurban areas. Sustainability 8. https://doi.org/10.3390/su8121238

Garlapati VK, Tewari S, Ganguly R (2019) Chapter 19 – Life cycle assessment of first-, second-generation, and microalgae biofuels. In: Hosseini M (ed) Advances in feedstock conversion technologies for alternative fuels and bioproducts. Woodhead Publishing, pp 355–371

Gholkar P, Shastri Y, Tanksale A (2021) Renewable hydrogen and methane production from microalgae: a techno-economic and life cycle assessment study. J Clean Prod 279:123726

Ghoroghi A et al (2022) Advances in application of machine learning to life cycle assessment: a literature review. Int J Life Cycle Assess 27(3):433–456

Giwa A et al (2018) Techno-economic assessment of the sustainability of an integrated biorefinery from microalgae and Jatropha: a review and case study. Renew Sust Energ Rev 88:239–257

Gnansounou E, Kenthorai Raman J (2016) Life cycle assessment of algae biodiesel and its co-products. Appl Energy 161:300–308

Gonçalves J et al (2023) Microalgae as biofertilizers: a sustainable way to improve soil fertility and plant growth. Sustainability 15. https://doi.org/10.3390/su151612413

Grierson S, Strezov V, Bengtsson J (2013) Life cycle assessment of a microalgae biomass cultivation, bio-oil extraction and pyrolysis processing regime. Algal Res 2(3):299–311

Grimaldi F et al (2021) Intensified production of zeolite A: life cycle assessment of a continuous flow pilot plant and comparison with a conventional batch plant. J Ind Ecol 25(6):1617–1630

Guieysse B, Plouviez M (2021) Chapter 13 – Sustainability of microalgae cultivation. In: Lafarga T, Acién G (eds) Cultured microalgae for the food industry. Academic Press, pp 343–365

Guo F et al (2016) Life cycle assessment of microalgae-based aviation fuel: influence of lipid content with specific productivity and nitrogen nutrient effects. Bioresour Technol 221:350–357

Gurreri L et al (2024) Microalgae production in an industrial-scale photobioreactors plant: a comprehensive life cycle assessment. Sustain Chem Pharm 39:101598

Hansen J et al (2022) Impact pathways from climate services to SDG2 ("zero hunger"): a synthesis of evidence. Clim Risk Manag 35:100399

Hariem Brundtland G (1985) World commission on environment and development. Environ Policy Law 14(1):26–30

Hellweg S et al (2023) Life-cycle assessment to guide solutions for the triple planetary crisis. Nat Rev Earth Environ 4(7):471–486

Hossain N, Zaini J, Indra Mahlia TM (2019) Life cycle assessment, energy balance and sensitivity analysis of bioethanol production from microalgae in a tropical country. Renew Sust Energ Rev 115:109371

Hosseinzadeh-Bandbafha H et al (2020) Life-cycle assessment (LCA) analysis of algal fuels. In: Spilling K (ed) Biofuels from algae: methods and protocols. Springer New York, New York, pp 121–151

Jorgensen SE (2016) Ecotoxicology and chemistry applications in environmental management. CRC Press

Jorquera O et al (2010) Comparative energy life-cycle analyses of microalgal biomass production in open ponds and photobioreactors. Bioresour Technol 101(4):1406–1413

Josa I, Garfí M (2023) Social life cycle assessment of microalgae-based systems for wastewater treatment and resource recovery. J Clean Prod 407:137121

Kalverkamp M, Helmers E, Pehlken A (2020) Impacts of life cycle inventory databases on life cycle assessments: a review by means of a drivetrain case study. J Clean Prod 269:121329

Ketzer F, Skarka J, Rösch C (2018) Critical review of microalgae LCA studies for bioenergy production. Bioenergy Res 11(1):95–105

Khan MI, Shin JH, Kim JD (2018) The promising future of microalgae: current status, challenges, and optimization of a sustainable and renewable industry for biofuels, feed, and other products. Microb Cell Factories 17(1):36

Lardon L et al (2009) Life-cycle assessment of biodiesel production from microalgae. Environ Sci Technol 43(17):6475–6481

Li G et al (2022) Cultivation of microalgae in adjusted wastewater to enhance biofuel production and reduce environmental impact: pyrolysis performances and life cycle assessment. J Clean Prod 355:131768

Li S et al (2023) Production of sustainable biofuels from microalgae with CO2 bio-sequestration and life cycle assessment. Environ Res 227:115730

Lopes Silva DA et al (2019) Why using different life cycle assessment software tools can generate different results for the same product system? A cause–effect analysis of the problem. Sustain Prod Consump 20:304–315

Maga D (2017) Life cycle assessment of biomethane produced from microalgae grown in municipal waste water. Biomass Convers Biorefinery 7(1):1–10

Magalhães IB et al (2021) Technologies for improving microalgae biomass production coupled to effluent treatment: a life cycle approach. Algal Res 57:102346

Marangon BB et al (2023) Emerging microalgae-based biofuels: technology, life-cycle and scale-up. Chemosphere 326:138447

Martinez-Porchas M, Martinez-Cordova LR (2012) World aquaculture: environmental impacts and troubleshooting alternatives. Sci World J 2012(1):389623

Merlo S et al (2021) Marine microalgae contribution to sustainable development. Water 13. https://doi.org/10.3390/w13101373

Morales M et al (2019) Chapter 20 – Life-cycle assessment of microalgal-based biofuel. In: Pandey A et al (eds) Biofuels from algae, 2nd edn. Elsevier, pp 507–550

Oele M, Dolfing R (2019) SimaPro. Retrieved from SimaPro: https://simapro.com/2019/whats-new-in-simapro-9-0

Ott D et al (2023) LCA as decision support tool in the food and feed sector: evidence from R&D case studies. Environ Syst Decis 43(1):129–141

Patil RA et al (2020) Chapter 27 – Process intensification applied to microalgae-based processes and products. In: Jacob-Lopes E et al (eds) Handbook of microalgae-based processes and products. Academic Press, pp 737–769

Pechsiri JS et al (2023) Comparative life cycle assessment of conventional and novel microalgae production systems and environmental impact mitigation in urban-industrial symbiosis. Sci Total Environ 854:158445

Pérez-López P et al (2014a) Life cycle assessment of the production of the red antioxidant carotenoid astaxanthin by microalgae: from lab to pilot scale. J Clean Prod 64:332–344

Pérez-López P et al (2014b) Life cycle assessment of the production of bioactive compounds from Tetraselmis suecica at pilot scale. J Clean Prod 64:323–331

Porcelli R et al (2020) Comparative life cycle assessment of microalgae cultivation for non-energy purposes using different carbon dioxide sources. Sci Total Environ 721:137714

Raghuvanshi S et al (2018) Comparative study using life cycle approach for the biodiesel production from microalgae grown in wastewater and fresh water. Procedia CIRP 69:568–572

Rajesh Banu J et al (2020) Microalgae based biorefinery promoting circular bioeconomy-techno economic and life-cycle analysis. Bioresour Technol 302:122822

Ramos Huarachi DA et al (2020) Past and future of social life cycle assessment: historical evolution and research trends. J Clean Prod 264:121506

Reap J et al (2008) A survey of unresolved problems in life cycle assessment. Int J Life Cycle Assess 13(4):290–300

Reijnders L (2020) Chapter 30 – Life cycle assessment of microalgae-based processes and products. In: Jacob-Lopes E et al (eds) Handbook of microalgae-based processes and products. Academic Press, pp 823–840

Renuka N et al (2018) Microalgae as multi-functional options in modern agriculture: current trends, prospects and challenges. Biotechnol Adv 36(4):1255–1273

Rybaczewska-Błażejowska M, Jezierski D (2024) Comparison of ReCiPe 2016, ILCD 2011, CML-IA baseline and IMPACT 2002+ LCIA methods: a case study based on the electricity consumption mix in Europe. Int J Life Cycle Assess 29:1799–1817

Sala S et al (2022) Toxicity impacts in the environmental footprint method: calculation principles. Int J Life Cycle Assess 27(4):587–602

Sangma CBK, Chalie-u R (2023) Chapter 8 – Life cycle assessment of wastewater treatment by microalgae. In: Bandh SA, Malla FA (eds) Valorization of microalgal biomass and wastewater treatment. Elsevier, pp 137–178

Saranya G, Ramachandra TV (2020) Life cycle assessment of biodiesel from estuarine microalgae. Energy Convers Manag: X 8:100065

Schade S, Meier T (2019) A comparative analysis of the environmental impacts of cultivating microalgae in different production systems and climatic zones: a systematic review and meta-analysis. Algal Res 40:101485
Schade S, Meier T (2020) Distinct microalgae species for food—part 1: a methodological (top-down) approach for the life cycle assessment of microalgae cultivation in tubular photobioreactors. J Appl Phycol 32(5):2977–2995
Schade S, Stangl GI, Meier T (2020) Distinct microalgae species for food—part 2: comparative life cycle assessment of microalgae and fish for eicosapentaenoic acid (EPA), docosahexaenoic acid (DHA), and protein. J Appl Phycol 32(5):2997–3013
Shimako AH et al (2016) Environmental assessment of bioenergy production from microalgae based systems. J Clean Prod 139:51–60
Sills DL et al (2020) The effect of functional unit and co-product handling methods on life cycle assessment of an algal biorefinery. Algal Res 46:101770
Silva AG et al (2015) Life cycle assessment of biomass production in microalgae compact photobioreactors. GCB Bioenergy 7(2):184–194
Silva TA et al (2022) Microalgae from food agro-industrial effluent as a renewable resource for agriculture: a life cycle approach. Resour Conserv Recycl 186:106575
Singh A, Olsen SI (2011) A critical review of biochemical conversion, sustainability and life cycle assessment of algal biofuels. Appl Energy 88(10):3548–3555
Smetana S et al (2017) Autotrophic and heterotrophic microalgae and cyanobacteria cultivation for food and feed: life cycle assessment. Bioresour Technol 245:162–170
Soh L et al (2014) Evaluating microalgal integrated biorefinery schemes: empirical controlled growth studies and life cycle assessment. Bioresour Technol 151:19–27
Sprecher B et al (2014) Life cycle inventory of the production of rare earths and the subsequent production of NdFeB rare earth permanent magnets. Environ Sci Technol 48(7):3951–3958
Sun C et al (2019) Life-cycle assessment of biohythane production via two-stage anaerobic fermentation from microalgae and food waste. Renew Sust Energ Rev 112:395–410
Suppipat S et al (2023) Designer's guide for life cycle assessment. In: Suppipat S et al (eds) Design for enhancing eco-efficiency of energy-related products: the integration of simplified LCA tools in industrial design education. Springer International Publishing, Cham, pp 15–29
Swathe Sriee AE et al (2024) A comparative assessment of microbial biodiesel and its life cycle analysis. Folia Microbiol 69(3):521–547
Taelman SE et al (2013) The environmental sustainability of microalgae as feed for aquaculture: a life cycle perspective. Bioresour Technol 150:513–522
Thielemann AK, Smetana S, Pleissner D (2021) Cultivation of the heterotrophic microalga Galdieria sulphuraria on food waste: a life cycle assessment. Bioresour Technol 340:125637
Togarcheti SC, Padamati RB (2021) Comparative life cycle assessment of EPA and DHA production from microalgae and farmed fish. Clean Technol 3:699–710. https://doi.org/10.3390/cleantechnol3040042
Tua C et al (2021) Integration of a side-stream microalgae process into a municipal wastewater treatment plant: a life cycle analysis. J Environ Manag 279:111605
Verdelho Vieira V et al (2022) Clarification of most relevant concepts related to the microalgae production sector. Processes 10. https://doi.org/10.3390/pr10010175
Verones F et al (2020) LC-IMPACT: a regionalized life cycle damage assessment method. J Ind Ecol 24(6):1201–1219
Wang X et al (2017) Effect of pretreatment on microalgae pyrolysis: kinetics, biocrude yield and quality, and life cycle assessment. Energy Convers Manag 132:161–171
Wei C et al (2023) Comparative s of various harvesting strategies for biogas production from microalgae: energy conversion characteristics and greenhouse gas emissions. Energy Convers Manag 289:117188
Wu W, Lei Y-C, Chang J-S (2019) Life cycle assessment of upgraded microalgae-to-biofuel chains. Bioresour Technol 288:121492
Yang J et al (2011) Life-cycle analysis on biodiesel production from microalgae: water footprint and nutrients balance. Bioresour Technol 102(1):159–165
Zaimes GG, Khanna V (2015) Chapter 8 – Life cycle sustainability aspects of microalgal biofuels. In: Klemeš JJ (ed) Assessing and measuring environmental impact and sustainability. Butterworth-Heinemann, Oxford, pp 255–276
Zhang Y, Kendall A (2019) Effects of system design and co-product treatment strategies on the life cycle performance of biofuels from microalgae. J Clean Prod 230:536–546
Zhang D et al (2022) Life cycle assessment of auto-tropically cultivated economic microalgae for final products such as food, total fatty acids, and bio-oil. Front Mar Sci 9:990635

Bioeconomy Analysis of Microalgal Products and Process: The Role of Microalgae Technology in the Blue Economy

23

Sergio Rossi

Abstract

The bioeconomy analysis of microalgal products and processes, particularly within the framework of the blue economy, provides valuable insights into the transformative potential of microalgae technology. The blue economy emphasizes sustainable economic activities derived from the oceans, seas, and other aquatic environments, aiming to balance resource use with ecological preservation. Microalgae, with their remarkable versatility and high growth rates, offer a promising solution for integrating renewable biological resources into the bioeconomy. These organisms can address a wide range of challenges, including energy production, food security, human health, and environmental sustainability. From biofuels and biodegradable plastics to high-value nutraceuticals and pharmaceutical compounds, microalgae's applications are diverse and multifaceted. Moreover, their capacity for bioremediation, carbon capture, and wastewater treatment highlights their role in mitigating environmental degradation and enhancing ecosystem health. The development of innovative cultivation techniques, coupled with advances in biotechnological processing, is expected to enhance the commercial viability of microalgae-based products significantly. These innovations will likely enable the scaling of microalgae applications, providing a robust foundation for sustainable industrial development. As such, microalgae technologies represent a key pillar in the global transition to a greener, more sustainable economy, positioning them as vital contributors to addressing contemporary global challenges such as climate change, resource depletion, and ecological degradation.

S. Rossi (✉)
DiSTeBA, University of Salento, Lecce, Italy

Institute of Marine Sciences (LABOMAR), Federal University of Ceará, Fortaleza, CE, Brazil

Underwater Gardens International, Barcelona, Spain
e-mail: sergio.rossi@unisalento.it

Keywords

Microalgae · Sustainable economy · Microalgal bioproducts · Innovative cultivation techniques · Biotechnological processing

23.1 Introduction

There is an increasing concern about the possibilities to adapt to the new socioeconomic conditions that are envisaged nowadays due to the shortage or depletion of certain resources. Food and energy are the most vulnerable in a world in which climate change is rapidly transforming the real possibilities to maintain our actual way of life. In this panorama, we should look for alternative solutions that help transition from one socioeconomic model to another, and microalgae may be part of the targets for both problems, food, and energy shortages. In fact, microalgae represent a highly versatile and renewable resource if it is cultivated efficiently in aquatic environments (Satyanarayana et al. 2011). Their lack of structural components (except the cell wall), their rapid growth, the quite affordable cultivation protocols, and their unique biochemical composition—rich in lipids, proteins, carbohydrates, and other bioactive compounds—make them ideal for a wide range of industrial applications (Dolganyuk et al. 2020). It is not new that these microorganisms have been considered part of the solutions explored, such as potential biofuels in the 1970 oil crisis (Chisti 2007). However, during the last two decades, they have been considered a proper solution for many different problems, rapidly increasing the number of scientific teams working on method optimization in freshwater and marine microalgal growth. For example, the high lipid content of microalgae, with respect to its dry weight, is especially advantageous for producing biodiesel and other biofuels (Merlo et al. 2021; Sajjadi et al. 2018). In such a framework, it has been demonstrated that, for the same amount of oil, microalgae can yield

D. U. Santos Ballardo, S. Rossi (eds.), *Microalgae as Promising Source of Commercial Bioproducts*, Developments in Applied Phycology 14, https://doi.org/10.1007/978-3-031-86433-9_23

significantly higher amounts per unit area with respect to land-based crops without competing with food production or requiring arable land (Chisti 2007).

The use of microalgae is now a reality also for food production. These photosynthetic microorganisms are now optimized in their cultivation protocols as a sustainable source of high-quality proteins, amino acids, and essential fatty acids (Okeke et al. 2022). These make them suitable as ingredients for aquaculture feed, livestock diets, and even human food products (Kusmayadi et al. 2021). There are many bioactive compounds in microalgae, such as pigments (e.g., chlorophyll, phycocyanin) and antioxidants (e.g., carotenoids, astaxanthin), that can be extracted in high quantities from the single cells, having significant health benefits (Abd El Baky and El-Baroty 2013). They are, in fact, increasingly used in supplements, functional foods, and therapeutic formulations, more and more common in the nutraceutical industry that has been changing some sources (e.g., fungi and plants) by these unicellular species. There is also a growing interest in the cosmeceutical industry. In this sense, the moisturizing, anti-aging, and UV-protective properties of several microalgal species make them valuable in cosmetic products (Al-Obaidi et al. 2021).

One of the most important parameters to account for microalgal successful projection in the industry is its ability to grow rapidly (doubling its cell number every 1–3 days) under diverse conditions, including saline, brackish, or wastewater environments (Patel et al. 2021; Gilani et al. 2024). This rapid growth and adaptability enhance their scalability and sustainability, making them relatively cheap and easy-to-handle sources of organic matter. Such characteristics align with the principles of the bioeconomy by reducing dependency on fossil resources and promoting circular industrial practices (Chisti 2007; Mata et al. 2010).

The biomass produced by microalgae may also be used in environmental applications, particularly bioremediation. In this context, they act as natural agents for water purification (Merlo et al. 2021; Dhanker et al. 2023), absorbing the excess nutrients like nitrogen and phosphorus, which are major contributors to eutrophication in aquatic ecosystems and potential disruptors of the aquatic ecosystem functioning. Microalgae can also accumulate heavy metals (e.g., cadmium, lead) and degrade organic pollutants, including hydrocarbons and pharmaceuticals, thus purifying wastewater and industrial effluents (Priya et al. 2022). Such a dual role in nutrient and pollutant accumulation makes these microorganisms an eco-friendly alternative for wastewater treatment processes, substantially reducing reliance on energy-intensive conventional methods (Pittman et al. 2011). Interestingly, the biomass generated during bioremediation may have a secondary value. It can be inoculated, harvested, and used for biofuel production, livestock feed, or as a raw material for bioplastics and fertilizers (Santos-Ballardo et al. 2016a, b; Merlo et al. 2021; Park and Lee 2023). This transforms waste into a resource, creating a sustainable loop in line with the bioeconomy cycle (Rawat et al. 2011).

23.2 Products from Microalgal Biomass in the Blue Economy

If we observe the latest data from the FAO Sofia report (2022), we will see that algae, generally, have been continuously growing as a suitable product for different targets. Macroalgae have a prominent role, as their use as low-energy associated products is more and more easy to introduce in the market (FAO 2022). Microalgae are also considered key products of the blue economy, recognized for their contributions to the production of nutritional supplements in the food and pharmaceutical industries (Selvam et al. 2024). For example, microalgae include omega-3 fatty acids, vitamins, and antioxidants, which are critical in promoting human health and well-being (Liu et al. 2022). They are also essential for aquaculture feeding, especially in the first development stages of fish (Bell and Sargent 1996).

In fact, we consider microalgae more sustainable and a plant-based alternative to traditional fish oil (e.g., coming from *Engraulis ringens*) as a source of omega-3 fatty acids, such as eicosapentaenoic acid (EPA) and docosahexaenoic acid (DHA) (Ackman 2004). These compounds are essential for cardiovascular health, brain function, and reducing inflammation in several species, not men. In terms of human consumption, they can address the rising consumer demand for eco-friendly and vegan nutritional options because of their low energy and land use requirements, which respect terrestrial crops (Spolaore et al. 2006).

Microalgae are also rich in vitamins (e.g., A, B12, and E) and antioxidants (e.g., carotenoids and astaxanthin), which help strengthen immune function in humans, but also in reared organisms like cattle or fish, protecting against oxidative stress, and improve the skin health (Del Mondo et al. 2020; Zhuang et al. 2022). Bioactive compounds are gaining popularity in preventing chronic diseases and enhancing overall well-being, reflecting the global shift toward natural and sustainable health solutions (Gouveia et al. 2008). The global market for microalgae-based supplements has seen significant growth, especially for food that can represent billions of dollars, currently with microalgae-based nutritional supplement products (Wang et al. 2020). By contributing to a balanced diet and preventing chronic diseases, these products align with the blue economy's broader goals, seamlessly integrating sustainability with human health priorities (Thilsted et al. 2016).

The high lipid content in microalgae positions them as a key resource for renewable energy production, particularly in the development of biofuels (Merlo et al. 2021). Moreover, microalgal lipids can be converted into biodiesel, a renewable energy source that offers a cleaner alternative to traditional fossil fuels. Biodiesel from microalgae reduces greenhouse gas emissions, contributing to climate change mitigation while addressing the global energy crisis (Chisti 2007). Microalgal biofuels represent a cornerstone of sustainable energy strategies, providing a pathway to decarbonization while reducing reliance on non-renewable resources (Santos-Ballardo et al. 2016a, b). In addition to energy production, the byproducts of microalgal biofuel processing, such as residual biomass, can be used for animal feed, fertilizers, or other bioproducts, fostering a circular economy model that maximizes resource efficiency (de Carvalho et al. 2022).

Microalgae serve as a foundational component in aquaculture, significantly enhancing the sustainability of this rapidly growing industry (Han et al. 2019; Napan et al. 2015). Microalgae are a natural feed for marine organisms, including fish larvae, shellfish, and crustaceans (Hemaiswarya et al. 2011). They are rich in essential nutrients such as proteins, lipids, vitamins, and fatty acids, ensuring the growth and health of farmed species (Pulz and Gross 2004). As previously highlighted, the use of microalgae reduces dependence on conventional feed ingredients like fishmeal and fish oil, as well as biomass coming from land crops, which are often sourced from overexploited wild fish stocks or occupy land that may be used for other purposes, including food production or reforestation programs. This contributes to conserving marine and land ecosystems, which are deeply threatened and have several biodiversity and functioning problems. Microalgae may represent one of the interesting alternatives supporting the expansion of sustainable aquaculture practices. It has also been suggested that microalgae cultivation in aquaculture systems may offer other benefits, such as the absorption of excess nutrients (e.g., nitrogen and phosphorus) from the polluted water. Together with Integrated Multitrophic Aquaculture practices in which we may use filter feeders to capture the excess of particles present in the water column (Borghese et al. 2025), the use of microalgae may absorb the excess of nutrients and diminish the Oxygen Biological Debt (OBD) of the system. In this way, these microorganisms may prevent eutrophication, reducing the environmental footprint of aquaculture operations. Integrating microalgae into aquaculture allows the industry to adopt more sustainable and ecologically responsible farming practices, aligning with global goals for sustainable food production and marine conservation (FAO 2020, 2022).

23.3 Process Optimization and Efficiency

Advancing the bioeconomy with microalgae-based products demands ongoing innovation and optimization in bioprocessing techniques. The most crucial issues to be overcome are enhancing the efficiency of the cultivation and harvesting process (i.e., looking for the more adequate species depending on the pursued industrial conditions, exploring methods to stimulate the bioproducts, contrasting indoor and outdoor cultivation conditions, etc.), scalability (i.e., calculate the Life Cycle Assessment (LCA) for the microalgal cultivation, understand the carbon footprint, make a realistic business plan, etc.), and sustainability (i.e., look for autochthonous microalgae, understand each cultivation species constraints in terms of potential impacts on the environment, etc.) of microalgae cultivation (Merlo et al. 2021). Among all these potential issues, the harvesting and extraction processes are probably the most essential to improving economic viability and meeting industrial demands (Zhu et al. 2024).

Maximizing biomass productivity and stimulating target compounds in the programmed yields relies on controlling critical growth parameters such as light intensity, temperature, nutrient concentration, and CO_2 levels (Fuentes-Grünewald et al. 2011; Ramanna et al. 2017). For example, Fuentes-Grünewald et al. (2011) limited the nutrient load (nitrogen compounds). They slightly increased the temperature (stressing the phytoplankton cells) to obtain a better lipid concentration per dry weight in order to increase biodiesel production. Another example comes from Wijffels and Barbosa (2010), in which they optimize CO_2 enrichment, boosting algal growth and contributing to carbon capture, aligning with sustainability goals. One of the optimization methods used to boost microalgal cultivation and harvesting is the photobioreactors. With these enclosed systems we have precise control over environmental conditions, enhancing productivity and reducing contamination risks (Koller 2015). However, enclosed systems are more expensive than open ponds in terms of construction, maintenance, and energetic requirements.

However, they allow the control of abiotic parameters, such as temperature, pH, and CO_2 injection. Consequently, the biomass obtained is of a higher quality and has a higher volumetric concentration (Carvalho et al. 2006). PBR may also have a high cost in terms of energy used for their construction, sometimes limited to good performance in the LCA (Sevigné Itoiz et al. 2012). A specific microalgae and biomass yield experiment compared plastic bags and methacrylate tubes (Santos-Ballardo et al. 2016a, b). In this comparison, the economic and carbon footprint was balanced toward the plastic bags, but the methacrylate definitively gave better harvesting performance. Other innovations, such as tubular or flat plate photobioreactors, improve light utilization and scalability and must be considered when an upscaling plan is made. The other possibility is the use of open ponds.

It has been demonstrated that they are cost-effective, but the most important problem is, with no doubt, the contamination and environmental variability to which they are subject (Qin et al. 2019). Hybrid systems that combine photobioreactor precision with open pond scalability are being explored to balance cost and productivity (Chiaramonti et al. 2013; Khor et al. 2022), and the use of outdoor photobioreactors has also been suggested as a good solution for algal biomass production (Fuentes-Grünewald et al. 2012). We may have 6–10 algal harvests in optimum conditions a year, with different methods to enhance biomass production (Wan et al. 2015). It has been stressed that microalgal harvesting accounts for a significant portion of production costs, requiring energy-efficient and cost-effective methods (Muhammad et al. 2021).

Centrifugation
Highly effective for achieving high recovery rates, but energy-intensive, highlighting the need for innovations in energy-efficient designs.

Filtration
It is effective for larger microalgal species but less efficient for smaller ones, emphasizing the need for filter materials and technological advancements.

Flocculation
Chemical or biological agents facilitate the aggregation of microalgae cells for easier collection, providing a cost-effective and energy-efficient alternative (Milledge and Heaven 2013). However, it must be considered that advanced extraction methods must be applied to isolate high-value compounds like lipids, proteins, and pigments (Mahato et al. 2019). In this case, several methods are proposed.

Solvent Extraction
Widely employed for lipid recovery, though solvent use poses environmental concerns, prompting the development of green solvents and solvent-free alternatives.

Supercritical Fluid Extraction
Employs high-pressure CO_2 to extract compounds with exceptional purity and minimal environmental impact, although initial costs are still significant.

Enzymatic Hydrolysis
Offers a selective, efficient, and eco-friendly method for extracting specific compounds, showing promise for protein and carbohydrate recovery.

Combining cultivation, harvesting, and extraction optimization with a careful study of its applicability and economic-environmental costs may reduce waste, enhance energy efficiency, and minimize operational complexity (Firoozi et al. 2024). As previously explained, coupling wastewater treatment with microalgae cultivation provides nutrient recycling and biomass production, supporting the well-known circular economy principles (Rawat et al. 2011; Merlo et al. 2021). Again, it is important to make a realistic approach to moving from laboratory-scale to industrial-scale production, as these barriers are more and more solved in microalgal production, implementing innovations in bioreactor design, harvesting technologies, and automation. Scaling up also requires addressing logistical challenges, such as supply chain management and regional suitability for large-scale cultivation, but also a clear roadmap regarding local possibilities and socioeconomic realities (Bello et al. 2024; Penloglou et al. 2024). Incorporating a participative process is essential for the upscaling project, as will be discussed later.

Optimizing each stage of the bioprocessing chain is crucial to reducing production costs. For instance, replacing energy-intensive harvesting techniques with gravity-driven or flotation methods can lead to substantial cost savings (Ghazvini et al. 2022). Similarly, process intensification, such as integrating cultivation with CO_2 capture, can boost profitability while advancing environmental objectives (Adamu et al. 2020).

An efficient bioprocessing roadmap is not only a financial imperative but also a key goal of sustainability. By minimizing energy use, reducing waste, and leveraging renewable resources, microalgae production aligns with global climate and sustainability objectives, such as the United Nations' Sustainable Development Goals (SDGs) (Merlo et al. 2021; Olabi et al. 2023).

23.4 A Needed Technological Innovation in Cultivation Systems

During the last two decades, there have been substantial advances in microalgal cultivation and harvesting protocols, which are essential to optimize productivity, reduce environmental impact, and enhance the sustainability of microalgae-based industries (Yap et al. 2021). The materials and processes involved in biomass production of different species are one of the most targeted improvements, in particular photobioreactors and closed-loop systems, which are at the forefront of this progress, transforming microalgae cultivation into a scalable and economically viable solution for the bioeconomy (Gurreri et al. 2023).

Photobioreactors (PBRs) offer highly controlled environments for microalgae cultivation, allowing precise regulation of key growth factors like light intensity, temperature, and nutrient availability (Paul Peter et al. 2022). They are far more convenient than transparent plastic bags to stimulate algal growth; their material performs better in light penetration even if it is more expensive in terms of economic and carbon print features (Santos-Ballardo et al. 2016b). Advanced PBR designs, including flat-panel, tubular, and vertical column systems, are engineered to optimize light distribution, enhancing photosynthetic efficiency.

Depending on factors like light incidence, quality, and exposition time, those systems must have good photosynthetic structural performance for the algae. By minimizing shading and ensuring uniform exposure, these designs can significantly boost biomass productivity while reducing energy consumption (Chisti 2007), taking into account that sometimes in outdoor conditions environmental factors may play a negative non-controllable role in algal biomass production (e.g., freezing by cold conditions, dying for heat waves) (Fuentes-Grünewald et al. 2012). The integration of automation and real-time monitoring systems further enhances efficiency by dynamically adjusting growth conditions to maintain optimal cultivation parameters (Ahmed et al. 2024).

Systematic use of PBRs ensures a consistent, high-quality product, which is crucial for industries like pharmaceuticals and nutraceuticals, where purity and reliability are of utmost importance (Richmond 2004). The last generation of PBRs incorporate different systems for recycling water and nutrients, reducing resource consumption and environmental impact, and also saving energy in bubbling, mixing, or in the routine input of nutrients (Dhanker et al. 2023). By capturing and reusing CO_2 and nutrients, these systems reduce operational costs while supporting sustainability goals through waste minimization.

Closed-loop systems are engineered to minimize resource inputs and maximize resource recovery, often incorporating water recycling, nutrient recovery, and CO_2 capture to establish a self-sustaining cultivation environment. Integrating various unit operations within a microalgal biorefinery framework further enhances resource recovery, process efficiency, and cost-effectiveness, aiming to produce a diverse range of valuable bioproducts. Additionally, circular process integration fosters self-sustainability and achieves zero environmental discharge, aligning with eco-friendly practices (Mohan et al. 2020). By reducing reliance on external resources, closed-loop systems significantly decrease the environmental footprint of microalgae production. These systems recycle essential inputs such as water, nutrients, and CO_2 within the cultivation process, minimizing waste and reducing the need for fresh supplies. This aligns with sustainability principles and lowers operational costs, making the production process more economically viable. By integrating resource-efficient practices, closed-loop systems pave the way for scalable and environmentally friendly microalgae cultivation, contributing to the broader goals of a sustainable bioeconomy (Koller et al. 2012).

One of the most promising features of closed-loop systems is their ability to repurpose waste streams as valuable inputs for the production process. These systems can effectively utilize byproducts such as industrial CO_2 emissions, wastewater, and agricultural residues to supply essential resources like carbon, water, and nutrients (Kara et al. 2022). For instance, industrial CO_2 emissions can be captured and used as a carbon source for microalgae cultivation, effectively reducing greenhouse gas emissions. Simultaneously, wastewater can serve as a rich source of nutrients such as nitrogen and phosphorus, essential for microalgal growth. This dual functionality addresses two major environmental challenges: mitigating pollution from industrial emissions and wastewater while repurposing these waste streams into valuable inputs. By converting waste into resources, this approach supports circular economy principles, enhances sustainability, and reduces the overall environmental footprint of the production process (Rawat et al. 2011).

The scalability of closed-loop systems makes them suitable for industrial applications across diverse sectors. With ongoing advancements, these systems are becoming increasingly cost-effective and adaptable, paving the way for large-scale implementation across various industries. In biofuel production, they provide a sustainable and efficient method for generating renewable energy sources. In aquaculture, they support the cultivation of high-quality feed while minimizing resource inputs. Additionally, these systems offer a dual benefit in wastewater treatment by purifying water and recovering valuable nutrients for reuse. As technology evolves, closed-loop systems are poised to play a transformative role in promoting sustainability and resource efficiency across these critical sectors (ElFar et al. 2021). The modularity of closed-loop designs also allows for specific solutions to meet specific local industrial requirements.

23.5 Economic and Environmental Sustainability

Microalgae represents a transformative opportunity within the blue economy, offering diverse applications that drive economic development while actively contributing to climate change mitigation. These versatile organisms provide sustainable solutions across various sectors, from renewable energy and food production to carbon capture and environmental restoration, positioning them as a key resource for advancing both economic and environmental goals (Carrasco et al. 2018; Olabi et al. 2023). The microalgae industry spans diverse roles throughout its value chain, from research and development to cultivation, processing, and product distribution. This multifaceted ecosystem fosters employment opportunities across skilled and unskilled labor sectors and bolsters economic resilience by contributing to sustainable industries (Collier et al. 2017). For example, professionals oversee microalgae farms, whether employing photobioreactors or open pond systems, ensuring optimal growth conditions and operational efficiency. Also, skilled personnel responsible for maintaining cultivation systems are required; monitoring nutrient levels and troubleshooting equipment issues are essential in the industrial application of biomass production.

Collect microalgae biomass using techniques such as centrifugation, filtration, and flocculation, ensuring minimal loss and energy efficiency. Employees also need to refine microalgae biomass into valuable products, such as biofuels, nutraceuticals, and pigments. In this framework, it is essential to count on researchers to develop innovative cultivation technologies, improve productivity, and explore new microalgae applications, from sustainable energy to pharmaceuticals. One of the key roles to consider is the product development specialist, who designs and commercializes microalgae-based products tailored to various industries' needs. The sales and marketing teams, after the work done with the quality control managers, are essential to selling microalgae products, expanding market reach, and educating consumers on their benefits. Finally, another source of hiring specific jobs in the microalgae production plant is the professionals that optimize cultivation practices to align with environmental goals, ensuring minimal ecological impact (Bourgougnon et al. 2021).

One of the most important things to take into account in the microalgal industry is their potential to boost local economies. The possibility of making controlled cultivations for different purposes has some difficulties. However, it can be overcome, and once the system is ready, the feasibility of giving the product is easier than in other industries (Merlo et al. 2021). For example, biomass production for energy consumption (biofuel) may be activated locally to give a certain independence in remote or poorly inhabited areas, giving independence from the mainstream fossil fuel markets. In a moment when we face a climatic and energetic crisis, this solution may be a transitional path that allows for democratizing the energy supply (Merlo et al. 2021). The cultivation facilities can drive local economic activity, especially in coastal and rural areas, by creating new jobs and attracting investments (FAO 2022). Also, the microalgae industry diversifies economies because of the different emerging field promotions such as renewable energy, sustainable aquaculture, and green technologies (Meyer 2017).

There is a long debate about the potential role of microalgae in climate change mitigation (Chisti 2007). Microalgae absorb the CO_2 that will return to the system if used for example as a biofuel. A recent proposition is using this huge amount of CO_2 captured and burying it in the sand in dry environments, collecting the nutrients from deep waters (more than 80 meters deep) to fuel the system (Schipper et al. 2019). The microalgae utilization of the photosynthetic mechanism to convert CO_2 into biomass, including lipids, proteins, and carbohydrates, is the key point to understand such a possibility in climate change mitigation. However, biomass production is never "net zero" because of several externalities needed to maintain the cultivation (Sevigné Itoiz et al. 2012). The higher growth rates with respect to terrestrial plants make microalgae more adaptable to sequester large amounts of CO_2 in relatively short periods, making them efficient agents for carbon capture (Chisti 2007). During the last decade, the microalgae cultivation systems were coupled with industrial facilities, directly capturing CO_2 emissions from power plants, cement factories, and other sources (Shoenung et al. 2019). Such an integration reduces atmospheric CO_2 while producing valuable biomass for downstream applications (Rawat et al. 2011). The advanced photobioreactors mentioned above maximize CO_2 uptake, enhancing the carbon-sequestration capacity of microalgae cultivation systems (Udaypal et al. 2024).

Using microalgae with a wide variety of products stimulates biomass production that can be adapted to the market needs (Singh and Ahluwalia 2013). For example, converting microalgae biomass into biochar offers a stable and long-lasting form of carbon immobilization, improving soil health when used as an amendment (Lehmann and Joseph 2015). In the aquatic systems, different species of microalgae (that may have very different growth rates and biomass productivity, Fuentes-Grünewald et al. 2011) contribute to the natural carbon cycle, supporting oceanic carbon sequestration and regulating atmospheric CO_2 levels (Prasad et al. 2021). Thus, marine microalgae are especially interesting in marine microalgae, which also have the advantage of using fewer freshwater resources (Maeda et al. 2018). Furthermore, microalgae sequester CO_2 and reduce reliance on fossil fuels by providing a sustainable biofuel feedstock, especially in small-medium areas where the demand may be satisfied with moderate land use (Merlo et al. 2021). By integrating microalgae cultivation into carbon capture and utilization frameworks, industries can adopt a circular approach to CO_2 management (Zieliński et al. 2023), supporting the United Nations' Sustainable Development Goals (SDGs) for climate action and sustainable industrialization.

23.6 Challenges and Considerations and the Need for Innovation

While microalgae offer immense opportunities in sectors like renewable energy, pharmaceuticals, and aquaculture, their large-scale production presents various challenges. These include issues related to cost efficiency, resource optimization, scalability, and environmental impact. To ensure the economic viability and widespread adoption of microalgae in the bioeconomy, it is crucial to address these challenges through technological advancements, process optimization, and sustainable practices.

One of the first issues to solve is understanding that setting up and operating the cultivation systems, whether photobioreactors (PBRs) or open ponds, is resource-consuming (Rasouli et al. 2018). Energy consumption for lighting, temperature control, mixing, and inputs such as CO_2 and nutrients are significant cost drivers in microalgae production. These factors contribute to the high operational expenses associated with large-scale cultivation (Sevigné Itoiz et al. 2012; Zhou and Yang 2023). Efficient designs and renewable energy integration can help mitigate these expenses.

Technologies for harvesting (e.g., centrifugation, flocculation, etc.) and processing microalgae into beneficial compounds (lipids for biofuels, proteins for food, etc.) are also a source of energy consumption and, indeed, one of the costliest operations in terms of final yield. Developing for example cost-effective and scalable processes, such as membrane technologies or low-energy extraction methods, is an important target for microalgal producers. The initial capital expenditure for infrastructure, including cultivation facilities, water management systems, and processing plants, presents a significant barrier to large-scale microalgae production (Dhakal and Chevalier 2017). The high upfront costs can deter potential investors and producers from entering the market. However, public-private partnerships and government incentives can play a crucial role in alleviating this financial burden. By offering funding, tax breaks, or subsidies, these collaborations can help lower the risk for producers, encourage innovation, and accelerate the adoption of microalgae-based technologies. Such support can make the industry more accessible, driving growth and fostering long-term sustainability (Dhakal and Chevalier 2017).

As with other products, the success of microalgae biomass production depends on the demand for end-products such as biofuels, omega-3 supplements, animal feed, and cosmetics (Tarannum et al. 2024). A robust market is needed to justify large-scale operations. Sometimes, this market is unavailable locally, so you have to make an extra effort to create a business plan for the product to be sold to other countries. However, engaging stakeholders and creating consumer awareness about the benefits of microalgae-derived products can expand market potential. Studying the necessities of the region or even the local potential consumers will be a good strategy, especially for biofuels, which may be a partial solution of the energetic local needs. In this case (biofuels), even if the fossil fuel market and providers are very well established, the microalgae-based products can be competitively priced against conventional alternatives if a good strategic plan is combined with the local stakeholders (Merlo et al. 2021). For instance, biofuels derived from microalgae must be competitive with fossil fuels and other biofuel sources in terms of cost while highlighting their additional environmental benefits, such as lower carbon emissions and reduced dependency on non-renewable resources. Achieving cost parity with conventional fuels is essential for widespread adoption, but the environmental advantages—such as carbon sequestration and a smaller ecological footprint—can serve as compelling selling points. By emphasizing these benefits, microalgae-based biofuels can attract consumers and policymakers seeking sustainable alternatives contributing to climate change mitigation (Singh and Gu 2010; Merlo et al. 2021).

One of the problems in putting the product on the market is, without a doubt, the regulatory frameworks. Compliance with regulatory standards is crucial for ensuring environmental safety, and maintaining food and pharmaceutical quality, for example, are some of the issues to be solved in the marketing process. These regulations safeguard public health and environmental integrity but can add complexity and cost to production processes. For example, nutraceuticals and cosmetics must meet stringent quality and safety requirements, such as purity, efficacy, and allergen testing, which demand rigorous testing protocols and certification. Navigating these regulatory landscapes requires careful planning and investment in compliance infrastructure, but it is essential for gaining consumer trust and accessing global markets.

To accelerate the potential obstacles to the product, policies promoting renewable energy, such as tax breaks, grants, or subsidies, can offset initial costs and drive growth. Also, a specific plan to invest in research and development (R&D) can accelerate innovation and technological adoption by the involved industry (Abdmouleh et al. 2015).

Developing low-energy, high-efficiency harvesting methods is a critical focus area for improving the economic viability of microalgae production. Innovative approaches, such as using magnetic nanoparticles for harvesting or enzymatic processes for extraction, can significantly reduce costs and energy consumption (de Lima Barizão et al. 2021). These advancements can enhance overall productivity. However, transitioning from laboratory-scale to industrial-scale production presents substantial challenges. Key factors include achieving uniform growth conditions across larger systems, maintaining consistently high productivity, and controlling operational costs during scale-up (Reisman 1993). Addressing these issues is vital for ensuring economic success and scalability, and, for this reason, combining technological innovation with effective scale-up strategies will be essential for a sustainable microalgae industry bioeconomy project.

23.7 Conclusions

The future of microalgae as a source of diverse products is not a distant vision but a present reality. The remarkable versatility of microalgae, particularly those species that live in seawater environments, offers significant potential for addressing a wide range of global challenges. These microorganisms are poised to play a clear role in the future of food security, renewable energy, and environmental sustainability. In the near future, microalgae could provide sustainable solutions for food production and biofuel generation, helping to alleviate some of the most pressing concerns related to resource depletion and global change.

For underdeveloped regions, the potential of microalgae is especially promising. Their ability to grow in marginal lands or saline environments, where traditional crops may not thrive, makes them an ideal candidate for bolstering local food systems and contributing to renewable energy production. Moreover, integrating microalgae-based technologies could serve as a transformative tool for economic development in these areas, empowering communities by providing access to sustainable and affordable resources.

The democratization of microalgae-based technologies—enabling economically challenged populations to harness the benefits of these organisms—represents a critical objective for global sustainability efforts. By focusing on the accessibility and scalability of microalgae applications, we can foster long-term resilience and self-sufficiency in regions most vulnerable to climate change and resource scarcity. In this way, microalgae could not only address the needs of the present but also lay the foundation for a more equitable and sustainable future for all.

References

Abd El Baky HH, El-Baroty GS (2013) Healthy benefit of microalgal bioactive substances. J Aquat Sci 1(1):11–22. https://doi.org/10.12691/jas-1-1-3

Abdmouleh Z, Alammari RAM, Gastli A (2015) Review of policies encouraging renewable energy integration & best practices. Renew Sust Energ Rev 45:249–262

Ackman RG (2004) The ocean supplies more EPA and DHA than we can use. OCL 11:112–115

Adamu A, Russo-Abegão F, Boodhoo K (2020) Process intensification technologies for CO_2 capture and conversion – a review. BMC Chem Eng 2(1):2. https://doi.org/10.1186/s42480-019-0026-4

Ahmed N, Zhang B, Deng L, Bozdar B, Li J, Chachar S, Chachar Z, Jahan I, Talpur A, Gishkori MS, Hayat F, Tu P (2024) Advancing horizons in vegetable cultivation: a journey from age-old practices to high-tech greenhouse cultivation—a review. Front Plant Sci 15:1357153. https://doi.org/10.3389/fpls.2024.1357153

Al-Obaidi JR, Alobaidi KH, Al-Taie BS, Wee DH-S, Hussain H, Jambari NN, Ahmad-Kamil EI, Ariffin NS (2021) Uncovering prospective role and applications of existing and new nutraceuticals from bacterial, fungal, algal and cyanobacterial, and plant sources. Sustain For 13(7):3671. https://doi.org/10.3390/su13073671

Bell MV, Sargent JR (1996) Lipid nutrition and fish recruitment. Mar Ecol Prog Ser 134:315–331

Bello M, Modu AK, Boryo DEA, Mahmoud AA, Ranganathan P (2024) Sustainable algal biorefinery: a review on current perspective on technical maturity, supply infrastructure, business and industrial opportunities. J Environ Manag 370:122208

Bourgougnon N, Burlot A-S, Jacquin A-G (2021) Algae for global sustainability? In: Jacquot J-P (ed) Advances in botanical research, vol 100. Academic Press, pp 145–212

Borghese J, Arduini D, Trani R, Anglano M, Doria L, Giangrande A, Aguilo-Arce J, Toso A, Putignano M, Rossi S (2025) Short-time variations in environmental factors of an innovative IMTA system in the Mar Grande of Taranto (Mediterranean Sea). Marine Pollution Bulletin 212:117580

Carrasco R, Fajardo C, Guarnizo P, Vallejo RA, Fernandez-Acero FJ (2018) Biotechnology applications of microalgae in the context of EU "blue growth" initiatives. J Microbiol Genet 10:2574–7371. JMGE-118. https://doi.org/10.29011/2574-7371.000018

Carvalho AP, Meireles LA, Malcata FX (2006) Microalgal reactors: a review of enclosed system designs and performances. Biotechnol Prog 22(6):1490–1506

Chiaramonti D, Prussi M, Casini D, Tredici MR, Rodolfi L, Bassi N, Chini Zittelli G, Bondioli P (2013) Review of energy balance in raceway ponds for microalgae cultivation: re-thinking a traditional system is possible. Appl Energy 102:101–111

Chisti Y (2007) Biodiesel from microalgae. Biotechnol Adv 25(3):294–306. https://doi.org/10.1016/j.biotechadv.2007.02.001

Collier ZA, Connelly EB, Polmateer TL et al (2017) Value chain for next-generation biofuels: resilience and sustainability of the product life cycle. Environ Syst Decis 37:22–33. https://doi.org/10.1007/s10669-016-9618-1

de Carvalho JC, Molina-Aulestia DT, Martinez-Burgos WJ, Karp SG, Manzoki MC, Medeiros ABP, Rodrigues C, Scapini T, Vandenberghe LPDS, Vieira S et al (2022) Agro-industrial wastewaters for algal biomass production, bio-based products, and biofuels in a circular bioeconomy. Fermentation 8(12):728. https://doi.org/10.3390/fermentation8120728

de Lima Barizão AC, de Oliveira JP, Gonçalves RF, Cassini ST et al (2021) Nanomagnetic approach applied to microalgae biomass harvesting: advances, gaps, and perspectives. Environ Sci Pollut Res 28:44795–44811. https://doi.org/10.1007/s11356-021-15260-z

Del Mondo A, Smerilli A, Sané E et al (2020) Challenging microalgal vitamins for human health. Microb Cell Factories 19:201. https://doi.org/10.1186/s12934-020-01459-1

Dhakal KP, Chevalier LR (2017) Managing urban stormwater for urban sustainability: barriers and policy solutions for green infrastructure application. J Environ Manag 203(1):171–178

Dhanker R, Khatana K, Verma K, Singh A, Heena K, R., & Mohamed, H. I. (2023) An integrated approach of algae-bacteria mediated treatment of industries generated wastewater: optimal recycling of water and safe way of resource recovery. Biocatal Agric Biotechnol 54:102936

Dolganyuk V, Belova D, Babich O, Prosekov A, Ivanova S, Katserov D, Patyukov N, Sukhikh S (2020) Microalgae: a promising source of valuable bioproducts. Biomol Ther 10(8):1153. https://doi.org/10.3390/biom10081153

ElFar OA, Chang C-K, Leong HY, Peter AP, Chew KW, Show PL (2021) Prospects of industry 5.0 in algae: customization of production and new advance technology for clean bioenergy generation. Energy Convers Manag: X 10:100048

FAO (2020) The State of World Fisheries and Aquaculture 2020. Sustainability in action. Rome. https://doi.org/10.4060/ca9229en

FAO (2022) The State of World Fisheries and Aquaculture 2022. Towards Blue Transformation. Rome, FAO. https://doi.org/10.4060/cc0461en

Firoozi AA, Tshambane M, Firoozi AA, Sheikh SM (2024) Strategic load management: enhancing eco-efficiency in mining operations through automated technologies. Results Eng 24:102890

Fuentes-Grünewald C, Garcés E, Alacid E, Sampedro N, Rossi S, Camp J (2011) Improvement of lipid production in the marine strains *Alexandrium minutum* and *Heterosigma akashiwo* by utilizing abiotic parameters. J Ind Microbiol Biotechnol 39:207–216

Fuentes-Grünewald C, Alacid E, Garcés E, Rossi S, Camp J (2012) Biomass and lipid production of dinoflagellates and raphidophytes in indoor and outdoor photobioreactors. Mar Biotechnol 15:37–47

Ghazvini M, Kavosi M, Sharma R, Kim M (2022) A review on mechanical-based microalgae harvesting methods for biofuel production. Biomass Bioenergy 158:106348

Gilani IE, Hosseini H, Al-Ghouti MA, Saadaoui I, Sayadi S (2024) Microalgal-based desalination brine remediation: achievements, challenges, and future research trends. Environ Technol Innov 34:103592

Gouveia L, Batista AP, Sousa I, Raymundo A, Bandarra NM (2008) Microalgae in novel food products. Innovative Food Sci Emerg Technol 9:72–84. https://doi.org/10.1016/j.ifset.2007.07.008

Gurreri L, Rindina MC, Luciano A, Lima S, Scargiali F, Fino D, Mancini G (2023) Environmental sustainability of microalgae-based production systems: roadmap and challenges towards the industrial implementation. Sustain Chem Pharm 35:101191

Han P, Lu Q, Fan L, Zhou W (2019) A review on the use of microalgae for sustainable aquaculture. Appl Sci 9(11):2377. https://doi.org/10.3390/app9112377

Hemaiswarya S, Raja R, Ravi Kumar R et al (2011) Microalgae: a sustainable feed source for aquaculture. World J Microbiol Biotechnol 27:1737–1746. https://doi.org/10.1007/s11274-011-0789-z

Kara S, Hauschild M, Sutherland J, McAloone T (2022) Closed-loop systems to circular economy: a pathway to environmental sustainability? CIRP Ann 71(2):505–528

Khor WH, Kang HS, Lim JW, Iwamoto K, Tang CHH, Goh PS, Quen LK, Shaharuddin NMRB, Lai NYG (2022) Microalgae cultivation in offshore floating photobioreactor: state-of-the-art, opportunities and challenges. Aquac Eng 98:102269

Koller M (2015) Design of closed photobioreactors for algal cultivation. In: Prokop A, Bajpai R, Zappi M (eds) Algal biorefineries. Springer, Cham, pp 131–153. https://doi.org/10.1007/978-3-319-20200-6_4

Koller M, Muhr A, Braunegg G (2012) Microalgae as versatile cellular factories for valued products. Algal Res 1(3):119–133. https://doi.org/10.1016/j.algal.2012.03.002

Kusmayadi A, Leong YK, Yen H-W, Huang C-Y, Chang J-S (2021) Microalgae as sustainable food and feed sources for animals and humans biotechnological and environmental aspects. Chemosphere 271:129800

Lehmann J, Joseph S (2015) Biochar for environmental management: science, technology, and implementation. Earthscan

Liu Y, Ren X, Fan C, Wu W, Zhang W, Wang Y (2022) Health benefits, food applications, and sustainability of microalgae-derived N-3 PUFA. Food Secur 11(13):1883. https://doi.org/10.3390/foods11131883

Maeda Y, Yoshino T, Matsunaga T, Matsumoto M, Tanaka T (2018) Marine microalgae for production of biofuels and chemicals. Curr Opin Biotechnol 50:111–120

Mahato N, Sinha M, Sharma K, Koteswararao R, Cho MH (2019) Modern extraction and purification techniques for obtaining high purity food-grade bioactive compounds and value-added co-products from citrus wastes. Food Secur 8(11):523. https://doi.org/10.3390/foods8110523

Mata TM, Martins AA, Caetano NS (2010) Microalgae for biodiesel production and other applications: a review. Renew Sust Energ Rev 14(1):217–232. https://doi.org/10.1016/j.rser.2009.07.020

Merlo S, Gabarrell X, Pedroso A, Rossi S (2021) Marine microalgae contribution to sustainable development. Water 13:1373. https://doi.org/10.3390/w13101373

Meyer R (2017) Bioeconomy strategies: contexts, visions, guiding implementation principles, and resulting debates. Sustain For 9(6):1031. https://doi.org/10.3390/su9061031

Milledge JJ, Heaven S (2013) A review of the harvesting of microalgae for biofuel production. Rev Environ Sci Biotechnol 12(2):165–178. https://doi.org/10.1007/s11157-012-9301-z

Mohan SV, Hemalatha M, Chakraborty D, Chatterjee S, Ranadheer P, Kona R (2020) Algal biorefinery models with self-sustainable closed loop approach: trends and prospective for blue-bioeconomy. Bioresour Technol 295:12212

Muhammad G, Alam MA, Mofijur M, Jahirul MI, Lv Y, Xiong W, Ong HC, Xu J (2021) Modern developmental aspects in the field of economical harvesting and biodiesel production from microalgae biomass. Renew Sust Energ Rev 135:110209

Napan K, Christianson T, Voie K, Quinn JC (2015) Quantitative assessment of microalgae biomass and lipid stability post-cultivation. Front Energy Res 3:15. https://doi.org/10.3389/fenrg.2015.00015

Okeke ES, Ejeromedoghene O, Okoye CO, Ezeorba TPC, Nyaruaba R, Ikechukwu CK, Oladipo A, Orege JI (2022) Microalgae biorefinery: an integrated route for the sustainable production of high-value-added products. Energy Convers Manag: X 16:100323

Olabi AG, Shehata N, Sayed ET, Rodriguez C, Anyanwu RC, Russell C, Abdelkareem MA (2023) Role of microalgae in achieving sustainable development goals and circular economy. Sci Total Environ 854:158689

Park YK, Lee J (2023) Achievements in the production of bioplastics from microalgae. Phytochem Rev 22:1147–1165. https://doi.org/10.1007/s11101-021-09788-8

Patel AK, Tseng Y-S, Singhania RR, Chen C-W, Chang J-S, Dong CD (2021) Novel application of microalgae platform for biodesalination process: a review. Bioresour Technol 337:125343

Paul Peter A, Koyande AK, Chew KW, Ho S-H, Chen W-H, Chang J-S, Krishnamoorthy R, Banat F, Show PL (2022) Continuous cultivation of microalgae in photobioreactors as a source of renewable energy: current status and future challenges. Renew Sust Energ Rev 154:111852

Penloglou G, Pavlou A, Kiparissides C (2024) Recent advancements in photobioreactors for microalgae cultivation: a brief overview. Processes 12(6):1104. https://doi.org/10.3390/pr12061104

Pittman JK, Dean AP, Osundeko O (2011) The potential of sustainable algal biofuel production using wastewater resources. Bioresour Technol 102(1):17–25. https://doi.org/10.1016/j.biortech.2010.06.035

Prasad R, Gupta SK, Shabnam N, Oliveira CYB, Nema AK, Ansari FA, Bux F (2021) Role of microalgae in global CO2 sequestration: physiological mechanism, recent developments, challenges, and future perspectives. Sustain For 13(23):13061. https://doi.org/10.3390/su132313061

Priya AK, Jalil AA, Vadivel S, Dutta K, Rajendran S, Fujii M, Soto-Moscoso M (2022) Heavy metal remediation from wastewater using microalgae: recent advances and future trends. Chemosphere 305:135375

Pulz O, Gross W (2004) Valuable products from biotechnology of microalgae. Appl Microbiol Biotechnol 65:635–648. https://doi.org/10.1007/s00253-004-1647-x

Qin L, Alam MA, Wang Z (2019) Open pond culture systems and photobioreactors for microalgal biofuel production. In: Alam MA, Wang Z (eds) Microalgae biotechnology for development of biofuel and wastewater treatment. Springer, Singapore, pp 69–92. https://doi.org/10.1007/978-981-13-2264-8_3

Ramanna L, Rawat I, Bux F (2017) Light enhancement strategies improve microalgal biomass productivity. Renew Sust Energ Rev 80:765–773

Rasouli Z, Valverde-Pérez B, D'Este M, De Francisci D, Angelidaki I (2018) Nutrient recovery from industrial wastewater as single cell protein by a co-culture of green microalgae and methanotrophs. Biochem Eng J 134:129–135

Rawat I, Kumar RR, Mutanda T, Bux F (2011) Dual role of microalgae: phycoremediation of domestic wastewater and biomass production for sustainable biofuels. Appl Energy 88:3411–3424. https://doi.org/10.1016/j.apenergy.2010.11.025

Reisman HB (1993) Problems in scale-up of biotechnology production processes. Crit Rev Biotechnol 13(3):195–253. https://doi.org/10.3109/07388559309041319

Richmond A (ed) (2004) Handbook of microalgal culture: biotechnology and applied phycology. Blackwell Science

Sajjadi B, Chen W-Y, Raman AAA, Ibrahim S (2018) Microalgae lipid and biomass for biofuel production: a comprehensive review on lipid enhancement strategies and their effects on fatty acid composition. Renew Sust Energ Rev 97:200–232

Santos-Ballardo DU, Rendón-Unceta MC, Rossi S, Vázquez-Gómez R, Reyes-Moreno C, Valdez-Ortiz A (2016a) Effects of outdoor cultures on the growth and lipid production of *Phaeodactylum tricornutum* using closed photobioreactors. World J Microbiol Biotechnol 32:128. https://doi.org/10.1007/s11274-016-2089-1

Santos-Ballardo DU, Rossi S, Reyes-Moreno C, Milán-Carrillo J, Valdez-Ortiz A (2016b) Microalgae potential as a biogas source: current status, restraints and future trends. Rev Environ Sci Biotechnol 15:243–264

Satyanarayana KG, Mariano AB, Vargas JVC (2011) A review on microalgae, a versatile source for sustainable energy and materials. Int J Energy Res 35:291–311. https://doi.org/10.1002/er.1695

Schipper K, Al Muraikhi M, Alghasal GSHS et al (2019) Potential of novel desert microalgae and cyanobacteria for commercial applications and CO_2 sequestration. J Appl Phycol 31:2231–2243. https://doi.org/10.1007/s10811-019-01763-3

Schoenung SM, Efroymson RA, Langholtz MH (2019) Considerations for the design of a gas transport system for co-location of microalgae cultivation with CO_2 sources. Environmental Sciences Division, OAK RIDGE NATIONAL LABORATORY, ORNL/TM-2019/1130, pp 25

Selvam J, Mal J, Singh S, Yadav A, Giri BS, Pandey A, Sinha R (2024) Bioprospecting marine microalgae as sustainable bio-factories for value-added compounds. Algal Res 79:103444

Sevigné Itoiz E, Fuentes-Grünewald C, Gasol CM, Garcés E, Alacid E, Rossi S, Rieradevall J (2012) Energetic balance and environmental impact analysis of marine microalgal biomass production for biodiesel generation in a photobioreactor pilot plant. Biomass Bioenergy 39:324

Singh UB, Ahluwalia AS (2013) Microalgae: a promising tool for carbon sequestration. Mitig Adapt Strateg Glob Chang 18:73–95. https://doi.org/10.1007/s11027-012-9393-3

Singh J, Gu S (2010) Commercialization potential of microalgae for biofuels production. Renew Sust Energ Rev 14:2596–2610. https://doi.org/10.1016/j.rser.2010.06.014

Spolaore P, Joannis-Cassan C, Duran E, Isambert A (2006) Commercial applications of microalgae. J Biosci Bioeng 101:87–96. https://doi.org/10.1263/jbb.101.87

Tarannum N, Patidar SK, Chaudhary N (2024) Promising microalgae for nutraceutical and food applications: solution for global food problems. In: Arya SK, Khatri M, Singh G (eds) Value added products from bioalgae based biorefineries: opportunities and challenges. Springer, pp 315–329. https://doi.org/10.1007/978-981-97-1662-3_16

Thilsted SH, Thorne-Lyman A, Webb P, Bogard JR, Subasinghe R, Phillips MJ, Allison EH (2016) Sustaining healthy diets: the role of capture fisheries and aquaculture for improving nutrition in the post-2015 era. Food Policy 61:126–131

Udaypal RKG, Mehariya S, Verma P (2024) Advances in microalgae-based carbon sequestration: current status and future perspectives. Environ Res 249:118397

Wan C, Alam MA, Zhao X-Q, Zhang X-Y, Guo S-L, Ho S-H, Chang J-S, Bai F-W (2015) Current progress and future prospect of microalgal biomass harvest using various flocculation technologies. Bioresour Technol 184:251–257

Wang A, Alam M, Xu JL, Wang Z (2020) Microalgae as a mainstream food ingredient: demand and supply perspective. In: Alam M, Xu JL, Wang Z (eds) Microalgae biotechnology for food, health and high value products. Springer, Singapore, pp 21–41. https://doi.org/10.1007/978-981-15-0169-2_2

Wijffels RH, Barbosa MJ (2010) An outlook on microalgal biofuels. Science 329:796–799. https://doi.org/10.1126/science.1189003

Yap JK, Sankaran R, Chew KW, Munawaroh HSH, Ho S-H, Banu JR, Show PL (2021) Advancement of green technologies: a comprehensive review on the potential application of microalgae biomass. Chemosphere 281:130886

Zhou Q, Yang Y (2023) Microalgae: an overview of their potential in biofuels and industrial biotechnology applications. Energy Convers Manag 276:113478

Zhu J, Wakisaka M, Omura T, Yang Z, Yin Y, Fang W (2024) Advances in industrial harvesting techniques for edible microalgae: recent insights into sustainable, efficient methods and future directions. J Clean Prod 436:140626

Zhuang D, He N, Khoo KS, Ng E-P, Chew KW, Ling TC (2022) Application progress of bioactive compounds in microalgae on pharmaceutical and cosmetics. Chemosphere 291(Part 2):132932

Zieliński M, Dębowski M, Kazimierowicz J, Świca I (2023) Microalgal carbon dioxide (CO2) capture and utilization from the European Union perspective. Energies 16(3):1446. https://doi.org/10.3390/en16031446

GPSR Compliance

The European Union's (EU) General Product Safety Regulation (GPSR) is a set of rules that requires consumer products to be safe and our obligations to ensure this.

If you have any concerns about our products, you can contact us on ProductSafety@springernature.com

In case Publisher is established outside the EU, the EU authorized representative is:

Springer Nature Customer Service Center GmbH
Europaplatz 3
69115 Heidelberg, Germany

Batch number: 10755966

Printed by Printforce, the Netherlands